AF527739

New Horizons in Insect Science: Towards Sustainable Pest Management

Akshay Kumar Chakravarthy
Editor

New Horizons in Insect Science: Towards Sustainable Pest Management

Editor
Akshay Kumar Chakravarthy
Department of Agricultural Entomology
College of Agriculture
Bangalore
Karnataka
India

ISBN 978-81-322-2088-6 ISBN 978-81-322-2089-3 (eBook)
DOI 10.1007/978-81-322-2089-3

Library of Congress Control Number: 2015931951

Springer New Delhi Heidelberg New York Dordrecht London

Printed on acid-free paper

Springer is part of Springer Science+Business Media (www.springer.com)

Foreword

It was sometime in mid-2012 that I received an invitation to deliver an inaugural key note for the International Conference on Insect Science (ICIS-2013) in Bengaluru, India. Despite preoccupations, it was impelling to attend ICIS-2013 because India is the centre of origin for the shoot and fruit borer, *Conogethes punctiferalis* that obsessed me scientifically for nearly three decades. Secondly, it marked 25 years of fruitful collaboration between Japan and India on *Conogethes*. Further, a global discussion was planned in ICIS-2013 on the crambid moth—*Conogethes*—that is undergoing speciation and is expanding geographically and also in its host range.

The material presented in this book deals with insect science and pest management that are intimately related. Practicable pest management programmes cannot be strategized without sound insect science. There were lively discussions on a wide variety of aspects of insect science and pest management involving a majority of species from the oriental region. However, all presentations have not found a place in this book. Of course, space is a limitation! Hot issues in Entomology like resistance management, food security, phytosanitory measures, pest risk analysis, molecular entomology, toxicology, management, biodiversity, biosystematics, conservation, climate change, ecology and behaviour have all been included.

Entomologists like in other spheres of research have become specialised and sophisticated. But an increasing extent of effort is required for extension services, social and environmental issues to implement pest management especially in developing and tropical countries. This is a challenging and daunting task given the stratified standards of living, trade barriers and societal concerns. Natural recourses and people's participation also need to be interjected and harnessed for implementing pest management strategies successfully as these are critically important. Equally important is organising conferences as ICIS-2013! Such formative scientific conferences provide a forum for expressing and developing important new ideas across a wide range of related disciplines. I look forward to more such conferences and books by well-organised teams of entomologists.

14 October 2013
Japan

Prof. Hiroshi Honda
Applied Entomology and Zoology
Faculty of Life and Environmental Sciences
University of Tsukuba, Ibaraki 305-8572, Japan

Preface

Great advances have been made during the last few decades in insect science and pest management. Undoubtedly, important factors contributing to the progress, firstly, have been the effective use of several sophisticated yet sensitive instruments like the GC/MS, gel electrophoresis, quantitative proteomics methods etc. More importantly, the present-day youth have not only increasingly become interested in insects and insect-related biological organisms but are also trying to understand them, their lives, and interactions, with multidisciplinary approach. Secondly, global trade and globalisation have led to rapid dissemination of information. As a result, new views, perspectives and interrelated sciences are evolving and emerging. Nowadays, the food growers and consumers have become more aware of the risks of pesticide residues in food, affecting market and trade. It is with this background that the Department of Entomology, UAS, Bengaluru and the Indian Society for Advancement of Insect Science, Ludhiana, Punjab conducted the International Conference on Insect Science (ICIS-2013) at Bengaluru, 14–17 February, 2013. ICIS-2013 was a mega, unique event drawing over 500 entomologists from 36 countries to a theme: New horizons in Insect Science with reference to molecular, climate change and pest management.

Papers on a number of relatively biodegradable, new molecules with a narrow spectrum of activity like anthrelinic diamide class of insecticides, cyazypyr and entomotoxic proteins namely lectins for the production of insect resistant transgenic crops were presented. These and other compounds have set new standards of efficacy and utility in plant protection. Some of these compounds represent landmarks for pest management for the present and the future. Presentations also focussed on pheromone tab with auto-confusing techniques, molecular-based genomic studies of pests, DNA barcoding of pests, pathogens, natural enemies and pollinators to ensure accurate identification, RNAi, impact of farming practices, pesticides and landscape management of wildlife in cultivated and wild habitats, identifying, conserving and declaring bee-rich areas as heritage sites, creating and marketing sustainable food goods that are safe, affordable and socially acceptable. A majority of the presentations were from developing countries focusing on pests that constrain generally small scale, sustainable, tropical agricultural production systems. The presentations reflected incredibly diverse aspects of insect science and pest management, and such coherent and up-to-date collection of views are often not available to readers. Manuscripts were received in two forms: original research material and research work with reviewed material. It is hoped that this book will be of value and use to insect-scientists, pest managers and students alike, worldwide.

Bangalore, Karnataka, India A. K. Chakravarthy

Acknowledgements

The ICIS-2013 Secretariat thanks all those who have helped in organizing and conducting ICIS-2013, a mega event, successfully. The honourable Vice Chancellor, Dr. K. Narayana Gowda, the Members of the Board of Regents, officers and Dr. M. A. Shankar, Director of Research, UAS, Bangalore, Dr. B. S. Dhillon, honourable Vice Chancellor, PAU, Ludhiana and Dr. B. V. Patil honourable Vice Chancellor, UAS® for their constant help, guidance and encouragement, and the Life members particularly Dr. Balwinder Singh, President and Professor and Head and Dr. Ramesh Arora, Secretary and Professor of Entomology of Indian Society for the Advancement of Insect Science (INSAIS), Punjab for their encouragement and support. I immensely thank Dr. S. Ayyappan (DG), Dr. N. K. Krishna Kumar DDG (Hort.), Dr. Swapan Kumar Dutta DDG (CC), Dr. T. P. Rajendran ADG (PP), Indian Council of Agricultural Research, New Delhi. The Director of Education, Director of Extension, Director of Students Welfare, Administrative Officer, Dean (Agri.), Dean (PGS), UAS, GKVK, Bangalore who deserve our gratitude for their constant help, guidance and encouragement. Dr. A. S. Siddu, Director, IIHR, Bangalore and Mr. A. N. Yellappa Reddy, IFS helped immensely in conducting ICIS-2013. I acknowledge all the foreign and Indian delegates who have participated and sent manuscripts. Prof. Sunil Natuyal, ISEC, Dr. H. Basappa, DOR, for untiring efforts for the success of ICIS-2013. I thank the staff of Holiday Palm Hotel, Bangalore for their constant help and services. The Secretariat, ICIS-2013 acknowledges the following for their valuable support: Agastya Foundation, Mr. Chandrashekar, Balur Estate, NBAII, NCBS, BCRL, IIHR, Biofuel Board, Canara Bank, Coffee Board, Dhanuka, DuPont, Excel Crop Care, Garuda Agrochemicals, Monsanto, NABARD, Rallies Pvt. Ltd., State Department of Horticulture, Sumitomo, Syngenta, IAT, Bangalore and Resolution Print Media. Staff & Students of Department of Entomology and Entomologists of UAS, Bangalore. Postgraduate students, faculty in particular Dr. N. G. Kumar, Dr. G. Govindaraj and supporting staff from UAS, Bangalore, Dr. P. K. Mehta, HPKV Palampur, Dr. H. C. Sharma ICRISAT and Dr. S. Sithanantham, Sun Agro Biotech, Chennai.

I greatly welcome the publication of ICIS-2013 proceedings in part by International Springer Group, New Delhi.

A. K. Chakravarthy

Prelude

New Horizons in Insect Science: Towards Sustainable Pest Management: An Introduction

A. K. Chakravarthy

This book deals with the kind of topics compiled on insect science and the way they have been arranged. These topics were presented at the International Conference on Insect Science, 14–17 February 2013 at Bengaluru, South India. The book addresses the topics of current and continuing significance. Also included in this book are areas in which effects of IPM on the environment, ecosystems and society are highly impacted. Emphasis is given to the role of integrated pest management (IPM), and new and evolving chemical and non-chemical pest suppression tools to reduce crop losses due to insect pests. An analysis is made of the current environmental problems associated with pesticide use and new insecticide molecules. The major objective of new branches in insect science has been to develop systems of pest management that optimize cost: benefit ratios on a long-term and sustainable basis for the farmer and the society at large. Thus, topics on insect science included in this book will be of tremendous benefit to farmers, policy makers, environmentalists, entomologists and agricultural scientists so that these could be further integrated into the entire crop production process.

The select topics have been grouped under seven broad sections. Part I includes Insect Taxonomy. As a group, insects are the most speciose of all living beings on earth. So, to identify and characterize them is the first and foremost important step. Sreedevi and others have discussed about the use of new integrative taxonomic approaches in precisely identifying and delineating species using molecular tools like DNA bar coding. Similarly, Jalali and others have shown ways to use DNA bar coding for identifications of insects important to agriculture, horticulture and forestry ecosystems.

Part II includes seven chapters on insect physiology. Shama Singh from New Delhi investigated the effects of changes in body size and colour on mating success of *Drosophila.* Dark strain showed significantly higher number of mated pairs and longer copulation duration as compared to the light strain. The longevity of big female *Trichopria sp.* (a tiny wasp) was significantly longer than small ones. Similarly, big females of the wasp produced significantly more progeny with higher sex ratio compared to small females, according to Veena and Manjunath. Vidhu and Evans from Kerala, India highlighted the presence of different forms and levels of formic acid in the behaviour and physiology of the red ant, *Oecophylla smaragdina*. The team of Mahfuza Khan from Bangladesh have researched on the impact of added bacteria in adult diets on the ovariole number of the pumpkin fruit fly, *Bactrocera tau*. Janardhana Jani and others have given the readers the effects of metabolites of *Psuedomonas sp.* on the health of the crop plants. Nowadays, *Bacillus thuringiensis* (Bt) transgenic crops are gaining a lot of importance. Leena Pathak and others have dealt with insect resistance in Bt crops.

The chapters selected on Insect Toxicology have been included in Part. III. Midgut and whole body extracts of the resistant *Cnaphalocrosis medinalis*, a pest on rice, showed differences in esterase-banding pattern with midgut producing three esterase bands, according to Ramesh Babu and Shashi Vemuri. Wei Qing Zheng and others from China carried out bioassays of Rongbao (active ingredients of calcium cyanamide) against housefly maggots. Nanoparticles and nanotechnology are going to be the *buzz* words in crop protection in the days to come. Chakravarthy and others showed that inorganic nanoparticles proved promising against *Spodoptera* and *Helicoverpa.* Alibabaie and Safaralizadeh from Iran demonstrated fumigant toxicity of nutmeg seed essential oil on cowpea weevil, *Callosobruchus maculatus.*

The next important new horizon of insect science is Insect Vectors compiled under Part. IV. Kumara and others from Srilanka discussed at length vectors of coconut leaf wilt disease. Chavan and Nagaraju have comprehensively reviewed the literature on plant viruses in South East Asia.

Molecular science has been the main stay of biologists today. Therefore, under Part. V, four important chapters on insect molecular science have been included. Ankit Patel and others have shared their research experience on molecular approaches for the improvement of *B. thuringiensis* against crop pests. Ponnuvel and others have deciphered in depth on diapause-related gene expression in eggs of *Bombyx mori*. Asokan and others have comprehensively dealt with the role of RNA interference in pest management. Shashank and coworkers have studied molecular characterization and management of shoot and fruit borer, *Conogethes punctiferalis.*

Another interesting, promising and newly emerging branch of Insect Science has been the Insect Chemical Ecology. Two interesting chapters on insect semiochemicals by a team of workers lead by Srinivasan Ramasamy from AVRDC Taiwan and Kamala Jayanthi from IIHR, Bengaluru have been included in Part. VI of this book.

In this book, insect science has embraced topics on basic and applied Entomology. So, included under Part. VII of the book are 15 chapters on Applied Entomology. Five chapters on host plant resistance as a means to suppress pest populations in cultivated ecosystems form an important part of this book. Vijayakumar et al. have elucidated the mechanism of resistance in rice gall midge. Besides inflicting direct injury to crops, insects also cause malady to crop plants by infecting disease causing organisms or debilitating the plant of nutrients. One such interesting case has been reported by Vijayakumar Ghante from Karnataka, South India on Bt cotton hybrids. Saravanaraman and coworkers have dealt with mutation breeding in shoot webber and capsule borer, *Antigastra catalaunalis* of sesamum. Selvanarayanan has elegantly addressed the redesigning research on crop resistance to insect pests.

Another novel and noble means of pest suppression has been through biocontrol agents. Ballal and Verghese have reviewed role of parasitoids and predators in pest management. Venkatesan and Jalali have gone a step further in improving efficiency of egg parasitoid *Trichogramma chilonis* against crop pests. Similarly, Srinivasmurthy et al. have dealt with reproductive alterations by Wolbachia in a braconid wasp. Magar and coworkers have addressed influence of herbivores on the biology of *Chrysoperla carnea.* There are six chapters in this book on insect pest management. Areawide integrated pest management in pigeonpea has come from Bhede and others. IPM for coconut pests has been addressed by Kumara and others. Sustainable management of tea mosquito bug on cashew is the issue dealt by Manja Naik and others. IPM for reducing pesticide residues in crops and natural resources has been the topic of Ranga Rao and others from ICRISAT, Hyderabad, India. One of the most important topics of interest today is climate change and the biological consequences of it on arthropod biodiversity and pest management has been discussed at length by H. C. Sharma from ICRISAT.

Contents

Contributors

Mahshid Alibabaie Department of Plant Protection, Agricultural Faculty West Azarbayjan province, Urmia University, Urmia, Iran

N. S. Aratchige Crop Protection Division, Coconut Research Institute, Lunuwila, Sri Lanka

Hosamani Arunkumar Main Agricultural Research Station, UAS, Raichur, Karnataka, India

R. Asokan Division of Biotechnology, Indian Institute of Horticultural Research, Bangalore, India

A. G. Badgujar Department of Entomology, Marathwada Krishi Vidyapeeth, Parbhani, Maharashtra, India

SK Bal National Institute of Abiotic Stress Management (ICAR), Pune, Maharashtra, India

K. Balaji Department of Entomology, Faculty of Agriculture, Annamalai University, Annamalainagar, Tamil Nadu, India

Chandish R. Ballal National Bureau of Agricultural Insect Resources (NBAIR) (erstwhile Project Directorate of Biological Control) (ICAR), Bangalore, Karnataka, India

G. Basana Gowda Department of Agricultural Entomology, Gandhi Krishi Vignan Kendra (GKVK), University of Agricultural Sciences, Bangalore, Karnataka, India

K. R. M. Bhanu Biocontrol Research Laboratories, Pest Control (India) Pvt. Ltd., Bangalore, Karnataka, India

B. V. Bhede Department of Entomology, Marathwada Krishi Vidyapeeth, Parbhani, Maharashtra, India

M. Bheemanna Main Agricultural Research Station, UAS, Raichur, Karnataka, India

S. P. Bhopale Department of Entomology, Dr. Panjabrao Deshmukh Krishi Vidyapeeth, Akola, India

B. B. Bhosle Department of Entomology, College of Agriculture, Marathwada Krishi Vidyapeeth, Parbhani, Maharashtra, India

A. K. Chakravarthy Department of Agricultural Entomology, Gandhi Krishi Vignan Kendra (GKVK), University of Agricultural Sciences, Bangalore, Karnataka, India

M. Chandrashekharaiah Department of Agricultural Entomology, Gandhi Krishi Vignan Kendra (GKVK), University of Agricultural Sciences, Bangalore, Karnataka, India

V. M. Chavan Pune, India

Hai Ying Chen Department of Vector Pest Control, Nanchang Center for Disease Control and Prevention, Nanchang, Jiangxi, China

Rajesh Chowdary Main Agricultural Research Station, UAS, Raichur, Karnataka, India

B. Doddabasappa College of Horticulture, Kolar, Karnataka, India

Le Duc Khanh Vietnam Academy of Agricultural Sciences, Hanoi, Vietnam

DA Evans Department of Zoology, University College, University of Kerala, Thiruvananthapuram, Kerala, India

Babasaheb B. Fand National Institute of Abiotic Stress Management (ICAR), Pune, Maharashtra, India

L. C. P. Fernando Crop Protection Division, Coconut Research Institute, Lunuwila, Sri Lanka

N. Murthy Geetha Seribiotech Research Laboratory, Bangalore, India

Vijaykumar N. Ghante Main Agricultural Research Station, UAS, Raichur, Karnataka, India

Yan Guo Department of Vector Pest Control, Nanchang Center for Disease Control and Prevention, Nanchang, Jiangxi, China

Thanh Hien Vietnam Academy of Agricultural Sciences, Hanoi, Vietnam

Hiroshi Honda Faculty of Life and Environmental Sciences, University of Tsukuba, Tsukuba, Ibaraki, Japan

Ritesh Inamake Biological Control Research Laboratory, Anand Agricultural University, Anand, Gujarat, India

S. K. Jalali National Bureau of Agricultural Insect Resources, Bangalore, Karnataka, India

Janardan Jani Biological Control Research Laboratory, Anand Agricultural University, Anand, Gujarat, India

Ankush L. Kamble National Institute of Abiotic Stress Management (ICAR), Pune, Maharashtra, India

Vasudev Kammar Department of Agricultural Entomology, Gandhi Krishi Vignan Kendra (GKVK), University of Agricultural Sciences, Bangalore, Karnataka, India

Subhash B. Kandakoor Department of Agricultural Entomology, Gandhi Krishi Vignan Kendra (GKVK), University of Agricultural Sciences, Bangalore, Karnataka, India

Vivek Kempraj Division of Entomology and Nematology, Indian Institute of Horticultural Research, Bangalore, Karnataka, India

Kuldeep Khatri Biological Control Research Laboratory, Anand Agricultural University, Anand, Gujarat, India

Chuanpit Khumsuwan AVRDC—The World Vegetable Center, East and Southeast Asia, Research and Training Station, Kasetsart University, Nakhon Pathom, Thailand

Mahesh Kumar National Institute of Abiotic Stress Management (ICAR), Pune, Maharashtra, India

N. K. Krishna Kumar ICAR, New Delhi, India

Ranjith Kumar Main Agricultural Research Station, UAS, Raichur, Karnataka, India

A. D. N. T. Kumara Crop Protection Division, Coconut Research Institute, Lunuwila, Sri Lanka

B. Ratna Kumari International Crops Research Institute for the Semi-Arid Tropics, Patancheru, Telangana State, India

Wei Min Li Department of Vector Pest Control, Nanchang Center for Disease Control and Prevention, Nanchang, Jiangxi, China

M-Y Lin AVRDC—The World Vegetable Center, Tainan, Taiwan

Vijay Kumar Lingaraj Department of Agricultural Entomology, College of Agriculture, UAS (B), Mandya, Karnataka, India

Xiao Qing Liu Department of Vector Pest Control, Nanchang Center for Disease Control and Prevention, Nanchang, Jiangxi, China

Hong Mei Ma Department of Vector Pest Control, Nanchang Center for Disease Control and Prevention, Nanchang, Jiangxi, China

Ravindra Mahadappa Aurade Division of Entomology and Nematology, Indian Institute of Horticultural Research, Bangalore, Karnataka, India

S. S. Madankar Department of Entomology, Dr. Panjabrao Deshmukh Krishi Vidyapeeth, Akola, India

P. N. Magar Department of Entomology, Dr. Panjabrao Deshmukh Krishi Vidyapeeth, Akola, India

M. Manamohan Division of Biotechnology, Indian Institute of Horticultural Research, Bangalore, India

Vu Manh Hai Vietnam Academy of Agricultural Sciences, Hanoi, Vietnam

D. Manjunath Department of Studies in Sericulture Science, University of Mysore, Mysore, India

M. K. Meegahakumbura Genetics and Plant Breeding Division, Coconut Research Institute, Lunuwila, Sri Lanka

Deepak Mehta Biological Control Research Laboratory, Anand Agricultural University, Anand, Gujarat, India

Naresh Meshram Division of Entomology, Indian Agricultural Research Institute, New Delhi, India

K. Srinivasa Murthy National Bureau of Agriculturally Important Insects, Bangalore, Karnataka, India

N. Nagaraju University of Agricultural Sciences GKVK, Bangalore, Karnataka, India

C. Manja Naik Department of Entomology, University of Agricultural Sciences, Bangalore, India

Prakash M. Navale Division of Biotechnology, Indian Institute of Horticultural Research, Bangalore, India

Rakshit Ojha National Bureau of Agricultural Insect Resources, Bangalore, Karnataka, India

A. S. Padmaja University of Agricultural Sciences GKVK, Bangalore, Karnataka, India

Deepak Panpatte Biological Control Research Laboratory, Anand Agricultural University, Anand, Gujarat, India

Noushad Parvez Biological Control Research Laboratory, Anand Agricultural University, Anand, Gujarat, India

Ankit Patel Biological Control Research Laboratory, Anand Agricultural University, Anand, Gujarat, India

Leena Pathak Biological Control Research Laboratory, Anand Agricultural University, Anand, Gujarat, India

Siddanagowda Ujjanagowda Patil Zonal Agricultural Research Station (ZARS), Bramhavar, Karnataka, India

Department of Agricultural Entomology, College of Agriculture, UAS (B), Mandya, Karnataka, India

Kamala Jayanthi PD Division of Entomology and Nematology, Indian Institute of Horticultural Research, Bangalore, Karnataka, India

L. Perera Genetics and Plant Breeding Division, Coconut Research Institute, Lunuwila, Sri Lanka

Kangayam M. Ponnuvel Seribiotech Research Laboratory, Bangalore, India

R. N. Pushpa University of Agricultural Sciences GKVK, Bangalore, Karnataka, India

S. L. Ramya National Bureau of Agriculturally Important Insects, Bangalore, Karnataka, India

G. V. Ranga Rao International Crops Research Institute for the Semi-Arid Tropics, Patancheru, Telangana State, India

CGP Rao Seribiotech Research Laboratory, Bangalore, India

Mohammad Hasan Safaralizadeh Department of Plant Protection, Agricultural Faculty West Azarbayjan province, Urmia University, Urmia, Iran

K. L. Sahrawat International Crops Research Institute for the Semi-Arid Tropics, Patancheru, Telangana State, India

M. Saravanaraman Department of Entomology, Faculty of Agriculture, Annamalai University, Annamalainagar, Tamil Nadu, India

Sirigineedi Sasibhushan Seribiotech Research Laboratory, Bangalore, India

N. S. Satpute Department of Entomology, Dr. Panjabrao Deshmukh Krishi Vidyapeeth, Akola, India

V. Selvanarayanan Department of Entomology, Faculty of Agriculture, Annamalai University, Annamalainagar, Tamil Nadu, India

Hari C. Sharma International Crops Research Institute for the Semi-Arid Tropics (ICRISAT), Patancheru, Telangana, India

O. P. Sharma Plant Protection, NCIPM, New Delhi, India

P. R. Shashank Division of Entomology, Indian Agricultural Research Institute, New Delhi, India

S. T. Shinde Department of Entomology, Marathwada Krishi Vidyapeeth, Parbhani, India

Shama Singh Department of Zoology, University of Delhi, Delhi, India

Department of Genetics, M. D. University, Rohtak, Haryana, India

K. Sreedevi Division of Entomology, Indian Agricultural Research Institute, New Delhi, India

R. Srinivasan AVRDC—The World Vegetable Center, Tainan, Taiwan

F-C Su AVRDC—The World Vegetable Center, Tainan, Taiwan

N. E. Thyagaraj Department of Entomology, University of Agricultural Sciences, Bangalore, India

Timmanna Department of Entomology, University of Agricultural Sciences, Bangalore, India

Henri EZ Tonnang Crop Management and Production Systems Division, International Potato Center (CIP), Lima 12, Peru

Ramesh Babu Veegala Crop Protection Division, CRIJAF, Barrackpore, West Bengal, India

N. Veena Forest Wood Protection Division, Institute of Wood Science and Technology, Bangalore, India

Shashi Vemuri AINP on Pesticide Residues, EEI Premises, ANGRAU, Rajendranagar, Andhra Pradesh, India

T. Venkatesan National Bureau of Agricultural Insect Resources, Bangalore, Karnataka, India

Abraham Verghese National Bureau of Agricultural Insect Resources (NBAIR) (erstwhile Project Directorate of Biological Control) (ICAR), Bangalore, Karnataka, India

VV Vidhu Department of Zoology, University College, University of Kerala, Thiruvananthapuram, Kerala, India

S. P. Wani International Crops Research Institute for the Semi-Arid Tropics, Patancheru, Telangana State, India

Sopana Yule AVRDC—The World Vegetable Center, East and Southeast Asia, Research and Training Station, Kasetsart University, Nakhon Pathom, Thailand

Wei Qing Zheng Department of Vector Pest Control, Nanchang Center for Disease Control and Prevention, Nanchang, Jiangxi, China

About the Editor

Dr. A. K. Chakravarthy is Professor, University of Agricultural Sciences, GKVK, Bangalore. With over three decades of experience of teaching, research and extension, Dr. Chakravarthy has been the investigator for over 30 research projects and has guided more than 20 students for post graduation. He has received a Ph.D. from Punjab Agricultural University and is a Fellow from IARI, New Delhi. He is a member of several National and International scientific academia, an advisor, a panelist, a referee, reviewer, editor and is associated with publication of 30 national and international journals worldwide. A field oriented, widely travelled biologist, he is actively working on novel approaches in Integrated Pest Management, Host Plant Interaction, Vertebrate Pest Management, Biodiversity and environmental conservation issues.

Part I

Insect Taxonomy

Insect Taxonomy—Basics to Barcoding

K. Sreedevi, Naresh Meshram and P. R. Shashank

Abstract

The integration of new knowledge and methods of population biology, phylogenetics, and other evolutionary disciplines into taxonomy is warranted (Sites and Marshall, Trends Ecol Evol 18:462–470, 2004). The analysis and interpretation of data used to delimit species have profound implications in taxonomic research. Integrative taxonomy gives priority to species delineation over the creation of new species names. The integration of all possible taxonomic approaches abridging the gaps of each in arriving at correct species delimitation is the need of the hour in the light of biodiversity inventory. Taxonomy needs to be pluralistic to improve species discovery and description, and to develop novel protocols to produce the much-needed inventory of life in a reasonable time. Insects, being vast and diverse on earth , need much more integrative taxonomic attention than other life systems. The unique characters of an organism that unravels the diagnostic character differences that delimit the species have to be assessed holistically.

Keywords

Evolutionary disciplines · Holistic approach · Population biology · Phylogenetics

Introduction

Insects, ever since their appearance 350 million years ago, are the dominant species in the biotic community widespread in all habitats of the earth. The various studies of insects prevail from ages to the modern era. However, the taxonomic studies originated in the eighteenth century with Carolous Linneaus work on *Systema Naturae*, first published in 1735. Aristotle was the first to recognize the hierarchical pattern in the diversity of animals but the scientific method of classification was put forth by Linneaus followed by workers like Latreille, Fabricius, etc. This paved way for the emergence of taxonomy and with Charles Darwin's theory of natural selection, systematics gained the momentum.

K. Sreedevi (✉) · N. Meshram · P. R. Shashank
Division of Entomology, Indian Agricultural Research Institute, New Delhi 110012, India
e-mail: kolla.sreedevi@gmail.com

A. K. Chakravarthy (ed.), *New Horizons in Insect Science: Towards Sustainable Pest Management*,
DOI 10.1007/978-81-322-2089-3_1, © Springer India 2015

Systematics forms the basis for any life science studies and advancement as it is important to all other fields of biology. It builds up the information on biodiversity of species. Taxonomy, a part of systematics is *the* basic scientific discipline of biology that gives the identity and background of the organism, on which all other related sciences rely. Systematics, in addition to classification and naming, also deals with the relationships and environmental adaptations, thus drawing attention to the evolution and phylogeny. The morphological, physiological, ecological, behavioral, geographic, and molecular characters of the organism, in aggregate, are considered for the holistic approach in systematics.

Insect Diversity

Now-a-days there is a good understanding of many things on the planet Earth and also the Universe but we are still lagging behind on knowledge of how many species or life exist on the earth. The term "biodiversity", coined by Wilson (1988) as a contraction of "biological diversity", represents the diversity of life at all levels including genetic, species, and ecosystem diversity and is the core of natural resources for sustainable development and biotic capital for sustenance of life-support system (Kim and Byrne 2006).

The fundamental unit of biodiversity—species—serves as focal point for studying the full panoply of life, allowing workers to zoom in and out along a scale from molecule to ecosystem. The species-centered view also provides a vital focus for conserving life forms and understanding the causes of declining biodiversity (Alder and Foottit 2009). The process of discovery and description of all species is at very slow pace. This now appears unlikely to resolve the question in the near future, if at all, for a variety of reasons such as the slow rate of description of new species, the high level of synonymy for most groups and the uneven distribution of taxonomic effort which results in deficits in the known number of species for many species-rich groups like insects (Stork 1993).

Most eukaryote species awaiting description are insects (Raven and Yeates 2007), which are the world's most diverse group of animals, making up more than 58 % of the known global biodiversity. The insects are known to be the most successful and diverse animals on earth and are closely associated with our lives and affect the welfare of humanity in diverse ways. At the same time, large number of insect species, including those not known to science, continue to become extinct or extirpated from local habitats worldwide. Our knowledge of insect biodiversity is far from complete. Insects are the most exuberant manifestation of earth's many and varied life forms. The members of the class Insecta, arranged in 29 orders with more than 1 million described species (Grimaldi and Engel 2005; Arillo and Engel 2006), deserve serious attention of the taxonomists. Four of these orders—the Coleoptera, Diptera, Hymenoptera, and Lepidoptera—account for 81 % of all the described species of living insects. A growing number of world checklists and catalogs are available online for various families and orders, yet many to find a place. Outfitted with search functions, they provide another tool for handling the taxonomic juggernaut of new species and nomenclatural changes. We can foresee a global registry of species in the near future that is updated with each new species or synonym, allowing real-time counts for any taxon. It is imminent that insect biodiversity research must take cognizance of its material, taxonomy, insect pest management related requirements in light of recent developments viz molecular taxonomy, bioinformatics, information technology, and other advancements (Ramamurthy 2003). Though 2010 has been designated International Year of Biodiversity by the Convention on Biological Diversity and the United Nations (Johns 2010), taxonomy, that strengthens the knowledge of biodiversity, is on constant decline (Wilson 1985; 2004).

Need for Insect Taxonomy

Huge is the biodiversity of insects and little is known from all the fronts. Species identity and information is the foremost step for advanced studies in any direction and that's where systematics has a big role to play in. Despite the ongoing biodiversity crisis, the number of new species described per scientist has not increased in the past 60–70 years, which has a huge impact on conservation science (Terry Sunderland 2012). Many species will become extinct before they are described and one will remain continually unaware of the total numbers of species that comprise global biodiversity and this is acknowledged by the Convention of Biodiversity and its signatories as a "taxonomic impediment" (Terry Sunderland 2012). Taxonomy enables the facilitation of certain conservation issues like endangered species, species richness estimates, etc. for sustainable management of the natural resources in a better way. An account on the role of taxonomy in species conservation was given by Mace (2004).

Taxonomy not only produces fascinating knowledge on the characteristics of life but also delivers basic and indispensible knowledge for many fields of human interest and contributes in many ways to the sustainability of our planet. Research in taxonomy and systematics involves the study of virtually all available specimens of a taxonomic group in order to ensure comprehensive treatment, and is dependent on the availability of well curated collections. In the course of these studies, species previously unrecognized are frequently discovered. A single holotype specimen designated for each species is the standard of definition for that species. Much of research in biology is ultimately dependent on the scientific name of the species.

Taxonomy is the pivotal but hidden service behind sectors ranging from conservation and management of biodiversity to food security, poverty reduction, health, biosecurity, new industrial product development, and ecotourism. Wherever evolutionary history is relevant to a problem, systematics provides the resources. Necessarily, then, systematics is at the leading edge of the study of evolutionary biology; and its central position is assured because new contributions of molecular and genetic research to understanding the evolution of species have to be related to the broader systematic concepts of the taxa concerned. Interpretation in biogeography depends substantially on an understanding of evolutionary history, and consequently is closely allied to systematics; the basic data for biogeography analysis are derived from specimens in collections. Global systematic work on insects has a great deal to contribute to the understanding of evolutionary and geological events in the distant past.

Impacts of taxonomy on society are often beneficial, sometimes in unpredictable ways. A case study related to the description of a new mealybug species in Africa on cassava reveals the importance of taxonomy (Smith et al. 2011). Cassava (manioc or tapioca; *Manihotesculenta*), a drought resistant, staple food crop for over 200 million people in sub–Saharan Africa was infested with new mealybug species, since described as *Phenacoccusmanihoti* in 1973. As a result of misidentification and misdirected pest eradication efforts, initial attempts to control this pest using natural enemies failed, severely impacting the livelihoods of millions of people. Correct taxonomic identification came into rescue in locating its natural enemy, *Anagyruslopezi*, which established successfully by 1990 in 25 African countries implying a cost/benefit ratio between 1:200 to over 1:600 (Smith et al. 2011). This is one of the examples where taxonomy has helped. In reality, many taxonomic works did not get highlighted as it does not deal directly in management of insect pests on crops, humans, and animals. Nevertheless, there will not be a proper applied research on an insect without taxonomic details.

The traditional taxonomy provides the most convenient and authentic classification based on the overall similarities, most visible characters between species. It is pivotal in species recognition (with identification keys) and management of biological collections. At the beginning, classification work was restricted to just taxonomic details of the organism without considering the degree of relatedness between species. Later in 1950s, the phylogenetic classification cropped up to take care of the evolutionary history of the organism. The different schools, (part of conventional taxonomy) that differ in their concepts of phylogenetic classification but still converge on the basis of morphological similarities between species, are presented hereunder.

Conventional Taxonomy

1. Evolutionary or traditional taxonomy
2. Phenetics or numerical taxonomy
3. Cladistics or phylogenetic systematics
4. Cladoendesis

Evolutionary Taxonomy

Evolutionary taxonomy, originated in early twentieth century, attempts to classify the organisms based on phylogenetic relationships coupled with degree of evolutionary changes. It takes taxon into consideration rather than a species. The characters differ in information content regarding phylogeny and hence have different weights. Both recency of phyletic splitting and rate of divergence are given importance. Evolutionary taxonomy evolved through the influence of theory of evolution on Linnean classification during post Darwinian period where the tree of life gained importance in scientific works with publication of *The Origin of Species*. In 1930s, few biologists developed a Mendelian framework for Darwinian evolutionary theory, result of which was the evolutionary synthesis (Marc Ereshefsky 2007). Theodore Dobzhansky, Ernst Mayr, and Gaylord Simpson were the few of evolutionary taxonomists. The school consists of two tenets, firstly all taxa being a genealogical lineage and secondly constructing classification that reflects both cladogenesis (branching) and anagenesis (divergence). In cladogenesis, speciation occurs with the selection pressure (genetic revolution) on the isolated population from the rest of the species where single lineage splits into two branches whereas in anagenesis, speciation occurs in a single lineage. As a result, evolutionary taxonomists see two types of taxa viz., monophyletic and paraphyletic taxa arising from the processes of speciation through cladogenesis and anagenesis, respectively. In brief, evolutionary taxonomists believe that classifications should highlight only genealogical taxa, and those taxa can be either monophyletic or paraphyletic (Marc Ereshefsky 2007).

Phenetics or Numerical Taxonomy

Phenetic systematics determines the relationships of organisms through a measure of similarity, considering plesiomorphies (ancestral traits) and apomorphies (derived traits) to be equally informative. It aims at natural classification using numeric algorithms like cluster analysis rather than using subjective evaluation of properties. A priori every character is given equal weight. From the twentieth century onward, it was superseded by cladistics, which considers plesiomorphies to be uninformative for an attempt to resolve the phylogeny of earth's various organisms through time. Today's systematists generally make extensive use of molecular biology and computer programs to study organisms. An alternative to these matrix methods in phylogenetics and systematics is cladoendesis.

Cladistics or Phylogenetic Systematics

Cladistics got conceptualized in second half of the twentieth century and is termed as Phylogenetic systematics by Willi Hennig (also the title of his 1966 book). In cladistics, classification is mainly based on common ancestry and hence, believes in cladogenesis, where two taxa originated in the same branching event have a common ancestor that is not shared by any other taxon. Thus, cladistics represents only monophyletic taxa in their classifications. Those who follow cladistics perceive that the concepts of phenotypic difference and adaptive zone are ambiguous and are applied inconsistently to different types of taxa (Hennig 1966; Eldredge and Cracraft 1980). Cladists believe that the concepts of phenotypic diversity and adaptive zone are too malleable and reject them as grounds for classifying taxa (Marc Ereshefsky 2007). A group of cladists developed the Phylocode—a phylogenetic code of biological nomenclature, which is considered alternative to the Linnaean system (Cantino and de Queiroz 2004). The widely used and popular phylogenetic approach is cladistics.

Cladoendesis—New Approach to Phylogenetic Construction

The term cladoendesis was introduced by N. J. Kluge in early twenty-first century, meaning "branch coupling" that pays more attention on the connection between apomorphies of each taxon and characteristics of higher taxa, so that the characters of all the taxa are, from the very beginning, considered to be interrelated within a certain hierarchy (Kluge 2012). It is a method of phylogenetic analysis opposed to various matrix methods. The phylogenetic trees are not built each time as new ones but reconstructed based on the previous results where each character of each taxon is compared with its ancestral condition in the ground plan of the higher taxon (Kluge 2012). Cladoendesis enables understanding of nature and evolution of metamorphosis in insects.

All the schools eventually yield phylogenetic trees, a visual representation of the fact that species are related by descent from a common ancestor. Depending upon the type of school used in the construction, it may be called as phenograms (arrived from phenetics) or cladograms (arrived from cladistics).

Molecular biology has taken the systematics towards a different turn. The convergence of unrelated in species under similar selection pressure and divergence of related in species under different selection pressure yields to surge toward molecular taxonomy as an additional tool in support of traditional taxonomy in diagnostics. Recent advances in molecular techniques have greatly helped to resolve the controversial classification schemes based either largely or entirely on morphological attributes (Viraktamath 2003).

Molecular Systematics

Early attempts at molecular systematics were also termed as chemotaxonomy that made use of proteins, enzymes, carbohydrates, and other molecules that were separated and characterized using techniques such as chromatography. These have been replaced in recent times largely by DNA sequencing, which produces the exact sequences of nucleotides or *bases* in either DNA or RNA segments extracted using different techniques.

The theoretical framework for molecular systematics was laid in 1960s which plunged into DNA–DNA hybridization during 1974–1986. The advantage claimed for using hybridization rather than gene sequencing was that it was based on the entire genotype, rather than on particular sections of DNA. Another application in molecular phylogeny is DNA barcoding, wherein the species of an individual organism is identified using small sections of mitochondrial DNA (mtDNA) or chloroplast DNA that demarcates species as lineages (Hebert et al. 2003).

Insect Mitochondrial DNQA and DNA Barcoding

Mitochondrial genes are often chosen for evolutionary studies as they have a number of positive characteristics like: (i) maternal inheritance with little or no recombination (ii) general conservation of gene order and composition (iii) small size compared with the nuclear DNA and (iv) the lack of introns (Gissi et al. 2008). mtDNA have proven to be informative in the study of species diversity and evolutionary processes because it is easy to isolate and contains conserved sequences that make it possible to use as universal primers (Otranto and Stewens 2002; Xie et al. 2006).

Extensive studies in the mtDNA of *Drosophila* species showed that the same genes are present in both mammals and invertebrates; however, their arrangements may differ (Clary and Wolstenholme 1985; Crozier and Crozier 1993). The protein coding genes are the most frequently sequenced mitochondrial genes for evolutionary studies and phylogenetic analysis. Protein coding genes commonly analyzed include; COI, COII, 16S, and 12S. In particular, the COI gene has been widely sequenced. Yet, the specific region chosen has varied from study to study (Lunt et al. 1996; Caterino et al. 2000). The first subunit of mtDNA CO gene, corresponding to nucleotides 1490–2198 of the *D. yakuba* sequence, has been identified as an area of interest for "DNA barcoding" (Hebert et al. 2003). This region has a rate of molecular evolution that is about three times that of 12S or 16S rDNA, its third position nucleotides showing a high incidence of base substitutions. The success of universal primers for this gene enables the analysis of amino acid substitutions to initially designate an unidentified organism to a higher taxonomic group before examining nucleotide substitutions to determine its species identity (Hebert et al. 2003).

Hebert et al. (2003) named this technique "DNA barcoding." Then, the Barcode of Life project was proposed to promote DNA barcoding as a global standard for sequence-based identification of eukaryotes. Recently, the Barcode of Life project entered a new phase with the launch of the International Barcode of Life project (IBOL; International Barcode of Life 2012). The IBOL is a huge international collaboration of 26 countries that aims to establish an automated identification system based on a DNA barcode library of all eukaryotes. In the first 5 years, the IBOL will focus mainly on developing a barcode library, including 5 million specimens of 500,000 species. The DNA sequences are used as genetic "barcodes" that may potentially be used as a bioidentification system for all animals and have proven to be a useful identification tool for vertebrates such as birds (Hebert et al. 2004), fish (Ward et al. 2005), and hexapod orders such as Lepidoptera (Hajibabaei et al. 2006), Coleoptera (Greenstone et al. 2005), Diptera (Smith et al. 2007), Hymenoptera (Smith et al. 2008), Ephemeroptera (Ball et al. 2005), and Hemiptera (Lee et al. 2011). DNA barcoding does not substitute but complement conventional taxonomical studies.

The three main taxonomic applications that DNA barcoding has been previously used are: (1) the identification of species previously defined by other criteria, including rapid identification, as well as linking specimens to established species that are unidentifiable by other means; (2) the description of new species by interpreting DNA diversity as an indicator of species diversity; (3) the definition of operational units for ecological studies (Rubinoff 2006).

DNA should be an excellent tool for inferring phylogenies: large number of homologous characters that should be less subject to convergent evolution than other characters that might lead to a confusion of grade and clade. A character can be phylogenetically informative when nucleotide changes are shared by two or more taxa. A character can be phylogenetically uninformative when all nucleotides are the same among taxa, or when only a single taxon has a different nucleotide.

According to Hillis et al. (1996) three main applications of molecular systematics are,

1. Reconstruction of phylogenetic relationships of organisms.
2. Studies of population structure, including geographic variation, mating systems, heterozygosity, and individual relatedness.
3. Identification of species boundaries including hybridization.

There are many methods to understand these molecular variations; few molecular methods with their applications were explained in Table 1

Molecular Markers Used in Systematics

A diverse range of novel molecular (DNA) markers are now available for taxonomic investigations. Both DNA and protein markers have revolutionized the biological sciences and have enhanced many fields of study, especially systematics. This has been possible because of the rapid advances in molecular biological methods and bench-level protocols for wider application. The utility of molecular markers as additional tools in systematic has led to "molecular systematics". Over a long time, significant contributions have been made in the field of insect systematics through morphometric traits, wherein a number of difficulties were encountered due to genotype–environment interactions. The limitations in using morphological, physiological, and cytological markers for assessing genetic diversity and population dynamics have been largely circumvented by the developments in DNA-based markers. Molecular markers, by nature, are neutral to the stage of development, physiological status, and environmental influences. Isozymes and other proteins as markers are often expressed codominantly and discriminate homozygous and

Table 1 Some methods presently employed in identification of the species. (Adopted from: Singh (2012))

Methods	Explanation
Hybridization	Genetic materials from two different species are subjected to hybridize. Closely related species show higher percentage of hybridization
DNA sequencing	DNA segments of two species are sequenced from one end to the other and the sequences of the two form the basis of establishing similarity or dissimilarity between them
Restriction mapping	Segments of DNA are isolated from different species and subjected to restriction mapping. Closely related species will have more similar restriction map
Chromosome banding	The chromosomes of different species are examined through microscope. Banding of chromosomes are also done for taxonomic purposes
Amino acid sequencing	Like DNA sequencing, protein sequencing is also done. The amino acid sequence of a given protein will be more similar between closely related species
Immunological methods	Antibodies that recognize specific macromolecules, usually on the cell surface are tested on different species. Antibodies that recognize macromolecules form one species will often recognize closely related species, but not from distantly related species

heterozygous individuals. However, the limited number of proteins and isozymes as markers and requirement of different protocols for each enzyme/protein limit their utility. Unlike morphological and protein-based markers, several DNA based markers are available to elicit the differences between individuals and populations, or they can be developed for each specific purpose. Although a large number of samples can be analyzed quickly, a number of other factors such as cost, speed, and requirements of technical skills are the major concern. DNA-based markers can generate large amount of high quality data compared to several biochemical marker systems, but degree of polymorphism detected and the statistical dependability of the results vary among marker systems.

Integrative Taxonomy

Will et al. (2005) a slight critique of DNA barcoding, used the term integrative taxonomy to mean a taxonomic process that was inclusive of all available data sources and not just mtDNA *COI* barcode data. Dayrat (2005) defined "Integrative taxonomy" as the science that aims to delimit the units of life's diversity from multiple and complementary perspectives. Thus, any study linking different kinds of data by mapping morphological diversity on to a molecular phylogeny is integrative (Yeates et al. 2011).

Integrative taxonomy gives priority to morphological characters because of their greater complexity and presumed multigenic origin, which are believed to constitute a more secure basis for separating species than small fragments of DNA sequence. The limitation of molecular systematics—being an essentially cladistic approach, is that it assumes that classification must correspond to phylogenetic descent, and that all valid taxa must be monophyletic also lead to integrative taxonomy. The recent discovery of extensive horizontal gene transfer among organisms also provides a significant complication to molecular systematics (indicating that different genes within the same organism can have different phylogenies) necessitating integration of both conventional and molecular taxonomy for holistic approach.

Molecular taxonomy, on other hand, can facilitate easy and rapid identification of the species provided the gene sequence has been deposited after authentic identification of the species morphologically supported by all metadata and photographs. Consistency index of molecular data is higher than that of morphological data. But accurate identification always comes from the morphological characters and hence addition of the new species to the list of fauna is encouraged by conventional taxonomy, which can be supported by molecular taxonomy. Both approaches have issues and limitations, which can supplement and complement for holistic species identity.

Schlick-Steiner et al. (2010) have offered the most detailed treatment of the process of integrative taxonomy. They stressed that integrative taxonomy does not replace traditional taxonomy, but that it uses complementarity among disciplines to improve rigor. Their integrative procedure relies on an agreement among three "conclusive" disciplines that is both proscriptive and restrictive, and disciplines and datasets are defined rather arbitrarily. Several studies cited as examples of integrative taxonomy by Schlick-Steiner et al. (2010) used correlative approaches to compare molecular data and morphology (Malhotra and Thorpe 2005; Yoder et al. 2005; Rissler and Apodaca 2007; Roe and Sperling 2007). The determination of origin and their evolution trajectories of a species will trigger the species delimitation in a better manner (Padial et al. 2010).

Yeates et al. (2011) surveyed the current taxonomic methods (inclusive of both morphological and molecular characters) employed in delimiting the species and identified two challenges, which if met, will provide a more complete toolkit and a more robust research program in integrative taxonomy and proposed alternatively the term "iterative taxonomy" for the practice of treating species boundaries as hypotheses that are to be tested with evidence. Delineating the species boundaries accurately is crucial to discovery of life's diversity (Dayrat 2005).

Role of Information and Communication Technologies (ICTs) in Systematics

A biggest challenge of a systematist while dealing with the larger taxa with numerous species having possessing complex variations at species and population levels is to look into the character evaluation (Ramamurthy 2003). Evaluation of character itself is a complex task that involves computational methods in screening the characters (Ramamurthy 2003). Computer simulation and modeling help in resolving certain phylogenetic issues in building up the tree with intercession. Potential evolutionary relationships are being evaluated with the help of computer assisted programs based on character data (let it be morphological in terms of numerical values or DNA sequences).

Another dimension of ICTs in systematics is in the field of automation of information with the taxonomic expertise in their respective areas. Lyal et al. (2008) stressed the need of accessibility of user-friendly identification tools. Digitization of the biological collections and descriptions helps in buildup of virtual information for the benefit of all researchers and scientists. The transformation of the processes of science and its outcomes is possible only through ICTs which are adept at dissemination. The databases created would largely help the present and future workers in respective line of work.

The taxonomists mandate should be built up of databases, online publications of new taxa and monographs, development of interactive keys and virtual repositories, generation of distribution maps and predictive models utilizing both the approaches, conventional and molecular for the benefit of speedy growth.

Conclusion

We attempted in this chapter to present an overview of the genesis and the growth of taxonomy, both as a science and as a tool, of species identification and documentation. The integration of new knowledge and methods of population biology, phylogenetics, and other evolutionary disciplines into taxonomy is warranted (Sites and Marshall 2004). The analysis and interpretation of data used to delimit the species have profound implications in taxonomic research (Yeates et al. 2011). Dayrat (2005) stated that "integrative taxonomy" gives priority to species delineation over the creation of new species names and also opined that a radical change in mentality is needed concerning the creation of names in order to achieve this integration and to prevent the overabundance of both synonyms and names of doubtful application from worsening.

The integration of all possible taxonomic approaches abridging the gaps of each in arriving at correct species delimitation is the need of hour in the light of biodiversity inventory. Padial et al. (2010) stated that taxonomy needs to be pluralistic to improve species discovery and description, and to develop novel protocols to produce the much-needed inventory of life in a reasonable time. Insects, being vast and diverse on earth, need much more integrative taxonomic attention than any other life system. The unique characters of an organism that unravels the diagnostic character differences that delimit the species (Sites and Marshall 2003) have to be assessed holistically from all directions and taxonomic approaches. Perusal of various concepts and views on taxonomical approaches reaffirms that the conventional mode of species identification still holds the relevance and undoubtedly serves as the foundation stone which supports the latest approaches.

References

Adler PH, Foottit RG (2009) Insect biodiversity: science and society. Blackwell Publishing Ltd, Garsington Road, Oxford, UK, pp 623. ISBN: 978-1-405-15142-9

Arillo A, Engel MS (2006) Rock crawlers in Baltic amber (Notoptera: antophasmatodea). Am Mus Novit 3539:1–10

Ball SL, Hebert PDN, Burian SK, Webb JM (2005) Biological identifications of mayflies (Ephemeroptera) using DNA barcodes. J N Am Benthol Soc 24:508–524

Cantino PD, de Queiroz K (2004) Phylocode: a phylogenetic code of biological nomenclature. http://www.ohiou.edu/phylocode. Accessed July 2013

Caterino MS, Cho S, Sperling FA (2000) The current state of insect molecular systematic: a thriving tower of babel. Ann Rev Entomol 45:1–54

Clary DO, Wolstenholme DR (1985) The mitochondrial DNA molecule of *Drosophila yakuba*: nucleotide sequence, gene organisation and genetic code. J Mol Evol 22:252–271

Crozier RH, Crozier YC (1993) The mitochondrial genome of the honeybee *Apis mellifera*: complete sequence and genome organization. Genetics 133:97–117

Dayrat B (2005) Towards integrative taxonomy. Biol J Linn Soc 85(3):407–415

Eldredge N, Cracraft J (1980) Phylogenetic patterns and the evolutionary process. Method and theory in comparative biology. Columbia University Press, New York, viii + 349 pp

Gissi C, Iannelli F, Pesole G (2008) Evolution of the mitochondrial genome of Metazoa as exemplified by comparison of congeneric species. Heredity 101:301–320

Greenstone MH, Rowley DL, Heimbach U, Lundgren JG, Pfannenstiel RA, Rehner SA (2005) Barcoding generalist predators by polymerase chain reaction: carabids and spiders. Mol Ecol 14:3247–3266

Grimaldi DA, Engel M (2005) The evolution of insects. Cambridge University Press, Cambridge

Hajibabaei M, Janzen DH, Burns JM, Hallwachs W, Hebert PDN (2006) DNA barcodes distinguish species of tropical Lepidoptera. Proc Natl Acad Sci U S A 103:968–971

Hebert PDN, Cywinska A, Ball SL, Dewaard JR (2003) Biological identifications through DNA barcodes. Proc R Soc Lond B 270:313–321

Hebert PDN, Stoeckle MY, Zemlak TS, Francis CM (2004) Identification of birds through DNA barcodes. PLoS Biol 2(10):1657–1663

Hennig W (1966) Phylogenetic Systematics. Translated by Davis D, Zangerl R. University of Illinois Press, Urbana, Illinois, iii+263pp

Hillis, David M, Moritz C, Mable BK (1996) Molecular systematics, 2nd edn. Sinauer Associates, Inc., Sunderland

Johns D (2010) The international year of biodiversity—from talk to action. Conserv Biol 24:338–340

Kim CK, Byrne LB (2006) Biodiversity loss and the taxonomic bottleneck: emerging biodiversity science. Ecol Res 21:794–810

Kluge NJ (2012) Cladoendesis and new look at the evolution of insect metamorphosis. Entomol Rev 92(6):622–632

Lee W, Kim H, Lim J, Choi H, Kim Y, Yang-Su K, Jeong-Yeon J, Foottit RG, Lee S (2011) Barcoding aphids (Hemiptera: Aphididae) of the Korean Peninsula: updating the global data set. Mol Ecol Resour 11:32–37

Lunt DH, Zhang DX, Szymura JM, Hewitt GM (1996) The insect *cytochrome oxidase* I gene: evolutionary patterns and conserved primers for phylogenetic studies. Insect Mol Biol 5:153–165

Lyal C, Paul K, Smith D, Smith R (2008) The value of taxonomy to biodiversity and agriculture. Biodiversity 9(1–2):8–13

Mace MG (2004) The role of taxonomy in species conservation. Phil Trans R Soc Lond B 359:711–719

Malhotra A, Thorpe RS (2005) Erratum to "A phylogeny of four mitochondrial gene regions suggests a revised taxonomy for Asian pitvipers (Trimeresurus and Ovophis)". Mol Phylogenet Evol 32:83–100

Marc Ereshefsky (2007) Species, taxonomy and systematics. In: Ruse M (ed) Philosophy of Biology. Prometheus Books. Amherst, New York, pp 403–428

Otranto D, Stewens JR (2002) Molecular approaches to the study of myiasis causing larvae. Int J Parasitol 32:1345–1360

Padial JM, Miralles A, Ignacio De la Riva, Vences M (2010) The integrative future of taxonomy. Front Zool 7:16

Ramamurthy VV (2003) Information technology in insect biodiversity research. In: B Subramanyam, VV Ramamurthy, VS Singh (eds) Proceedings of the national symposium on frontier areas of entomological research. IARI, New Delhi

Raven PH, Yeates DK (2007) Australian biodiversity: threats for the present, opportunities for the future. Aust J Entomol 46:177–187

Rissler LJ, Apodaca JJ (2007) Adding more ecology into species delimitation: ecological niche models and phylogeography help define cryptic species in the black salamander (*Aneides flavipunctatus*). Syst Biol 56:924–942

Roe AD, Sperling FAH (2007) Population structure and species boundary delimitation of cryptic Dioryctria moths: an integrative approach. Mol Ecol 16:3617–3633

Rubinoff D (2006) DNA barcoding evolves into the familiar. Conserv Biol 20:1548–1549

Schlick-Steiner BC, Steiner FM, Seifert B, Stauffer C, Christian E, Crozier RH (2010) Integrative taxonomy: a multisource approach to exploring biodiversity. Annu Rev Entomol 55:421–438

Singh AK (2012) Molecular taxonomy: use of modern methods in the identification of a species. Ind J Life Sci 2(1):143–147

Sites JW, Marshall JC (2003) Delimiting species: a Renaissance issue in systematic biology. Trends Ecol Evol 18:462–470

Sites JW, Marshall JC (2004) Operational criteria for delimiting species. Annu Rev Ecol Evol Syst 35:199–227

Smith MA, Wood DM, Janzen DH, Hallwachs W, Hebert PDN (2007) DNA barcodes affirm that 16 species of apparently generalist tropical parasitoid flies (Diptera, Tachinidae) are not all generalists. Proc Natl Acad Sci U S A 104(12):4967–4972

Smith MA, Rodriguez JJ, Whitfield JB, Andrew RD, Janzen DH, Winnie H, Hebert PDN (2008) Extreme diversity of tropical parasitoid wasps exposed by iterative integration of natural history, DNA barcoding, morphology, and collections. Proc Natl Acad Sci U S A 105:12359–12364

Smith R, Rassmann K, Davies H, King N (2011) Why. taxonomy.matters. http://www.bionet†intl.org/why. BioNET-INTERNATIONAL, Egham, UK. Accessed 25 Oct 2013

Stork NE (1993) How many species are there? Biodivers Conserv 2:215–232

Terry Sunderland (2012) http://blog.cifor.org/8746/why-taxonomy-is-important-for-biodiversity-based-science#.VG2sqjSUfSM. Accessed June 2013

Viraktamath CA (2003) Insect molecular systematics. In: B Subramanyam, VV Ramamurthy, VS Singh (eds) Proceedings of the national symposium on frontier areas of entomological research. IARI, New Delhi

Ward RD, Zemlak TS, Innes BH, Last PR, Hebert PDN (2005) DNA barcoding Australian fish species. Philos Trans R Soc Lond B Biol Sci 360:1847–1857

Wilson EO (1985) The biological diversity crisis: a challenge to science. Issues Sci Technol 2:20–29

Wilson EO (1988) The current state of biological diversity. In: Wilson EO, Peters FM (eds) Biodiversity. National Academy Press, Washington, DC, pp 3–18

Wilson EO (2003) The encyclopedia of life. Trends Ecol Evol 18:77–80

Wilson EO (2004) Taxonomy as a fundamental discipline. Philos Trans R Soc Lond B Biol Sci 359:739

Xie L, Hong XY, Xue XF (2006) Population genetic structure of the two spotted spider mite (Acari: Tetranychidae) from China. Ann Entomol Soc Am 99:959–965

Yeates DK, Seago A, Nelson L, Cameron SL, Joseph L, Trueman JWH (2011) Integrative taxonomy or iterative taxonomy. Systematic Entomol 36:209–217

Yoder AD, Olson LE, Hanley C, Heckman KL, Rasoloarison R, Russell AL, Ranivo J, Soarimalala V, Karanth KP, Raselimanana AP, Goodman SM (2005) A multidimensional approach for detecting species patterns in Malagasy vertebrates. Proc Natl Acad Sci U S A 102:6587–6594

DNA Barcoding for Identification of Agriculturally Important Insects

S. K. Jalali, Rakshit Ojha and T. Venkatesan

Abstract

Molecular characterization and DNA barcoding is a taxonomic method that uses a short genetic marker in an insect DNA to identify a species, including an unknown species. The DNA barcode method of identification includes, for example, identifying insect species from any developing stage and part; otherwise, identifying insects morphologically generally depends on adult stage and male genitalia. The utility of DNA barcoding for these purposes is subject to debate; however, in insects at least, it is approximately 650 bp of the mitochondrial gene, cytochrome oxidase I (*COX I*). The approximate number of described insect species in India is 59,000; however, the number of barcodes generated from India is 4.6% of known species, while the corresponding global scenario is about 16% of described species; a lot of emphasis is required to catch up with the world scenario. In order to speed up taxonomic identification, DNA barcoding is now been considered as an alternative tool for insect biodiversity identification in India and the world. The present chapter deals with the use of barcode in the identification of insects belonging to different orders and families, using the neighbour-joining approach with the bootstrapping method and the Kimura-2 parameter to obtain a clear phylogenetic signal. In a neighbour-joining tree for all sequences, two clades were obtained: the first cluster consisting of hymenopteran insects and the other consisting of other orders. This phylogeny also agrees with the traditional phylogeny of insects. The present results, thus, favour DNA barcoding as a decisive tool in quick and reliable identification of insects.

Keywords

COX 1 · DNA barcoding · India · Phylogeny

Introduction

Insects are the most abundant of all life on earth and have evolved into a tremendous range of different forms. It took nearly 200 years for

S. K. Jalali (✉) · R. Ojha · T. Venkatesan
National Bureau of Agricultural Insect Resources, Post Bag No. 2491, H. A. Farm Post, Hebbal Bellary Road, Bangalore, Karnataka 560024 India
e-mail: jalalisk1910@yahoo.co.in

A. K. Chakravarthy (ed.), *New Horizons in Insect Science: Towards Sustainable Pest Management*,
DOI 10.1007/978-81-322-2089-3_2, © Springer India 2015

taxonomists to describe 10 % of the total number of species estimated. In this context identification of insects has been a monumental task where it calls for the availability of more number of specialists and funding. To catalogue the vast numbers of species, naturalists came up with the idea of classifying living beings on the basis of taxonomy, which is a branch of science that helps us to describe a living being on the basis of morphological features. After 250 years of Darwin and Linnaeus, a new method called DNA barcoding, a tool of DNA-based taxonomy is in current use to identify known and unknown species on the basis of the pattern of nucleotide arrangement in a fragment of DNA of a particular species (Novotny *et al.* 2002). Several researchers have suggested the use of DNA barcoding in taxonomy as a method to achieve rapid species descriptions in the context of the current biodiversity crisis (Hebert *et a*l. 2003a, b; *Ball* and *Armstrong* 2006). DNA barcoding is the use of a short standardized DNA sequence (in insects, a 658 bp fragment of the mitochondrial cytochrome *c* oxidase (*COX I*) gene) to identify and assign unknown specimens to species besides facilitating the discovery of new species. Wilson (2012) observed that library barcodes gain their value due to an intimate association, through voucher specimens from where they came, with other data, particularly, Linnaean names, collection localities, and morphology in the form of digital images. This tool is widely accepted all over the globe from hard-core taxonomists' to graduate molecular biologists and also well received by governmental and nongovernmental organizations to catalogue all the species on our planet. With the advent of molecular biology and molecular tools, identification of life forms, including insects has become quick, precise, and easy.

India is one of the world's most biodiverse regions, with a total land area of about 3,287,263 km^2, covering a variety of ecosystems ranging from deserts to high mountains and tropical to temperate forests. Insects are the most abundant of all life forms on earth. India with about 2 % of the global land area is among the top 12 mega biodiversity nations in the world accounting for 7.10 % of the world insect fauna. It is estimated that over 900,000 species of insects are known across the globe with over 60,000 species described from India with nearly as many species remaining to be named. Barcode of Life Datasystem (BOLD) Systems is populated with nearly 142,398 insect species barcodes out of which India has only 2758 barcodes; NBAII had 110 barcodes as on November 2013 (in six different insect orders; Fig. 1, Table 1).

To catalogue such vast diversity a simple, rapid, and accurate method is the current need. DNA barcoding is a tool that fulfils all the above said criteria to identify a specimen to species level. The rDNA internal transcribed spacers region 2 (ITS-2) (Ashok Kumar et al. 2009), cytochrome c oxidase subunit 1 (*COX 1*), NADH dehydrogenase subunit 1 (nadh1), and cytochrome b (cytb) markers used in recent molecular analysis have substantially increased our understanding of the phylogenetic relationships between insect species. However, cytochrome c oxidase subunit 1 (*COX 1*) has been used extensively by molecular biologists across the globe to discriminate insect species. In the present chapter, the major focus is on insect pests and parasitoids of agricultural importance in India.

DNA barcoding is an emerging tool, therefore, a reliable database has to be built by performing *COX 1* sequencing on specimens previously identified by a taxonomist. Therefore, a prerequisite for genetic investigations in this study will be the technical step of constructing a database of insect pests and natural enemies in India and the world.

Molecular identification and phylogeny using species identification markers using *COX 1* of the mitochondrial region is regarded as efficient. The main advantage of DNA barcoding is the rapid acquisition of molecular data (Monaghan et al. 2005). Mitochondria are energy-producing organelles, found in nearly every cell in nearly every plant and animal species. The mitochondrial genome in particular has turned out to be exceedingly useful in tracing evolutionary history, as it is present in all eukaryotic organisms, evolves rapidly as compared to nuclear DNA. Nuclear and mitochondrial genomes exhibit different patterns of inheritance (Behura 2006).

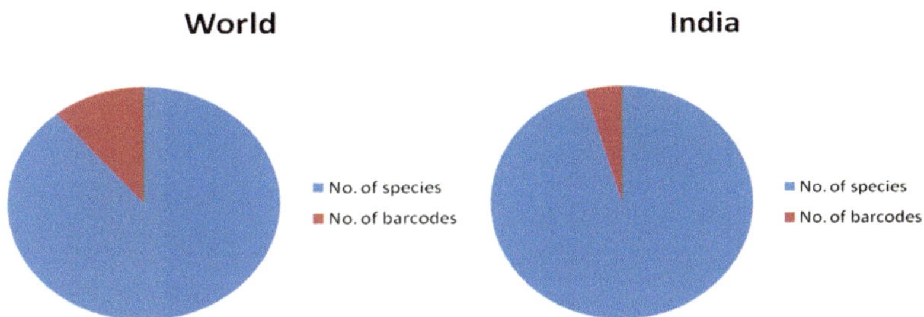

Fig. 1 Number of insect species and barcodes in India and the world

Table 1 Species barcodes in six different orders submitted to BOLD Systems by NBAII, Bangalore

Insect order	Number of insects barcoded
Lepidoptera	15
Diptera	14
Hemiptera	4
Coleoptera	26
Neuroptera	1
Hymenoptera	50
Total	*110*

Mitochondrial markers are used for revealing phylogenetic relationships among related groups, because mtDNA is maternally inherited, it evolves fairly rapidly, and most of the nucleotide substitutions occur at neutral sites. With respect to this genetic marker the intra- and inter-phylogenetic relationships have been studied using the sequence data obtained from the *COX 1* marker gene amplification. Relative homogeneity is maintained by concerted evolution, where mutations rapidly spread to all members of the gene family even if there are arrays located on different chromosomes (Arnheim 1983; Gerbi 1985; Tautz et al. 2002). However, these advantages are associated with a major drawback; while mitochondrial DNA was considered to be a neutral marker that reflects the history of the species, Ballard and Whitlock (2004) and Bazin *et al.* (2006) have recently argued that mitochondria are in fact often under strong selection and evolve under unusual evolutionary rules when compared with other genomes. Hurst and Jiggins (2005) suggested that selection can act directly on the mtDNA itself, but it can also arise indirectly from disequilibrium with other maternally transmitted DNA.

While morphological data are usually time-consuming and need specialists, DNA barcoding techniques are a uniform and practical method of species identification of insects and can be used for the identification of all developmental stages of insects, their food webs and biotypes and this may not be possible with morphology-based taxonomy. Tree-based taxon clustering as well as statistical taxon separation analysis indicates that molecular evidence does coincide with morphological hypotheses. Hence, species identification based on DNA sequence analysis proved to be feasible for the analysed taxa. DNA barcoding has the potential of being a valuable tool to biologists. It has helped to evolve many advanced tools as species diagnostics like species-specific primers developed for tea mosquito bugs (Rebijith et al. 2012) and mini barcodes for archival specimens.

Insect Mitochondrial Genome

The arrangement of genes in mt genomes has been studied in more insects than in any other group of invertebrates. So far, 15 species of insects have had their mt genomes sequenced completely; the mitochondria of insects contain their own double-stranded circular genomes (Fig. 2), which range from 14,503 bp (Beckenbach and Joy 2009) to 19,517 bp in size (Lewis et al. 1995).

The arrangement of genes in the mt genomes of insects studied so far are conserved since all species, except the wallaby louse, have the same arrangement of protein-coding and rRNA genes and most tRNA genes. Only the positions of a few tRNA genes differ, in particular, those in

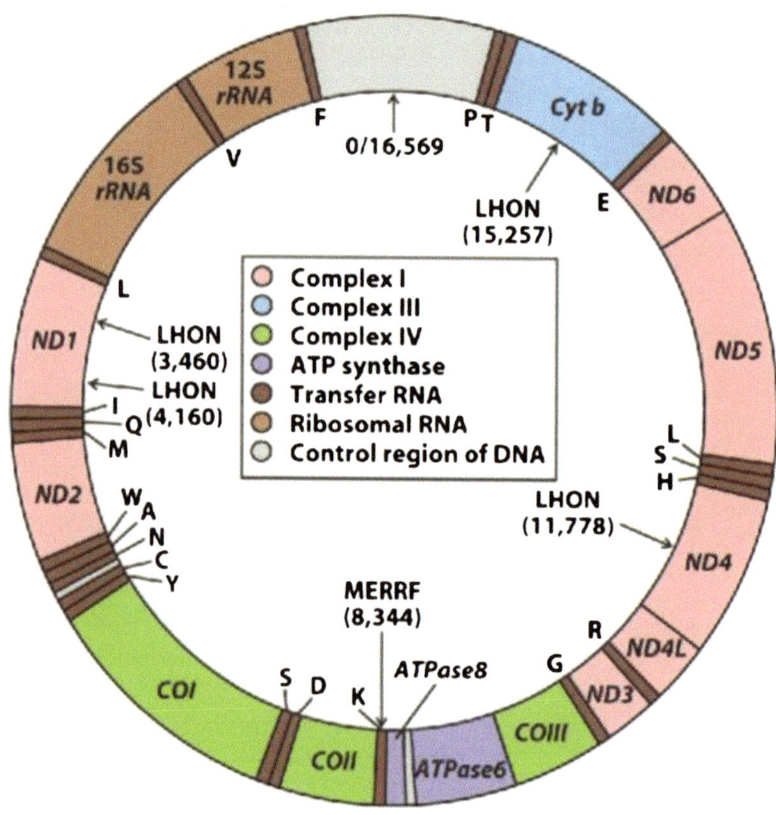

Fig. 2 Organization of insect mitochondrial genome (Source: http://chimerasthebooks.blogspot.in/2011_12_01_archive.html)

"hot spot" regions (Dowton and Austin 1999), e.g., near the control region, and in the two clusters of tRNA genes, *trnK-trnD* and *trnA-trnR-trnNtrnS1-trnE-trnF*. The most common arrangement of the 37 genes in the mt genome, which is present in the fruit fly *Drosophila yakuba,* the bug *T. dimidiata,* and many other species, is inferred to be ancestral for insects (Boore and Brown 1998; Crease 1999).

Agriculturally Important Insects

Insects have been used in landmark studies in biomechanics, climate change, developmental biology, ecology, evolution, genetics, paleolimnology, and physiology. Because of their diversity and many roles, they are familiar to the general public. However, their conservation is a challenge. The goal of this chapter is to document agriculturally dominant insect pests, their natural enemies, pollinators, and veterinary insect pests.

Importance of DNA Barcoding in Agricultural Entomology

Insect pests are a major concern for farmers across the world and more than 10,000 species of insects have been recorded damaging crops (Dhaliwal et al. 2007). Sometimes the yield loss by insects reaches as high as 60–70 % and it is reported that Indian agriculture is currently suffering an annual loss of about ₹ 86.39 million due to insect pests (Dhaliwal et al. 2007). An automated DNA-based system will free taxonomists from routine identifications, allowing them to direct their efforts to new collections, descriptions, and assessments of taxonomic relationships. In 2003, Paul D.N.

Hebert from the University of Guelph, Ontario, Canada, proposed the compilation of a public library of DNA barcodes that would be linked to named specimens. This library would "provide a new master key for identifying species, whose power will rise with increased taxon coverage and with faster, cheaper sequencing." The goal of a DNA barcoding library is the construction of an enormous, online, freely available sequence database. Participants in the DNA barcode initiative come in many configurations, including consortia, databases, networks, labs, and projects that range in size from local to global. An extensive survey is required to catalogue all the insect pests for better understanding of their habits and habitat so that proper measures can be used to control them. In recent years a new taxonomic approach called "DNA barcoding" has been proposed to aid the determination of species. This method suggests that large-scale surveys of DNA variation would accelerate studies on ecology, biodiversity, and conservation planning of poorly studied ecosystems and groups of organisms. Recently, several museums, herbaria, universities, biodiversity inventory agencies, and commercial experts have created the international consortium Barcode of Life (CBOL). The use of DNA sequence for species recognition, assessment, and taxonomic description, these include taxonomic accuracy, low cost, ease of application in diverse contexts (including by non-specialists), portability, routine and immediate access to information, and utility across a broad phylogenetic and taxonomic spectrum of organisms, including many species new to science.

Materials and Methods

Collection and Identification

Insects were collected from various ecosystems across India. At each site, insects were collected by brush and cotton wool pad and sweeping net methods and transferred to collection tubes containing 95 % alcohol. The specimens were identified to species level and distributed into their respective families to obtain a clear phylogenetic signal (Table 2) immediately upon their collection.

The specimens, thus collected and morphologically identified, were used for *COX 1* barcoding at the National Bureau of Agriculturally Important Insects (NBAII) Bangalore, India.

Genetic Analysis

The DNA was extracted from somatic tissue rich in mitochondria (e.g., thorax and upper abdominal region) using Qiagen DNeasy® Kit, following the manufacturers' protocols. The remaining parts of each of the insects and respective individuals were kept as voucher specimens at NBAII. The extracts were subjected to PCR amplification of a 658 bp region near the 5′ terminus of the *COX 1* gene following standard protocols. Primers used were forward primer: (LCO 1490 5′-GGTCAACAAATCATAAAGATATTG G-3′) and reverse primer: (HCO 2198 5′-TAAACTTCAGGGTGACCAAAAAATCA-3′).

PCR reactions were carried out in 96-well plates with 50 µl reaction volume containing GeNeiTM Taq buffer 5 µl, 1 µl of GeNeiTM 10 mM dNTP mix, 2.5 µl of (20 pmol/µl) forward primer, 2.5 µl of (20 pmol/µl) reverse primer, 1 µl of GeNeiTM Taq DNA polymerase (1 U/µl), DNA (50 ng/µl) 2 µl, and sterile water 36 µl. Thermocycling consisted of an initial denaturation of 94 °C for 5 min, followed by 30 cycles of denaturation at 94 °C for 1 min, annealing at 55 °C for 1 min, extension at 72 °C for 1 min. PCR was performed using a C1000™ Thermal Cycler. The amplified product was analysed on a 1.5 % agarose gel electrophoresis as described by Sambrook and Russell (2001). The amplified products were sent to commercial sequencing company, M/s Eurofins Pvt Ltd. India. Each species was bidirectionally sequenced and checked for quality by Bioedit 7.0.2 software and homology, insertions and deletions, stop codons, and framshifts by using NCBI BLAST. All sequences were uploaded to GenBank and the BOLD (http://www.boldsystems.org).

Table 2 Distribution of insect species into their respective subfamilies and tribes on the basis of classification

Order	Family	Insect	Genbank accession numbers	Barcode ID
Hemiptera	Agromyzidae	*Phytomyza orobanchia*	KC732453	AGIMP017–13
	Aphididae	*Xylocoris flavipes*	KF365462	Not available
	Anthocoridae	*Blaptostethus pallescens*	KF365463	Not available
Diptera	Tephritidae	*Acroceratitis histrionica*	KF471502	*Not available*
		Bactrocera correcta	KF289766	AGIMP022–13
		Bactrocera dorsalis	KF289767	AGIMP023–13
		Bactrocera zonata	KF289768	AGIMP024–13
Lepidoptera	Pyralidae	*Chilo auricilius*	KC306949	AGIMP003–12
		Chilo partellus	KC911712	AGIMP007–13
		Chilo sacchariphagus indicus	KC306951	AGIMP005–12
		Conogethes punctiferalis	KF114864	AGIMP012–13
		Galleria mellonella	KF289770	AGIMP026–13
		Polyocha depressella	KC306950	AGIMP004–12
		Scirpophaga excerptalis	KC306948	AGIMP002–12
	Noctuidae	*Bombyx mori*	JX025640	BMSW002–12
		Corcyra cephalonica	KF289769	AGIMP025–13
		Helicoverpa armigera	KC911713	AGIMP008–13
	Plutellidae	*Plutella xylostella*	KC911716	AGIMP011–13
	Galleriidae	*Sesamia inferens*	KC911715	AGIMP010–13
	Bombycidae	*Spodoptera litura*	KC911714	AGIMP009–13
Coleoptera	Scolytidae	*Euwallacea fornicates*	KC590061	AGIMP027–13
	Anobiidae	*Stegobium panecium*	KF471501	Not available
Hymenoptera	Formicidae	*Anoplolepis gracilipes*	JN987860	ANIND016–11
		Aphaenogaster beccarii	JN886031	ANIND005–11
		Camponotus compressus	JN886027	ANIND001–11
		Camponotus compressus GR-17	JN987857	ANIND013–11
		Camponotus irritance	JN886033	ANIND007–11
		Camponotus parius	JN886032	ANIND006–11
	Braconidae	*Chelonus blackburni*	KF365461	Not available
	Formicidae	*Leptogenys chinensis*	JN886030	ANIND004–11
		Monomorium scabriceps	JN987858	ANIND014–11
		Myrmicaria brunnea	JN886029	ANIND003–11
		Oecophylla samaragdina	JN886035	ANIND009–11
		Paratrechina longicornis	JN886034	ANIND008–11
		Pheidologeton diversus	JN987859	ANIND015–11
		Plagiolepis sp.	JN886037	ANIND011–11
		Solenopsis geminate	JN886028	ANIND002–11
		Tapinoma melanocephalum	JN886036	ANIND010–11
		Technomyrmex albipes	JN886038	ANIND012–11
	Trichogrammatidae	*Trichogramma achaeae*	KF234139	AGIMP021–13
		Trichogramma chilonis	KF234137	AGIMP019–13
		Trichogramma japonicum	KF234138	AGIMP020–13

Data Analysis

The pairwise analysis of 43 sequences was conducted using neighbour-joining bootstrap method and Kimura-2 parameter in MEGA5. The number of base substitutions per site was analysed between all sequences. Codon positions included were 1st+2nd+3rd+non-coding. All positions containing gaps and missing data were eliminated from the dataset. The A, T, G, C, AT, and GC content of all sequences was obtained using a computer program designed in the bioinformatics lab at NBAII, Bangalore, India. The AT% at three codon positions was calculated using the same program.

Sequences were aligned using the Mega 5 software package. Residue and pairwise distances were estimated using the Clustal W tool of MEGA 5 software with default settings of gap opening penalty ten, and a gap-extension 0.1 in pairwise and 0.05 in multiple alignments, sequence divergences were calculated and an NJ tree of distances was created to provide a graphic representation of the among-species divergences (Tamura et al. 2011). Sequences and other specimen information are available in the project "Ants of India," project code: ANIND in the campaign section "Ants of the World" and in "General Projects, Agriculturally Important Insects of India," project code: AGIMP in BOLD Systems at website (www.barcodinglife.org). Sequences and other specimen information are available at BOLD Systems (www.barcodinglife.org) and GenBank.

Results and Discussion

All mt CO1 sequences were submitted to the NCBI-GenBank under accession numbers provided in Table 3, PCR products from different ant species were easily produced and checked for low-quality bases at the ends.

The sequences were edited accordingly and aligned using Bioedit 7.0.2 software. The visualized PCR product contained no double bands on agarose gel, thus indicating that sequences obtained were mitochondrial DNA and not nuclear pseudogenes. The CO1 region in almost all the samples was in the range of 500–658 bp. A total of 42 insect species were studied; there were a total of 540 positions in the final datasets (software generated) according to the full K2P/NJ tree (Fig. 3). All 42 species could be differentiated by CO1 barcoding. Most of the amplified sequences were up to 658 bp in length. In general, the lengths suggest that nuclear DNA sequences originally from mitochondrial DNA sequences (NUMTs) were not sequenced because NUMTs are generally smaller than 600 bp. Although the *COX 1* region is highly conserved, there were differences in the length and sequence of the regions flanking *COX 1*. Previous phylogenetic studies have showed the utility of *COX 1* for the identification of genetic variability. Insect species were also subjected for analysis of nucleotide composition in all the different specimens, AT, GC%, and AT content at the first, second, and third codon positions were calculated, the difference was attributed to the AT content of the 3rd codon, AT_3, (25.58) which ranged from 32.07–5.78%. The AT content at the first and second codon positions was nearly invariant (Table 3). As expected, AT content was significantly found higher by 69.07% than the GC content of 30.92%. Average genetic distances among the different groups of insects used in this study showed higher values at the 3rd codon position, indicating that detailed study of the 3rd codon position for insects might reveal possible evolutionary information among this closely related group of organisms. Sequences were heavily AT-biased due to this 3rd codon position, which is expected in insect mtDNA. A phylogeny tree constructed using the N–J method revealed two clusters (Fig. 3); two clades were obtained, the first cluster consisting of orders lepidoptera, diptera, hemiptera, and coleoptera, whereas another clade showing relationship between hymenopteran insects. All the sequences obtained were submitted to GenBank. Sequence and barcode information is also available at BOLD Systems (www.barcodinglife.org) in the General Project section as Agriculturally

Table 3 AT% at the first, second, and third codon positions and AT and GC% of 42 different insect species

Insect species	First	Second	Third	AT%	GC%
Acroceratitis histrionic	17.21	19.11	30.80	67.14	32.85
Anoplolepis gracilipes	20.96	27.58	19.51	68.1	31.9
Aphaenogaster beccarii	20.44	24.91	20.27	65.6	34.4
Bactrocera correcta	16.40	18.75	27.50	62.65	37.34
Bactrocera dorsalis	17.18	18.75	27.96	63.90	36.09
Bactrocera zonata	17.03	14.68	5.78	62.5	37.5
Blaptostethus pallescens	19.47	18.53	30.99	69.00	30.99
Bombyx mori	25.90	18.88	24.88	69.66	30.33
Camponotus compressus	30.74	22.26	20.49	73.5	26.5
Camponotus compressus GR17	30.39	21.58	20.51	72.5	27.5
Camponotus irritans	20.49	24.95	19.83	65.3	34.7
Camponotus parius	25.22	20.51	20.36	65.5	34.5
Chelonus blackburni	24.18	21.59	30.35	76.13	23.86
Chilo auricilius	20.45	19.14	31.26	70.86	29.13
Chilo partellus	20.63	19.72	30.19	70.56	29.43
Chilo sacchariphagus indicus	20.13	19.31	30.27	69.72	30.27
Conogethes punctiferalis	18.84	18.99	30.69	68.54	31.45
Corcyra cephalonica	19.68	19.06	31.09	69.84	30.15
Euwallacea fornicates	17.62	19.14	27.81	64.58	35.41
Galleria mellonella	19.37	19.06	31.25	69.68	30.31
Helicoverpa armigera	21.09	19.27	30.19	70.56	29.43
Leptogenys chinensis	21.32	30.90	20.82	73.1	26.9
Monomorium scabriceps	25.07	19.90	20.51	65.5	34.5
Myrmicaria brunnea	20.66	26.11	20.00	66.8	33.2
Oecophylla smaragdina	33.13	22.18	20.97	76.3	23.7
Paratrechina longicornis	28.57	20.66	21.12	70.4	29.6
Pheidologeton diversus	24.20	19.78	20.49	64.5	35.5
Phytomyza orobanchia	19.30	30.24	18.23	67.78	32.21
Plagiolepis sp.	28.11	20.82	21.12	70.1	29.9
Plutella xylostella	20.63	19.27	29.74	69.65	30.34
Polyocha depressella	20.62	19.47	32.07	72.17	27.82
Scirpophaga excerptalis	19.63	19.47	30.60	69.72	30.27
Sesamia inferens	20.18	19.11	30.34	69.65	30.34
Solenopsis geminate	25.22	19.75	20.66	65.7	34.3
Spodoptera litura	20.96	19.46	29.61	70.04	29.95
Stegobium paniceum	19.56	18.93	31.29	69.79	30.20
Tapinoma melanocephalum	24.46	20.82	20.51	65.8	34.2
Technomyrmex albipes	26.89	20.51	20.82	68.2	31.8
Trichogramma achaeae	23.41	20.48	31.54	75.44	24.55
Trichogramma chilonis	23.08	20.00	32.03	75.12	24.87
Trichogramma japonicum	23.18	20.45	31.56	75.20	24.79
Xylocoris flavipes	17.55	18.19	28.50	64.25	35.74
Mean	22.12	20.76	25.58	69.07	30.92

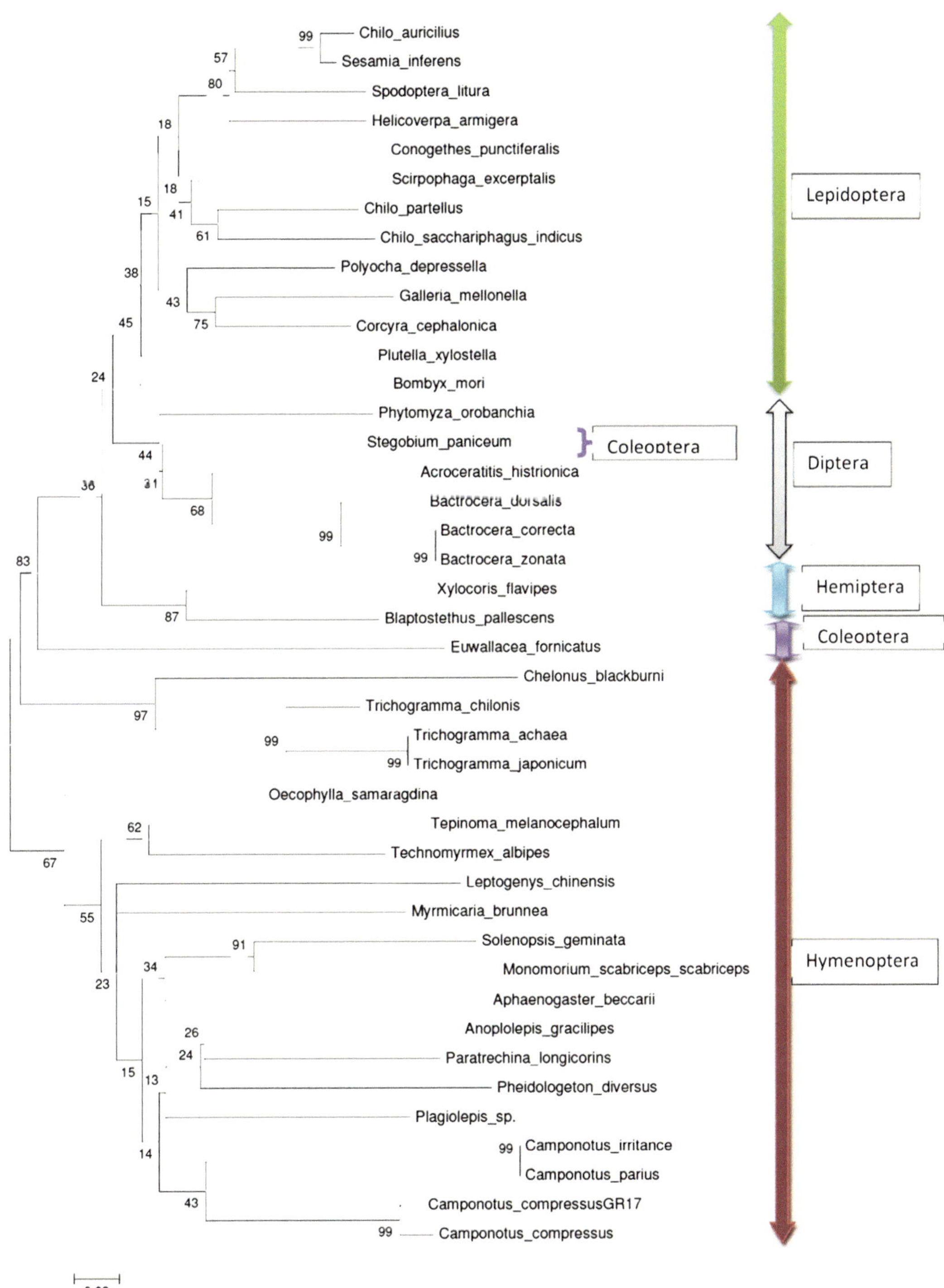

Fig. 3 Phylogenetic tree for 42 insect species (5 orders) depicting genetic relationships derived from CO1 sequences. Note: Bootstrap consensus tree generated by Bootstrap Test Phylogeny using neighbour-joining (N–J) method of MEGA 5 Software. All the 43 species are from 5 orders, which are distributed into two main clades that are 61 % similar

Important Insects in India (AGIMP) and in Ants of the World section as Ants of India (ANIND).

Limitations of DNA Barcoding

A short standardized DNA sequence originating a fragment of the mitochondrial gene has emerged as the standard barcode region for animals for unknown species and an aid in the discovery of new species. Mitochondrial DNA genes are maternally inherited which sometimes may result in interspecific hybridization or endosymbiont infections that generate transfer of mitochondrial genes outside the species, therefore DNA barcoding requires an expertise at the analysis level, and one must be trained on the analytical part of it. The proper knowledge about different kinds of bioinformatic tools enables one to analyse DNA sequence. Pseudogenes commonly known as nuclear mitochondrial DNA (NUMTs), originating from mitochondrial nucleus is one of the major obstacles in discriminating species on the mitochondrial DNA basis. Their integration into the nuclear genome was originally associated with transposable elements or short dispersed repeats, but close examination of many different NUMTs loci reveals a lack of common features at integration sites (Bensasson et al. 2001). Whitworth et al. (2007) observed that the patterns of mitochondrial variability can be confounded by the spread of maternally transmitted bacteria that cosegregate with mitochondria. They further reported that here, the performance of barcoding in a sample comprising 12 species of the blow fly genus *Protocalliphora*, known to be infected with the endosymbiotic bacteria *Wolbachia* showed very limited success: assignment of unknown individuals to species is impossible for 60 % of the species, while using the technique to identify new species would underestimate the species number in the genus by 75 %. In another study, Smith et al. (2012) analysed > 2 million insect *COX I* trace files on the BOLD and reported that *Wolbachia COX I* was present in 0.16 % of the cases. It is possible to generate *Wolbachia COX I* using standard insect primers; however, that amplicon was never confused with the *COX I* of the host. *Wolbachia* alleles recovered were predominantly Super group A and were broadly distributed geographically and phylogenetically and it was concluded that the presence of the *Wolbachia* DNA in total genomic extracts made from insects is unlikely to compromise the accuracy of the DNA barcode library and suggested that regular assays for *Wolbachia* presence and type can, and should, be adopted by large-scale insect barcoding initiatives.

Future Perspectives

DNA barcoding will greatly facilitate and complement taxonomic studies; the sequencing data coupled with traditional taxonomy is a model that can be applied on various disciplines and will allow analytical needs to be scaled to match the enormity of the current biodiversity crisis. It will help in the identification and conservation of the evolutionary processes that generate and preserve biodiversity. For groups in which identification can be difficult, the potential utility of DNA barcoding is immense. In this study, we showed that DNA barcoding allows the rapid identification of important functional units of hyper diverse arthropods in the rapid manner needed by conservation groups responding to habitat destruction and degradation. Insect diversity, measured via DNA barcoding in collaboration with taxonomists, should provide the essential fine-scale maps for assessing biodiversity at a scale at which conservation decisions are made. Our results reveal that *COX 1* barcoding will permit the unambiguous identification of insect species of India. Taxonomists, equipped with modern tools and collaborations, have a chance to move systematically to the forefront of conservation. DNA barcoding is not a perfect approach, but it has immense impact on the scientific community, becoming a widely used approach, characterized by many relevant aspects of uniformity and generalization. A critical knowledge of the method is essential for a proper use of it. In recent past many online resources help researchers to upload and retrieve DNA sequence and specimen data across insect orders for phylogenetic and barcoding studies.

Acknowledgements The authors wish to express their sincere thanks to Dr. Abraham Verghese, Director, NBAII, Bangalore, for financial assistance and providing the necessary facilities required for our research. The authors are also thankful to Dr. PDN Hebert of Biodiversity Institute of Ontario, Canada, for his useful suggestions and advice, and Drs. CA Viraktamath and TM Mushtak Ali of University of Agricultural Sciences, GKVK Campus, Bangalore, for their useful suggestions, corrections, and proper classification of ant species.

References

Arnheim N (1983) Concerted evolution in multigene families. In: Nei M, Koehn R (eds) Evolution of genes and proteins. Sinauer, Sunderland, pp 38–61

Ashok Kumar G Jalali SK Venkatesan T Stouthamer R Niranjana P Lalitha Y (2009) Internal transcribed spacer-2 restriction fragment length polymorphism (ITS-2-RFLP) tool to differentiate some exotic and indigenous trichogrammatid egg parasitoids in India. Biol Control 49(3):207–213

Ball SL, Armstrong KF (2006) DNA barcodes for insect pest identification: a test case with tussock moths (Lepidoptera: Lymantriidae). Can J Forest Res 36(2):337–350

Ballard JW, Whitlock MC (2004) The incomplete natural history of mitochondria. Mol Ecol 13(4):729–744

Bazin E, Glemin S, Galtier N (2006) Population size does not influence mitochondrial genetic diversity in animals. Science 312(5773):570–572

Beckenbach AT, Joy JB (2009) Evolution of the mitochondrial genomes of gall midges (Diptera: Cecidomyiidae): rearrangement and severe truncation of tRNA genes. Genome Bio Evol 1:278–287

Behura SK (2006) Molecular marker systems in insects: current trends and future avenues. Mol Ecol 15(11):3087–3113

Bensasson D, Zhang D-X, Hartl DL, *Godfrey MH* (2001) *Mitochondrial pseudogenes: evolution's misplaced witnesses*. Trends Ecol Evol 16(6):314–321

Boore JL, Brown WM (1998) Big trees from little genomes: mitochondrial gene order as a phylogenetic tool. Curr Opin Genet Dev 8(6):668–674

Crease TJ (1999) The complete sequence of the mitochondrial genome of *Daphnia pulex* (Cladocera: Crustacea). Gene 233(1–2):89–99

Dhaliwal GS, Dhawan AK, Singh R (2007) Biodiversity and ecological agriculture: issues and perspectives. Indian J Ecol 34:100–109

Dowton M, Austin AD (1999) Evolutionary dynamics of a mitochondrial rearrangement "hot spot" in the Hymenoptera. Mol Biol Evol 16(2):298–309

Gerbi SA (1985) Evolution of ribosomal DNA. In: MacIntyre RJ (ed) Molecular evolutionary genetics. Plenum, New York, pp 419–517

Hebert PDN, Cywinska A, Ball SL, deWaard JR (2003a) Biological identifications through DNA barcodes. Proc R Soc B 270:313–321

Hebert PDN, Ratnasingham S, deWaard JR (2003b) Barcoding animal life: cytochrome c oxidase subunit 1 divergences among closely related species. Proc R Soc B 270(Suppl. 1):96–99

Hurst GD, Jiggins FM (2005) Problems with mitochondrial DNA as a marker in population, phylogeographic and phylogenetic studies: the effects of inherited symbionts. Proc R Soc B 272:1525–1534

Lewis DL, Farr CL, Kuiken LS (1995) *Drosophila melanogaster* mitochondrial DNA: completion of the nucleotide sequence and evolutionary comparisons. Insect Mol Biol 4(4):263–278

Monaghan MT, Balke M, Gregory TR, Vogler AP (2005) DNA-based species delineation in tropical beetles using mitochondrial and nuclear markers. Philos Trans R Soc Lond B Biol Sci 360(1462):1925–33

Novotny V, Basset Y, Miller SE, Weiblen GD, Bremer B, Cizek L, Drozd P (2002) Low host specificity of herbivorous insects in a tropical forest. Nature 416:841–844. (Proc Biol Sci 274(1619):1731–1739)

Rebijith KB, Asokan R, Krishna V, Krishna Kumar NK, Ramamurathy VV (2012) Development of species-specific markers and molecular differences in mitochondrial and nuclear DNA sequences of Aphis gossypii and Myzus persicae (Hemiptera: Aphididae). Florida Entomol 95:674–682.

Sambrook J, Russell DW (2001) Molecular cloning: a laboratory manual, (N.Y), 3rd edn. Cold Spring Harbor Laboratory Press, New York

Smith MA, Bertrand C, Crosby K, Eveleigh ES, Fernandez-Triana J, Fisher BL, Gibbs J, Hajibabaei M, Hallwachs W, Hind K, Hrcek J, Huang DW, Janda M, Janzen DH, Li Y, Miller SE, Packer L, Quicke D, Ratnasingham S, Rodriguez J, Rougerie R, Shaw MR, Sheffield C, Stahlhut JK, Steinke D, Whitfield J, Wood M, Zhou X (2012) *Wolbachia* and DNA barcoding insects: patterns, potential, and problems. PLoS ONE 7(5):e36514. doi:10.1371/journal.pone.0036514

Tamura K, Peterson D, Peterson N, Stecher G, Nei M, Kumar S (2011) MEGA5: molecular evolutionary genetics analysis using maximum likelihood, evolutionary distance, and maximum parsimony methods. Mol Biol Evol 28(10):2731–2739

Tautz D, Arctander P, Minelli A, Thomas RH, Vogler AP (2002) DNA points the way ahead in taxonomy. Nature 418(6897):479

Whitworth TL, Dawson RD, Magalon H, Baudry E (2007) DNA barcoding cannot reliably identify species of the blowfly genus *Protocalliphora (Diptera: Calliphoridae)*. Proc Biol Sci 274(1619):1731–1739

Wilson JJ (2012) DNA barcodes of insects. Methods Mol Biol 858:17–46

Part II

Insect Physiology

Changes in Body Melanisation and Not Body Size Affect Mating Success in *Drosophila immigrans*

Shama Singh

Abstract

Wild flies and laboratory strains of *Drosophila immigrans* were investigated to study effects due to changes in body size and body melanisation on mating success (mated pairs, mating latency (ML), and copulation duration(CD)). Accordingly, the hypothesis, whether changes in body size or body melanisation are correlated with mating success in *D. immigrans,* was tested. A comparison of copulating and noncopulating flies of *D. immigrans* in the field showed contrasting differences due to body melanisation but not with body size. Similar results were found in copulating flies with early- vs. late-mating propensity. Isofemale lines varying either in body melanisation or in body size, but not in both the traits were selected. Laboratory data on dark vs. light isofemale lines showed a significant effect of body melanisation on mating success. By contrast, small vs. large body size strains did not differ in their mating success. Further, laboratory-selected dark strain showed a significantly higher number of mated pairs and longer CD as compared with light strain; whereas ML was longer for light strain and much shorter for dark strain. Thus, changes in mating success are associated with body melanisation and not with body size in *D. immigrans*.

Keywords

Body melanisation · Body size · Copulation duration · *Drosophila immigrans*

S. Singh (✉)
Department of Zoology, University of Delhi, Delhi 110007, India
Department of Genetics, M. D. University, Rohtak, Haryana 124001, India
e-mail: ssingh.genetics@gmail.com

Introduction

Mating success is an important component of fitness affected by genetic as well as environmental factors (Long et al. 1980; Markow 1996). In some *Drosophila* species, females have been shown to choose their mating partner based on body size.

A. K. Chakravarthy (ed.), *New Horizons in Insect Science: Towards Sustainable Pest Management,*
DOI 10.1007/978-81-322-2089-3_3,

Only a few studies in *Drosophila* species have shown larger males to have higher mating success than smaller males, e.g. *D. melanogaster* (Taylor and Kekic 1988); *Drosophila simulans* and *Drosophila mojavensis* (Markow and Ricker 1992); *Drosophila testacea* (James and Jaenike 1992). By contrast, smaller males have shown more matings than larger males in *Drosophila subobscura* (Steele and Partridge 1988) and *Drosophila montana* (Aspi and Hoikkala 1995). For *Drosophila pseudoobscura*, two different studies have shown contradictory results, e.g. higher mating success of larger males (Partridge et al. 1987) whereas no difference in the mating success of males of different sizes (Markow and Ricker 1992). Further, in *Drosophila willistoni*, body size differences in males do not impact mating success (Da Silva and Valente 2001). Such studies suggest that the targets of sexual selection might vary in different *Drosophila* species, or body size differences within a species may not be important to female choice. The relationships between male mating success and body size are inconsistent across different *Drosophila* species, but these studies have not considered effects due to changes in body melanisation.

Studies have shown that changes in body melanisation affect mating behavior in diverse insect taxa. In ladybird beetles, dark-coloured individuals have shown increased mating success and earlier emergence times in spring (Brakefield 1984; Brakefield and Larsen 1984). Further, changes in body melanisation impact behavioural patterns, e.g. mate preference in butterfly (Kronforst et al. 2006); alternative body colour/behaviour morphs in grasshoppers *Schistocerca gregaria* (Leo et al. 2005); aggressive mating behaviour of body colour morphs in the fish *Gambusia holbrooki* (Horth 2003) and *Tetrix undulata* (Forsman et al. 2002); courtship display in two *Drosophila* species (Yeh et al. 2006). Genetic analysis of *D. melanogaster* mutants affecting body colour also differ in mating success (de Magalhaes and Rodrigues-Pereira 1976; Mizugushi and Almeida 1983). For wild populations of *Drosophila* species, associations between body colour morphs and copulation duration (CD) have been shown for *Drosophila elegans* (Hirai et al. 1999). Recent data on *D. melanogaster* has shown that changes in body melanisation are correlated with mating-related traits (Parkash et al. 2011; Dev et al. 2013) and not with body size.

Mating speed and CD are fitness-related traits in diverse insect taxa (Prakash 1967; Spiess 1970; Pitnick 1991; Markow 1996; Lefrance and Bundgaard 2000). Both these traits are under polygenic control and respond to laboratory selection. Such experiments have resulted in strains with faster and slower mating speeds (Spiess and Langer 1964; Brncic and Koref-Santibanez 1964; Kessler 1969) and longer and shorter CDs (Macbean and Parsons 1966; Gromko et al. 1991). However, laboratory selection experiments have not considered correlated changes in body size or body melanisation. Thus, it is not clear whether there are direct or correlated selection responses for mating-related traits in *Drosophila* species.

Drosophila immigrans is a cold-adapted cosmopolitan species characterised by larger body size and lack of sexual dimorphism in body melanisation; and a much longer CD of about 45 min (Markow and O'Grady 2006) in comparison to other *Drosophila* species. Accordingly, this species is suitable for comparing the impact of body size vs. abdominal melanisation on mating success. The present study involves addressing the following questions: (a) Whether field collected copulating pairs and noncopulating flies differ in body size or body melanisation? (b) Do heritable strains for body size (large vs. small) or body melanisation (dark vs. light) differ in mating success? (c) Do laboratory-selected dark and light strains influence mating success in *D. immigrans?* Body size and body melanisation in both the sexes; mating latency (ML), and CD in different strains of *D. immigrans* were analyzed. Laboratory strains for mating propensity under no choice experiments were compared in crosses for small- and large-sized flies of both the sexes for differences in ML, CD and body melanization of homo- and hetero-specific matings. Finally, mating success of laboratory selected dark and light strains in *D. immigrans* was compared.

Materials and Methods

Field Collections Copulating pairs and noncopulating flies of *D. immigrans* from decaying fallen fruits in a highland locality (winter of 2007–2008) were aspirated. Flies were collected in the morning (6:00–8:00 a.m.) when they were actively involved in mating. The copulating pairs were collected in food vials with aspirator. After 90 min, the unsuccessful flies (male and female both) that is, the one that did not copulate were aspirated in separate food vials (noncopulating flies).

D. immigrans (n=120–180) were also collected from three altitudinal sites (a low-, mid- and highland locality) covering a range of 761–2708 m in October–November (2007–2008) using banana bait traps and net sweeping. The sampling sites were characterized by their latitude, altitude, average annual temperature (T_{ave}) and relative humidity (RH) on the basis of climatological tables obtained from Indian Institute of Tropical Meteorology (IITM; www.tropmet.res.in). For each population, 45 isofemale lines were set up. Wild populations are adapted to different climatic conditions, but between population differences can be analysed on the basis of common garden experiments. Thus, flies of each population were reared in biological oxygen demand (BOD.) incubators at 21 ± 0.2 °C. The density was controlled by limited egg laying period (6–8 h) on cornmeal–yeast–agar medium (70–80 eggs/vial, 37 × 100 mm).

Culture Conditions For each population, 2–3 replicates of isofemale lines were maintained for simultaneous analysis of body melanisation, wing length and mating related traits. All experiments were initiated soon after collections and performed with successive generations. The experimental conditions were made uniform with temperature controlled room set at 21 °C. Further, for each season, G_1 and G_2 (generations 1 and 2) of the wild caught flies were analysed to avoid possible effects of laboratory adaptations or inbreeding effects.

Trait Measurements Lines were selected on the basis of body size (wing length) and body melanisation, i.e. (a) lines similar in body size but differing in body melanisation (b) lines similar in body melanisation but varying in body size from the 45 isofemale lines prepared. Ten randomly chosen individuals per isofemale line were investigated. Body size (wing length) and body melanisation were simultaneously measured in wild caught and laboratory reared individuals. For each fly, wing length (in millimetre) was measured from the thorax articulation to the tip of third longitudinal vein under Olympus SZ-11 microscope fitted with a micrometer.

Melanisation was estimated from dorsal as well as lateral views of the abdomen giving values ranging from 0 (no pigment) to 10 (complete darkness) for each of the six visible abdominal tergites. For visual scoring of melanisation under stereo zoom microscope (Olympus, www.olympus.com) the melanisation score (0–10) for each tergite was multiplied with its relative size. Since the abdominal tergites (2nd to 7th) differ in size, the total area of each tergite was transformed into its relative size, i.e. 0.66, 0.88, 1.0, 0.88, 0.66 and 0.33 for 2nd to 6th tergites, respectively, for both the sexes. However, the size of 7th tergite differed between sexes, i.e. 0.22 for males and 0.33 for females. The abdomen of each fly minus viscera was mounted on a slide and total body melanisation per fly was also estimated through Biowizard image analysis software (Dewinter Optical Inc., www.dewinterindia.com). All experiments were done on assorted dark and light flies (melanisation difference ~20–25 %) and assorted small and large flies (wing length difference ~0.50 mm) of isofemale lines from a highland population. Laboratory selection experiment for body melanisation was carried out in a midland population (Solan) at 21 °C for 18 generations and different traits were analysed.

Analyses of Mating-Related Traits For mating-related traits (ML, CD, ovariole number and fecundity), assorted isofemale lines were analysed. Mating experiments were conducted when flies were more active: early in the morning (6:00–8:00 a.m.). During experiments, two main categories of behaviours were observed: (i) Mating/courtship latency (ML), which is the time taken by the male to initiate courtship of female after its introduction into the chamber (ii) CD, the amount of time till the female forces or pushes out the male. Observations were recorded for 90 min with a stopwatch.

Six-day-old virgin females and males were introduced into cylindrical Tarsons plastic food vials (37 × 100 mm) with cotton plug. All observations were made in morning (6.00–8.00 a.m.) in a thermo-controlled room at 21 °C. For all the observed matings, ML and CD were recorded with a stop watch started at 0 s. Aspirated mated pairs were used for estimating fecundity in egg-laying chambers. Everyday the flies were transferred to fresh egg-laying chambers and fecundity was monitored under the Olympus stereo-zoom microscope SZ-11, Japan. This was continued for 15 successive days (7th–21st) as this period coincided with highest egg production and the data were shown as daily fecundity. Total ovariole number per fly was counted from ovaries dissected in a saturated solution of potassium dichromate.

Statistical Analyses For all traits, isofemale line and population means ± SD were used for illustrations and tabular data. Since all the traits showed higher repeatability values across generations, data were pooled. Mean trait values for body melanisation and wing length in copulating and noncopulating wild flies were compared with *t*test. Mating related traits in wild caught flies of *D. immigrans* in early vs. late mating propensity groups were also compared using *t* test. Further, the trait values for overall four types of matings (homo- and hetero-specific) on the basis of body melanisation and size were subjected to contingency χ^2 test to find whether matings are random or not. Correlation values (*r*) of mating related traits (ML, CD and daily fecundity) with body size and with melanisation in assorted isofemale lines were calculated. Data on within-population analyses were used for calculation of correlation values of different traits as a function of body melanisation. Statistica (Statsoft Inc., Release 5.0, Tulsa, OK, USA) was used for all calculations as well as illustrations.

Results

Mean ± SD for field observations on body size and body melanisation in copulating and non - copulating flies of *D. immigrans* are shown in Table 1. For each day observation, *t* test was performed to compare wing length and body melanisation of the copulating and noncopulating categories both for males as well as females. No significant (ns) differences were observed for wing length, whereas quite high significant values (***$p<0.001$) were obtained for body melanisation (Table 1). The copulating flies show higher body melanisation irrespective of wing length compared with noncopulating flies. These results favour the role of body melanisation and not body size on mating success in wild flies of *D. immigrans*.

No difference was found in body size (wing length) among early and late mating groups of flies (Table 2). Results showed high percentage melanisation, mated pairs, CD and fecundity; and lower ML in early-mating propensity flies than late-mating propensity flies (Table 2; significance level: $p<0.001$).

Isofemale lines varying either in body melanisation or in body size but not in both the traits were selected. Data showed that isofemale lines of *D. immigrans* similar in body size but varying in body melanisation (dark vs. light) differ significantly in ML and CD (Fig. 1). By contrast, isofemale lines similar in body melanisation but varying in body size (large vs. small) exhibit no difference in ML and CD in *D. immigrans* (Fig. 2). The results are illustrated in Figs. 1 and 2.

No choice matings were performed to confirm the role of body melanisation in mating success in *D. immigrans*. Data on no-choice tests based on isofemales varying only in body melanisation (dark—D and light—L) but not in body size; and varying in body size (small—Sm and large—La) but not in body melanisation are given in Table 3. Isofemale lines differing in body melanisation

Table 1 Field observations on body size and body melanisation in aspirated copulating and noncopulating flies (at 9:00 a.m. daily for 7 days) of *D. immigrans*. Traits were compared with *t* test

Day	Sex	Copulating pairs			Noncopulating flies			*t* test b/w copulant vs. noncopulant	
		N	WL	% melanisation	*n*	WL	% melanisation	WL	% melanisation
1	M	13	3.57±0.11	71.1±3.01	11	3.51±0.10	21.6±2.32	ns	***
	F		3.90±0.13	76.0±4.12		4.01±0.09	27.4±2.10	ns	***
2	M	7	3.48±0.09	62.4±3.06	9	3.50±0.12	30.2±2.45	ns	***
	F		3.52±0.10	68.2±3.50		3.55±0.11	36.4±2.74	ns	***
3	M	11	3.84±0.14	56.4±2.72	19	3.80±0.13	24.2±2.23	ns	***
	F		3.49±0.08	52.4±3.04		3.46±0.08	29.6±2.15	ns	***
4	M	12	3.70±0.10	73.5±3.56	6	3.74±0.13	28.2±3.06	ns	***
	F		3.94±0.15	70.4±3.61		3.96±0.15	31.7±2.55	ns	***
5	M	4	3.85±0.11	63.8±3.79	8	3.89±0.10	27.6±2.41	ns	***
	F		4.08±0.09	69.2±3.20		4.00±0.09	33.1±2.50	ns	***
6	M	14	3.26±0.13	61.2±3.15	7	3.21±0.12	34.0±3.02	ns	***
	F		3.77±0.12	64.3±3.98		3.83±0.11	37.5±2.86	ns	***
7	M	21	3.69±0.09	74.3±3.10	15	3.76±0.13	28.3±2.91	ns	***
	F		4.02±0.07	75.0±2.09		4.07±0.12	32.9±2.46	ns	***

ns nonsignificant, *M* male, *F* female, *WL* Wing Length, *b/w Between*
***$p<0.001$

Table 2 Data on mating related traits in wild-caught copulating flies of *D. immigrans* in early- vs. late-mating propensity (MP) groups

Traits	Sex	Early MP	Late MP	*t* test
Wing length (mm)	M	3.25±0.09	3.26±0.11	ns
	F	3.57±0.10	3.54±0.12	ns
Melanisation (%)	M	71.0±3.20	32.1±2.41	***
	F	75.8±3.37	36.3±2.60	***
Mating latency (min)	–	7.21±3.46	21.0±3.02	***
Mated pairs (%)	–	67.4±1.42	32.5±1.16	***
Copulation duration (min)	–	64.2±3.75	41.1±3.51	***
Fecundity (no. of eggs/day)	–	59.5±2.88	40.7±2.19	***

ns nonsignificant, *M* male, *F* female
***$p<0.001$

and not body size show difference in mated pairs, ML and CD for light and dark homo-specific (L × L; D × D) as well as hetero-specific matings (L × D; D × L) on the basis of contingency χ^2 test (Table 3). By contrast, flies of isofemale lines varying in body size and not in body melanisation, under no choice mating conditions, exhibited no difference in mated pairs, ML and CD (Table 3). Thus, laboratory results (similar to wild data) on dark vs. light isofemale lines show significant effect of body melanisation on mating success. By contrast, small vs. large body size lines did not differ in mating success.

Between and Within Population Variations Basic data on body melanisation, wing length, mating latency, copulation duration, ovariole number and fecundity in a lowland and a highland population grown at 21 °C are shown in Table 4. Variations in these traits show parallel changes along altitude. There is no difference for melanisation in each sex. Further, differences in melanisation (dark and light flies) in both the populations correspond to changes in reproductive traits (ML, CD, and fecundity; Table 4). Statistical comparisons show that within population differences are quite significant for all traits

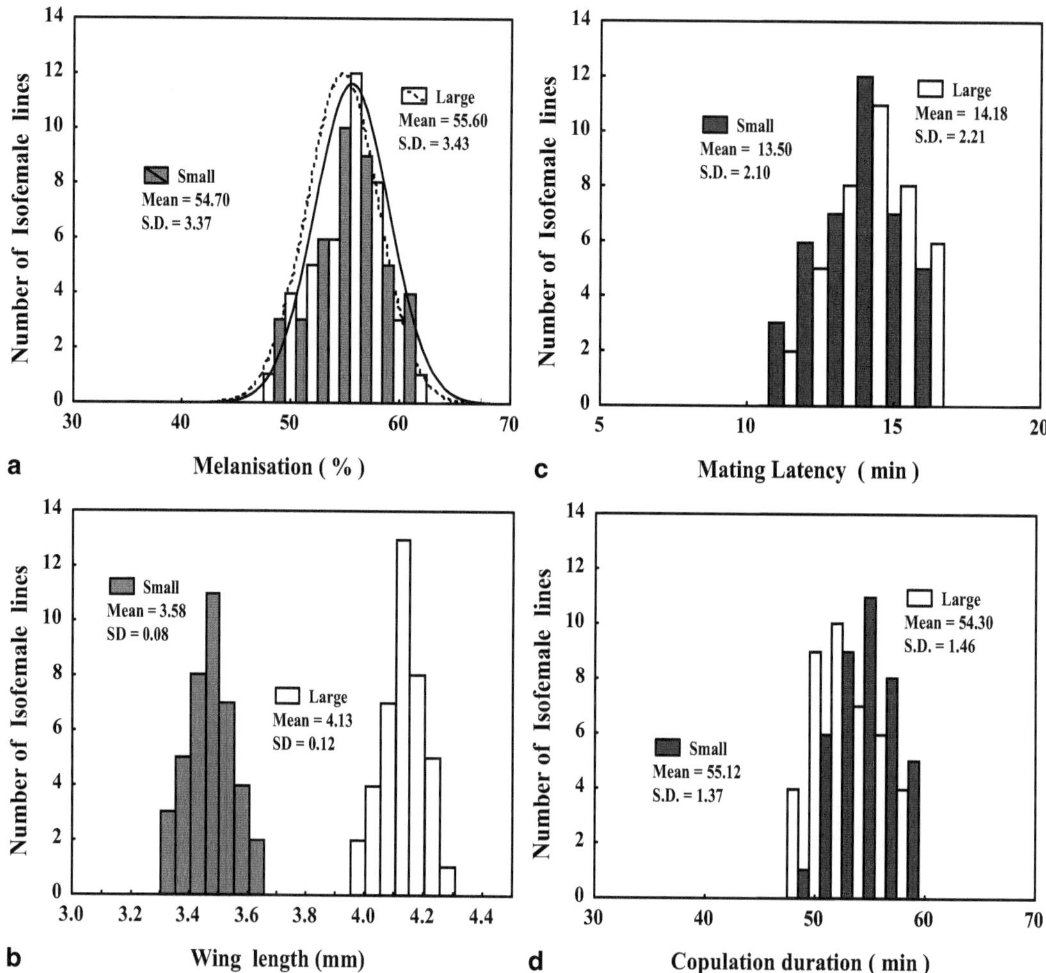

Fig. 1 Isofemales of *D. immigrans* similar in body size (**a**), but varying in body melanization (*dark* vs. *light*) (**b**), differ significantly in mating latency (**c**), and copulation duration (**d**)

(except body size:wing length) compared to between population differences (Table 4).

Correlations Between Traits Based on isofemale lines data, within population analyses for correlation of mating related traits (mating latency, copulation duration and fecundity) with body melanisation are shown in Fig. 3 and Table 4. There are significant correlations for ML (r=−0.90); CD (r=0.81), and fecundity (r=0.95) with assorted dark and light flies on the basis of body melanisation (Fig. 3). Small- and large-sized flies assorted on the basis of body size (wing length) show lack of correlations with ML, CD and fecundity (r=ns). Thus, within population variation in body size show a lack of correlation with body melanisation as well as mating related traits (Fig. 3). This analysis further supports correlated changes in body melanisation (not body size) and mating related traits in *D. immigrans*.

Selection: Supporting Evidence Results of laboratory selected dark (DSS) and light (LSS) body color strains are illustrated as bars (Fig. 4). For body melanisation, darker selected strain showed about 1.6-fold increase while lighter strain exhibited ~2 fold decrease in comparison to control population after 18 generations of selection (Fig. 4). The selected strains varied

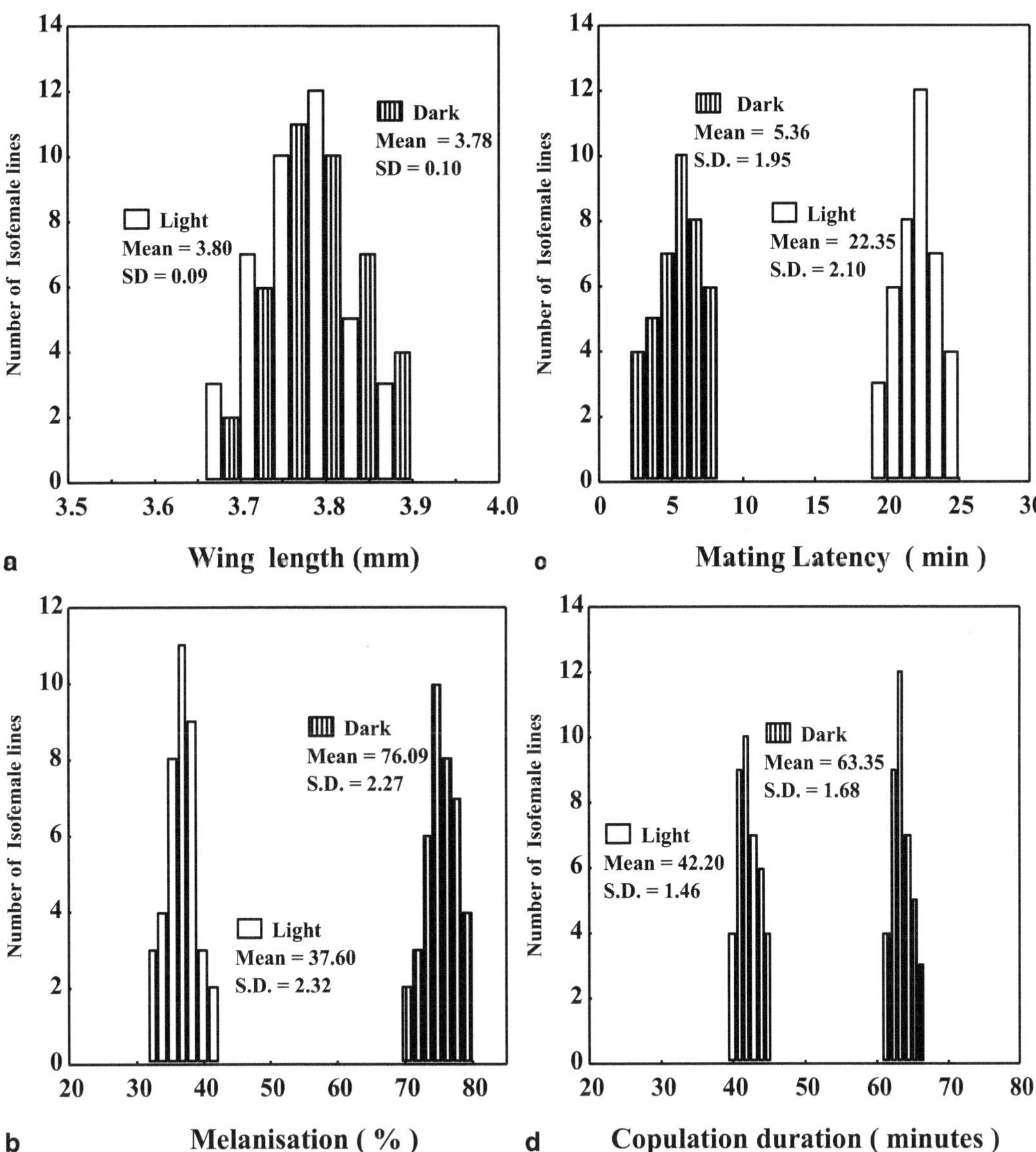

Fig. 2 Isofemales similar in body melanisation (**a**) but differing in body size (*large* vs. *small*; **b**) show lack of differences in mating latency (**c**) and copulation duration (**d**) in *D. immigrans*

Table 3 Data on no-choice tests based on isofemales varying only in body melanisation (dark—D and light—L) but not in body size (A), and varying in body size (small—Sm and large—La) but not in body melanisation (B). For each mating trait results of contingency χ^2 test are also shown

Traits	(A) Body melanisation					(B) Body size				
	L×L	D×D	L×D	D×L	Cont. χ^2	Sm×Sm	La×La	Sm×La	La×Sm	Cont. χ^2
% MP	10.3±2.05	40.0±2.84	23.8±3.01	25.9±2.73	***	10.3±2.50	11.2±2.83	12.0±2.66	12.5±3.14	ns
ML	23.4±2.92	5.5±1.72	16.5±2.10	14.3±1.60	***	23.4±2.92	24.0±2.78	25.3±3.14	21.2±3.01	ns
CD	38.0±2.61	64.5±2.42	47.0±2.56	53.2±2.44	***	39.0±2.61	41.5±3.20	44.5±2.13	40.1±2.46	ns

MP mated pairs, *ML* mating latency, *CD* copulation duration, *ns* nonsignificant
***$p < 0.001$

Table 4 Data on mating related traits in dark and light isofemale lines ($n=30$) from lowland and highland populations of *D. immigrans* grown at 21 °C. Correlations of different traits with body melanization are also shown

Traits	Sex	Lowland (761 m)			Highland (2708 m)		
		Dark	Light	*r* with melanisation	Dark	Light	*r* with melanisation
% melanisation	F	50.0±3.06	26.3±3.15	–	76.0±3.27	38.6±3.32	–
	M	47.5±2.91	24.1±2.80	–	71.1±3.15	32.4±3.10	–
Wing length (mm)	F	3.57±0.10	3.54±0.09	0.09±0.12 (ns)	4.00±0.12	3.92±0.14	0.12±0.18 (ns)
	M	3.21±0.12	3.26±0.13	0.12±0.18 (ns)	3.51±0.16	3.57±0.19	0.16±0.20 (ns)
Mating latency (min)	–	12.4±2.40	28.5±2.36	−0.88±0.07***	5.26±2.57	22.3±2.62	−0.90±0.07***
Copulation duration (min)	–	50.0±2.44	31.2±2.51	0.85±0.11***	65.3±2.54	43.2±2.63	0.81±0.11***
Ovariole no.	–	48.2±2.30	46.0±2.45	0.24±0.17 (ns)	56.5±2.40	59.1±2.78	0.22±0.21 (ns)
Fecundity (no. of eggs/day)	–	45.1±1.83	33.0±1.90	0.92±0.06***	64.0±2.01	48.0±2.13	0.95±0.06***

ns nonsignificant
***$p<0.001$

significantly in all the three fitness related traits except body size (wing length), suggesting that evolutionary response to laboratory selection did not affect body size. For darker selected strain, a significant increase in CD (Fig. 4) and fecundity (data not shown) and decrease in ML was observed. By contrast, for lighter selected strain, a decrease in CD and fecundity (data not shown); and an increase in ML were found in comparison to control population (Fig. 4). Thus, laboratory selection results support that changes in body colour are correlated with mating related traits in *D. immigrans*, similar to wild and laboratory data.

Discussion

Large size seems to be at an advantage in a wide variety of ecological contexts in the flies' mating system. Research has shown that larger males and females exhibit higher mating success as compared to smaller individuals. Large males generally experience greater success in aggressive competition than smaller ones especially in species, in which there is direct male–male competition for access to mates (Thornhill and Alcock 1983; Otranen 1984; Day and Butlin 1987; Cook 1988). The relationship between large body size and success also holds for fighting, in everything from minute insects like aphids and thrips, to elephant seals (Alcock 1993). In particular, large females often have greater longevity and higher fecundity and larger males have enhanced mating success (Butler and Day 1984). A male preference for larger and hence more fecund females has been demonstrated in several species of insects (Hieber and Cohen 1983; Johnson and Hubbell 1984).

Several studies have shown the role of body size in mating success in insects as well as *drosophilids* such as: in Japanese beetle *Popillia japonica,* males prefer larger females due to their egg characteristics (Saeki et al. 2005); in *drosophilid Phorticella striata* female preferred males with long wings. Other studies (Markow 1988; Partridge et al. 1987a, b; Hedge and Krishna 1997) also favour role of body size. In *D. willistoni*, no significant differences in wing length were observed between copulating and noncopulating flies (Da Silva and Valente 2001). The present study in *D. immigrans* shows no effect of body size (wing length) on ML, CD and fecundity.

Ecological significance of body melanisation has been investigated in melanic and typical morphs in butterflies and beetles (Majerus 1998; True 2003). In ladybird beetle, *Coccinella septempunctata*, the melanic morphs with higher elytral pigmentation showed greater fecundity than lesser pigmented individuals in relation with radiant heat levels (Rhamhalinghan 1999).

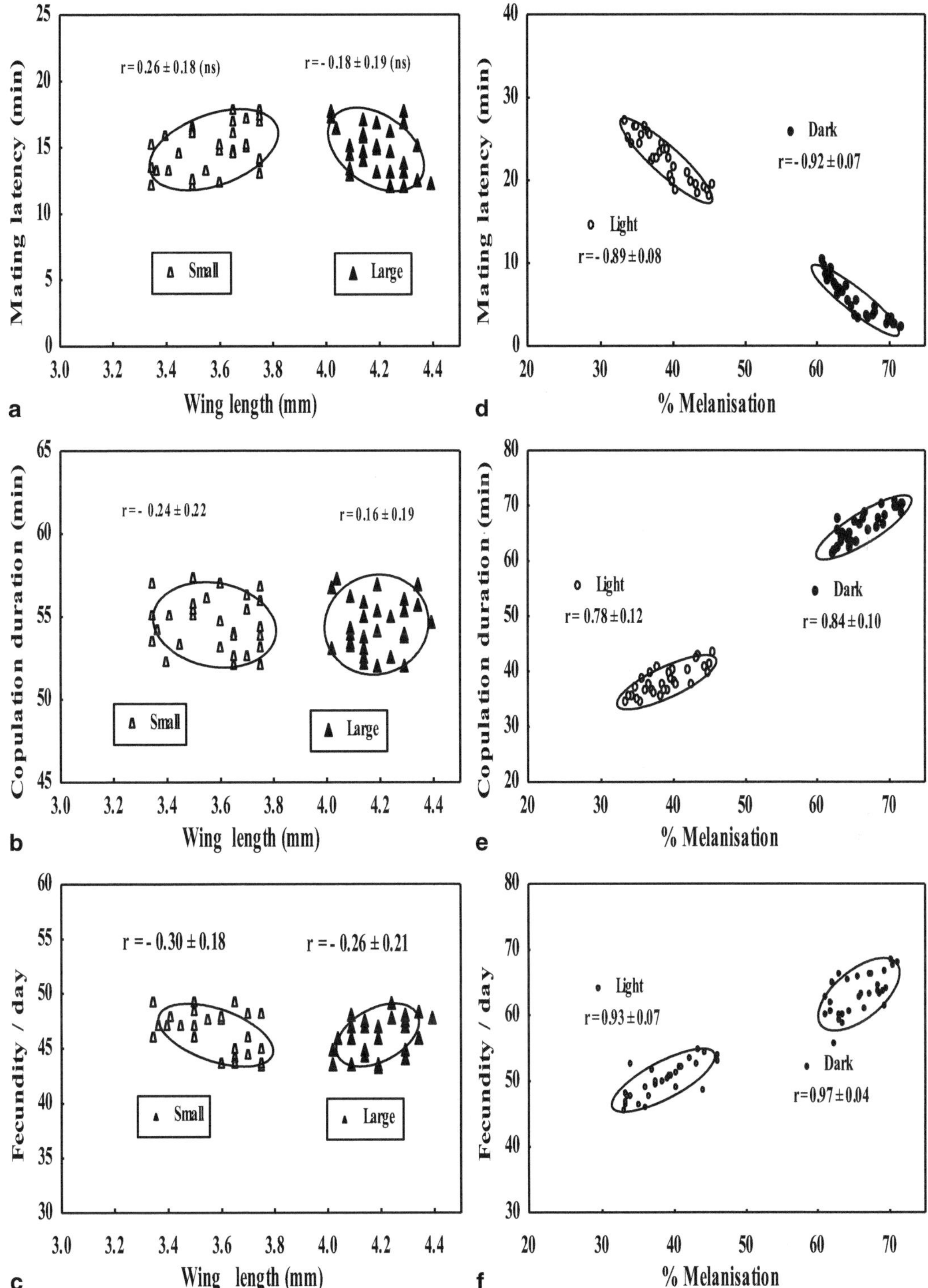

Fig. 3 Correlations of mating related traits (mating latency, copulation duration and daily fecundity) with body size (*large* vs. *small*; **a**–**c**) and with body melanisation (**d**–**f**) in assorted isofemale lines of *D. immigrans*

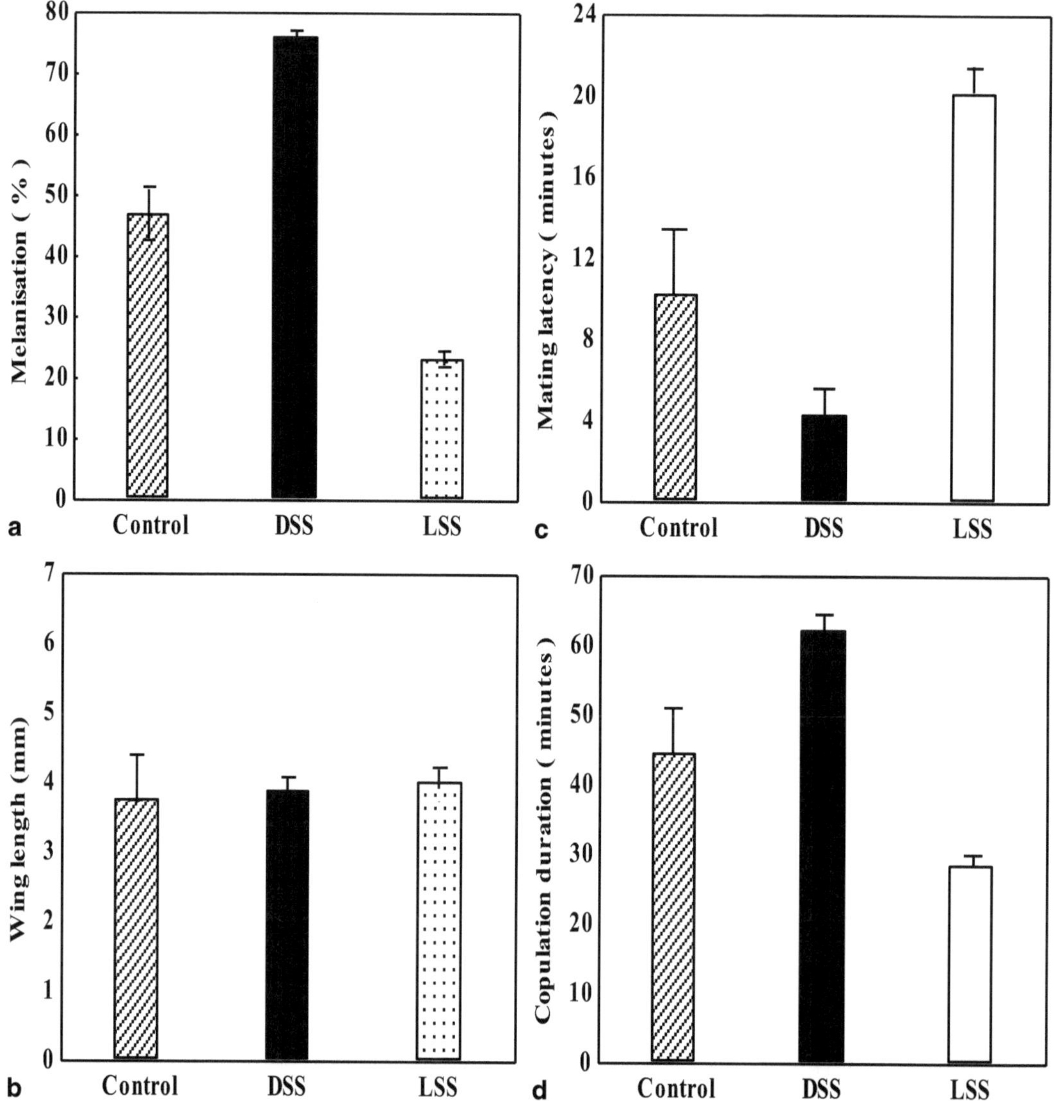

Fig. 4 Results of laboratory selected *dark* (DSS) and *light* strains (LSS) (**a**) of *D. immigrans* on body size (*wing length*) (**b**), mating latency (**c**), and copulation duration (**d**). *Darker* (DSS) and *lighter* (LSS) selected strains vary in mating success but not in body size in *D. immigrans*. Trait values for control (*base*) population are also shown

In *Adalia bipunctata* (coleopteran) and *Ephestia kuehniella* (lepidopteran), melanics display higher mating success compared with typicals (Verhoog et al. 1998). However, less attention has been paid to most visible body colour variations (Hirai et al. 1999; Brakefield 1984) which show pleiotropic effects on mating behaviour (True 2003). A few recent studies have shown the possible consequences of changes in body melanisation for fitness traits in *drosophilids* (Rajpurohit et al. 2008; Singh et al. 2009; Parkash et al. 2011; Dev et al. 2013). However, there is no detailed study that has considered the impact of changes in body size as well as body melanisation simultaneously on mating success in *Drosophila* species.

In diverse insect taxa, pleiotropic effects of body melanization have been observed for various behavioural, physiological and developmental traits (True 2003). For example, in *D. elegans*, CD is shorter in brown morph as compared with the black morph (Hirai et al. 1999). In ladybird beetle (*Adalia bipunctata*) dark colour individuals benefit from increased mating success (Brakefield 1984). In another study, on *Coccinella septempunctata* higher fecundity was correlated with increase in body melanisation (Rhamhalinghan 1999). Pleiotropy or genetic linkage is responsible for generation of such correlated behaviour traits. Indirect evidences of such pleiotropy can be evidenced from correlated phenotypes in laboratory selection experiments. In the present work, we performed selection experiment for body melanisation till 18 generations to study its correlated effect on mating related traits, i.e. ML and CD in *D. immigrans*. Selection results show an increase for body melanisation, CD and fecundity in dark selected strain and almost similar decrease for all the traits in light selected strain irrespective of body size—small or large.

Present work documents the effect of body size as well as body melanisation on mating success in wild and laboratory (isofemale line approach) of *D. immigrans*. Laboratory-selected dark strain obtained through selection experiment showed significantly higher number of mated pairs and longer CD as compared with light strain. ML was longer for light and shorter for dark strain. Thus, the current study supports the role of body melanisation in mating success in *D. immigrans*.

References

Alcock J (1993a) The effects of male body size on territorial and mating success in the landmark-defending fly *Hermetia comstocki*, (Stratiomyidae). Ecol Entomol 18:1–6

Alcock J (1993b) Animal behaviour. An evolutionary approach. Sinauer, Sunderland

Aspi J, Hoikkala A (1995) Male mating success and survival in the field with respect to size and courtship song characters in *Drosophila littoralis* and *D. montana* (Diptera—*Drosophilidae*). J Insect Behav 8:67–87

Brakefield PM (1984) Selection along clines in the ladybird *Adalia bipunctata* in the Netherlands: a general mating advantage to melanics and its consequences. Heredity 53:37–49

Brakefield PM, Larsen TB (1984) The evolutionary significance of dry and wet season forms in some tropical butterflies. Biol J Linn Soc 22:1–12

Brncic D, Koref-Santibanes (1964) Mating activity of homo and hetero-karyotypes in *D. pavani*. Genetics 49:585–591

Butler RK, Day TH (1984) The effect of larval competition on development time and adult size in seaweed fly, *Coelopa frigida*. Heredity 54:107–110

Cook DF (1988) Sexual selection in dung beetles. II. Female fecundity as an estimate of male reproductive success in relation to horn size and alternative behavioural strategies in *Onthophagus binodis* (Scarabeidae: Onthophagini). Aust J Zool 36:521–532

Da Silva LB, Valente VLS (2001) Body size and mating success in *Drosophila willistoni* are uncorrelated under laboratory conditions. J Genet 80:77–81

Day TH, Butlin RK (1987) Non-random mating in natural populations of the seaweed fly, *Coelopa frigida*. Heredity 58:213–220

de Magalhaes LE, Rodrigues-Pereira MAQ (1976) Frequency dependent mating success among mutant ebony of *D. melanogaster*. Experientia 32:309–310

Dev K, Chahal J, Parkash R, Kataria SK (2013) Correlated changes in body melanisation and mating traits of *Drosophila melanogaster*: A seasonal analysis. Evol Biol 40(3):366–376

Forsman A, Ringblom K, Civantos E, Ahnesjo J (2002) Coevolution of color pattern and thermoregulatory behavior in polymorphic pygmy grasshoppers *Tetrix undulata*. Evolution 56:349–360

Gromko MH, Briot A, Jensen SC, Fukui HH (1991) Selection on copulation duration in *Drosophila melanogaster*: Predictability of direct response versus unpredictability of correlated response. Evolution 45:69–81

Hegde SN, Krishna MS (1997) Size-assortative mating in *Drosophila malerkotliana*. Anim Behav 54:419–426

Hieber CS, Cohen JA (1983) Sexual selection in the lovebug, *Plecia nearctica*: the role of male choice. Evolution 37:987–992

Hirai Y, Sasaki H, Kimura MT (1999) Copulation duration and its genetic control in *Drosophila elegans*. Zool Sci 16:211–214

Horth L (2003) Melanic body colour and aggressive mating behavior are correlated traits in male mosquitofish (*Gambusia holbrooki*). Proc Roy Soc Lond B 270:1033–1040

James AC, Jaenike J (1992) Determinants of mating success in wild *Drosophila testacea*. Anim Behav 44:168–170

Johnson LK, Hubbell SP (1984) Male choice, experimental demonstration in a brentid weevil. Behav Ecol Sociobiol 15:183–8

Kessler S (1969) The genetics of *Drosophila* mating behaviour II. The genetic architecture of mating speed in *D. pseudoobscura*. Genetics 62:421–433

Kronforst MR, Young LG, Kapan DD, McNeely C, O'Neill, RJ, Gilbert LE (2006) Linkage of butterfly mate preference and wing color preference cue at the genomic location of wingless. Proc Natl Acad Sci U S A 103:6575–6580

Lefrance A, Bundgaard J (2000) The influence of male and female body size on copulation duration and fecundity in *Drosophila melanogaster*. Hereditas 132:243–247

Leo Lester R, Grach C, Paul Pener M, Simpson SJ (2005) Stimuli inducing gregarious colouration and behavior in nymphs of *Schistocerca gregaria*. J Insect Physiol 51:737–747

Long CE, Markow TA, Yager P (1980) Relative male age, fertility and competitive mating success in *Drosophila melanogaster*. Behav Genet 10:163–170

MacBean IT, Parsons PA (1967) Directional selection for duration of copulation in *Drosophila melanogaster*. Genetics 56:233–239

Majerus MEN (1998) Melanism: evolution in action. Oxford University Press, Oxford

Markow TA (1988) Reproductive behavior of *Drosophila* in the laboratory and in the field. J Comp Psychol 102:169–174

Markow TA (1996) The evolution of mating systems. Evol Biol 29:73–106

Markow TA, O'Grady P (2006) A guide to species identification and use. Elsevier/Academic, New York

Markow TA, Ricker JP (1992) Male size, developmental stability and mating success in natural populations of three *Drosophila* species. Heredity 69:122–127

Mizugushi Y, de Almeida JR (1983) Effect of yellow mutant on mating speed and duration of copulation in *D. melanogaster*. Annu Soc Entomol Braz 12:145–150

Otranen M (1984) Male contests for territories and female in the fly *Dryomyza anilis*. Anim Behav 32:882–90

Parkash R, Sharma V, Chahal J, Lambhod C, Kajla B (2011) Impact of body melanization on mating success in *Drosophila melanogaster*. Entomol Exp Appl 139:47–59

Partridge L, Ewing E, Chandler A (1987a) Male size and mating success in *Drosophila melanogaster*. The role of female and male behavior. Anim Behav 35:555–562

Partridge L, Hoffmann A, Jones JS (1987b) Male size and mating success in *Drosophila melanogaster* and *D. pseudo-obscura*, under field conditions. Anim Behav 35:468–476

Pitnick S (1991) Male size influences mate fecundity and remating interval in *Drosophila melanogaster*. Anim Behav 41:735–745

Prakash S (1967) Association between mating speed and fertility in *Drosophila robusta*. Genetics 57:653–663

Rajpurohit S, Parkash R, Ramniwas S, Singh S (2008) Variations in body melanisation, ovariole number and fecundity in highland and lowland populations of *Drosophila melanogaster* from the Indian subcontinent. Insect Sci 15:553–561

Rhamhalinghan M (1999) Variations in the realized fecundity of melanics of *Coccinella septempunctata* L in relation to degree of elytral pigmentation and radiant heat level. J Insect Sci 12:22–26

Saeki Y, Kruse KC, Switzer PV (2005) Male preference for large females and female reproductive condition in the Japanese beetle, *Popillia japonica* Newman (Coleoptera: Scarabaeidae). J Kans Entomol Soc 78:13–19

Singh S, Ramniwas S, Parkash R (2009) Fitness consequences of body melanization in *Drosophila immigrans* from montane habitats. Entomol Res 39:182–191

Spiess EB (1970) Mating propensity and its genetic basis in *Drosophila*. In: Hecht MK, Sreere WC (eds) Essays in evolution and genetics in honour of Theodosius Dobzhansky. Springer, Berlin

Spiess EB, Langer B (1964) Mating speed control by gene arrangement in *D. persimilis*. Evolution 18:430–444

Steele RH, Partridge L (1988) A courtship advantage for small males in *Drosophila subobscura*. Anim Behav 36:1190–1197

Taylor CE, Kekic V (1988) Sexual selection in a natural population of *Drosophila melanogaster*. Evolution 42:197–199

Thornhill R, Alcock J (1983) The evolution of insect mating systems. Harvard University Press, Cambridge

True JR (2003) Insect melanism: the molecules matter. Trend Ecol Evol 18:640–647

Verhoog MD, Breuker CJ, Brakefield PM (1998) The influences of genes for melanism on the activity of moth *Ephestia kuehniella*. Anim Behav 56:683–688

Yeh SD, Liou SR, True JR (2006) Genetics of divergence in male wing pigmentation and courtship behavior between *Drosophila elegans* and *D. gunungcola*. Heredity 96:383–395

Do Size and Age of Female *Trichopria* Sp. Influence Longevity, Reproductive Performance, and Sex Ratio?

N. Veena and D. Manjunath

Abstract

The present study was conducted to record the effect of female size and age on the longevity and reproductive performance of *Trichopria* sp. (Hymenoptera: Diapriidae), an endo-pupal, gregarious parasitoid of *Exorista bombycis* Louis (Diptera: Tachinidae). Being an endo-larval parasitoid of the silkworm *Bombyx mori* L. (Lepidoptera: Bombycidae), *E. bombycis* causes a cocoon yield reduction of 10–20 % in south India. Two-day-old females of the wasp were segregated into big and small based on their size. The longevity of big and small females was recorded at 24 h intervals by feeding 30 % honey. The progeny production and sex ratio were assessed by allowing big and small gravid parasitoid females as well as 1–8 day-old females of the wasp to parasitize 3 day-old puparia of *E. bombycis* for 2 days at a wasp: host ratio of 1:4. Number of host puparia parasitized, parasitoid developmental duration, progeny production, and sex ratio were recorded. The longevity of big parasitoid female wasps was significantly longer than small ones. Similarly, big females of the parasitoid produced significantly more progenies with higher sex ratio compared to small females. With regard to parasitoid age, young females (1–3 day-old) produced significantly more progenies in comparison to old ones (7–10 day-old).

Keywords

Female longevity · Parasitoid size · Progeny

N. Veena (✉)
Forest Wood Protection Division, Institute of Wood Science and Technology, Malleswarm, Bangalore 560003, India
e-mail: veena_daya@rediffmail.com

D. Manjunath
Department of Studies in Sericulture Science, University of Mysore, Manasagangotri, Mysore 570006, India
e-mail: manjunath4591@yahoo.co.in

Introduction

One of the important components of integrated pest management is the use of natural enemies as biocontrol agents. This has many advantages over the traditional chemical control (Scholler and Flinn 2000). The natural enemies are self-perpetuating, effective on a long term, and

A. K. Chakravarthy (ed.), *New Horizons in Insect Science: Towards Sustainable Pest Management*, DOI 10.1007/978-81-322-2089-3_4, © Springer India 2015

economical. With regard to parasitoids as biocontrol agents, they assume immense importance as they account for 87% suppression of pest insects in nature in comparison to 12% by predators and 1% by pathogens (van Lenteren 1983). To realize their effectiveness as potent biocontrol agents, they need to be released in the field subsequent to their mass production in the insectary employing appropriate protocol. Further, the mass-produced parasitoid individuals, especially females are required to possess superior fitness characteristics in order to accomplish the desired goal in the field. In addition, it would be ideal if the population of the mass-produced parasitoid adults is female-biased. It is a well established fact that fitness traits as well as sex ratio of parasitoids are influenced by various host, parasitoid, and environmental factors. Fitness is often correlated with adult size and it correlated with enhanced longevity, temperature tolerance, effective host searching ability, and greater reproductive potential (Ji et al. 2004; Aruna 2007; Gangadhar 2009). The factors that contribute to parasitoid fitness include host age and size, parasitoid age and size, temperature, mating, and adult diet.

Parasitoid age influenced developmental characteristics. In *Trichogramma chilonis* and *Trichogramma astriniae*, female progenies were the highest on the first day of adult life (Hirashima et al. 1990). Progeny production and fecundity were higher in young parasitoid females than in older ones (Guang and Oloo 1990). The sex ratio of older females was male-biased and young ones were female-biased in *Trichogramma minutum* (Leatemia et al. 1995). Oviposition rate decreased with the parasitoid age in *Anagyrus pseudococci* Girault (Avidov et al. 1967), *Apoanagyrus lopezi* De Santis (Odebiyi and Bokonon-Ganta 1986), *Tetrastichus incertus* Ratzeburg (Pitcoirn and Gutierrez 1992), and *Catolaccus grandis* (Morales-Ramos and Cates 1992). In *Ceratogramma etiennei* Delvare, the percent parasitization increased in young females while it decreased with female age (Amalin et al. 2005). In *Ascogaster reticulatus* Watanabe (Honda 1998) and *Chelonus* sp. nr. *curvimaculatus* (Hentz 1998) young parasitoid females were more fecund than older ones.

Parasitoid size is known to influence fecundity with large females consistently laying more eggs than small females. In some species, they produce more eggs of larger size (Speight 1994). Reproduction rate of large females remains higher and can be twice as high as small females regardless of the size of the males they mate with (Ji et al. 2004). Size fitness relationship of a parasitoid has been demonstrated in *Aphaereta minuta* (Nees) by Visser (1994) and *Achrysocaroides zwoelferi* by West et al. (1996) who observed that large females were more fecund with larger eggs, high longevity, and greater host searching efficiency than smaller ones. Similar results were obtained with *Ascobara tabida* (Ellers et al. (1998). Further, it was also observed that the fitness of the parasitoid increased linearly with size. Sagarra et al. (2001) reported that, *Anagyrus kamali*, large females have higher longevity, higher reproductive longevity, higher fecundity, higher oviposition rate, and a large number of progenies.

Trichopria sp. is a 1.5–2 mm long, gregarious parasitoid wasp of *Exorista bombycis*. Being an endo-larval parasitoid of *Bombyx mori*, *E. bombycis* is a serious problem while raising silkworm for cocoon production causing 10–20% mortality of larvae. Unlike the other parasitoids of *E. bombycis*, *Trichopria* sp. is observed to prevail and parasitize the puparia of *E. bombycis* throughout the year in most cocoon markets located in traditional districts of Karnataka, thereby indicating its potential as an agent of biological suppression of *E. bombycis*. It is proposed to exploit *Trichopria* sp. as a biocontrol agent of *E. bombycis* alongside *Nesolynx thymus*, which is currently the recommended biocontrol agent of this pest.

Material and Methods

Host Culture

The puparia were obtained after allowing the post-parasitic maggots collected from silkworm (*B. mori*) cocoon markets to pupate in the laboratory (23–28 °C and 60–80% RH) of the Department

Table 1 Impact of age of *Trichopria* sp. on reproductive efficiency

Parasitoid age (days)	No. of puparia parasitized@	Developmental duration (days)	Progeny production (no.)			Sex ratio (♀/♂)
			Male	Female	Total	
1	2.00±0.00	22.20±0.13	81.00±1.12[bc]	123.80±1.13[a]	204.80±1.61[ab]	1.51±0.02[a]
2	2.00±0.00	22.30±0.15	91.81±1.15[a]	128.40±1.27[a]	222.20±1.65[a]	1.40±0.03[b]
3	1.90±0.10	22.40±0.16	81.10±4.82[bc]	107.20±5.76[b]	187.50±9.90[bc]	1.33±0.04[b]
4	2.00±0.00	22.30±0.15	88.00±1.15[a]	94.80±1.17[c]	180.00±1.66[cd]	1.10±0.02[c]
5	1.90±0.10	22.30±0.15	87.50±4.50[a]	94.30±4.74[c]	182.10±9.18[cd]	1.09±0.03[c]
6	1.90±0.00	22.30±0.15	87.50±4.50[a]	94.30±4.74[c]	182.10±1.18[cd]	1.09±0.03[c]
7	2.00±0.00	22.40±0.16	67.80±1.57[e]	92.10±5.19[c]	159.90±1.25[e]	1.36±0.05[b]
8	1.90±0.10	22.40±0.16	74.80±4.15[d]	87.30±5.20[c]	162.10±9.05[de]	1.16±0.04[c]
F test	NS	NS	*	*	*	*

Data are the means of 10 replications (mean ± SE)
Mean values followed by the same superscript in columns are not significantly different ($P > 0.05$)
NS nonsignificant
*Significant at 1 %

of Studies (DOS) in Sericulture Science, University of Mysore, Manasagangothri, Mysore. The maggots were kept in a single layer in a perforated tray that was placed over a non-perforated tray in such a way so as to create some space between the two trays. This allowed the healthy maggots to pass through the perforations and pupate in the lower tray. The maggots pupated in about 24 h.

Parasitoid Culture

Nucleus cultures of the parasitoid were obtained from the Central Sericulture Training and Research Institute (CSTRI), Mysore. They were maintained at 23–28 °C and 60–80 % RH by feeding 30 % honey.

The impact of the female *Trichopria* sp. on its progeny production was assessed by allowing 2-day-old big and small parasitoid females to parasitize the 3-day puparia of *E. bombycis* at a ratio of 1:4. The influence of parasitoid age on reproductive efficiency was evaluated by offering 3 day-old puparia of *E. bombycis* to 1–8 day-old *Trichopria* sp females at 4:1. The parameters considered for the purpose included the number of host puparia parasitized, parasitoid developmental duration, progeny production, and sex ratio. To record the effect of body size on the longevity of the adult females of the parasitoid, they were categorized into big and small based on body length using an ocular micrometer and only those exhibiting significant variation in body length were considered for the study. Parasitoid females of both size were fed 30 % honey *ad libitum* to record the longevity. The data were analyzed statistically using ANOVA followed by Duncan's multiple range test (DMRT) with SPSS ver. 10.00 (SPSS Inc., Chicago, IL, USA), and the accepted level of significance was 5 %.

Results and Discussion

Significantly more number ($P < 0.01$) of host puparia (3.00 ± 0.00) was parasitized by big parasitoid females compared to small females (2.10 ± 0.31). While producing the progenies, the parasitoid females of either size took identical duration (2.30 ± 0.15 days) for completing the life cycle. With regard to male, female, and total progeny production by each of the females, the values were significantly superior ($P < 0.01$) with big females (56.00 ± 0.94, 233.40 ± 1.07, and 288.40 ± 1.35, respectively) to small females (36.60 ± 1.40, 122.70 ± 0.89, and 159.30 ± 1.70, respectively). However, the progeny sex ratio (number of females/male) from big (3.65 ± 0.37) and small females (3.37 ± 0.12) remained statistically identical (Table 1). In so far as the longevity of adult females was concerned (Fig. 1), significantly longer ($P < 0.01$) survival of 12.31 ± 0.34

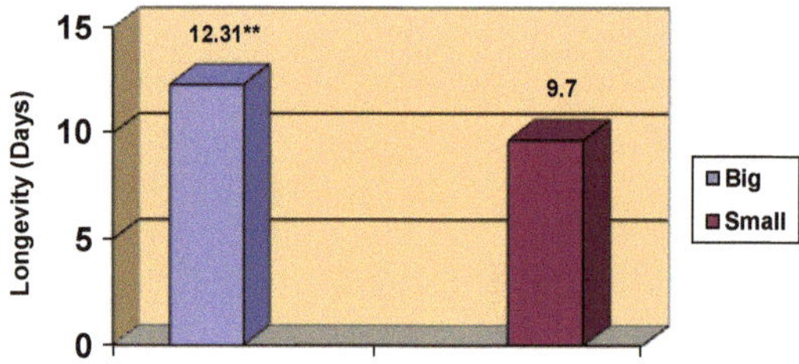

Fig 1 Impact of size of *Trichopria* sp. female on its longevity; data are the means of 10 replications (mean ± SE). **Significant at 1 %

days was observed with big parasitoid females in comparison with the smaller ones (9.70 ± 0.23 days) (Fig. 1).

No significant variation in the rate of parasitism of host puparia by *Trichopria* sp. was noticed when the puparia of 1–8 day-old (at an age difference of 1 day) were offered, with the results among the treatments varying between 2.00 ± 0.00 (1, 2, 4, and 7 day-old) to 1.90 ± 0.10 (3, 5, 6, and 8 day-old). The parasitoid's life cycle too revealed statistical similarity among the treatments. Considering the male, female, and total progenies produced by parasitoid females of different ages, the values for these parameters decreased almost gradually and consistently with increase in female age (Table 2). Statistical analysis of data revealed that the total progenies produced by 1 (204.80 ± 1.61) and 2 day-old (222.20 ± 1.65) females of *Trichopria* sp. were significantly superior to other treatments with the values 187.50 ± 9.90–159.90 ± 1.25 being at par with each other. The data for sex ratio of the parasitoid progenies generated from the females of different ages ranged between 1.51 ± 0.02 (1 day-old) and 1.09 ± 0.03 (6 day-old) without showing significant variation.

The impact of the female size of *Trichopria* sp. on its progeny production was assessed by allowing 2 day-old big and small parasitoid females to parasitize 3 day-old puparia of *E. bombycis* at 1:4. It can be understood that the performance of big parasitoid females has been superior to small females with respect to most parameters related to progeny production. The superiority of the big parasitoid female has been found with respect to the number of puparia parasitized and progeny production that included males, females, and the total population. Obviously, the higher efficiency of big females in terms of the rate of parasitism and progeny production can be considered as attributes associated with parasitoid fitness/quality. Big parasitoid females parasitized nearly 50 % more host puparia than small females. Similar effect was noticed with regard to total progeny produced by big females. Though both big and small parasitoid females have revealed an identical sex ratio in the progenies, significant increase in the rate of parasitism of host with concomitant enhancement in the progeny production, including females, underlines the importance of parasitoids in effective parasitism. Enhancement in female progeny production by big parasitoid females following parasitism of more number of host puparia can be considered advantageous for biological suppression.

Large parasitoid females of *Aphytis melinus* and *Aphytis Lignanensis* (Opp and Luck 1986), *Diglyphus begini* (Heinz 1991), *A. minuta* (Visser 1994), and *A. kamali* (Sagarra et al. 2001) were found to be more fecund than small females. In addition, big females produced more female progenies in *Dinarmus basalis* (Waage and Ming 1984), *Trichogramma* sp. (Waage and Ng 1984; Greenberg et al. 1998), *Trichogramma pretiosum* (Kazmer and Luck 1995) and *Anisopeteramulus calandrae* (Ji et al. 2004). The findings of the present investigation with *Trichopria* sp. are in consonance with those of the above investigators. But, these authors have not presented any account of the sex ratio of the parasitoids studied. Admittedly, they have gathered information on female longevity and host searching efficiency/host encounter, with big parasitoid females scoring over small females with reference to these parameters. Observations related to the influence of parasitoid female size on the sex ratio of progenies are scanty. Nevertheless, Sagarra et al. (2001) recorded that the sex ratio of progenies produced by small females of *A. kamali* was higher than that in the progenies of big females. Our results differ from those of the above authors (Sagarra et al. 2001), with the sex ratio being higher in the progenies produced by big females of *Trichopria* sp.

Table 2 Effect of size of *Trichopria* sp. female on its reproductive performance

Parasitoid size (mm)[a]	No. of puparia parasitized[b]	Developmental duration (days)	Progeny production (no.)			Sex ratio (♀/♂)
			Male	Female	Total	
Big (1.997±0.01)	3.00±0.00	22.30±0.15	56.00±0.94	233.40±1.07	288.40±1.35	3.65±0.37
Small (1.271±0.00)	2.10±0.31	22.30±0.15	36.60±1.40	122.70±0.89	159.30±1.70	3.37±0.12
t-test	*	NS	*	*	*	NS

Data are the means of 10 replications (mean±SE)
NS nonsignificant
* Significant at 1%
[a] Big and small parasitoid females differed significantly in their body length
[b] Out of four puparia provided for parasitism

A considerable enhancement in the parasitoid female longevity was recorded in big females indicating that big parasitoid females have an inherent potential to live longer than small females. This was also observed in *A. zwaelferi* (West et al. 1996), *A. minuta* (Visser 1994), and *A. kamali* (Sagarra et al. 2001). The possession of potential by *Trichopria* sp. to live longer could be considered as one of the fitness characteristics of the parasitoid.

The influence of parasitoid age on reproductive efficiency was evaluated by offering 3-day-old puparia of *E. bombycis* to 1–8-day-old females of *Trichopria* sp. at 1:4. It was found that a number of puparia parasitized by each of the parasitoid females remained similar irrespective of the age of the parasitoid females, indicating that the age of the parasitoid does not influence the rate of parasitism. However, the parasitoid age did have an influence on progeny production with female and total progeny production decreasing as parasitoid age increased. The reason for decrease in progeny production with parasitoid age seems to be associated with parasitoid fecundity, with older ones being less fecund compared to the younger ones. A negative relationship between the parasitoid sex ratio and parasitoid age also has been observed based on production of relatively more female progenies by younger parasitoid females and this decreased relative to male progenies as the parent female age advanced. *Trichopria* sp. is a proovigenic parasitoid with short adult life span. It is quite possible that the parasitoid female's ability to contract the spermatheca to ensure the passage of sperms into the common oviduct for fertilizing the eggs has declined.

Three distinct effects in the rate of parasitism vis-a-vis parasitoid female age have been reported. Singh et al. (1997) using *Trichomelopsis apanteloctena* and Kumar et al. (1990) working on *N. thymus* have recorded no significant effect of parasitoid age on the rate of parasitism. In the second category of effect, Amalin et al. (2005) have demonstrated an increase in the rate of parasitism in young females of *C. etiennei*. In the third category of effect, the rate of parasitism by *Glyptapanteles flavicoxis* decreased as the parasitoid age advanced (Hu et al. 1986). The observations are that the rate of parasitism by *Trichopria* sp. remained almost similar and are in agreement with those documented by Kumar et al. (1990) and Singh et al. (1997).

There are quite a few reports available on the influence of parasitoid age on progeny production and sex ratio. That the parasitoid females are more fecund and/or efficient in progeny production when they are young has been recorded in *T. chilonis* (Guang and Oloo 1990), *C. grandis* (Greenberg et al. 1995), *C. curvimaculatus* (Hentz 1998), and *A. reticulatus* (Honda 1998). In contrast, Kumar et al. (1990) observed nonsignificant influence of parasitoid age on progeny production in *N. thymus*. Significantly superior sex ratio has been realized in the progenies produced by younger parent females of *B. intermedia* (Barbosa and Frongillo 1979), *B. lasus* (Simser and Coppel 1980), *T. chilonis* and *T. astriniae* (Hirashima et al. (1990) and *T. minutum* (Leatemia et al. 1995). The observations recorded with reference to progeny production in the present study employing *Trichopria* sp. fall in line with those reported by Greenberg et al. (1995), Guang and Oloo (1990), Hentz (1998), and Honda (1998) and those regarding sex ratio corroborate the findings of Hirashima et al. (1990) and Leatemia et al. (1995).

Acknowledgments The senior author expresses sincere gratitude to the chairman, Department of Studies in Sericulture Science, University of Mysore.

References

Amalin DA, Pena JE, Duncan RE (2005) Effects of host age, female parasitoid age, and host plant on parasitism of *Ceratogramma etiennei* (Hymenoptera: Trichogrammatidae). Fla Entomol 88(1):77–82

Aruna AS (2007) Developmental dynamics of a eulophid ectopupal parasitoid (*Nesolynx thymus*) on some dipteran hosts. Ph.D Dissertation. University of Mysore, Mysore, India

Avidov Z, Rossler Y, Rosen D (1967) Studies of an Israel strain of *Anagyrus pseudococci* (Girault) (Hym: Encyrtidae). II. Some biological aspects. Entomophaga 12:111–118

Barbosa P, Frongilla E (1979) Host parasitoid interactions affecting reproduction and oviposition by *Brachymeria intermedia*. Entomophaga 24:139–143

Eller J, Van Alphen JK, Svenster G (1998) A field study of size fitness relationship in the parasitoid *Ascobara tabida*. J Anim Ecol 67:316–324

Gangadhar B (2009) Biology and evaluation of *Tetrastichus howardi* (Olliff) (Hymenoptera: Eulophidae) with reference to some Insect hosts. PhD Dissertation. University of Mysore, Mysore, India

Greenberg SM, Morales-Ramos JA, King EG, Summy KR, Rojas Mg (1995) Biological parameters for propagation of *Catolaccus grandis* (Hymenoptera: Pteromalidae). Env Entomol 24:1322–1327

Greenberg SM, DA Nordlund, King EG (1998) Influence of rearing host on adult size and ovipositional behavior of mass produced female *Trichogramma minutum* Riley and *T. pretiosum* Riley (Hymenoptrea: Pteromalidae). Biol Control 11:43–48

Guang LQ, Oloo GW (1990) Host Preference studies in *Trichogramma* Sp. nro Mwanzai. Schulten and Feijen (Hym: Trichogrammatidae) in Kenya. Insect Sci Appl 11:757–763

Heinz KM (1991) Sex-specific reproduction consequences of body size in the solitary ectoparasitoid, *Diglyphus begini*. Evolution 5:1511–1515

Hentz MG (1998) Development, longevity, and fecundity of *Chelonis* sp. nr. *Curvimaculatus* (Hymenoptera: Braconidae), an egg-larval parasitoid of pink bollworm (Lepidoptera: Gelechiidae). Environ Entomol 27:443–449

Hirashima J, Miura K, Miura T (1990) Studies on the biological control of the diamondback moth, *Plutella xylostella* (Linnaeus). IV. Effect of temperature on the development of egg parasitoids *Trichogramma chilonis* and *Trichogramma ostriniae*. Science Bulletin of the faculty of agriculture. Kyushu University 44:81–87

Honda T (1998) Age-related fecundity and learning ability of the egg-larval parasitoid *Ascogaster reticulates* Watanabe (Hymenoptera: Braconidae). Biol Control 13:177–181

Hu Cui, Barbora P, Martinat PJ (1986) Influence of age of female *Glyptapanteles flavicoxis* (March) (Hym: Broconidae) and its periodicity of Oviposition activity on levels of parasitism and reproductive output. Ann Entomol Soc Am 79:280–282

Ji J, Choi WI, Ryoo MI (2004) Fitness and sex allocation of *Anisopteramalcus calandrae* (Hym: Pteromalidae): relative fitness of large females and males in a multipatch system. Ann Entomol Soc Am 97(4):825–830

Kazmer DJ, Luck RF (1995) Field tests of the size-fitness hypothesis in the egg parasitoid *Trichogramma pretiosum*. Ecology 76:412–425

Kumar P, Kumar A, Sengupta K (1990) Parasitoids of uzi fly, *Exorista sorbillans* Widemann (Diptera: Tachinidae) IX. Effect of host and progeny production of *Nesolynx thymus*. Indian J Seri 29(2):208–212

Leatemia JA, Laing JE, Corrigan JE (1995) Effects of adult nutrition on longevity, fecundity, and offspring sex ratio of *Trichogramma minutum* Riley (Hymoenoptera: Trichogrammatidae). Can Entomol 127:245–254

Morales-Ramos JA, Cates JR (1992) Laboratory determination of age-dependent fecundity, and development (Hymenoptera: Pteromalidae). Ann Entomol Soc Am 85:469–476

Odebiyi JA, Bokonon-Ganta AH (1986) Biology of *Epidiniocarsis [= Apoanagyrus] lopezi* (Hymenoptera: Encyrtidae) an exotic parasite of cassava mealybug, *Phenacoccus manihoti* (Homoptera: Pseudococcidae) in Nigeria. Entomophaga 31:251–260

Opp SB, Luck RF (1986) Effects of host size on selected fitness components of *Aphytis melinus* DeBach and *A. lingnanensis* Compere (Hymenoptera: Aphelinidae). Ann Entomol Soc Am 79:700–704

Pitcoairn MJ Gutierrez AP (1992) Influence of adult size and age on the fecundity and longevity of *Tetrastichus incertus* (Hymenoptera: Eulophidae). Ann Entomol Soc Am 88:53–57

Sagarra LA, Vincent C, Stewart RK (2001) Body size as an indicator of parasitoid quality in male and female *Anagyrus kamali* (Hymenoptera: Encyritidae). Bull Entomol Res 91:363–367

Scholler M, Flinn PW (2000) Parasites and predators. In: Subramanyam B, Hagstrum D (eds), Alternatives to pesticides in stored-product IPM, Kluwer, Boston, pp 229–272

Simser DH, Coppel HC (1980) Courtship and mating behaviour of *Brachymeria lasus* (Hym: Chalcididae), an imported gypsy moth parasitoid. Entomophaga 25:349–356

Singh J, Brar KS, Bakhelia DRC, Shenhmar M (1997) Effect of storage on the emergence, sex ratio and parasitization efficiency of *Trichogramma chilonis* Ishii. Third Agricultural Science Congress Punjab Agricultural University, Ludhiana, March 12–15, 249 p

Speight HR (1994) Reproductive capacity of horse chestnut scale insect *Pulvinaria regalis*. Z Angew Entomol 118:59–67

Van Lenteren JC (1983) The development of host discrimination on the preventing of superparasitism in the parasite *Pseudocoila bochei* Weld. Agric Ecosystem Environ 10:143

Visser ME (1994) The importance of being large: the relationship between size and fitness in females of the parasitoid *Aphaereta minuta* (Hymenoptera: Brachonidae). J Anim Ecol 6:963–978

Waage HK, Ming NS (1984) The reproductive strategy of the parasitic wasp. I. Optimal progeny and sex allocation in *Trichogramma evanescens*. J Anim Ecol 53:401–415

Waage JK, Ng SM (1984) The reproductive strategy of a parasitic wasp. I. Optimal progeny and sex allocation in *Trichogramma evanescens*. J Anim Ecol 53:401–415

West SA, Flanagan KE, Godfray, HCJ (1996) The relationship between parasitoid size and fitness in the field study of *Achrysocarioides zwoelferi* (Hymenoptera: Eulophidae). J Anim Ecol 65:631–639

Impact of Adult Size and Sib, Conspecific, and Random Mating in *Trichopria* Sp. on Progeny Production and Sex Ratio

N. Veena and D. Manjunath

Abstract

The gregarious endo-pupal parasitoid, *Trichopria* sp. (Hymenoptera: Diapriidae) was evaluated for progeny production and sex ratio as influenced by adult size and sib, conspecific and random mating. *Trichopria* sp. is a parasitoid of the tachinid fly, *Exorista bombycis* (Louis), which causes a reduction of 10–20% in silkworm (*Bombyx mori* L) cocoon yield in south India. Two-day-old virgin females and unmated males of *Trichopria* sp. were categorized into big and small and were allowed to mate in the following manner: (1) big female × big male, (2) big female × small male, (3) small female × big male, and (4) small female × small male. With regard to sib, conspecific, and random mating, the following mating combinations were set up: (1) mating among the progeny of the same mother (sib mating), (2) mating between the female progeny of one mother (A) and male progeny of another mother (B) (♀A × ♂B) and vice versa (♀B × ♂A), and (3) mating between the progenies of several mothers (random mating). Observations were made on the number of pupae parasitized, parasitoid developmental duration, progeny production, and sex ratio. Both big and small females of *Trichopria* sp. produced significantly more progenies when they mated with big males. Random mated females produced significantly higher progenies than those from sib and conspecific mating.

Keywords

Trichopria sp · Parasitoid size · Progeny production · Mating

N. Veena (✉)
Forest Wood Protection Division, Institute of Wood Science and Technology, Malleswarm, Bangalore 560003, India
e-mail: veena_daya@rediffmail.com

D. Manjunath
Department of Studies in Sericulture Science, University of Mysore, Manasagangotri, Mysore 570006, India
e-mail: manjunath4591@yahoo.co.in

Introduction

Use of parasitoids in pest management programs is bound to pay rich dividends in the long run as these agents can perpetuate under natural conditions with concomitant suppression in the pest

DOI 10.1007/978-81-322-2089-3_5,

populations while rendering the pest management program eco-friendly, long lasting, and cost effective. In biological control program, hymenopteran parasitoids are often mass produced before they are released to field. During mass multiplication, emphasis is laid on the production of high-quality parasitoid with female-biased population. The quality of mass-produced parasitoids are often influenced by the host, parasitoid, and environmental factors. Therefore, it is important to identify the factors and their influence on the production of quality females. These factors include host age and size, host density, host quality, parasitoid age, size and density, mating, adult diet, and temperature, etc.

Literature pertaining to factors influencing parasitoid quality and sex ratio is rather scanty. In nature, most hymenopteran parasitoids reproduce by arrhenotokous parthenogenesis, which is the dominant mode of sex determination, where an unfertilized egg develops into a male and a fertilized one into a female (Heimpel and de Boer 2008). For a female, finding a mate is only the first step towards production of an optimal sex ratio throughout her life. Finding a mate does not always guarantee that the female will produce progenies with optimal sex ratio. The quality of male and his sperms impact the production of daughters at an optimal rate (Boivin 2013). Sib mating is commonly found among gregarious parasitoids that might select multiple mating ways thus diversifying their progenies through multiple males (Ridley 1993).

Trichopria sp. is one among the 20 hymenopteran parasitoids that have been identified to parasitize Tachinid (uzi) fly, *Exorista bombycis*. Being an endo-larval parasitoid of mulberry silkworm, *Bombyx mori*, *E. bombycis* has been known to inflict 10–20% reduction in cocoon yield in the premier silk-producing states of India since 1980. Unlike other parasitoids of *E. bombycis*, *Trichopria* sp. assumes considerable importance because the parasitoid prevails and parasitizes the puparia of *E. bombycis* throughout the year in the cocoon markets located in the traditional districts of Karnataka, thereby underlining its potential as an agent of biological suppression of *E. bombycis*. The present investigation is undertaken to generate information about how adult size and mating condition (sib, conspecific, and random) influenced progeny production and sex ratio in *Trichopria* sp., an idiobiontic pupal parasitoid of a number of lepidopteran and a few dipteran pests. The information thus generated would be of help in understanding a few mechanisms among many others involved in enhancing the efficiency of the mass production unit of the parasitoid.

Material and Methods

Culture of the Host

The hosts, *E. bombycis* puparia, were obtained after allowing the post-parasitic maggots, collected from the government cocoon markets at Ramanagaram/Kollegal (Karnataka), to pupate in the laboratory (23–28 °C and 60–80 % RH) of the Department of Studies (DOS) in Sericulture Science, University of Mysore.

Culture of the Parasitoid

The stock cultures of the parasitoid, *Trichopria* sp., cultured in the pupae of *E. bombycis*, were procured from the Central Sericultural Research and Training Institute (CSRTI) and maintained at 23–28 °C and 60–80 % RH with feeding 30 % honey.

Two-day-old virgin females and unmated males of the parasitoid were used for study. The effect of the adult male and female size of *Trichopria* sp. on mating efficiency and reproductive performance was determined by allowing 2 day-old virgin females and unmated males to mate and offering them 3 day-old puparia of *E. bombycis* for oviposition at 1:4. The following mating combinations were used to record the reproductive performance.

Influence of Parasitoid Size

Mating combinations were: (i) B♀ × B♂, (ii) B♀ × S♂, (iii) S♀ × B♂, and (iv) S♀ × S♂.

Table 1 Effect of size of *Trichopria* sp. on mating efficiency and progeny production

Type of mating	No. of puparia parasitized[a]	Developmental duration (days)	Progeny production (no.)			Sex ratio (♀♀/♂)
			Male	Female	Total	
B♀ × B♂	3.00±0.00[b]	22.37±0.18	45.87±1.00[c]	127.00±1.87[b]	172.87±1.60[b]	2.77±0.09[b]
B♀ × S♂	2.00±0.00[c]	22.25±0.16	25.75±1.33[e]	56.37±1.43[e]	82.12±2.06[d]	2.23±0.13[c]
S♀ × B♂	2.00±0.00[c]	22.37±0.18	68.62±2.13[b]	106.37±1.33[c]	175.0±2.82[b]	1.55±0.04[d]
S♀ × S♂	2.00±0.00[c]	22.25±0.16	39.87±2.09[d]	77.62±0.92[d]	117.50±2.52[c]	2.00±0.10[c]
F test	*	NS	*	*	*	*

Data are the means of ten replications (mean±SE)
Mean values followed by the same superscript in columns are not significantly different from each other $P>0.05$
B♀ × B♂—big female × big male, B♀ × S♂—big female × small male
S♀ × B♂—small female × big male, S♀ × S♂—small female × small male
NS nonsignificant
*Significant at 1%
[a]Out of four puparia provided for parasitism

Impact of Sib, Conspecific, and Random Mating

Mating combinations were: (i) mating between the progeny individuals of a single mother (sib mating), (ii) mating between female progeny of mother A and male progeny of mother B (♀A × ♂B) and vice versa (♀B × ♂A), and (iii) mating between male and female progenies of several mothers.

The number of pupae parasitized, parasitoid developmental duration, progeny production, and sex ratio were obtained in all combinations. The data were analyzed using one-way Analysis of Variance (ANOVA) followed by Duncan's multiple range test (DMRT). The analysis was carried out using SPSS ver. 10.00 (SPSS Inc., Chicago, IL, USA).

Results and Discussion

The mean number of host puparia parasitized by *Trichopria* sp. were significantly different ($P<0.01$) among all the mating combinations involving big and small males and females of the parasitoid which ranged from 3.00±0.00 (B♀ × B♂) to 2.00±0.00 (Table 1). The duration of the parasitoid life cycle in the treatments did not significantly differ among treatments and varied between 22.37±0.18 (B♀ × B♂) and 22.25±0.16 days (S♀ × S♂). With reference to male, female, and total progenies produced by *Trichopria* sp., the mean numbers were significantly different among the treatments ($P<0.01$) and varied from 68.62±2.13 (S♀ × B♂) to 25.75±1.33 (B♀ × S♂), from 127.00±1.87 (B♀ × B♂) to 56.37±1.43 (B♀ × S♂), and from 175.00±2.82 (B♂ × S♀) to 82.12±2.06 (B♀ × S♂), respectively. The number of parasitoid female/male (sex ratio) of the parasitoid in the progeny was significantly higher (2.77±0.09) in mating combination involving B♂ × B♀ compared to S♀ × S♂ combination (1.55±0.04). The sex ratio of the progenies from B♂ × B♀ was significantly greater ($P<0.01$) than the progenies from other mating combinations (Table 1).

Significantly higher ($P<0.01$) number of host puparia (4.00±0.00) was parasitized by *Trichopria* sp., when random mating than the parasitoid females mated with sibs (2.00±0.00) and conspecifics (3.00±0.00). The developmental duration of the parasitoid among various mating combinations was not significantly different among treatments and ranged from 22.37±0.18 (random and ♀B × ♂A) to 22.12±0.12 d (sib mating). Regarding progeny production, significantly ($P<0.01$) higher number of male progeny was obtained from the conspecific mating combinations compared to sib and random mating. Similar result ($P<0.01$) was seen with regard to female (240.12±0.74) and total progenies (333.25±0.97) generated from the random-mated females. Among the mating combinations, the results of the sib mating were significantly inferior ($P<0.01$) with respect to male (51.25±1.43), female (63.50±1.55), and total

Table 2 Effect of sib, conspecific, and random mating on the progeny production of *Trichopria* sp

Mating combinations	No. of puparia parasitized[a]	Developmental duration (days)	Progeny production (no.)			Sex ratio (♀♀/♂)
			Male	Female	Total	
Sib mating	2.00±0.00[d]	22.12±0.12	51.25±1.43[d]	63.50±1.55[e]	114.75±1.97[d]	1.25±0.14[c]
♀A × ♂B	3.00±0.00[c]	22.25±0.16	135.62±1.23[b]	176.75±0.88[c]	312.37±1.76[c]	1.29±0.01[c]
♀B × ♂A	3.00±0.00[c]	22.37±0.18	136.62±0.88[b]	172.50±1.51[d]	309.12±0.85[c]	1.25±0.01[c]
Random mating	4.00±0.00[b]	22.37±0.18	93.12±1.04[c]	240.12±0.74[b]	333.25±0.97[b]	2.57±0.03[b]
F test	*	NS	*	*	*	*

Data are the means of ten replications (mean±SE)
Mean values followed by the same superscript in columns are not significantly different from each other ($P>0.05$)
Sib mating—mating among the progeny of same mother
Conspecific: ♀A × ♂B—mating of female progenies of mother A and male progenies of mother B
Conspecific: ♀B × ♂A—mating of male progenies of mother A and female progenies of mother B
Random mating—mating between male and female progenies of several mothers
NS nonsignificant
*Significant at 1%
[a]Out of four puparia provided for parasitism

progenies (114.75±1.97) when compared with other mating combinations. The progeny sex ratio was significantly higher ($P<0.01$) in random mating (2.57±0.03) in comparison to other combinations (Table 2).

A significantly higher rate of parasitism, progeny production, and sex ratio resulted in B♀ × B♂ mating combination among other combinations, i.e., S♀ × B♂ and S♀ × S♂. Of the remaining mating combinations, though the rate of parasitism remained at par, there was considerable fluctuation in progeny production and sex ratio. Considering female progeny production, which is understood to be an important aspect in mass production of parasitoids, the combination involving S♀ × B♂ scored over others though the treatment remained inferior in terms of sex ratio. Though in terms of sex ratio the combinations of B♀ × S♂ and S♀ × S♂ were significantly superior, which underlines their importance at the present juncture, they do not assume importance in view of significantly fewer female progenies. Therefore, considering the importance of female progeny production, the mating combinations that assume significance were in the following order: B♀ × B♂>S♀ × B♂>S♀ × S♂>B♀ × S♂.

The probable reason for lower progeny production in the mating combination of B♀ × S♂ could be that big females generally show considerable reluctance to accept small males for mating. In the absence of adequate mating, the urge for mating continues with female and, obviously, the urge for oviposition remains at low level. In a separate study, we have observed that the virgin female delays oviposition for a considerable period of time (1–2 days) by which time its progeny production ability seems to reduce considerably. But when mating is allowed, the female oviposits readily. It is, therefore, obvious that for the production of more progenies the female needs to get her mating urge fulfilled first.

Random-mated females of parasitoid have exhibited greater vigor as evidenced by significantly more number of host pupae parasitized, higher number of progenies produced, and the highest sex ratio compared to the females in sib and conspecific mating combinations. These parameters, except sex ratio, with conspecific mating combinations too are significantly superior to those with sib mating. Decline in sex ratio among the conspecifics is due to reduced production of females relative to males.

The observations recorded from random and conspecific mating combinations clearly establish the fact that outbreeding is undoubtedly superior to inbreeding in *Trichopria* sp. from the viewpoint of progeny production and progeny sex ratio. Nevertheless, significant increase in sex ratio in the progenies produced by random-mated females can be construed as a distinct advantage while undertaking mass production of the parasitoid under a biocontrol program. Therefore, considering the importance of female population in the suppression of host (pest) population in bio-control programs, the superiority of random mating among the progenies of several mothers requires special mention.

There is a dearth of information as to the role of mating among the progenies produced by a single mother (sib) and those produced by a few (conspecific) as well as several mothers (random). Fisher (1930) reported that the parasitoids that resort to random mating (panmictic) invest equally in sons and daughters as against Hamilton (1967) who noted that parasitoids in the nature produce female-biased population when their progenies were allocated in host populations when available in patches. Based on his in-depth studies, he proposed a model that explains that all the parental females are equally fecund, the females of the offspring of these females are inseminated before their dispersal due to random mating, and mating groups are limited to the offspring emerging from the host patch parasitized by the parental females. In the present investigation, the fact that the random-mated females of *Trichopria* sp. have invested more in daughters compared to sons is in consonance with the hypothesis propounded by Hamilton (1967).

When big and small males as well as big and small females were confined together, it was observed that big males mated with big females invariably. In some replications, it was observed that when small males mounted on big females, the big males disturbed the mating pair and succeeded in warding off the small male mating with the big female. In the treatment where small females were confined with big and small males, the small males were found to mate with big females although in some replications big males also mated with small females. It can, therefore, be understood that big females prefer to mate with big males and small females accept both big and small males. Although in both treatments, the rate of parasitism remained identical, progeny production, including the females, was significantly higher in the treatment involving mating of big females with big males as against the mating between small females and small males. The total progeny production and female progenies from the mating combination of small females and small males though have declined considerably, the sex ratio remained at par with mating between big females and big males as increase in female production in the former and decrease in female number in the latter relative to male number has remained identical. However, keeping in view the importance of female numbers in the population of a mass-produced parasitoid, any mating combination resulting in the enhancement of female production is desirable. Nonetheless, in the study, we have realized such a performance in the mating combination consisting of big females and big males of *Trichopria* sp.

Ji et al. (2004) observed that in *Anisopteramalcus calandrae* parasitoid did not show distinct preference for mating with males of specific size. The reproduction rate of larger females in this parasitoid remained higher than smaller ones and could be twice as high as smaller females. Shimamoto et al. (2006) working on *Colletes perforator* found that bigger males mate more often than smaller males. Influence of male body size related mate choice has been reported in *Megalothynnus klugii* and *Macrothynnus* sp. (Alcock and Gwynne 1987) with larger males more likely to get mates. Our findings with *Trichopria* sp. are in conformity with those of Shimamoto et al. (2006) and Alcock and Gwynne (1987).

Acknowledgments The help rendered by way of extending research facilities to the senior author during the course of the Ph.D. program, Department of Studies in Sericulture Science, University of Mysore is gratefully acknowledged.

References

Alcock J, Gwynne DT (1987) Courtship feeding and mate choice in thynnine wasps (Hymenoptera: Tiphiidae). Aust J Zool 35:39–47

Anonymous (1999) Natural enemies of Uzifly. Annual Report, Central Sericultural Research and Training Institute, Mysore, p 28

Boivin G (2013) Sperm as a limiting factor in mating success in Hymenoptera parasitoids. Entom Exp et App 146(1):149–155

Fisher RA 1930) The genetical theory of natural selection. Oxford University Press, Oxford

Hamilton WD (1967) Extraordinary sex ratios. Science 156:477–488

Heimpel GE, De Boer JG (2008) Sex determination in Hymenoptera. Ann Rew Entom 53:209–230

Ji J, Choi WI, Ryoo MI (2004) Fitness and sex allocation of *Anisopteramalcus calandrae* (Hym: Pteromalidae): relative fitness of large females and males in a multi-patch system. Ann Entomol Soc Am 97(4):825–830

Ridley M (1993) Clutch size and mating frequency in parasitic hymenoptera. Am Nat 142:893–910

Shimamoto K, Kasuya E, Yasumoto AA (2006) Effects of body size on mating in solitary bee *Colletes perforato*r (Hymenoptera: Colletidae). Ann Entomol Soc Am 99(4):714–717

Importance of Formic Acid in Various Ethological States of *Oecophylla smaragdina* (Fabricius)

VV Vidhu and DA Evans

Abstract

In polymorphic colony of *Oecophylla smaragdina* Fabricus, the content of formic acid (ΓA) in major workcrs was 9.7⊥0.7 mg/gm ti33uc which was one forth of the content of minor workers, one third of the content of intermediate workers and 10 times higher than non-biting reproductive forms. Larvae and pupae had no traces of FA in their body. During the 24 h cycle, the major workers showed a chemical rhythm in content of FA with a peak at noon and low amounts in morning, evening and night. The workers aggressively defend enemies by painful bite with sharp mandibles and simultaneous spray of FA and some of them continued biting until they died, at which time the FA content decreased to non-detectable amounts and the body weight was reduced to 60% of the original weight. Continuous disturbance caused an increase in the content of FA in workers. Continuous bite for 30 min resulted in a significant increase in the content of protein together with a decrease in the content of free amino acids (FAA) and complete stoppage of activity of transaminase enzymes together with a sharp decline in uric acid and urea in both head and thorax. Slight but significant decrease in glucose together with no significant change in the contents of glycogen and lactic acid after 30 min of biting highlights the efficiency of insect tracheal system in supplying O_2. A sharp increase in acetylcholine (Ach) in head and thorax may be the reason for the observed

VV. Vidhu (✉) · DA. Evans
Department of Zoology, University College, University of Kerala, 34, Thiruvananthapuram, Kerala, India
e-mail: vidhuvijayan08@gmail.com

DA. Evans
e-mail: drevansda@gmail.com

A. K. Chakravarthy (ed.), *New Horizons in Insect Science: Towards Sustainable Pest Management*, DOI 10.1007/978-81-322-2089-3_6, © Springer India 2015

immobility or lack of coordination in movements in ants after biting. The observed hyper-proteinemia in ants within 30 min was evidenced by additional bands in electrophorogram.

Keywords

Altruism · Formic acid · *Oecophylla smaragdina*

Introduction

The Asian arboreal weaver ant is a dominant, highly aggressive and predaceous insect. They form the most elaborate communicative organization in the insect world. They form complex organizations through their simple intelligence to emerge as an elaborate social structure. These ants are living as a polymorphic colony possessing industrious workers, reproductive males, females and developing stages (Holldobler and Wilson 1977). The worker ants are known for their painful irritating bite and so they are called as the 'living pesticides' utilized in biological control of the world earliest record in China (Konishi and Ito 1973; Huang and Yang 1987). The workers aggressively defend territory and prey on any organism, which can be over powered by them with sharp mandibles and a simultaneous spray of formic acid (FA) from abdominal glands. Observation in the field and bioassay in the laboratory revealed that FA plays an important role in the life of these worker ants. Biting the enemy or intruder and going on biting until death is a peculiar feature seen in this particular species of worker ants and the biochemistry behind this altruistic behaviour also form the subject matter of this chapter.

Material and Methods

The ant nests were collected in glass jar from the garden plants from the University College Campus, Trivandrum with minimum disturbance and were anesthetized using chloroform. The colony individuals were separated, weighed and the FA content was estimated (Colowick and Kaplan 1963). The FA content of colony individuals and head, thorax and abdomen of worker ants was estimated. The FA content at different time intervals of 24 h such as morning, afternoon, evening and night (7 am, 1 pm, 6 pm and 11 pm) was also estimated. Periodicity studies were conducted during sunny days of April and May 2010. The ants were disturbed for 15 min by gentle tapping on the nests and after that FA content was estimated. The major workers were allowed to bite at a particular site on the body for more than 30 min or till death and after that, FA content was estimated. Standard biochemical techniques were used for the quantitative estimation of protein (Lowry et al. 1951), free amino acids (FAA) (Spies 1957), glycogen (Carrot et al. 1956), lactic acid (Colowick and Kaplan 1963) and acetylcholine (Ach; Augustinson 1957). The activities of aspartate aminotransferase (AsAT) and alanine aminotransferase (AlAT) were estimated as described by Reitman and Frankel (1957). The content of glucose, urea and uric acid were estimated by enzyme kits (Span Diagnostics, India Ltd.).

Abdomen of major workers were dipped in ice-cold distilled water for 2 min, after that fine surgical needle was inserted in to the tip of abdomen, the glands were exposed by gentle pull and viewed under dissection microscope. Photos of poison gland and Dufour's were taken using Sony cyber-shot 4× optical zoom camera. The protein profile of the head and thorax of worker ants, before and after 30 min of continuous bite was done by standard electrophoretic techniques (SDS PAGE). Statistical analysis was done as described by Daniel (2006).

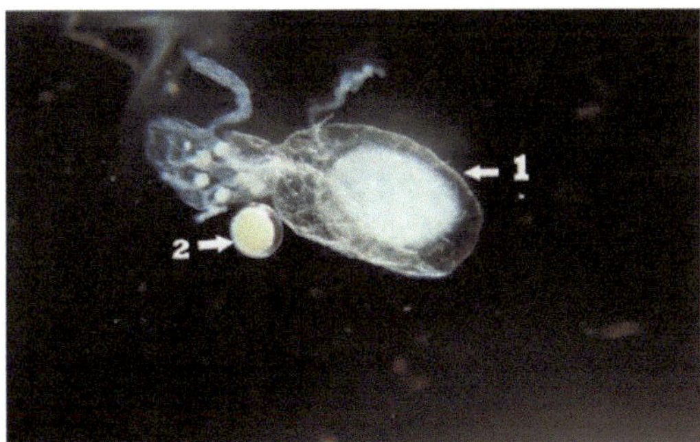

Fig. 1 *1* Poison gland, *2* Dufour's gland

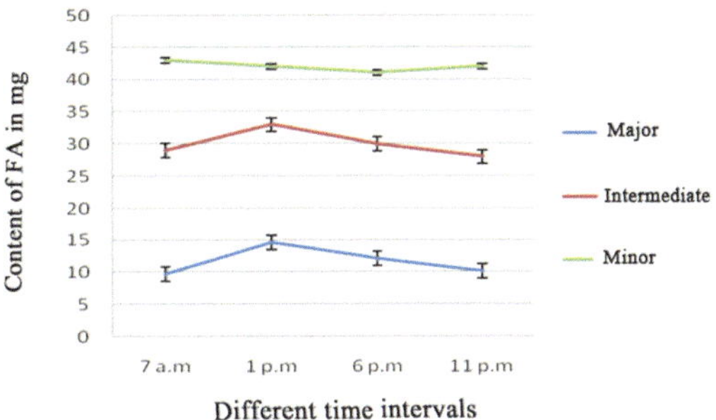

Fig. 2 Fluctuations of FA content of different worker castes of *O. smaragdina* in 24 h cycle (All values are mean ± SD, $n=4$. Values are expressed in mg/gm fresh tissue, Elevation of FA content in major and intermediates at 1 pm was significantly different from FA content at morning. $p<0.01$, $n=4$)

Table 1 The content of formic acid in different individuals in *O. smaragdina*

Individuals		Formic acid
Workers	Major	9.7 ± 0.7*
	Intermediate	29.1 ± 2.3*
	Minor	43.7 ± 3.1*
Major worker	Head	0.63 ± 0.02
	Thorax	0.11 ± 0.01
	Abdomen	42.9 ± 3.6
Winged forms	Males	1.27 ± 0.12
	Females	1.23 ± 0.11
Developing forms	Larva	Nil
	Pupa	Nil

All values are mean ± SD of 4 observations ($n=4$). Values are expressed in mg/gm fresh tissue
*indicates the values are significantly different. ($p<0.01$)

Results and Discussion

The observation on an exposed nest revealed that the polymorphic colony of *Oecophylla smaragdina* possessed winged males, winged females and three types of apterous workers (major, intermediate and minor). FA is present in the poison gland seen in association with Dufour's gland in workers (Fig. 1).

Among different individuals of the colony, worker castes possessed high content of FA but the winged males and females possessed negligible quantities of FA. Among the worker castes, minor workers possessed very high content of FA followed by intermediate and major workers. The major share of FA of workers was in abdomen but the head and thorax possessed negligibly low content of FA. Head and thorax after repeated washing in distilled water and subsequent estimation of FA did not show any detectable quantity. The developing stages such as larvae and pupae had no traces of FA in their body (Table 1).

On estimating FA content of worker castes at different time intervals, it was found that FA of major workers showed a strong periodicity with physical activity. In major workers, FA during afternoon (1 pm) showed a 50 % elevation compared with the contents at morning (7 am). The amount of FA gradually decreased at evening and reached at its initial amount (the content at morning) at night. The intermediate workers also showed a slight but significant elevation of FA content at 1 pm but its fluctuation were not prominent as in major worker (Fig. 2). The minor workers did not show any significant fluctuation in the content of FA during different time intervals of 24 h cycle.

When the major workers were allowed to bite on the body, most of them remained detached from the site of bite and searched for new sites. After three or four bites, each lasting less than 15 s, they did not show any interest in biting at a new site. A small proportion of ants (less than 25 %) did not detached from the site of bite and continued on biting for more than 25 min and after that they became immobile. During the course of bite, they intermittently constricted the mandibles with a simultaneous spray of FA and it was accomplished through forward bend of abdomen. These ants when forcefully detached

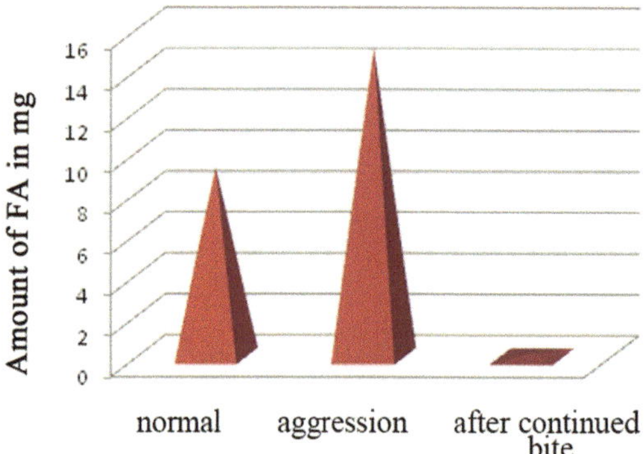

Fig. 3 Variation in the content of FA in relation to different physiological states of major workers of *O. smaragdina* (Values are expressed as mg/gm fresh tissue. All values are mean ± SD, $n=4$)

from bite site showed lack of co-ordination in movements and finally died. The body weight of a major worker with body weight of 11.8 ± 0.8 after 25–30 min of prolonged biting was reduced to 60% of its original weight and it was 6.6 ± 0.4 mg and at that state there was no detectable quantities of FA in the body (Fig. 3). Continuous disturbance made on the nest, resulted in a increase in FA in major workers.

After prolonged bite of 25 min total protein content of head and thorax showed a sharp elevation at the same time the total free amino acid contents showed a sharp decrease. Acetylcholine after 25 min of biting showed a significant elevation both in head and thorax. The content of lactic acid did not show either significant increase or decrease (Table 2).

Prolonged bite resulted in sharp decline in the content of urea and uric acid in the head region but in thorax the amount of the above constituents did not show any significant change (Table 2). Glucose content of head and thorax showed significant decrease after 25 min of biting, but the glycogen content did not show any change both in head and thorax. Activities of both transaminases AsAT and AlAT showed sharp decline in head and thorax after 25 min of continuous bite. Protein profile of both head and thorax showed additional bands when the ants were allowed to bite continuously for 20 min (Fig. 4). Thorax region (T2) of dead ants showed formation of new protein bands in between 30–66 KDa and 25–30 KDa.

Among the worker castes, minor workers possessed very high content of FA followed by intermediate and major workers (Table 1). Minor workers have an important role in nursing the brood (Holldobler and Wilson 1990) and hence, the FA present in them may have role in protection of brood from microbes. It was reported that FA has antimicrobial activity and it is used as an antibacterial agent in poultry industry and in live stock management (Thomson and Hinton 1997). Larvae and pupae had no traces of FA in body. Winged males and females also showed some amount of FA in body (Table 1). Adult males and females did not show any tendency to bite while handling. It was reported that the winged males and females at the time of rain make a nuptial flight and after that the female shed their wings and start a new colony (Sudd 1967). Traces of FA present in body of reproductive females help from predation during their reproductive phase as a solitary, apterous organism.

The minor workers did not show any significant fluctuation in the content of FA during different time intervals of 24 h cycle. It was reported that in *O. smaragdina* temperature has profound influence on the foraging activity (Rossy and Narendran 1988). Hence, it can be concluded that the FA content in the body of major workers is directly proportional to the physical activities such as predation, maintenance of territory and scavenging. It was observed that ants became active in sunny days especially after one or two days of rain, at which time the temperature and humidity will be high. The minor workers were mostly confined within the nest itself and thus not encountering with prey's and intruders and hence FA content did not show any fluctuation during the 24 h cycle.

Poison gland secretion of formicine ants is believed to be the simultaneous secretion of both of these glands. Poison gland secretes FA and Dufour's gland secretes mixtures of hydrocarbons and long chain aliphatic compounds (Hermann and Blum 1968). It was reported that in workers of formicine ant *Formica rufa* L., the formic

Table 2 Changes in Biochemical Profile in different physiological states of major workers of *O. smaragdina*

Biochemical Parameters		Normal		After Prolonged Bite	
		Head	Thorax	Head	Thorax
Protein*		3.54±0.36	3.25±0.29	9.75±0.19	11.1±0.47
Free amino acids*		4.0±0.85	3.27±0.31	2.72±0.40	2.43±0.50
Glucose*		0.39±0.21	0.57±0.01	0.2±0.008	0.06±0.001
Glycogen		0.24±.016	0.19±0.016	0.19±0.08	0.20±0.02
Urea[a]		8.0±0.20	4.0±0.15	1.0±0.003	4.0±0.04
Uric acid[a]		7.0±0.31	1.0±0.001	1.0±0.001	1.0±0.003
Lactic acid[a]		2.10±0.02	3.02±0.001	2.01±0.003	3.03±0.012
Acetylcholine*		3.08±0.45	4.0±0.93	5.62±0.47	5.87±0.39
Transaminases*	AsAT	21.33±1.83	21.41±1.0	1.94±0.02	0.24±0.08
	AlAT	247.63±3.2	230.03±5.3	1.62±0.01	0.32±0.005

Values are expressed in mg/100 mg fresh tissue
All values are mean±SD, $n=4$
*Control and Test values are significantly different at $p<0.01$
[a]Values are expressed in μg/100 mg fresh tissue

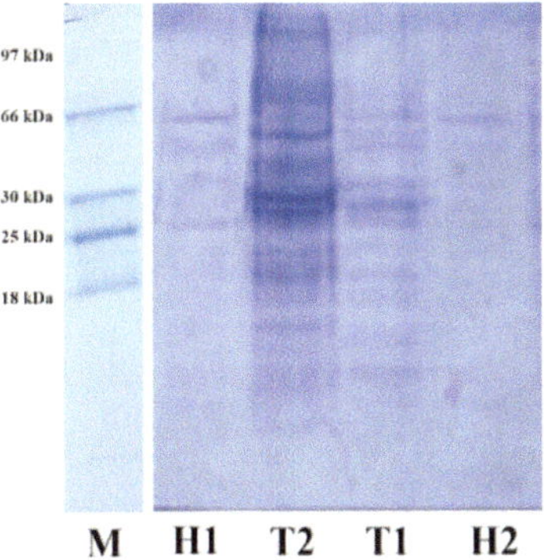

Protein Profile Indicates

M -Marker

H1-Head of normal ants

T1-Thorax of normal ants

H2-Head of ants after 30 min. of prolonged bite

T2-Thorax of ants after 30 min. of prolonged bite

Fig. 4 Electrophorogram-SDS PAGE

acid and some hydrocarbons have a combined effect and their relative concentrations regulate the intensity and duration of the alarm behaviour. In *O. smaragdina* the FA in the body of workers were related to behaviour (Lofqvist 2003). It was observed that when there was continuous aggression there will be a sharp elevation in the content of FA with transfer of it from abdominal end to mandibles together with the increased production of FA. The worker ants on prolonged biting showed absolute loss of FA and at the same time the body weight of a major worker with body weight of 11.8±0.8 was reduced to 60% of its original weight, and it was 6.6±0.4 mg (Fig. 2). Even though FA is a secondary metabolite, in the absence of that defensive chemical the existence of ant itself appeared as immaterial.

In a comparative study on the amount of various bio chemicals in head and thorax of ants, before and after 30 min of prolonged bite, revealed that protein was doubled with a sharp decrease in FAA at both sites. In larvae of *Oryctes rhinoceros*, it was observed that exposure of high temperature of 45 °C for 2 h or after the infection of *Bacillus thuringiens* resulted in sharp elevation of haemolymph protein content (Adhira et al. 2011).

Total prevention of protein catabolism through almost complete stoppage of both aminotransferases can contribute to protein elevation. Sharp decline in both urea and uric acid in thorax and abdomen within 30 min of extreme stress (continuous biting) can also be suggested as an index of protein catabolism. The normal transaminase activities of insects are very much higher than that in mammalian body (Evans and Kaleysaraj 1992; Subhramanyam et al. 1998).

Within a period of 30 min of continuous bite, the glucose content of head and thorax of *O. smaragdina* showed a slight decrease but was significantly different from control and at the same time no significant change on the contents of glycogen was observed. This may be due to the difference in energy metabolism between insects and mammals. The content of lactic acid in head and thorax also showed no significant difference from control. In mouse, 30 min of swimming exercise caused a sharp elevation in the content of lactic acid together with sharp decline in the contents of both glycogen and glucose (Evans et al. 2002). In insects, all the tissues are richly supplied with direct supply of oxygen by elaborate tracheal system with a pair of lip-like openings called spiracles, which lead to a tree-like series of tubes called tracheoles, which repeatedly divide into numerous microscopic ends and penetrate into individual body cells. Unlike vertebrate systems, this gives every insect cell a continuous airline to the atmosphere and hence chances of anaerobic glycolysis will be negligibly low (Candy and Kilby 1959) and that also might be the reason for no accumulation of lactic acid in the worker ants. Zebe and McShan (1957) have studied the lactic acid metabolism in different orders of Class Insecta and showed that glycolysis is unimportant in insects due to the peculiarity of respiratory system.

The content of acetylcholine showed a sharp elevation in head and in thorax after biting. Insects on treatment with organophosphorus insecticides also showed a sharp elevation of acetylcholine and subsequent paralysis (Evans and Kaleysaraj 1991). The observed immobility or lack of co-ordination in movements of ants after continuous bite may be due to accumulation of acetylcholine in body. Enzymes are involved in all the chemical reactions within living organisms. Respiration, growth, muscle contraction, digestion, nerve conduction and so on, all are enzyme mediated reactions. The AsAT and AlAT are the principal enzymes concerned with non-essential amino acid metabolism and gluconeogenesis and interrelates carbohydrate and protein metabolism by catalyzing the incorporation of pyruvate and oxaloacetate from alanine and aspartate, respectively (Wilson 1973; Ortem and Neuhas 1982). Sharp decrease in content of total FAA together with decrease in content of nitrogenous waste materials and almost complete suppression of transaminase activity strongly supports the formation of new protein. Hyper protienemia within a short period of 30 min of continuous biting was evidenced by additional protein bands in electrophorogram (Fig. 4).

Aggressive defence strategies observed in social Hymenoptera, social aphids and termites demonstrate that defensive behaviour are usually detrimental to the defenders but facilitate the survival of the reproductives and brood within the nest. In honey bees, the barbed sting of worker bees eviscerate and kills the workers after stinging, suggesting that the cost of altruism to individual is high but trivial to the entire colony (Vincent and Ring 2003). All the above mentioned biochemical mechanisms acted in favour of worker ants to eliminate the intruder and to sacrifice their life for the sake of the colony.

References

Adhira MN, Syamala DG, Evans DA (2011) Effect of bacillus *thuringiensis israelensis* infection on the biochemical and haematological profile of the hemolymph of *Oryctes rhinoceros* (L.) (Coleoptera:Scarabaeidae) larva. Entomon 35(4):1–5

Augustinson BK (1957) In: Glick D (ed) Methods of biochemical analysis, vol 5. Interscience publishers, New York, pp 1–63

Candy DJ, Kilby BA (1959) Site and mode of trehalose biosynthesis in the locust. Nature 183:1594–1595

Carrot NV, Longly RW, Roe JH (1956) The determination of glycogen in liver and muscle by the use of anthrone reagent. J Biol Chem 220:583–593

Colowick SP, Kaplan NO (1963) Methods in enzymology, vol 3. Academic, London, pp 285–291

Daniel WW (2006) In: Biostatistics – A Foundation for Analysis in Health Sciences, 7th edn, Georgia State University, Wiley and Sons (Asia) Pvt. Ltd., pp. 295–308.

Evans DA, Kaleysaraj R (1991) Quassin: a mosquito larvicide with selective toxicity. J Ecotoxicol Environ Monit 4:243–249

Evans DA, Kaleysaraj R (1992) Total protein, amino acid profile and certain related enzymes in adults and developing stages of Culex quinque fasciatus and effect of quassin. Indian J Comp Anim Physiol 10:46–54

Evans DA, Subramaniam A, Rajasekharan S, Pushpangadan P (2002) Effect of Trichopin zeylanicum leaf extract on the energy metabolism of mice during exercise and at rest. Indian J Pharmacol 34:32–37

Hermann HR, Blum MS (1968) The Hymenopterus poison apparatus-VI. *Camponotus pensilvanicus* (Hymenoptera: Formicidae). Psyche 75:216–227

Holldobler B, Wilson EO (1977) Weaver ants. Sci Am 237:146–154

Holldobler B, Wilson EO (1990) The ants. Belknap Press of Harvard University, Cambridge

Huang HT, Yang P (1987) Ancient cultured citrus ant used as biological control agent. Bioscience 37(9):665–671

Konishi M, Ito Y (1973) Early entomology in East Asia. In: Smith RF, Mittler TE, Smith N (eds) History of entomology. Annual Review Inc., Palo Alto

Lofqvist J (2003) Formic acid and saturated hydrocarbons as alarm pheromones for the ant Formica rufa. J Insect Physiol 22:1331–1346

Lowry OH, Roseborough NT, Farr A, Randall LRJ (1951) Protein measurement with the Folin phenol reagent. J Biol Chem 173:263–275

Orten JM, Neuhas OW (1982) Hormones; Chemical regulators of metabolism. In: Human Biochemistry,10th edn. (Orten JM, Neuhaus OW., Eds), The C.V. Mosby. Co, St. Louis, pp.592-648

Reitman S, Frankel S (1957) A colorimetric method for the determination of serum glutamic oxaloacetic acid and glutamic pyruvic transaminases. Am J Clin Pathol 28:56–63

Rossy PA, Narendran TC (1988) Effects of soil surface of temperature on the foraging activity of *Oecophylla smaragdina* (Hymenoptera: formicidae) of India. J Zool Res 1(2):151–152

Spies JR (1957) Colorimetric procedures for amino acids. In: Colowick SP, Kaplan NO (eds) Methods in enzymology, vol 3. Academic, New York

Subhramanyam A, Evans DA, Rajasekharan S, Pushpangadhan P (1998) Hepato protective activity of *Trichopus zeylanicas* extreact against paracetamol induced hepatic damage in rats. Indian J Exp Biol 38:385–389

Sudd JH (1967) An introduction to the behavior of ants. Edward Arnold Ltd., London

Thompson JL, Hinton M (1997) Antibacterial activity of formic and propionic acids in the diet of hens on *Salmonellas* in the crop. Br Poult Sci 38(1):59–65

Vincent HR, Ring TC (2003) Encyclopedia of Insects, Academic Press, California, USA; p.1048

Wilson RP (1973) Nitrogen metabolism in channel cat fish, *Ictalurus punctatus*. Tissue distruction of aspartate and alanine amino transferases and glutamic dehydrogenases. Comp Biochem Physiol 466:617–624

Zebe EC, Shan MWH (1957) Lactic acid and α-glycerol phosphate dehydrogenases in insects. J Gen Physiol 40:5

Metabolites of *Pseudomonads*: A New Avenue of Plant Health Management

Janardan Jani, Noushad Parvez and Deepak Mehta

Abstract

Biotic threat in the form of insect pests is a major cause for yield loss in agricultural systems and an important factor affecting the structure and productivity of crop plant communities. However, bacteria antagonistic to plant pathogens and harmful insects are known to reduce plant contagion. These bacteria have been extensively studied in agricultural systems where they significantly contribute to soil suppressiveness, which is the natural potential of soils to inhibit plant pathogens. The genus *Pseudomonas* has been reported extensively not only for preventing infectious diseases but also promoting plant growth. Many *Pseudomonas spp.* have been reported for the presence of genes that are responsible for construct, and produce an array of imperative metabolites such as indole acetic acid (IAA), 2–4 di-acetyl phluoroglucinol (DAPG), HCN, phenazines, lipodepsipeptide, pyrrolnitrin, pyoverdin (Pvd) and pyochelin, etc. for such twofold and significant tasks. Improved *Pseudomonas* strains for their potential genes and control over the transformation of proteins responsible for the formation of such metabolites and also their desirable expression in the plant vicinity are nowadays a major concern throughout the world. Loss of biodiversity of such *Pseudomonas spp.* is proved likely to reduce the resistance of plant communities to soil borne diseases and highlight that the interrelationships between plants and such microorganisms need closer consideration to understand the functioning of ecosystems and to manage agricultural systems in an environmentally friendly and sustainable way.

Keywords

Metabolites · *Pseudomonads* · Plant Health Management

J. Jani (✉) · N. Parvez · D. Mehta
Biological Control Research Laboratory, Anand Agricultural University, Anand, Gujarat 388110, India
e-mail: janchait01@yahoo.com

A. K. Chakravarthy (ed.), *New Horizons in Insect Science: Towards Sustainable Pest Management*,
DOI 10.1007/978-81-322-2089-3_7, © Springer India 2015

Introduction

Biotic threats for plants, such as insects, soil-borne pathogens, attack plants and cause important damage to crop health. Ecological health concerns confine the use of chemicals in soil to control root diseases. The treatment of soil or planting material with certain strains of plant-beneficial, root-colonizing *Pseudomonas* spp. is a promising alternative to control biotic threats. Strains of *Pseudomonas* spp. are also recognized for plant growth promotion activities (Kumar 1998; Kaur et al. 2007; Nashwa et al. 2008, Mahesh et al. 2010; Sharma et al. 2011).

Pseudomonas genus comprises more than hundred species with different lineages, groups and subgroups based on multilocus sequence analysis. Many of the plant associated strains belong to *Pseudomonas fluorescens* group, which currently includes more than 50 named species (Yamamoto et al. 2000; Mulet et al. 2010). *Pseudomonas* spp. can utilize a variety of organic compounds (Frias et al. 1994; Olalemi and Arotupin 2012) as energy sources, and produce an array of secondary metabolites foremost as 2, 4-diacetylphloroglucinol (DAPG), lipopeptides, phenazines, pyrrolnitrine, pyochelin and hydrogen cyanide (Keel et al. 1992; Haas and Defago 2005). Certain strains live in a commensal relationship with plants, protecting them from infection by pathogens that would otherwise cause disease. Control of root diseases by beneficial bacteria involves a blend of possible mechanisms that may complement each other. Direct antagonism against the pathogen by production of diffusible or volatile antibiotic compounds or by inactivation of virulence traits of the pathogen is considered to be a primary mechanism of bio control (Diby et al. 2005; Dikin et al. 2007). Another important mechanism is the indirect inhibition of the pathogen by bacterial stimulation of defence responses in the plant host. As such, *Pseudomonas* spp. functions as key components of ecological processes that suppress plant diseases in agricultural and natural environments.

Metabolites in Pseudomonas: Biosynthetic Pathway and Genetic Organization

The genes responsible for the synthesis of antibiotics in *Pseudomonas* spp. are highly conserved. Phenazines are unusual nitrogen-containing heterocyclic molecules, of which over 60 different derivatives have been identified in nature. Phenazines may play important roles in both symbiotic and pathogenic microbe-microbe and microbe-host interactions. Phenazines have been classified as broad spectrum antibiotics and have been shown to inhibit a wide variety of plant pathogenic organisms (Poritsanos et al. 2006). Pyrrolnitrin or 3-chloro-4- (2′-nitro-chloro-phenyl)-pyrrole (Prn) is produced by a diverse number of *Pseudomonas* spp. such as *Pseudomonas aureofaciens*. It inhibits growth of plant parasitic fungi and bacteria (Angayarkanni et al. 2005).

Pyrrolnitrin is synthesized from tryptophan through a biochemical pathway determined primarily by radio labeling studies using tryptophan analogs. In recent years, it has been identified that biosynthetic and regulatory loci are required for the production of these antibiotics. Besides pathway-specific regulators, a number of global regulatory elements are involved in the control of the biosyntheses of these compounds, among them the sigma factors RpoD, RpoS and RpoN and a two-component system composed of the sensor kinase GacS and the response regulator GacA (Bjornlund et al. 2009; Neidig et al. 2011; Lalaouna et al. 2012). In addition, numerous biotic and abiotic signals may influence production of these antipathogenic compounds, including different mineral and carbon sources and metabolites released by microorganisms and plants. Recently, in *P. fluorescens*, CHA0, which keeps the antibiotics DAPG and PLT at a fine-tuned balance can be affected by microbial and plant phenolics. By the help of reporter system based on autofluorescent green (GFP) and red (DsRed) proteins to monitor changes in the balance of DAPG and PLT expression in the rhizosphere of healthy and pathogen-attacked plants.

Gene activation (GacS/GacA) signal-transduction pathway is one of the proposed models to explain the post-transcriptional control of metabolite in *Pseudomonas* spp. and it has been proved in *P. fluorescens* CHAO (Bjornlund et al. 2009; Reddy 2009). GacS/GacA pathway plays an important role in contributing the antipathogenicity against *Pythium ultimum* and it also induces stress in nematode (Haas and Defago 2005). It confirmed that *Pseudomonas* spp. control over nematode infection (Siddiqui et al. 2006) and GacS\GacA pathway, DAPG is the most important factor in stress enhancement in *Caenorhabditis elegans* (Neidig et al. 2011).

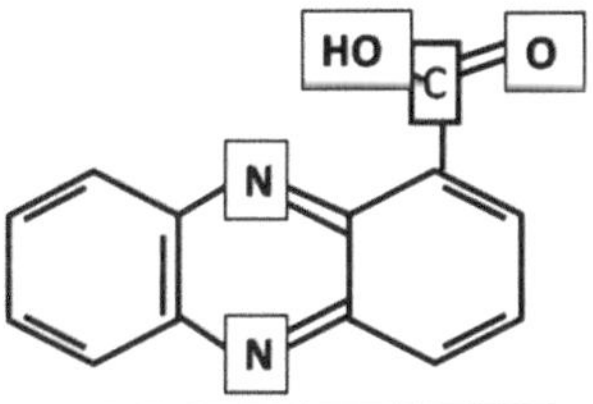

Phenazine 1- carboxylate

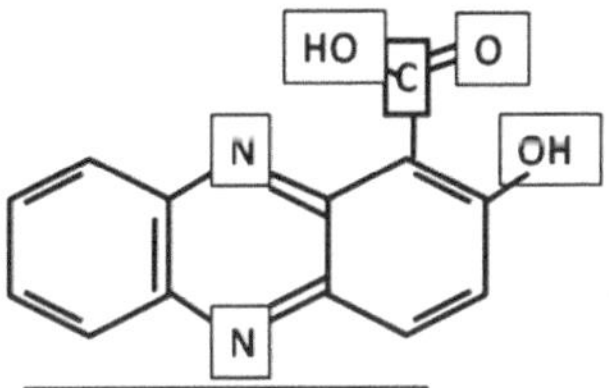

2-hydroxyphenazine 1- carboxylate

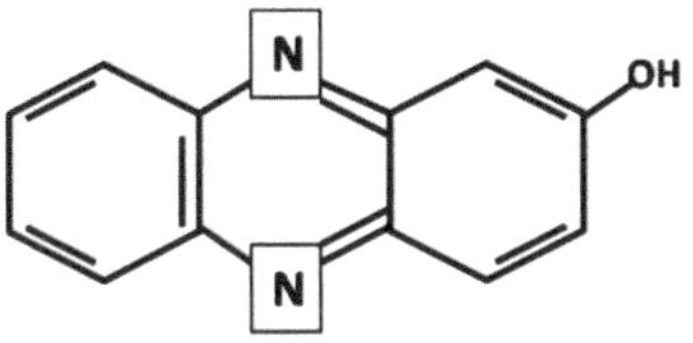

2-hydroxyphenazine

Pyrrolnitrin (Prn)

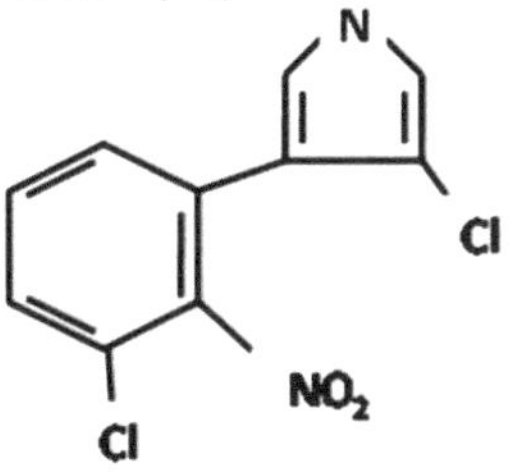

Role of Secondary Metabolites in Bacterial supremacy

Pseudomonads are ubiquitous soil microflora playing an important role in plant protection. The suspicion benefits of these depends on various factors, such as their ability to efficiently exploit root exudates and to withstand predation by nematodes and protozoa (de Mesel et al. 2004; Jousset et al. 2008; Pedersen et al. 2009; Rosenberg et al. 2009). Bacteria have evolved an array of anti-predatory resistance mechanisms, such as toxicity, and in soil, unpalatable or toxic strains again competitive advantage in presence of predators (Ronn et al. 2001; Jousset et al. 2008). Extracellular metabolites of *Pseudomonas* spp. drive complex interactions with predators, affecting their physiology and behaviour. Secondary metabolite works specifically on predators, acting as repellents, stressors or toxics. Production of secondary metabolites by biocontrol bacteria serves multiple functions, and metabolites protecting plants against pathogens improve bacterial resistance against predator (Bjornlund et al. 2009; Pedersen et al. 2009; Neidig et al. 2011).

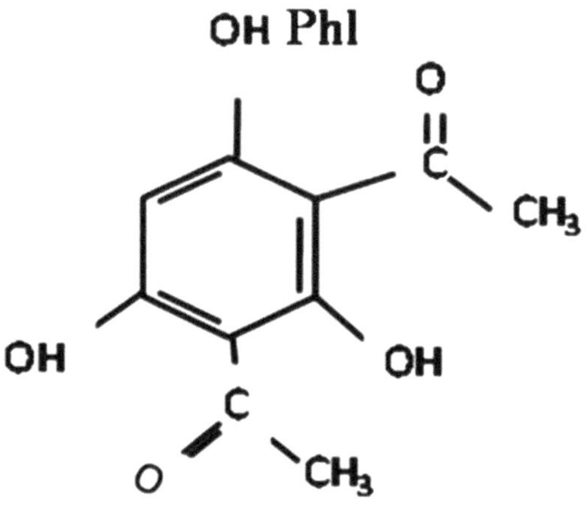

OH Phl
O
C
CH_3
OH
OH
C
O
CH_3

Potentiality, Mechanisms and Factors Of Bacterial Metabolites

The antibiotics pertain to polyketides, heterocyclic nitrogenous compounds and lipopeptides have broad-spectrum action against several plant pathogens, affecting crop plants. Secondary metabolites reduce various pathogens in vitro and some of these metabolites have been detected in the rhizosphere (Leon et al. 2009). Role of biocontrol bacterial agents and their antagonistic properties was evidenced from the comparison of wild-type strains, non-producing insertion or deletion mutants, and complemented derivatives (Raaijmakers and Weller 1998; Haas and Defago 2005). Moreover, shuffling of specific genes encoding biocontrol trait to less efficient strains in respective of particular trait may confer or enhance biocontrol potential to non-producing *Pseudomonas* respectively. Metabolites play roles in different plant-protection mechanism, e.g. pyoverdine in ISR and competition, and DAPG in ISR and antagonism (Maketon et al. 2012). In addition, certain strains display multiple plant-beneficial traits, such as some exhibits ACC deaminase activity and produce the phytohormone indole-acetic acid, pyoverdine, DAPG and hydrogen cyanide. The modes of action of these secondary metabolites are partly understood.

The phenazines, which are analogues of flavin coenzymes, inhibit electron transport and are known to have various pharmacological effects on pathogenic cells (Byng et al. 1979). In the presence of ferripyochelin, phenazines catalyze the formation of hydroxyl radicals, which damage lipids and other macromolecules. 2, 4-diacetylphloroglucinol (DAPG), Phl is the best-known phloroglucinol compound in a family of related. Phenazines Phl causes membrane damage to *Pythium* spp. and is particularly inhibitory to zoospores of this oomycet. Pyrrolnitrin has been described as an inhibitor of fungal respiratory chains. Synthetic analogues of pyrrolnitrin have been developed for use as agricultural fungicides (Siddiqui et al. 2006). Cyclic lipopeptides, which include biocontrol-active substances have surfactant properties, and are able to insert into membranes and perturb their function (Groboillot et al. 2011). Cyanide ion derived from HCN is a potent inhibitor of many metalloenzymes, especially copper-containing cytochrome c oxidases. Tryptophan-dependent IAA synthesis involved various enzymes of *Pseudomonas* that help in the development of crop plants. Apart from antibiotic metabolites, it also secretes organic acids and enzymes responsible for mineral solubilization and antipathogenic activity (Fig. 1).

Siderophores

Antagonism and competition may concern the attainment of organic substrates released by seeds and roots (Rudresh et al. 2004; Kamilova et al. 2005) and micronutrients such as soluble iron. Iron acquisition entails the production of siderophores, perceptibly fluorescent pyoverdines. Once it complexes to ferric iron in soil or the root zone, the siderophores are then taken up using outer membrane receptors. In a perspective of biological control, competition for iron involves the synthesis of siderophores of higher affinity compared with siderophores used by phyto-pathogens (Lemanceau et al. 1992). Proper upholding of trait and conditions can achieve the control of pathogens by competition where siderophore mediated iron competition may also prevent growth of *Escherichia coli* O157:H7 on food human pathogenic strain of *E. coli* (McKellar 2007).

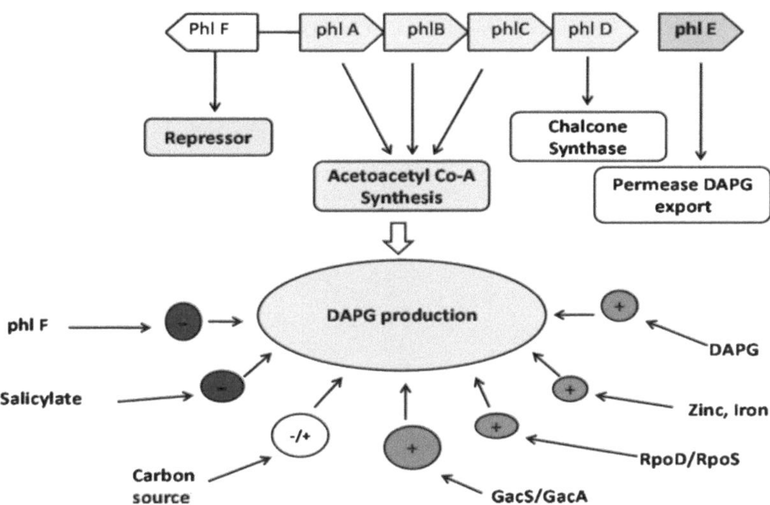

Fig. 1 Factors and genes (phl A-F) cascade which control level of 2-4 Di acetyl phluoroglucinol (*DAPG*)

Phyto-hormones

Biocontrol agents in vicinity, makes the plant more efficient in fighting back against phyto-pathogens. *P. fluorescens* WCS417r, ISR involves the jasmonate and ethylene as signals. They activate plant genes involved in defence mechanisms. ISR can be triggered by contact of the plant to certain cell surface components of biocontrol strains, such as lipopolysaccharides and flagella, or exposure to biocontrol metabolites including pyoverdine and DAPG (Wang et al. 2000; Pieterse et al. 2003; Bakker et al. 2007). Deamination of the ethylene precursor 1-aminocyclopropane-1-carboxylate (ACC) can diminish the quantity of plant ACC for ethylene synthesis and introduction of ACC deaminase locus into *Pseudomonas* spp. improved suppression of phyto-pathogens (Glick 2005; Blaha et al. 2006; Rezzonico et al. 2007; Weller 2007) such as *Pythium* damping-off. *Pseudomonas* biocontrol strains produce extracellular lytic enzymes (Diby et al. 2005), but genetic evidence for an actual role in biocontrol is lacking. Although Prn was first identified in 1964, little is known regarding mechanism of action. Prn inhibits growth but does not kill the target organism (Imanaka et al. 1965; Table 1).

Based solely upon its similarity to other phenyl pyrroles, it seems likely that Prn may interfere with an aspect of normal membrane function in target organisms. Reduced pyocyanine is hypothesized to interfere with the normal membrane function in target organisms resulting in increased levels of toxic intracellular active oxygen products, such as superoxide (O^{2-}), hydrogen peroxide (H_2O_2), hydroxyl radical (OH−). Prn has been described as an inhibitor of fungal respiratory chains (Arima et al. 1965). Synthetic analogues of Prn have been developed for use as agricultural fungicides. Cyclic lipopeptides, which include biocontrol-active substances have surfactant properties, and are able to insert into membranes and perturb their function (Groboillot et al. 2011). Cyanide ion derived from HCN is a potent inhibitor of many metalloenzymes, especially copper-containing cytochrome c oxidases (Hagins et al. 2009). Tryptophan-depen-

Table 1 Prominent Siderophores and Producer *Pseudomonas* spp

Siderophore	*Pseudomonas* species
Pseudobactin	*Pseudomonas putida*
Pyoverdin (pvd)	*Pseudomonas aeruginosa, Pseudomonas fluorescence*
Cepabactin	*Pseudomonas chlororaphis*
Chrysobactin	*Pseudomonas cepacia*
Ornibactin	*Pseudomonas aeruginosa*
Desferrioxamine b&e	*Pseudomonas fluorescence, Pseudomonas cepacia*

Table 2 List of metabolites of *Pseudomonas* acquiesced for plant health

Metabolites	Characteristics	Name	Promising strains
Class I			
Antifungal antibiotic	Non volatile	Phenazine Pyrrolnitrin Pyoluteorin 2,4 Diacetylphloroglucinol (DPAG, Phl)	*Pf* CHAO *Pseudomonas chlororaphis* *Pseudomonas aureofaciens* *Pf-5*
	Cyclic peptides	Lokisin	*Pf dr54*
	Volatile	HCN	*Pf* CHAO
Antinematode		2,4 diacetylphloroglucinol (DPAG, Phl) HCN	*Pf* CHAO
Antibacterial antibiotic	Secondary metabolites	Pseudomonic acid Oomycin Pyrrolnitrin	*Pseudomonas putida*
Antiviral antibiotic	Secondary metabolite	Karalicin	*Pseudomonas rotkzi*
Class II			
Siderophores	Coloured pigment Ion chelating	Pseudobactin Pyochelin (Cu^{++} and Zn^{++}) Pyoverdin	*Pseudomonas aeruginosa* *Pseudomonas fluorescence*
Phyto-hormones	Auxins	IAA	*Pf-5*
Antifungal enzymes		Chitinases β-1,3-glucanases β-1,4-glucanases Lipase (EC.3.1.1.3)	*Pf-5*
Acids	Phosphate solubilization	Gluconic acids	*Pseudomonas cepacia*

dent IAA synthesis involved various enzymes of *Pseudomonas*, which help in the development of crop plants. Apart from antibiotic metabolites, it also secretes organic acid and enzymes, which are responsible for mineral solubilization and antipathogenic activity.

Insecticidal Activity in Root Associated Pseudomonads

It is reported indisputably that some of the root-associated *Pseudomonas* strains express chitinase and/or chitinase like metabolite and show potent insecticidal activity during in vitro studies (Mette and Jan 1999; Nandakumar et al. 2007; Ramarathnam et al. 2011). Moreover anti-insect activity is due, in part, to the production of a novel protein toxin (Ramarathnam et al. 2011) that is related to insect toxins produced by entomopathogenic bacteria associated with insect-invading nematodes. However, *P. stutzeri* and *P. fluorescens* have been found to produce chitinolytic activity in batch growth cultures, but the enzyme activity has not been characterized in detail. *P. aeruginosa* has been reported fleetingly for the chitinase activity. It posses two isozymes (FI and FII) for endochitinase activity against insect (Wang and chang 1997; Mette and Jan 1999) (Table 2).

Interactions in Rhizosphere

Pseudomonas spp. is adapted to survival in soil and colonization of plant roots (Kiely et al. 2006; Arima et al. 1964). Biological control of threats by pseudomonas depends on biochemical compound that attack the parasites. Such mechanism principally affected by effective colonization capability of the bioagent in rhizosphere and phyllosphere. Effective colonization on the plant system as well as inside root tissues system is a specific trait and may noticeably aid biocontrol activity. Notz (2001) checked the effects of four different crops as biotic factor on the expression of Phl genes and colonization of *P f* CHAO. Maximum expression of Phluoroglucinol and colonization of culture was exhibited in maize plant. On the other side in absence of adequate colonization posses diminish biocontrol activity. It is well reported an inversely proportional relation between the numbers of bacteria present on the wheat root and the number of take-all lesions seen on the plant (Schippers et al. 1987; Bull et al. 1993). However, reports also claim absence of any effects of colonization on biocontrol activities of bacteria (Roberts et al. 1994).

It is also desired to document the impact of the deviated expression of such metabolites on native rhizospheric microflora. During the study of impact of the wild type *P. fluorescens* strain CHA0-Rif and CHA0-Rif/pME3424 which is a Phl^- and Pyoluteorin overproducing derivative on the indigenous culturable bacterial and fungal populations in the cucumber rhizosphere has been investigated. Compared with untreated plants, Natsch et al. (1998) demonstrated that neither CHA0-Rif nor CHA0-Rif/pME3424 affected the frequency of dominant bacterial groups, whereas Girlanda et al. (2001) observed a detectable influence on the culturable fungal population.

Among all metabolites of *Pseudomonas* spp. Pyoverdin, Pyoluteorin and Phluoroglucinol (Phl) studied as major metabolites to control fungal disease on crops. Phluoroglucinol efficiently contributes in control of *Caenorhabditis elegans* and other nematodes. Expression of the metabolite depends on biotic and abiotic factors/conditions. Biocontrol *Pseudomonads* are not specific for one plant. The composition of root exudates is species and cultivar specific and the differences in their quantity and/or quality can modulate the production of metabolites. In addition, the role of intra-species and inter-species signalling is the fruitful aspect as a scientific research with equally relevant applications. For example, increasing understanding of the role of N-acyl homoserine lactone (NAHL) signal molecules in antifungal metabolite production and the identification of promoters that can be induced in the rhizosphere is providing new approaches for the development of novel biocontrol agents.

References

Angayarkanni T, Kamalakannan A, Santhini E, Predeepa D (2005) Identification of biochemical markers for the selection of *Pseudomonas fluorescens* against *Pythium* spp. In: Asian conference on emerging trends in plant-miceobial interactions. University of Madras, Chennai

Arima K, Imanaki H, Kousaka M, Fukuta A, Tamura G (1964) Pyrrolnitrin, a new antibiotic substance, produced by Psneudomonas. Agric Biol Chem 28:575–576

Arima K, Imanaki H, Kousaka M, Fukuda A, Tamura G (1965) Studies on pyrrolnitrin a new antibiotic. I. Isolation and properties of pyrrolnitrin. J Antibiot A 18:201–204

Bakker PAHM, Pieterse CMJ, van Loon LC (2007) Induced systemic resistance by fluorescent *Pseudomonas* spp. Phytopathology 97:239–243

Bjornlund L, Ronn R, Pechy-Tarr M, Maurhofer M, Keel C, Nybroe O (2009) Functional GacS in *Pseudomonas* DSS73 prevents digestion by *Caenorhabditis elegans* and protects the nematode from killer flagellates. ISME J 3:770–779

Blaha D, Prigent-Combaret C, Mirza MS, Moenne-Loccoz Y (2006) Phylogeny of the 1-aminocyclopropane-1-carboxylic acid deaminase encoding gene acdS in phyto beneficial and pathogenic Proteobacteriaand relation with strain biogeography. FEMS Microbiol Ecol 56: 455–470

Bull CT, Weller DM, Thomashow LS (1993) In situ measurement of root respiration and soluble C-concentrations in the rhizosphere. Soil Biol Biochem 25:1189–96

Byng G, Eustice DC, Jensen RA (1979) Biosynthesis of Phenazine pigments in mutant and wild-type cultures of *Pseudomonas aeruginosa*. J bacteriol 13:884–86

De Mesel I Derycke S Moens T Van Der Gucht K Vincx M Swings J (2004) Top-down impact of bacterivorous nematodes onthe bacterial community structure: a microcosm study. Environ Microbiol 6:733–744

Diby P, Saju K., Jisha PJ, Sharma YR, Kumar A, Anandraj M (2005) *Pseudomonas fluorescens* mediated vigour in black pepper (*Piper nigrum* L). Ann of Micro 55:129–133

Dikin A, Sijam K, Kadir J, Seman IA (2007) Mode of action of antimicrobial substances from *Burkholderia multivorans* and *Microbacterium testaceum* against *Schizophyllum commune* Fr. Int J Agri Biol 9:311–314

Frias JE, Flores E, Herrero A (1994) Requirement of the regulatory protein NtcA for the expression of nitrogen assimilation and heterocyst development genes in the cyanobacterium *Anabaena* sp. PCC 7120. Mol Microbiol 14:823–32

Girlanda M, Perotto S, Moenne-Loccoz Y, Bergero R, Lazzari A, Defago G, Bonfante P, Luppi AM (2001) Impact of biocontrol *Pseudomonas fluorescens* CHA0 and a genetically modified derivative on the diversity of culturable fungi in the cucumber rhizosphere. Appl Environ Microbiol 67:1851–1864

Glick BR (2005) Modulation of plant ethylene levels by the bacterial enzyme ACC deaminase. FEMS Microbiol Lett 25:1–7

Groboillot A, Portet-Koltalo F, Derf F, Feuilloley MJG, Orange, Poc CD (2011) Novel application of cyclolipopeptide amphisin: feasibility study as additive to remediate polycyclic aromatic hydrocarbon (PAH) contaminated sediments. Int. J Mol Sci 12:1787–1806

Haas D, Defago G (2005) Biological control of soil-borne pathogens by fluorescent *Pseudomonads*. Nat Rev Microbiol 3:307–319

Hagins JM, Locy R, Silo-Suh L (2009) Isocitrate lyase supplies precursors for hydrogen cyanide production in a cystic fibrosis isolate of *Pseudomonas aeruginosa*. J Bacteriol 191:6335–6339

Imanaka H, Kousaka M, Tamura G, Arima K (1965) Studies on pyrrolnitrin, a new antibiotic. III Structure of pyrrolnitrin. J Antibiot A 18:207–210

Jousset A, Scheu S, Bonkowski M (2008) Secondary metabolite production facilitates establishment of rhizobacteria by reducing both protozoan predation and the competitive effects of indige-nous bacteria. Funct Ecol 22:714–7194

Kamilova F, Validov S, Azarova T, Mulders I, Lugtenberg B (2005) Enrichment for enhanced competitive plant root tip colonizers selects for a new class of biocontrol bacteria. Environ Microbiol 7:1809–1817

Kaur R, Kaur J, Singh RS, Alabouvette C (2007) Biological control of *Fusarium oxysporum* f. sp. ciceri by non-pathogenic *Fusarium* and *fluorescent Pseudomonas*. Int J Bot 3:114–117

Keel C, Schnider U, Maurhofer M, Voisard C, Laville J, Burger U, Wirthner P, Haas D, Defago G (1992) Suppression of root diseases by *Pseudomonas fluorescens* CHA0: importance of the bacterial secondary metabolite 2, 4-diacetylphloroglucinol. Mol Plant Microbe Interact 5:4–13

Kiely PD, Haynes JM, Higgins CH, Franks A, Mark GL, Morrissey JP, O'Gara F (2006) Exploiting new systems-based strategies to elucidate plant-bacterial interac-tions in the rhizosphere. Microb Ecol 51:257–266

Kumar BSD (1998) Disease suppression and crop improvement through fluorescent *Pseudomonads* isolated from cultivated soils. World J Microbiol Biotechnol 14:735–741

Lalaouna D, Fochesato S, Sanchez L, Schmitt-Kopplin P, Haas D, Heulin T, Achouak W (2012) Phenotypic switching in *Pseudomonas brassicacearum* involves GacS- and GacA dependent Rsm small RNAs. Appl Environ Microbiol. 78(6):1658–65

Lemanceau P, Bakker PAHM, de Kogel WJ, Alabouvette C, Schippers B (1992) Effect of pseudobactin 358 production by *Pseudomonas putida* WCS358 on suppression of *Fusarium* wilt of carnation by nonpathogenic *Fusarium oxysporum* Fo 47. Appl Environ Microbiol 1 58:2978–2982

Leon M, Yaryura PM, Montecchia MS, Hernandez AI, Correa OS, Pucheu NL, Kerber NL, Garcia AF (2009) Antifungal activity of selected indigenous *Pseudomonas* and *Bacillus* from the soybean rhizosphere. Int J Microbiol 10:1–9

Mahesh M, Saifulla M, Sreenivasa S, Shashidhar KR (2012) Integrated management of pigeon pea wilt caused by *Fusarium udum* butler. EJBS 2:1–7

Maketon C, Fortuna AM, Okubara PA (2012) Cultivar-dependent transcript accumulation in wheat roots colonized by *Pseudomonas fluorescens* Q8r1–96 wild type and mutant strains. Biol Control 60:216–224

McKellar RC (2007) Role of nutrient limitation in the com-petition between *Pseudomonas fluorescens* and *Escherichia coli* O157:H7. J Food Protect 70:1739–1743

Mette NN, Jan S (1999) Chitinolytic activity of *Pseudomonas fluorescens* isolates from barley and sugar beet rhizosphere. FEMS Microbiol Ecol 30:217–227

Mulet M, Lalucat J, García-Valdes E (2010) DNA sequence-based analysis of the *Pseudomonas* species. Environ Microbiol 12:1513–1530

Nandakumar R, Babu S, Raguchander T, Samiyappan R (2007) Chitinolytic activity of native *Pseudomonas fluorescens* strains. J Agric Sci Technol 9:61–68

Nashwa MA, Sallam KAM, Abo-Elyousr, Hassan MAE (2008) Evaluation of *Trichoderma* species as biocontrol agents for damping-off and wilt diseases of *Phaseolus vulgaris* L. and efficacy of suggested formula. Egypt J Phytopatho 36(1–2):81–93

Natsch A, Keel C, Hebecker N, Laasik E, De fago G (1998) Impact of *Pseudomonas fluorescens* strain CHA0 and a derivative with improved biocontrol activity on culturable resident bacterial community on cucumber roots. FEMS Microbiol Ecol 27:365–380

Neidig N, Paul RJ, Scheu S, Jousset A (2011) Secondary metabolites of *Pseudomonas fluorescens* CHA0 drive complex non-trophic interactions with Bacterivorous nematodes. Microb Ecol 61:853–859

Notz D (2001) Prozesse der Meereisentwicklung in der sommerlichen Arktis, Diploma thesis, Meteorol. Inst. Univ. Hamburg, Hamburg, Germany.

Olalemi AS, Arotupin DJ (2012) Effect of refined petroleum products contamination on bacterial population and physicochemical characteristics of cultivated agricultural soil. J Microbiol Biotechnol Food Sci 2(2):684–700

Pedersen A, Nybroe O, Winding A, Ekelund F, Bjornlund L (2009) Bacterial feeders, the nematode *Caenorhabditis elegans* and the flagellate *Cercomonas longicauda*, have different effects on outcome of competition among the *Pseudomonas* biocontrol strains CHA0 and DSS73. Microb Ecol 57:501–509

Pieterse CMJ, Van Pelt JA, Verhagen BWM, Ton J, Van Wees SCM, Leon-Kloosterziel KM, Van Loon LC (2003) Induced systemic resistance by plant growth promoting rhizobacteria. Symbiosis 35:39–54

Poritsanos N, Selin C, Fernando WGD, Nakkeeran S, de Kievit TR (2006) A GacS deficiency does not affect *Pseudomonas chlororaphis* PA23 fitness when growing on canola, in aged batch culture or as a biofilm. Can J Microbiol 52:1177–1188

Raaijmakers JM, Weller DM (1998) Natural plant protection by 2,4 diacetylphloroglucinol—producing *Pseudomonas* spp. in take-all decline soils. Mol Plant-Microbe Interact 11:144–152

Ramarathnam R, Fernando WGD, de Kievit T (2011) The role of antibiosis and induced systemic resistance, mediated by strains of *Pseudomonas chlororaphis*, *Bacillus cereus* and *B. amyloliquefaciens*, in controlling black leg disease of canola. Bio Control 56:225–235

Reddy BP, Reddy MS, Kumar KV (2009) Characterization of antifungal metabolites of *Pseudomonas fluorescens* and their effect on mycelial growth of *Magnaporthe grisea* and *Rhizoctonia solani*. Int J PharmTech Res 1:1490–1493

Rezzonico F, Zala M, Keel C, Duffy B, Moenne Loccoz Y, Defago G (2007) Is the ability of biocontrol fluorescent *Pseudomonads* to produce the antifungal metabolite 2, 4-diacetylphloroglucinol really synonymous with higher plant protection? New Phytol 173:861–872

Roberts DP, Short NMJ, Maloney AP, Nelson EB, Schaff DA (1994) Role of colonization in biocontrol: studies with *Enterobacter cloacae*. Plant Sci 101:83–89

Ronn R, Grunert J, Ekelund F (2001) Protozoan response toaddition of the bacteria *Mycobacterium chlorophenolicum* and *Pseudomonas chlororaphis* to soil microorganisms. Biol Fertil Soils 33:126–131

Rosenberg K, Bertaux J, Scheu S, Bonkowski M (2009) Soil amoeba rapidly change bacterial community composition in *Arabidopsis thaliana* rhizosphere. ISME J 3:675–684

Rudresh DL, Shivaprakash MK, Prasad RD (2004) Effect of combined application of *Rhizobium*, phosphate solubilizing bacterium and *Trichoderma* spp. on growth, nutrient uptake and yield of chickpea (*Cicer aritenium* L.). Appl Soil Ecol 28:139–146

Schippers B, Bakker AW, Bakker PAHM (1987) Interactions of deleterious and beneficial rhizosphere microorganisms and the effect on cropping practices. Annu Rev Phytopathol 25:339–58

Sharma N, Sharma KP, Gaur RK, Gupta VK (2011) Role of chitinase in plant defense. Asian J Biochem 6: 29–37

Siddiqui IA, Shaukat SS, Sheikh IH, Khan A (2006) Role of cyanide production by *Pseudomonas fluorescens* CHA0 in the suppression of root-knot nematode, *Meloidogyne javanica* in tomato. World J Microbiol Biotechnol 22:641–650

Wang SL, Chang WT (1997) Purification and Characterization of two bifunctional chitinases/lysozymes extracellularly produced by *Pseudomonas aeruginosa* K-187 in a shrimp. Appl Environ Microbiol 63:380–386

Wang C, Knill E, Glick BR, Defago G (2000) Effect of transferring 1-aminocyclopropane-1-carboxylic acid (ACC) deaminase genes into *Pseudomonas fluorescens* strain CHA0 and its derivative CHA96 on their growth-promoting and disease-suppressive capacities. Can J Microbiol 46:1–10

Weller DM (2007) Pseudomonas biocontrol agents of soil-borne pathogens: looking back over 30 years. Phytopathology 97:250–256

Yamamoto S, Kasai H, Arnold DL, Jackson RW, Vivian A (2000) Phylogeny of the genus *Pseudomonas*: intrageneric structure reconstructed from the nucleotide sequences of *gyrB* and *rpoD* genes. Microbiology 146:2385–2394 (Reading, England)

Chitinase Expressed as an Inducible Trait in *Pseudomonas aeruginosa* Schröter P-15

Noushad Parvez, Leena Pathak, Ankit Patel, Deepak Panpatte, Ritesh Inamake, Kuldeep Khatri and Janardan Jani

Abstract

Extracellular chitinases play a significant role to stipulate the bioefficacy level of native *Pseudomonas* isolates against *Fusarium udum* and *Spodoptera litura* Fab. Present exertion is to confirm whether extracellular chitinase production is considered as an inducible biocontrol trait in *Pseudomonas* spp. *Pseudomonas aeruginosa* p-15 showed maximum chitinase activity in King's B broth 3.75 U/ml among all seven chitinase positive native isolates from Anand, Gujarat. Enhanced chitinase activity by 28.0, 15.2, and 8.0 % was observed after 84 h when broth was amended with 1.5 % of colloidal chitin, dry powder of tyndallized *Fusarium udum* and *Spodoptera litura* (third instar), respectively. The components of King's B medium were modified and used for chitinase expression as optimized using response surface methodology (RSM) for the responses such as colony count, chitinase activity, and biocontrol activity. These efforts suggest chitinase as an inducible trait in *Pseudomonas aeruginosa* p-15.

Keywords

Chitinase · Inducible trait · Pseudomonas aeruginosa

Introduction

Conventional synthetic chemical pesticides have long served as agents for reducing the incidence of plant disease; however, they are costly, can cause environmental pollution, and may induce pathogen resistance. Microbial products are especially valuable because their toxicity to nontarget animals and humans is extremely low. They are safe for both the user and consumers of treated crops compared to other commonly used chemicals (Prakob et al. 2009). Fluorescent pseudomonads, particularly *Pseudomonas aeruginosa*, *P. putida*, and *P. fluorescens*, which are commonly isolated from the plant rhizosphere, have been shown to protect plants from nematodes, insect attack, and fungal infection. *P. aeruginosa* strains

N. Parvez (✉) · L. Pathak · A. Patel · D. Panpatte · R. Inamake · K. Khatri · J. Jani
Biological Control Research Laboratory, Anand Agricultural University, Anand 388 110, Gujarat, India
e-mail: shahriak@gmail.com

A. K. Chakravarthy (ed.), *New Horizons in Insect Science: Towards Sustainable Pest Management*,
DOI 10.1007/978-81-322-2089-3_8, © Springer India 2015

are reported well as potential agent against many biotic threats like *Drosophila melanogaster* (Lau et al. 2003), *Galleria mellonella* (Bulla et al. 1975; George et al. 2000; Mostakim et al. 2012), *Meloidogyne javanica* (Siddiqui et al. 1999a, 2000b, 2001c), *Fusarium udum* (Badri and Sariah 2009), and *Pythium* (Buysens et al. 1994).

Chitinase and such metabolites work as an important biocontrol trait in *Pseudomonas*. *P. aeruginosa* is well reported for the production of chitinase enzyme by chitinase genes like *chi A* and *chi C*. The expression level of such chitinolytic enzymes can be increased by involving inducers and optimal media conditions.

Material and Methods

Microorganism and Cultural Conditions

Present work was carried out using one of the seven native *Pseudomonas* isolates isolated from rhizospheric soils collected from wilt (*Fusarium*) suppressive soil of Anand, Gujarat. Before its biochemical identification as *P. aeruginosa* p-15, it was confirmed positive for in vitro chitinase, Indole acetic acid production and proved nonpathogenic for pigeon pea (*Cajanus cajan* var. ICP 2376), wheat (*Triticum aestivum* var. GW173), and rice (*Oryza sativa* var. Gurjari) plants in greenhouse studies. Bacterial culture was multiplied overnight in King's B (KB) at pH 7 and 29 °C. For fungal powder preparation and inhibition assay *Fusarium udum* Butler was obtained from ICRISAT, Hyderabad, which was cultured for mycelium on potato dextrose agar (Himedia Labs, Mumbai, India) for 7 days at 28 °C. Lignocellulose agar sporulation media was used to harvest spores after 10 days in 0.5 M tris HCl.

Media Component Preparation Chitin Powder

Chitin, insect, and mycelia powders were prepared for the alternative media preparation by modified practices. One gram of chitin powder was added slowly into 18 ml of concentrated HCl under vigorous stirring and the mixture was then added to 100 ml of ice-cold ethanol with rapid stirring, kept overnight at 25 °C. The precipitate was collected by centrifugation at 8000× g (Bekman Coulter Allegra 64 R) for 10 min at 4°C and washed with sodium phosphate buffer until it was neutralized (pH 7) stored at −20 °C and used for further applications (Hackman 1962; Hackman and Goldberg 1965a, 1974b).

Insect and Mycelia Powder

Harvested mycelia and insect were dried in oven at 45 °C until constant weight. They were subjected to grinded and subsequently 0.5 % tris HCl (pH 7.5) added to it and allowed to stand overnight at room temperature. The mixture was subjected for centrifugation at 8000× g for 10 min at 4 °C. The supernatants were described and pellets of insect and mycelia powder were tantalized in 0.5 % tris HCl pH 7.5. The procedure was repeated for three consecutive days followed by centrifugation. The pellets were dried, stored at −20 °C, and used for further application.

Crude Glycerol By-Product from Jatropha Biodiesel Plant

Modification of the basic King's B media was practiced as alternative, significant, and economical production processes. Crude glycerol was used as alternative carbon source in chitinase production in place of purified glycerol. Crude glycerol, a by-product of biodiesel production from Jatropha (*Jatropha curcas* L.), was obtained from biodiesel production plant, Department of Food Processing and Bioenergy, Anand Agricultural University. It was autoclaved and analyzed for the specifications like glycerol percentage, moisture, color, etc. (Table 1).

Quantification Assays (Responses)

To check optimal growth, chitinase assay and antipathogenicity, two sets of experiments having

Table 1 Crude glycerol specifications: an alternative carbon source

Parameter	Extent
Water %	8
Glycerol %	15
Methanol %	2
pH	6.1
Color	Dark brown
Density (gcm^{-3})	1.2

four subsets each, were designed. The first set utilized KB media with pure glycerol while in second set pure glycerol was replaced by crude glycerol. Each set was supplemented with colloidal chitin, fungal powder, and insect powder and inoculated with native strain of *P. aeruginosa* p-15. All the subsets were checked for growth, chitinase, and antifungal activity.

Chitinase Assay (Uml^{-1})

Chitinase (N-acetyl-β- glucosaminidase) activity was checked in different media system by a modified approach described by Roberts and Selitrennikoff (1988). Each sample was centrifuged at 8000× g for 5 min and the supernatant was used for enzyme activity. One millileter supernatant was used for direct estimation of N-acetyl-d-glucosamine (GlcNAc). In second part, 1 ml of supernatant was incubated with 1 ml of 1% colloidal chitin in a 0.05 M phosphate buffer, pH 7.0 at 37 °C for 1 h, centrifuged at 10,000× g for 15 min. The amount of N-acetyl-d-glucosamine released in the supernatant was determined by the standard method (Lingappa and Lockwood 1962) using GlcNAc as a standard.

Microbial Biomass (Log CFU/ml)

Culture growth at 29 °C at 72 and 82 h was compared. After centrifugation at 10,000× g for 15 min pellets were washed with sterile distilled water and resuspended in 5 ml King's B basal medium. Optical density of concentrated (4×) cultured broth was determined by UV-visible spectrophotometer (Bekman Coulter DU® 730) at 600 nm and used for determination of the colony forming unit (CFU) by comparing the standard graph of OD verses Log CFU ml^{-1}. The colonies were calculated and transformed to logarithmic values (Thompson 1996) for quantification.

Antipathogenic Activity (mm)

The antifungal activity was also assayed in vitro for cultured broth supernatant by inhibition of the growth of *F. udum* Butler on Saboaraud's Dextrose Agar (SDA) medium (Dennis and Webster 1971; Velusamy et al. 2011). Fungal pathogen was preinoculated by pouring 5 ml of soft agar with 10^3 spores/ml. 0.2 ml supernatant (filtered with 0.2 μm Advavantec® cellulose acetate membrane filter) of 72 and 84 h culture grown broth was added in 10 mm well. Sterile distilled water was added to the wells of control plates. Diameter of the zone of inhibition was measured and expressed in millimeter (mm) after 7 days of incubation at 26 °C.

Effects of Amendment and Media Alteration on Responses

KB media with 2% crude glycerol (KBCG) and second set with 1% of pure glycerol were used. Twenty millileter broth was prepared for chitinase production in 50 ml sugar tubes. Both media with and without supplementation of 1.5% colloidal chitin, fungal powder and insect powder (eight different treatments and three repetitions) were inoculated with 0.2 ml overnight grown culture and assayed for chitinase activity after 84 h of incubation. Results were statistically analyzed using completely randomized design (CRD).

Modified Media Components and Experimental Designs

KBCG amended with colloidal chitins was used in significant component's screening study. Chitinase activities were analyzed by inoculating 0.2 ml overnight grown culture and incubating it for 84 h in 20 ml media system.

Table 2 Variables showing process parameters used in Plackett–Burman design

Variables	Medium components	(+) Values (g/lt or ml/lt) (H)	(–) Values (g/lt or ml/lt) (L)
X_1	Crude glycerol	5	50
X_2	Colloidal chitin	3	30
X_3	Peptone	5	50
X_4	Tryptone	5	50
X_5	K_2HPO_4	0.5	5
X_6	$MgSO_4.7H_2O$	0.5	5

Table 3 Plackett–Burman design matrix of six process variables (X_1–X_6) and five dummy variables (D_1–D_5) along with observed response (Chitinase activity)

Run No.	X_1	X_2	X_3	X_4	X_5	X_6	D_1	D_2	D_3	D_4	D_5	Chitinase activity (Uml^{-1})
1	H	H	L	H	H	H	L	L	L	H	L	2.63
2	L	H	H	L	H	H	H	L	L	L	H	2.71
3	H	L	H	H	L	H	H	H	L	L	L	1.11
4	L	H	L	H	H	L	H	H	H	L	L	2.08
5	L	L	H	L	H	H	L	H	H	H	L	1.35
6	L	L	L	H	L	H	H	L	H	H	H	0.51
7	H	L	L	L	H	L	H	H	L	H	H	1.00
8	H	H	L	L	L	H	L	H	H	L	H	2.84
9	H	H	H	L	L	L	H	L	H	H	L	3.00
10	L	H	H	H	L	L	L	H	L	H	H	2.06
11	H	L	H	H	H	L	L	L	H	L	H	1.49
12	L	L	L	L	L	L	L	L	L	L	L	0.35

Plackett–Burman design

It is an efficient way to identify the important factors among a large number of variables (Stanbury et al. 1986; Abedin and Taha 2008) used to screen important variables that significantly influenced the response like chitinase production by *P. aeruginosa* p-15 (Plackett and Burman 1946). Total number of trials to be carried out according to Plackett–Burman is $k+1$, where k is the number of variables. For the selection of the key ingredients significantly affecting the responses, six nutrient factors considered for the design were crude glycerol, colloidal chitin, peptone, tryptone, K_2HPO_4, and $MgSO_47H_2O$, which were designated as X_1, X_2, X_3, X_4, X_5, and X_6, respectively (Table 2). Apart from these, five dummy variables were also used denoted as D_1 to D_5, respectively.

In this study, 11-run Plackett–Burman design was applied to evaluate selected factors. The entire variable had two levels of the lower (L) and the higher (H) concentration. The number of H and L per trial was $(k+1)/2$ and $(k-1)/2$, respectively.

The main effect figure with an H indicates that high concentration of this variable is nearer to optimum and an L indicates that the low concentration of this variable is nearer to optimum. The principal effect of each on chitinase production was estimated as the difference between both averages of measurements made at the higher level and at the lower level. The main effect of each variable (Table 3) was determined with the following equation:

$$Exi = (\Sigma Mi^{+} - \Sigma Mi^{-}) / N,$$

where Exi is the variable main effect, Mi^{+} and Mi^{-} are the responses value in trails; the independent variable (xi) was present in high and low concentrations, respectively, and N is the number of trails divided by 2. Using Microsoft Excel, statistical t-values for equal unpaired sample

Table 4 Central composite design (CCD) matrix of independent variables and the corresponding experimental and predicted values

Run No.	Factors						Response	
	A: crude glycerol (ml/l)		B: colloidal chitin (g/l)		C: peptone (g/l)		Chitinase activity (Uml^{-1})	
	Coded value	Actual value	Coded value	Actual value	Coded value	Actual value	Observed value	Predicted value
1	0.00	55.00	0.00	27.50	0.00	28.75	2.18	2.29
2	−1.00	32.50	1.00	38.75	1.00	39.38	2.42	2.47
3	0.00	55.00	0.00	27.50	0.00	28.75	2.29	2.29
4	−1.00	32.50	1.00	38.75	−1.00	18.13	2.69	2.72
5	1.00	77.50	−1.00	16.25	−1.00	18.13	2.31	2.25
6	0.00	55.00	−1.68	8.58	0.00	28.75	2.30	2.39
7	1.00	77.50	1.00	38.75	1.00	39.38	2.20	2.23
8	−1.68	17.16	0.00	27.50	0.00	28.75	1.55	1.53
9	0.00	55.00	0.00	27.50	0.00	28.75	2.29	2.29
10	0.00	55.00	0.00	27.50	0.00	28.75	2.50	2.29
11	0.00	55.00	0.00	27.50	0.00	28.75	2.32	2.29
12	1.00	77.50	−1.00	16.25	1.00	39.38	2.06	2.02
13	0.00	55.00	1.68	46.42	0.00	28.75	3.55	3.47
14	0.00	55.00	0.00	27.50	0.00	28.75	2.18	2.29
15	0.00	55.00	0.00	27.50	−1.68	10.88	2.00	2.03
16	−1.00	32.50	−1.00	16.25	−1.00	18.13	1.68	1.64
17	1.00	77.50	1.00	38.75	−1.00	18.13	2.40	2.41
18	−1.00	32.50	−1.00	16.25	1.00	39.38	1.37	1.35
19	1.68	92.84	0.00	27.50	0.00	28.75	1.80	1.83
20	0.00	55.00	0.00	27.50	1.68	46.62	1.63	1.62

were calculated for determination of variable significance. Experimental error was estimated by calculating the variance among the dummy variables as follows:

$V_{eff}=\Sigma(E_d)^2/n$, where V_{eff} is the variance of the concentration effect, E_d is the concentration effect of dummy variable, and n is the number of dummy variables.

The standard error (S.E.) of the concentration effect was the square root of the variance of an effect and the significance level (P-value) of each concentration effect was determined using student's test:

$t_{(xi)}=E_{(Xi)}/\text{S.E}$, where $E_{(Xi)}$ is the effect of variable X_i.

Response Surface Methodology (RSM) To describe the nature of response surface in the experimental region and to elucidate the optimal concentrations of the most significant independent variables, a central composite design (CCD; Box and Behnken 1960) was applied, which is an RSM. Factors of highest confidence levels, namely, crude glycerol (A), peptone (B), and colloidal chitin (C) were tested at five levels (−α, +α, 0, +1, and −1). According to the design, 20 treatments trial combinations were executed (Table 4). For predicting the optimal point, the following second-order polynomial model was fitted to correlate relationship between independent variables and three responses:

$$Y = \beta o + \Sigma \beta i x i + \Sigma \beta i j x i x j + \Sigma \beta i i x i^2$$

where Y is the predicted response, β_0 is a constant, βi is the linear coefficient, βii is the squared coefficient, βij is the cross product coefficient, and x_i is the dimensionless coded value of (X_i). The above equation was solved by using the software design expert (Version 7.0.2, State ease inc., USA). A 2^5 factorial design with five replicates at the center point with a total number of 20 trials were employed (Table 3).

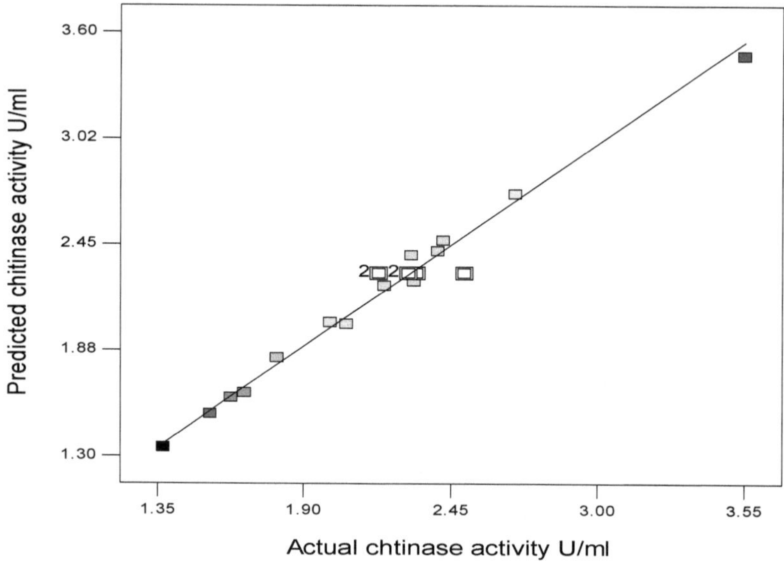

Fig. 1 Deviations in predicted and observed response

Validation of Optimization of Media

For validation studies, periodic analysis of chitinase activity up to 5 days was performed by taking six separate 20 ml fermentation systems for unoptimized fermented KBCG broth with 1.5 % colloidal chitin v/v at 29 °C and compared with periodical responses in optimized media provided by the design. Log CFU/ml and antifungal activity of matured broth at the time period of maximum chitinase activity were also analyzed.

Results and Discussion

Effects of crude glycerol and inducers chitinase assay (Uml^{-1}) were performed from supernatants of cultured broth for 84 h. Among all the eight, which had amendments in basal KB media and KB media with crude glycerol (KBCG), supernatant of cultured broth in presence of colloidal chitins showed significantly higher 3.51and 2.72 Uml^{-1} chitinase activity, respectively. Inhibition zone and microbial biomass were higher in case of colloidal chitins with KB and KBCG, respectively (Table 5). Colloidal chitin proved to be the most significant inducer for chitinase production in *P. aeruginosa* p-15 after 84 h. In presence of altered carbon source as crude glycerol as carbon source in KB media showed higher chitinase activity and inhibition diameter and exhibited significantly higher microbial biomass (Fig 1).

Screening of Important Medium Components via Plackett–Burman Design

To enhance the production of chitinase, Plackett–Burman design was employed as a statistical approach for the screening of suitable medium components. Table 2 represents the independent variables and their respective high and low concentrations used in the optimization study, while Table 3 shows the Plackett–Burman experimental design for 12 trials with two levels of concentrations for each variable and the corresponding chitinase activity in terms of (Uml^{-1}) of the culture medium. The variables X_1–X_6 represent the experimental variables, whereas D_1–D_5 represent the dummy variables. Table 6 represent the effect, S.E., $t_{(xi)}$, and *P*-values of each process variable for chitinase production.

The significant process variables were screened at probability value ($P \leq 0.05$). The probability value of crude glycerol, colloidal

Table 5 Effects of media amendments on chitinase production, growth and in vitro antipathogenic activity against *F. udum* Butler

Treatments	Chitinase activity (U/ml)	Microbial biomass (Log CFU/ml)	Inhibition zone (mm)
KB	2.73	8.00	2.69
KB+FP	3.16	11.00	11.00
KB+CC	3.51	9.00	20.00
KB+IP	2.96	8.00	9.00
KBCG	2.00	8.00	6.00
KBCG+FP	2.10	8.50	14.00
KBCG+CC	2.72	13.00	13.00
KBCG+IP	2.23	10.00	10.00
SEM	0.08	0.32	0.06
CD	0.25	0.95	0.20
CV %	5.64	5.82	4.27

Table 6 Statistical analysis of process parameters for chitinase activity

Factors	Fermentation parameters	Chitinase activity (Uml^{-1})			
		Effect (*Exi*)	S.E.	$t(_{xi})$	P-value
X_1	Crude glycerol	12.07	0.66	4.58	0.00593
X_2	Colloidal chitin	15.32		14.48	0.00003
X_3	Peptone	11.72		3.52	0.01695
X_4	Tryptone	9.88		−2.09	0.09131
X_5	K_2HPO_4	11.26		2.12	0.08785
X_6	$MgSO_4.7H_2O$	11.15		1.78	0.13487

chitin, and peptone ($P<0.05$) for chitinase production was considered as significant positive variables. However, tryptone, K_2HPO_4, and $MgSO_4.7H_2O$ showed no significant influence on chitinase production.

To obtain above results, Plackett–Burman design is considered a powerful tool for identifying factors that have significant influence on chitinase production. The optimal concentration of the individual factor was further determined by the subsequent CCD experiment.

Optimization of Screened Medium Components Using CCD

CCD was used to determine the optimal concentration (level) of the medium components. A total of 20 experiments with three variables (components of the medium) and five coded levels (five different concentrations) were performed. Model was run for optimization of significant media components screened via Plackett–Burman experimental design without any transformation for chitinase activity values. Chitinase response ranged from 1.37 to 3.55 Uml^{-1} and the ratio of maximum to minimum was found 2.59.

Based on the results obtained from Plackett–Burman design, we selected three variables, namely, crude glycerol, colloidal chitin, and peptone were selected. As they were positive influencing, increased concentrations were studied for optimization. The other components of the production medium were nonsignificant and hence, their concentrations were set at middle level in CCD. The other conditions were: temperature 29 °C, 0.2 % inoculum size, agitation speed (140 rpm), incubation period 84 h and pH 7.2. (Fig. 2)

Table 7 represents the experimental design matrix for CCD along with the experimental results of predicted responses for chitinase activity in broth. The experimental values for the regression coefficient were obtained by quadratic polynomial equation, where only significant

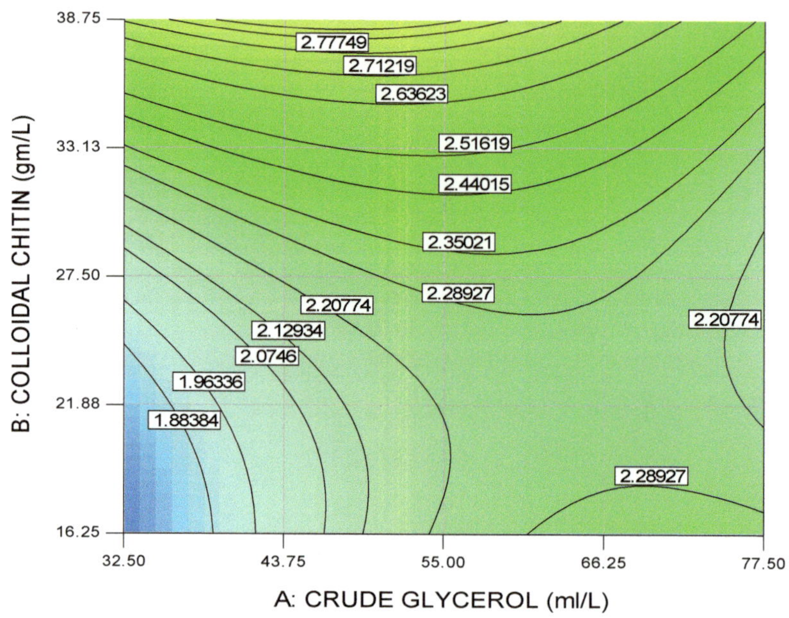

Fig 2 Interaction effects of crude glycerol and colloidal chitin on chitinase activity

Table 7 Analysis of variance (ANOVA) for the experimental results of the CCD

Source	Sum of squares	Degree of freedom	*F*-value	*p*-value Prob *F*	Remark
Model	4.119	9	47.0175	<0.0001	Significant
A—crude glycerol	0.110	1	11.3875	0.0071	
B—colloidal chitin	1.412	1	145.1031	<0.0001	
C—peptone	0.199	1	20.5334	0.0011	
AB	0.418	1	42.9998	<0.0001	
AC	0.002	1	0.2169	0.6513	
BC	0.001	1	0.1040	0.7537	
A^2	0.672	1	69.0752	<0.0001	
B^2	0.735	1	75.5607	<0.0001	
C^2	0.399	1	41.0462	<0.0001	

coefficients ($P<0.05$) were considered (Table 3). The smaller *P*-values indicate the higher significance of the corresponding coefficient. The insignificant coefficients were not omitted from the equations, since it was a hierarchical model. The predicted response Y (Table 4) for the chitinase activity was obtained as follows:

where Y is the chitinase activity (Uml^{-1}) and A, B, and C are coded values of the independent variables (crude glycerol, colloidal chitin, and peptone, respectively). The statistical significance of the quadratic model for the experimental responses was evaluated by the analysis of vari-

$$Y = 2.29 + 0.090^{*}A + 0.32^{*}B - 12^{*}C - 0.23^{*}A^{*}B + 0.016^{*}A^{*}C + 0.011^{*}B^{*}C - 0.22^{*}A2 + 0.23^{*}B2 - 0.17^{*}C2$$

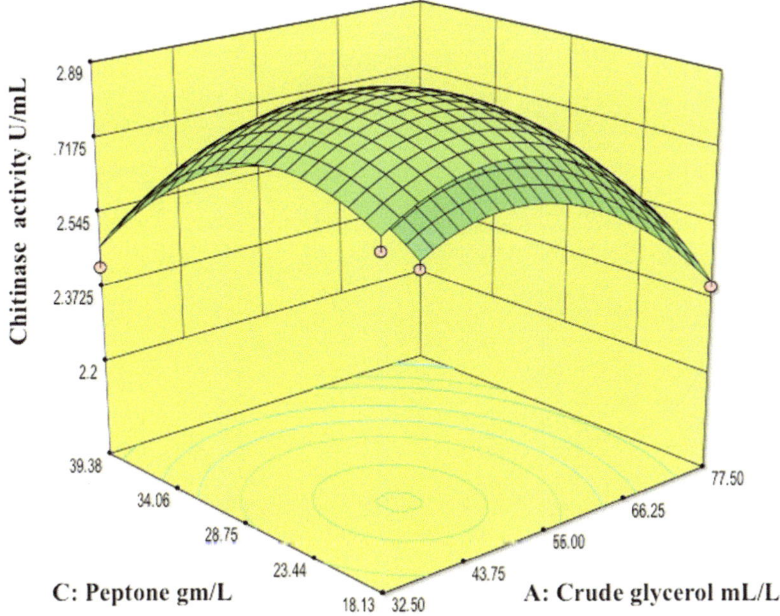

Fig. 3 Effects of peptone and crude glycerol

ance (ANOVA). According to ANOVA (Table 7), the model was significant with an F-test of a very low probability $[(P>F)<0.0001]$. The goodness of fit for the model was expressed by coefficient of determination R^2 and the value was 0.9769. The value of R^2 indicated that the experimental values were in agreement with the predicted values and suggested that the model is suitable and practicable.

Model run showed that the predicted and obtained response values were with minimum alteration as shown in Fig. 1. The 3-D surface and contour plots illustrate the response over a region of interesting factor levels, the relationship between the response and experimental levels of each variable and the type of interactions between the test variables in order to deduce the optimal composition of the culture medium. But at higher concentration of crude glycerol, the effect of different concentrations of peptone was little on chitinase production. It is evident from fig. 3 that at low concentration of colloidal chitin, production of chitinase increased with decrease in concentration of peptone and on increasing crude glycerol. When colloidal chitin was higher, higher chitinase production was obtained at lower values of both the variables. There was maximum production of chitinase at low peptone value and highest colloidal chitin (Fig. 4). We can determine the optimal concentration of the medium components from the data obtained from the 3-D surface and contour plots and the equations obtained from the multiple regression analysis.

The model predicts that the chitinase production (2.88 Uml^{-1}) is located at the actual values: crude glycerol—47.70 ml/l, colloidal chitin—38.75 g/l, and peptone—25.59 g/l. The predicted values of the response obtained and their corresponding concentration of medium components vary accordingly. Thus, graphical optimization of the overall desirability function was performed to determine the best possible combination for each response simultaneously.

To validate optimized media components for chitinase activity, periodical production curve with 0.2% culture inoculation up to 120 h was carried out and compared with both unoptimized KBCG and KB in presence of 1.5% of colloidal chitin. It was found that optimal media components were responsible for premature production peak (72 h) with 3.75 Uml^{-1} which was 12 h earlier as well as 46.4, 30.20, and 5% higher than

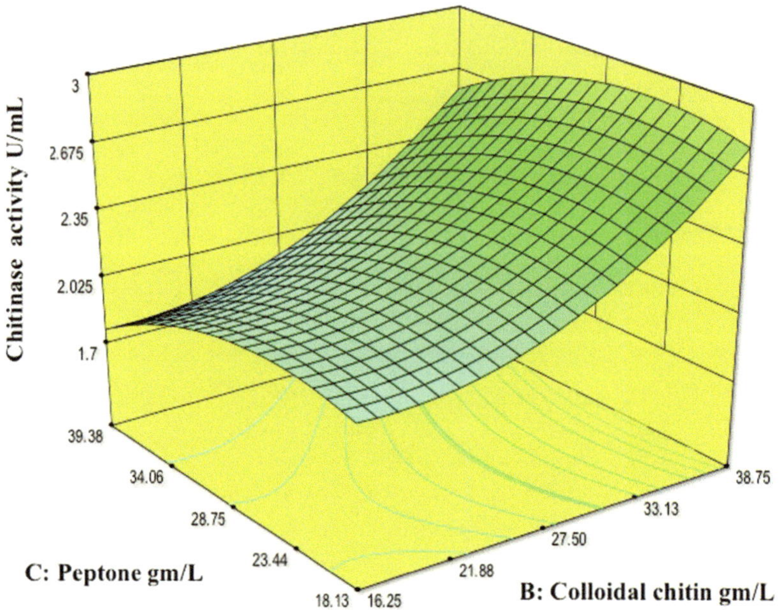

Fig. 4 Effects of peptone and colloidal chitin

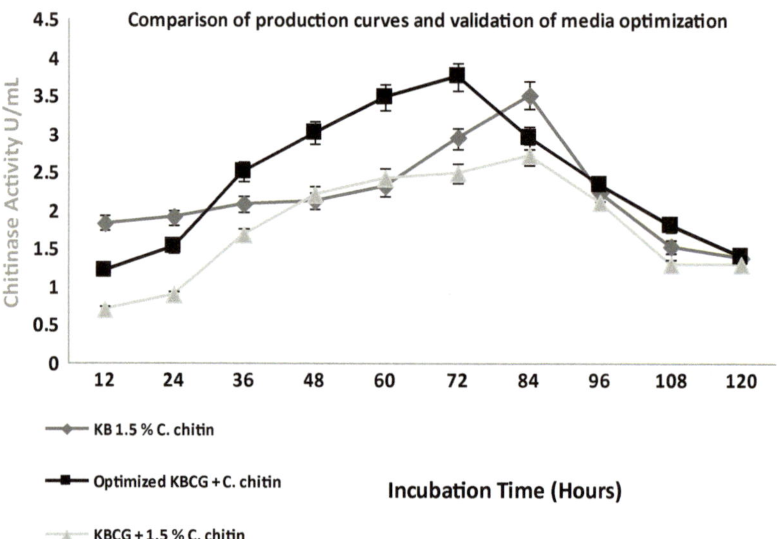

Fig. 5 Validation of optimization for difference responses

unoptimized KBCG, predicted optimized response and KB, respectively (Fig. 5).

Further effects on other responses such as Log CFU/ml and *Fusarium* growth inhibition (mm) were performed at 72 and 84 h, and found Log CFU value (14 Log CFU/ml) and inhibition (14 mm) value were higher at 84 h as compared to 13 Log CFU/ml and 11 mm, respectively, at 72 h. These results also suggest that inhibition of *F. udum* may be a contribution of combination of other metabolites of *P. aeruginosa* p-15 also.

There are several reports on the optimization of medium composition for the production improvement using statistical approaches as it was found a reliable methodology to obtain reproducible results (Li et al. 2007; Ghanem et al. 2010; Dong et al. 2012). *Pseudomonas* is a potential producer of chitinolytic and other biocontrol enzymes (Fallahzadeh et al. 2010). The optimal concentration ranges of the three factors were optimized using CCD in RSM. As a result, a quadratic model was found to fit for chitinase production and the optimal medium composition was determined as follows: crude glycerol—47.70 ml/l, colloidal chitin—38.75 g/l, peptone—25.59 g/l, and chitinase activity reached 3.75 Uml^{-1} when compared with the predicted value of 2.88 Uml^{-1}. A successful and significant improvement (46.48%) in the production of chitinase by *P. aeriginosa* p-15 was accomplished using cheaper carbon source and inducer. The optimized medium established in this work might result in a significant reduction in the cost of medium constituents. These preliminary results also suggested employing crude glycerol in media optimization studies for chitinase production. Future trends in biocontrol research will unite fundamental biology with the quest for solutions that will make biocontrol integral to the safe and wise management of every agroecosystem.

References

Abedin RMA, Taha HM (2008) Antibacterial and antifungal activity of cyanobacteria and green microalgae. Evaluation of media components by Plackett-Burman design for antimicrobial activity of *Spirulina platensis*. Global J Biotechnol Biochem 3:22–31

Badri FA, Sariah M (2009) Molecular characterization of *Pseudomonas aeruginosa* UPM P3 from oil palm rhizosphere. Am J Appl Sci 6:1915–1919

Box GEP, Behnken DW (1960) Some new three level designs for the study of quantitative variables. Technometrics 2:455–475

Bulla LA, Rhodes RA, St. Julian G (1975) Bacteria as insect pathogens. Annu Rev Microbiol 29:163–190

Buysens S, Poppe J, Hofte, M (1994) Role of siderophores in plant growth stimulation and antagonism by *Pseudomonas aeruginosa* 7NSK2. In: Ryder MH, Stephens PM, Bowen GD (eds) Improving plant productivity with rhizosphere bacteria, vol 1. CommonwealthScientific and Industrial Research Organisation, Adelaide pp 139–141

Dennis C, Webster J (1971) Antagonistic properties of species groups of *Trichoderma* I. Production of non-volatile antibiotics. Trans Br Mycol Soc 57:25–39

Dong Y, Zhang N, Lu J, Lin F, Teng L (2012) Improvement and optimization of the media of *Saccharomyces cerevisiae* strain for high tolerance and high yield of ethanol. Afr J Microbiol Res 6:2357–2366

Fallahzadeh V, Ahmadzadeh M, Sharifi R (2010) Growth and pyoverdine production kinetics of *Pseudomonas aeruginosa* 7NSK2 in an experimental fermentor. J Agric Technol 6:107–115

George J, Laurence GR, Frederick MA (2000) Positive correlation between virulence of *Pseudomonas aeruginosa* mutants in mice and insects. J Bacteriol 182:3843

Ghanem KM, Al-Garni SM, Al-Makishah NH (2010) Statistical optimization of cultural conditions for chitinase production from fish scales waste by *Aspergillus terreus*. Afr J Biotechnol 9:5135–5146

Hackman RH (1962) Studies on chitin 5. Action of mineral acids on chitin. Aust J Biol Sci 15:526–532

Hackman RH, Goldberg M (1965a) Studies on chitin—nature of alpha—and beta—chitins. Aust J Biol Sci 8:935–941

Hackman RH, Goldberg M (1974b) Light-Scattering and infrared-spectrophotometric studies of chitin and chitin derivatives. Carbohydr Res 38:35–45

Lau GW, Goumnerov BC, Walendziewicz CL, Hewitson J, Xiao W, Mahajan-Miklos S, Tompkins RG, Perkins LA, Rahme LG (2003) The *Drosophila melanogaster* toll pathway participates in resistance to infection by the Gram-negative human pathogen *Pseudomonas aeruginosa*. Infect Immun 71:4059–4066

Li Y, Liu Z, Cui F, Liu Z (2007) Application of Plackett- Burman experimental design and Doehlert design to evaluate nutritional requirements for Xylanase production by *Alternaria mali* ND-16. J Appl Microbiol Biotechnol 77:285–291

Lingappa Y, Lockwood JL (1962) Chitin media for selective isolation and culture of *Actinomycetes*. Phytopathol 52:317–323

Mostakim M, Soumya E, Mohammed IH, Ibnsouda SK (2012) Biocontrol potential of a *Pseudomonas aeruginosa* strain against *Bactrocera oleae*. Afr J Microbiol Res 6:5472–5478

Plackett RL, Burman JP (1946) The Design of optimum multifactorial experiments. Biometrika 33(4):305–325

Prakob W, Nguen-Hom J, Jaimasit P, Silapapongpri S, Thanunchai J, Chaisuk P (2009) Biological control of lettuce root-knot disease by the used of *Pseudomonas aeruginosa*, *Bacillus subtilis* and *Paecilomyces lilacinus*. J Agril Technol 5:179–191

Roberts WK, Selitrennikoff CP (1988) Plant and bacterial chitinases differ in antifungal activity. J Gen Microbiol 134:169–176

Siddiqui MR (2000b) Parasites of plants and insects, 2nd edn. CABI Publishers, Wallingford

Siddiqui IA, Ehteshamul-Haque S, Ghaffar A (1999a) Root dip treatment with *Pseudomonas aeruginosa* and *Trichoderma* spp., in the control of root rot-root knot disease complex in chili (*Capsicum annuum* L.). Pak J Nematol 17:67–75

Siddiqui IA, Ehetshamul-Haque S, Shahid-Shaukat S (2001c) Use of Rhizobacteria in the control of root rot and root knot disease complex of mung bean. J Phytopatholol 149:337–346

Stanbury PF, Whitaker A, Hall SJ (1986) Media for industrial fermentations. In: Principles of fermentation technology, vol 1. Pergamon, Oxford, pp 93–122

Thompson DC (1996) Evaluation of bacterial antagonist for reduction of summer patch symptoms in Kentucky blue grass. Plant Dis 80:856–862

Velusamy P, Ko HS, Kim KY (2011) Determination of antifungal activity of *Pseudomonas* sp. A3 against *Fusarium oxysporum* by high performance liquid chromatography (HPLC). Agric Food Anal Bacteriol 1:15–23

Insect Resistance to *Bacillus thuringiensis* (Bt) Transgenic Crops and Its Management

Leena Pathak, Noushad Parvez, Ankit Patel and Janardan Jani

Abstract

Transgenic crops producing *Bacillus thuringiensis* (Bt) toxins for insect–pest control have been successful and started paying lucrative returns to the farmers in terms of increased production due to low pest damage, savings in cost of pesticides and manpower involved for pest control. However, their efficacy shows a reducing trend due to evolution of resistance among the target pests is a significant environmental risk. Resistance is a genetically based decrease in susceptibility of a population to an insecticide. To date, field-evolved resistance to Bt crops has been documented in only three insect species: *Helicoverpa zea* Boddie, *Spodoptera frugiperda* Smith, and *Busseola fusca* Fuller to Bt cotton and Bt corn producing Cry1Ac, Cry1F, and Cry1Ab, respectively. Scientists in the industry, government, and academia now recognize evolution of resistance to Bt in pests as a great threat to the continued success of Bt. Insect Resistance Management (IRM) strategies begin with resistance risk assessment. Phenotypic monitoring methods are best studied for low-dose events and genic methods are best suited for high-dose events. Resistance risks are real. But they can be managed. Resistance issues are associated with first-generation technologies and incomplete or compromised IRM programs. Next generation technologies (NGS) with multiple pyramided modes of action are needed.

Keywords

Insect · Resistance · Bt transgenic

Introduction

Even though, *Bacillus thuringiensis* (Bt) has been widely exploited, some critical questions left behind by the insect species have proved to be an uphill task for agriculturists to answer. Perhaps the most serious threat to the durability of this novel insect control technology is the potential of insect populations to develop resistance to Bt Cry proteins. Despite many efforts over the successful exploitation of Bt, some insects have developed resistance to Bt-toxin. This situation contributed to the development of biological con-

L. Pathak (✉) · N. Parvez · A. Patel · J. Jani
Biological Control Research Laboratory, Anand Agricultural University, 388110 Anand, Gujarat, India
e-mail: leenapandya0411@yahoo.com

A. K. Chakravarthy (ed.), *New Horizons in Insect Science: Towards Sustainable Pest Management*,
DOI 10.1007/978-81-322-2089-3_9, © Springer India 2015

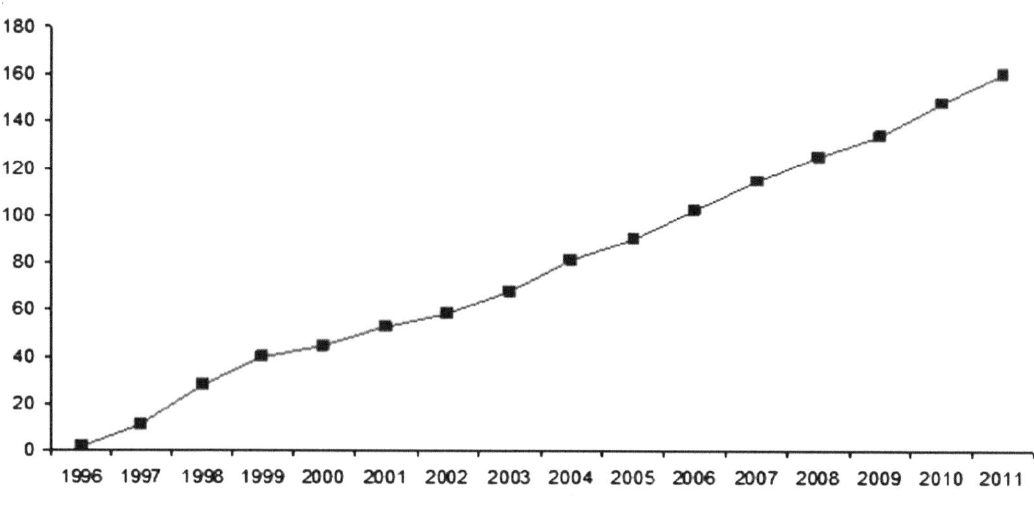

Fig. 1 Global area of biotech crops, 1996–2011 (mha)

trol strategies on a wider basis. Several biological insecticides have been developed ranging from bacterial, fungal, viral, etc., for the efficient pest control in formulations suitable for agroclimatic conditions. Among these agents, Bt is the most widely studied and exploited worldwide (Estada and Ferre 1994).

Bt occurs over a wide range of habitats such as soil, insect hosts, treated habitats, phylloplane, stored products, etc. (Uribe et al. 2003). The ability of Bt to produce many insecticidal crystalline toxins during sporulation has been exploited and δ-endotoxin, the main principle of insect toxicity is successfully utilized in pest-control programmes. Despite many efforts over the successful exploitation of Bt, insects have developed resistance to Bt-toxins. The first report on insect resistance to Bt was published by McGaughey 1985 in Indian-meal moth. *Plutella xylostella* was the first insect to develop resistance in field (Tabashnik et al. 1994) while other species have the genetic potential to develop resistance in the near future. First, Bt commercial formulation was made available for field in 1958. Whiteley and Schnepf cloned a Bt-toxin gene in 1981. Monsanto developed first transgenic cotton plant in 1990. According to ISAAA Executive Report, the global area of biotech crops continued to increase for the 16th year at a sustained growth rate of 8 % or 12 million ha (30 million acres), reaching 160 million ha or 395 million acres (Fig. 1). Biotech crops have set a precedent in that the biotech area has grown impressively every single year for the past 16 years, with almost a remarkable 94-fold increase since commercialization began in 1996. Thus, biotech crops are considered as the fastest adopted crop technology in the history of modern agriculture (James 2011a).

Action of Bt is very specific. Different strains of Bt are specific to different receptors in insect gut wall. Bt toxicity depends on recognizing receptors, damage to the gut by the toxin occurs upon binding to a receptor. Each insect species possesses different types of receptors that will match only certain toxin proteins, like a lock to a key. However, the history of insecticide resistance informs us that adaptation by insects could diminish the long-term efficacy of Bt crops and the associated economic, health, and environmental benefits. To date, field-evolved resistance to Bt crops has been documented in only three insect species: *Helicoverpa zea*, *Spodoptera frugiperda*, and *Busseola fusca* to Bt cotton and Bt corn producing Cry1Ac, Cry1F, and Cry1Ab, respectively (Tabashnik 2008). This chapter summarizes the current status of resistances in Bt crops, the principles of IRM for Bt crops and what they mean for the design of IRM programs.

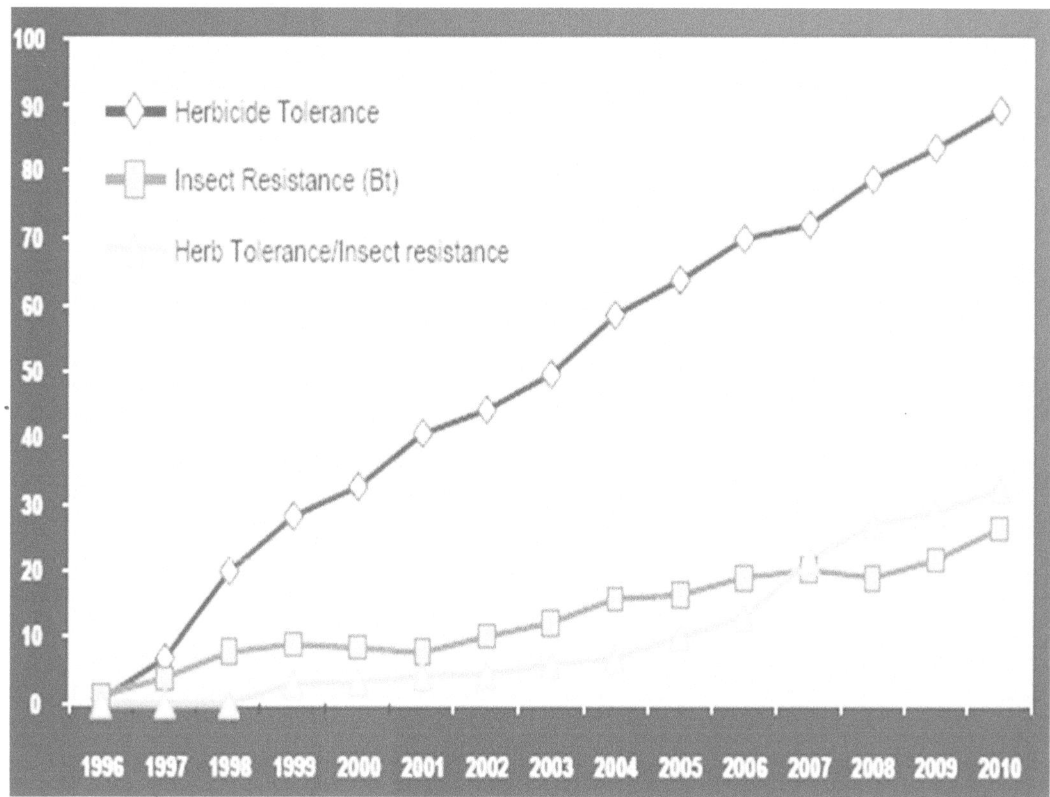

Fig. 2 Global area under GM crops by traits (mha)

Table 1 Response of DBM from parental susceptible (LAB-P) and resistant (NO-QA) strains to four toxins. (Tabashnik et al. (1997))

Strain	Toxin	10 mg/lit		100 mg/lit	
		n	% mortality	*N*	% mortality
LAB-P	Cry1Aa	86	98	197	99
	Cry1Ab	118	97	61	100
	Cry1Ac	50	94	109	100
	Cry1F	91	100	220	97
NO-QA	Cry1Aa	86	11	168	4
	Cry1Ab	130	2	164	1
	Cry1Ac	140	10	212	3
	Cry1F	91	0	120	0

Resistance and Its Mechanism

Resistance is the phenotype of an individual that gives the individual ability to survive on the transgenic insecticidal plants from egg to adult and produce viable offspring. Resistance is caused by gene in the target insect that reduces susceptibility to a toxin and is a trait of an individual. Nine different insect like diamondback moth, pink bollworm, *Helicoverpa armigera*, *Heliothis virescens*, *Earias vittella*, etc., have already shown resistance to Bt toxins worldwide. The global status of various traits incorporated in GM crops summarized in ISAAA Executive Report, 2011, (Fig. 2); it is showing a logarithmic increasing trend in the usage. However, insect resistance gene incorporation among various crops has declined while traits like herbicide tolerance or mixture of herbicide tolerance and insect resistance shows an increasing trend (James 2011b).

Table 2 Field-evolved resistance to Bt toxins is prays and transgenic cotton. (Tabashnik et al. (2008))

Insect	Bt spray or toxin	Resistance ratio			
		>10	>100	>1000	Maximum
P. xylostella	Dipel	2	0	0	36
T. ni	Dipel	23	2	0	160
H. zea	Cry1Ac	54	14	2	>1000

Table 3 Geographical variability in susceptibility of *H. armigera* to Bt toxin across northern Karnataka cotton ecosystem. (Yenagi et al. (2010))

Location	LC_{50} mg/ml	95 % FL		RF
		Lower	Upper	
Dharwad	0.149	0.042	0.220	1.00
Haveri	0.710	0.507	2.154	4.77
Raichur	0.828	0.570	1.021	5.66
Bijapur	0.186	0.090	0.159	1.25
Belgaum	0.174	0.047	0.259	1.17

Tabashnik and Liu (1997) studied the response of diamondback moth larvae from parental susceptible and resistant strain (Table 1) to four toxins and observed that at concentration of 10 and 100 mg/lit toxin killed 90–100 % larvae of susceptible strain but only 0–11 % of resistant counterpart. Gould et al. (1992) studied the generation verses resistance ratio (RR) between selected and control strain of *H. virescens* against Cry1Ac and reported that after ten generations the ratio of LC_{50} of selected strain to control was 10 and RR suddenly increased to 50 after 17th generation of selection.

Tabashnik et al. (2008) submitted the report on field-evolved resistance to the Bt toxin in sprays and transgenic cotton which say that *H. zea* had maximum RR (> 1000 fold) to Cry1Ac while field population of diamondback moth and cabbage looper showed the 36 RR in Hawaii and 160 RR in British Columbia, respectively, against Bt spray (Table 2). He also conducted a laboratory diet bioassay on field population of *H. zea* resistant to Cry2Ab and observed that the number of resistant populations to the total population was 0 of 8 in 2002, 1 of 25 in 2003, 1 of 24 in 2004, and 5 of 10 in 2005. Means proportion of resistant population was significantly higher in 2005 so, he concludes that continues use of Bt toxin may develop resistance simultaneously in insect species. Rajagopal et al. (2009) studied development of resistance by *H. armigera* to Cry1Ac on an artificial diet and observed that after ten generations the larvae were able to tolerate 72-fold more Cry1Ac than susceptible strain. The resistance dose is reflected more in the LC_{90} values than in the LC_{50} values, suggesting that the larvae become much more adopted to survive an ever increasing dose of the selection pressure. Yenagi et al. (2010) studied the resistance development in *H. armigera* in response to Delfin for different Bt cotton ecotypes in northern Karnataka, namely, Dharward, Haveri, Raichur, Bijapur, and Belgaum (Table 3). The populations from Raichur and Haveri were found tolerant to Bt toxin. LC_{50} values resulting from 0.149 to 0.828 mg/ml. The Dharward strain was the most susceptible.

Belgaum and Bijapur population were similar to each other at RF. This would suggest that under field condition tolerant individual is likely to persist and may subsequently contribute to the resistant pool. Studies carried out by CICR (Kranthi 2012b) showed that there was a decline in the proportion of susceptible populations. These results were obtained by studying LC_{50} (median lethal concentration) and IC_{50} (median growth inhibitory concentration) which were expressed in terms of μg Cry1Ac/ml of diet (Table 4) and further they compared RR value with reference to susceptible strains.

However, the LC_{50} values ranged from 0.02 to 0.54 μg Cry1Ac/ml of diet in 2002 and 0.246–5.10 μg Cry1Ac/ml of diet in 2011–2012. The IC_{50} values ranged from 0.003 to 0.034 μg

Table 4 Resistance monitoring to Cry1Ac toxin of Bt cotton on cotton bollworm populations collected from various locations in India. (Kranthi (2012b))

Year	Sites	Highest IC_{50}	Resistance ratio	Highest LC_{50}	Resistance ratio
1999–2000	10	0.034	2	0.67	7
2002–2003	45	0.043	2	0.54	5
2003–2004	20	0.023	1	0.38	4
2004–2005	21	0.104	5	0.74	7
2005–2006	39	0.166	9	0.72	7
2006–2007	27	0.195	10	0.79	8
2007–2008	49	0.201	11	1.15	12
2008–2009	26	0.58	31	3.12	31
2009–2010	31	0.59	31	3.14	31
2010–2011	27	0.24	13	3.26	33
2011–2012	17	0.36	19	5.10	51

Table 5 Cross-resistance of CryIC resistant *S. exigua* to other Cry proteins. (Moar et al. (1995))

Generation	Toxin	*n*	LC50 (μg of protein/g diet)	RR
20	Cry IAb	92	Nd	20
Susceptible		86	25.3	
22	Cry IAb	96	5866	93
Susceptible		96	63.2	
34	Cry IIA	258	10,731	73
Susceptible		282	147	
34	Cry IH	53	80	12
Susceptible		409	6.6	

All proteins used were toxins except Cry IIA (protoxin)
Nd not determined

Cry1Ac/ml of diet in 2002 and 0.036–0.363 μg Cry1Ac/ml of diet in 2011–2012. Ninety percent of the populations showed typical susceptible response, RRs of 31-fold were recorded in one or two locations during 2008–2009 to 2010–2011 and 51-fold in one location during 2011–2012.

Apart from that, Bt cotton was found effective in bollworms in some districts where the RRs were 51-fold. Thus, it appears that the data did not indicate levels of resistance in the populations that may be adequate for significant survival of the populations under field conditions in any of the populations tested. However, the data indicated that there was a clear decrease in the proportion of susceptible populations. Bt produces δ- endotoxin which comes in contact with insect gut enzymes; gets solublized and is converted from protoxin to active toxin to which insect is susceptible. If there is any change in configuration the gut enzyme/s may be by mutation or selection, the protoxin does not get converted to active toxin and in turn it is unable to bind to the receptors on brush border membrane which result into inactivation of toxin and insect becomes resistant. Tabashnik et al. (1994) studied cross resistance to Cry1B, Cry1F between resistant and susceptible strain of DBM. Resistant strain of Cry1Ab and Cry1F showed RROF 750 and 240, respectively, at the same concentration. Moar et al. (1995) studied the cross-resistance of CryIC resistant *S. exigua* to other Cry proteins and observed that at 22nd generation of resistant strain (Table 5) showed highest RR of 93 against Cry1Ab followed by an RR of 73 against CryIIA at 34th generation.

Biochemical and Genetic Basis of Resistance

Evolution of resistance to Bt toxin has necessitated proper understanding of biochemistry of toxin-insect interaction. Some biochemical mechanisms responsible for resistance development are altered proteolytic processing, modification

Table 6 Toxicity of commercial formulations and purified Cry toxins to susceptible laboratory and resistant field populations of *P. xylostella*. (Sayyed et al. (2004))

Population	Toxin used	LC50 (μg/ml)	Resistance ratio
LAB-UK (susceptible)	Dipel	0.0039	–
	MVP	0.027	–
	Cry1Ac	0.007	–
	Cry1Ab	0.47	–
	Cry1Aa	0.039	–
	Cry1Fa	0.20	–
Karak (resistant field)	Dipel	2.97	770
	MVP	9800	363,000
	Cry1Ac	>40	>5710
	Cry1Ab	77.0	164
	Cry1Aa	32.9	845
	Cry1Fa	82.7	414

of binding site, absence of specific receptors, hindrance in pore formation, and removal of a highly conserved glycosylation pathway (Heckel 1994). Candas et al. (2002) studied the changes in the levels of proteins identified in the midgut epithelium of Indian-meal moth larvae resistant to Bt and found that the levels of aminopeptidase N, vacuolar ATPase subunit B, phosphopyruvate hydratase, cytochrome oxidase subunit I, NADH dehydrogenase subunit V, 3-Dehydroecdysone reductase, F_1F_0-ATPase/resistant variant, and GSH transferase showed significant increase. Ferre and van Rie (2002) proposed a model for binding site modification of Cry proteins to brush border membrane of midgut cells of *P. xylostella* larvae and found conformational changes in binding site can alter the degree of specificity of toxin binding. Aroian et al. (2003) proposed an oligosaccharide receptor-based model for *bre* genes in intestinal cells of *Caenorhabditis elegans* and reported that *bre* genes synthesize an oligosaccharide which serves as receptor for crystal toxins. Loss of any of the *bre* enzymes leads to high level of resistance to Cry5B suggesting that oligosaccharide is the major receptor for Cry5B. In the case of Cry14A, the reduced level of resistance conferred by mutants suggests that other receptors can partly compensate in the absence of *bre* oligosaccharide.

Genetic reasons for resistance development can be intraspecific variation in baseline susceptibility, frequency of resistance gene allele, mode of inheritance of resistance, and stability of resistance. Sayyed et al. (2004) studied the toxicity of commercial formulations and purified Cry toxins to susceptible laboratory and resistant field populations of *P. xylostella* and reported that among the various toxins tested, susceptible populations were less sensitive to Cry1Ab, Cry1Aa, and Cry1Fa, respectively. While among the resistant field populations tested, they developed severe resistance especially to Cry1Ac and MVP (Table 6) along with resistance development to other toxins tested. This suggests that there is an intraspecific variation in the baseline susceptibility among susceptible populations.

Kranthi (2008) has been assigned job to monitoring the shifts in baseline susceptibility (development of tolerance/resistance) in the *H. armigera* against Cry1Ac toxin in various cotton growing regions of the country. Yu Cheng Zhu et al. (2009) studied frequencies of resistance alleles to Bt cotton in field population of *H. armigera*. In 1999, the allele frequency estimated for population (0.0058) significantly fluctuated during 2003–2005. F1 and F2 screens conducted in 2006 and 2007 revealed >3-fold increase of resistant gene frequency compared to the levels of 2003–2005, >18-fold increases over 1999 in same population.

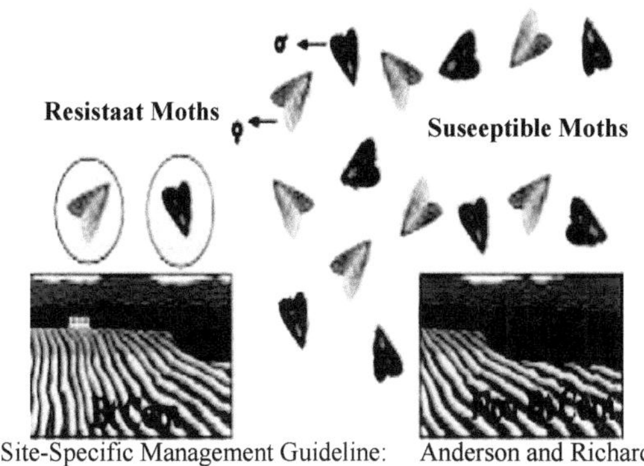

Fig. 3 Refuge strategy

Insect Resistance Management (IRM) Strategy

The ultimate goal of IRM programs for Bt crops—as with IRM programs for any insect-control technology—is to slow the rate at which insect-resistance evolves. IRM programs cannot be expected to prevent resistance, but they should be designed to maximize the effective life of a Bt crop. The economic benefits of this strategy are obvious and prolonged product life increases the likelihood that next-generation products can be developed and commercialized in a timely manner, creating a paradigm of continuous improvement in technologies rather than sequential replacement to keep up with resistance (Head and Greenplate 2012). Resistance management strategies try to prevent or diminish the selection of the rare individuals carrying resistance genes and hence to keep the frequency of resistance genes sufficiently low for insect control. Strategy development generally relies heavily on theoretical assumptions and on computer models simulating insect population growth under various conditions (Alstad and Andow 1994). Proposed strategies include the use of multiple toxins (stacking or pyramiding), crop rotation, high or ultrahigh dosages, and spatial or temporal refugia (Tabashnik 1994; McGaughey 1992). It is expected that each pest-crop complex may require a specific implementation of certain resistance management strategies that may have to address the use of both *B. thuringiensis* sprays and transgenic crops. Experience with transgenic crops expressing *cry* genes grown under different agronomic conditions is essential to define the requirements of resistance management. It is equally important to design a resistance management strategy acceptable to everyone involved: technology suppliers, seed companies, extension workers, crop consultants, regulators, and, most of all, growers Kennedy and Whalon (1995). To prevent the loss of this valuable management tool, IRM guidelines have been established to delay or stop the development of ECB resistance.

The refuge strategy, which is mandated in the USA and elsewhere, is based on the idea that most of the rare resistant pests surviving on Bt crops will mate with abundant susceptible pests from nearby refuges of host plants without Bt toxins. If inheritance of resistance is recessive, the hybrid progeny from such matings will die on Bt crops, substantially slowing evolution of resistance. This approach is sometimes called the "high-dose refuge strategy" because it works best of the dose of toxin ingested by insects on Bt plants is high enough to kill all or nearly all of the aforementioned hybrid progeny. In principle, if high-dose is achieved, resistance can be delayed by increasing refuge abundance which lowers proportion of the population selected for resistance to compensate for survival of hybrid progeny of Bt plants (Tabashnik and Gould 2004). Thus, the US environmental protection agencies guidelines for high-dose specified that Bt plants should kill at list 99.99 % of susceptible insects in the field. As an example, Anderson and Hellmich (2005) described the refuge strategy of corn in the Site-Specific Management Guideline that Refuge plants, or nontransgenic corn, are an important component of IRM. The purpose of planting a refuge is to dilute resistance genes by supplying an abundance of susceptible ECB moths that can mate with the rare resistant moths that have survived exposure to Bt corn (Fig. 3). Offspring from these matings are likely to be susceptible to Bt corn.

Refuge strategy: reduce chances that resistant moths mate with each other by providing large numbers of susceptible moths from the refuge, non-Bt corn. Offsprings of these moths are susceptible to Bt. Recommended amounts of refuge to delay resistance is 20 % or more located within one-half mile of the Bt corn.

If transgenic plants can express a *cry* gene at doses high enough to kill even homozygous resistant insects, that crop will become a nonhost. While such an ultrahigh dose might be impractical with a sprayable product due to high cost, incomplete coverage, toxin breakdown, and plant growth, it may be possible with toxin-engineered plants, taking into account the currently attainable levels of Cry expression in planta (Jansens et al. 1997). For example, a Colorado potato beetle population 100-fold resistant to a Cry3A-containing *B. thuringiensis* spray could not survive on potato plants expressing the same protein (Altre et al. 1996). It remains to be seen if a combination of toxins with ultrahigh expression can overcome all homozygous resistance alleles, changing the crops into nonhost plants. Metz et al. (1995) demonstrated that F1 larvae from a cross between a susceptible laboratory *P. xylostella* colony and a field-resistant colony did not survive on transgenic broccoli expressing Cry1Ac. It has been reported that the inclusion of refuge plants in cages with transgenic broccoli plants resulted in slower evolution of resistance in populations of *P. xylostella*. Supporting evidence also comes from selection experiments using *B. thuringiensis* subsp. *aizawai* and a diamondback moth population that had evolved resistance to Cry1Ab and Cry1Ca in the field. In these studies, a 10 % refuge delayed resistance over a nine-generation test (Luiand Tabashnik 1997).

A specific planting strategy that has been recommended to reduce selection is the use of seed mixtures of toxin-expressing and toxin-free plants to provide prepackaged refugia. The seed mix strategy, still controversial, would probably only be effective for insect species whose larvae move very little between plants (Mallet and Porter 1992). Another valuable option for resistance management, in combination with the use of refugia, is the expression of multiple Cry proteins in crops or incorporation of multiple proteins in *B. thuringiensis* sprays, provided these toxins have different modes of action with respect to the insect's mechanism of resistance. Cry toxins that recognize different receptors in the same target species could be deployed in this strategy, since they are less prone to cross-resistance. As noted above, diamondback moth populations resistant to field applications of Cry1A-containing *B. thuringiensis* formulations showed minimal cross-resistance to other crystal proteins such as Cry1Ba, Cry1Bb, Cry1Ca, Cry1Da, Cry1Ia, Cry2A, and Cry9Ca, while they were cross-resistant to Cry1Fa and Cry1Ja (Lambert et al. 1996). For many insect species, multiple Cry1A proteins would not be an appropriate choice, since some of these proteins share binding sites with one another and even with other toxins of the Cry1 class. Yet for other insects, Cry1A proteins have been shown, at least on ligand blots, to recognize different binding proteins. Additionally, *B. thuringiensis* Cry toxins could be combined with other insecticidal proteins.

The multiple-attack strategy assumes that within a population, if insects homozygous for one resistance gene are rare, then insects homozygous for multiple resistance genes are extremely rare. Crops or sprays deploying multiple toxins would still control even insects homozygous for one or two resistance genes yet heterozygous for another gene. A critical condition for the success of this strategy is that each of the insecticides on its own should have high mortality for susceptible homozygote. An example is *O. nubilalis*, in which Cry1Ab and Cry1Ba, both highly active, bind to different receptors. A strong argument for the utility of multiple-gene pyramiding is found in the recent results of Georghiou and Wirth (1997). Their field-collected *C. quinquefasciatus* populations readily developed resistance in the laboratory to a single *B. thuringiensis* subsp. *israelensis* toxin (Cry11A) but remained remarkably sensitive when selection was with the full complement of toxins from this variety.

Due to the urgent need for a more complete understanding of the parameters of effective resistance management, companies developing *B. thuringiensis* biopesticidal sprays and transgenic plants formed the *B. thuringiensis* Management Working Group to promote research on the judicious use of *B. thuringiensis* products. It is hoped that an increased understanding of the complex interplay among Cry toxins, their bacterial hosts, their target organisms, and the ecosystems they share will allow for the long-term, effective use of Cry toxins for pest management.

Evolution of resistance in target pests to transgenic insecticidal crops is a significant environmental risk that could affect multiple stakeholders including those outside of agriculture. Reduction in binding is a major mechanism of resistance in all cases of field-evolved resistance to Bt products or Cry proteins in *P. xylostella*, except for resistance against Cry1Ca. In Pink bollworm and other pests, resistance could evolve via selection for alleles that are not recessive, have lower fitness cost or completely overcome disadvantage on Bt plants. A number of studies have reported the ability of insects resistant to Cry1Ac to acquire cross-resistance to structurally similar insecticidal proteins such as Cry1Aa, Cry1Ab, Cry1C, etc. Of all of the various strategies and tactics considered for IRM, the high-dose/refuge strategy is by far the most widely considered and used. Refuges can delay pest resistance to Bt crops, especially when resistance is recessive and refuges are abundant.

References

Alstad DN, Andow DA (1994) Managing the evolution of insect resistance to transgenic plants. Science 268:1894–1896

Altre JA, Grafius EJ, Whalon ME (1996) Feeding behavior of CryIIIA-resistant and susceptible Colorado potato beetle (Coleoptera: Chrysomelidae) larvae on *Bacillus thuringiensis tenebrionis*-transgenic CryIIIA-treated and untreated potato foliage. J Econ Entomol 89:311–317

Anderson PL, Hellmich RL (2005) Bt Corn and insect resistance management: what are they? The site specific management guideline, SSMG-33. Department of Entomology Iowa State University Ames, Iowa.

Aroian RV, Griffitts JS, Huffman DL, Whitacr JL, Barrows BD, Marroquin LD, Mullers R, Brown JR, Hennet T, Esko JD (2003) Resistance to a bacterial toxin is mediated by removal of a conserved glycosylation pathway required for toxin-host interactions. J Biol Chem 278(46):45594–45602

Candas M, Loseva O, Oppert B, Kosaraju P, Bulla LA (2002) Insect resistances to *Bacillus thuringiensis*-alterations in the Indianmeal moth larval gut proteome. Mol Cellul Proteomics 2(1):19–28

Estada U, Ferre J (1994) Binding of insecticidal crystal proteins of *Bacillus thuringiensis* to the midgut brush border of the cabbage looper, *Trichoplusia ni* (Hubner) (Lepidoptera: Noctuidae) and selection for resistance to one of the crystal proteins. Appl Environ Microbiol 60(10):3840–3846

Ferre J, Van Rie J (2002) Biochemistry and genetics of insect resistance to *Bacillus thuringiensis*. Annu Rev Entomol 47:501–533

Georghiou GJ, Wirth MC (1997) Influence of exposure to single versus multiple toxins of *Bacillus thuringiensis* subsp. *israelensis* on development of resistance in the mosquito *Culex quinquefasciatus* (*Diptera: Culicidae*). Appl Environ Microbiol 63:1095–1101

Gould F, Martinez-Ramirez A, Anderson A, Ferre J, Silva FJ, Moar WJ (1992) Broad-spectrum resistance to Bt toxins in *Heliothis virescens*. Proc Natl Acad Sci U S A 88:7986–7990

Head GP, Greenplate J (2012) The design and implementation of insect resistance management programs for *Bt* crops. GM Crops Food 3(3):144–153

Heckel DG (1994) The complex genetic basis of resistance to *Bacillus thuringiensis* toxin in insects. Biocontrol Sci Technol 4:405–417

James C (2011a) Global status of commercialized biotech/GM crops. ISAAA executive report brief No 37

James C (2011b) Global Status of Commercialized Biotech/GM crops. ISAAA Brief No. 43. ISAAA: Ithaca, NY

Jansens S, Van Vliet C Dickburt, Buysse L, Piens C, Saey B, De Wulf A, Gossele V, Paez A, Goebel E, Peferoen M (1997) Transgenic corn expressing a Cry9C insecticidal protein from *Bacillus thuringiensis* protected from European corn borer damage. Crop Sci 37:1616–1624

Kennedy GG, Whalon ME (1995) Managing pest resistance to *Bacillus thuringiensis* endotoxins: constraints and incentives to implementation. J Econ Entomol 88:454–460

Kranthi KR (2008) Monitoring for shifts in baseline susceptibility (development of tolerance/resistance) in the cotton bollworms (*Helicoverpa armigera*, and *Earias vittella* against Cry1A(c) toxin in various cotton growing regions of the country. Annual progress report, Central Institute of Cotton Research

Kranthi KR (2012b) *Bt* cotton Central Institute for Cotton Research, Nagpur. Indian society for cotton improvement (isci) Mumbai, pp 1–71

Lambert BL, Buysse C Decock, Jansens, Piens C, Saey B, Seurinck J, Van Audenhove K, Van Rie J, Van Vliet A, Peferoen MA (1996) *Bacillus thuringiensis* insecticidal protein with a high activity against members of the family Noctuidae. Appl Environ Microbiol 62:80–86

Lui YB, Tabashnik BE (1997) Experimental evidence that refuges delay insect adaptation to *Bacillus thuringiensis*. Proc R Soc Lond Ser B 264:605–610

Mallet J, Porter P (1992) Preventing insect adaptation to insect resistant crops: are seed or refugia the best strategy? Proc R Soc Lond Ser B 250:165–169

Metz TD, Roush RT, Tang JD, Shelton AM, Earle ED (1995) Transgenic broccoli expressing a *Bacillus thuringiensis* insecticidal crystal protein: implications for pest management strategies. Mol Breed 1:309–317

Moar WJ, Pusztai-Carey M, Van Faassen H, Bosch D, Frutos R, Rang C, Luo K, Adang MJ (1995) Development of *Bacillus thuringiensis* Cry1C resistance by *Spodoptera exigua* (Hubner) (Lepidoptera: Noctuidae). Appl Environ Microbiol 61(6):2086–2092

Rajagopal R, Arora A, Sivakumar S, Rao NGV, Nimalkar SA, Bhatnagar RK (2009) Resistance of *Helicoverpa armigera* to Cry1Ac toxin from *Bt* is due to improper processing of the protoxin. Biochem J 419:309–316

Sayyed AH, Raymond B, Ibiza-Palacios MS, Escriche B, Wright DJ (2004) Genetic and biochemical characterization of field evolved resistance to *Bacillus thuringiensis* toxin Cry1Ac in Diamondback moth, *Plutella xylostella*. Appl Environ Microbiol 70:7010–7017

Tabashnik BE, Liu YB (1997) Experimental evidence that refuges delay insect adaptation to *Bacillus thuringiensis*. Proc R Soc Lond B 264:605–610

Tabashnik BE, Gould F (2004) Delaying evolution of insect resistance to transgenic crops by decreasing dominance and heritability. J Evol Biol 17:904–912

Tabashnik BE, Finson N, Groeters FR, Moar WJ, Johnson MW, Luo K, Adang MJ (1994) Reversal of resistance to *Bacillus thuringiensis* in *Plutella xylostella*. Proc Natl Acad Sci U S A 91:4120–4124

Tabashnik BE, Gassmann AJ, Crowder DW, Carriere Y (2008) Field-evolved insect resistance to transgenic *Bt* crops. ISB News Report

Uribe D, Martinez W, Ceron J (2003) Distribution and diversity of *Cry* genes in native strains of *Bacillus thuringiensis* obtained from different ecosystems from Colombia. J Invertebr Pathol 82:119–127

Yenagi BS, Patil VS, Biradar DP, Khadi BM (2010) Geographical variability in susceptibility of cotton bollworm (*Helicoverpa armigera*) to *Bacillus thuringiensis* toxin across northern Karnataka cotton ecosystem. J Ent Res 34:61–63

Yu Cheng Zhu, Liu F, Xu Z, Shen J (2009) Monitoring of resistance development to Bt cotton in field populations of H. armigera. Nanjing Agricultural University, China, pp 674–680

Part III

Insect Toxicology

Carboxylesterase and Glutathione-S-Transferase Quantification Mediating Resistance in Populations of Rice Leaf Folder, *Cnaphalocrocis medinalis* (Guenee)

Ramesh Babu Veegala and Shashi Vemuri

Abstract

A study was taken up for cross resistance between *Bacillus thuringiensis* Cry toxins and synthetic insecticides in larvae for two field populations of *Cnaphalocrocis medinalis* collected from the Directorate of Rice Research (DRR) and ICRISAT using the leaf-dip bioassay method. Bioassays with third instar larvae were carried out for two selected rice leaf folders, *C. medinalis* field populations with monocrotophos. The bioassay results showed a twofold difference in the resistance ratios for the two populations. Qualitative and quantitative changes of carboxylesterase (CarE) and glutathione-s-transferase (GST's) were worked out with midgut extracts of the two *C. medinalis* populations of α-napthyl acetate and chloro dinitro benzene substrates. The results revealed a 1.35-fold and 2.245-fold more variation in CarE and GST's in midgut homogenates of the two populations. Midgut and whole body extracts of the resistant *C. medinalis* showed difference in the esterase banding pattern with midgut producing three esterase bands. Inhibitor studies of the esterase isozymes separated under native PAGE with specific esterase inhibitors DDVP, eserine, and serine sulphate at different concentrations revealed esterase isozymes are B type esterases and were associated with carboxyl esterase activity.

Keywords

Carboxylesterase · *Cnaphalocrocis medinalis* · Glutathione-s-transferase · Isozymes

R. B. Veegala (✉)
Crop Protection Division, CRIJAF, Barrackpore, West Bengal, 700120 India
e-mail: veegalaramesh@gmail.com

S. Vemuri
AINP on Pesticide Residues, EEI Premises, ANGRAU, Rajendranagar, Andhra Pradesh, 500030 India
e-mail: Sash_3156@Yahoo.Co.In

Introduction

The rice leaf folder (RLF), *Cnaphalocrocis medinalis* (Guenee) (Lepidoptera: Pyralidae) is an important rice pest, widely distributed in many rice growing areas of Asia (Cheng 1996).There have been frequent and serious outbreaks of this

A. K. Chakravarthy (ed.), *New Horizons in Insect Science: Towards Sustainable Pest Management*,
DOI 10.1007/978-81-322-2089-3_10, © Springer India 2015

pest in many countries, namely India, Korea, Japan, China, Malaysia, Sri Lanka, and Vietnam (Dale 1994). The over-use of broad-spectrum insecticides, such as methyl parathion, monocrotophos, and endosulfan has been cited as a major cause of outbreaks of *C. medinalis* because excessive spraying of insecticide disrupts natural control (Dale 1994). Insecticides still remain the major control tactics against leaf folder. Carbofuran and fenthion (Chandramohan and Jayaraj 1976), bendiocarb, acephate and carbosulfan, quinolphos, monocrotophos, and phosphamidon (Raju et al. 1990) and fenvalerate (Ramaraju and Natarajan 1997) were the common insecticides used against rice leaf folder.

Genetically modified rice lines expressing *B. thuringiensis* insecticidal crystal proteins have been produced that are highly tolerant to lepidopteran pests. In China, three GM rice lines transformed with *cry1Ac/cry1Ab* genes(GM Minghui 63), *cry1Ac/CpTI* genes (GM Minghui 86) and *cry1Ab* genes(GM Kemingdao) effective against *Scirpophaga incertulus*, *C. medinalis*, and *Chilo suppressalis* have been tested both at field and laboratory levels and are on the verge of commercialization (Tu et al. 2000; Ye et al. 2003; Han et al. 2007). Though economic and environmental benefits of GM crops are well established, a matter of concern is the possibility of the target insect pest developing resistance to *B. thuringiensis* insecticidal toxins. Though a couple of resistance mechanisms have been reported for conferring resistance to *B. thuringiensis* viz. reduced binding of crystal toxins to the brush border membrane vesicles (BBMVs) of midgut epithelium and alteration in the midgut proteases that cleave the protoxin to active toxin. A new resistance mechanism to *B. thuringiensis* Cry toxins is identified and is associated with increased activity of midgut carboxylesterase activity (Gunning et al. 2005). The involvement of glutathione-S-transferase, carboxylesterase, and microsomal mono-oxygenase in insecticide resistance has been reported in insecticide-resistant strains of many insect species. Carboxylesterases (CES, EC 3.1.1.1) are members of a superfamily of serine hydrolases that hydrolyze ester, amide, and carbamate bonds. Several different carboxylesterase (CarE) genes exist with evidence of multiple gene duplication in insects. Esterases hydrolyze ester bonds from various substrates with a carboxylic ester. Esterases are frequently implicated in the resistance of insects to organophosphorus, carbamates, pyrethroids, neonicotinoids, and to many other new classes of insecticides through gene amplification, upregulation, coding sequence mutations, or a combination of these mechanisms (Li et al. 2007).

Material and Methods

Two field populations of *C. medinalis* were collected from the Directorate of Rice Research (DRR), Rajendranagar and ICRISAT, Patancheru. *C. medinalis* adults were collected from rice fields during the boot leaf stage in rabi 2011. The collected adults were released into pots containing TN-1 plants for egg-laying and were covered with a muslin cloth for aeration, 20 % honey solution was also provided for feeding. Ten pairs of *C. medinalis* adults were released into each TN-1 pot. *C. medinalis* populations from different locations were reared separately and after larval hatching the third instar larva was used for bioassay.

The leaf-dip bioassay method was used. Three–four long tender leaves from TN-1 were used. The leaves were first washed with distilled water and were then dipped in monocrotophos solution and thoroughly air-dried for about 10 min, different concentrations of the insecticide were prepared, and bioassays were carried out first at tenfold variation. Based on 20–80 % mortality, concentrations were prepared in a narrow range of fivefold for further bioassays. Six concentrations were tested with ten third instar larvae per treatment and replicated thrice. Larvae were allowed to feed on insecticide treated leaves for 24 h and mortality was recorded for 24 h after treatment. Control treatments with larval mortality more than 20 % were discarded and bioassays were repeated. Statistical analysis for calculating the LC_{50} values for the bioassay was estimated using maximum likelihood program MLP 3.01 (Ross 1987). The corrected percentage mortality was calculated using Abbott's formula (Abbott 1925).

Preparation of Enzyme Homogenate

Fifth instar larvae *C. medinalis* were used for enzyme preparation. Larval midguts were excised with replicated samples and were homogenized in 500 µl homogenization buffer (50 mM sodium phosphate buffer, pH 7.4). After centrifugation at 10,000 rpm for 20 min, the clear supernatant was collected and used as enzyme sources for analysis. All the operations were carried out on ice and centrifugation was carried out at 4 °C to minimize losses of enzyme activity. The protein content of enzyme extract was estimated by Coomassie Brilliant Blue G-250 dye binding method using bovine serum albumin as the standard (Bradford 1976).

Carboxylesterase Assay

Carboxylesterase activity was determined following van Asperen (1962) with necessary modifications and α-naphthyl acetate as a substrate. A 0.3 mM substrate solution of 1-naphthyl acetate was prepared in acetone. The assay mixture contained 15 µl of enzyme preparation, 0.5 ml of 50 mM sodium phosphate buffer pH 7.4, and 800 µl of 0.3 mM substrate solution. The mixture was incubated at 30 °C for 30 min. Finally, 200 µl of 0.1 % tetrazotized*o*-dianisidine (Fast blue B) in 3.5 % Sodium Dodecyl Sulphate (SDS) was added and incubated for 20 min at room temperature in the dark. The α-naphthol formation was measured at 590 nm. Enzyme activity was calculated from α-naphthol standard curve.

Esterase Isozyme Studies

Native Polyacrylamide Gel Electrophoresis (PAGE) with 10 % resolving gel was performed to separate esterase isozymes. Qualitative changes in the esterase-banding pattern were performed using F_2 generation larvae reared after surviving the insecticidal bioassay with monocrotophos. On the native PAGE, 5 µg protein concentrations of the midgut homogenate per well were loaded and run at a constant voltage of 90 for 45 min. Gels were stained briefly for esterase activity with freshly prepared 0.05 % (w/v) α-napthyl acetate and 0.1 % (w/v) fast blue B in 50 mM phosphate buffer pH 7.4. For inhibition studies, gels were cut into strips and incubated in 10^{-4} and 10^{-6} M serine sulphate each and 10^{-4} M DDVP individually in 50 mM phosphate buffer pH 7.4 for 30 min at 28 °C with occasional shaking. Control gels were incubated for 30 min in buffer alone. All the gel strips were stained and incubated for 30 min in α-napthyl acetate substrate diazonium mixture for confirming the esterase activity.

Glutathione-S-Transferase Activity Estimation

Glutathione-S-transferase assay was performed using reduced glutathione (50 mM), midgut homogenate supernatant (10,000 × *g*) from the F_2 generation larvae reared after surviving the insecticidal bioassay with monocrotophos, chlorodinitro benzene (CDNB) 50 mM, sodium phosphate buffer (pH 6.5, 100 mM), and Ethylenediaminetetraacetic acid (EDTA) (1 mM). The assay mixture contained 50 µl of 50 mM CDNB, 150 µl of reduced glutathione, and 2.77 ml of 100mM, pH 6.5 phosphate buffer containing 1mM of EDTA. To the above assay 30 µl of enzyme (midgut homogenate) was added and the contents shaken gently and incubated for 2–3 min at 25 °C then the contents were transferred to 4 ml cuvettes and absorbance was recorded for 6–7 min at 340 nm. Based on the increase in absorbance over 5 min the enzyme activity was calculated in µmol/min/mg/protein.

Results and Discussion

Bioassay results for the two selected *C. medinalis* populations, DRR and ICRISAT, Patencheru revealed LC_{50} of 60 ppm for ICRISAT *C. medinalis* population and showed twofold resistance ratio over the DRR population which showed LC_{50}30 ppm against monocrotophos (Table 1).

Qualitative and quantitative changes of CarE and GST's for DRR population in α-napthyl acetate and CDNB revealed a

Table 1 Toxicity of monocrotophos to third instar larvae *C. medinalis* 24 HAT

Location	LC_{50} (ppm)	RR	SLOPE ± SE	'F' limits		χ^2 (degrees of freedom)
				Lower	Upper	
DRR population	30	1.00	1.29±0.21	0.0018	0.0057	4.0(4)
ICRISAT population	60	2.00	1.14±0.21	0.002	0.011	2.31(4)

RR resistance ratio over one generation, *F Limits* fiducial limits, *HAT* hours after treatment

Table 2 Carboxylesterase and GST activity of third instar *C. medinalis* larvae

Location	CarE (μmols/min/mg protein)	CarE folds	GST (μmols/min/mg protein)	GST folds
DRR population	114.39	1.00	5.660	1.00
ICRISAT population	155.2	1.35	12.598	2.245

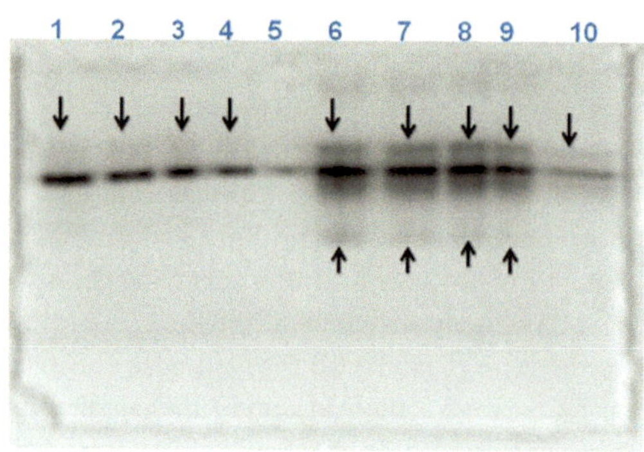

Fig. 1 Carboxylesterase, profiles from whole body and midgut homogenates of ICRISAT, *C. medinalis* population. Lanes 1–5 are whole body homogenates showing two bands of esterase isozymes, Lanes 6–10 are midgut homogenates showing three bands of esterase isozymes

titre of 114.3 μmols min^{-1} mg^{-1} protein and 5.66 μmols min^{-1} mg^{-1} protein while ICRISAT population showed 155.2 μmols min^{-1} mg^{-1} protein and GST titre of 12.59 μmols min^{-1} mg^{-1} protein, respectively. The results revealed that ICRISAT *C. medinalis* population had 1.35-fold greater carboxyl esterase and 2.245-fold more GST's in its midgut homogenates over DRR population (Table 2). Similar findings were reported by Mohan and Gujar (2003a), where 1.2–1.8-fold increased CarE activity was observed in *P. xylostella* for monocrotophos, cartap and fipronil resistant populations. Yamamoto et al. (2008) reported that GST from RLF, *C. medinalis*, was inhibited by fenitrothion, permethrin, and deltamethrin, suggesting GST may be involved in metabolizing organophosphorus and pyrethroid insecticides.

Esterase activity visualized in native PAGE following incubation in substrate solution (0.05 % α-naphthyl acetate in 0.1 % fast blue ® revealed two isozyme bands. The major band with diffused esterase activity was relatively of more molecular mass than the other esterase activity band (Fig. 1). Difference in esterase banding pattern in ICRISAT *C. medinalis* population was observed when using midgut and whole body extracts. The midgut produced all three types of esterase bands while whole body homogenates produced only two bands (Fig. 2). Inhibitor studies with the esterase isozymes separated under native PAGE when subjected to inhibition by incubating with class-specific esterase inhibitors in buffers containing dichlorvos (DDVP)10^{-4} M, eserinesulphate, 10^{-6} and 10^{-4} M concentrations indicated that these two esterase isozymes are B type esterases as eserinesulphate, a specific inhibitor of cholinesterase did not inhibit the esterase activities at 10^{-6} and 10^{-4} M and esterases were characterized to have carboxylesterase activity (Fig. 2).

In insects, the esterase bands separated electrophoretically under native condition are classified into three types by the substance which inhibits their activity (Aldridge 1953; Van Asperen 1962). The results of the present study are in conformity with cypermethrin resistant *Plutella xylostella* strain, wherein the carboxylesterase levels from the first to the fourth instar, pupa and adult showed 2.64-, 3.16-, 2.61-, 3.04-, 2.93-, and 2.75-fold higher carboxylesterase activity in comparison to *P. xylostella* susceptible strain (Moharil et al. 2008) and also reported DDVP

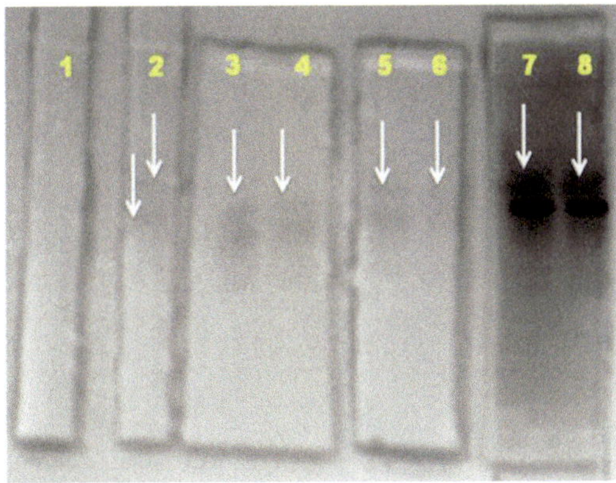

Fig. 2 Characterization of *C. medinalis* carboxylesterase with specific substrates. Lane: 1—DDVP (10^{-4}M), Lanes: 2–4—Eserine (10^{-4}M), Lanes: 5, 6—Eserine (10^{-6}M), Lanes: 7 and 8—Control

to be the best inhibitor amongst chlorpyriphos, monocrotophos, profenofos, quinalphos, and phenthoate, as it showed 54.34, 72.54, 78.71, 80.82, and 82.94 % inhibition of carboxylesterase titres associated with cypermethrin resistance at 0.01, 0.1, 1, 5, and 10 mg/ml and suggested that DDVP is the best synergist to mitigate cypermethrin resistance in *P. xylostella*. Rashad (2008) reported high titres of esterase activity in brain, foregut, midgut, and ovary of 2-day old adults of *Schistocerca gregaria* (Forsskal) while in 13-day-old adults the hindgut exhibited high esterase levels. Inhibitory studies with EDTA and profenofos depicted high levels of both carboxylesterase and phosphotriesterases in the brain tissues of two ages, attributed to play a role in insecticide resistance.

In the present study, correlation between LC_{50} value of monocrotophos of the third instar larvae and carboxylesterase activity was similar to that obtained by Kranthi et al. (1997) who studied the seasonal dynamics of metabolic mechanisms responsible for pyrethroid resistance in *H. armigera* and assigned it to the involvement of microsomal oxidase and esterases. Young et al. (2005) reported pyrethroid resistance in *H. armigera* and attributed to overproduction of esterase isoenzymes that metabolize and sequester pyrethroid insecticides and found out that pyrethroid-resistance-associated esterases were inhibited by piperonylbutoxide (PBO) and maximum inhibition achieved 3–4 h after dosage and again restored by 24 h.

Esterase zymogram in the present study showed an additional fast moving band by midgut extracts in comparison with whole body homogenates; this is in conformity with the findings in DBM (Mohan and Gujar 2003a).

In the present study, the carboxylesterase present in ICRISAT population was calibrated in vitro to be 155.2 μmols min^{-1} mg^{-1} protein and the esterase zymograms showed intensely stained bands depicting resistance association. The median lethal concentration, LC_{50}, for this strain was 60 ppm though more than the discriminating dose for monocrotophos, 0.35 μg per larvae (Anandan and Regupathy 2007). The base-line LC_{50} estimate for the ICRISAT population with Cry1Ab toxin is 0.50 μg/ml, the matter of concern in this regard is that indiscriminate usage of insecticides for control of *C. medinalis* may further bring an elevation in the esterase titre which may sequester the toxin before it reaches the target site as exemplified in the case of 'Silver strain' *H. armigera* towards Cry1Ac expressed by transgenic cotton Ingard ® in Australia, where sequestration by esterases was recognized as a potential resistance mechanism apart from previous resistance mechanisms viz. reduced binding by the Cry toxin to BBMVs of midgut epithelium and alteration in midgut proteases that cleave protoxin to active toxin (Gunning et al. 2005).

This chapter emphasizes the importance of monitoring carboxylesterase enzyme titre and its characterization in RLF larvae that may also mediate resistance to Cry toxins along with other resistance mechanisms when transgenic rice gets commercialized.

References

Abbott WS (1925) A method of computing the effectiveness of an insecticide. J Econ Entomol 18(2):265–267

Aldridge WN (1953) Serum esterases—1: two types of esterase (A and B) hydrolysing p-nitrophenyl acetate, propionate and butyrate, and a method for their determination. Biochem J 53:110–117

Anandan GK, Regupathy A (2007) Assessment of acute toxicity of insecticides for monitoring insecticide resistance in rice leaf folder, *Cnaphalocrocis medinalis* (Guenee) in Tamil Nadu, India. Resist Pest Manag Newslett 16(2):3–5.

Bradford MM (1976) A rapid and sensitive method for the quantification of microgram quantities of protein utilizing the principle of protein–dye binding. Anal Biochem 74:248–254.

Chandramohan N, Jayaraj S (1976) Evaluation of certain granular and foliar insecticides against rice leaf folder, *Cnaphalocrocis medinalis* (Guenee). Madras Agric J 63:364–366

Cheng JA (1996) Rice pests. China Agricultural Press, Beijing

Dale D (1994) Insect pests of rice plant—their biology and ecology. In: Heinrichs EA (ed) Biology and management of rice insects. Wiley Eastern Limited, New Delhi

Gunning RV, Dang HT, Kemp FC, Nicholson IC, Moores GD (2005) New resistance mechanism in *Helicoverpaarmigera* threatens transgenic crops expressing *Bacillus thuringiensis* Cry1Ac. Appl Environ Microbiol 71(5):2558–2563

Han LZ, Wu KM, Peng YF, Wang F, Guo YY (2007) Efficacy of transgenic rice expressing Cry1Ac and CpTI against the rice leaf folder, *Cnaphalocrocis medinalis* (Guenee). J Invert Pathol 96(1):71–79

Kranthi KR, Armes NJ, Rao NGV, Raj S, Sundaramurthy VT (1997) Seasonal dynamics of metabolic mechanisms mediating pyrethroid resistance in *Helicoverpa armigera* in central India. Pestic Sci 50(2):91–98

Li X, Schuler MA, Berenbaum MR (2007) Molecular mechanisms of metabolic resistance to synthetic and natural xenobiotics. Ann Rev Entomol 52:231–253

Mohan M, Gujar GT (2003) Local variation in susceptibility of the diamondback moth, *Plutella xylostella* (Linnaeus) to insecticides and role of detoxification enzymes. Crop Prot 22(3):495–504

Moharil M, Prasad P, Nagarjuna AW, Rao GV, Sachin NT, Rai MK, Nimbalkar SA (2008) Detection of a carboxylesterase-mediated resistance mechanism in *Plutella xyloestella* (L.) by diagnostic microplate assay. Res J Agric Biol Sci 4(6):623–629

Raju N, Gopalan M, Balasubramanian G (1990) Ovicidal action of insecticides, mould inhibitors and fungicides in the eggs of rice leaf folder and stem borer. Ind J Plant Prot 18(1):5–9

Ramaraju K, Natarajan K (1997) Control of leaf folder under extreme weather conditions. Madras Agric J 66:252–254

Rashad EM (2008) Esterase activity and detection of carboxylesterase and phosphotriesterase in female desert locust *Schistocerca gregaria* (Forskal) in relation to tissues and ages. Egypt Acad J Biol Sci 1(2):135–143

Ross GES (1987) Maximum likelihood programme. The Numerical Algorithms Group, Rothamsted Experimental Station, Harpenden

Tu J, Zhang G, Datta K, He Y, Zhang Q, Khush GS, Datta SK (2000) Field performance of transgenic hybrid rice expressing *Bacillus thuringiensis* delta-endotoxin. Nat Biotechnol 18:1101–1104

Van Asperen K (1962) A study of house fly esterase by means of a sensitive colorimetric method. J Insect Physiol 8:401–416

Yamamoto K, Teshiba S, Aso Y (2008) Characterization of glutathione S-transferase of the rice leaf-folder moth, *Cnaphalocrocis medinalis* (Lepidoptera: Pyralidae): comparison of its properties of glutathione S-transferases from other lepidopteran insects. Pestic Biochem Physiol 92(3):125–128

Ye GY, Yao HW, Shu QY, Cheng X, Hu C, Xia YW, Gao MW, Altosaar I (2003) High levels of stable resistance in transgenic rice with cry1Ab gene from *Bacillus thuringiensis* Berliner to rice leaf folder, *Cnaphalocrocis medinalis* (Guenee) under field conditions. Crop Prot 22(2):171–178

Young SJ, Gunning RV, Graham DM (2005) The effect of piperonylbutoxide on pyrethroid resistance associated esterases in *Helicoverpaarmigera* Hubner) (Lepidoptera: Noctuidae). Pest Manage Sci 61(4):397–401

Bioassay of Rongbao (Active Ingredients of Calcium Cyanamide) Against Housefly Maggots

Wei Qing Zheng, Yan Guo, Wei Min Li, Hong Mei Ma, Xiao Qing Liu and Hai Ying Chen

Abstract

Calcium cyanamide (CC) has been extensively reported as a plant growth regulator and N-resource fertilizer. CC phytotoxicity and cytotoxicity have also been showed on organisms such as microorganisms, schistosomes, mollusks, mammals and plants. But to date, little is known about its bioactivity against filth flies. In this study, housefly maggots, *Musca domestica* (L.), were used to measure biological activity of CC. Laboratory bioassays showed that CC (at a dosage of approximately 2 % (W/W)) resulted in average mortality rate of 37.67 % on day 1 and 100 % on day 2, compared with the average mortality rate of 34.51 % on day 1, 89.80 % on day 2 and 100 % on day 3 at dosage approximately 1 % (W/W). This outcome demonstrated that the application of CC could be used as a part of the filth fly-integrated management. In the relationship between CC concentrations and pH value evaluation, the performance of CC and non-CC larval rearing medium differed within pH. The positive relationship of concentration of CC dissolved in water with pH was supported by the formula $y=43.33x+6.61$ ($R^2=0.95$, $p<0.01$). The results showed that housefly maggots' death had a positive relationship with the pH of larval rearing medium.

Keywords

Calcium cyanamide · Houseflies · Larvicidal efficacy · pH value

Introduction

Flies are a large class of insects of the order Diptera (Hardy 1981; Hardy and Delfinado 1980). They have widespread food choices ranging from decayed plants to decayed bodies of animals, including human beings (Beadle et al. 1938; Fabre et al. 2003; Greathead 1969; Hanski 1977; Kumara et al. 2012; Landolt and John 1992; Liquido

W. Q. Zheng (✉) · Y. Guo · W. M. Li · H. M. Ma · X. Q. Liu · H. Y. Chen
Department of Vector Pest Control, Nanchang Center for Disease Control and Prevention, Honggutan, Nanchang, 330038 Jiangxi, China
e-mail: zhengweiqing2001@163.com

H. Y. Chen
e-mail: chy@nccdc.org.cn

A. K. Chakravarthy (ed.), *New Horizons in Insect Science: Towards Sustainable Pest Management*,
DOI 10.1007/978-81-322-2089-3_11,

et al. 1994; Lysyk and Axtell 1986; Syamsa et al. 2012). They also suck plant juice and animal blood (Perlman 1962; Schofield and Torr 2002). Flies cause major damage in agriculture, forestry, animal husbandry and food-processing industry (Bonnefoy et al. 2008; Chapin et al. 1989; Ekesi 2010; Schofield and Torr 2002; Yadav et al. 2006). The global economic loss cannot be measured every year because of the damage and infestation caused by flies (Bonnefoy et al. 2008). In China, field flies mainly infest citrus fruit, vegetables, cowhide and silkworm, and seriously degrade the commercial value of these products (Huang and Huang 2012; Li et al. 2000; Liang et al. 2012; Liu et al. 2012; Ma et al. 2008). Of the thousands of species of flies, only a few are common in and around the home. Some of the most common flies in China are housefly (*M. domestica*), false stable fly (*Muscina stabulans*) and the lesser house fly (*Fannia canicularis*). The most commonly found flies are the common green bottle fly (*Lucilia sericata*), the oriental latrine fly (*Chrysomya megacephala*), *Aldrichina grahami*, *Anthomyia illocata*, *Helicophagella melanura* and the flesh fly (*Boettcherisca peregrina*) (Jia and Wu 2008). These pests breed in pathogen-laden animal wastes, carrion and decaying organic material from where they can pick up bacteria and viruses. A single filth fly can carry over 1 million pathogens that can be a threat to public health (Barin et al. 2010; Bonnefoy et al. 2008; Graczyk et al. 2005). Around 60 different diseases can be transmitted by filth flies, from salmonella infection to dysentery (Bonnefoy et al. 2008).

The most commonly used control measures for flies are environmental management, physical control, chemical control and biological control, but in some instances integrated fly control has also been implemented (Collier and Van Steenwyk 2003; Cook et al. 2007). (1) *Environmental management:* Most filth flies can reproduce quickly in warm weather. To keep fly population under control, we should limit the places where they can breed and feed. Good sanitation practices can largely reduce fly population. There is nothing a filth fly who likes less than proper sanitation. Adult flies need a place to lay their eggs, and a clean home and yard just won't appeal to mama fly (Powell 1993; Rozendaal 1997). (2) *Physical control:* Following all the sanitation rules above will significantly reduce, but not completely eliminate flies in or around our yard and home. Physical control steps can be adopted to further limit and exclude the number of these unhealthy pests using barriers and traps to keep doors, windows and vents closed, and by screening and sealing these and other fly entry points. Automatic door closing devices and air curtains that blow air away from doorways can also be installed to supplement an integrated fly management program. In addition to fly swatting, mechanical fly control includes trapping. Sticky fly paper is one type of fly trap. Ultraviolet light traps are another type, often used to supplement fly control in commercial buildings (Kaufman et al. 2005; Powell 1993; Rozendaal 1997). (3) *Chemical control:* Pesticides are available to control flies but should never be used as the first line of defence against them. This type of control provides only temporary relief. Even worse, the indiscriminate and improper use of chemical insecticides has given rise to many well-known and serious problems such as the risk of developing insect resistance and residual insecticides for humans and the environment. These problems have been coupled with acute neurotoxicity for human beings and domesticated animals (Brogdon and McAllister 1998; Elliott et al. 1978; Metcalf 1989). (4) *Biological control:* This type of control measure is generally accepted due to advantages such as environmental friendliness and harmlessness to non-target organisms. Biological control can also be subdivided into release of parasitic wasp, introduction of natural enemies, use of pathogenic microorganisms and application of secondary compound derived from plants or animals.

Parasitic Wasp Release

A few datum available indicate that *Spalangia* and *Muscidifurax* species are commercially available parasites most likely to attack both house fly and stable fly pupae in feedlots. *Muscidifurax* species are parasites of house flies and *Spalangia* species are parasites of stable flies (Aluja et al. 2003; Axtell and Arends 1990; Feder 1995; Ratcliffe et al. 2002; Rivers et al. 2012).

Introduction of Natural Enemies

Predators sometimes have a decisive effect on controlling fly populations in an orchard, vegetable garden and human-inhabited area. Invertebrate predators may include spiders, ants, carabid beetles, assassin bugs, staphylinid beetles among others. Vertebrate predators, such as birds, domestic fowls and primates, can consume flies, resulting in marked reduction in their numbers (Bruns 1960; Geden et al. 1988; Kaufman et al. 2000; Klopsch et al. 2012; Prasad and Snyder 2004; Redford 1984; Urbaneja et al. 2006; Wright et al. 1960).

Pathogenic Microorganisms for Flies

Bacillus thuringiensis (Bt) is best known as a pathogen of lepidopterous larvae, which was first isolated at the beginning of the twentieth century from diseased silkworms in Japan. Commercial production of pesticides based on Bt began in the USA in 1958 (Alfazairy et al. 2013; Farrar Jr et al. 2009; Hughes et al. 1986; La Lacey et al. 2001). Today, Bt products are the most widely used biological agents for the control of lepidopteran pests on food crops and forest trees, including dipteran pests such as flith flies. Isolation of mosquito pathogenic strains of *Bacillus sphaericus* (Bs) predated the isolation of Bt by some 10 years, but these early strains showed low toxicity. Following an intensive isolation and screening program organized by the WHO, more toxic strains of Bs were recovered and these, together with several isolated strains in other research project, have considerable potential as control agents. On the other hand, an important attribute of Bs seems to be its residual effect in the environment following application and activity in heavily polluted areas which have promoted its use as a biocontrol agent for insect pests of public health importance (Charles et al. 1996; La Lacey et al. 2001; Priest 1992).

Plant or Animal Secondary Compound Application

Natural products, especially plant or animal secondary compounds, are well known to have a range of useful biological properties against insect pests. Botanical pesticides with azadirachtin, rotenone, pyrethrin, hot-pepper wax, garlic oil, citrus oil and herbal essential oils as their active ingredients are commercially applied in crop protection, nuisance and diseases control and prevention in human and livestock (Al-Doghairi and El Hag 2003; Dorman and Beasley 1991; Ho et al. 1996; Liu et al. 2009; Madanlar et al. 2000; Miller and Chamberlain 1989; Park and Shin 2005; Shalaby et al. 1998; Singh et al. 2003; Singh and Singh 1998; Witt et al. 1999). According to research, other plant-based materials, such as basil oil, catnip oil, turpentine, rosin, camphor, cineole, have been proved to have the toxicity and/or repellant efficacy to control insect pests (Al-Doghairi and El Hag 2003; Bernier et al. 2005; Chang et al. 2009; Dube et al. 1989; Isman 2000; Lee et al. 2004; Obeng-Ofori et al. 1998; Park and Shin 2005; Peterson et al. 2002; Prates et al. 1998; Witt et al. 1999).

Although both environmental management and physical control have been recently emphasized as basic fly-control measures all over the world, it is important to stress that environmental management and physical control cannot operate alone; they must work alongside other control strategies in an integrated fashion. Chemical control remains the first consideration for fly control, because its killing effect can immediately solve the insect pest problem. However, chemicals harm the environment. Biological control is the best recommended fly-control measure as it is eco-friendly and is harmless to non-target organisms. It is urgent to find an alternative way to complement those measures.

Calcium cyanamide (CC) has been identified as an N-fertilizer resource, vegetable product improver, soil ameliorant, microorganism control, snail and schistosome control and herbicide control (Bourbos et al. 1997; Donald et al. 2004; Kaushal et al. 2002; Shi et al. 2009; Weeks et al. 2000; Wei et al. 2009; Xu 2009). Meanwhile, it has gained increased attention as a contributor to alcohol control and treatment (Ferguson 1956; Jones et al. 1988). The present chapter reports the results of CC killing efficacy tests on houseflies (*M. domestica*) larvae (maggot), and reveals the relationship between maggot mortality and pH.

Material and Methods

Test Maggots

Housefly maggots, *M. domestica*, were nurtured in laboratory at 25–27 °C and 60–80 % relative humidity (RH). The maggots were reared using mixed materials whose formula was 100 parts of wheat bran every 5 parts of dried milk powder; dechlorinated water was added to the mixture to the extent of water saturation but not dripping. Eggs laid on mixed material from female houseflies were hatched into larvae 1 for 2–3 days and larvae 2 for 4–5 days. Randomly mixed houseflies larvae 1 and 2 were chosen for bioassay.

Test Chemicals

Rongbao® dust is a dark gray powder made by Ningxia Darong Industrial Group Co., Ltd., the People's Republic of China. Rongbao® also known as CC is a secondary product of metallurgy. N resource fertilizer product Rongbao® in this study contains approximately 50 % CC, and other inert fillers.

Killing Efficacy

The killing efficacy of CC was evaluated using chemical-feed-maggot complex touch technique. The testing period lasted up to 72 h depending on the efficacy. The timing of the tests depended on whether the target maggots were killed by movement when touched. Evaluations were carried out in laboratory, at 26 ± 1 °C and 60 ± 10 % RH. Maggots (larvae 1 and 2 in combination) together with their larval rearing medium were prepared and placed into a 100 ml beaker. Before 10 g complex was separated and introduced into a 25 ml beaker, maggots and larval rearing medium were mixed gently and evenly using glass rod. 0.2/0.4 g Rongbao® dust on the basis of weight were applied to the mixed larval rearing medium and gently mixed again. The 25 ml beaker with larval rearing medium was covered with a gauze fastened tightly by a rubber band. The treatment group was set to two groups ($n=3$). A series of concentration was prepared from each treated groups; one group for 0.2 g Rongbao® dust with its dosage of approximately 1 %, the other group 0.4 g with approximately 2 % dosage. Dead maggots in the complex were selected, counted and recorded everyday until 72 h of deadline that was previously defined. As a blank control, nothing was placed on larval rearing medium with the same process as the test group.

Data Analysis

Formulated CC (Rongbao® dust) was assessed for insecticidal activity to *M. domestica* maggots at dilutions of approximately 1 % and 2 % in larval rearing medium. Maggots in control were not treated with any chemical compound. The rate of larvicidal activity of CC was calculated as the percentage reduction in maggots' mortality rate. Mortality rate (MR, %) was calculated as:

$$MR = (S - L) / L \times 100\%, \qquad (1)$$

where S is the number of the total insects in the treated (control) beakers and L is the number of live insects in the treated (control) beakers.

Mortality data was corrected for control mortality when mortality rate of maggots in control beakers was less than 20 %. Adjusted mortality rate (AMR, %) was calculated as:

$$AMR = (T - C) / (1 - C) \times 100\%, \qquad (2)$$

where C is the mortality rate of maggots in control beakers and T is the mortality rate of maggots in the treated beakers.

Microsoft Excel (version 2003) software was applied to record primary data and calculate mortality rate and adjusted mortality rate according to formulas 1 and 2 listed above. Data for mean, standard deviation (SD) and analysis of variance (ANOVA) were calculated and analyzed using SPSS 20 software.

Results and Discussion

Effect of CC on Maggots

To determine the effect of CC on maggots, mortalities following two concentrations of CC exposure were compared to an equivalent control group. Cumulative mortality rates of groups exposed to two different concentrations of CC on day 1, day 2 and day 3 were corrected for mortalities of corresponding control groups, whereas for the other treatments no corrections were made since their control mortalities measured on corresponding day did not exceed the 5% level. CC-induced cumulative mortality measured on day 1, day 2 and day 3 was higher in groups exposed to 2% concentration compared with groups exposed to 1% concentration on the corresponding day. The cumulative mortality rate increased from 34.51 to 37.67% on day 1, 89.80 to 100% on day 2 and 100 to 100% on day 3 as CC concentrations increased from 2 to 4%, respectively. Groups exposed to 2% CC on day 1 demonstrated insignificantly higher mortality rates than 1% counterpart (chi square=0.51, p=0.48). However, once the period of maggots treatment had proliferated for two days, exposure to 2% CC induced significantly higher mortality rates (chi square=23.21, p=0) compared to 1% concentration. One percent CC-induced mortality rates measured on day 3 reached the complete 100% level the same as 2% CC-induced mortality rates measured on day 2. As for blank control groups, null-mortality rates in housefly maggots were observed with correction of mortality rates in the treated groups in vain (Table 1).

Relationship Between CC Concentration and pH Value

The performance of CC and non-CC larval rearing medium differed within pH value. The pH values of larval rearing medium were 7.0 and 7.5 after treatment with 1 and 2% CC, respectively, whereas that of the control group was 6.63. The non-CC larval rearing medium had less pH value than CC 1%, and then the 1% concentration had less pH value compared to 2% concentration (Table 2). The positive relationship of the concentration of CC dissolved in water with the pH value was supported by the formula $y=43.33x+6.61$ ($R^2=0.95$, $p<0.01$) (Fig. 1).

Common species of filth fly in and around the home breed between manure and human and livestock food from a wide range of human and domestic animal (Bonnefoy et al. 2008). They are not only a group of pests affecting growth, feed intake and feeding efficiency of livestock and poultry, but also pose a major threat to human health (Bonnefoy et al. 2008; Schofield and Torr 2002; Yadav et al. 2006). The filth fly is primarily a nuisance to human, livestock and poultry, and also transmits pathogens and acts as a vector for more than 100 human and animal intestinal diseases (Bonnefoy et al. 2008; Coffey and Maier 1950; Round 1961). As a consequence, it has now become a major focus of medical and veterinary research and its control is an ongoing battle. Pyrethroids, carbamates, organophosphates and organochlorines-based products are insecticides applied as aerosols, baits, granules or sprays for larval and adult fly control (Bonnefoy et al. 2008; Coffey and Maier 1950; Powell 1993; Rozendaal 1997; WHO 2006). However, the repetitive and inappropriate use of synthetic pyrethroids has been reported with different levels of resistance or cross resistance in Chinese field-captured population. Such is often the case with carbamates and organophosphates (Ma et al. 2004).

CC is a calcium compound commonly used as N-resource fertilizer, first synthesized in 1898 by Adolph Frank and Nikodem Caro (van Der Ploeg et al. 2001). It is commercially known as Rongbao® in China, and used as vegetable product improver and inducer, micro-organisms con-

Table 1 Killing efficacy of different concentrations of CC against housefly maggots in larval rearing medium 72 h after each touch

Contents	Time	1			2			3			Mean		
		T	*N*	*M* (%)	*T*	*N*	*M* (%)	*T*	*N*	*M* (%)	*T*	*N*	*M* (%)
1%	Day 1	71	13	18.31	78	22	28.21	106	53	50	255	88	34.51
	Day 2		67	94.37		62	79.49		100	94.34		229	89.80
	Day 3		71	100		78	100		106	100		255	100
2%	Day 1	69	28	40.58	100	44	44	46	9	19.57	215	81	37.67
	Day 2		69	100		100	100		46	100		215	100
	Day 3		69	100		100	100		46	100		215	100
Control	Day 1	55	0	0	35	0	0	43	0	0	133	0	0
	Day 2		0	0		0	0		0	0		0	0
	Day 3		0	0		0	0		0	0		0	0

T is the total number of tested maggots, *N* is the cumulative number of the dead maggots, *M* is the cumulative mortality rate of the dead maggots

Table 2 Relationship between CC concentrations and pH value of larval rearing medium after CC dissolved in the complex

Concentration	Time	pH value
1%	Day 1	7.00
	Day 2	7.00
	Day 3	7.00
	Mean ± SD	7.00 ± 0
2%	Day 1	7.50
	Day 2	7.50
	Day 3	7.50
	Mean ± SD	7.50 ± 0
Blank control	Day 1	6.50
	Day 2	6.80
	Day 3	6.60
	Mean ± SD	6.63 ± 0.15

trol, snail control and herbicide (Shi et al. 2009; Auchmoody and Wendel 1973; Lu et al. 2006; Xu et al. 2010). The larvicidal activity of CC against housefly maggots was significantly higher in 2% concentration than in 1%. The optimum killing activity concentration for CC was 2%. The optimum length of exposure to determine the effectiveness of treatment was 2 days. The test showed that CC had great effect on killing housefly maggots. In the study of effect of CC on killing eggs of *Schistosoma japonicum* and maggots, all death of maggots from liquid homogenized excreta 3 days after 1% (W/V) CC exposure was found and the effect lasted for 15 days (Wei et al. 2009). This finding is in agreement with the observation by Wei et al. The present data indicated that CC could be suitable as a promising alternative housefly larvicide for medical use.

CC is soluble in water and is hydrolyzed forming a soluble acidic salt $Ca(HCN_2)_2$ and calcium hydroxide. Continuous carbonation of CC in water results in calcium hydroxide and formation of hydrogen cyanamide (H_2NCN) referred to as cyanamide (Miller and Bann 1956). When applied to moist soil, CC undergoes hydrolysis as a result of which calcium hydroxide and urea ($CO(NH_2)_2$) are formed, inducing pH value increase in soil environment. Soil pH value directly affects nutrient availability for plant growth and development. The common practice of making soils less acidic is to apply a material that contain some form of lime or CC to adjust the soil pH value. CC not only raises soil pH value, but also provides abundant nitrogenous fertilizer (Oh et al. 2006). Similarly, when CC is in contact with water in larval rearing medium, it decom-

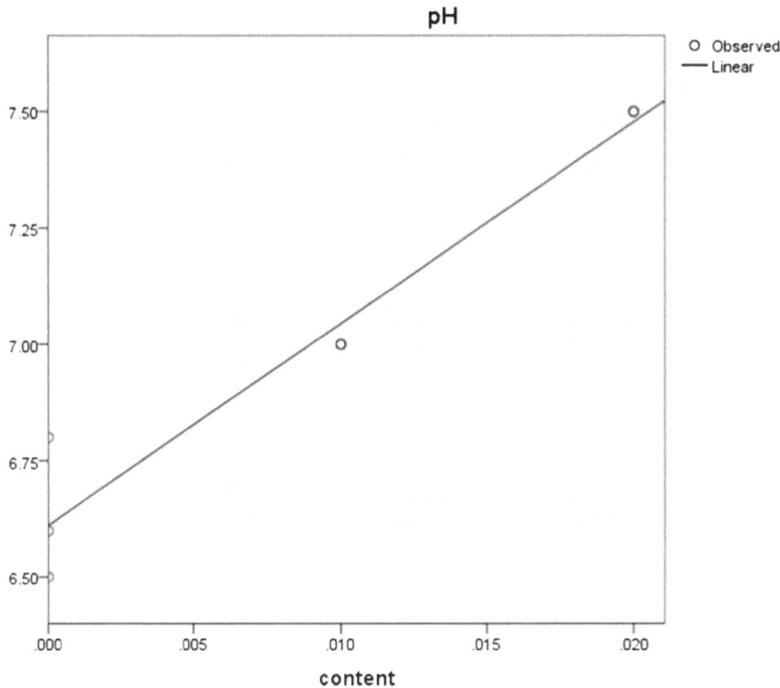

Fig. 1 Effect of contents on pH values of larval rearing medium after CC dissolved in water. *Little circles* represent observed value. *Line* represents the expected relationship between contents and pH. The relationship of contents and pH value is described as the formula $y=43.33x+6.61$ ($R^2=0.95, p<0.01$)

poses and liberates more hydroxyl ion to raise pH value in the mixed system. Our simple study design, based on pH value test in the larval rearing medium, absolutely confirms our hypothesis presented above. In our study, results from analyses of relation between CC and pH value demonstrated a positive lineal pattern of increase in pH value in response to the increasing concentration of CC. Data from blank control group evidenced background pH value in larval rearing medium. pH value occurred in blank control group among three repeats fluctuated in small size among 3 days. This difference is interpreted as due to the influence of larval rearing medium environment upon the pH value of blank control group. However, no pH value fluctuations in treatment group during the experimental period was observed.

Nutrient-rich substrates such as animal manure and human excreta provide an excellent developmental substrate for maggots' development and growth. However, recent research suggests that the acidity of rearing substrates may also affect the development of larval flies. The application of sodium bisulfate to calf bedding material indirectly reduced housefly larval survival. Acidification as a mode of fly management can be beneficial to many areas of a dairy facility, especially for areas housing newborn calves. By reducing larval density, calf housing areas can potentially serve as less attractive sites to adult flies for laying eggs (oviposition) (Calvo et al. 2010). When caring for larval rearing medium, pH level should be kept moderate in order to maintain a healthy environment for housefly maggots. If the pH balances are broken, it could mean death for maggots. This viewpoint is supported by our test result. In our test, the application of CC to larval rearing medium has shown to increase pH value and maggots survival was significantly affected after the chemical addition. But, it is still unknown whether alkalization was responsible for the reduction in maggots survival, thus, further study is required to verify this phenomenon.

After one day of larval rearing medium treatment by CC, the substrates were in the condition of compaction, appearing rigid and turning gray. Compaction is the process of compacting materials. It means to compress, condense or consolidate. Severe compaction reduces materials' pore space, thereby increasing bulk density, expressed as grams per cubic centimeter (g/cm^3) (Horn et al. 1995). In soil, compaction has impact on microorganisms, growth of plant and soil animal, just because soil compaction induces several stresses which negatively influences physical properties of soil (bulk density and pore space) and may interact simultaneously, including increased soil strength, decreased aeration and reduced hydraulic conductivity (Tracy et al. 2011). Therefore, soil compaction may consequently limit soil microfungi, which are significant for nutrient bioavailability. Non-normal microbe population further heads off plant and soil animal development (Kara and Bolat 2007). These problems such as aeration decrease and hydraulic conductivity shrinkage demonstrated in soil compaction may happen in larval rearing medium. Larval rearing medium compaction due to chemical or physical reaction after CC addition can have profound effects on water communication and air exchange and, hence, can have a detrimental effect on maggots' normal metabolism. We performed a study to observe the effect of different CC concentrations on larval rearing medium compaction. The effect was predominantly obvious, higher concentration CC having heavier compaction compared with the lower concentration counterpart. Our observed results indicated some negative effects of larval rearing medium compaction focusing on minimizing water and air exchange between medium and maggots, and then heading off normal metabolism, inducing maggots mortality at last. Compared to null maggots in control group, approximately one third of the whole maggots sneaked up on side wall of the beakers in treatment counterpart. This phenomenon strongly and further confirms decrease in aeration and hydraulic conductivity shrinkage in larval rearing medium after its compaction.

CC and its break-down products hydrogen cyanamide (HC) exhibit some bioactivity to other organisms involving fungi, mollusk and nematode. The mode of action of CC or HC on kinds of plants has extensively been exposed. HC stimulates the fermentative pathway, inhibits respiration to release buds from dormancy and restricts plant growth (Vergara et al. 2012). The bud response to HC is showed, more condensed and stronger, as reflected by a higher number of regulated genes and a higher intensity of regulation. HC perturbed mitochondrial activity, developed oxidative stress and established a situation that resembled hypoxia (Ophir et al. 2009). Oxidative stress or hypoxia induced expression of Ca^{2+}-ATPase and then evoked an increase in $[Ca^{2+}]$cyt. Similar induction was confirmed for calmodulin, calmodulin-binding protein, and calcium-dependent protein kinase (CDPK). HC induced-mechanism of Ca^{2+} signaling led to release of bud dormancy. HC also changed the interplay between abscisic acid (ABA) and ethylene metabolism. It temporarily caused an increase of acetaldehyde, ethanol and ethylene which enhanced bud break (Ophir et al. 2009). Most genes identified following HC application appeared to be associated with the reactivation of growth. Groups of genes that were rapidly up-regulated in response to HC were the glutathione S-transferase (GST) class of genes. Phylogenetic analysis of these GSTs showed that they clustered into sub-clades, suggesting a strongly correlation between their expression and bud-break across species (Walton et al. 2009). HC also induced early bud break and at the same time down-regulated *PpDAM5* and *PpDAM6* expression, two of the six peach (*Prunus persica*) dormancy-associated MADS-box genes (Yamane et al. 2011). In the short term, HC induced transiently the expression of hypoxic responsive genes (HRG) and flowering locus T(VvFT), a transcription factor related to dormancy release in Vitis, and hastened the sprouting of endodormant grapevine-buds. In the long term, along with the advancement of bud-break, the expression of these genes moved forward in treated buds, suggesting that this second induction that occurred just before bud-break was developmentally regulated (Vergara et al. 2012). On the other hand, HC restricted growth of plant in a dose-dependent manner. Cytological observations of root tip cells suggested HC disturbance in cell division. The process changes were reduction of mitotic cells, inhibition of proliferation of meristematic cell and cell cycle, and modification of cytoskeleton arrangement. The phytotoxic effect of HC was manifested by modifications in expansion of gene expression, especially in expansions responsible for cell wall remodeling after the cytokinesis (*LeEXPA9*, *LeEXPA18*) (Soltys et al. 2011).

Recently, the mode of action of CC on plants has been demonstrated. However, little information is available on the mode of action of CC or its break-down products HC on filth fly maggots. CC initiated hypoxia round the body of maggots because of compaction in larval rearing medium, when it was reacted with water from larval rearing medium and in maggots' skin. Hypoxia developed oxidative stress, perturbed mitochondrial activity and then disturbed respiration. The hypothesis of mechanism of action on maggots should be conducted through series of physiological and cytological experiments for confirmation. Other possible mode of action of CC on maggots and its related genes and enzymes should be further explored.

References

Al-Doghairi M, El Hag E (2003) Effect of several biopesticides on Alfalfa Weevil Larvae, *Hypera bruneipennis* (Boheman). Pak J Biol Sci 6:777–781

Alfazairy AA, El-Ahwany AM, Mohamed EA, Zaghloul HA, El-Helow ER (2013). Microbial control of the cotton leafworm *Spodoptera littoralis* (Boisd.) by Egyptian *Bacillus thuringiensis* isolates. Folia Microbiol 58(2):155–162.

Aluja M, Rull J, Sivinski J, Norrbom A, Wharton R, Macías-ordónez R, Díaz-fleischer F, López M (2003) Fruit flies of the Genus *Anastrepha* (Diptera: Tephritidae) and associated native parasitoids (Hymenoptera) in the tropical rainforest biosphere reserve of Montes Azules, Chiapas, Mexico. Environ Entomol 32:1377–1385

Auchmoody LR, Wendel GW (1973) Effect of calcium cyanamide on growth and nutrition of planted yellow-poplar seedlings. USDA Forest Service Research Paper 1–14

Axtell R, Arends J (1990) Ecology and management of arthropod pests of poultry. Annu Rev Entomol 35:101–126

Barin A, Arabkhazaeli F, Rahbari S, Madani S (2010) The housefly, *Musca domestica*, as a possible mechanical vector of Newcastle disease virus in the laboratory and field. Med Vet Entomol 24:88–90

Beadle G, Tatum EL, Clancy C (1938) Food level in relation to rate of development and eye pigmentation in *Drosophila melanogaster*. Biol Bull 75:447–462

Bernier U, Furman K, Kilne D, Allan S, Barnard DR (2005) Comparison of contact and spatial repellency of catnip oil and N, N-Diethyl-3-methylbenzamide (Deet) against mosquitoes. J Med Entomol 42:306–311

Bonnefoy X, Kampen H, Sweeney K (2008) Flies: public health significance of urban pests. World Health Organization, Europe, pp 209–237

Bourbos V, Skoudridakis M, Darakis G, Koulizakis M (1997) Calcium cyanamide and soil solarization for the control of Fusarium solani f.sp. cucurbitae in greenhouse cucumber. Crop Prot 16:383–386

Brogdon W, McAllister J (1998) Insecticide resistance and vector control. Emerg Infect Dis 4:605–613

Bruns H (1960) The economic importance of birds in forests. Bird Study 7:193–208

Calvo MS, Gerry AC, McGarvey JA, Armitage L, Mitloehner FM (2010) Acidification of calf bedding reduces fly development and bacterial abundance. J Dairy Sci 93:1059–1064

Chang C, Cho I, Li Q (2009) Insecticidal activity of basil oil, trans-anethole, estragole, and linalool to adult fruit flies of *Ceratitis capitata, Bactrocera dorsalis*, and *Bactrocera cucurbitae*. J Econ Entomol 102:203–209

Chapin J, Grant J, Sullivan M (1989) Hessian fly (Diptera: Cecidomyidae) infestation of wheat in South Carolina. J Agricul Entomol 6:137–146

Charles J, Nielsen-LeRoux C, Delécluse A (1996) *Bacillus sphaericus* toxins: molecular biology and mode of action. Ann Rev Entomol 41:451–472

Coffey J, Maier P (1950) Fly control techniques. Am J Public Health 40:561–566

Collier T, Van Steenwyk R (2003) Prospects for integrated control of olive fruit fly are promising in California. Calif Agric 57:28–32

Cook S, Khan Z, Pickett J (2007) The use of push-pull strategies in integrated pest management. Ann Rev Entomol 52:375–400

Donald E, Lawrence J, Porter I (2004) Influence of particle size and application method on the efficacy of calcium cyanamide for control of clubroot of vegetable brassicas. Crop Prot 23:297–303

Dorman D, Beasley V (1991) Neurotoxicology of pyrethrin and the pyrethroid insecticides. Vet Hum Toxicol 33:238–243

Dube S, Upadhyay P, Tripathi S (1989) Antifungal, physicochemical, and insect-repelling activity of the essential oil of Ocimum basilicum. Can J Bot 67:2085–2087

Ekesi S (2010) Combating fruit flies in eastern and southern africa (COFESA): Elements of a Strategy and Action Plan for a Regional Cooperation Program. International Centre of Insect Physiology and Ecology, Nairobi, Kenya: 9–11.

Elliott E, Janes N, Potter C (1978) The future of pyrethroids in insect control. Annu Rev Entomol 23:443–469

Fabre F, Ryckewaert P, Duyck P, Chiroleu F, Quilici S (2003) Comparison of the efficacy of different food attractants and their concentration for melon fly (Diptera: Tephritidae). J Econ Entomol 96:231–238

Farrar R Jr, Shepard B, Shapiro M, Hassell R, Schaffer M, Smith C (2009) Supplemental control of lepidopterous pests on Bt transgenic sweet corn with biologically-based spray treatments. J Insect Sci 9:1–10

Feder J (1995) The effects of parasitoids on sympatric host races of *Rhagoletis pomonella* (Diptera: Tephritidae). Ecology 76:801–813

Ferguson J (1956) A new drug for the treatment of alcoholism. Can Med Assoc J 74:793–795

Geden C, Stinner R, Axtell B (1988) Predation by predators of the house fly in poultry manure: effects of predator density, feeding history, interspecific interference, and field conditions. Environ Entomol 17:320–329

Graczyk T, Knight R, Tamang L (2005) Mechanical transmission of human protozoan parasites by insects. Clin Microbiol Rev 18:128–132

Greathead D (1969) A study in East Africa of the bean flies (Diptera, Agromyzidae) affecting *Phaseolus vulgaris* and of their natural enemies, with the description of a new species of Melanagromyza Hend. Bull Entomol Res 59:541–561

Hanski I (1977) Biogeography and ecology of carrion flies in the Canary Islands. Ann Entomol Fennicae 43:101–107

Hardy DE (1981) Diptera: Cyclorrhapha IV, series Schizophora section Calyptratae. In: Zimmerman EC (ed) Insects of Hawaii. University Press of Hawaii, Hawaii, pp 5–6

Hardy DE, Delfinado M (1980). Diptera: Cyclorrhapha III, series Schizophora section Acalyptera, exclusive of family Drosophilidae. In: Zimmerman EC (ed) Insects of Hawaii. University Press of Hawaii, Hawaii, pp v–vi

Ho S, Koh L, Ma Y, Huang Y, Sim K (1996) The oil of garlic, *Allium sativum* L. (Amaryllidaceae), as a potential grain protectant against *Tribolium castaneum* (Herbst) and *Sitophilus zeamais* Motsch. Postharvest Biol Technol 9:41–48

Horn R, Domżżł H, Słowńska-Jurkiewicz A, van Ouwerkerk C (1995) Soil compaction processes and their effects on the structure of arable soils and the environment. Soil Tillage Res 35:23–36

Huang Z, Huang K (2012) Comparative study on distribution, harm and morphological characteristics of important insects in bactrocera in fruits and vegetables. Acta Agric Jiangxi 24:73–75

Hughes P, van Beek N, Wood H (1986) A modified droplet feeding method for rapid assay of Bacillus thuringiensis and baculoviruses in noctuid larvae. J Invert Path 48:187–192

Isman M (2000) Plant essential oils for pest and disease management. Crop Prot 19:603–608

Jia D, Wu G (2008) Fly prevention and control (part two):filth fly ecology and the common species in China. Chin J Hyg Insecti Equi 14:128–132

Jones A, Neiman J, Hillbom M (1988) Concentration-time profiles of ethanol and acetaldehyde in human volunteers treated with the alcohol-sensitizing drug, calcium carbimide. Br J Clin Pharmcol 25:213–221

Kara O, Bolat I (2007) Influence of soil compaction on microfungal community structure in two soil types in Bartin Province, Turkey. J Basic Microb 47:394–399

Kaufman P, Long S, Rutz D, Glenister C (2000) Prey-and Density-Mediated Dispersal in *Carcinops pumilio* (Coleoptera: Histeridae), a predator of house fly (Diptera: Muscidae) eggs and larvae. J Med Ent 37:929–932

Kaufman P, Rutz D, Frisch S (2005) Large sticky traps for capturing house flies and stable flies in dairy calf greenhouse facilities. J Dairy Sci 88:176–181

Kaushal T, Suganuma T, Fujikake H, Ohtake N, Sueyoshi K, Takahashi Y, Ohyama T (2002) Effect of deep placement of calcium cyanamide, coated urea, and urea on soybean (*Glycine max* (L.) Merr.) seed yield in relation to different inoculation methods. Soil Sci Plant Nutr 48:855–863

Klopsch C, Kuhlmann H, Barth F (2012) Airflow elicits a spider's jump towards airborne prey. I. Airflow around a flying blowfly. J Royal Soc Interface 9:2591–2602

Kumara T, Abu HA, Che Salmah M, Bhupinder S (2012) Larval aggregation on a burned human remain. Trop Biomed 29:197–199

La Lacey, Frutos R, Kaya H, Vail P (2001) Insect pathogens as biological control agents:do they have a future? Biol Control 21:230–248

Landolt PJ, John S (1992) Effects of time of day, adult food, and host fruit on incidence of calling by male Caribbean fruit flies (Diptera: Tephritidae). Environ Entomol 21:382–387

Lee, BH, Annis PC, Tumaalii FA, Choi WS (2004) Fumigant toxicity of essential oils from the Myrtaceae family and 1, 8-cineole against 3 major stored-grain insects. J Stored Prod Res 40:553–564

Li H, Ye H, Lv J (2000) Damages and distributions of fruit Fly (*Bactrocera dorsalis* Hendel) in Yunnan Province, Southern China. J Yunnan Uni (Natural Sciences) 22:473–475

Liang C, Xu J, Gui G, Zhu M, Fang Q, Liang S (2012) Survey and integrated prevention of silkworm myiasis in Xiushui County, Jiangxi Province, China. Newslet Seric Tea 37:17–18

Liquido NJ, Harris EJ, Dekker L (1994) Ecology of bactrocera latifrons (Diptera: Tephritidae) populations: host Plants, natural enemies, distribution, and abundance. Ann Entomol Soc Am 87:71–74

Liu Y, Jin H, Peng X, Loeschcke V (2009) Effects of rotenone on locomotor activity and gene expression related to dopamine synthesis of fruit fly *Drophila melanogaster*. Acta Entomol Sinica 52:769–774

Liu H, Liu C, Li N, Zeng L (2012) Damage of leafminer Liriomyza trifolii in vegetable fields in Zhongshan city. Guangdong Agricul Sci 48:92–96

Lu GL, Wei WY, Li GP, Xia M, Yang GF, Zhao ZY (2006) Obervation on long-term field molluscidal of Rongbao. Pract Prev Med 13:1462–1465

Lysyk T, Axtell R (1986) Movement and distribution of house flies (Diptera: Muscidae) between habits in two livestock farms. J Econ Entomol 79(4):993–998

Ma Y, Jiang Z, Han Z (2004) Analysis of current situation and control measure of house fly resistance to insecticides in China. Chin J Hyg Insecti Equi 10:277–280

Ma Z, Shao L, Jie F (2008) Brief introduction of cowhide myiasis damage and prevention. Anim Husb Vet Med 74:47

Madanlar N, Yoldas Z, Durmusoglu E (2000). Laboratory investigations on some natural pesticides for use against pests in vegetable greenhouses. Bulletin OILB/SROP 23(1):281–288

Metcalf R (1989) Insect resistance to insecticides. Pesti Sci 26:333–358

Miller SA, Bann B (1956) Cyanamide. J Appl Chem 6:89–93

Miller J, Chamberlain W (1989) Azadirachtin as a larvicide against the horn fly, stable fly, and house fly (Diptera: Muscidae). J Econ Entomol 82:1375–1378

Obeng-Ofori D, Reichmuth CH, Bekele A, Hassanali A (1998) Toxicity and protectant potential of camphor, a major component of essential oil of Ocimum kilimandscharicum against four stored product beetles. Int. J Pest Manage 44(4):203–209

Oh K, Kato T, Li ZP, Li FY (2006) Environmental problems from tea cultivation in Japan and a control measure using calcium cyanamide. Pedosphere 16:770–777

Ophir R, Pang X, Halaly T, Venkateswari J, Lavee S, Galbraith D, Or E (2009) Gene-expression profiling of grape bud response to two alternative dormancy-release stimuli expose possible links between impaired mitochondrial activity, hypoxia, ethylene-ABA interplay and cell enlargement. Plant Mol Biol 71:403–423

Park I, Shin S (2005) Fumigant Activity of Plant Essential Oils and Components from Garlic (Allium sativum) and Clove Bud (Eugenia caryophyllata) Oils against the Japanese Termite (Reticulitermes speratus Kolbe). J Agric Food Chem 53:4388–4392

Perlman F (1962) Insects as allergen injectants. Calif Med 96:1–10

Peterson C, Nemetz L, Jones L, Coats J (2002) Behavioral activity of catnip (Lamiaceae) essential oil components to the German cockroach (Blattodea: Blattellidae). J Econ Entomol 95:377–380

Powell PK (1993) Household pest control: house flies. West Virginia University, Virginia, pp 3

Prasad R, Snyder W (2004) Predator interference limits fly egg biological control by a guild of ground-active beetles. Biol Control 31:428–437

Prates H, Santos J, Waquil J, Fabris J, Oliveira A, Foster J (1998) Insecticidal activity of monoterpenes against *Rhyzopertha dominica* (F.) and *Tribolium castaneum* (Herbst). J Stored Prod Res 34:243–249

Priest F (1992) Biological control of mosquitoes and other biting flies by Bacillus sphaericus and Bacillus thuringiensis. J Appl Bacteri 72:357–369

Ratcliffe S, Robertson H, Jones C, Bollero G, Weinzierl A (2002) Assessment of parasitism of house fly and stable fly (Diptera: Muscidae) pupae by pteromalid (Hymenoptera: Pteromalidae) parasitoids using a polymerase chain reaction assay. J Med Entomol 39:52–60

Redford K (1984) The relationship between frugivory and insectivory in primates. Primates 25:433–440

Rivers D, Kaikis A, Bulanowski D, Wigand T, Brogan R (2012) Oviposition restraint and developmental alterations in the ectoparasitic wasp, Nasonia vitripennis, when utilizing puparia resulting from different size maggot masses of *Lucilia illustris*, *Protophormia terraenovae*, and *Sarcophaga bullata*. J Med Entomol 49:1124–1136

Round M (1961) Observations on the possible role of filth flies in the epizootiology of bovine cysticercosis in Kenya. J Hyg (Lond) 59:505 513

Rozendaal J (1997) Houseflies. In: Rozendaal J (ed), vector control: methods for use by individuals and communities. World Health Organization, Geneva, pp 308–310

Schofield S, Torr S (2002) A comparison of the feeding behaviour of tsetse and stable flies. Med Vet Entomol 16:177–185

Shi K, Wang L, Zhou Y, Yu J (2009) Effects of calcium cyanamide on soil microbial communities and Fusarium oxysporum f. sp. cucumberinum. Chemosphere 75:872–877

Singh NP, Kumari V, Chauhan D (2003) Mosquito larvicidal properties of the leaf extract of a herbaceous plant, Ocimum canum (Family: Labiatae). J Commun Dis 35(1):43–45

Singh S, Singh R (1998) Neem (Azadirachta indica) seed kernel extracts and azadirachtin as oviposition deterrents against the melon fly (Bactrocera cucurbitae) and the oriental fruit fly (Bactrocera dorsalis). Phytoparasitica 26:191–197

Soltys D, Rudzińska-langwald A, Kurek W, Gniazdowska A, Sliwinska E, Bogatek R (2011) Cyanamide mode of action during inhibition of onion (Allium cepa L.) root growth involves disturbances in cell division and cytoskeleton formation. Planta 234:609–621

Syamsa R, Ahmad F, Zuha R, Khairul A, Marwi M, Shahrom A, Omar B (2012) An occurrence of Synthesiomyia nudiseta (Wulp) (Diptera: Muscidae) from a human corpse in a high-rise building in Malaysia: a case report. Trop Biomed 29:107–112

Tracy SR, Black CR, Roberts JA, Mooney SJ (2011) Soil compaction: a review of past and present techniques for investigating effects on root growth. J Sci Food Agri 91:1528–1537

Urbaneja A, GarcíaMarí F, Tortosa D, Navarro C, Vanaclocha P, Bargues L, Castanera P (2006) Influence of ground predators on the survival of the Mediterranean fruit fly pupae, Ceratitis capitata, in Spanish citrus orchards. Biocontrol 51:611–626

Van Der Ploeg, Schweigert RP, Bachmann RJ (2001) Use and misuse of nitrogen in agriculture: the German story. Sci World 1:737–744

Vergara R, Rubio S, Pérez FJ (2012) Hypoxia and hydrogen cyanamide induce bud-break and up-regulate hypoxic responsive genes (HRG) and VvFT in grapevine-buds. Plant Mol Biol 79:171–178

Walton EF, Wu RM, Richardson AC, Davy M, Hellens RP, Thodey K, Janssen BJ, Gleave AP, Rae GM, Wood M, Schaffer RJ (2009) A rapid transcriptional activation is induced by the dormancy-breaking chemical hydrogen cyanamide in kiwifruit(Actinidia deliciosa) buds. J Exp Bot 60:3835–3848

Wei W, Zhu H, Liu Z, Lv G, Ding L, Gu, J (2009) Effect of calcium cyanamide on killing eggs of Schistosoma japonicum and maggots. Chin J of schistosomiasis control 21:127–129

Weeks JT, Koshiyama KY, Maier-Greiner U, Schaäeffner T, Anderson OD (2000) Wheat transformation using cyanamide as a new selective agent. Crop Sci 40: 1749–1754

World Health Organization (WHO) (2006) Dept. of Communicable Disease Prevention, Control and Eradication WHO Pesticide Evaluation Scheme (6th ed). Pesticides and their application: for the control of vectors and pests of public health importance. World Health Organization, Gevena, pp 36–46

Witt J, Warren S, Ranney T, Baker J (1999) Biorational and conventional plant protectants reduce feeding by adult Japanese beetles. J Environ Horti 17:203–206

Wright D, Hughes R, Worrall J (1960) The effect of certain predators on the numbers of cabbage root fly (*Erioischia brassicae* (Bouche)) and on the subsequent damage caused by the pest. Ann Appl Biol 48:756–763

Xu Z (2009) The effect of soil decontaminant Ronbao on the yield and economic returns of broccoli, tower cauliflower and Chinese cabbage. Bull Sci Tech 25:456–459

Xu J, Xu ZX, Li QS (2010) Effect of soil decontaminant Rongbao on the yields of tomato, green pepper and corn. Chin J Trop Agri 30:26–30

Yadav A, Khajuria J, Soodan J (2006) Warble fly infestation in goats of Jammu. J Vet Parasit 20:149–152

Yamane H, Ooka T, Jotatsu H, Hosaka Y, Sasaki R, Tao R (2011) Expressional regulation of PpDAM5 and PpDAM6. peach (Prunus persica) dormancy-associated MADS- box genes, by low temperature and dormancy-breaking reagent treatment. J Exp Bot 62:3481–3488

Nanomaterials: A Review of Their Action and Application in Pest Management and Evaluation of DNA-Tagged Particles

M. Chandrashekharaiah, Subhash B. Kandakoor, G. Basana Gowda, Vasudev Kammar and A. K. Chakravarthy

Abstract

Nanotechnology, a new field of research, prompted scientists to work on a wide range of aspects. Nanoparticles relating to pest management include formulation for herbicides and pesticides. The potential uses of nanotechnology in insect–pest management include the slow release, efficient dosage of insecticides, and provide diagnostic tools for early detection. Application of nanaoparticles also includes development of nanodispensers, nanogels, and nanocapsules. Effects of different inorganic nanoparticles against selected insects were evaluated under laboratory conditions. DNA-tagged nanogold caused 30.50, 57.50, and 75.00% mortality on third, fourth, and fifth instar *Spodoptera litura* larvae, respectively. CdS nanoparticle caused highest *S. litura* larval mortality of 21.41–93.79% at 150 and 2400 ppm, respectively. The nano-TiO2 showed maximum of 73.79% *S. litura* larval mortality at 2400 ppm and the least was 18.50% at 150 ppm. Nano-Ag caused maximum 56.89% *S. litura* mortality at 2400 ppm followed by 46.89 and 33.44% mortality at 1200 and 600 ppm, respectively. Nanoparticles coated with ecdysteroid analogues like tebufenozide and halofenozide were tested against *Corcyra cephalonica*. The treated eggs did not hatch due to arrest of embryonic development. Tebufenozide and halofenozide caused maximum larval mortality at 80 ppm. These two compounds at 80 ppm reduced fecundity and fertility in adults. Tebufenozide against *Helicoverpa armigera* larvae reduced the larval weight. Tebufenozide at 5.00 ppm was reduced the larval weight significantly (14.23 ± 1.43 and 112.35 ± 0.29, respectively) compared to control. Histopathological effects of tebufenozide at the light microscopic level showed vacuolation and inhibition of imaginal buds. At electron microscopic level,

A. K. Chakravarthy (✉) · M. Chandrashekharaiah
S. B. Kandakoor · G. B. Gowda · V. Kammar
Department of Agricultural Entomology, Gandhi Krishi Vignan Kendra (GKVK), University of Agricultural Sciences, Bangalore, Karnataka 560065 India
e-mail: chakravarthyakshay@gmail.com

A. K. Chakravarthy (ed.), *New Horizons in Insect Science: Towards Sustainable Pest Management*, DOI 10.1007/978-81-322-2089-3_12, © Springer India 2015

peritropic membrane was completely disrupted in the larval stages and dearrangement of columnar cells was observed.

Keywords

Corcyra cephalonica · Halofenozide · Nanotechnology · *Spodoptera litura* · Tebufenozide

Introduction

Nanoscience is concerned with the unique properties of matter at its nanolevel and exploits them to create novel structures, devices, and systems for a variety of different uses. Nanoparticles are 1 billionth of a meter. These have strikingly different properties due to their small size and thus are found useful in many applications. With the development of nanotechnology systems delivering pesticides appropriately can contributes to improve pest-management practices. Nanoparticles and nanotechonology can be used in more than one way in pest management. Select ways of use of nanaoparticles in countries like India is only discussed here. This chapter initially reviews the possibilities of using nanoparticles in then results of evaluation trials on moult using growth regulators are furnished.

Nanoparticles possess insecticidal property due to novel characteristics like extraordinary strength, chemical reactivity, and electrical conductivity. They have distinct physical, biological, and chemical properties associated with their atomic strength (Leiderer and Dekorsy 2008). Nanoparticles are agglomerated atom by atom, and their size/shape may be maintained specifically (Roy 2009) and particles can be arranged into ordered layers (Ulrich et al. 2006). Such self-assembly is due to forces such as hydrogen bonding, dipolar forces, hydrophilic and hydrophobic interactions, surface tension, and gravity. Since, inorganic nanoparticles have unique properties owing to the quantum size effects and the large number of unsaturated atoms, polymeric films containing inorganic nanoparticles exhibit novel catalytic, magnetic and optical properties.

Nanoparticles in natural ecosystems have different biological responses than those observed in laboratory cell-based toxicity assays. Properties of nanoparticles can be exploited in the production of new insecticides (Owolade et al. 2008). These particles are released slowly but efficiently to a particular host plant against an insect pest (Scrinis and Lyons 2007). Seema Singh (2012) and a team of researchers from IIT Madras have developed nanoparticles from gold, silver, copper, and several other metallic oxides that have been found effective against insect pests.

Recently, nanotechnology embraced the world of pesticides and pest control and has the potential to revolutionize modern-day agriculture. Recently, smart polymer or nanocapsules with crop protection agents like insecticides, herbicides, fungicides, pheromones, repellents, and allomones are being used in pest control (Perez-de-Luque and Rubiales 2009; Racuciu et al. 2009; Matsumoto et al. 2009; Roy et al. 2010). Several nanoparticles like nanoporous zeolites, nanocapsules, and nanosensors may be used in insect–pest suppression (Hallberg 2010). Certain carbon nanotubes (1 nm) have the tremendous potential to protect host plants from insect pests (Yao 2010). Nanomaterials possess important properties of self-assembly, stability, specificity, encapsulation, and biocompatibility (Ehdaie 2007). Nanobiotechnology can be used to enhance the yield and nutritional values of crops as well as increase the plant's ability to resist insect pests (Bhattacharyya and Debnath 2008; Bhattacharyya et al. 2010). Nanotechnological tools are useful in detecting host plant diseases, types of viral infections, and crop pathogens. Moreover, a tremendous loss of crops due to attack by insect pests can be prevented. A recent approach to the control of insect pests is the use of DNA-tagged nanoparticles, insecticide-coated nanoparticles, and hormonal blended nanoparticles. Chowdappa and Shivakumar Gowda (2013) reviewed the literature on the status and scope of nanopartechnology in crop protection.

Nanodispensers

Nanoparticle with good adsorption capacity of pheromone molecules, slow and controlled release of molecules during the entire period of insect activity provide an opportunity to exploit such technology for pest management. Many kinds of such materials were exploited as pheromone dispensers, viz., filter paper, nylon mesh, polymers, paraffin wax, zeolites, and glass ampoules among others (Beroza et al. 1971; Shorey et al. 1972, Bradley et al. 1995). The commercially available dispensers are mainly made from different kinds of polymers. Among several kinds of silica-based porous glasses, porous vycor glass (PVG) is a commercial transparent porous material obtained by acid leaching of a phase-separated alkaline borosilicate glass (Aline Tiboni et al. 2008). The soluble borate phase is dissolved, leaving an open porous structure of essentially pure silica with interconnecting pores, a narrow pore size distribution, and a pore volume of nearly 28 %. The pore surface contains slightly acidic silanol groups. The nanometric pores in PVG have been used to incorporate several compounds (conducting polymers, oxides, semiconductors, amorphous carbon, and carbon nanotubes), with the purpose of obtaining novel functional nanocomposite materials. PVG was impregnated with pheromones of *Grapholita molesta* (Lepidoptera: Tortricidae) and *Leucoptera coffeella* (Lepidoptera: Lyonetiidae), the main pests of apple and coffee plantations in Brazil, respectively (Aline Tiboni et al. 2008). The release rate was dependent on the interaction between the molecules of the impregnated pheromone and the surface of the PVG pores. A good performance of the porous glass was observed, similar to that of rubber septa commercially used as pheromone dispensers (Aline Tiboni et al. 2008).

Nanogel is a synthetic polymer or biopolymer which is chemically or physically cross linked. Nanogels are usually in the tens to hundreds of nanometers in diameter. The pores in nanogels can be filled with small molecules or macromolecules and properties such as swelling, degradation, and chemical functionality can be controlled (Bencherif et al. 2009). A nanogel pheromone, methyl eugenol is stable at ambient conditions that results in slow release rate of pheromone. Nanogelled pheromone brought about an effective management of *Bactrocera dorsalis*, a prevalent harmful pest for a number of fruits including guava (Bhagat et al. 2013).

Hydrogel is a network of polymer chains that are hydrophilic, sometimes found as a colloidal gel in which water is the dispersion medium. Hydrogels are highly absorbent natural or synthetic polymers. Hydrogels also possess a degree of flexibility similar to natural tissue, due to their significant water content. These hydrogels can be used in diverse areas of agriculture such as dryland/rainfed agriculture, hi-tech horticulture and floriculture, nursery raising in soil-less media, soil reclamation, agroforestry, artificial lawns and landscapes, terrace gardening, etc. Heat-tolerant species of entomopathogenic nematode, *Steinernema thermophilum*, incorporated and immobilized into water insoluble superabsorbent hydrogel matrix, has been developed for the first time by a simple method that comprises swelling of the xerogel in appropriate volume of aqueous suspension of the nematode. The formulation obtained has improved shelf life ranging from few hours to 36 months at storage temperatures varying from 5 to 50 °C. The formulation is not infected by any microorganism and hence does not require any antimicrobial or antifungal chemicals.

Nanoparticles—A Revolution in Future

Agricultural productivity enhancement by development of slow-release formulation, with efficient dosage insecticides, biopesticides, hormones, and nanosensors for pest detection (Scrinis and Lyons 2007; Liu and Du 2004) has been effected. Nanoparticles help to produce new pesticides, insecticides, and insect repellants

(Owolade et al. 2008). Nanoencapsulation is a process through which a chemical or an insecticide is slowly but efficiently released to a particular system for efficient pest control. Nanoencapsulation with nanoparticles in the form of pesticide allows for proper absorption of the chemical into the plants unlike the case of larger particles (Scrinis and Lyons 2007). This process can also deliver DNA and other desired chemicals into plant tissues for protection of host plants against insect pests (Torney 2009). Release mechanisms of nanoencapsulation include diffusion, dissolution, biodegradation, and osmotic pressure with specific pH (Vidhyalakshmi et al. 2009; Ding and Shah 2009). Nanoparticles loaded with garlic essential oil are efficacious against *Tribolium castaneum* Herbst (Yang et al. 2009). It is also known that aluminosilicate-filled nanotube can stick to plant surfaces while nanoingredients of nanotube have the ability to stick to the surface hair of insect pests and ultimately enter the body and influence certain physiological functions. Sukul et al. (2009) have reported that potentized drugs significantly increased plant growth, chlorophyll, protein, and water content in the leaves as compared to the control. *CCC* 30 (nano) was found more effective than *CCC* 30. Potentized drugs are thought to initiate action on the integral membrane proteins of leaves and modulate cell physiology towards growth. Barik et al. (2008) opined that more ambitious uses of nanoparticles are bioremediation of contaminated environments, biocides and antifungals on textiles.

Surface-modified hydrophobic as well as lipophilic nanosilica could be effectively used as novel drugs for treatment of nuclear polyhedrosis virus (BmNPV), a scourge in the silkworm industry. Also, research on silkworm *Bombyx mori*, L. race Nistari, clearly demonstrates that nanoparticle could stimulate more production of fibroin protein which can help in producing carbon nanotube in future (Bhattacharyya et al. 2008; Bhattacharyya 2009). This highlights the putative effects of nanoparticles on insects, as these small particles are present in their body. Research on nanoparticles and insect control should be geared toward introduction of faster and ecofriendly pesticides in future (Bhattacharyya et al. 2007). Nanopesticide formulations increase the solubility of poorly soluble active ingredient and help in releasing the active ingredients slowly. Rotenone, a water-soluble botanical insecticide used to control aphids, thrips, and acari from decades; however, its effective utilization has limited due to its poor water solubility, stability, degradation, and isomerization when exposed to sunlight. Lao et al. (2010) synthesized nanoparticles by loading rotenone into a nanoparticle increasing their effectiveness by several times than free rotenone in water-soluble one. It is high time, therefore, that leading chemical companies to focus on formulation of nanoscale pesticides for delivery into the target host tissue through nanoencapsulation. At present, the toxicological and ecotoxicological risks linked to this expanding technology ("emerging technology") cannot be assessed yet. While nanotechnology is increasingly moving into the center of public attention, it is currently not yet linked to any great degree to concerns about health and the environment.

Environmental Risks

Several environmental scientists working with nanoscale structures, natural weathering of minerals such as iron oxides and silicates, silicates and microorganisms produce nanoscale colloids, which include dispersion of nanosized particles in media with special properties that can be important. Anthropogenic and natural colloids of solids and liquids in gasses are commonly encountered in the environment. However, nanotechnology is not just about the size of very small things. It is about structure and the ability to work—observe, manipulate, and build at the atomic or molecular level. The nano when it is micro then the systems that often exhibit novel and significantly changed physical, chemical, and biological properties due to their size and structure. This new property has the abilities to improve catalysis, tunable wavelength sensing ability, and increase mechanical strength. The basic structures of particles include nanoparticles, nanolayers, and nanotubes. Nanoparticles or nanopolymers have the ability to prevent pollution at source and other practices that efficiently use raw materials, energy, water or other resources to reduce or eliminate creation of wastes. This strategy also includes using less toxic and renewable reagents and processing materials, where possible, and the production of more environmentally benign manufactured products. Nanotechnology could play a key role in pollution-prevention technologies. For example, nanotechnology-based home lighting can reduce energy consumption. It can consider as the silicon nanowires that detect pH of the soil, carbon nanotubes, small organic molecules, and biomolecules are examples of nanoscale materials, devices, and circuits that could be used for pollutants sensing, prevention, and treatment. Nanotechnology applications also help to create benign substances that replace toxic materials. For example, nontoxic energy-efficient computer monitors are replacing those made of cathode ray tubes, which contain many toxic materials. Newer liquid crystalline displays are smaller, do not contain lead, and consume less power than CRT computer monitors. Using carbon nanotubes in computer displays may further diminish the environmental impacts by eliminating toxic heavy metals and drastically reduce material and energy use requirements, while providing enhanced performance for consumer needs. Only few examples of beneficial and toxic effects of nanoparticles are mentioned here. As it is a new technology, there are several pros and cons of this technology. Possible effects of the nanoparticles in the environment have to be studied. The early impact of nanotechniology research has been mostly in remediation and end-of-pipe treatments of pollutions in the environment. Thus, nanotechnology could substantially enhance environment quality and substainability through pollution prevention, treatment, and remediation.

DNA-Tagged Particles

DNA-tagged particles are oligofunctional DNA-gold nanoparticle conjugated and highly functionalized reagents can be produced from gold nanoparticles containing up to seven different DNA oligionucleaotide sequence. The individual oligomers are orthogonally addressable and reveal an efficient reactivity that is comparable to the analogous monofunctional conjugates.

The electrochemical properties of gold nanaoparticles (AuNps) have led to their widerspread use as DNA labels. This fact has improved the design strategies of the electrochemical detection modes that are based on either AuNP detection after dissolving or the direct detection of the AuNP/DNA conjugates anchored onto the genosensor surface. Various enhancement strategies have been reported so as to improve the detection limit. Most are based on catalytic deposition of silver onto AuNP. Other strategies on the use of AuNPs as carrier/amplifier of other labels will also be revised. The developed techniques are characterized by sensitivities and specificities that enable further applications of the developed DNA sensors in several fields.

Material and Methods

In order to evaluate the efficacy of inorganic nanoparticles like DNA-tagged nanogold on *Spodoptera litura* Fab., CdS, Nano-Ag, and Nano-TiO2 against *Spodoptera litura* (Fabricius), tebufenozide—RH-5992, and halofenozide—RH-0345) on the development of *Corcyra cephalonica* (Stainton), and tebufenozide on *Helicoverpa armigera* Hubner was tested under laboratory conditions.

Activated DNA-tagged gold nanoparticles were prepared following Chakravarthy et al. (2012a) method. A 200, 300, 400, and 500 ppm were prepared and 10 μl of the suspension was mixed with chickpea (*Cicer aritinum*)-based semisynthetic diet filled in 5 ml glass vials. Second instar *S. litura* larvae were released onto the diet 20 min after surface treatment. Twenty *S. litura* larvae of were exposed to each concentration of DNA-tagged gold nanoparticle for 30 s. A control with ten untreated larvae was also maintained.

Studies were also conducted to evaluate the nanopartiles coated tebufenozide (RH-5992) and halofenozide (RH-0345) against *C. cephalonica*. The effect of 5, 10, 20, 40, and 80 ppm of the

Table 1 Effect of different concentrations of DNA-tagged gold nanoparticle on second instar *S. litura* 3, 4, and 5 days after treatment

Treatment (ppm)	Percentage of larval mortality days after treatment[a]		
	3rd	4th	5th
200	10.0 (16.0)b	27.5 (31.4)c	35.0 (36.0)b
300	22.5 (28.2)a	42.5 (40.7)b	62.5 (52.3)a
400	25.0 (29.7)a	55.0 (47.9)ab	72.5 (58.6)a
500	30.0 (33.2)a	57.5 (49.6)a	75.0 (60.5)a
Control	0.0 (0.6)c	0.0 (0.6)d	0.0 (0.6)c
SEM ±	2.92	2.87	3.05
CD at 5 %	8.81	8.66	9.20

[a] Mean of 30 larvae/replication/treatment; figures in the parentheses are angular transformed values; means followed by same letters in each column are not significantly different by LSD at 5 %

above compounds on eggs, larvae, and adults was evaluated under laboratory conditions.

The potential effect of CdS, Nano-Ag, and Nano-TiO2 nanoparticles was evaluated on *S. litura*. The preparation of the nanoparticles was as per Chakravarthy et al. (2012b). Serial dilutions from 150 to 2400 ppm were prepared in the geometric order. The nanoparticle suspensions were then subjected to ultrasonication. Ten, second instar larvae were released on the castor leaf discs (4 × 4 cm) treated with desired concentrations of the solutions of nanoparticles (leaf dip method bioassay technique). Each leaf disc served as replicate, 3 per treatment and leaf discs treated with water served as control. The larvae after release on the castor leaf discs were incubated at 25 ± 1 °C with a specific 70 % relative humidity (RH) in a BOD incubator. The numbers of larvae killed were recorded at 24 h intervals up to 9 days and the percent larval mortality was computed.

An attempt was also made to evaluate the effect of tebufenozide (RH-5992) against fruit borer, *H. armigera*. A stock solution of tebufenozide (RH-5992) in acetone was first made, from which the different test concentrations were prepared and dispensed with the synthetic diet. Neonate larvae were fed on diet treated with tebufenozide at 0.01, 0.25, 0.50, 1.00, 2.50, and 5.00 ppm for 96 h. The fourth instar larvae were fed on diet treated with 1.00, 2.50, 5.00, 10.00, and 20.00 ppm of tebufenozide for 96 h. Tebufenozide was dissolved in acetone and thoroughly mixed with the diet and control diet treated with acetone alone was kept for allowing it to evaporate. The synthetic diet was prepared according to Nagarkatti and Prakask (1974). Each experiment was replicated five times with 50 larvae per replicate. All data recorded from the experiments were analyzed using software package R. version 2.10.0 (R development Core Team 2008) for statistical analysis.

Observations on developmental processes, behavior and mortality were recorded on the larvae till adults emerged. Histopathological observations: newly born and fourth instar larvae of *H. armigera* reared in laboratory were utilized for histopathological study. Treated and untreated group of larvae were fixed in Bovin's solution overnight and rinsed in 70 % ethanol four times prior to dehydration. The newly and fourth instar larvae were embedded in paraffin and cross sections (4–6 µm thickness) were stained with haematoxvlin-eosi. Cross sections of midgut of untreated (control) larvae were also examined under the microscope.

Results

Effect of DNA-Tagged Nanogold Against *Spodoptera litura* Fab.

As the concentration and days after treatment increased, the larval mortality of second instar *S. litura* larvae also increased significantly ($p < 0.05$). The maximum mortality of 30.0, 57.5, and 75.0 was observed at 500 ppm on 3rd, 4th, and 5th day, respectively (Table 1). DNA-tagged gold nanoparticle has a devastating effect on the larval tissue of *S. litura*. These metal nanoparticles were the better alternative to synthetic insecticides, in addition to being a toxicant that inhibits the biological and physiological systems of insects (Biju 2007). The high binding affinity of noble metals (e.g., gold) for proteins is well documented (Joshi et al.

Table 2 Effect of select concentrations of RH-5992 on eggs, larvae, and adults of *C. cephalonica*

Treatment PPM	Egg	Second instar larvae	Third instar larvae	Fourth instar larvae	Adults
5	77.1	56.3	30	45.3	57.6
10	77.2	65	45	54	62.4
20	80.1	66.1	53	54.7	70.8
40	89.2	72.8	56.3	60.3	72.6
80	96.4	74.3	66.3	67.5	89.3
Control	0	0	0	0	0
SEm±	1.68	0.96	1.63	0.51	0.94
CD@5%	3.80	2.86	4.85	1.54	2.88

2004; Love et al. 2005) and colloidal gold is commonly used to stain proteins in gels or blotted on membranes. The high binding affinity of noble metal like gold to proteins was attributed to the affinity of the thiols for the amino groups (Joshi et al. 2004; Love et al. 2005; Duchesne and Fernig 2007; Whaley et al. 2000; Zhu et al. 2008a, b). It has also been reported that gold nanoparticles can bind with SH-group containing amino acids of proteins (Duchesne et al. 2008). Gold metal is a conductor of heat, but, when its size is reduced by 1 billion, it becomes insulator so that it can be used to deliver the toxicant to the target cells directly. This unique property will be of immense value in medical, entomological, and allied sciences.

Peculiar symptoms of nanoparticles were larval death and larval tissue lysis. It has been reported that gold nanoparticles have the ability to stimulate different physiological enzymes (Biju 2007). DNA-tagged gold nanoparticles can affect phosphorylation in relation to kinase activity that helps to inhibit the indirect effect of DNA functions and thus lysis of the insect–pest tissue leads to death of the *S. litura*. DNA-conjugated-gold nanoparticles have an effect on kinase activity (Brennan et al. 2006; Wang et al. 2006; Ehrlich et al. 2008). In this context, DNA-tagged nanopolymer has a tremendous potential to locate the specific damaged part of the tissues in relation to its application in different physiological processes (Stadler et al. 2007; Liedl et al. 2010). Thus, DNA-tagged gold nanoparticles may stimulate the chitinases activity for disruption of the larval tissue of *S. litura* from 3rd day onwards.

Effect of Inorganic Nanoparticles CdS, Nano-Ag, and Nano-TiO2 Against *S. litura*

The CdS exhibited negligible larval mortality, as the days advanced, larval mortality increased (Table 2). On 3rd day after treatment (DAT), the larval mortality ranged from 4.50 to 13.79%, whereas on 9th DAT the cumulative larval mortality was ranged from 21.41 to 93.79% in 150 and 2400 ppm, respectively. Maximum mortality (93.79%) was recorded in 2400 ppm and followed by 1200 and 600 ppm (73.79 and 51.72%, respectively). In case of other treatments, the larval mortality was the least (150 and 300 ppm recorded 24.41 and 38.40% mortality, respectively). Nano-Ag caused only 56.89% mortality at 2400 ppm (Table 2) followed by 46.89 and 33.44% mortality at 1200 and 600 ppm, respectively. Least mortality (22.50 and 15.45%) was observed in 300 and 150 ppm. Nano-TiO2 showed very low effect at 3rd and 5th days after treatment, the mortality ranged from 2.50 to 13.79% and 7.84 to 33.79 per ppm, respectively. After 7 days, the mortality further increased and maximum of 53.79% larval mortality was recorded in 2400 ppm and the lowest was 13.50% in 150 ppm. During 9th DAT, resulted 73.79% of cumulative larval mortality in 2400 ppm and the least was 18.50% in 150 ppm. Another peculiar observation recorded was in this treatment discharge of larval inner contents due to ruptured midgut during 7th and 9th days after treatment.

The CdS, Nano-Ag, and Nano-TiO2 ceased active movement larvae, the skin and entire body became stiff and hard and oozing of the

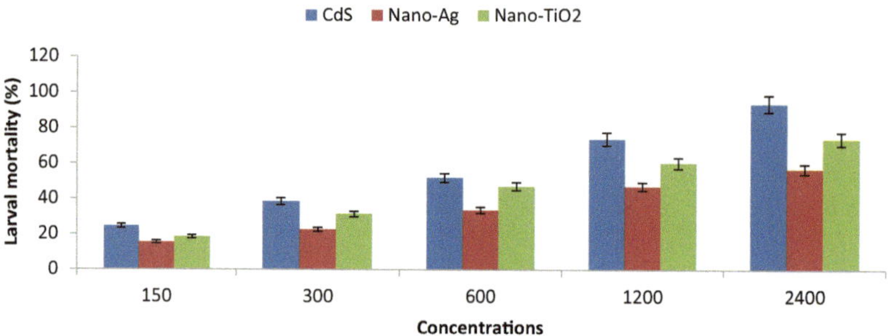

Fig. 1 Killing effect of artificially synthesized CdS, Nano-Ag, and Nano-TiO_2 nanoparticles on *S. litura*

body contents (lysis) was observed. Further, the body became swollen, pulpy, and fragile and body turned dark brown. The larvae showed premature moulting and larvae attained pupal shape, all the internal contents oozed out, and eventually death occurred.

Nanosilver particles possess insecticidal properties due to morphological and structural features brings about physiological changes (Nel et al. 2006). Stadler et al. (2010) successfully tested nanoalumina against two stored grain pests *Sitophilus oryzae* Linn. and *Rhyzopertha dominica* (F.). Glutathione-coated CdS quantum dots (GSH-CdS) exhibited an absorption peak at 366 nm, indicative of 2.4 nm core size. It also interacts with more than one protein molecule and affinity of GSH-CdS for proteins was tested (Gabellieri et al. 2011). This study demonstrated that CdS nanoparticle has adverse effects on *S. litura* larvae and could be a better alternative to synthetic insecticides, in addition to being a toxicant that inhibits biological and physiological systems of insects and also essential components of new biosensors and self-assembled nanodevices.

The acute toxicity of silver is dependent on its chemistry and free ions. Research has shown that aqueous concentration of 1–5 mg/l was sensitive to aquatic insects, trout, and flounder (Bryan and Langston 1992; Wood et al. 1994). Eisler (1997) indicated that the accumulation of silver has lead to adverse effects on growth, because of their different physico–chemical properties and free-ions released from nanosilver. Asharani et al. (2007) reported that silver nanoparticles have the potential to cause chromosomal aberrations and DNA damage and are capable of inducing cell proliferation in cell lines of zebrafish. Further, it was shown that these particles have the capability to enter cells and cause cellular damage (Hussain et al. 2005; Ji et al. 2007). Indeed, several lines of evidence support the enhanced efficiency of silver nanoparticles on antimicrobial activity and are highly reactive as they generate Ag^+ ions while metallic silver is relatively unreactive (Morones et al. 2005). It was also shown that the nanoparticles efficiently penetrate into microbial cells, which implies lower concentrations of nanosized silver would be sufficient for microbial control. This would be efficient, especially for some organisms that are less sensitive to antibiotics due to the poor penetration of antibiotics into cells (Samuel and Guggenbichler 2004; Fig. 1).

Effect of Tebufenozide- and Halofenozide-Coated Nanoparticles on *Corcyra cephalonica* (Stainton)

Mortality of eggs increased significantly with an increase in concentration. The two ecdysteroids progressively caused egg mortality (Tables 2 and 3). When the unhatched eggs were observed under stereo binocular microscope, there was no or negligible embryonic development. Administration of Juvinile Hormone (JH) mimics to *Cydia pomonella* (L.) eggs during incubation period inhibited egg embryogenesis and showed deformities in embryogenesis (Gelbic and Sehnal 1973). Larvae of *Malacosoma californicum pluviale*

Table 3 Effect of select concentrations of RH-0345 on eggs, larvae, and adults of *C. cephalonica*

Treatment PPM	Egg	Second instar larvae	Third instar larvae	Fourth instar larvae	Adults
5	73.5	58.1	33.7	42.5	60.5
10	77.4	68.8	51	55	65.5
20	86.2	74.3	53.8	55.7	70.8
40	89.2	75.5	61	60.6	80.6
80	98.6	76.3	71.5	69	92.4
Control	0	0	0	0	0
SEm±	1.92	0.96	1.63	0.51	0.94
CD@5%	2.34	2.86	4.85	1.54	2.1

(Dyar) do not hatch if the yolk content in the eggs is decreased (Wellington and Maelzer 1967).

Similar trend of mortality was also observed when 2nd, 3rd, and 4th instar larvae treated with tebufenozide and halofenozide (Tables 2 and 3). At higher concentrations, maximum of 74.3, 66.3, and 67.5% mortality was observed when RH-5992 was treated on 2nd, 3rd, and 4th instar larvae, respectively, whereas 76.3, 71.5, and 69.00% mortality was observed when 2nd, 3rd, and 4th instar larvae were treated with RH-0345. Adults' mortalities at higher concentration were 89.30 and 92.40% on RH-5992 and RH-0345, respectively.

It has been reported that several insecticides reduce *Corcyra* sp. infestation (Jalali et al. 2007). Introduction of ecdysteroid analogues in agricultural field is not usual (Jenson 2008). Several insecticides have the ability to interfere with the cuticular enzymes like phenoloxidase, quinone methide isomerase, and DOPA decarboxylase and these help in cuticle sclerotization in insects (Tsubota et al. 2010). The successful effect of ecdysteroid analogues in *C. cephaplonica* denoted that these substances have tremendous effect in the cuticular cell DNA. The modified insect protoxin nucleic acid sequences can also protect the plant host, plant cells, and seeds of plants from insect–pest attack (Abad et al. 2004). Moreover, tebufenozide and halofenozide (RH-5992; RH-0345) can induce the feeding mechanism leading to incomplete moult and thus exhibit death of the larvae. The fecundity of treated females was reduced by 60% normal value and hatchability of laid eggs ranged 0–40%, as compared to 96.80% in control. Due to decrease in fecundity and hatchability, the females treated with 80 ppm of the two compounds produced no offspring and suffered mortality. In some insects, the decrease in fecundity after treatment with a mimic was shown to be due to the derangements in the differentiation of oogonia and follicular cells (Metwally et al. 1973; Rohdendorf and Sehnal 1972).

It is clear that tebufenozide and halofenozide also act through the receptor binding mechanisms and transactivates the expression of upregulated genes. But, because of its persistence, the down-regulated genes that are normally expressed in the mild presence of 20E do not express themselves. It has been established now that tebufenozide (RH-5992) is a potential insecticide (Retnakaran et al. 2001; Kreutzweiser et al. 2011). Tebufenozide possesses several alkyl groups such as CH3, *n*-C3H7, *i*-C3H7, *n*-C4H9, and *n*-C5H11 at the *para*-position of the benzene ring and also possess *tert*-butyl group. Tebufenozide is more active because it possesses the 1- and 2-naphthyl derivatives and these are very active moieties in its structure (Nakagawa et al. 2000; Nakagawa 2007).

Other types of nanoparticles for storage pest are nanopesticides and nanoencapsulated pesticides, which are expected to reduce the volume of application and slowdown the fast release kinetics (Elibol et al. 2003, Niemeyer and Doz 2001, Leiderer and Dekorsy 2008). Mode of actions of nanoparticles includes destruction of the cuticle layers, the waxy layer of the cuticle results in the desiccation of arthropods. Since there is no chemical alteration of the absorbed lipids the mode of action can be described as physisorption (Leiderer and Dekorsy 2008). Stadler et al. (2010) showed that nanoalumina can be successfully used to control stored grain pests. Nanocarriers are designed to reduce the volume of application and slowdown the release kinetics of agrochemicals (Perez de Luque and

Table 4 Effect of feeding on tebufenozide (RH-5992) treated diet on the developmental responses of neonate larvae of *H. armigera* for 48 h

Tebufenozide (conc. ppm)	Average single larval weight after 7 days (mg) (X±SE)	Average larval weight (mg) (X±SE)	Day 100% larval mortality occurred	Average single pupal weight (mg) (X±SE)
Control	78.52±2.61a	506.51±11.50a	–	208.52±12.12a
0.10	64.46±2.82b	325.63±7.16b	–	197.47±15.36a
0.25	60.57±2.43b	309.47±15.60b	–	163.51±18.20b
0.50	52.51±5.01c	270.54±11.76c	–	124.63±8.56c
1.00	30.27±1.73d	264.32±20.48c	16	–
2.50	27.64±1.84d	146.57±14.28d	14	–
5.00	14.23±1.43e	70.51±7.36e	12	–

Experiment terminated 48 h after treatment; means followed by the same letter are not significantly different ($p<0.01$; one-way ANOVA followed by Tukey–Kramer test)

Rubiales 2009). Sabbour (2012) evaluated nano aluminium oxide (Al_2O_3) and titanium dioxide (TiO_2) against rice weevil *Sitophilus oryzae* (Coleoptera: Curculionidae). Nano Al_2O_3 found highly effective against *S. oryzae* and nano TiO_2 has low to moderately effective against *S. oryzae*. Silica nanoparticle (SNP) was evaluated Debnath et al. (2011) against *S. oryzae* and its efficacy was compared with bulk-sized silica (individual particles larger than 1 lm). Amorphous SNP was highly effective against this insect pest causing more than 90% mortality.

Effect of Tebufenozide—A Moult Inducing Growth Regulator on *Helicoverpa armigera* Larvae

Effect of tebufenozide on the neonate and fourth instar larvae of *H. armigera* (Lepidoptera: Noctuidae) was recorded through feeding for 96 h and then transferred the larvae to normal diet. At lower concentrations of IGR, i.e., at 1.0 and 2.5 ppm, behavior of the larvae was normal on the 1st DAT. Larvae showed taxis towards the source and palpated on the diet. However, at higher concentrations of IGR, i.e., at 10 and 20 ppm, larvae exhibited kinesis (undirected orientation and movements), larvae turned sluggish and treated diet surface inhibited feeding in larvae. After 18 days, the compound resulted in 100% larval mortality. Study clearly denoted significant change in the developmental processes of newly borne larvae *H. armigera* larvae. The sensitivity was observed after about 60 h of feeding on tebufenozide. This insect growth regulator disrupted the normal development of neonate larvae subsequently in a more or less dose-dependent manner (Table 4). At 0.10 ppm, average of single neonate larval weight after 7 days from the start of the experiment was 64.46±2.82 mg. In control, the neonate larval weight after 7 days from the start of the experiment was 78.52±2.61 mg. At the lowest concentration of 0.10 ppm, tebufenozide failed to exert influence on the pupal weight (Table 4). However, with subsequent increase in concentration pupal weights (163.51±18.20 mg and 124.63±8.56 mg) differed significantly from control (208.52±12.12 mg) (Table 4). Concentration gradient effect of tebufenozide was also reflected in the neonate larval weights taken 7 days after the experiment. The average pupal weight at 0.25 ppm was 163.51±18.20 mg and at dose 0.50 ppm, it was 124.63±8.56 mg, respectively.

Fourth instar larvae of fruit borer fed on synthetic diet treated with tebufenozide for 96 h disrupted the development in a dose-dependent manner. This disruption of growth was reflected in the maximum average larval weight as well as average pupal weight (Table 2). Average pupal weight in all concentrations differed significantly from control (Table 2). There were no significant differences in average fourth instar larval weight after 7 days. Data denoted that at 10 and 20 ppm pupation did not take place since 100% larval mortality occurred on day 7 (Table 5). Both first and fourth instars larvae of *H. armigera* after tebufenozide reduced intake of food, growth and caused rupture of the cuticle, resulting in malformation and mortality.

Table 5 Effect of feeding on tebufenozide (RH-5992) treated diet on growth and development of IV instar larvae of *H. armigera* for 48 h

Tebufenozide (conc. ppm)	Average single larval weight after 7 days (mg) (X±SE)	Average larval weight (mg) (X±SE)	Day 100% larval mortality occurred	Average single pupal weight (mg) (X±SE)
Control	113.51±0.21	568.45±21.4a	–	209.53±11.9a
1.00	109.46±0.60	468.55±19.6b	–	188.45±15.4b
2.50	110.57±0.74	428.37±14.3b	–	178.55±13.8b
5.00	112.35±0.29	220.51±18.5c	–	162.35±17.5c
10.00	111.53±0.35	145.57±11.3d	6	–
20.00	112.57±0.31	138.41±20.4d	6	–

Experiment terminated 48 h after treatment; means followed by the same letter are not significantly different ($p<0.01$; one-way ANOVA followed by Tukey–Kramer test)

The purpose of this work was to evaluate whether this specific hormone analog RH-5992 has potential effect in insect hormonal alteration, which then can be introduced as ecofriendly insecticide. The long-term effect of the above-said product might alter the hormone feedback mechanisms in the physiology of *H. armigera*. The tebufenozide RH-5992, a nonsteroid agonist, exhibited an effective disruption in the insect development, which included moulting and metamorphosis controlled by the steroid hormone 20-hydroxyecdysone (20E).

Naturally, it may be assumed that tebufenozide RH-5992 can bind ecdysterone receptors to cause a precocious and incomplete moulting that is lethal to larvae. RH-5992 has the ability to bind with lepidopteran ecdysterone receptor (Retnakaran et al. 2001; Hu et al. 2004). *H. armigera* undergoes six larval instars and it is from the third instar that active feeding on tomato fruits begin. The life cycle is completed in 30 to 35 days on tomato crop, depending on the conditions (David and Ramamurthy 2011). Tomato fruit loss can be prevented in early instar larvae are prevented from feeding on fruits without any adverse effects on environment, human health, and other nontarget species.

The biochemical alterations lead to death of the larvae. It has already been reported that insect chitinases were found in the integument, gut, and fat bodies but were absent in the haemocytes which belong to glycosylhydrolases. These glycosylhydrolases have been observed in moulting fluid and midgut tissues. Moreover, these chitinases can synthesize chitin through peritrophic membrane (PM) of the midgut. Genes and cDNAs encoding insect chitinases have been identified and characterized from several lepidopteran insects (Zhu et al. 2008). Further observations have established that certain endotoxins may bind with the midgut epithelial cells and can inhibit the synthesis of proteases and aminopeptidases causing lysis of the midgut epithelial cells (Sanjay et al. 2001).

Cross section of the midgut of the newly born and fourth instar treated larvae revealed morphological differences in the treated compared to untreated larvae. In newly born larvae, the epidermal cells were dearranged and the imaginal buds failed to show any growth and development. The cement, epicuticle, and exocuticle layers were dearranged. In the fourth instar larvae, the epicuticle was partially detached from the exocuticle and cells were dearranged. The cells showed damage to varying degrees. Vacuolation of epithelial cells, destructured cytoplasmic, and protoplasmic organelles were evident from the cross section of the midgut.

Tebufenozide (RH-5992) at 10 to 20 ppm disrupted growth and development of *H. armigera* larvae in a week. Limited knowledge on the mode of action of this IGR in the physiology of neonate larvae of *H. armigera* will of value in the future. Tebufenozide (RH-5992) showed consistent effects in disrupting the gene functions. Thus, the study revealed that tebufenozide potentially can be used as ecofriendly biopesticide.

Acknowledgment The investigators are thankful to the Director of Research, University of Agricultural Sciences, GKVK, Bangalore 650 065. Tebufenozide [RH-5992] was gifted from Dr. Daniel Doucet of Canada. This study was also facilitated by the National Bureau of Agriculturally important Insects (NBAII) and IIHR, Bangalore.

References

Abad A, Flannagan RD, Herrmann R, Lu AL, McCutchen BF, Presnail JK, Rice JA, Wong JF, Yu CG (2004) Plant activation of insect toxin. Application number: 11/021,115, Publication number: US 2005/0166284 A1, International Classification, A01H001/00;C12N015/82; A01H005/00

Asharani PV, Nair G, Zhiyuan H, Manoor P, Valiyaveettil S (2007) Potential health impacts of silver nanoparticles. Abstracts of Papers, 234th ACS National Meeting, Boston, MA, USA, August 19–23, 2007. pp:TOXI–099

Barik TK, Sahu B, Swain V (2008). Nanosilica-from medicine to pest control. Parasitol Res 103(2):253–258

Bencherif SA, Siegwart DJ, Srinivasan A, Horkay F, Hollinger JO, Washburn NR, Matyjaszewski K (2009) Nanostructured hybrid hydrogels prepared by a combination of atom transfer radical polymerization and free radical polymerization. Biomaterials 30:5270–5278

Beroza M, Bierl BA, Knipling EF, Tardif JGR (1971) The activity of the gypsy moth sex attractant disparlure vs that of the live female moth. J Econ Entomol 64:1527–1529

Bhagat D, Samanta SK, Bhattacharya S (2013) Efficient management of fruit pests by pheromone nanogels. Sci Rep 3(1294):1–8

Bhattacharyya A (2009). Nanoparticles-from drug delivery to insect pest control. Akshar 1(1):1–7

Bhattacharyya A, Debnath N (2008). Nano Particles-A Futuristic Approach in Insect Population. In Proceedings on UGC Sponsored National Seminar On Recent Advances In Genetics and Molecular Biology, Biotechnology and Bioinformatics, 21st and 22nd November, 2008. Jointly Organized by Department of Zoology and Botany, Vidyasagar College, Kolkata-700006.West Bengal, Indian

Bhattacharyya A, Bhaumik A, Usha Rani P, Suvra Mandal S, Epidi TT (2010) Nano- particles - A recent approach to insect pest control. Afr. J. Biotechnol. 9(24):3489–3493

Bhattacharyya A, Barik B, Kundu P, Mandal DN, Das A, Sen P, Rao CV, Mandal S (2007) Bioactivity of nanoparticles and allelochemicals on stored grain pest-Sitophilus oryzae (L) (Coleoptera: Curculionidae). Accepted in 27th Annual Session of The Academy of Environmental Biology and National Symposium of Biomarkers of Environmental Problems, October 26–28, 2007, Jointly organized by Department of Zoology and Department of Environmental Sciences, Charan Singh University, Meerut, U. P., Indian

Bhattacharyya A, Gosh M, Chinnaswamy KP, Sen P, Barik B, Kundu P, Mandal S (2008) Nano-particle (allelochemicals) and Silkworm Physiology. In: Chinnaswamy KP, Vijaya Bhaskar Rao A (eds) Recent trends in seribiotechnology. Bangalore, pp. 58–63

Biju V (2007) Quantum dot-insect neuropeptide conjugates for fluorescence imaging, transfection, and nucleus targeting of living cells. Langmuir 23:10254–10261

Bradley SJ, Suckling DM, McNaughton KG, Wearing CH, Karg G (1995) A temperature-dependent predictive model for polyethylene tubing pheromone dispenser release rates. J Chem Ecol 21:745–760

Brennan JL, Hatzakis NS, Tshikhudo TR, Dirvianskyte N, Razumas V, Patkar S, Vind J, Svendsen A, Nolte RJM, Rowan AE, Brust M (2006) Bionanoconjugation via click chemistry: the creation of functional hybrids of lipases and gold nanoparticles. Bioconjugate Chem 17(6):1373–1375

Bryan G, Langston W (1992) Bioavailability, accumulation and effects of heavy metals in sediments with special reference to United Kingdom estuaries: a review. Environ Pollut 76:89–131

Chakravarthy AK, Bhattacharyya A, Shashank PR, Epidi TT, Doddabasappa B, Swapan K (2012a) DNA-tagged nano gold: a new tool for the control of the armyworm, *Spodoptera litura* Fab. (Lepidoptera: Noctuidae). Afr J Biotechnol 11(38):9295–9301

Chakravarthy AK, Chandrashekharaiah, Kandakoor SB, Bhattacharya A, Dhanabala K, Gurunatha K, Ramesh P (2012b) Bio efficacy of inorganic nanoparticles CdS, Nano-Ag and Nano-TiO2 against *Spodoptera litura* (Fabricius) (Lepidoptera: Noctuidae). Curr Biotica 6(3):271–281

Chowdappa P, Shivakumar Gowda (2013) Nanotechnology in crop protection: status and scope. Pest Manag Hortic Ecosyst 19(2):131–151

David BV, Ramamurthy VV (2011) Elements of economic entomology, 6th edn. Namrutha Publications, Chennai, p 386

Debnath N, Das S, Seth D, Chandra R, Somesh Ch, Bhattacharya S Ch, Goswami A (2011) Entomotoxic effect of silica nanoparticles against *Sitophilus oryzae* (L.). J Pest Sci 84:99–105

Ding WK, Shah NP (2009) Effect of various encapsulating materials on the stability of probiotic bacteria. J Food Sci 74(2):M100–M107

Duchesne L, Fernig DG (2007) Silver and gold nanoparticle-coated membranes for femtomole detection of small proteins and peptides by Dot- and Western blot. Anal Biochem. 362(2):287–289

Duchesne L, Wells G, Fernig DG, Sarah A, Harris SA, Levy RI (2008) Supramolecular domains in mixed peptide self assembled monolayers on gold nanoparticle. Chem Biol Chem 9:2127–2134

Ehdaie B (2007) Application of Nanotechnology in Cancer Research: Review of Progress in the National Cancer Institute's Alliance for Nanotechnol. Int. J. Biol. Sci. 3:108–110

Ehrlich H, Janussen D, Simon P, Bazhenov VV, Shapkin NP, Erler C, Mertig M, Born RE, Heinemann S, Hanke T, Worch H, Vournakis JN (2008) Nanostructural organization of naturally occurrin composites-part II: silica-chitin-based biocomposites. J Nanomaterials 5:1–8

Eisler R (1997) Silver hazards to fish, wildlife and invertebrates: A synoptic review. Washington, DC, US Department of the Interior, National Biological Service, 44 pp. (Biological Report 32 and Contaminant Hazard Reviews Report 32).

Elibol OH, Morisette DD, Denton JP, Bashir R (2003) Integrated nanoscale silicon sensors using top-down fabrication. Appl Phys Lett 83:p. 4613

Gabellieri E, Cioni P, Balestreri E, Morelli E (2011) Protein structural changes induced by glutathione-coated CdS quantum dots as revealed by Trp phosphorescence. Eur Biophys J 40(11):1237–1245

Gelbic I, Sehnal F (1973) Effects of juvenile hormone mimics on the codling moth Cydia pomonella (L.) (Lep. oletreutidae). Bull Ent Res 63:7–16

Hu W, Cook BJ, Ampasala DR, Zheng S, Caputo G, Krell PJ, Retnakaran A, Arif BM, Feng Q (2004) Morphological and molecular effects of 20-Hydroxyecdysone and its agonist tebufenozide on CR-203, a midgut-derived cell line from the spruce budworm, Choristoneura Fumiferana. Arch Insect Biochem Physiol 55(2):68–78

Hussain SM, Hess KL, Gearhart JM, Geiss KT, Schlager JJ (2005) In vitro toxicity of nanoparticles in BRL 3A rat liver cells. Toxicol in vitro 19:975–983

Jalali SK, Venkatesan T, Murthy KS, Rabindra RJ, Lalitha Y (2007) Vacuum packaging of *Corcyra cephalonica* (Stainton) eggs to enhance shelf life for parasitisation by the egg parasitoid *Trichogramma chilonis*. Biol Control 41(1):64–67

Jenson AE (2008) Survival of egg, neonate and wandering-phase larvae of the Indian meal moth (Plodia interpunctella (Hubner)) exposed to surface and aerosol applications of methoprene. An abstract of a dissertation submitted in partial fulfillment of the requirements for the degree Doctor of Philosophy, Department of Entomology, Collegeof Agriculture Kansas State University Manhattan, Kansas. 1–186

Ji JH, Jung JH, Kim SS, Yoon JU, Park JD, Choi BS, Chung YH, Kwon IH, Jeong J, Han BS, Shin JH, Sung JH, Song KS, Yu, IJ (2007) Twenty-eight-day inhalation toxicity study of silver nanoparticles in Sprague-Dawley rats. Inhal Toxicol 19:857–871

Joshi H, Shirude PS, Bansal V, Ganesh KN, Sastry M (2004) Isothermal titration calorimetry studies on the binding of amino acids to gold nanoparticles, J Phys Chem B 108:11535–11540

Kreutzweiser DP, Gunn JM, Thompson DG, Pollard HG, Faber MJ (2011) Zooplankton community responses to a novel forest insecticide, tebufenozide (RH-5992), in littoral lake enclosures. Can J Fish Aquat Sci 55:639–648

Lao SB, Zhang ZX, Xu HH, Jiang GB (2010) Novel amphiphilic chitosan derivatives: Synthesis, characterization and micellar solubilisation of rotenone. Carbohydr Polym 82:1136–1142

Leiderer P, Dekorsy T (2008) Interactions of nanoparticles and surfaces Tag der mÄundlichenPrÄufung: 25. April. URL: http://www.ub.unikonstanz.de/kops/volltexte/2008/5387/; URN: http://nbn-resolving.de/urn:nbn:de:bsz:352-opus-53877

Liedl T, Hogberg B, Tytell J, Donald E, Ingber DE, Shih WM (2010) Self-assembly of three-dimensional prestressed tensegrity structures from DNA. Nature Nanotechnol. Published online: 20 June 2010 | doi:10.1038/nnano.2010.107

Liu XY, Du N (2004) Zero-sized effect of nano-particles and inverse homogeneous nucleation. J Biol Chem 279(7):6124–6131

Love JC, Estroff LA, Kriebel JK, Nuzzo RG, Whitesides GM (2005) Selfassembled monolayers of thiolates on metals as a form of nanotechnol. Chem Rev 105:1103–1169

Matsumoto S, Christie RJ, Nishiyama N, Miyata K, Ishii A (2009) Environment- responsive block copolymer micelles with a disulfide cross-linked core for enhanced siRNA delivery, Biomacromology, 10:119–127

Metwally MM, Sehnal F, Landa V (1973) Reduction of fecundity and the control of the Khapra beetle by the juvenile hormone mimics, J Econ Ent 65:1603–1605

Morones JR, Elechiguerra JL, Camacho A, Holt K, Kouri JB, Ramirez JT, Yacaman MJ (2005) The bactericidal effect of silver nanoparticles. Nanotechnology 16:2346–2353

Nagarkatti S, Prakash S (1974) Rearing Heliothis armigera (Hubn.) on artificial diet. Commonw Inst Biol Contr Tech Bull 17:169–173

Nakagawa Y (2007) Structure–activity relationship and mode of action study of insect growth regulators. J Pestic Sci 32(2):135–136

Nakagawa Y, Hattori K, Minakuchi C, Kugimiya S, Ueno T (2000) Relationships between structure and molting hormonal activity of tebufenozide, methoxyfenozide, and their analogs in cultured integument system of *Chilo suppressalis* Walker. Steroids 65(3):117–123

Nel A, Xia T, Mädler L, Li N (2006) Toxic potential of materials at the nanolevel. Science 311:622–627

Niemeyer CM, Doz P (2001) Nanoparticles, proteins, and nucleic acids: biotechnology meets materials science. Angew Chem Int Edn 40(22):4128–4158

Owolade OF, Ogunleti DO, Adenekan MO (2008) Titanium dioxide affects disease development and yield of edible cowpea. EJEAF Chem 7(50):2942–2947

Perez-de-Luque A, Rubiales D (2009) Nanotechnology for parasitic plant control. Pest Manage. Sci. 65:540–545

R Development Core Team (2008) R: a language and environment for statistical computing. R Foundation for Statistical Computing, Vienna, Austria. ISBN 3–900051-07–0, URL http://www.R-project.org

Racuciu M, Miclaus S, Creanga D (2009) The response of plant tissues to magnetic fluid and electromagnetic exposure. Rom. J. Biophys. 19:73–82

Retnakaran A, Gelbic I, Sundaram M, Tomkins W, Ladd T, Primavera M, Feng Q, Arif B, Palli R, Krell P (2001) Mode of action of the ecdysone agonist Tebufenozide (RH-5992), and an exclusion mechanism to explain resistance to it. Pest Manage Sci 57(10):951–957

Rohendorf EB, Sehnal F (1972) The induction of ovarian dysfunction in *Thermobia domestica* by the cercopia juvenile hormones. Experientia 28:1099–1101

Roy SC (2009) There's plenty of holes at the bottom: The other side of Nano. Sci. Cult. 75(1-2):1–3

Roy D, Cambre JN, Sumerlin BS (2010) Future perspectives and recent advances in stimuliresponsive materials. J. Prog. Polym. Sci. 35:278–301

Sabbour MM (2012) Entomotoxicity assay of two nanoparticle materials 1-(Al2O3and TiO2) against *Sitophilus oryzae* under laboratory and store conditions in Egypt. J Nov Appl Sci 1(4):103–108

Samuel U, Guggenbichler JP (2004) Prevention of catheter-related infections: the potential of a new nano-silver impregnated catheter. Int. J Antimicrob Agents 23(1):75–8

Sanjay SI, Trivedi N, Prasad R, Kuruvilla J, Rao KK, Chhatpav HS (2001) Aminopeptidase-N from the *Helicoverpa armigera* (Hubner) brush border membrane vesicles as a receptor of *Bacillus thuringiensis* Cry1 A-endotoxin. Current Microl 43:255–259

Scrinis G, Lyons K (2007). The emerging nano-corporate paradigm: nanotechnology and the transformation of nature, food and agrifood systems. Int J Sociol Food Agric 15(2):1–23

Shorey HH, Kaae RS, Gaston LK, McLaughlin JR (1972) Sex pheromones of Lepidoptera. XXX. Disruption of sex pheromne communication in Trichoplusia ni as a possible means of mating control. Environ Ent 1:641–645

Singh S (2012) IIT Madras develops nanoparticles to fight pesticide residues. http://www.livemint.com/2008/01/07133544/IIT-Madrasdevelopsnanoparticic.html

Städler B, Solak HH, Frerker S, Bonroy K, Frederix F, Voros J, Grandin MH (2007). Nanopatterning of gold colloids for label-free biosensing. Nanotechnology 18(15):155–306

Stadler T, Butelerb M, Weaver DK (2010) Novel use of nanostructured alumina as an insecticide. Pest Manag Sci 66:577–579

Sukul NC, Singh RK, Sukul S, Sen P, Bhattacharyya A, Sukul A, Chakrabarty R (2009). Potentized drugs enhance growth of pigeon pea. Environ Ecol 26(3):1115–1118

Tiboni A, Coracini MDA, Lima ER, Zarbin PHG, Zarbin AJG (2008) Evaluation of porous silica glasses as insect pheromone dispensers. J Braz Chem Soc 19(8):1634–1640

Torney F (2009) Nanoparticle mediated plant transformation. Emerging technologies in plant science research. Interdepartmental Plant Physiology Major Fall Seminar Series. Phys. p. 696

Tsubota T, Minakuchi C, Nakakura T, Shinoda T, Shiotsuki T (2010) Molecular characterization of a gene encoding juvenile hormone esterase in the red flour beetle. *Tribolium castaneum*. Insect Mol Biol 19(4):527–535

Ulrich C, Mewis I, Goswami A, Chatterjee SD, Banerjee SP, Adhikary S, Bhattacharyya, A (2006) Biodiversity-Macro and Micro: To be Nano or Not to be!. Everyman's Science 11(6):433–443

Vidhyalakshmi R, Bhakyaraj R, Subhasree RS (2009) Encapsulation the future of probiotics-A Review. Adv Biol Res 3(3–4):96–103

Wang J, Profitt JA, Pugia MJ, Suni II, (2006) Aunanoparticle conjugation for impedance and capacitance signal amplification in biosensors, Anal. Chem 78:1769–1773

Wellington WG, Maelzer DA (1967) Effects of farnesyl methyl ether on the reproduction of the western tent caterpillar, *Malacosoma pluvial*: some physiological, ecological and practical implications. Can Ent 99:249–263

Whaley SR, English DS, Hu EL, Barbara PF, Belcher AM (2000) Selection of peptides with semiconductor binding speci®city for directed nanocrystal assembly. Nature 405:665–668

Wood C, Munger S, Galvez F, Hogstrand C (1994) The physiology of silver toxicity in freshwater fish. In: Andren, A. and Bober, T., editors. Transport, fate, and effects of silver in the environment. Proceedings of the 2nd international conference. 11_14 September 1994. Madison, WI, University of Wisconsin Sea Grant Institute. pp 109–114

Yang FL, Li XG, Zhu F, Lei CL (2009) Structural Characterization of Nanoparticles Loaded with Garlic Essential Oil and Their Insecticidal Activity against *Tribolium castaneum* (Herbst) (Coleoptera:Tenebrionidae). J. Agric. Food Chem. 57(21):10156–10162

Zhu Q, Arakane Y, Beeman RW, Kramer KJ, Muthukrishnan S (2008a) Functional specialization among insect chitinase family genes revealed by RNA interference. Proc Natl Acad Sci U S A 105(18):6650–6655

Zhu X, Yang Q, Huang J, Suzuki I, Li G (2008b) Colorimetric study of the interaction between gold nanoparticles and a series of amino acids. J Nanosci Nanotechnol 8:353–357

Fumigant Toxicity of Nutmeg Seed Essential Oil (*Myristica fragrans* Houtt.) (MF, Myristicaceae) on Cowpea Weevil, *Callosobruchus maculatus* F. (Coleoptera: Bruchidae)

Mahshid Alibabaie and Mohammad Hasan Safaralizadeh

Abstract

Unlike conventional pesticides, usually these natural aromatic products, present less risk to humans and the environment. Essential oils are recognized as alternatives to fumigants. The biological activity of essential oil extracted from nutmeg seed, *Myristica fragrans* against adults of *Callosobruchus maculatus* was investigated in a series of laboratory experiments carried out at 27 ± 1 °C and 60 ± 5 % relative humidity (RH), in dark condition. Dry seeds were subjected to hydrodistillation, using a modified Clevenger-type apparatus. The oil was applied against 1–3 day old adults of *C. maculatus*. The LC_{50} of *M. fragrans* on *C. maculatus* was 4.232 µL/L air. The relationship between exposure time and oil concentration on mortality showed that mortality increased as oil concentration and exposure time increased. The concentration of 30 µL/L and exposure time of 24 h was sufficient to obtain 100 % kill of insects. *C. maculatus* was significantly susceptible to this essential oil. Bioassays conducted in air-tight glass chambers showed fumigant toxicity as per the filter-paper method. The experiment was conducted in four replications using a completely randomized design. The results suggest that *M. fragrans* can be used to control *C. maculatus*.

Keywords

Biological activity · Essential oils · Nutmeg seed · Weevil

M. Alibabaie (✉) · M. H. Safaralizadeh
Department of Plant Protection, Agricultural Faculty
West Azarbayjan province, Urmia University, Urmia, Iran
e-mail: m.alibabaie68@gmail.com

Introduction

Insects are among major pests of stored cereals often causing an important economic damage amounting to 5–10 % in the temperate zone and to 20–30 % in the tropical one (Haque et al. 2000).

A. K. Chakravarthy (ed.), *New Horizons in Insect Science: Towards Sustainable Pest Management*, DOI 10.1007/978-81-322-2089-3_13, © Springer India 2015

In the countries where modern storage technologies have been introduced, bean pests are usually controlled either by contact insecticides or by fumigation with an insecticidal gas. However, residual toxicity, resistant insect strains, worker's safety, and high cost of the treatment call for new systems for their control (Yildirim et al. 2001).

Fumigant Toxicity

Fumigation is one of the major chemical methods to control stored-product insect infestations worldwide. Currently, phosphine and methyl bromide are the products most widely used (Bond 1984; Fields and White 2002; Lee et al. 2004; Emekci 2010). Carbon dioxide and sulfuryl fluoride are also registered for fumigation of stored grain in several countries. Fumigation is the method of choice for many stored-grain managers because it is effective against all life stages, inexpensive, rapid, and leaves minimal residues (van Someren Graver 2004). However, there are some concerns about the current fumigants. Methyl bromide has largely been phased out in developed countries, and it is slated to be phased out in the rest of the world by 2015, because it is an ozone-depleting substance (TEAP 2000; Fields and White 2002). Phosphine is not effective against insect populations in India, Australia, and Brazil because of resistance (Bell and Wilson 1995; Benhalima et al. 2004; Collins et al. 2005; Pimentel et al. 2009). Carbon dioxide requires high temperatures and high concentrations to control insect populations (Soderstrom et al. 1992). Sulfuryl fluoride is used as a replacement for methyl bromide, but eggs require high doses or long durations to be effective (Kenaga 1957; Baltaci et al. 2009). However, for small subsistent farmers in developing countries these fumigants are not available or too costly to use. Several essential oils have antiparasitical, bactericidal, fungicidal, virucidal, and insecticidal properties (Bakkali et al. 2008; Rajendran and Sriranjini 2008). The essential oils are rich in monoterpenes and cause death of insects by inhibiting acetylcholinesterase activity in nervous system (Houghton et al. 2006).

Control of stored-product insects primarily depends upon the continuing application of liquid–gaseous insecticides (White and Leesch 1995; Ren et al. 2008). Although effective their repeated use for several decades has disrupted biological control system by natural enemies and led to outbreaks of insect pests, widespread development of resistance, undesirable effects on nontarget organisms, environmental and human health concerns (Champ and Dyte 1977). The increasing concern about its adverse effects has highlighted the need for the development of selective insect-control alternatives. Plants may provide potential alternatives to currently used insect-control agents because they constitute a rich source of bioactive chemicals (Regnault-Roger 1997; Weaver and Subramanyam 2000; Isman 2006; Negahban et al. 2007). Much effort has, therefore, been focused on plant derived materials for potentially useful products as commercial insect-control agents (Rajendran and Sriranjini 2008). Sometimes it has been claimed that monoterpenoids have comparable fumigant action to that of methyl bromide (Shaaya et al. 1997; Dunkel and Sears 1998). Many aromatic plant species are indigenous to Iran (Naghibi et al. 2005), but essential oils have scarcely been evaluated (Negahban et al. 2007).

Essential Oil

Nutmeg (*Myristica fragrans*), whose seed is widely used as a spice, is a tropical, dioeciously evergreen tree native to Moluccas or Spice Island of Indonesia. Nutmeg has a characteristic pleasant fragrance and has slightly warm taste. It is used to flavor many kinds of baked goods, confections, puddings, meats, sausages, saucers, vegetables, and beverages (Panayotopoulos and Chisholm 1970).

The hypnotic, analgesic, and hypotensive activities of *M. fragrans* have also been reported (Grover et al. 2002). With the recent gain in popularity of herbal medicine the world over it is also possible to misuse *M. fragrans* because of its medicinal properties. It has been reported that the spice can be toxic when ingested in large quantities (1–3 nutmegs) (Forrest and Heacock 1972).

The utility of nutmeg as a spice has been known since ancient times in Indonesia. Nutmeg has aromatic, stimulant, narcotic, carminative, astringent, aphrodisiac, hypolipidaemic, antithrombotic, antiplatelet aggregation, antifungal, antidysenteric, and anti-inflammatory activities. The spice is used as a remedy for stomach ache, rheumatism, and vomiting during pregnancy (Olajide et al. 1999; Sonavane et al. 2002). This chapter describes a laboratory study to assess the potential fumigant effects of essential oils extracted from the fruits of *M. fragrans* on adults of *Callosobruchus maculatus*.

Material and Methods

Cowpea (*Vigna unguiculata*) Blackeye, a variety susceptible to *C. maculatus* (Baker et al. 1989), was stored in a freezer at 18 °C for a week and subsequently dried in a stove at 60 °C for about a week to guarantee the absence of viable insects without having to use chemicals. The beans were stored in airtight plastic containers at room temperature before use. Only visually uninfested beans were used for the experiments.

C. maculatus was collected in Urmia on local cowpea. The beetles were reared on cowpea in laboratory for a year (≈ 12 generations) prior to the experiments. The rearing was done in a climate chamber at 30 ± 1 °C with a 12-h photoperiod and 50–80 % relative humidity (RH) For the tests, newly emerged (1–1.5 h) insects were used. For repellence tests, female beetles were used that had been kept for 1 h with a surplus of newly emerged males and were assumed to have mated.

In the experiments, the day of death of the adults was determined as the day the antennae and legs did not move upon gentle disturbance with forceps. Unhatched eggs recognized by the color of the egg. In the cowpea variety that produced big yellowish seeds, the tunnel that is dug by the developing larva could be seen a few days before emergence of the beetle as a bluish spot under the seed coat. From the size and the clearness of this spot, the stage of development could be estimated, but further investigation was avoided to keep the beans intact. If no beetles emerged from the beans for 5 days, the larvae were considered dead.

Plant Materials

Plants were collected and dried in Urmia (Northwest Iran). The plant samples were stored in plastic bags (1 L volumes) in the dark at 4 °C. Shortly before use, after warming up to room temperature, the dry plant material was isolated.

Isolation and Extraction of Essential Oil

The essential oil of nutmeg seeds was isolated with a yield of 6.85 % w/w. The essential oils were extracted by hydrodistillation of dried plant material (100 g of each sample in 500 ml of distilled water) using a modified Clevenger-type apparatus for 4 h. The oils were dried over anhydrous sodium sulfate and stored in sealed glass vials at 4–5 °C prior to analysis. Yields were averaged over four experiments and calculated according to dry weight of the plant materials.

Gas Chromatography Mass Spectrometry (GC/MS) Analysis

Measurements were performed using a QP-5050A (Shimadzu) gas chromatograph coupled to a VG autospect mass spectrometer at 70 eV, 40–550 amu with a fused silica capillary column (DB-5MS, 30 m × 0.25 mm) using helium as a carrier gas and with temperature programming from 60 °C/5 min to 300 °C/2 min (10 °C/min) for essential oils (Adams, 1995). The MS was operated using an interface temperature of 240 °C, and an electron impact ionization of 70 eV with a scan mass range of 40–350 m/z (sampling rate of 1.0 scan/s).

Table 1 Chemical composition of essential oil of nutmeg seeds

No.	Retention time	LRI	Compounds	Percentage
1	5.967	920	α-thujene	0.78
2	6.246	931	α-prinene	10.23
3	6.583	943	Camphene	0.16
4	7.508	978	Sabinene	21.38
5	7.792	989	α-myrcena	2.38
6	8.493	1017	α-terpinene	2.72
7	8.843	1032	Limonene	5.57
8	9.142	1045	β-ocimene	0.03
9	9.525	1061	y-terpinene	3.98
10	9.728	1070	trans-sabinene hydrate	0.03
11	10.097	1085	Terpinolene	1.62
12	10.393	1098	Linalool	0.75
13	10.821	1119	Fenchyl alcohol	0.05
14	11.603	1158	*cis*-sabinene hydrate	0.06
15	12.306	1193	4-terpineol	13.92
16	12.492	1203	α-terpineol	3.11
17	12.67	1212	Citronellol	0.77
18	13.297	1247	Linalyl acetate	0.06
19	13.949	1282	Bornyl acetate	0.24
20	14.158	1293	Safrole	4.28
21	15.76	1392	Methyl eugeunol	0.77
22	16.568	1447	Isoeugeunol	1.74
23	17.773	1530	Myristicin	13.57
24	18	1551	Elimicin	1.42
25	18.573	1586	Metoxyeugeunol	0.1
26	18.742	1595	β-asaron	0.03
27	20.625	1767	Myristic acid	0.11
28	21.01	1789	Ethyl miristate	0.04
29	21.352	1716	Palmitic acid	0.03
30	22.946	1954	Ethyl palmitate	0.07
31	25.103	2181	Stearic acid	0.01
32	25.183	2193	Ethyl oleate	0.01

LRI linear retention indices

Test on Toxicity

All tests were done in a climate chamber at 27±1°C with a 12-h photoperiod at ambient RH Untreated beans were used as control for every experiment. Different amounts of essential oils, at 0, 1.5, 2.21, 3.31, 4.95, and 7.80 μL/L in air were placed on to Whatman No. 1 filter-paper disks of 3 cm diameter. Each filter-paper disk was then air dried for 2 min and placed on the underside of the screw cap of a glass vial (300 mL). Caps were screwed tightly on the vials, each of which contained 20 unsexed adults (1–2 days old). Cap was screwed tightly and the lid was sealed with parafilm. Each concentration and control was replicated four times. Insect mortality was checked after 24 h. Test insects were considered dead if appendages did not move when prodded with a fine brush. Each treatment was replicated three times.

Statistical Analyses

Using SPSS (SPSS Inc. 1989–2002), data were analyzed by one-way analysis of variance (ANOVA) and means were compared using Tukey test at $P<0.05$. (If significant differences between the treatments were found, they were tested by Tukey's tes.) Mortality of adults was arcsine square root transformed prior to analysis. The number of dead and live insects was counted and probit analysis was used to estimate LC_{50} and LC_{95} values with their Wducial limits by SAS 8.0 (SAS Institute 2000). One-way ANOVA using Statistica (Statsoft 1998) was performed on the data. A Duncan test was applied to the means to detect significant differences of repellency among concentrations and oils at the 0.05% level. Data are presented in tables as means with standard errors. A median lethal concentration (LC_{50}) was considered significantly different when the respective 95% fiducial limits did not overlap.

Results and Discussion

The essential oil of nutmeg seeds was isolated with a yield of 6.85% w/w. Thirty-two components were identified. The major compounds in the oil were sabinene (21.38%), 4-terpineol (13.92%), and myristicin (13.57%). On the other hand, allylbenzene and propylbenzene derivatives (myristicin, safrole, eugenol, and derivatives thereof) were the predominant compounds in nutmeg seeds. The data describing select compounds extracted from nutmeg seeds are presented in Table 1.

Table 2 Fumigant toxicity of *M. fragrans* essential oil against *C. maculates*

Insect species	LC_{50}	LC_{95}	Slope ± SE	Degrees of freedom	Chi-square
C. maculates	4.232	19.158	2.72±0.26	6	6.502
	(3.52–4.81)	(13.09–32.28)			

Units LC_{50} and LC_{95}=μL/L, applied for 24 h at 27 °C; 95 % lower and upper fiducial limits are shown in parenthesis

This research showed that the essential oil of nutmeg seeds is of good quality as it has ideal percentages of essential oil and active compounds. The standard of Materia Medica Indonesia (MMI) requires that the essential oil content in nutmeg seeds be 5–10 %, and it was found to be 6.85 % in this study. Additionally, the content of myristicin and safrole measured in this study exceeded MMI requirements of 5–10 %. *M. fragrans* oil showed strong fumigant activity against *C. maculates* adults at different concentrations and exposure times. In all of the concentrations, the oil yielded more than 90 % mortality after 12 h, so the effective time for mortality could be 12 h.

Concentration of 30 μL/L in air and exposure time of 24 h were enough to obtain 100 % kill of the insects. At the lowest concentration, the oil yielded nearly more than 50 % after 9 h against *C. maculates*. Probit analysis showed that for this essential oil the LC_{50} is 4.232 and LC_{95} is 19.158 (Table 2).

This is the first report on the insecticidal activity of *M. fragrans* in the world. It has been recognized that plant-derived insect-control agents could be developed into products suitable for integrated pest management because they are selective to pests, have no or little harmful effect against nontarget organisms or the environment (Arnason et al. 1989; Schmutterer 1992; Hedin et al. 1997; Isman 2000). The essential oil extracted from Myristaceae family has shown strong effect on stored-product insects such as *C. maculates*. In this study, the fumigant activity of *M. fragrans* was attributed especially to sabinene, 4-terpineol and myristicin as major constituents of the oil.

In this study, *M. fragrans* was characterized by a rapid knockdown effect, hyperactivity, convulsion and paralysis and dead. These effects show that this essential oil could act like traditional fumigants. *M. fragrans* naturally produce many volatile compounds that are not only important for aromatic and flavor characteristics, but also have the additional benefit of offering for the management of stored-product pests.

Therefore, essential oil of *M. fragrans* could have potential as an alternative to methyl bromide for postharvest control of insect pests in storage. Regardless of potent activity of essential oils as fumigants, however, there are several barriers to commercialization for botanical insecticides including availability of sufficient quantities of plant material to produce the pesticides, the standardization and refinement of pesticide products. While some essential oils are currently available in large quantities, essential oils from rare plants may be difficult to obtain in sufficient quantities (Isman et al. 2007).

It may be possible to produce botanical insecticides by phytopharming through genetic engineering of an existing field crop to produce high-value natural products. Due to rapid volatilization and low persistence of the essential oils in the environment, it is unlikely to use them in field crops; however, this property is conductive to use them in stored-product pest-controlled conditions.

The insecticidal constituents of many plant extracts and essential oils are monoterpenoids. Due to their high volatility, they have fumigant activity that might be of importance for controlling stored-product insects (Coats et al. 1991; Konstantopoulou et al. 1992; Regnault-Roger and Hamraoui 1995; Ahn et al. 1998).

In general, the longer insects are exposed to a fumigant, the lower the dose that is required to control insects. The rate of decline is described by the concentration-time (CT) product; concentration, usually the LD_{50} or the LD_{95}, multiplied by time. For some fumigants, such as methyl bromide, the CT is constant with time, in other words the dose can be reduced by half if the duration is doubled (Estes 1965; Bell and Glanville 1973; Bond 1984).

Fumigants, such as phosphine, are much more effective for longer durations (Bell and Glanville 1973). For example, the CT of phosphine reduces by 50% when the exposure goes from 2 to 24 h (Lindgren and Vincent 1966), whereas other fumigants, such as sulfuryl fluoride, have higher CT with longer durations (Kenaga 1961). For many of the pure products we tested the LD_{50} remained constant with time, causing the CT to increase with time. For products such as camphor the LD_{50} declined with time, the CT increased with time. Understanding the factors that affect CT of essential oils will be important in predicting mortality under field conditions, where concentrations varied due to loss of gas from leakage and absorption by commodities (Bond 1984). The amount of fumigant needed to control insects depends upon a number of factors: stage, species, duration, temperature, and commodity (Bond 1984).

This study reports that the volatile compounds detected in blood samples such as myristicin, 4-terpineole, and safrole were associated with fumigant toxicity in *C. maculates*. It is suggested that the fumigant toxicity by nutmeg seed essential oil is due, at least in part, to the direct pharmacological action of one or more of its constituents. Although the essential oils and their constituent compounds tested as fumigants were not as active as commercial fumigants, they, unlike commercial fumigants, do act as repellents and have contact toxicity (Bakkali et al. 2008).

References

Adams RP (1995) Identification of essential oil components by gas chromatography/mass spectroscopy. Allured Publishing Corporation, Carol Stream, pp 469

Ahn YJ, Lee SB, Lee HS, Kim GH (1998) Insecticidal and acaricidal activity of carvacrol and b-thujaplicine derived from *Thujopsis dolabrata* var hondai sawdust. J Chem Ecol 24:1–90

Arnason JT, Philogene BJR, Morand P, Imrie K, Iyengar S, Duval F, Soucy-Breau C, Scaiano JC, Werstiuk NH, Hasspieler B, Downe AER (1989) Naturally occurring and synthetic thiophenes asphotoactivated insecticides. In: Arnason JT, Philogene BJR, Morand P (eds) Insecticides of plant origin. ACS Symposium series no 387, Washington DC, pp 164–172

Baker TA, Nielsen SS, Shade RE, Singh BB (1989) Physical and chemical attributes of cowpea lines resistant and susceptible to *Callosobruchus maculatus* (F.) (Coleoptera: Bruchidae). J Stored Prod Res 25:1–8

Bakkali F, Averbeck S, Averbeck D, Idaomar M (2008) Biological effects of essential oils. Food Chem Toxicol 46:446–475

Baltaci D, Klementz D, Gerowitt B, Drinkall MJ, Reichmuth C (2009) Lethal effects of sulfuryl fluoride on eggs of different ages and other life stages of the warehouse moth *Ephestia elutella* (Hübner). J Stored Prod Res 45:19–23

Bell CH, Glanville V (1973) The effect of concentration and exposure in tests with methyl bromide and with phosphine on diapausing larvae of Ephestia elutella (Hübner) (Lepidoptera, Pyralidae). J Stored Prod Res 9:165–170

Bell CH, Wilson SM (1995) Phosphine tolerance and resistance in *Trogoderma granarium* everts (Coleoptera: Dermestidae). J Stored Prod Res 31:199–205

Benhalima H, Chaudhry MQ, Mills KA, Price NR (2004) Phosphine resistance in stored-product insects collected from various grain storage facilities in Morocco. J Stored Prod Res 40:241–249

Bond EJ (1984) Manual of fumigation for insect control. Food and agricultural organization of the United Nations. Rome: 1984, pp 298–306

Champ BR, Dyte CE (1977) FAO global survey of pesticide susceptibility of stored grain pests. FAO Plant Prot Bull 25:49–67

Coats JR, Karr LL, Drewes CD (1991) Toxicity and neurotoxic effects of monoterpenoids in insects and earthworms. In: Hedin PA (ed) Naturally occurring pest bioregulators. American Chemical Society, Washington DC, pp 305–316

Collins PJ, Daglish GJ, Pavic H, Kopittke RA (2005) Response of mixed-agecultures of phosphine-resistant and susceptible strains of lesser grain borer, *Rhyzopertha dominica*, to phosphine at a range of concentrations and exposure periods. J Stored Prod Res 41:373–385

Dunkel FV, Sears LJ (1998) Fumigant properties of physical preparations from mountain big sagebrush, *Artemisia tridentate* Nutt. ssp. *vaseyana* (Rydb.) beetle for stored grain insects. J Stored Prod Res 34:307–321

Emekci M (2010) Quo Vadis the fumigants. In: Carvalho OM, Fields PG, Adler CS, Arthur FH, Athanassiou CG, Campbell JF, Fleurat-Lessard F, Flinn PW, Hodges RJ, Isikber AA, Navarro S, Noyes RT, Riudavets J, Sinha KK, Thorpe GR, Timlick BH, Trematerra P, White NDG (eds) Proceedings of The10th International Working Conference on stored product protection. Julius-Kühn-Archiv, Portugal, pp 303–313

Estes PM (1965) The effects of time and temperature on methyl bromide fumigation of adults of *Sitophilus granarius* and *Tribolium confusum*. J Econ Entomol 58:611–614

Fields PG, White NDG (2002) Alternatives to methyl bromide treatments for stored-product insect and quarantine insect. Annu Rev Entomol 47:331–359

Forrest JE, Heacock RA (1972) Nutmeg and mace; the psychotropic Spices from Myristica Fragrans, Lloydia 35:440–449

Grover JK, Khandkar S, Vats V, Dhunnoo Y, Das D (2002) Pharmacological studies on Myristica fragrans--antidiarrheal, hypnotic, analgesic and hemodynamic (blood pressure) parameters. Experimental Clinical Pharmacol 24(10):675–80

Haque MA, Nakakita H, Ikenaga H, Sota N (2000) Development inhibiting activity of some tropical plants against *Sitophilus zeamais* Motschulsky (Coleoptera Curculionidae). J Stored Prod Res 36:281–287

Hedin PA, Hollingsworth RM, Masler EP, Miyamoto J, Thompson DG (1997) Phytochemicals for pest control, ACS Symp series no. 658. American Chemical Society, Washington DC

Houghton PJ, Ren Y, Howes MJ (2006) Acetylcholinesterase inhibitors from plants and fungi. Nat Prod Rep 23:181–199

Isman MB (2000) Plant essential oils for pest and disease management. Crop Prot 19:603–608

Isman MB (2006) Botanical insecticides, deterrents, and repellents in modern agriculture and an increasingly regulated world. Annu Rev Entomol 51:45–66

Isman MB, Machial CM, Miresmailli S, Bainard LD (2007) Essential oil-based pesticides: new insights from old chemistry. In: Ohkawa H, Miyagawa H, Lee P (eds) Pesticide chemistry. Wiley, Weinheim, pp 201–209

Kenaga EE (1957) Some biological, chemical and physical properties of sulfuryl fluoride as an insecticidal fumigant. J Econ Entomol 50:1–6

Kenaga EE (1961) Time, temperature and dosage relationships of several insecticidal fumigants. J Econ Entomol 54:537–542

Konstantopoulou LL, Vassilopoulou L, Mavragani-Tsipidou P, Scouras ZG (1992) Insecticidal effects of essential oils. A study of the effects of essential oils extracted from eleven Greek aromatic plants on *Drosophila auraria*. Experientia 48:616–619

Lee BH, Annis PC, Tumaalii FA, Choi WS (2004) Fumigant toxicity of essential oils from the Myrtaceae family and 1, 8-cineole against 3 major stored-grain insects. J Stored Prod Res 40:553–564

Lindgren DL, Vincent LE (1966) Relative toxicity of hydrogen phosphide to various stored-product insects. J Stored Prod Res 2:141–146

Naghibi F, Mosaddegh M, Mohammadi MS, Ghorbani A (2005) Labiatae family in folk medicine in Iran, from ethnobotany to pharmacology. Iranian J Pharm Res 2:63–79

Negahban M, Moharramipour S, SeWdkon F (2007) Fumigant toxicity of essential oil from *Artemisia sieberi* Besser against three stored product insects. J Stored Prod Res 43:123–128

Olajide OA, Ajayi FF, Ekhelar AI, Awe SO, Makinde JM, Alada AR (1999) Biological effects of *Myristica fragrans* (nutmeg) extract. Phytother Res 13:344–345

Panayotopoulos DJ, Chisholm DD (1970) Hallucenogenic effect ofNutmeg. Brithish medical J 1:754–760

Pimentel MAG, Faroni LRDA, Guedes RNC, Sousa AH, Tótola MR (2009) Phosphine resistance in Brazilian populations of *Sitophilus zeamais* Motschulsky (Coleoptera: Curculionidae). J Stored Prod Res 45:71–74

Rajendran S, Sriranjini V (2008) Plant products as fumigants for stored-product insect control. J Stored Prod Res 44:126–135

Regnault-Roger C (1997) The potential of botanical essential oils for insect pest control. Integr Pest Manage Rev 2:25–34

Regnault-Roger C, Hamraoui A (1995) Fumigant toxic activity and reproductive inhibition induced by monoterpenes on *Acanthoscelides obtectus* (Say) (Coleoptera), a bruchid of kidney bean (*Phaseolus vulgaris* L.). J Stored Prod Res 31:291–299

Ren YL, Mahona D, Gravera JVS, Head M (2008) Fumigation trial on direct application of liquid carbonyl sulphide to wheat in a 2500 t concrete silo. J Stored Prod Res 44:115–125

SAS (2000) SAS Institute, SAS/STAT user's guide, version 8. SAS, Cary, p 3, 884

Schmutterer H (1992) Control of diamondback moth by application of neem extracts. In: Talekar NS (ed) Diamondback moth and other crucifer pests. Proceedings of Second International Workshop Asian Vegetable Research and Development Center, Taipei. Taiwan, pp 325–332

Shaaya E, Kostjukovski M, Eilberg J, Sukprakam C (1997) Plant oils as fumigant and contact insecticides for the control of stored product insects. J Stored Prod Res 33:7–15

Soderstrom EL, Brandl DG, Mackey B (1992) High temperature combined with carbon dioxide enriched or reduced oxygen atmospheres for control of *Tribolium castaneum* (Herbst) (Coleoptera: Tenebrionidae). J Stored Prod Res 28:235–238

Sonavane GS, Sarveiya VP, Kasture VS, Kasture SB (2002) Anxiogenic activity of *Myristica fragrans* seeds. Pharmacol Biochem Behav 71:239–244

SPSS Inc (1989–2002) SPSS for windows. Chicago, Illinois, USA

Statsoft, (1998) STATISTICA for Windows (Computer program electronic manual). StatSoft Inc, Tulsa

TEAP (2000) UNEP Technology and Economic Assessment Panel (TEAP). Report of the Technology and Economic Assessment Panel, April 2000. UNEP, Nairobi

Van SGJE (2004) Guide to fumigation under gas-proof sheets. Food and Agricultural Organization of the United Nations, Rome

Weaver DK, Subramanyam B (2000) Botanicals. In: Subramanyam B, Hagstrum DW (eds) Alternatives to pesticides in stored-product IPM. Kluwer, Massachusetts, pp 303–320

White NDG, Leesch JG (1995) Chemical control. In: Subramanyam B, Hagstrum DW (eds) Integrated management of insects in stored products. Marcel Dekker, New York, pp 287–330

Yildirim E, Ozbek H, Aslan I (2001) Pests of stored products. Ataturk University Agricultural Faculty Press, No. 191, Erzurum, p 117

Part IV
Insect Vectors

Identification of Putative Vectors of Weligama Coconut Leaf Wilt Disease in Sri Lanka

A. D. N. T. Kumara, L. Perera, M. K. Meegahakumbura, N. S. Aratchige and L. C. P. Fernando

Abstract

Weligama coconut leaf wilt disease (WCLWD) is phytoplasma borne and reported in the southern part of Sri Lanka. The disease debilitates the palm, resulting in a drastic yield loss. It is well known that phytoplasma is transmitted from plant to plant by insect vectors, particularly phloem-feeders in the order Homoptera, i.e. leafhoppers and plant hoppers. A survey was conducted in the disease affected areas to collect homopterans present on the coconut palms, other plant species in the vicinity and in the environment. Insects were collected by hand, aspirator, sticky and light traps. The most abundant species were subjected to the nested polymerase chain reaction (PCR) using universal phytoplasma specific primers, P1/P7 and Pc399/P1694 to detect phytoplasma DNA present in their bodies. Thirty two homopteran and a few hemipteran species were collected from coconut plantations. Eight homopteran species, *Goniagnathus (T.) punctifer, Recilia dorsalis* Motschulsky, *Kolla ceylonica* (Melichar), *Idioscopus clypealis* (Lethierry), *Proutista moesta* (Westwood), *Proutista* sp., *Nisia nervosa* (Motschulsky) and an unknown Cixiid and a hemipteran species, *Stephanitis typica* (Distant) gave positive bands at 1280 bp. The DNA sequence of these bands was similar to WCLWD phytoplasma sequence (Gene Bank: EU635503), suggesting them as putative vector species of WCLWD.

Keywords

Coconut · Leaf Wilt Disease · Phytoplasma · Putative Vectors · Weligama

A. D. N. T. Kumara (✉) · N. S. Aratchige · L. C. P. Fernando
Crop Protection Division, Coconut Research Institute, Lunuwila 61150, Sri Lanka
e-mail: adnthissakumara@yahoo.com

L. Perera · M. K. Meegahakumbura
Genetics and Plant Breeding Division, Coconut Research Institute, Lunuwila 61150, Sri Lanka

Introduction

Coconut (*Cocos nucifera* L) is a major economically important plantation crop widely grown in Sri Lanka. Most plantations are distributed in the north western region called 'coconut triangle'. The southern region of Sri Lanka is the

A. K. Chakravarthy (ed.), *New Horizons in Insect Science: Towards Sustainable Pest Management*, DOI 10.1007/978-81-322-2089-3_14, © Springer India 2015

important coconut growing area found at about more than 100 km away from the coconut triangle and covering 13,498 ha. A debilitating yellowing syndrome of coconut plantation appeared in 2006, and it was identified as a phytoplasma-borne disease (Wijesekara et al. 2008). The disease was named as Weligama coconut leaf wilt disease (WCLWD) (Wijesekara et al. 2008; Perera et al. 2010). It was confirmed that WCLWD phytoplasma belong to the 16SrXI *Candidatus Phytoplasma* oryzae group and is most similar to the sugarcane white leaf phytoplasma as well as root wilt (Kerala wilt) phytoplasma in India (Perera et al. 2012). Symptoms of the disease are flattening and ribbing of the leaflets called flaccidity, marginal necrosis of leaflets and extensive yellowing of leaflets. These symptoms are more similar to the root wilt disease symptoms in India, but the symptoms of inflorescence rotting and yellowing of midworld fronds are not common in WCLWD. Both of these maladies act as predisposing factors for leaf rot disease, and palm become unproductive in a short period.

Phytoplasmas are prokaryote organisms of the class Mollicutes, affecting more than 700 plant species from tropical to temperate countries (Jones 2002). The phytoplasma cannot be cultivated invitro; therefore, molecular methods like PCR, RFLP, etc are the best approach for their detection, identification and characterization. Based on PCR-amplified ribosomal DNA (16S rgene) of phytoplasmas are classified using restriction fragment length polymorphism (RFLP) and DNA sequence analysis (IRPCM 2004). Phytoplasma diseases are transmitted through insect vectors, vegetative propagation of disease shoot to the healthy stocks, a vascular connection of healthy plant and disease plant by parasitic plants like Dodder (Dale and Kimks, 1969; Weintraub and Beanland 2006). Recently another mode of transmission of phytoplasma through seed/embryo has also been confirmed (Cordova et al., 2003; Khan et al., 2002; Oropeza, pers. comm.). However, the major mode of transmission is through insect vectors. In general, phloem-feeding insects, mainly in the suborder Auchenorrhyncha of order Homoptera, i.e. leafhoppers (Cicadellidae) and plant hoppers (Maixner 2005; Weintraub and Beanland 2006).

Most of identified phytoplasma-born disease vectors are belongs to the family Cicadellidae to date. Within Cicadellidae, subfamily Deltocephalinae contains more than 75 % of all confirmed phytoplasma vector species (Nielson 1979; Weintraub and Beanland 2006). Insects in the subfamily Deltocephalinae are either monophagous or polyphagous and are able to transmit single or multiple phytoplasma species (Nielson 1979; Weintraub and Beanland 2006). Four families of plant hoppers namely Cixiidae, Delphacidae, Derbidae and Flatidae are well-known vectors of phytoplasma diseases (Weintraub and Beanland 2006). In addition, few species of Hemiptera have been reported as vectors of phytoplasma-borne diseases. *Stephanitis typica* (Hemiptera: Tingidae) has been identified as a vector of Kerala (root) wilt disease of India (Mathen et al. 1990). Brown marmorated stink bug, *Halyomorpha halys* (Hemiptera: Pentotomidae), responsible for transmission of the witches' broom phytoplasma to *Paulownia* trees in Asia (Hiruki 1999). Most of phytoplasma vectors are non-destructive feeders of plant and both nymphs and adult feed on same plant parts while living in the same habitat (Weintraub and Beanland 2006). In India, three insect vector species namely *S. typica* (Hemiptera: Tingidae), *Proutista moesta* (Westwood) (Homoptera: Derbidae), *Sophinia greeni* (Homoptera: Cicadellidae) have been confirmed in Kerala (root) wilt disease (Mathen et al. 1990; Solomon et al. 1998). The vector of lethal yellowing disease of coconut in the Caribbean and Central America was identified as a *Mindus crudus* Van Duzee (Homoptera: Cixidiidae) (Howard et al. 1983); whereas, in the African region, vector of lethal yellowing is not yet identified. The putative vectors of plant hoppers and leafhoppers were identified phytoplasma diseases, i.e. Kalimanthan wilt of Indonesia and lime decline phytoplasma in Saudi Arabia (Alhudaib et al. 2009; Warokka et al. 2006). This chapter reports abundance of phloem-feeding insect fauna associated with WCLWD, their seasonal variation, and PCR-based phytoplasma detection.

Material and Methods

Field Collection of Insects

Monthly surveys were conducted during 2010–2011 at ten locations in the disease-affected area in south Sri Lanka. The rainfall and maximum and minimum temperature data were obtained from the climatology department of Sri Lanka. Initially, preliminary observations and insect collections were made for the selection of suitable collection methods and insect distribution in the palms and surroundings using sticky traps, hand collection, aspirator and sweep netting. Thereafter, the survey was conducted in young palms and surrounding areas in the affected plantations based on the observations of preliminary survey. It was conducted in severely, moderately and mildly affected areas. Collected insect samples were freeze-dried and transported for further studies to the laboratory at the Coconut Research Institute. Collected insects were categorized according to the rainfall pattern as two rainy seasons and two off-rainy seasons, respectively, i.e. March–May, September–November, June–August and December–February. All the collected insects were separated and stored in the freezer for DNA extractions. Initially, insects were classified up to their genus level and sent to the Faculty of Agriculture, University of Peradeniya for further classification.

DNA Extraction and PCR Analysis

The stored insects were used for DNA extraction. The insect total nucleic acids were extracted from preserved insects using a protocol slightly modified by Mpunami (1997). CTAB extraction buffer (2% Cetyltrimethylammonium bromide); 1.4 M NaCl; 20 mM EDTA pH 8.0; 1% (wt/v) PVP–40; 0.2% (v/v) 2-mercaptoethanol) was used for DNA extraction. Pre-warmed extraction buffer—300 μl—was added into 1.5 ml sterile microcentrifuge tubes. Three insects from each species were placed in tubes and crushed using a pestle. This pestle was made from a sterile 1000 μl micropipette tip while sealing it at the distal end using a flame. Each sample was crushed and incubated at 65 °C in a water bath for 30 min for the lysis of the insect DNA. Then the tubes were allowed to cool at room temperature and were extracted by adding an equal volume of chloroform: isoamyl alcohol (24:1). The solution was gently mixed by inversion for 5 min, and was centrifuged at 12,000 rpm for 10 min to separate the phases. Then, the upper aqueous phase was transferred to a clean sterile tube. This chloroform: isoamyl alcohol (24:1) extraction method was repeated again to get pure DNA. Thereafter, separated upper aqueous phase was transfered into clean new tube and added the 0.6 volume of ice cold isopropanol into it. The mixture was inverted slowly in few minutes and was kept for few minutes at −20 °C for DNA precipitation. Tubes were then centrifuged at 12,000 rpm for 10 min for forming DNA pellet. DNA pellet was isolated by removing the liquid portion and then washing with 70% (v/v) ethanol. Washed pellet was air-dried and it was dissolved in 25 μl sterile distilled water or TE buffer and stored DNA at 4 °C for analysis.

The amplification of 16S ribosomal DNA was performed using a nested PCR and the phytoplasma universal primer pairs P1 (5′–AAG AGT TTG ATC CTG GCT CAG GAT T–3′ (Deng and Hiruki 1991) and P7 (5′–CGT CCT TCA TCG GCT CTT–3′ (Smart et al. 1996) used for the first PCR. The universal phytoplasma primers PC 399 (5′–GAA ACG ACT GCT AAG ACT GG–3′) and P1694 (5′–TGA CGG GCG GTG TGT ACA AAC CCC G–3′) (Skrzeczkowski et al. 2001) were used for the second PCR. PCR was performed in 20 μl reaction volumes in 0.25 ml micro tubes, and this reaction mixture contained 20–30 ng of template DNA, 0.5 μM of each primer, 200 μM each of the four dNTPs, 2 mM $MgCl_2$, 10x polymerase buffer/PCR buffer, sterile water and 0.4 μl of *Taq* DNA polymerase (Go Taq polymerase, USA). Thermo-cycling parameters for primers P1/P7 and PC 399/P1694 followed the same procedure previously described by Smart et al. 1996; Heinrich et al. 2001; Skrzeczkowski et al. 2001; Perera et al. 2012. The primary PCR (first PCR) products were then diluted: 2 μl in 38 μl sterile distilled water, and 4 μl of dilution was used for the nested PCR/2nd PCR. In nested PCR, the primers P1/P7 were used followed by universal primers Pc399/P1694, which amplify a 1280 bp DNA fragment (Lee et al. 1993; Skrzeczkowski et al. 2001; Perera et al. 2012). This PCR amplified template DNA of suspected insects collected from the diseased area, the insects collected from disease-free areas, diseased coconut leaf midribs and sugar cane white leaf disease DNA (positive control) and sterile distilled water (negative control). Amplified PCR products were subjected to electrophoresis in 1% agarose gel (using TBE buffer) by staining it with ethidium bromide (5 μg/ml). DNA bands were visualized with a UV transilluminator. The amplified DNA bands were cut and purified using Wizard ® SV Gel and PCR Cleanup System in accordance with the manufactured protocol. The purified products were sent for DNA sequencing to the University of Nottingham, UK and Gene Tech lab (Pvt.), Colombo.

Data Analysis

The mean number of each putative vector insects in each season was analyzed using ANOVA, and compared regression analysis of each insect group with rainfall and average atmospheric temperature using IBM SPSS version 19.

Result and Discussion

Relative Abundance

Preliminary observations revealed that both adult and young palms were associated with similar species of sucking insects. Lethal decline disease of coconut in Tanzania and lethal yellowing disease in the Caribbean, where disease symptoms appear in both young and old palms, and most of the insects captured were on young palms (Mpunami 1997). The insect-collection methods facilitated the trapping of a large number of airborne adult insects, and allowed collection of insects potentially moving between coconut trees or from ground-cover plants to and from coconut. Sticky traps were used to collect the insects living and flying around the coconut canopy; sweep nets caught those species living under the canopy and on grass; light traps gave an indication of species that were abundant at the vicinity of plantations as well as insects with nocturnal habit. Insects were collected from both the palms with symptoms and without symptoms in the selected plantations. In the survey, 32 Homopteran and one Heteropteran were collected. Insects belonged to seven families, i.e. Cicadellidae, Cixiidae, Delphacidae, Derbidae, Meenoplidae, Membracidae and Tingidae (Table 1). Among these, 22 species belonged to Cicadellidae. However, considering the abundance of individuals, the highest number of individuals in the affected area was from *Stephanitis typica* (Tingidae), *Proutista moesta*, *Proutista* sp. (Derbidae), *Recilia dorsalis*, *Nephotetreix virescens* (Cicadellidae) and *Nisia nervosa* (Meenoplidae) (Fig. 1). *Schwartziella typica* completes its life cycle on the coconut palm and the life stages on the coconut leaves, while adults of *P. moesta* and *Proutista* sp. were found only on the coconut palms, and their eggs and nymphal stages were found on the decaying organic material especially palm waste like decaying fronds, inflorescence etc. *N. nervosa* and *K. ceylonica* were mostly collected from surrounding vegetation, and few individuals were noticed on the coconut palm.

PCR and DNA Sequencing

According to the nested PCR test, the 1280 bp bands were given DNA samples of nine insect species collected from disease areas, four samples of the disease-affected coconut leaf midrib DNA, sugarcane white leaf disease sample. The DNA sample of insects (*P. moesta* and *S. typica*) collected from disease-free areas, sterile water samples and the three diseased coconut samples collected from diseased area were not given the required size of bands (Fig. 2a, b). The insects given to them that gave positive results among the 33 insect species were *P. moesta*, *Proutista* sp., *S. typica*, *N. nervosa*, *R. dorsalis*, *Goniangnathus (T.) punctifer* (Cicadelidae), *Idioscopus clypealis*

Table 1 Insect species collected in a survey from the WCLWD affected palms in Weligama

Order	Suborder	Family	Species
Homoptera	Auchenorhyncha	Cicadellidae	*Goniagnathus (T.) punctifer* *Recilia dorsalis* *Kolla ceylonica* *Kolla paulula* *Scaphoideus morosus* *Idioscopus clypealis* *Nephotetrix virescens* *Nephotetrix nigropictus* *Exitianus indicus* *Exitianus* sp. *Hishimonus* sp. *Platyrectus marginatus* *Stirellus* sp. 1 *Stirellus* sp. 11 *Balclutha* sp. *xvi. Hecalus porrectus* xvii. 5, Unknown species
Homoptera	Auchenorhyncha	Derbidae	*Proutista moesta* *Proutista sp1* Unknown sp.
Homoptera	Auchenorhyncha	Meenoplidae	*Nisia nervosa*
Homoptera	Auchenorhyncha	Cixidae	Unknown sp.
Homoptera	Auchenorhyncha	Delphacidae	Unknown sp. Unknown sp. Unknown sp. Unknown sp. Unknown sp.
Homoptera	Auchenorhyncha	Membracidae	Unknown sp.
Hemiptera		Tingidae	*Stephanitis typica*

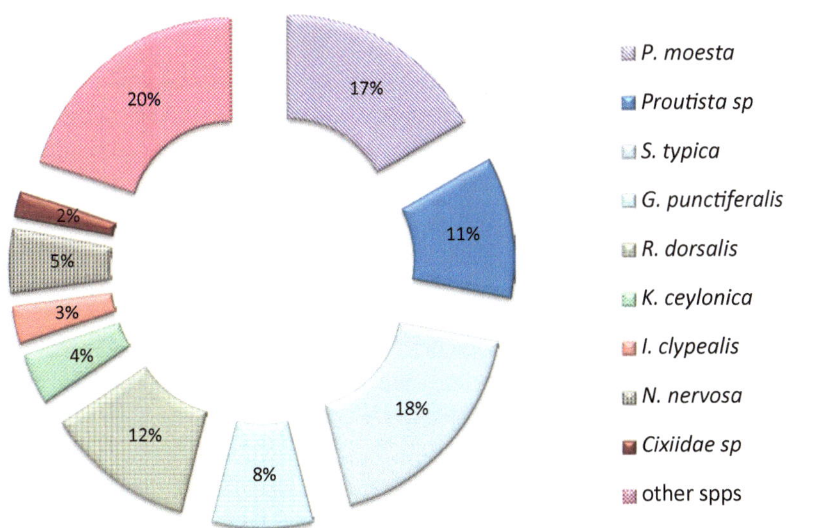

Fig. 1 Relative abundance of phloem feeding insect association in the WCLWD-affected coconut plantations

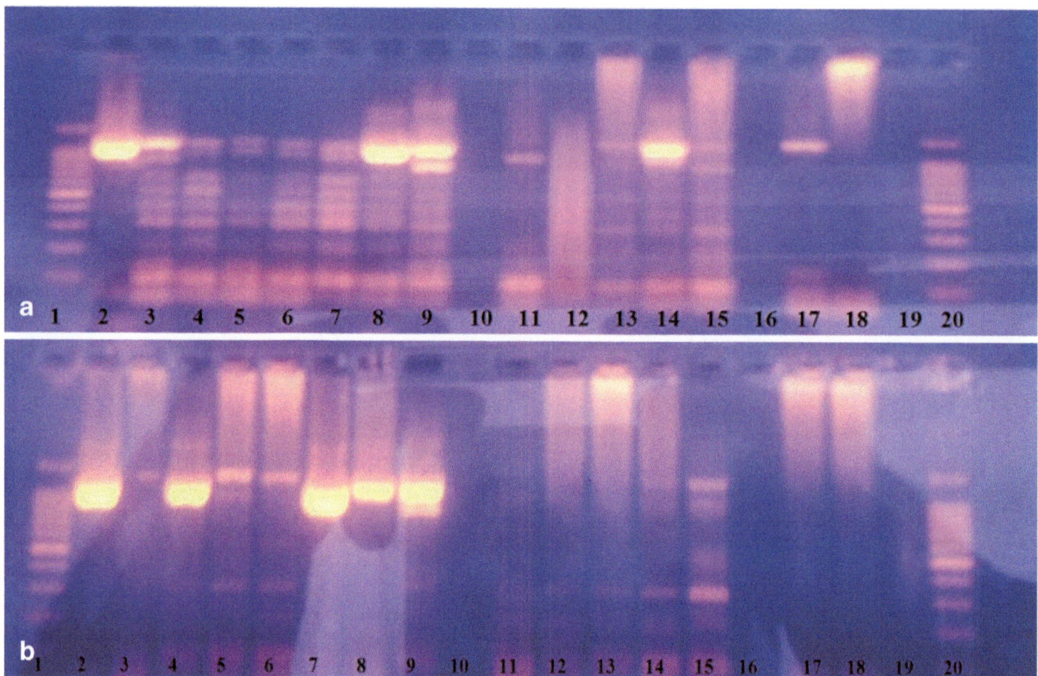

Fig. 2 Representative gels showing phytoplasma rDNA amplified from the DNA of putative insect vectors WCLWD with P1/P7 followed by nested PCR with primers PC 399/P1694; A Lane=1 & 20, 1 kB ladder; 2–10, insect samples11 and 12 insect DNA collected from disease free area; 13, 14 and 15 infected coconut sample; 17 sugar cane white leaf sample, 18 sterile distilled water; *B lane*=1 & 20, 1 kB ladder; 2–10, insect samples collected infected area; 10 and16 collected infected area; 10, 16 and 19 blank well, blank well, 11 and 12 insect DNA collected from disease-free area; 13, 14 and 15 infected coconut sample; 17 and 18 sterile distilled water

(Cicadelidae), *Kolla ceylonica* (Cicadelidae) and an unknown Cixiid sp. (Fig. 2 a&b). The sequencing and blasting of the positive DNA bands indicated that these sequences are matched with the gene bank deposited WCLWD sequence (Gene Bank accession number: EU635503). Among the tested insects, the highest percentage of positive results was given by *P. moesta* and *S. typica*, while other insects relatively were given low percentage of positive results. The results varied with time; due to a low titre of phytoplasma DNA present in the plant tissues especially in woody plants like coconut tree, the phytoplasma DNA was comparatively in very low concentrations in coconut tree tissues (Perera et al. 2012). However, the phytoplasma DNA concentration within the insect body was comparatively higher than in the plant due to multiplication of phytoplasma within the insect body before transmission (Weintraub and Beanland 2006). Comparatively, results showed on the DNA collected from *P. moesta*, *Proutista* sp., *S. typica* were consistent while the other six species showed a low constancy. It may be due to their feeding habit that adult insects always feed on the coconut leaf phloem sap. Their body has been filled every time with infected sap. *R. dorsalis* relatively is a more abundant insect but the percentage of positive results given was relatively low (Fig 3). The nine insect species have a wide host range and were not specific to coconut or family Palmae.

All phloem-feeding insects can acquire phytoplasma by feeding phloem sap. However, the insects are acquired the phytoplasma from plant sap through feeding, it was not translocated within the all phloem feeding insects. Within the insect digestive tract of most insects, the acquired phytoplasma was digested within short period of time before translocated. But few insects were allowed to multiple and translocation of phyto-

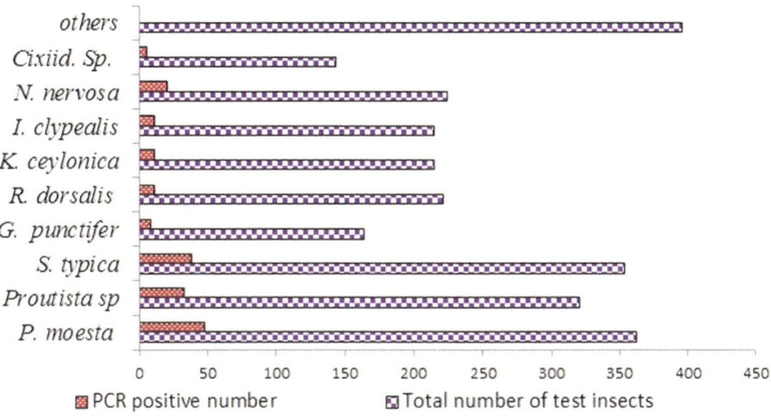

Fig. 3 Total number of each insect who were subjected to the nested PCR, and the number of those insects have given positive results

plasma in their body and they act as vectors. According to the Weintraub and Beanland (2006), vector species are found mainly in four families of Fulgorids: Cixiidae, Delphacidae, Derbidae and Flatidae. The first three families have at least one species that transmit a phytoplasma in the coconut lethal yellows group (16SrIV). Several species in these families also transmit phytoplasmas from the stolbur (Sr16XII) group. A Flatid vector, *Metcalfa pruinosa* (Say), transmits aster yellows (AY) (group Sr16I). This study was identified nine insect species as the putative vectors of WCLWD in Sri Lanka and five species of among them were recorded as vectors of several other phytoplasma diseases. The subfamily Deltocephalinae has the most highly derived lineages. More than 75% of all confirmed phytoplasma vector species are found in this subfamily. The feeding habits of the species within the Deltocephalinae range from monophagous to polyphagous, and members of this group can transmit one or more different phytoplasma taxa (Weintraub and Beanland 2006). The four species identified belong to this subfamily, viz. *R. dorsalis, K. ceylonica, I. clypealis* and *G. punctifer*. All the homopterans are non-destructive feeders (Mitchell 2004; Okuda et al. 1998), and among these nine species of insects, eight species are non-destructive feeders except *S. typical*. Hence, the feeding marks were also not visible on the coconut palms.

The presence of phytoplasma DNA within the body of these insects can be suspected as putative vectors of the WCLWD in Sri Lanka. This disease is more similar to root wilt disease of India, and they confirmed *P. moesta*, *S. typica* and *Sophonia greeni* (distant) are vectors of disease through electron microscopic studies, but they have not confirmed the disease using molecular techniques (Mathen et al. 1990). Some insects act as the dead-end host of phytoplasma and they acquire phytoplasma into body but do not transmit them. The alternate host plants also act as a source of inoculum without showing the symptoms (Weintraub and Beanland 2006). The alternative host plant present in the WCLWD-affected area surrounding the coconut plantations and insect may acquire the phytoplasma directly from that plant. This study may not clearly reflect the percentage of positive insects, as one sample contains DNA of three insects (Fig. 3).

Phytoplasma pass through insect generations via transovarian transmission, and if the infected female lays eggs, the phytoplasma goes to the next generation. Therefore, the insect, before feeding on the affected plant, may have phytoplasma in their body (Mitsuhashi et al. 2002; Alma et al. 1997). The identification and confirmation of the vector is not an easy task; however, it is important to identify the putative or possible vector before going to transmission studies.

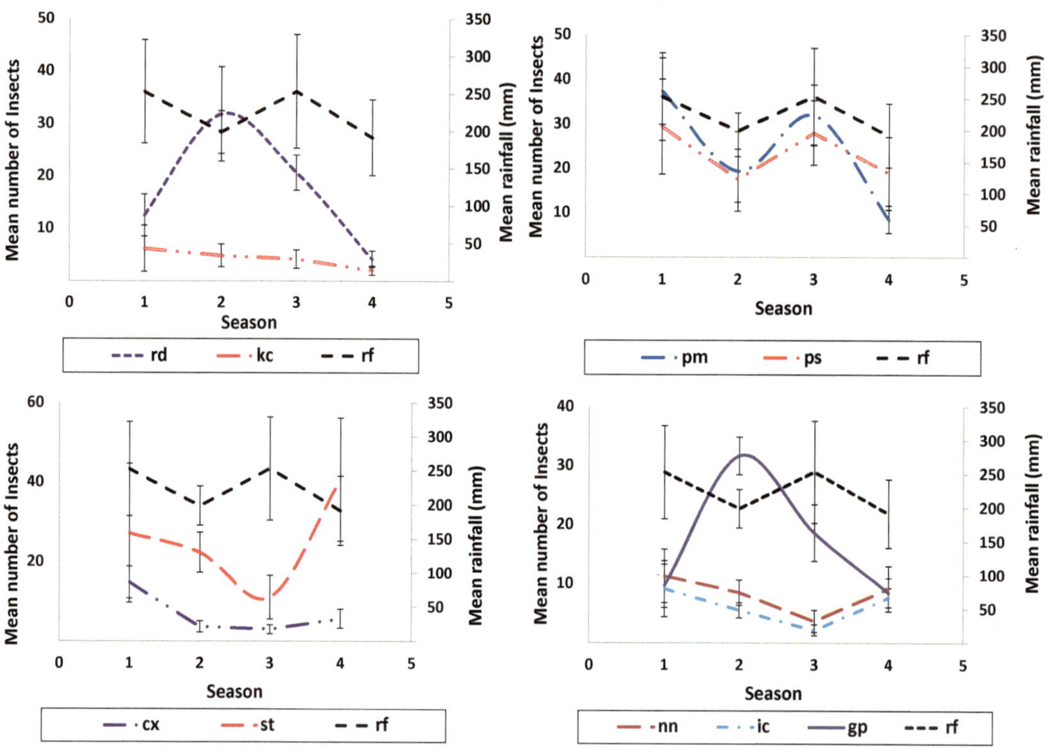

Fig. 4 Relationship insect vectors with rainfall pattern according to the season (within 2 years); season *1* December–February, *2* March–May, *3* June–August, *4* September–November. *Rf* Mean rainfall pattern, *pm P. moesta*, *ps Proutista* sp., *cx* cixiid sp., *st S. typica*, *rd R. dorsalis*, *kc K. ceylonica*, *nn N. nervosa*, *ic I. clypialis*, *gp G. punctifer*

Relationship with Climatic Factors

The 2-year survey indicated that the insect species and their abundance varied throughout the year. Compared to the mean number of insects with the season, *P. moesta* (F=3.505, p=0.034), *G. punctifer* (F=8.653, p=0.001), *R. dorsalis* (F=5.015 p=0.009), *Cixiid* sp. (F=4.732, p=0.012) were significantly different in the four seasons. Their abundance was generally higher in rainy seasons, and the number gradually declined with the reduction of rainfall (Fig. 3). However, only *P. moesta* and *Proutista* sp. followed the rainfall pattern (Fig. 4) (Pearson Correlation coefficient=0.460, p=0.012, R^2=0.47). The average atmospheric temperature is not significantly different in the four seasons, and there is no correlation with insect and the atmospheric average temperature (F=0.264, p=0.613). *S. typica* showed negative relationship with rainfall but this relationship is not significantly correlated (Fig. 4). In addition, there is a significant correlation of abundance pattern between insect species. The similar trends were observed in an individual species with others of their abundance in the field i.e. *R. dorsalis* and *G. punctifer* (Pearson Correlation coefficient = 0.460, p=.012, R^2=0.47) also, *K. ceylonica and Proutista* sp. (Pearson Correlation coefficient = 0.455, p=0.025, R^2=0.41), *N. nervosa* correlate with (Fig. 4) *I. clypialis* (Pearson Correlation coefficient =0.744, p=0.001) and Cixiid species (Pearson Correlation coefficient =0.712, p= 0.001), and *Proutista* sp. also correlated with the *N. nervosa,* (Pearson Correlation coefficient = 0.368, p=0.038) *I. clypialis* (Pearson Correlation coefficient =0.376, p=0.033) as well as *K. ceylonica* (Pearson Correlation coefficient =0.455, p=0.013). The above data indicated that all putative vectors followed the same trend in the field (Fig. 4) and their abundance is related to the rainfall pattern, but all the data were not significantly related due to the higher variation. It can be concluded that all the nine species are putative vectors of WCLWD in Sri Lanka, and *P. moesta, S. typica* and *Proutista* sp. are the more abundant ones.

Acknowledgements We thank Dr. (Mrs.) Raji Ganeswaram (Faculty of Science University of Jaffna), Dr. Hemachandra (Faculty of Agriculture, University of Peradeniya), Dr. M. Dickinson (University of Nottingham, UK), Dr. R Wejesekara, Mrs. N. I. Suwadharathne, Mr. D. Hettiarachchi, Mr. C. Munasinghe, Mr. C. Senarathne, Mrs. S. Fernado, Miss R. S. Godage, Miss Anupama, Mr. Danuska and Mr. N. G. Premasiri.

References

Alhudaib K, Arocha Y, Wilson M, Jones P (2009) Molecular identification potential vectors and alternative hosts of the phytoplasma associated with a lime decline disease in Saudi Arabia. Crop Prot 28:13–18

Alma A, Arzone A, Bosco D (1993) Grapevine MLO transmission by insects. Extended Abstracts 11th Meeting ICVG, Montreux, Switzerland, 6–9 Sept. 1993, 84–85

Corodova I, Jones P, Harrison NA, Oropeza C (2003) *In situ* PCR detection of phytoplasma DNA in embryos from coconut palms with lethal yellowing disease. Mol Plant Pathol 4:99–108

Dale JL, Kim KS (1969) Micoplasma like bodies in dodder parasitizing aster yellows infected plants. Phytopathology 59:1765–66

Deng S, Hiruki C (1991) Amplification of 16S rRNA genes from culturable and non culturable mollicutes. J Microbiol Methods 14:53–61

Heinrich M, Botti S, Caprara L, Arthofer W, Strommer S, Hanzer V, Katinger H, Bertaccini A, Machalo MLC. (2001) Improved detection methods for fruit tree phytoplasmas. Plant Molecular Biology Reporter, (19):169–179

Hiruki, C. (1999) Paulownia witches'-broom disease important in East Asia. Acta horticulturae (496):63–68

Howard FW, Norris RC, Thomas DL (1983) Evidence of transmission of palm yellowing agent by a plant hopper, *Mindus crudus* (Homoptera; Cixidae). Tropical Agric (Trinidad) 60:168–171

IRPCM Phytoplasma/Spiroplasma Working Team-Phytoplasma Taxonomy Group (2004) '*Candidatus Phytoplasma*', a taxon for the wall-less, non-helicaln prokaryotes that colonize plant phloem and insects. Int J Syst Evol Microbiol 54:1243–1255

Jones P (2002) Phytoplasma plant pathogens. In: Waller M, Lenne´ JM, Waller SJ (eds) Plant pathologists pocketbook. CAB International, Part 12. Oxford University Press, USA, pp 126–139

Khan AJ, Botti S, Paltrinieri S, Al-Subhi AM, Bertaccini AF (2002) Phytoplasmas in alfalfa seedlings: infected or contaminated seeds? In: Abstracts of the14th international organization of mycoplasmology conference, Vienna, Austria, p 148

Lee IM, Hammond RW, Davis RE, Gundersen DE (1993) Universal amplification and analysis of pathogen 16S rDNA for classification and identification of mycoplasmalike organisms. Phytopathology 83:834–842

Maixner M (2005) Risks posed by the spread and dissemination of grapevine pathogens and their vectors. In: Plant protection and plant health in Europe: introduction and spread of invasive species, Symposium Proceedings No. 81. The British Crop Production Council, Alton, Hampshire, UK, pp 141–146

Mathen K, Rajan P, Nair CPR, Sasikala M, Gunasekaran M, Govindankutty MP, Solomon JJ (1990) Transmission of root (wilt) disease to coconut seedlings through *Stephanitis typica* (Distant) (Heteroptera: Tingidae). Tropical Agric 67(1):69–73

Mitchell PL (2004) Heteroptera as vector of plant pathogens. Neotrop Entomol 33:519–545

Mitsuhashi W, Saiki T, Wei W, Kawakita H, Sato M. (2002) Two novel strains of Wolbachia coexist in both species of mulberry leafhoppers, Hishimonoides sellatiformis and Hishimonus sellatus which are vectors of mulberry dwarf phytoplasma. Insect Mol Biol (11):577–584

Mpunami AA (1997) Molecular methods for detecting the coconut lethal disease (LD) phytoplasma in Tanzania. Dissertation, University of Nottingham, 222 pp

Nielson MW (1979) Taxonomic Relationships of leafhopper vectors. In: Maramorosch K, Harris KF (ed) Leafhopper vectors and plant disease agents. Academic, London

Okuda S, Naakano Y, Goto T, Natsuaki T (1998) 16Sr DNAs of paulownia wiches' broom phytoplasma transmitted by *Halymota mista*. Presented at the 7th int. Congress Plant Patho., Pap. No. 3.7.33. Edingburgh, Scotland

Perera L, Meegahakumbura MK, Wijesekara HTR, Kalani NGA, Munasing CE, Fernando WBS, Dickinson MJ (2010) Detection of Weligama coconut leaf wilt disease in Sri Lanka by polymerase chain reaction. In: George VT, Krishnakumar V, Augustine JB, Niral V, Josephrajkumar A (eds) Proceedings of the International Conference on Coconut Biodiversity for Prosperity. Niseema Printers and Publishers Kochi–18, India, p 146

Perera L, Meegahakumbura MK, Wijesekara HTR, Fernando WBS, Dickinson MJ (2012) A Phytoplasma is associated with the Weligama Coconut Leaf Wilt Disease in Sri Lanka. J Plant Pathol 94(1):205–209

Skrzeczkowski LJ, Howell WE, Eastwell KC, Cavileer TD (2001) Bacterial sequences interfering in detection of phytoplasma by PCR using primers derived from the ribosomal RNA operon. Acta Hort 550:417–424

Smart CD, Schneider B, Blomquist CL, Guerra LJ, Harrison NA, Ahrens U, Lorenz KH, Seemüller E, Kirkpatrick BC (1996) Phytoplasma-specific PCR primers based on sequences of the 16–23 rRNA spacer region. Appl Evol Microbiol 62:2988–2993

Solomon JJ, Nair CPR, Rajan P, Srinivasan N, Govindankutty MP, Mathen K, Pillai NG (1998) Etiology-Phytoplasma In: Nampoothiri KUK, Koshy PK (ed) Coconut root (wilt) disease. Central Plantation Crops Research Institute, India

Warokka WA, Jones P, Dickinson M (2006) Detection of phytoplasmas associated with Kalimanthan wilt disease of coconut by Polymerase chain reaction. J Littri 12:154–160

Weitraub PG, Beanland L (2006) Insect vectors of phytoplasmas. Annu rev Entomol 51:91–111

Wijesekara HTR, Perera L, Wickramananda IW, Herath I, Meegahakumbura MK, Fernando WBS, De Silva PHPR (2008) Preliminary investigation on Weligama coconut leaf Wilt disease: a new disease in Southern Sri Lanka. In: Nainanayake NPAD, Everard JMDT (ed) Proceedings of the Second Symposium on Plantation Crop Research—Export competitiveness through quality improvements. Coconut Research Institute, Lunuwila

Plant Virus Disease Spread Through Insect Vectors and Their Management

V. M. Chavan

Abstract

Arthropod-borne plant viruses are among the most important complex and extensively distributed plant disease agents. The vectors of any plant virus are always restricted to one of the major taxa, such as aphids, leafhoppers, whiteflies, thrips, mealy bugs, beetles, mites, or nematodes. The first vectors recognized as being associated with a plant virus disease were insects, i.e., leaf hoppers (*Deltocephalis dorsalis* Motschulsky) transmitting rice stunt virus disease. Different terms or categories used for describing the relationship between vectors and viruses give some indication of the behavior of the virus in the vector during transmission or mechanism of virus transmission by vector. The most threatening property of the insect vector is its wide host range. The knowledge of virus–vector relationship is essential to devise suitable measures against vector-borne diseases. Most approaches to control the vectors of virus diseases are aimed at eradicating or altering one or more of the primary participants in the transmission process (vector, virus, and host plant) or at preventing their coming together. Broadly, the management of vectors of viral diseases can be done by adopting methods of cultural control, biological control, and chemical control, e.g., an integrated management strategy consisting of appropriate transplanting time, use of barrier crops, and periodical application of synthetic and botanical chemicals for management of aphid vectors of papaya ring spot virus (PRSV) on papaya has been successfully developed at I. A. R. I. Regional Station, Pune.

Keywords

Arthropod · Integrated management · Plant virus · Vector

V. M. Chavan (✉)
I.A.R.I. Regional Station, Survey No. 125A, Baner Phata,
I.T.I. Road, Aundh, Pune-411007, India
e-mail: vmcpune@yahoo.com

A. K. Chakravarthy (ed.), *New Horizons in Insect Science: Towards Sustainable Pest Management*,
DOI 10.1007/978-81-322-2089-3_15, © Springer India 2015

Introduction

Most of the devastating plant diseases are attributed to viruses transmitted by vectors. In the absence of vectors, these diseases would be of little importance. Therefore, the control measure is dependent on vectors. Arthropod-borne plant viruses are among the most important, most complex, and most extensively distributed plant disease agents in the world. They feed on plants and move from one plant to the other, during which viruses get effectively transmitted. The vectors of any one plant virus are always restricted to one of the major taxa, such as the aphids, the leafhoppers, the whiteflies, the thrips, the mites, or the nematodes. The first vectors that are recognized as being associated with a plant virus disease were insects when Takata (1895) confirmed observation by a rice farmer that the rice stunt virus disease was found transmitted by leafhoppers *Deltocephalis dorsalis* Motschulsky. Later on, *Nephottetix apicalis* Motschulsky was found to be very efficient vector of rice stunt viral disease. Subsequently, aphids were also identified as vectors of plant viruses (Doolittle 1916). The first authentic report of plant hoppers, *Peregrinus madis* (Ashmead) as vectors of corn mosaic virus was given by Kunkel (1922). The first report of whitefly *Bemisia tobaci* (Gennadius) as virus vector transmitting mosaic disease of abutilon was demonstrated by Orlando and Silberschmidt (1946) in Brazil. Not all vectors are insects, and some viruses are transmitted by leaf and bud-feeding eriophyoid mites, e.g., *Phytoptus ribis* Nal. transmitted current reversion agent (Amos et al. 1927). More recently, vectors of viruses have been found among soil-inhabiting organisms, a dorylaimid nematode by Hewitt et al. (1958) and a chytrid fungus by Teakle (1960). All these vectors transmit more than 283 viruses and other similar pathogens (Harris 1981). More than half of the nearly 550 vectors transmitted virus species recorded are disseminated by aphids (55 %), 11 % by leafhoppers, 11 % by beetles, 9 % by whiteflies, 7 % by nematodes, 5 % by fungi and plasmodiophorids, and the remaining 2 % by thrips, mites, mirids, or mealybugs (Astier et al. 2001).

All above plant viruses' vectors show a common feature i.e., they all penetrate unwounded plant cells, usually when feeding on the plants and thus have an opportunity to acquire virus from an infected plant and inoculating it. These processes occur in very different ways with different vectors and viruses, making a fascinating series of biological adaptation. The most threatening property of the insect vector is its wide host range. Natesan et al. (1996) studied the host range, vector relation, and serological relationship of cotton leaf curl virus in South India and found that the virus was transmitted by whitefly, *B. tabaci* to 24 plant species in 6 families. The major hosts included were bean, cotton, tobacco, tomato, and several other weed hosts of the insect.

Relationship Between Plant Viruses and Vectors

Kennedy et al. (1962) introduced terms which are descriptive of the relationship between vectors and viruses and give indication of the location and route followed by the virus–vector relationship.

Terms Used in Virus–Vector Relationship

Acquisition Access Period

The time for which, initially, a virus-free vector is allowed to access a virus source and could desire to feed on that source.

Acquisition Feeding Period or Acquisition Threshold

The time for which, initially, a virus-free vector actually feeds on the virus source.

Inoculation Access Period

The time for which the virus-carrying vector is allowed to access a virus-free plant and feed on it.

Inoculation Feeding Period

The time for which the virus-carrying vector appears to be feeding on virus-free plants.

Latent Period

The time for the beginning of acquisition feeding until the vector can infect healthy plants with the virus. It is also called as preinfection period or incubation period.

Transmission Threshold Period

The minimum total time that a vector needs to acquire a virus and inoculate it to a virus-free plant.

Infective Capacity or Retention Period of Vector

This is the period for which an insect carries/retains/transmits the virus to host plants and remains viruliferous.

Persistence

The time for which a vector remains infective after leaving a virus source. Persistence is further divided into three main but integrating categories.

A. Nonpersistence Virus persisting for few (usually for few seconds, minutes, or less than an hour) hours at about 20 °C.

B. Semipersistence Virus persisting for 10–100 h.

C. Persistent Persistence for more than 100 h, in some instances for the life of the vector.

Recently, there has been an attempt to replace these somewhat arbitrary categories of persistence by categories based on the behavior of the virus in the vector during transmission or mechanism of virus transmission by vector.

Mechanism of Virus Transmission

Stylet-Borne Viruses

These are in fact nonpersistent viruses which are carried in the stylets of the vector to the site of inoculation. The term was proposed by Bradley (1964) based on a series of experiments. Stylet-borne viruses adhered to tips of stylets, are immediately acquired by vector during feeding and are transmitted by vectors soon after acquisition. Viruliferous insects can transmit these viruses for a limited period, rarely for a few days but never forever or after moulting. Thus, they can infect only a limited number, sometimes only one or two plants. These viruses are transmitted by aphids. They are acquired mainly from the epidermis within the first few seconds of probing. Many viruses are stylet-borne viruses and a great majority of them induce mosaic symptoms and are sap transmitted, e.g., *Cucumovirus*, *Carlavirus*, and *Potyvirus* are aphid stylet-borne viruses.

All aphids and leaf hoppers feed on plants and possess piercing and sucking mouth parts, consisting of pairs of stylets, a labium, a large slender rigid organ with a deeply concave anterior surfaces forming the channel of beak, a labrum, mandibles, and maxillae. The two pairs of stylets form a compact bundle or fascicle with slides in the groove of labium and constitute the piercing organ. Piercing organ has two channels; through one (salivary channel), saliva is injected into a plant and through the second (food channel), plant sap is sucked. Mean length of 75 single stylets of adult *Myzus persicae* (Sulzer) is 496 ± 17 μm (Forbes and Mac Carthy 1969).

Aphids and leafhoppers secrete two types of the salivary sheath material, which coagulate rapidly and form a salivary sheath or stylet sheath in the path of stylets and water-soluble saliva (Miles 1968). Most of the aphids and leafhoppers form this sheath around their stylets during the feeding process. Stylet sheath is laid by the vector in tissue of the host and it stays there even after the withdrawal of stylets. It shows the intercellular (in majority of aphids) or intracellular (in majority of leaf hoppers) path that the stylet follows during its passage through the tissues and also the point where it terminates. Stylets move fairly rapidly within this sheath but are subsequently extended beyond the sheath for ingestion of food material from host cells. Most aphids are surface feeders and feed in parenchyma and mesophyll and hence, induce mosaic symptoms while most leafhoppers feed on phloem.

Foregut-Borne Viruses

An ingestion–egestion mechanism for transmission of both nonpersistently and semipersistently transmitted viruses was proposed by Harris (1977), where virus enters the foregut, attaches to the lining of the anterior portion of the alimentary canal, and is inoculated when the vector egests while probing a plant. Foregut-borne viruses have no latent period in their vectors, cannot recover from the vectors hemolymph and cannot be transmitted after injection into the vector's hemocoel. Infectivity is retained only a few days and it is lost after ecdysis, e.g., Rice tungro spherical virus.

Circulative Viruses

Kennedy et al. (1962) used the term circulative for these viruses that after ingestion, pass through the gut wall into hemolymph, and then pass through the salivary glands to be discharged with the salivary secretion. Circulative viruses can be recovered experimentally from the vector's hemolymph, can be transmitted after injection into the vector's hemocoel, are not lost after ecdysis and may be transmitted for weeks, sometimes for the life of the vector (Sylvester 1980). These viruses are not transmitted mechanically or immediately after acquisition but become transmissible only after a latent or incubation period within the body of insects. The latent period may be several hours or days and is temperature-dependent. No circulative virus is transovarially transmitted from ineffective females to their progeny. Circulative viruses are mainly transmitted by aphids, e.g., Barley yellow dwarf, beet curly top, and beet western yellows.

Propogative Viruses

These viruses propogate or multiply within their insect vector which transmits them for a long time but vary upon as they live. These viruses possess incubation period (one or more weeks) which is presumed to be the time needed for them to multiply and to reach a definite concentration to become transmissible. Thus, they have a definite biological relationship with their vectors. Some propogative viruses are transovarially transmitted from a viruliferous female to its progeny, e.g., Maize rayado fino virus, rice stripe virus.

Transmission by Aphids

Aphids form the largest group of insect vector both because of large number of species, about 370 involved and the large number of viruses, about 300 they transmit. *M. persicae* alone is estimated to transmit about 100 viruses while other aphids transmit more than 30 viruses each. Some aphids on the contrary can transmit only one virus each. Following three types of virus transmission are observed in aphid vectors.

Nonpersistent Viruses

The nonpersistent viruses transmitted by aphids include *Potyviruses*, *Carlaviruses*, and alfalfa mosaic virus. Specificity between the virus and species of vector is not well developed and usually one virus is transmitted by several or many aphid species. The transmission threshold period can be as short as 2 min, and the virus can be efficiently acquired and inoculated during probes of only 10–30 s indicating that the virus is acquired and inoculated to cells of the epidermis. Transmission of nonpersistent virus is favored by making the aphids fast before giving acquisition feeding. Experimental results show that nonpersistent viruses are generally stylet borne. Some workers have suggested that a virus is held in the food canal and still others opined that it is carried in a plug of gelled saliva where the stylet punctures the leaf.

Semipersistent Viruses

A small heterogeneous group of viruses is acquired by aphids or other insects during feeding times from several minutes to several hours but not usually by probing. Efficiency of transmission increases with increase in both acquisition feeding and inoculation feeding time but there is no definite incubation period and fasting before acquisition feeding has no effect on efficiency of transmission. There is no good evidence on how or where these viruses are held with the aphids but they seem not to be retained through moult.

Persistent Viruses

Persistent aphid transmitted viruses have minimum acquisition access and inoculation access period of 10–60 min usually, have latent period of 12 h or more and can be retained by the aphids for at least a week. Infected plants mainly show leaf rolling and yellowing symptoms and many viruses are probably concentrated in the phloem of the plant.

Transmission by Leafhopper and Plant Hopper

Leafhopper and plant hopper make up the second most important group of virus vectors. Most of the vector species are leafhopper (Cicadellidae), but nearly 20 are plant hoppers (Fulgoridae) and the tomato pseudo curly top agent is transmitted by the treehopper, *Micratalis* sp.

The viruses transmitted by leaf and plant hoppers cause yellowing and/or leaf rolling and only a few are sap transmissible. The vector mainly feeds from the phloem of plants. Most commonly, the viruses multiply in their vectors and persist for a long periods and several of them are transmitted through the eggs to the progeny.

Transmission by Whiteflies

Of the 1100 identified species of whiteflies in the world, only three are recognized as vectors of plant viruses. *B. tabaci* is considered as the most common and important whitefly vector of plant viruses.

None of the whitefly–transmitted (WFT) viruses is transovarially or seed transmissible. Only a few of these viruses induce mosaic symptom. Whiteflies generally acquire viruses more rapidly from young leaves and recently infected plants. Females are more efficient vectors of virus transmission than males. Muniyappa (1980) divided WFT virus diseases into four groups, i.e., yellow mosaic, yellow vein mosaic, leaf curl, and mosaic diseases. Transmission efficiency of whiteflies increases with longer feeding periods, a short incubation period occurs in most cases, viruses retain in vectors from few days up to 20 days and serial transmission is generally intermittent and inefficient. Hence, relationship of WFT viruses to their vectors is of circulative type. However, Brown (1994) stated that Gemini viruses are transmitted in a persistent manner by whitefly as well as some WFT gemini viruses were experimentally transmitted by sap or mechanical inoculation.

Transmission by Thrips

Thrips tabaci (L.) and three species of *Frankliniella* are the only known vectors of tomato spotted wilt virus (TSWV). *T. tabaci* is a cosmopolitan species feeding on at least 140 species in 40 families of plants. It reproduces parthenogenitically and feeds by sucking the content of the subepidermal cells of the host plant. Only larvae can acquire TSWV with a minimal acquisition time of 15–30 min and with incubation period of 4–18 days. Before completion of incubation period, the larvae become ineffective. Virus may be retained for life with erratic transmission. No transversal transmission. Probably the virus multiplies in the vector (Matthews 1970).

Transmission by Mealy Bugs

Several species of mealy bugs are recorded as vectors of isolates of cacao swollen shoot virus. The virus can be acquired in 1 h and inoculated in 15 min. This virus persists for 3–4 days in *Planococcoides njalensis* (Laing) and is retained through the moult (Gibbs and Harrison 1976).

Transmission by Beetles

Nearly 40 viruses are transmitted by beetles which are sap-transmissible, relatively stable, highly antigenic, and icosahedral except TMV and PSTV. They always cause mosaic diseases.

Beetles have biting mouthparts and transmit virus mechanically. Viruses are acquired by vectors within short acquisition feeding period of few minutes, have no incubation period and are always transmitted immediately after acquisition period. Some of the viruses are retained by vectors for short time usually ranging from 24–48 h, while others are retained by vectors for prolonged periods ranging from 7 to 20 days. Virus vector relationship of these viruses is of nonpersistent type except that some viruses are retained for many days.

Transmission by Mites

Some eriophyid mites are firmly established as virus-vector. Mites possess mouthparts that are spherically adopted for piercing and sucking process. Relationship of wheat streak virus with its mite vector is of persistent type but not circulative because of long acquisition, long inoculation, and retention period of virus in its vector through moulting. Virus particles mostly exist in densely packed bundles in greatest concentration in the lumen of mid-gut and rectum-like sac of hind-gut. Plants could possibly get infected by the virus by its back flow from gut to mouth parts during feeding or by introduction of defected virus into plant cells through feeding punctures or abrasions caused by anal setae or anal suckers.

Transmission by Nematodes

About 25 viruses belonging to two groups are transmitted by three genera of soil inhabiting plant parasitic nematodes. The nepoviruses, which have isometric particles about 30 nm in diameter, and the tobra viruses, which are straight tubular ones. *Nepoviruses* are transmitted by species of *Xiphinema* or *Longidorus* (Dorylaimidae), whereas vectors of tobra viruses are *Trichodorus* species (Trichodoridae). The three species are cosmopolitan in distribution and are considered as obligatory but migratory ectoparasite getting their food mainly from root tips. Both adults and larval stages of most nematode vector species can transmit viruses equally efficiently and there is no transovarial transmission.

Important Aspects of Virus Transmission by Vectors

Sequential Spread of Viruses in Insects

Propogative viruses occur systematically in almost all body parts of vector. They follow a definite route and sequence to spread systematically within the body of a vector after ingestion. The virus is sucked into the gut which is generally the primary site of virus multiplication, passes into hemolymph through some part of alimentary track, enters the blood stream, is carried to salivary glands by circulating blood and is finally injected into healthy plant with salivary liquid during feeding.

Transovarial Spread of Viruses

Propogative viruses occur systematically in internal parts including presumably, ovaries and eggs of vector. Progeny of viruliferous females will be infected. This is called as transovarial passage of viruses.

Dependent Transmission

Aphids sometimes transmit virus from an infected plant only if the later is also infected by second virus. In this dependent transmission, the presence of second virus, i.e., *helper virus* is essential, e.g., potato virus C is transmitted only in presence of potato virus Y.

Epidemiology of Virus Transmission

Spread of aphid-borne nonpersistent viruses follows the seasonality of vectors; especially the dominant species (Mora-Aguilera et al. 1993; Basky 1986). Alate aphids land on available plants regardless of species and are unable to distinguish host plants from nonhost prior to landing. Host selection occurs after arriving on plant surface and after ingestion of plant sap. Virus transmission occurs if the aphid is carrying specific virus, even though the aphid does not colonize the plant. The acceptance or rejection of plant host by a vector with piercing-sucking mouth parts is performed by a series of brief probes into multiple plant epidermal cells, which is sufficient to inoculate the nonpersistent viruses. If brief feeding probes designate the plant as an acceptable host or food source, the vector is likely to initiate prolonged feeding. However, in case of nonregular host plant, there is little chance of initiating prolonged feeding and in this case the number of efficient vectors involved decides the epidemiology of virus incidence. Thus, in papaya ring spot virus (PRSV), the rapid spread of virus occurs within a short period when the vector population is high and it appears that the fresh incidence of PRSV coincided with number of aphids trapped 2 weeks prior to infection suggesting a strong link between aphid vectors and PRSV incidence (Krisna Kumar et al. 2010). Similarly, to understand ecological factors mediating the spread of insect-borne plant pathogens, vector species for these pathogens need to be identified, e.g., different strains of grapevine leaf roll virus were found to be borne by two mealybug, *Planococcus ficus* Signoret and *Pseudococcus longispinus* (Tsai et al. 2010).

Transmission Specificity of Plant Viruses by Vectors

Virus transmission by a vector is often characterized by some degree of specificity. Numerous studies suggest the involvement of a virus–ligand interaction in transmission specificity. The coat protein (CP) and its derivatives and nonstructural proteins, such as a helper component (HC) or a transmission factor, are major viral determinants of transmission specificity. The CP or its derivatives, in the case of *Luteoviruses*, *Cucumber mosaic virus* (CMV), *Cucumber necrosis virus* (CNV), and Grape vine fan leaf virus (GFLV), and a HC or a transmission factor in *Potyviruses*, *Caulimoviruses*, and *Waikaviruses* have a profound role in transmission specificity (Andret-Link and Fuchs 2005).

Cross Protection

A strain of propogative virus already present in a vector may cross protect that vector against acquisition and transmission of second strain., e.g., Aster yellows and corn stunt.

Ingestion–Egestion Mechanism of Transmission

According to Harris (1977), the transmission process is based in epidermal and intracellular zones of the host tissue. The insect vectors during their sap sampling and host selection routine, take in the cell sap or protoplasm to their fore alimentary canal and in case the host is infected, the virus-laden material (cell sap or protoplasm) contaminate the free alimentary canal. This is the ingestion step. The virus gets transmitted and the transmission cycle is completed in the next step, all or part of the virus-infected material is egested during subsequent sap sampling probes by the same vector in healthy plants.

Horizontal and Vertical Transmission

Spread of viruses by vector in a field or in different areas is horizontal transmission while transversal transmission is the vertical transmission of viruses.

Effect of Viruses on Vectors

Harmful Effects

Deleterious effects induced by viruses and mollicute like organism (MLO) are identical and can be grouped under four categories.

Cytopathological and Histological Effects

Fat bodies take up different shapes and their cells contain scanty cytoplasm, decreased carbohydrates and numerous vacuoles. Their nuclei become stallate or sharply irregular, sometimes they first enlarge and then shrink and amount of their DNA and RNA may damage. Mycetome undergoes premature degeneration in *Colladonus montanus* infected with western X MLO and become abnormally hardened in *Nephotettix* sp. infected with rice dwarf virus.

Effect on Age of Vectors

Cytopathological and histopathological effects are liable to reduce life span of infected vectors. Mean adult longevity of *N. cincticeps* infected with rice dwarf virus decreased from 29 days life of healthy insects to 20 days. Simultaneously, nymph mortality increased from 21.7 to 35 %.

Effect on Reproductive Capacity of Vectors

Viruses and MLO cause partial to total reduction in their egg laying capacity. *N. cincticeps* infected with rice dwarf virus lays only 32–72 % of eggs laid by healthy leaf hoppers.

Metabolic Effects

Viruliferous insects may show increased or decreased respiration of *N. cincticeps* infected with rice dwarf virus.

Beneficial Effects

Virus-infected plant serves as better host for aphids than healthy plant. Aphids fed on infected plant in certain cases have longer life span, greater egg laying capacity, early attainment of adulthood and rapid breeding. *A. fabae* breeds more rapidly on sugar beet plant infected with beet yellow mosaic virus than on healthy beets. Similarly, maturation of aphids as winged forms (alatae) was favored when cereal grain aphids, *Sitobion avenae* F. and *Rhopalosiphum padi* (L.), were reared from birth on barley yellow dwarf virus (BYDV)-infected oats or barley (Gildow 1983).

Management of Vectors of Virus Diseases

Most approaches to control the vectors of virus diseases are aimed at eradicating or altering one or more of the primary participants in the transmission process (vector, virus, and host plant) or at preventing their coming together. Knowledge of virus–vector relationship is essential to devise suitable measures against vector-borne diseases (Watson and Plumb 1972; Basu and Handa 1987). Broadly, management of vectors of viral diseases can be done by adopting cultural control, biological control, and chemical control.

Cultural Control Methods

The nonchemical methods of control are becoming increasingly popular due to limitations of pesticide in preventing disease spread, growing problem of insect resistance, and recent awareness of pollution problems. The control of vectors of plant virus diseases by cultural practices is not new. These methods include:

1. Use of disease-free seeds, seedings and tubers for initial sowing/planting
2. Removal of weeds, volunteer crops, and crop residues being alternate source for virus and vector
3. Vector avoidance by crop rotation, crop isolation or growing barrier crops
4. Use of reflective surfaces as crop mulches to deter vectors from alighting on the crops
5. Manipulating plant distribution, density and field size to minimize vector population and there by checking disease spread

6. Suitable adjustment of planting and harvesting dates to avoid high-density populations of vectors.
7. Breeding vector-resistant cultivars

Biological Control Methods

Although biological control by introduction of predators and parasites of vectors is attractive for economic and environmental reasons, it is not at present very successful in virus control.

Chemical Control Methods

Use of chemicals remains the basic control measure for vectors of virus diseases but for maximum economic returns and in order to use them only when needed, the epidemiology of the virus and its vector needs detailed study.

Insecticides, particularly the systemic ones, can sometimes efficiently control the spread of circulative viruses because long acquisition and inoculation feedings are required for transmission. On the other hand, pesticides hardly affect the spread of NP viruses and may rather aggravate the situation by enhancing vector management and more sap sampling.

Mineral and vegetable oils and milk lipids might inhibit NP transmission by modifying the probing behavior responsible for transmission. There, the physico- and electrochemical properties might insulate the secondary transduction systems of insect feeding apparatus and inhibit sap sampling (Simmons et al. 1977).

To prevent virus spread by insect, integrated programmes have to be developed based on an understanding of the biology and phenology of the vectors and means of perpetuation of viruses, adopting cultural control measures and minimum application of the proper pesticidal formulations at carefully chosen times.

Integrated Management of Aphid Vectors *vis-a vis* PRSV (P)

The field experiment consisting of eight treatments including control as well as synthetic and plant-based chemicals with maize as border crop was laid out in randomized block design at I. A. R. I. Regional Station, Pune. The border crop of maize was raised 15–20 days before transplanting of papaya seedling. Starting 15 days after transplanting till commencement of flowering of plants, four applications of treatments as foliar sprays were carried out at a fortnight interval. Observations were recorded on 3rd, 7th and 14th day after each application for number of migratory aphids on five randomly selected plants per plot from which mean number of aphids per plot was worked out separately for each application as well as pooled data. Similarly, numbers of plants showing PRSV (P) symptoms were counted in each plot at weekly interval to get the percentage of disease incidence after each application as well as pooled data.

Management of Aphid Vectors on Papaya

After all the spray applications (Table 1), except second application, the different treatments showed significant effect in checking the aphid incidence on papaya plants as compared to control plot receiving no spray application. Plants in the plots with border crop of maize and treated with fortnightly alternate sprays of dimethoate (0.05 %) and azadirachtin (Nimbicidin 0.03 EC 4 ml/lt) were most promising in reducing the aphid population on papaya plants. The average effect of all the four sprays indicated significant reduction in aphid incidence by different treatments as compared to untreated control having 1.33 numbers of aphids per plant. The most significant treatment of maize as a border crop with

Table 1 Management of aphid vectors on papaya

Treatments	Mean number of survival aphids per plant				
	1st spray	2nd spray	3rd spray	4th spray	Pooled
Thiamethoxam 0.005 %	0.73	0.31	0.76	0.44	0.56
Imidacloprid 0.01 %	0.87	0.20	0.58	0.33	0.51
Acetamiprid 0.005 %	0.56	0.33	0.69	0.24	0.46
Buprofezin 0.025 %	0.91	0.65	1.36	0.33	0.81
Dimethoate 0.05 %	1.60	0.38	0.87	0.36	0.80
Azadirachtin 4 ml/L (Nimbicidin 0.03 EC)	0.89	0.42	0.76	0.29	0.59
Border crop of maize + fortnightly alternate spray of dimethoate 0.05 % and azadirachtin 4 ml/L (Nimbicidin 0.03 EC)	0.29	0.20	0.27	0.13	0.22
Untreated	1.67	0.73	1.76	1.16	1.33
S.E.m.±	0.18	0.13	0.14	0.12	0.07
C.D. at 5 %	0.54	N.S	0.44	0.35	0.21

NS nonsignificant

Table 2 Management of PRSV (P) on papaya

Mean incidence of PRSV (P) (%)					
Treatments	1st spray	2nd spray	3rd spray	4th spray	Pooled
Thiamethoxam 0.005 %	8.33	25	64.58	80	44.48
	(16.60[a]	(29.22)	(54.44)	(63.93)	(43.01)
Imidacloprid 0.01 %	6.67	31.67	61.64	76.48	44.12
	(14.76)	(33.24)	(51.71)	(61.96)	(42.08)
Acetamiprid 0.005 %	5.00	28.33	63.33	76.67	43.33
	(7.60)	(30.38)	(53.39)	(66.15)	(41.75)
Buprofezin 0.025 %	11.67	43.33	73.15	82.92	52.77
	(19.90)	(41.07)	(59.73)	(66.34)	(47.42)
Dimethoate 0.05 %	10	35	68.33	81.48	48.70
	(18.05)	(35.59)	(56.33)	(65.25)	(44.44)
Azadirachtin 4 ml/L (Nimbicidin 0.03 EC)	10	33.33	66.67	81.67	47.92
	(18.05)	(34.92)	(55.01)	(65.00)	(44.22)
Border crop of maize + fortnightly alternate spray of dimethoate 0.05 % and azadirachtin 4 ml/L (Nimbicidin 0.03 EC)	5.00 (10.45)	11.67 (16.45)	25.00 (28.45)	48.33 (44.03)	22.50 (28.14)
Untreated	11.67	51.67	90	100	64.58
	(23.85)	(45.90)	(72.53)	(90)	(54.75)
S.E.m. ±	3.90	7.11	6.28	4.77	2.71
C.D. at 5 %	N.S.	N.S.	19.05	14.47	8.22

NS non significant

[a]Figures in parentheses are arc sine transformed values

alternate sprays of dimethoate and azadirachtin (0.22 aphids/plant) was followed by other effective treatments of acetamiprid (0.005 %), imidacloprid (0.01 %), thiamethoxam (0.005 %), and azadirachtin (4 ml/lt).

Management of Incidence of PRSV (P)

As observed after each spray application (Table 2), all the treatments showed significant reduction in disease incidence caused by PRSV on papaya plants in the field as compared to control plot. The pooled observations of all the four sprays indicated significant reduction in the incidence of disease by different treatments as compared to the untreated control showing 64.88 percent disease incidence. Plants in the plots having maize as a border crop and subsequently sprayed with dimethoate (0.05 %) and azadirachtin (4 ml/lt) alternately at fortnightly interval recorded lowest incidence of the virus, (22.50 %) and thus was significantly superior over all other treatments in checking the disease incidence. The other remaining treatments in order of their effectiveness were acetamiprid (0.005 %), imidacloprid (0.01 %), thiamethoxam (0.005 %), azadirachtin (4 ml/lt), dimethoate (0.05 %), and buprofezin (0.025 %) showing 43.33, 44.12, 44.48, 47.92, 48.70, and 52.27 percent incidence of disease caused by PRSV (P). Kalleshwarswamy et al. (2009) also reported that aphid vector management using timely chemical intervention as a component has a place in integrated management of PRSV on papaya.

References

Amos J, Hatton RG, Knight RC, Massee AM (1927) Experiments in the transmission of reversion in black currants. Ann Rep East Malling Res Station (II Suppl):, 1925 pp 126–150

Andret-Link P, Fuchs M (2005) Transmission specificity of plant viruses by vectors. J Plant Path 87(3):153–165

Astier S, Albouy J, Maury Y, Lecoq H (2001) Principes de Virologie Végétale. INRA, Paris, France

Basky Z (1986) Relationship between activity of virus vector aphids and virus incidence in seed cucumber fields. Acta Phytopathologica et Entomologica Hungarica 21 (1-2):29–33

Basu AN, Handa A (1987) Aphid transmission of plant viruses in India. J Aphidology 1(1&2):5–17

Bradley RHE (1964). Aphid transmission of stylet-borne viruses. In Corbett MK, Sisler HD (eds) Plant virology. University of Florida Press, Gainesville , pp 141-174

Brown JK (1994) Current status of *Bemisia tabaci* as a plant pest and virus vector in agro-ecosystem worldwide. FAO Plant Prot Bull 42(1/2):3–31

Doolittle SP (1920) The mosaic disease of cucurbits. USDA Bull 879, pp 69

Forbes AR, Mac Carthy HR (1969) Morphology of the Homoptera, with emphasis on virus vectors. In: Maramarosch K (ed) Viruses, vector and vegetation. Interscience, New York, pp 211-234

Gibbs A, Harrison B (1976) Plant virology: the Principles. Edward Arnold Pub. London , UK

Gildow FE (1983) Influence of barley yellow dwarf virus–infected oats and barley on morphology of aphid vectors. Phytopathology 73(8):1196–1199

Harris KF (1977) An ingestion-egestion hypothesis of non circulative virus transmission. In: Harris KF, Maramosch K (ed) Aphids as virus vectors. Academic Press, New York pp 167-227

Harris KF (1981) Arthopods and nematodes vectors of plant viruses. Ann Rev Phytopath 19:391–420

Hewitt WB, Raski DJ, Goheen AC (1958) Nematode vector of soil-borne fanleaf virus of grapevines. Phytopathology 48:586–595

Kalleshwarswamy CM, KrishnaKumar NK, Dinesh MR, Chandrashekar KN, Manjunatha M (2009) Evaluation of insecticides and oils on aphid vectors for the management of *Papaya ring spot virus* (PRSV). Karnataka J Agric Sci 22(3 Spl issue):552–553

Kennedy JS, Day MF, Eastop VF (1962) A conspectus of aphids as vectors of plant viruses. Commonwealth Institute of Entomology, London , pp 114

Krishna Kumar NK Singh HS Kalleshwaraswamy CM (2010) Aphid (Aphididae: Homoptera), vectors of *Papaya ring spot virus* (PRSV), bionomics, transmission efficiency and factors contributing to epidemiology. Acta Hort 851:434–442

Kunkel LO (1922) The corn mosaic of Hawali distinct from sugarcane mosaic. Phytopathology 17:41

Matthews REF (1970) Relationship between plant viruses and invertebrates. In Plant virology. Academic Press, New York, pp 554-555

Miles PW (1968) Insect secretions in plants. Ann Rev Phytopathol 6:137–164

Mora-Aguilera G, Nieto Angel D, Campbell CL (1993) Development of a prediction model for Papaya ringspot in Vera Cruz, Mexico. Plant Dis 77:1205–1211

Muniyappa V (1980) Viral diseases transmitted by whiteflies. In: Harris KF, Maramosch K (eds) Vectors of plant pathogens. Academic Press, New York

Natesan HM, Muniyappa V, Swaanson MM, Harrison BD (1996) Host range, vector relation and serological relationship of cotton leafcurl virus from South India. Ann Appl Bio 2:233–244

Oralando A, Silberschmidt K (1946) Studies on the natural dissemination of the virus on infectious chlorosis of the Malvaceae (Abutilon virus 1 Baur.) and its relation with the insect vector, *Bemisia tabaci* (Genn.) (Homoptera-Aleyrodidae). Arquiv Inst Biol (Sao-Paulo) 17:1–36

Simmons JN, Mc Lean DL, Kinsey MG (1977) Effects of mineral oil on the probing behavior and transmission of stylet borne viruses by *Myzus persicae*. J Econ Entolmol 49:301–315

Sylvester ES (1980) Circulative and propogative virus transmission by aphids. Ann Rev Entomol 30:71–78

Takata K (1895) Results of experiments with stunts disease of rice. J Jpn Agric Sci 171:1–4

Teakle DS (1960) Association of olpidium-brassicie and tobacco necrosis virus. Nature 188(13):431–432

Tsai C.-W., Rowhani A Golino DA Daane KM Almeida RPP (2010) Mealybug transmission of grapevine leaf roll viruses: an analysis of virus–vector specificity. Phytopathology 100(8):230–234

Watson MA, Plumb RT (1972) Transmission of plant-pathogenic viruses by aphids. Ann Rev Entomol 17:425–452

Vectors of Plant Viruses of Crop Plants in Southeast Asia

N. Nagaraju, A. S. Padmaja, G. Basana Gowda and R. N. Pushpa

Abstract

Plant viruses are transmitted by aphids, whiteflies, leafhoppers, plant hoppers, thrips, mites, fungi, and nematodes mainly from one host plant to another. These have been classified as non-persistent, semi-persistent, and persistent, depending on the length of the period the vector that can harbour infectious particles, which can range from minutes to hours (non-persistent) to days (semipersistent) and to lifetime and even inheritance by the insect progeny (persistent). Aphids are the vectors for the nonpersistent viruses. Some aphids, leaf hoppers, thrips, and whiteflies are the vectors for different viruses that fit in the persistent category.

Keywords

Insects· Plant viruses · Vectors · Virus–vector relationship

Introduction

Many plant viruses cannot spread to other plants without the help of seeds, tubers, pollen, and other propagating materials. However, importantly other major external factor that spread viruses is a vector. Vectors feed on the infected plant, become contaminated or infested with the virus, then move to healthy plants and infect them by feeding. Arthropods (insects and mites) are the most common plant virus vectors, though nematodes and primitive soil microorganisms can also transmit these pathogens (Table 1). Insect and other noninsects transmit 76 % of all the diseases. However, hemipteran insects transmit a majority of the vectored viruses (55 %).

Watson and Roberts (1939) coined the terms non-persistent and persistent viruses, as a first attempt to categorize and understand plant virus vector transmission relationships. The nonpersistent viruses had very short retention times (>12 h) in vectors in contrast to persistent viruses where retention time was 12 h to indefinite. Nonpersistent viruses were efficiently transmitted after relatively brief (<5 min) acquisition access period (AAP) and inoculation access period (IAP), while persistent viruses required longer AAP and IAP, and optimum transmission efficiencies were associated with feeding (Table 2). Some of the

N. Nagaraju (✉) · A. S. Padmaja · G. Basana Gowda · R. N. Pushpa
University of Agricultural Sciences GKVK, 560 065 Bangalore, Karnataka, India
e-mail: nagaraju63kgere@yahoo.co.in

A. K. Chakravarthy (ed.), *New Horizons in Insect Science: Towards Sustainable Pest Management*,
DOI 10.1007/978-81-322-2089-3_16,

Table 1 Vectors and plant viruses in relation to transmission. (Hogenhout et al. 2008; referred to the International Committee on Taxonomy of Viruses (ICTV), http://phene.cpmc.Columbia.edu/Ictv/index.htm)

Vector taxa	Vector groups	Virus groups					
		Icosahedral particles RNA genome	Rod-shaped particles RNA genome	DNA genome	Enveloped particles RNA genome	Total	%
Hemiptera	Aphids	26	153[a]	13	5	197	28
	Whiteflies	–	13	115[b]	–	128	18
	Leaf hoppers	8	–	15	3	26	4
	Plant hoppers	10	4[c]	–	4	18	3
	Other Hemiptera	–	8	5	–	13	2
Thysanoptera	Thrips	2	–	–	14	16	2
Coleoptera	Beetles	50	1	–	–	51	7
Acari	Mites	10	9	–	–	10	1
Nematoda	Nematodes	45	3	–	–	48	7
Mycota	Fungi	8	16	–	–	24	3
	No identified vectors	84	60	19	3[d]	166	24
	Total	233	268	167	30	697	
	%	33	39	24			

[a] Includes 110 virus species of the genus Potyvirus, family Potyviridae
[b] Virus species of the genus Begomovirus, family Geminiviridae; these are all tenuiviruses that have multiple shapes
[c] Tenuiviruses that have multiple shapes
[d] These viruses probably have insect vectors

Table 2 Transmission characteristics and timing of plant viruses transmitted by hemipteran insects. (Hogenhout et al. 2008; referred to the International Committee on Taxonomy of Viruses (ICTV), http://phene.cpmc.Columbia.edu/Ictv/index.htm)

Biological characteristics	Nonpersistent stylet borne	Semipersistent foregut—borne	Persistent circulative	Persistent propagative
AAP and IAP	Seconds, minutes	Minutes, hours	Hours, days	Hours, days
Latent period	None	None	Hours, days	Days, weeks
Retention time in the vector	Minutes, lost after moulting	Hours, lost after moulting	Days, weeks	Lifespan of the vector
Presence in vector's hemolymph	No	No	Yes	Yes
Multiplication in vector	No	No	No	Yes
Transovarial transmission	No	No	No	Often

AAP acquisition access period, *IAP* inoculation access period

important and major viral diseases can be identified through the visual symptoms (Fig. 1).

Plant viruses cause severe yield losses to the cereal, vegetable, fruit, floral industries and substantially lessen the quality of crop products. Due to virus infection, losses of over US$1.5 billion are reported in rice in Southeast Asia (Hull 2002), and estimates of losses have been calculated at US$63 million in apple in the USA (Cembali et al. 2003), and over US$20 million in potato in the UK (Hull 2002). Tomato spotted wilt virus (TSWV) alone is responsible for losses of over US$1 billion in vegetable and ornamental crops. TSWV is transmitted by thrips and has the largest host range of any plant virus infecting more than a thousand plant species from 84 families (Parella et al. 2003).

Most plant viruses are absolutely dependent on a vector for plant-to-plant spread. Although a number of different types of organisms are vector for different plant viruses, phloem-feeding hemipterans are the most common and transmit

Fig. 1 Important plant viral diseases

the great majority of plant viruses. The complex and specific interactions between the hemipteran vectors and viruses depend on two general strategies, the capsid and helper strategies. Both strategies are required for transmission by aphids in a nonpersistent and semipersistent manner (Ng and Falk 2006).

More than 200 plant viruses are transmitted by hemipteroid insects beginning a few hours or days after acquisition and for upto life of the insects, i.e., in a persistent-circulative or persistent-propagative mode. These viruses move through the insect vector, from the gut lumen into hemolymph or other tissues and finally into the salivary glands, from which these viruses are introduced back into plant host during insect feeding. The movement and/or replication of the viruses in the insect vectors require specific interactions between viruses and vector components (Hogenhout et al. 2008).

Many plant viruses are transmitted from plant to plant in nature by invertebrate vectors and the plant virus vector interactions are very specific (Muniyappa and Veeresh 1986; Hohn 2007). Members of the class, Insecta and Arachnida from the phylum Arthropoda and the members of the order Dorylaimida from of the phylum Nematoda are the major vectors of plant viruses. The *Homopterans* feed by sucking sap from plants and are numerically the most important suborder containing plant virus vectors. The tomato yellow leaf curl virus tomato yellow leaf curl virus (TYLCV) and its whitefly vector *Bemisia tabaci* Genn. have been of increasing importance recently in many regions with tropical, subtropical and arid Mediterranean climates due to a rapid expansion in geographic distribution and host range of the virus and its vector (Pico et al. 1996).

In this review, vectors that transmit viral diseases on important crops are discussed (Table 3). Viruses cause serious diseases of crop plants reducing both quality and quantity of final produce and keeping quality of plant produce. Some of the important viral diseases are: mosaics in chilli and cucumber, tomato mosaic, tomato spotted wilt, potato virus Y in potato, papaya ringspot virus in papaya and cucurbits, citrus tristiza, chilly leaf curl, banana bunchy top, etc. Aphids constitute the largest vector group transmitting plant viruses compared to whitefly, thrips leafhoppers, psyllids and mites. The number of transmission patterns that have evolved are unique in their variety compared to those found with other vector groups. In India, the viruses transmitted by aphids, mainly *Aphis craccivora, Myzus persiacae, Aphis gossypii,* are more important in transmitting many viruses compared to other South Asian country infecting varied crop plants ranging from cereals, vegetables, fruit, and ornamental crops. mMst of them are major viruses which are not so important in countries like Pakistan, China, and neighboring countries.

Additionally *Rhopalosiphum padi, R. maidis and Aphis gossypii* are found important in transmitting sugarcane mosaic and chilli veinal mottle virus (ChiVMV). The occurrence of *M. persicae* in many South Asian countries on transmission of several viruses in nonpersistent manner can be observed. However, viruses like tomato leaf curl virus, yellow mosaics in legumes, Bhendi are transmitted through whitefly in a semipersistent manner. *The tomato yellow leaf curl virus* (TYLCV) and *tomato yellow leaf curl China virus* (TYLCCNV) transmit through *B. tabaci* are major viral diseases in China and Nepal, and differences can be found in transmission, symptom expression, and molecular description compared to Indian ToLCV. In recent years, tomato spotted wilt (Tospo) virus in tomato, capsicum and lettuce has become serious. In general, viruses do not exist and survive in nature without another living organism. In Thailand, *Ceratothripoides claratris* is a major thrip vector which is transmitting capsicum chlorosis virus in capsicum and other solanaceous crops causing huge loss. Leaf hoppers are found to be successful transmitters of many viral diseases in cereals compared to vegetables particularly phytoplasma diseases viz. southern rice black-streaked dwarf virus, rice gall dwarf virus, grassy stunt virus in a persistent propagative manner. Mites *Aceria tulipae* and *Aceria tosichella* are also important virus vectors, transmitting wheat streak mosaic and pigeon pea sterility mosaic virus. Plant viruses can be transmitted by insects in many ways. These have been classified as nonpersistent, semipersistent, and persistent, depending on the vector that can harbor infectious particles, ranging from minutes to hours (non-persistent) to days (semipersistent) and to life-time and even inheritance by the insect progeny (persistent) (Hohn 2007).

Aphids Aphids are among the most destructive insect pests on cultivated plants in temperate regions. The damage they cause to plants has made them enemies of farmers and gardeners the world over, though from a zoological standpoint they are a highly successful group of organisms. Their success is due in part to the asexual reproductive capabilities of some species. About 4400 species of 10 families are known. Historically, far fewer families were recognized, as most species were included in the family Aphididae. Around

Table 3 Vectors of plant viruses on important crop plants of South Asian countries

Vector species	Crops	Viruses	Major	Minor	References
Aphids					
India					
Pentalonia nigronervosa	Cardamom	Cardamom mosaic	Major	–	Deshapande et.al. 1972
Aphis craccivora Koch and *M. persicae Sulz*	Beans	Common bean mosaic	Major	Minor	Muniyappa 1976
Toxoptera citricida	Citrus	Citrus tristeza virus		Minor	Balaraman and Ramakrishnan, 1977a, 1977b
Aphis gossypii and *Myzus persicae*	Bell pepper	Pepper vein banding virus	Major		Nagaraju and Reddy 1980, 1981
A. craccivora, *A. gossypii,* and *M. persicae*	Chilli	Cucumber mosaic, pepper vein banding, pepper veinal mottle, potato virus Y and tobacco etc.	Major		Bidari and Reddy 1990
M. persicae	Chilli, brinjal, tomato	Cucumber mosaic virus			Kiranmai et al. 1997
M. persicae and *A. craccivora*	Chenopodium album	Garlic mosaic virus–D isolate		Minor	Ghosh and Ahlawat 1997
A. gossypii, *A. craccivora,* and *M. persicae*	Sunflower	Mosaic		Minor	Nagaraju et al. 1997
A. craccivora, *A. gossypii,* and *M. persicae*	Cowpea	Cucumber mosaic virus		Minor	Nagaraju and Keshavamurthy 1994; Mahalakshmi et al. 2008; Nagaraju and Keshavamurthy 1997
Myzus persicae	Soybean	Soybean mosaic virus		Minor	Balgude and Sawant 2012
M. persicae, *A. gossypii,* and *A. craccivora*	Papaya	Papaya ring spot virus	Major		Lakshminarayana Reddy et al. 2007; Sreenivasa Rao Gude et al. 2008; Kalleshwaraswamy and Krishna kumar 2008; Krishnakumar et al. 2010; Singh Vimla and Singh Devendra 2010
A. craccivora, *A. gossypii*, *M. persicae,* and *M. Nicotianae*	Gherkin	Papaya ring spot virus strain P (PRSV-P) and cucumber green mottle virus (CGMV)	Major		Rashmi et al. 2005a and 2005b
M. persicae (Sulz.)	Marrow	Watermelon mosaic virus		Minor	Raychaudhuri and Varma 1977
A. craccivora and *A. gossypii*	Banana	Cucumber mosaic virus	Major		Dheepa and Paranjothi 2010
China					
Rhopalosiphum padi	Barley	Barley yellow dwarf virus	Major		Du et.al. 2007
Pakistan					

Table 3 (continued)

Vector species	Crops	Viruses	Major	Minor	References
A. gossypii	Chilli and tobacco	Chilli veinal mottle virus (ChiVMV)		Minor	Shah et al. 2008
A. gossypii	Chilli	Pakistani isolate of chilli veinal mottle potyvirus	Major		Shah et al. 2008
Rhopalosiphum padi and *R. maydis*	Corn	Sugarcane mosaic virus	Major		Hasan et al. 2003
Nepal					
M. persicae	Sweet bean in Nepal	Mosaic		Minor	Pudashini et al. 2013
Malayasia					
Aphis gossypii	Citrus	Citrus tristeza virus (CTV)		Minor	Ayzapour 2013
Japan					
M. persicae and *A. craccivora*	Chenopodium album	Garlic mosaic virus–D isolate	Major		Noda and Inouye 1989
Asia and Pacific					
Pentalonia nigronervosa Coq	Banana	Bunchy top	Major		Magnaye and Valmayor 1995
M. persicae and A. gossypii	Potato	Potato virus Y	Major		Sadeghi et al. 2008
M. persicae and *A. gossypii*	Cowpea	Cowpea mottle virus		Minor	Hajiabadi 2012
M. persicae	Egg plant	Eggplant blister mottled virus (EBMV)		Minor	Al-Ani et al. 2011
Whitefly					
India					
B. tabaci Genn	Tomato	Leaf curl	Major		Muniyappa 1980; Muniyappa and Veeresh 1984. Saikia and Muniyappa 1989; Muniyappa et al. 2000; Banks et al. 2001; Rekha et al. 2005; Shankarappa et al. 2007
	Dolichos	Dolichos yellow mosaic virus	Major		Manjunath and Muniyappa 1995; Varma and Malathi 2003; Maruthi et al. 2006
	Bhendi	Yellow vein mosaic	Major		Khan and Mukhopadhyay 1986; Bhagabati and Goswami 1972; Jose et al. 2003; Karri and Acharyya 2012
	Malvastrum coromandelianum, a weed host	Mosaic		Minor	Harrison et al. 1991
	Cassava	Mosaic	Major		Mathew and Muniyappa 1991

Table 3 (continued)

Vector species	Crops	Viruses	Major	Minor	References
	Cotton	Leaf curl		Minor	Nateshan et al. 1996; Khan and Ahmad 2005
	Croton	Croton yellow vein mosaic		Minor	Mandal and Muniyappa 1991
	Hibiscus	Leaf curl	Major		Rajeshwari et al. 2005
	Pumpkin	Yellow vein mosaic			Jayashree et.al. 1999; Muniyappa et al. 2003
	Jatropha	Mosaic		Minor	Aswatha Narayana et al. 2007
	Tobacco	Leaf curl	Major		Valand and Muniyappa 1992
	Cowpea	Mild mottle virus		Minor	Muniyappa and Reddy 1983
	Sunflower	Leaf curl	Major		Govindappa et al. 2011
	Chilli	Leaf curl	Major		Senanayake et al. 2012
	Mungbean	Mungbean yellow mosaic virus	Major		Manjunath et al. 2012
	Pole bean	Yellow mosaic	Major		Jyothi et.al. 2013
China *B. tabaci* Genn	Tomato	TYLCV and tomato yellow leaf curl China virus (TYLCCNV)	Major		Wang et al 2010; Pan et al. 2012; Pan et al. 2013; Liu et al. 2013
Pakistan *B. tabaci* Genn	Cotton	Leaf curl virus	Major		Briddon and Markham 2000
Nepal *B. tabaci* Genn	Tomato	TYLCV	Major		Ghimire et al. 2001; Ajlan et al. 2007
Bangladesh *B. tabaci* Genn	Jute	Jute leaf mosaic virus (JLMV)	Major		Dastogeer et.al. 2012
Indonesia *B. tabaci* Genn	Tobacco	Leaf curl	Major		Aidawati et al. 2002
Middle East *B. tabaci* Genn	Tomato	Leaf curl virus	Major		Cohen and Harpaz 1964
Thrips					
India					
Thrips palmi	Groundnut	Bud necrosis	Major		Vijaylakshmi 1994 and Wightman et al. 1995
Thrips palmi	Chilli, tomato, onion	Necrosis	Major		Adkins et al. 2010
Thailand					
Ceratothripoides claratris	Capsicum	Capsicum chlorosis virus	Major		Premachandra et al. 2005

Table 3 (continued)

Vector species	Crops	Viruses	Major	Minor	References
Iran					
Thrips tabaci	Soybean	Tomato yellow ring virus		Minor	Ali Raza Golnaraghi et al. 2007
IV. Plant hopper					
Iran *Agallia vorobjevi*	Egg plant	Eggplant mottled dwarf virus (EMDV)		Minor	Babaie and Izadpanah 2003
Pakistan					
Orosius albicinctus	Sesamum	Phyllody	Major		Akthar et al. 2009
	Chick pea	Chickpea chlorotic dwarf virus	Major		Akthar et al. 2011
Nepal					
Nephotettix virescens and *N. nigropictus*	Rice	Rice virus	Minor		Dahal 1998
Taiwan					
Laodelphax striatellus	Rice	Rice stripe virus		Minor	Hsieh 1974
China					
Sogatella furcifera Nilaparvata lugens, Laodelphax striatellus	Rice	Southern rice black-streaked dwarf virus		Minor	Lingling Pu et al. 2012; Zhou 2013
Recilia dorsalis	Rice	Rice gall dwarf virus	Major		Chen et al. 2013
Vietnam					
Nilaparvata lugens	Rice	Grassy stunt virus		Minor	Ta et al. 2013
V. Psyllids					
Taiwan					
Diaphorina citri *Trioza erytreae*	Citrus	Citrus greening		Minor	Huang et al. 1990; Batool et al. 2007
VI. Mite					
Aceria tosichella Keifer	Wheat	Wheat streak mosaic		Minor	Murugan et al. 2011
Aceria cajani	Pigeon pea	Sterility mosaic virus	Major		Kulkarni et al. 2002

250 species are serious pests for agriculture and forestry as well as an annoyance for gardeners.

Virus Transmission and Virus–Vector Relationship Aphids constitute the largest vector group transmitting plant viruses. Cardamom mosaic (Katte) is an important virus affecting the production of cardamom in Karnataka. The virus was transmitted by aphid, *Pentalonia nigronervosa*. There was a negative correlation between aphid population and the disease spread. The highest spread of disease occurred when aphid population was low. The spread of the disease was more associated with aphid activities rather than aphid population (Deshapande et.al. 1972).

The seed borne nature of common bean mosaic virus and its transmission by *Aphis craccivora* Koch, *A. fabae sub-sp solanella*, *A. gossypii* Glov. *Liphaphis erysimi,* and *M. persicae* Sulz. were reported by Muniyappa (1976). Citrus tristeza virus (CTV) and a virus disease resembling Hassaku dwarf in acid lime (*Citrus aurantifolia*) were transmitted efficiently by aphid, *Toxoptera citricida* Kirkaldy (Balaraman and Ramakrishnan1977a, 1977b).

Marrow mosaic virus, a strain of watermelon mosaic virus, is transmitted by *M. persicae* (Sulz.) to marrow plants in a typical stylet-borne manner. Preacquisition starving of vectors was not essential for transmission, but did increase the transmission rate. Single aphid could transmit the virus, more than 5/plant were required for 100 % transmission. A feeding period of only 30 s was needed for acquisition of the virus, but one of 2 min resulted in maximum percentage transmission (Raychaudhuri and Varma 1977).

Pepper vein banding virus on bell pepper was reported to be transmitted by two aphid species, *A. gossypii* and *M. persicae* with the transmission efficiency of 62.0 % and 75.0 %, respectively (Nagaraju and Reddy 1981). Similarly, pepper veinal mottle virus and cucumber mosaic virus also reported on bell pepper and were transmitted by the same aphids (Nagaraju and Reddy 1980).

In the absence of reliable virus-free garlic plants, *Chenopodium album* local lesion host of garlic mosaic virus–D isolate was successfully transmitted nonpersistently through *M. persicae* and *A. craccivora* (Noda and Inouye 1989). Banana bunchy top virus transmission is through its insect vector, the banana aphid (*Pentalonia nigronervosa* Coq) with minimum feeding period of 1.5–2 h on susceptible plants. An average of about 25 days incubation is necessary for the development of banana bunchy top symptoms. Nymphs are more effective vectors than mature aphids (Magnaye and Valmayor 1995).

Cucumber mosaic virus (CMV), pepper vein banding, pepper veinal mottle, potato virus Y, and tobacco etch viruses infected chilli crop through aphids namely *A. craccivora, A. gossypii,* and *M. persicae* in a nonpersistent manner as reported by Bidari and Reddy (1990). Likewise, mosaic disease was reported in sunflower crop for the first time from Karnataka (India) and they were transmitted successfully by three aphid species *viz.*, *A. gossypii, A. craccivora* and *M. persicae*. *A. craccivora* and *M. persicae transmitted* the mosaic virus in garlic nonpersistently to *Chenopodium album* (Ghosh and Ahlawat 1997).

The efficient transmitter *M. persicae* was able to transmit three cucumber mosaic virus (CMV) isolates in a nonpersistent manner to solanaceous crops. The efficiency of transmission was ranging from 67 to 70 % for CMV in brinjal, 81–85 % for CMV in chilli, and from 77–81 % for CMV in tomato (Kiranmai et al. 1997). Transmission of CMV infecting cowpea was through *A. craccivora, A. gossypii,* and *M. persicae* in a nonpersistent manner (Nagaraju and Keshavamurthy 1994; Mahalakshmi et al. 2008). Single aphid *M. persicae* was able to transmit the CMV with 3 min AAP and 5 min IAP. The transmission efficiency of vector increased with 1 h preacquisition starvation (Nagaraju and Keshavamurthy 1997).

PRSV was reported to be transmitted by several aphid species tested with transmission efficiency of more than 90 % (Lakshminarayana Reddy et al. 2007). A strain of PRSV was reported infecting several cucurbitaceous crops around Bangalore through *M. persicae, A. gossypii,* and *A. craccivora* (Sreenivasa Rao Gude et al. 2008).

Gherkin (*Cucumis anguria*) crop was found infected with two viral diseases namely, PRSV strain P (PRSV-P) and cucumber green mottle virus (CGMV) (Rashmi et al. 2005a, 2005b).

PRSV-P is transmitted by four species of aphids namely, *A. craccivora, A. gossypii, M. persicae,* and *M. nicotianae.* Whereas, CGMV could not be transmitted by any of the above aphid species. Shah et al. (2008) found efficient transmission of ChiVMV in chilli and tobacco plant when one hour starved *A. gossypii* was allowed for 2–3 min of acquisition-feeding period with 5–10 viruliferous aphids.

Single aphid of *M. persicae and A. gossypii* was able to transmit PRSV with a transmission of 56% and 53%, respectively, compared to *A. craccivora* (38%). The time required for the initiation of the first probe on inoculated test plants was significantly shorter compared to *A. craccivora.* There was a perceptible decline in transmission efficiency as the sequestration period increased, although *M. persicae* successfully transmitted PRSV after 30 min of sequestration (Kalleshwaraswamy and Krishna kumar2008; Krishnakumar et al.2010).

The natural spread of PRSV disease in the eastern Uttar Pradesh region (India) occurs by aphid vectors transmiting the disease through wounds created during sucking of sap for feeding. Five aphid vectors *viz. A. craccivora, A. gossypii, A. citricola, M.persicae,* and *Rhopalosiphum maidis*, were common in the surveyed areas. The virus–vector relationship with *M. persicae* was the most efficient transmitting, 70% of diseases within 12 days after inoculation. Aphids could acquire the virus without any preacquisition fasting, and showed a decline in transmission after 4 h of preacquisition fasting. The aphid could readily transmit PRSV after 2 min of infection feeding with an optimum transmission after 6 min. of infection feeding. The virus was totally inactivated at 4 h of postacquisition fasting. *M. persicae* ceased to be infective very soon and could infect not more than two plants in succession revealing "nonpersistent nature" stylet borne nature of PRSV (Singh Vimla and Singh Devendra 2010).

Two aphid species *A. craccivora* and *A. gossypii* are efficiently transmitted the CMV in a nonpersistent manner within 20 min of acquisition and inoculation of 10–15 min. Thirty plants of robusta were inoculated with infective *A. gossypii* and 30 plants of cowpea and commulina inoculated with infected *A. craccivora.* (Dheepa and Paranjothi 2010).Two aphid species, *M. persicae* and *A. gossypii*, transmitted the strain of potato virus Y causing egg plant mosaic in southern Iran between Turkish tobacco plants (Sadeghi et al. 2008).

Five to ten viruliferous aphids, *A. gossypii* transmitted successfully Pakistani isolate, *chilli veinal mottle virus* (ChiVMV) when they were starved for 1 h with acquisition-feeding period of 2–3 min and IAP of 1 h (Shah et al. 2008).

Symptoms characteristic of the virus eggplant blister mottled virus (EBMV) appeared when healthy eggplants exposed to group of aphids *M. persicae* precaged on virus-infected plants for 1/2 and 1 min. *A. fabae* and *B. tabaci* failed to transmit the virus. The same period is also required for inoculation of healthy plants. This indicated that *M. persicae* transmitted the virus in a nonpersistent manner (Rakib et al. 2011). The enzyme-linked immunosorbent assay (ELISA) tests showed that two aphid species, *M. persicae* and *A. gossypii,* carried cowpea mottle virus. These tests also showed weak reaction with *T. tabaci* colonies collected from infected fields, but no positive reaction was observed with *T. tabaci* colonies collected from non-infected fields (Hajiabadi 2012).

Citrus triteza virus (CTV) is a member of genus Closterovirus with long flexuous virions, monopartite genome and is vectored by aphids. Efficiency of CTV transmissibility is affected by the species of aphids, the source plant at acquisition feeding and the CTV isolate. The Brown citrus aphid (BrCA), i.e., *A. gossypii* is the most efficient vector of CTV in major citrus growing areas of Peninsular Malaysia, all citrus varieties and its hybrids including Fortunella sp., *Citrofortunella microcarpa* and Citromelo were infected with CTV in a high rate. The survey of CTV vector(s) in Peninsular Malaysia revealed that atleast there is one efficient vector *(A. gossypii)* of CTV in citrus growing areas (Ayzapour 2013).

In Nepal, sweet bean (*Lablab purpureus* L.) is an important legume crop. During 2010, sweet bean plants showed mottling and leaf deformation, severe mosaic, necrosis, downward curling of leaves, and reduction in leaf size in addition to malformation of leaves and pods with 60–70% incidence. It is transmitted in a nonpersistent manner by *M. persicae* and is also reported as seed-transmitted (Pudashini et al. 2013).

Whitefly Whitefly, *Bemisia tabaci* (Hemiptera: Aleyrodidae) transmitted viruses (WTVs) occur predominantly in the tropics where their vectors are most abundant. Several crops in Karnataka, India are infected with WTVs causing enormous losses (Muniyappa 1980; Muniyappa and Veeresh 1984. Saikia and Muniyappa 1989; Banks et al. 2001.

B. tabaci originated in the tropics and subtropics, has rapidly spread as a consequence of the international trade in flowers and other nursery stock. Because of its wide host range, rapid propagation, and superior ability to transmit virus, *B. tabaci* has become one of the most important pests in field crops worldwide. *B. tabaci* is a complex of numerous genetically distinct populations, previously referred to as biotypes and now recognized as cryptic species. There are about 24 cryptic species of *B. tabaci*, including the two most widely distributed and invasive biotypes, B and Q, hereafter referred to as B and Q whiteflies. *B. tabaci* is the only known vector of TYLCV, which seriously reduces tomato production and quality. TYLCV is a single-stranded DNA (ssDNA) plant virus in the genus begomovirus, family Geminiviridae that originated in the Middle East (Cohen and Harpaz 1964; Varma and Malathi 2003) Begomoviruses are transmitted by *B. tabaci* in a circulative manner and persist in the whitefly vector. Transmission of tomato leaf curl virus was successful when *B. tabaci* (Genn) were allowed to feed on leaf curl infected tomato for 24 h AAP and IAP with 10–15 whiteflies.

As early as 1947, mosaic on common weed, *Malvastrum coromandelianum,* was reported, and now it is whitefly transmitted begomovirus (Harrison et al. 1991). Bhendi yellow vein mosaic (Harrison et al. 1991), Cassava mosaic (Mathew and Muniyappa 1991), Cotton leaf curl (Nateshan et al. 1996), croton yellow vein mosaic (Mandal and Muniyappa 1991), Hibiscus leaf curl (Rajeshwari et al. 2005), Pumpkin yellow vein mosaic (Muniyappa et al. 2003), Jatropha mosaic (Aswatha Narayana et al. 2007), Tobacco leaf curl (Valand and Muniyappa1992), and Tomato leaf curl (Muniyappa et al. 2000) begomoviruses were transmitted by *B. tabaci* in a semipersistent manner (circulative manner).

For the first time, Cowpea mild mottle virus (CMMV) has been demonstrated transmitted by *B. tabaci* in a nonpersistent manner. *B. tabaci* adults acquired CMMV in 10 min transmitted it within 5 min to soybeans. Starvation before acquisition had no effect upon transmission, but starvation after acquisition decreased transmission frequency (Muniyappa and Reddy 1983). The Bhendi yellow mosaic virus (BYVMV) disease is caused by the whitefly (*B. tabaci*) transmitted virus complex consisting of a monopartite begomovirus BYVMV and a betasatellite molecule. The causal virus and its associated betasatellite molecule infect the crop at all the stages (Jose et al. 2003).

The infection of BYVMV to Bhendi under natural field conditions was depending on the environmental parameters, crop characteristics, and efficient vector whitefly (*B. tabaci*) population (Khan and Mukhopadhyay 1985; Bhagabati and Goswami 1972). Susceptibility of cultivars encourages its incidence in the field in the presence of the active vectors. Considering it as one of the major constraints of okra cultivation, it is essential to gather basic information to understand the nature of infection, source and gradual increase with the increase of plant age survival capacity of the virus and mode of spread among different varieties in a cropping season (Karri and Acharyya 2012).

High incidence of TYLCV in most tomato-growing pockets and yield losses of 40 % or even higher in Risingpatan, Tanahun and Kudule, of western hills of Nepal is recorded. The TYLCV vector, whitefly (*B. tabaci*), was found active throughout the crop-growing period in commercial tomato-growing pockets (*Sita et al.* 2001*)*.

In China, TYLCV and t*omato yellow leaf curl China virus* (TYLCCNV) were shown to be horizontally transmitted by both B and Q biotypes of whitefly, but transmission frequency was low (Wang et al. 2010). In 2006, TYLCV was introduced into China, approximately 10 years after the introduction of an invasive whitefly, B. tabaci B biotype. After the introduction of Q biotype into China in 2003, the prevalence and spread of

TYLCV started to accelerate and found that the two biotypes might not be equally competent vectors of TYLCV. B. tabaci B and Q biotypes needed 48 and 12 h AAP, respectively, to reach their respective maximum viral loads (Pan et al. 2012). Vector salivation is essential for TYLCV transmission.

The transmission efficiency of TYLCV by the whitefly vector, *B. tabaci* is high when it associated with the bacterial symbiont *Hamiltonella* with a high frequency with close association of bacterial symbiont *Hamiltonella* with high frequency. The *Hamiltonella* helps in retention and transmission efficiency of TYLCV helps in retention and transmission efficiency of TYLCV by the whitefly vector (Pan et al. 2013). Tomato leaf curl virus (ToLCV-Ban4) was transmitted by single whitefly, but five insects were necessary to achieve 100 % transmission. Minimum AAP and IAP were 10 min and 20 min, respectively. A latent period of 6 h was required for *B. tabaci* to efficiently infect tomato test plants. Following a 24 h AAP, the insect retained its ability to infect tomato test plants for 12 days, but not for its entire life. In one insect/one plant inoculation tests, female whiteflies were more efficient (~95 %) than males (~25 %) in transmitting the virus (Muniyappa et al. 2000). Similar virus vector relationships were found in whitefly transmitted viruses such as Croton yellow vein mosaic virus (CYVMV) (Mandal and Muniyappa 1991); ICMV (Mathew and Muniyappa 1991) and pumpkin yellow vein mosaic virus (PYVMV) (Muniyappa et al. 2003). The existence of a short incubation period and persistence of virus for 12 days in the vector suggests that the virus is circulative but not propagative. The long persistence of ToLCV in vector and ability of single whitefly to cause disease, together with intensive cultivation of tomato throughout the year and abundance of weed hosts are important factors for very high incidence of ToLCV in nature (Muniyappa et al. 2000). At least 25 *B. tabaci* adults were required for the transmission of hibiscus leaf curl virus (HLCV) from hibiscus to hibiscus with minimum AAP and IAP of 12 and 24 h, respectively (Rajeshwari et al. 2005). The minimum AAP and IAP were 30 min (10 % transmission) and 15 min (10 % transmission), respectively, and female *B.tabaci* are able to transmit TYLCV. The efficiency of transmission increased by increasing both AAP and IAP, as well as insect numbers (Ajlan et al. 2007).

The B-biotype *B. tabaci* was first observed in Kolar, Karnataka in India, during 1999 on tomato. High population of B-biotype was responsible for outbreak of tomato leaf curl virus which resulted in total failure of tomato crop (Banks et al. 2001). The B-biotype was more common than the indigenous *B. tabaci*, in locations where it had been present more than 2 years. It was found in most districtrs of Karnataka, Andhra Pradesh and Tamil Nadu, South India. (Rekha et al. 2005; Shankarappa et al. 2007).

Virus causing yellow vein disease on *Calendula officinalis* L.was transmitted from naturally infected *C.Officinalis* to healthy seedlings of *C. officinalis* through whiteflies. inoculation. A single whitefly could transmit the virus and showed 20.55 % infection and 15 whiteflies were required for 100 % transmission. The minimum acquisition access feeding period and minimum inoculation access feeding period for the present virus were 10 min and 30 min, respectively. Acquisition access feeding period and inoculation access feeding period was 100 % at 6 h and 3 h, respectively. A pre-acquisition access feeding period of 3 h or more gave 100 % transmission. Post acquisition access feeding period of whiteflies did not have any effect in increasing transmission of CYVV. As post acquisition access feeding period increased, the transmission of CYVV was gradually decreased (Bushra Afreen 2000)

[PYVMV was transmitted to healthy pumpkin plants only by *B. tabaci* and not through sap inoculation. A single viruliferous whitefly was able to cause 21.67 % infection but 100 % infection was obtained when 15 whiteflies were used per plant. The acquisition threshhold and the inoculation threshhold periods for the whitefly were 6 h and 3 h, respectively. The percentage of transmission increased with increase in both acquisition and inoculation feeding periods. A preacquisition starvation period of 3 h gave 100 % transmission, but postacquisition starvation was found to reduce the transmission efficiency (Jayashree et al. 1999).

Dolichos yellow mosaic virus (DoYMV) was transmitted poorly by *B. tabaci*. Whitefly acquired DoYMV in 6 h and transmitted in 1 h. A minimum of four *B. tabaci* adults were required for the transmission of virus with 1.6% efficiency which increased to 18.3% using 50 adults/plant after 24 h AAP and IAP (Manjunath and Muniyappa 1995; Maruthi et al. 2006). The rate of transmission of Jatropha mosaic virus (JMV) by *B. tabaci* was low (4%) when 5 or 10 adults were used for inoculation. About 25 adults that were given 24 h AAP and 48 h IAP transmitted the virus to an extent of 40% (Aswatha Narayana et al. 2007).

A single whitefly was capable of transmitting chilli leaf curl virus, and eight or more whiteflies/plant resulted in 100% transmission. The minimum AAP and IAP were 180 and 60 min, respectively. The virus persisted in whiteflies for up to 5 days post acquisition (Senanayake et al. 2012). The infected plants develop partial or complete brilliant golden yellow mosaic on leaves, stunted growth, less number of pods with reduced pod size. The rate of transmission of virus with 10 viruliferous whiteflies (*Bemisia tabaci)* was 100% at 24h acquisition and inoculation access, and latent period for symptom expression was between 7 and 15 days after inoculation. Single whitefly per plant could transmit the disease up to 40%. (Jyothi et al. 2013)

Thrips Thrips are small to minute insects with an adult body size ranging in most species from about 0.5 to 5 mm and adults usually have four slender wings. About 5000 species of thrips have been recognized represents a little over 60% of the world total. As plant disease perspective all known vectors of tospoviruses are only in two genera of thirps: *Franklieniella* and *Thrips.* Among several thrips, *Thrips palmi* Karny commonly known as melon thrips is believed from Southeast Asia (Mound 1996). *T. palmi* Karny is a phytophagous pest, especially of species in the Cucurbitaceae and Solanaceae, such as sweet pepper, tobacco, cucumber, watermelon, melon, squash, and pumpkin and potato. However other crops such as cotton, cowpea, pea, common bean, soybean, sunflower, and sesame are also hosts. *T. palmi* lifecycle from egg to adult lasts 17.5 days at 25 °C. The adults emerge from pupae in the soil and migrate to leaves and flowers of host plants, where they can be found in pockets, cracks or crevices. Eggs are laid on the host. The second larva goes into the soil, where it develops and pupates, thus completing the life cycle. *T. palmi* transmits Groundnut Bud Necrosis Virus (GBNV) (Vijaylakshmi 1994; Wightman et al. 1995. In India Adkins et al. (2010) reported major thrips species viz., *T. palmi, T. tabaci, F. schultzei, Scirtothrips dorsalis* and *T. hawaiiensis* in crops like tomato, chilli, pepper and onion.

In Thailand, a new vector of capsicum chlorosis virus, i.e., adult *Ceratothripoides claratris* showed 69% transmission to tomato after acquiring the CaCV (isolate AIT) as freshly emerged (<1 h) first-instar larvae. However, when just molted (<1 h) second-instar larvae acquired the virus, the percentage of adult transmitters significantly decreased (48%). The transmission efficiency of up to 47% was detected with second instar larvae of *C. claratris* which had acquired the virus as freshly emerged first-instar larvae (Premachandra et al. 2005). Transmission efficiency did not significantly differ between adult males and females, irrespective of the larval stage at which the virus was acquired. Highest transmission efficiency for CaCV was recorded in adult *C. claratris* derived from second-instar larvae collected from infected tomato plants in a greenhouse. Lowest transmission efficiency was observed in adults directly collected from infected tomato plants in the greenhouse. The spread of CaCV on tomato plants in greenhouses showed a close association with thrips infestations.

Members of the subfamily Deltocephalinae include many vector species transmitting pathogens among economically important crops. Species of ten genera belonging to seven tribes namely, *Exitianus ball, Nephotettix matsumura, Deltocephalus Burmeister, Hecalus Stål, Balclutha Kirkaldy, cicadulina China, Hishimonus ishihara, Orosius distant, Changwhania Kwon and Doratulina melichar* were detected in a survey carried out during 2007–2008 in Mid Country associated with the vegetable ecosystems of Sri Lanka. An illustrated dichotomous key based on the morphology and genitalia characters is presented here for easy identification of field collected leafhoppers at generic level (Gnaneswaran et al. 2010).

After 3 weeks of exposition to a colony of *Thrips tabaci* previously grown on tomato yellow ring virus infected soybeans, some tobacco plants (2/12, 16.7 %) showed leaf chlorosis and diffuse necrosis symptoms. In another test, 7 out of 20 tobacco seedlings (35.0 %) exhibited the same symptoms 21 days after the inoculation with larval thrips (Ali Raza Golnaraghi et al. 2007).

Leaf Hopper Transmission of eggplant mottled dwarf virus (EMDV), a plant rhabdovirus, was achieved by the agallian leafhopper *Agallia vorobjevi.* Symptoms began to appear 20 days after inoculation. Viruliferous leafhoppers were also captured in the field (Babaie and Izadpanah 2003). Chickpea chlorotic dwarf virus (CpCDV), genus Mastervirus, family Geminiviridae), is the most common viral disease of chickpea in Pakistan. Transmission results showed that leafhopper O. *albicinctus* successfully transmitted CpCDV from diseased to healthy chickpea plants. (Akthar et al. 2011). The phytoplasma that causes phyllody disease in Sesamum was successfully transmitted from infected to healthy plants via leafhopper *O.albicinctus.* The causative agent was successfully transmitted to ten healthy plants, producing disease symptoms within 25–35 days in all the plants with 60 % of healthy plants (Akhtar et al. 2009).

Psyllids Two species of citrus psyllid, *Diaphorina citri* Kuwayama (Asiatic psylla) and *Trioza erytreae* Del Guerico, (African psylla) can transmit the greening pathogen. Acquisition feeding period is 30 min or longer. The pathogen remains latent for 3–20 days. Inoculation feeding period is 1 h or more (Huang et al. 1990; Batool et al. 2007).

Management

Khan and Mukhopadhyay (1985) found that soil application of methyl phosphoro dithioate (furatox-104) at 15 kg ha/ha followed by four sprays of matasystax 25 EC at 0.03 % at 15 days intervals recorded reduced yellow vein mosaic incidence in Okra with 23.26 % compared to control (81.22 %). The average whitefly population in treated plot was 59.66 compared to 231 in control with enhanced yield up to 59.45 q/ha over 23.8 in control. Three sprays of phosphomidon (0.02 %) or metasystax (0.02 %) or soil application of phorate 15 kg ha^{-1} or early sowing or intercropping okra with cowpea or mungbean recorded less number of *B. tabaci* with decreased BYVMV incidence and increased yield (Singh and Singh 1989). Neem-based products showed no effect in controlling whitefly and the disease (Pun et al. 2000).

As reported by Shankarappa (2002) the seed treatment with imidacloprid 70WG (3 g/kg seed) followed by three sprays of imidacloprid 200SL (0.03 ml/L) in tomato recorded lowest incidence of ToLCV disease (30.0 %) with increased yield of 51.17 t/ha over control (14.94 t ha). Seed treatment with imidacloprid 70WG (5 g kg seed) was effective in controlling YVMV in okra transmitted by with highest C:B ratio of 5.9 (Shivapuri et al. 2004).

Yellow mosaic virus (YMV) in okra could effectively managed by growing maize as border crop, the disease incidence was 30.0 % compared to 50.06 % in control with yield of 46.90 and 2.73 t ha, respectively (Pun et al. 2005). The lowest YMV incidence was due to obstruction in the movement of viruliferous whiteflies from outside. Profenophos 40 EC (500 g a.i/ha) and thiomethoxam 25 % WDG (25 g a.i/ha) effectively controlled whitefly population with reduced leaf curl incidence and increased yield in tomato (Rajashri et al. 2009).

The management of leaf curl disease, by plant products showed that neem seed kernal extract (5 %) was most effective than karanj and tumba seed extract. Management by insecticides, imidacloprid 17.8 SL (0.003 %) was most effective than spinosad 48 EC (0.02 %), malathion 50 EC (0.05 %), acephate 75 SP (0.1%) and methyl-demeton 25 EC (0.025 %). Management of chilli leaf curl was done by seed extract of plants and insecticides at different concentrations. (Pandey et al. 2010).

The pole bean yellow mosaic disease could be effectively managed by integrated module viz., border crop with African tall maize, seed treatment with of imidacloprid 70 % WS at 5.0 g kg, use of reflective mulch, spraying triazophos 40 EC at 0.175 % at 30 days after sowing (DAS) and with thiomethoxam 25WDG at 0.05 % at 45 DAS. The integrated module resulted in reduced disease incidence (7.4 %) and whitefly count (1.0), highest yield of 32.2 t ha with cost:benefit ratio of 1:3.17 compared to untreated control recording mean incidence of 33.0 %, white fly count 4.73 and 9.70 t ha yield (Jyothi et al. 2013).

Vectors play a main role in the epidemiology of the virus diseases. Most of the viruses are transmitted by aphids and leafhoppers. Once pathogens are identified, molecular diagnostic techniques could be established, and this would help in epidemiology and identification of durable resistant sources. Potential transmission of viruses with other vector groups (fungi, nematodes, and other insect species) has to be identified. Biodiversity of both viruses and vectors should be studied for better management of viruses. In future, genetic engineering techniques facilitate with resistant genes from wild species into cultivated genotypes to combat viruses.

References

Adkins S, Karthikeyan G, Damayanthi T, Kodetham G, David JR, Naidu R (2010) IPM CRSP project on tospoviruses and thrips vectors in south and south East Asia. J Insect Sci 10:2

Afreen B, Baghel G, Khan AA, Naqvi QA (2000) Virus–vector relationship of yellow vein disease of calendula and the whitefly. Plant Virol 20:1–6

Aidawati N, Hidayat SH, Suseno R, Sosromarsono S (2002) Transmission of an Indonesian Isolate of *Tobacco leaf curl virus* (*Geminivirus*) by *Bemisia tabaci* Genn. (Hemiptera: Aleyrodidae). Plant Pathol J 18(5):231–236

Ajlan AM, Ghanem GAM, Adilsalam (2007) Tomato yellow leaf curl virus (TYLCV) in Saudi Arabia: identification, partial characterization and virus-vector relationship. Arab J Biotech 10(1):179–192

Akhtar KP, Sarwar G, Dickinson M, A hmad M, Haq MA, Hameed S, Iqbal MJ (2009) Sesame phyllody disease: its symptomatology, etiology, and transmission in Pakistan. Turk J Agric For 33:477–486

Akhtar KP, Ahmad M, Shah TM, Atta BM (2011) Transmission of Chickpea chlorotic dwarf virus in Chickpea by the Leafhopper *Orosius albicinctus* Distant in Pakistan. Plant Prot Sci 47(1):1–4

Al-Ani RA, Adhab MA, Kareem AH, Ismail (2011) Eggplant blister mottled virus (EBMV): a possible new potyvirus characterized from Iraq. J Gen Mol Virol 3(3):049–052

Aswatha Narayana DS, Rangaswamy KT, Shankarappa KS, Maruthi MN, Lakshminarayana Reddy CN, Rekha AR, Keshava Murthy KV (2007) Distinct begomoviruses closely related to cassava mosaic viruses cause Indian jatropha mosaic disease. Int J Virol 3:1–11

Ayzapour K, Sijam K, Vadmalai G, Jaafar H (2013) Preliminary identification of Citrus tristeza virus (CTV) vectors in Peninsular Malaysia. Afr J Agric Res 8(30):4218–4133

Babaie GH, Izadpanah K (2003) Vector transmission of eggplant mottled dwarf virus in Iran. J Phytopathol 151:679–682

Balaraman K, Ramakrishnan K (1977a) Studies on strains and strain interaction in citrus tristeza virus. Booklet published by the University of Agricultural Sciences, Bangalore, pp 39

Balaraman K, Ramakrishnan K (1977b) A virus disease of acid lime [*Citrus aurantifolia* (Christm.)Swing.] resembling Hassaku dwarf. Mysore J Agric Sci 11:185–188

Balgude YS, Sawant DM (2012) Relationship of Soybean Mosaic Virus with its Aphid Vectors. Bioinfolet 9(1):61–65

Banks GK, Colvin J, Chowda Reddy RV, Maruthi MN, Muniyappa V, Venkatesh HM, Kiran Kumar M, Padmaja AS, Beitia FJ, Seal SE (2001) First report of the *Bemisia tabaci* B biotype in India and an associated tomato leaf curl virus disease epidemic. Plant Dis 85:231–235

Batool A, Iftikharl YS, Mughal M, Khan MM, Jaskani MJ, Abbas MI, Khan A (2007) Citrus greening disease—a major cause of citrus decline in the world—a review. Hort Sci (Prague) 34(4):159–166

Bhagabati KN, Goswami BK (1972) Incidence of yellow vein mosaic virus disease of Okra in relation to white fly population under different sowing dates. Indian J Virol 8(1):37–39

Bidari VB, Reddy HR (1990) Identification of naturally occurring viruses on commercial cultivars of chilli. Mysore J Agric Sci 24:45–51

Briddon RW, Markham PG (2000) Cotton leaf curl virus disease. Virus Res 71(1–2):151–159

Cembali T, Folwell RJ, Wandschneider P, Eastwell KC, Howell WE (2003) Economic implications of a virus prevention program in deciduous tree fruits in the US. Crop Prot 22:1149–1156

Chen H, Zheng L, Jia D, Zhang P, Chen Q, Liu Q, Wei T (2013) Rice Yellow dwarf exploit tubules to facilitate viral spread among cultured insect vector cells derived from leaf hopper *Recilia dorsalis*. Front Microbiol 4:1–6

Cohen S, Harpaz I (1964) Periodic, rather than continual acquisition of a new tomato virus by its vector, the tobacco whitefly (*Bemisia tabaci* Gennadius). Entomol Exp Appl 7:155–166

Dahal G (1998) Leafhopper and plant hopper vectors of rice viruses and their management in Nepal. Integr Pest Manag Rev 2(3):139–150

Dastogeer KMG, Ashrafuzaman M, Ali MA (2012) The virus-vector relationship of the jute leaf mosaic virus (JLMV) and its vector, *Bemisia tabaci* Gen. (Hemiptera: Aleyrodidae). Asian J Agric Sci 4(3):188–192

Deshapande RS, Siddappaji C, Viswanath S (1972) Epidemiological studies on Katte disease of cardamom (*Elettaaria cardamomum* Maton). Mysore J Agric Sci 6:4–9

Dheepa R, Paranjothi S (2010) Transmission of Cucumber Mosaic Virus (CMV) infecting banana by aphid and mechanical methods. Emir J Food Agric 22(2):117–129

Du ZQ, Li L, Liu L, Wang XF, Zhou (2007) Evaluation of Aphid transmission abilities and vector transmission phenotypes of Barley Yellow Dwarf Virus in China. J Plant Pathol 89(2):251–259

Ghimire SR, Subedi PP, Green SK (2001) Status of Tomato Yellow Leaf Curl Virus in tomato in the western hills of Nepal. Nepal Agric Res J 4 and 5:1–4

Ghosh DK, Ahlawat YS (1997) Filamentous viruses associated with mosaic disease of garlic in India. Indian Phytopathol 50(2):266–276

Gnaneswaran G, Hemachandra KS, Wijayagunasekara HNP, Ahangama D (2010) A key for identification of deltocephaline genera (hemiptera: cicadellidae: deltocephalinae) associated with vegetable ecosystem in Sri Lanka. Trop Agric Res 21(4):421–426

Golnaraghi AR, Pourrahim R, Farzadfar S, Ohshima K, Shahraeen N, Ahoonmanesh A (2007) Incidence and distribution of tomato yellow fruit ring virus on soybean in Iran. Plant Pathol J 6:14–21

Govindappa MR, Shankergoud, Shankarappa KS, Wickramaarachchi WART, Reddy BA, Rangaswamy KT (2011) Molecular detection and partial characterization of begomovirus associated with leaf curl disease of sunflower (*Helianthus annuus*) in Southern India. Plant Pathol J 10:29–35

Gude SR, Nagaraju, Pankaja NS, Ravi Kumar B, Anjula N (2008) Occurrence of a strain of papaya ring spot virus on cucurbitaceous crops around Bangalore. Env Ecol 26(3):1068–1069

Hajiabadi AM, Asaei F, Abdollahi M, Rastgou M (2012) Natural incidence of tomato viruses in the North of Iran. Phytopathol Mediterr 51(2) 390–396

Harrison BD, Muniyappa V, Swanson MM, Roberts IM, Robinson DJ (1991) Recognition and differentiation of seven whitefly transmitted geminiviruses from India and their relationships to African cassava mosaic and Thailand mungbean yellow mosaic viruses. Ann Appl Biol 118:299–308

Hasan MU, Sahi GM, Wakil W, Imanat Y (2003) Aphid transmission of sugarcane mosaic virus (SCMV). Pak J Agric Sci 40(1–2):74–76

Hogenhout SA, Ammar El-D, Whitefield AE, Redinaugh MG (2008) Insect vector interactions with persistently transmitted viruses. Annu Rev Phytopathol 46:327–359

Hohn T (2007) Plant virus transmission from the insect point of view. Proc Natl Acad Sci U S A 104(46):17905–17906

Hsieh CY (1974) Transmission of rice stripe virus by *Laodelphax striatellus* fallen in Taiwan. Plant Prot Bull (Taiwan) 15(4):153–162

Huang CH, Liaw CF, Chang L, Lant (1990) Incidence and spread of citrus likubin in relation to the population fluctuation of *Diphorina citri*. Plant Prot Bull (Taiwan, Roc) 32:167–176

Hull R (2002) Matthew's plant virology. Academic Pres, New York

Jayashree K, Pun KB, Sabitha Doraiswamy (1999) Virus—vector relationships of yellow vein mosaic virus and whitefly (*Bemisia tabaci*) in pumpkin. Indian Phytopathol 52(1):1–5

Jose J, Usha R (2003) Bhendi Yellow Vein Mosaic Disease in India is caused by association of a DNA Beta satellite with a Begomovirus. Virology 305:310–317

Jyothi V, Nagaraju N, Padmaja AS, Rangaswamy KT (2013) Transmission, detection and management of yellow mosaic virus in polebean (*Phaseolus vulgaris* L.) 43(2):229–236

Kalleshwaraswamy CM, Krishna kumar NK (2008) Transmission efficiency of Papaya Ring Spot Virus by three aphis species. Phytopathol 98:541–546

Karri SR, Acharyya P (2012) Incidence of Yellow Vein Mosaic Virus disease of Okra (Abelmoschus esculentus (L.) Moench] under summer and rainy environment. Int J Cur Res 4:018–021

Khan JA, Ahmad (2005) Diagnosis, monitoring and transmission characteristics of Cotton leaf curl virus. Curr Sci 88(11):1803–1809

Khan MA, Mukhopadhyay S (1985) Studies on the effect of same alternative cultural methods on the incidence of yellow vein mosaic virus disease of okra (*Abelmoschus esculentus*). Indian J Virol 1:69–72

Kiranmai G, Sreenivasulu P, Nayudu MV (1997) Characterization of Cucumber mosaic virus isolates naturally infecting three solanaceous vegetable crops in Andhra Pradesh. Indian Phytopathol 50(3):416–425

Krishna kumar NK Singh HS Kalleshwaraswamy CM (2010) Aphid (Aphididae: Homoptera) vectors of Papaya Ring Spot Virus (PRSV), bionomics, transmission efficiency and factors contributing to epidemiology Proc..II IS on Papaya. In: Kumar N et al. (eds). Acta Hortic 431–442

Kulkarni NK, Lava Kumar P, Muniyappa V, Jones T, Reddy DVR (2002) Transmission of Pigeon pea sterility mosaic virus by the Eriophyid Mite, *Aceria cajani* (Acari: Arthropoda). Plant Dis 86(12):12997–1302

Lakshminarayana Reddy CN Nagaraju Prasanna Kumar MK Venkataravanappa V (2007) Transmission and host range of the papaya ring spot virus (PRSV). J Plant Dis 2(1):9–13

Lingling Pu Guohua Xie Chunyan Ji Ling B Zhang M Xu D Zhou G (2012) Transmission characteristics of Southern rice black-streaked dwarf virus by rice Plant hoppers. Crop prot 41:71–76

Liu B, Preisser EL, Chu D, Pan H, Xie W, Wang S, Wu Q, Zhou X, Zhang Y (2013) Multiple forms of vector manipulation by a plant-infecting virus: *Bemisia tabaci* and tomato yellow leaf curl virus. J Virol 87(9):4929–4937

Magnaye LV, Valmayor RV (1995) BBTV, CMV and other viruses affecting banana in Asia and the Pacific. Food and Fertilizer Technology Centre data base

Mahalakshmi B, Reddy BMR, Nagaraju (2008) Characterization of a mosaic virus disease occurring on cowpea, *Vigna unguiculata* (L.) Walp. Mysore J Agric Sci 42(2):278–282

Mandal B, Muniyappa V (1991) Transmission of croton yellow vein mosaic virus by *Bemisia tabaci*: virus—vector relationships. Fitopatol Brasileir 16:260–262

Manjunath B, Muniyappa V (1995) Purification and transmission of Dolichos yellow mosaic geminivirus. Indian J Mycol Plant Pathol 25:91–92

Manjunath B, Prameela HA, Neetha J, Muniyappa V (2012) Transmission efficiency of *Mungbean yellow mosaic virus* by indigenous and B-biotype whiteflies. J Food Legumes 25(2):135–138

Maruthi MN, Manjunatha B, Rekha AR, Govindappa MR, Colvin J, Muniyappa V (2006) Dolichos yellow mosaic virus belongs to a distinct lineage of old world begomoviruses; its biological and molecular properties. Ann Appl Biol 148:1–9

Mathew AV, Muniyappa V (1991) Transmission of Indian cassava mosaic virus by *Bemisia tabaci*. Fitopatol Brasileria 16:46–49

Mound LA (1996) The Thysanoptera vector species of tospoviruses. Acta Hortic 431:298–307

Muniyappa V (1976) Studies on a mosaic disease of French bean (*Phaseolus vulgaris* Linn.). Madras Agric J 63:69–70

Muniyappa V (1980) Whiteflies, pp 39–85. In: Harris KF, Maramorosch K (eds) Vectors of Plant Pathogens. Academic, NY, pp 467

Muniyappa V, Reddy DVR (1983) Transmission of cowpea mild mottle virus by *Bemisia tabaci* in a non—persistent manner. Plant Dis 67: 391–393

Muniyappa V, Veeresh GK (1984) Plant virus diseases transmitted by whiteflies in Karnataka. Proc Indian Acad Sci (Anim Sci) 93:397–406

Muniyappa V, Veeresh GK (1986) Arthropod vectors of virus and mycoplasma diseases of plants in India. Colemania 2:19–38

Muniyappa V, Venkatesh HM, Ramappa HK, Kulkarni RS, Zeidan M Tarba CY, Ghanim M, Czosnek H (2000). Tomato leaf curl virus from Bangalore (ToLCV-Ban4): sequence comparison with Indian ToLCV isolates, detection in plants and insects, and vector relationships. Arch Virol 145:1583–1598

Muniyappa V, Maruthi MN, Babitha CR, Colvin J, Bridon RW, Rangaswamy KT (2003) Characterisation of pumpkin yellow vein mosaic virus from India. Ann Appl Biol 142:323–331

Murugan M, Sotelo Cardona P, Duraimurugan P, Whitfield AE, Schneweis D, Starkey S, Smith CM (2011) Wheat curl mite resistance: interactions of mite feeding with wheat streak mosaic virus infection. J Econ Entomol 104(4):1406–1414

Nagaraju N, Keshavamurthy KV (1994) Studies on a sap transmissible virus associated with mosaic disease of cowapea. Indian Phytopathol 47:23–26

Nagaraju N, Keshavamurthy KV (1997) Studies on the relationship between cowpea mosaic virus and its vector *Myzus persicae* Sulz. Mysore J Agric Sci 31:170–174

Nagaraju N, Reddy HR (1980) Natural occurrence of pepper veinal mottle virus on bell pepper. Indian Phytopathol 33:282–284

Nagaraju N, Reddy HR (1981) Natural occurrence of pepper vein banding virus on bell pepper. Mysore J Agril Sci 15(1):50–53

Nateshan HM, Muniyappa V, Swanson MM, Harrison BD (1996) Host range, vector relations and serological relationships of cotton leaf curl geminivirus from southern India. Ann Appl Biol 128:233–244

Ng JCK, Falk BW (2006) Virus-vector interactions mediating nonpersistent and semipersistent plant virus transmission. Ann Rev Phytopathol 44:183–212

Noda C, Inouye N (1989) Leek yellow stripe virus isolated from Allium plant in Japan. Ann Phytopathol Soc Jpn 55:208–215

Pan H, Chu D, Yan W, Su Q, Liu B, Wang S, Wu Q, Xie W, Jiao X, Li R, Yang N, Yang X, Xu B (2012) Rapid spread of tomato yellow leaf curl virus in china is aided differentially by two invasive whiteflies. PLoS One 7(4):34817

Pan SQ, Liu B, Chu D, Xie W, Wu Q, Wang S, Xu B, Zhang Y (2013) Insect symbiont facilitates vector acquisition, retention, and transmission of plant virus. Sci Rep 3:1367

Pandey SK, Mathur AC, Srivastava M (2010) Management of Leaf Curl Disease of Chilli (*Capsicum annuum* L.). Int J Virol 6:246–250

Parella G, Gognalons P, Gebre-Selassi e WK, Vovlas G, Marchoux G (2003) An update of the host range of Tomato spotted wilt virus. J Plant Pathol 85:227–264

Pico B, Diez MJ, Nuez F (1996) Viral diseases causing the greatest economic losses to the tomato crop. II. The Tomato yellow leaf curl virus—a review. Sci Hortic 67:151–196

Premachandra WTSD, Borgemeister C, Maiss E, Knierim D, Poehling HM (2005) *Ceratothripoides claratris*: a new vector of a Capsicum chlorosis virus Isolate Infecting tomato in Thailand. Phytopathol 95:659–663

Pudashini BJ, Shahid MS, Natsuaki KT (2013) First report of bean common mosaic necrosis virus (BCMNV) infecting sweet bean in Nepal. Plant Dis 97(2):290

Pun KB, Doraiswamy S, Jeyarajan R (2000) Screening of virus inhibitory chemicals and neem products against okra yellow vein mosaic virus. Indian Phytopathol 53:95–96

Pun KB, Sabita Doraiswamy, Jeyranjan R (2005) Management of Okra yellow vein mosaic virus disease and its whitefly vector. Indian J Virol 16:32–35

Rajasri M, Lakshmi K, Vijaya, Prasada Rao RDVJ, Loka RK (2009) Management of whitefly transmitted Tomato leaf curl virus using guard crops in tomato. Indian J Plant prot 3:101–103

Rajeshwari R, Chowda Reddy RV, Maruthi MN, Colvin J, Seal SE, Muniyappa V (2005) Host range, vector relationships and sequence comparison of a begomovirus infecting hibiscus in India. Ann of Appl Biol 147:15–25

Rashmi CM, Lakshminarayana Reddy CN, Praveen HM, Nagaraju (2005a) Natural occurrence of Cucumber Green Mottle Mosaic Virus (CGMMV) on Gherkin (*Cucumis anguria* L.). Env Ecol 23(Spl 4):781–784

Rashmi CM, Krishna Reddy M, Praveen HM, Nagaraju (2005b) Occurrence of Papaya Ring spot Virus (PRSV-p) on Gherkin (*Cucumis anguria* L.). Env Ecol 23(Spl.4):785–788

Raychaudhuri M, Varma A (1977) Virus-vector relationship of marrow mosaic virus with *Myzus persicae* Sulz. Indian J Entomol 37(3) 247–250

Rekha AR, Maruthi MN, Muniyappa V, Colvin J (2005) Occurrence of three genotypic clusters of *Bemisia tabaci* and the rapid spread of the B-biotype in South India. Entomol Exp et Appl 117:221–233

Sadeghi MS, Behjatnia SAA, Masumi M, Izadpanah (2008) Characterisation of a strain of Potato virus Y causing eggplant mosaic in southern Iran. Aust Plant Pathol 37(1):76–86

Saikia AK, Muniyappa V (1989) Epidemiology and control of tomato leaf curl virus in southern India. Trop Agric (Trinidad) 66:350–354

Senanayake DMJB, Varma A, Mandal B (2012) Virus–vector relationships, host range, detection and sequence comparison of *chilli leaf curl virus* associated with an epidemic of leaf curl disease of chilli in jodhpur. India. J Phytopathol 16(3):146–155

Shah H, Yasmin T, Fahim M, Hameed S, Haque MI (2008) Transmission and Host range studies of Pakistani isolate of Chilli Veinal Mottle Virus. Pak J Bot 40(6):2669–2681

Shankarappa KS (2002) Spread of Tomato leaf curl virus and *Bemisia tabaci* Genn. in open pollinated tomato varieties. M. Sc. (Agri.) Dissertation, University of Agricultural Sciences, Bangalore, pp 96

Shankarappa KS, Rangaswamy KT, Aswatha Narayana DS, Rekha AR, Raghavendra N, Lakshminarayana Reddy CN, Chancellor TCB, Maruthi MN (2007) Development of silver leaf assay, protein and nucleic acid-based diagnostic techniques for quick and reliable detection and monitoring of biotype B of the whitefly, *Bemisia tabaci* (Gennadius). Bull Entomol Res 97:1–11

Shivapuri A, Bhargava KK, Chhipa AP, Ghasolia RP (2004) Management of yellow vein mosaic virus of okra. J Mycol Plant Pathol. 34:353–355

Singh BR, Singh M (1989) Control of yellow vein mosaic of okra by checking its vector whitefly through adjusting dates of sowing, insecticidal application and crop barrier. Indian J Virol 5:61–66

Singh V, Singh D (2010) Studies on natural transmission of papaya ring spot virus disease in Eastern Uttar Pradesh. Ann Plant Prot Sci 18(1):188–192

Ta HA, Nguyen DP, Causse S, Nguyen TD, Ngo VV, Hébrard E (2013) Molecular diversity of *Rice grassy stunt virus* in Vietnam. Virus Genes 46(2):383–386

Valand GB, Muniyappa V (1992) Epidemiology of tobacco leaf curl virus in India. Ann Appl Biol 120:257–267

Varma A, Malathi VG (2003) Emerging geminivirus problems: a serious threat to crop production. Ann Appl Biol 142:145–164

Vijaylakshmi K, Reddy DVR, Ranga Rao GV, Wightma JA (1994) Transmission of tospovirus, bud necrosis virus by thrips. In: Inernatinal Conference of Virology in Tropics, Lucknow, 2–6 Dec, pp 74

Wang J, Zhao H, Liu J, Jiu M, Qian YJ (2010) Low frequency of horizontal and vertical transmission of two begomoviruses through whiteflies exhibits little relevance to the vector infectivity. Ann Appl Biol 157:125–133

Watson MA, Roberts FM (1939) A comparative study of the transmission of Hysogamus virus 3, Potato Virus Y and Cucumber virus 1 by the vector *Myzus persicae* (Sulz), M circumflex (Buckton) and Macrosiphum gei (Koch). Proc Roy Soc London Ser B 127:543–576

Watson MA, Roberts FM (1940) Evidence against the hypothesis that certain plant viruses are transmitted mechanically by aphids. Ann Appl Biol 227–233

Wightman JA, Ranga Rao GV, Vijaya Lakshmi K (1995) Thrips palmi, general pest and vector of some tospoviruses in Asia. Pages 11–15 in Recent studies on peanut bud necrosis disease: proceedings of a Meeting, 20 Mar 1995, ICRISAT Asia Center, India (Buiel, A.A.M., Parlevliet, J.E., and Lenne, J.M., eds.). Patancheru 502 324, Andhra Pradesh, India: International Crops Research Institute for the Semi-Arid Tropics; and P O Box 386, 6700 AJ Wageningen, The Netherlands: Department of Plant Breeding, Agricultural University of Wageningen. 80 pp

Zhou G, Xu D, Xu D, Zhang M (2013) Southern rice black-streaked dwarf virus: a white-backed planthopper-transmitted fijivirus threatening rice production in Asia. Front Microbiol 4:270

Part V

Insect Molecular Biology

Molecular Approaches for the Improvement of *Bacillus thuringiensis* Against Pests

Ankit Patel, Leena Pathak, Noushad Parvez, Deepak Panpatte, Kuldeep Khatri and Janardan Jani

Abstract

Multlple pest management tools including biological approaches, are generally recognized as possible solutions. Bacillus thuringiensis (*Bt*) Berliner has been used as a biopesticide in agriculture. The insecticidal properties of this bacterium are due to insecticidal proteins. Genetic improvement of *Bt* natural strains, in particular *Bt* recombination, offers a promising means of improving efficacy and cost-effectiveness of *Bt*-based bioinsecticide products to develop new biotechnological applications; which describes site-specific recombination, including transposition and transduction. Applications of homologous recombination include disruption of *cry* and *cyt* genes to assess their contribution to pesticidal activity and inactivation of protease producing genes to increase crystal production and stability. Conjugation-like system is used to transfer *cry* encoding plasmids within strains but most *cry* genes are not readily transmissible by this process. Nevertheless, a number of transconjugant and naturally occurring strains producing Cry proteins distinct from those of *Bt* HD-1 subsp. *kurstaki*, including strains of *Bt* subsp. *aizawai* and *Bt* subsp. *morrisoni*, have been registered with Environmental Protection Agency (EPA). Engineering *Bt* and *B. cereus* came in 1989 through electroporation to transform vegetative cells with plasmid DNA. Genes conferring resistance to insects have been inserted into crop plants.

Keywords

Bacillus thuringiensis · Molecular approaches · Pest · Resistance

A. Patel (✉) · L. Pathak · N. Parvez · D. Panpatte · K. Khatri · J. Jani
Biological Control Research Laboratory, Anand Agricultural University, Anand, Gujarat, 388110 India
e-mail: aspatel1990@yahoo.co.in

Introduction

Since World War II, control methods of insect disease have relied heavily on broad-spectrum synthetic chemical insecticides. However, many governments have restricted the usage of chemical insecticides due to their negative environmental effects. As a result, the use of biopesticides, as a component of an integrated pest management (IPM), has been gaining worldwide acceptance. *Bacillus thuringiensis* (*Bt*) has been used as a biopesticide. Its advantages are specific toxicity against target insects, non-polluting residues and safety to non-target organisms. *Bt* produces crystals, composed of protein/s known as insecticidal crystal protein/s (ICPs) during sporulation, which are selectively toxic to different species of several invertebrate phyla. In addition, some *Bt* strains secrete vegetative insecticidal proteins (VIPs) during vegetative growth that are toxic to many lepidopteran pests, causing lysis of midgut epithelial cells and gut paralysis. Some *Bt* strains also secrete proteases, chitinases, and other virulence factors conferring insecticidal activity.

Genetic improvement of *Bt* strains for the development of novel biopesticides entails increasing their potency against target insects, broadening the insecticidal spectra for specific crop applications, improving persistence on plants and optimizing fermentation production and the most important of all meeting the resistance development needs. This review focuses on the developments of genetic manipulation for improving *Bt* strains.

Genetic Manipulation of *BT*

Transduction

The first genetic exchange system available in *Bt* was generalized transduction mediated by the *Bacillus cereus* Frankland and Frankland phages CP-51 or CP-54. This system allows the transfer of chromosomal markers. Since then, various phage-based systems for mapping chromosomal genes have been developed for several groups. The generalized transducing phage CP-55, obtained by Thorne's laboratory, was able to realize inter-varietal transduction between *Bt* subspecies *galleriae* and *israelensis*. TP-13, a broad host-range temperate phage, which infected 18 of 19 tested *Bt* subspecies, is useful in scanning large segments of chromosome. TP-18, a narrower host-range phage, which infected 9 of 21 *Bt* subspecies and had a head size seven times smaller than that of TP-13, was effective for ordering markers that were too closely linked. CP-51 could also be used to transduce plasmid molecules such as pBC16 and pC194 into strains of *B. cereus, B. anthracis* and *Bt*, at frequencies as high as 10^{-5}. Transduction has allowed the transfer of recombinant plasmids carrying crystal protein genes between *Bt* strains at detectable frequencies. This is of particular value for biotechnological applications because introduction of cloned crystal protein genes into various *Bt* subspecies, including both Cry^- and Cry^+ strains can expand the insecticidal spectrum of each strain, allowing native strains to have better insecticidal activity against more insect species. Transduction is an efficient means of gene transfer yielding high levels of gene expression, but not all strains are amenable to phage transduction. Furthermore, the stability of the introduced plasmids depends mainly on the host.

Conjugation

The second important advance in genetic exchange is the discovery of a conjugation process where the donor and the recipient strains are grown together for a few generations in nutrient broth, and the mixture is plated onto an appropriate medium that is selective for the recipient cells. The transcipient cells are scored by analyzing the plasmid profiles of the resulting recipient colonies by electrophoretic analysis. Using the cell-mating method, both large and small plasmids were found to transfer equally well between strains *Bt*, *B. cereus* and *B. anthracis*. Subsequently, there were many reports on the use of this conjugation-like process for transferring specific plasmids in *Bt*. A recombinant plasmid from *B. subtilis* (containing a cloned *Bt* toxin gene) transferred readily into *Bt* subsp. *kurstaki*, *israelensis* and *sotto*. This mating system can also be used for *Bt* and other gram-positive species. Plasmid pC194 from *Bt* strain 0016 and pBC16 from *B. cereus* could be transferred to *Bt* subsp. *israelensis*. Plasmid pBC16 transfers from *B. subtilis* to *B. anthracis*, *B. cereus*, *B. licheniformis*, *B. megaterium*, *B. pumilus*, *B. subtilis*, and *Bt* using the conjugation-like process. In soil microcosm, the conjugation-like process has been poorly

Table 1 Bioassay of transgenic plant against *Trichoplusia ni* and *Autographa nigrisigna*

Insects	*n*	Insect mortality (%)[a]	
		Control	Cry1Ac
Trichoplusia ni	25	0	100
Autographa nigrisigna	30	10	93.3

[a] Data were recorded 7 days after feeding. (Bao et al. 2009)

described (Vilas Boas et al. 2000). This is very important from a biotechnological point of view because many biopesticides can be created. For example, scientists in Ecogen Corporation (USA) have developed a broad spectrum *Bt* biopesticide foil for controlling lepidopteran, coleopteran and many plant pests through conjugal transfer and plasmid curing. An important advantage of the conjugational approach is that the transconjugants are treated as "non-genetically engineered" and are, thus, subject to relatively simple regulatory registration. Strain improvement may also entail elimination of any undesirable activities through plasmid curing.

Transformation

Initially, transformation of *Bt* was only possible using a protoplast, which yielded low transformant frequencies. This technique is largely inefficient, complex and can probably not be applied to most *Bt* strains. The preferred method of gene transfer employs the electroporation technique, for which numerous protocols are available, where a high voltage electric discharge through a cell suspension results in a transient increase in permeability of the cell membrane and hence DNA enters the cells and transformants can be obtained within 24 h. Two plasmids can be transformed simultaneously into a recipient using this technique. The transformation frequencies are in the range of 10^2–10^5 colony forming units (cfu)/µg, depending on the strain or the replicon used. This technique is useful for the introduction of cloned toxin genes, in either their native or modified form, into a variety of host strains, including acrystalliferous strains. This allows to the broadening of the insecticidal activity spectrum of many strains, which can have biotechnological applications against many plant pests. Bao et al. (2009) applied *Agrobacterium*-mediated genetic transformation to produce transgenic plants of spinach (*Spinacia oleracea*) resistant to two pest species, *Trichoplusia ni* (Hubner) and *Autographa nigrisigna.* They checked the effect of Cry1Ac toxin on the development of *Autographa nigrisigna* larvae and observed that larvae fed with non-transgenic plant developed into pupal stage while the larvae fed with transgenic plants died within 1 week (Table 1, Fig. 1).

Mehrotra et al. (2011) modified *cry*1Ab and *cry*1Ac insecticidal genes of *Bt* under the control of two different constitutive promoters and introduced into chickpea (*Cicer arietinum* L.) by Agrobacterium-mediated transformation. One hundred and eighteen stable transformed T0 plants as independent transformation events were obtained expressing individual *cry*1Ab, *cry*1Ac, or both pyramided genes for their co-expression driven by either cauliflower mosaic virus 35S promoter with duplicated enhancer (CaMV35S) or synthetic constitutive promoter (Pcec) and their combinations. Integration and inheritance of transgenes in T0 and T1 population of transgenic chickpea plants were determined by PCR, RT-PCR and Southern hybridization. The performed insect bioassay with modified transgenic plant showed relatively higher toxicity for plants expressing Cry1Ac protein as compared to Cry1Ab to *Helicoverpa armigera* (Hubner). Pyramided transgenic plants with moderate expression levels (15–20 ng mg^{-1} of TSP) showed high-level of resistance and protection against pod borer larvae of *H. armigera* as compared to high level expression of a single toxin. These results have shown the significance of pyramiding and co-expression of two Cry toxins for efficient protection against lepidopteran pests of chickpea. Ibrahim et al. (2008) studied Cry1Ag-expressing *BT*4 *Bt* strain caused mortality of *S. littoralis* larvae only slightly (the LC_{50} was 104 ppm) and observed that in the presence of only Cry1C, the LC_{50} was 64 ppm. In the presence of Cry1C co-expressed with Cry1Ag, the LC_{50} decreased to 2.2 ppm. Thus, a combination of the Cry proteins 1C and 1Ag

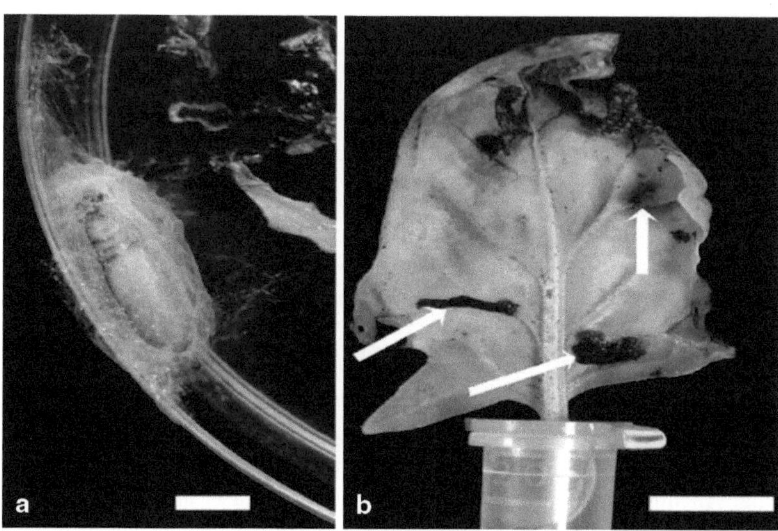

Fig. 1 Effect of Cry1Ac toxin on development of *Autographa nigrisigna* larvae. A larva developed into pupal stage when fed with leaves of control non-transgenic plant (**a**), whereas the larvae fed with leaves of transgenic plants died within 1 week of feeding (**b**). *Arrows* indicated dying larvae. (Bao et al. 2009)

could result in effective insect control. With this approach, a combination of Cry proteins can be designed rather than discovered (Fig. 2, Table 2).

Genetic Recombination

To avoid segregation or structural instability of plasmids in the recombinant strain, strategies for the integration of genes into resident plasmids or into the chromosome of the desired strain by in vivo homologous recombination were developed with the help of integration vectors. Since 1991, genetic recombination has been used as a technique to improve *Bt* strains. The integration via single or double crossover can occur by homologous recombination between the resident plasmid and the *Bt* subsp. *israelensis*. DNA fragments flanking the erm gene harbored in the integration plasmid. A very low frequency of recombination was observed for this event; this is not surprising, since the integration of the nonreplicative plasmid is the result of both a transformation and a recombination event between two different plasmids. Due to low frequency, the insecticidal host range of *Bt*, transformation and recombination were uncoupled using a thermo sensitive plasmid as integrative vector.

Another tool for the recombination in *Bt* strains was the deployment of a site-specific recombination (SSR) system, to selectively delete ancillary or foreign DNA elements from recombinant plasmids after their introduction into a *Bt* host. The SSR system is useful for engineering strains with unique combinations of *cry* genes, resulting in new active ingredients with improved insecticidal properties. This system consists in Tn5401, which is a transposable element indigenous to the *Bt* subsp. *morrisoni* strain EG2158. This transposon encodes a transposase protein (TnpA), a recombinase protein (TnpI) and a site-specific recombination site, or internal resolution site (IRS) which is required for Tn5401 transposition in *Bt* (Baum et al. 1999). Transposon Tn4430 was used to eliminate in vivo unwanted DNA sequences from transforming vectors harboring two IRSs. The transposon vector pEG922, containing transposon Tn5401 was used to disrupt the spo0F gene. In addition, a native *Bt* plasmid replicon was combined with an indigenous site specific recombination system that allowed for the selective removal of foreign DNA from the recombinant bacterium after introduction of the Cry-encoding plasmid vector. In this way, a coleopteran-active strain, approved as the native ingredient for Raven OF bioinsecticide, was constructed. The lethal concentration 50% (LC_{50}) of G033A against *S. exigua, P. xylostella*, and *H. amigera* was 4.26, 0.86, and 1.76 µg/ml, respectively (Table 3).

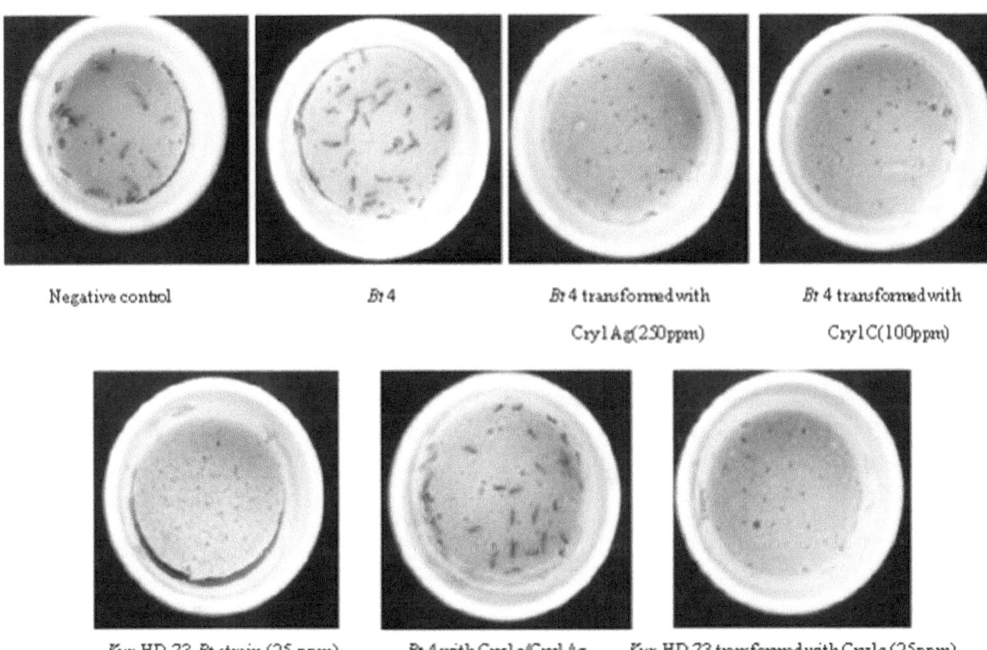

Fig. 2 The effect of toxicity of Bt Bt4 *cry*⁻ and Bt4 transformed with *cry*1C, *cry*1Ag and *cry*1C/1Ag on larvae of *S. littoralis* compared with the effect of Bt strain *kur*-HD73 and its transformed one (*kur*-HD73 harboring *cry*1C; Ibrahim et al. 2008)

Table 2 The LC_{50} values in ppm of toxins and combination of toxins used against the cotton leaf worm *S. littoralis*

Strain/toxin[a]	LC_{50}[bcd] (ppm)	95 % confidential limits	Slope/SE
Kur-HD73 Cry1Ac	197.42	(150.42–321)	1.79±0.36
Bt/NC3 Cry1C	63.23	(221.15–22.07)	1.67±0.391
Bt/NAg Cry1Ag	103.69	(122.71–87.71)	2.94±0.40
Bt/N1C1Ag Cry1C & Cry1Ag	2.216	not determined	0.87±0.38
Entomocidus Cry1C	41.48	not determined	3.9±1.016
Transformed-HD73 Cry1C & Cry1Ac	6.65	not determined	1.89±1.059
Mixture of HD73 and *entomocidus*	31.32	(33.49–29.23)	9.13±0.926
Mixture of BtN1Ag and BtN1C	6.6	not determined	1.89±1.059

SE standard error

[a] Bioassays were performed on spore-crystal preparations from T3 liquid cultures (Ibrahim et al. 2008)

[b] LC_{50} is a concentration of toxin required to kill 50 % of 1st instar larvae

[c] LC_{50}s were calculated by probit analysis

[d] Probit model is $Y = a + b * x$; where Y = probit value, a = intercept, probit value for $x = 0$, b = slope, regression coefficient of y on x, x = log (dose)

Yue et al. (2005a) obtained two integrative recombinant *Bt* strains; BMB1520-E, and BMB1520-F. In recombinant BMB1520-F, the *cry*1C gene was expressed stably at a significant level and did not reduce the expression of endogenous crystal protein genes. Bioassay results indicated that BMB1520-E and BMB1520-F showed a higher level of activity against *S. exigua* third-instar larvae than did their parent strains, in addition to the high toxicity to *Plutella xylostella* third-instar later larvae (Table 4).

Table 3 Insecticidal assay results of lyophilized spore-crystal mixtures of the strains

Strains	Insecticidal activity (LC_{50}; 95 % confidence interval)			
	P. xylostella (μg ml^{-1})	*H. amigera* (μg ml^{-1})	*S. exigua* (μg ml^{-1})	*P. aenescens* (mg ml^{-1})
G03	0.71(0.49–1.07)[a]	1.48(1.2–1.77)[a]	2.98(1.85–3.98)[a]	NA
G033A	0.86(0.56–1.26)[a]	1.76(1.29–2.22)[a]	4.26(2.74–6.51)[a]	0.35(0.24–0.57)[b]
Bt22	NA	NA	NA	0.22(0.12–0.36)[b]

NA no activity

[a] Only for the 130 kDa components contained in the dry powder of tested stain (Wang et al. 2006)

[b] Only for the 67 kDa components contained in the dry powder of tested stain

Table 4 Bioassays of wildtype strain and integrative recombinants. (Yue et al. 2005a)

Test pests	Strains	Regression equation of toxicity	Correlation coefficient	LC_{50} value (μL/mL)
P.xylostella	YBT1520	Y=−1.348+2.073X	0.94	1.066
	BMB1520–B	Y=3.382+0.538X	0.99	1.203
	BMB1520–E	Y=−1.165+2.312X	0.98	1.186
	BMB1520–F	Y=2.797+1.122X	0.97	1.019
S.exigua	YBT1520	Y=2.684+0.681X	0.97	1.597
	BMB1520–B	Y=2.793+0.925X	0.98	1.308
	BMB1520–E	Y=−1.331+2.074X	0.93	1.048
	BMB1520–F	Y=1.668+1.125X	0.98	0.856

Table 5 Insecticidal activity assay of lyophilized spore crystal mixtures of the strains

Strains	Insecticidal activity (LC_{50}; 95 % confidence interval)	
	Plutella xylostella (μg ml^{-1})	*Leptinotarsa decernlineata* (mg ml^{-1})
UV17	18.21 (16.0–20.0)	>500.0
UV173A	18.03 (13.0–22.0)	0.19 (0.09–0.33)
Bt22	NA	0.44 (0.25–0.80)

NA no activity (Wang et al. 2008)

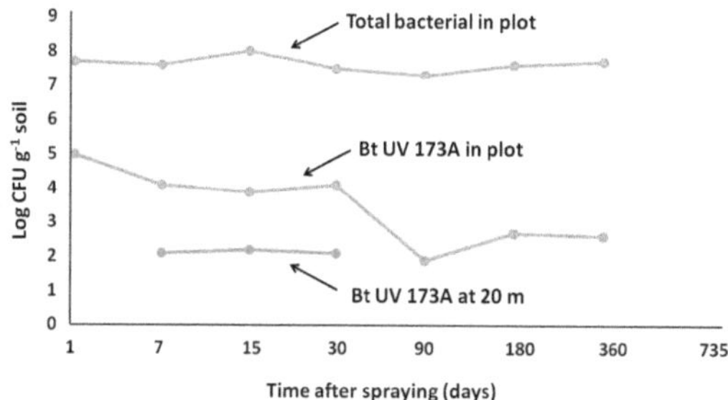

Fig. 3 Re-isolation of the released strain UV173A inside and outside the release plot. Included in graphs are concentrations of total bacteria within the plot (log CFU g^{-1} soil dry weight), the released strain UV173A within the plot (log CFU g^{-1} soil dry weight), as well as at 20 m from the plot. All samples were analyzed in duplicate (Wang et al. 2008)

Wang et al. (2008) studied the cry-type gene of wild *Bt* strain UV17, identified and a novel *cry*1Ba gene was cloned. The *cry*3Aa7 gene, which was highly toxic to coleopteran pests, was introduced into UV17, and a recombinant strain designated as UV173A was obtained. Bioassay results showed that UV173A was not only highly toxic against *Plutella xylostella* (50 % lethal concentration [LC_{50}] = 18.03 μg ml^{-1}), but also against coleopteran *Leptinotarsa decernlineata* (LC_{50} = 0.19 μg ml^{-1}). The recombinant strain was then tested in field trials to monitor its spatial variation of population and to investigate the impact on non-target invertebrates. A recombinant *Bt* strain, UV173A with broad insecticidal spectrum was obtained, and it did not cause adverse effects on the population of non-target organisms (Table 5, Fig. 3).

Table 6 Advantages and disadvantages of each genetic method for improvement of Bt strains. (Sansinenea et al. 2010)

Method of genetic manipulation	Advantages	Disadvantages
Transduction	Is an efficient means of gene transfer yielding high levels of gene expression	Not all strains are amenable to phage transduction and the stability of the introduced plasmids depends mainly on the host
Conjugation	(1) Both large and small plasmids were found to transfer equally well between strains (2) Transconjugants are treated as "non-genetically engineered" and are, thus, subject to relatively simple regulatory registration	(1) Plasmid incompatibility, as strains are often limited in their capacity to act as donors or recipients (2) Location of *cry* genes on non-transmissible plasmids (3) A plasmid may carry additional undesirable genes (4) Eventual segregational loss of plasmid in transconjugant strains (5) Unintended transfer of plasmids from one to another strain may occur
Transformation by electroporation	Rapid, simple and effective method of introducing plasmid DNA intro strains of Bt	(1) The transformation depends on the strain or the replicon used (2) Segregational or structural instability of some plasmids in the recombinant strain
Genetic recombination	Avoids segregational or structural instability of plasmids in the recombinant strain integrating genes into resident plasmids or into the chromosome	The low frequency of recombination makes this approach unsatisfactory for poorly transformable strains

This review has shown that all contributions are focusing on the main biotechnological application of *Bt* as bioinsecticide, although they are using different genetic techniques. However, all techniques have advantages and disadvantages that make them different from each other and true to the purpose of each research as summarized in Table 6.

References

Bao JH, Chin DP, Fukami M, Ugaki M, Nomura M, Mii M (2009) *Agrobacterium*-mediated transformation of spinach (*Spinacia oleracea*) with *Bacillus thuringiensis cry*1Ac gene for resistance against two common vegetable pests. Plant Biotech J 26:249–254

Baum JA, Gilmer AJ, Mettus AML (1999) Multiple roles for TnpI recombinase in regulation of Tn5401 transposition in *Bacillus thuringiensis*. J Bacteriol 18:6271–6277

Ibrahim NAGAA, Abdallah SO, Salama MS, Madkour MA (2008) Construction of a potent strain of *Bacillus thuringiensis* against the cotton leaf worm *Spodoptera littoralis*. Agric For Res 58:111–123

Mehrotra M, Singh AK, Sanyal I, Altosaar I, Amla DV (2011) Pyramiding of modified *cry*1Ab and *cry*1Ac genes of *Bacillus thuringiensis* in transgenic chickpea (*Cicer arietinum* L.) for improved resistance to pod borer insect *Helicoverpa armigera*. Euphytica 182:87–102

Sansinenea E, Vazquez C, Ortiz A (2010) Genetic manipulation in *Bacillus thuringiensis* for strain improvement. Biotechnol Lett 32:1549–1557

Vilas Boas LA Vilas Boas GFLT Saridakis H Lemos MVF Lereclus D Arantes OMN (2000) Survival and conjugation of *Bacillus thuringiensis* in a soil microcosm. FEMS Microbiol Ecol 31:255–259

Wang G, Zhang J, Song F, Wu J, Feng S, Huang D (2006) Engineered *Bacillus thuringiensis* GO33A with broad insecticidal activity against *lepidopteran* and *coleopteran* pests. Appl Microbiol Biotechnol 72:924–930

Wang G, Zhang J, Song F, Gu A, Uwais A, Shao T, Huang D (2008) Recombinant *Bacillus thuringiensis* strain shows high insecticidal activity against *Plutella xylostella* and *Leptinotarsa decemlineata* without affecting non-target species in the field. J Appl Microbiol 105:1536–1543

Yue C, Sun M, Yu Z (2005) Improoved production of insecticidal protein in *Bacillus thuringiensis* strain carrying an additional Cry 1C gene in its chromosome. Biotechnol Bioeng 92:1–7

Diapause-Related Gene Expression in Eggs of Multivoltine *Bombyx mori* L. Silkworm Races

Kangayam M. Ponnuvel, Sirigineedi Sasibhushan, N. Murthy Geetha and CGP Rao

Abstract

The diapause hormone has been identified as a major factor inducing diapause in silkworm, *Bombyx mori* L. However, the molecular mechanism controlling egg diapause in multivoltine silkworm remains unclear. In this study, an attempt is made to decipher molecular events during embryonic diapause in multivoltine silkworm. Using suppressive subtractive hybridization (SSH) 186 cDNA clones were isolated and sequenced from both diapause and non-diapause eggs wherein 29 diapause-related genes were identified and classified into six functional groups. The qPCR analysis confirmed the expression of 11 of these genes, wherein, three were upregulated during diapause and another seven during non-diapause, while, one gene remained unchanged. The gene expression profiles of diapause-induced and non-diapause eggs of multivoltine silkworm were also investigated at 18 and 30 h after oviposition using oligonucleotide microarrays containing 24,924 probes. Data analysis revealed upregulation of 638 genes and downregulation of 1136 genes at 18 h after oviposition. Further at 18 and 30 h after oviposition, 115 genes were stably upregulated, while 117 were stably downregulated. The annotated genes showing fold change of one and above were classified into seven functional groups, viz. immune, metabolism, stress, signal transduction, cell cycle, transcription, and apoptosis. The overall genes expressed at 18 h was higher than that at 30 h under both diapause and non-diapause conditions indicating that crucial biological processes for initiation or termination of diapause mechanism occur during 18 h time period.

Keywords

Bombyx mori · Diapause · Gene expression · Multivoltine

K. M. Ponnuvel (✉) · S. Sasibhushan · N. M. Geetha · CGP. Rao
Seribiotech Research Laboratory, Carmelaram Post, Kodathi Bangalore, Karnataka 560035, India
e-mail: kmpvel@yahoo.com

A. K. Chakravarthy (ed.), *New Horizons in Insect Science: Towards Sustainable Pest Management*, DOI 10.1007/978-81-322-2089-3_18, © Springer India 2015

Introduction

Among the factors contributing to the abundance of insects, a major element is the adaptation to survive under unfavourable environmental conditions through diapause in different developmental stages leading to suppressed metabolism. It has long been known that the major cues that insects use to enter diapause are photoperiodic, sensing a reduction in day length and thermoperiodic, sensing a reduction in temperature. Suppression of metabolism enables the insect to stretch its food reserves to bridge the unfavourable period. Survival during diapause may also be enhanced by coupling suppression of metabolic activities with synthesis of polyols, other cryoprotective agents, and heatshock proteins that reduce injury at low temperatures. Upon termination of diapause, the metabolic rate rapidly increases, which initiates development. Thus, the diapause and non-diapause phase of the insect's life cycle represent striking contrasts, and these differences at the molecular level remains largely unknown.

The silkworm, *Bombyx mori,* is a holometabolous insect that has four distinct life stages, including embryo, larva, pupa, and moth. It is a model organism for Lepidoptera in molecular genetics as well as functional genomics (Denlinger 2002). Diapause hormone (DH), one of the neurohormones, has been identified as a major factor inducing diapause in the resulting embryos (Hwang et al. 2005). The expression of *DH* mRNA in the early pupal stage correlates to the incidence of diapause (Sato et al. 1993; Xu et al. 1995). Although these findings clearly show that this hormone regulates the induction of embryonic diapause, it is still unidentified whether the individual gene expression profile regulates the stage of initiation or termination of diapause.

Several classes of diapause upregulated genes such as stress response genes, developmental arrest genes, and genes involved in regulating specific physiological activities that are unique to diapause have been noted. Although some genes are turned on at the onset of diapause and remain upregulated until diapause has been broken, others are uniquely expressed only in early or late diapause.

Earlier, Flannagan et al. (1998) identified genes differentially regulated during diapause in the flesh fly (*Sarcopahaga crassipalpis* Macquart). The study resulted in significant observations like downregulation of proliferating cell nuclear antigen during diapause. Another interesting discovery was of the ribosomal protein P0, expressed in a cyclic pattern throughout diapause (Craig and Denlinger 2000). However, the characterization of genes upregulated during the egg diapause of *Bombyx mori* is still quite limited.

Although the molecular mechanism controlling egg diapause remains obscure, it has recently been reported that *Heat shock protein 70a, samui, sorbitol dehydrogenase (SDH a & b)* and *glycerol kinase* genes play important roles in the mediation of egg diapause in silkworm, *B. mori* (Kihara et al. 2011). However, the role of each gene in diapause regulation is unclear. Further, to decipher the molecular events, it is essential to identify additional genes upregulated during diapause. In the present study, efforts have been made to use the suppressive subtractive hybridization (SSH) technique for identifying genes that are differentially expressed in the diapause-induced multivoltine silkworm eggs of *B. mori.* Using this technique, several genes have been identified that are expressed during diapause and non-diapause. These genes offer insight into the molecular mechanisms contributing to egg diapause.

Material and Methods

Insect Culture

The multivoltine strain MW13 (Indian origin) was selected for the study. The larvae were reared by the standard rearing method of Krishnaswami (1978) up to the third instar. The late stage (fourth and fifth instars) larvae were then reared under low temperature (18 °C) and photoperiod (6L: 18D) up to the cocooning stage and the moths were made to lay eggs at normal room temperature (25 °C).

RNA Isolation

After oviposition, the diapause and non-diapause egg samples were collected from 6–48 h at every 6 h time interval. The total RNA was extracted from the diapause and non-diapause eggs using TRIzol reagent (invitrogen) and quantified by UV absorbance at 260 or 280 nm. The RNA sample was denatured in formaldehyde, formamide, and electrophoresed in 2.0 % agarose gels.

Construction of Subtracted cDNA Library Through Suppressive Subtractive Hybridization

The SSH was performed using Clonetech PCR-Select™ cDNA Subtraction Kit to select genes that are upregulated and downregulated during diapause and non-diapause. The forward selection of SSH consisted of cDNA from non-diapausing eggs as tester and diapause-induced eggs as driver and the reverse selection had diapause-induced eggs as tester and non-diapausing eggs as driver.

The forward and reverse subtracted libraries were cloned using InsTAclone™ PCR Cloning Kit (Fermentas). Transformed plasmids were inserted into competent *Escherichia coli* cells and grown overnight on Luria-Bertani (LB) plates containing ampicillin. Over 100 colonies were isolated from each library and grown overnight in LB-ampicillin broth at 37 °C. Colonies were then purified with GeneJET™ Plasmid Miniprep Kit (Fermentas), run on a 1 % agarose gel to determine concentration and all subtracted clones were subjected to sequencing using M13 primer.

Basic Local Alignment Search Tool Analysis

The sequences were edited and assembled using MEGA version 5. Putative sequence homologies were determined by BLASTn and BLASTx algorithms in Silkbase database (http://www.silkdb.org) and GenBank (http://www.ncbi.nlm.nih.gov/).

Real-Time PCR Analysis

The total RNA isolated from diapause and non-diapause eggs from 6–48 h at each 6 h time interval was DNase-treated and reverse transcribed in a 20 μl reaction using M-MLV Superscript III reverse transcriptase (Invitrogen). One μl of the first strand cDNA was used in a 20 μl reaction mixture using the specific primers designed for real-time PCR (qPCR) (Table 1). The reactions were conducted on a STRATAGENE Mx 3005P real-time PCR system. The experiment was performed in triplicate and results were standardized to the expression level of the constitutive β actin gene. A non-template control (NTC) sample was also run to detect contamination if any.

Microarray Experiment and Data Analysis

A genome wide oligonucleotide microarray containing 24,924 probes were used to investigate the gene expression profiles of diapause-induced and non-diapause eggs of multivoltine silkworm *B. mori* at 18 and 30 h after oviposition. The complete sets of raw and normalized data from this study have been deposited in the NCBI Gene Expression Omnibus (GEO) repository (accession number GSE35622).

Results and Discussion

Diapause and Non-diapause Subtraction

Two SSH experiments were carried out under which, 186 cDNA clones each (94 non-diapause, 92 diapause) were specifically identified from forward (non-diapause) and reverse (diapause) subtraction. The plasmids were isolated from all 186 clones and run on a 0.8 % agarose gel. Based on size variations among the clones amplified by using M13 primers, 40 clones were selected from forward- and reverse-subtracted libraries and sequenced. The sequences obtained were subjected to BLAST analysis to know their

Table 1 Primers used in real-time qPCR for confirmation of differentially expressed genes

Gene	Forward primer	Reverse primer
Propanediol utilization protein	5′TGTCTACCATCGTGCCAAAG3′	5′CATGCATTCGTCAAACGAGA3′
Ubiquitin family protein	5′GCCATCTTCAAGCTGTTTGC3′	5′GAGGTGGCATGCAGATCTTT3′
Translationally controlled tumor protein homolog	5′ATATTCCATCATGGCAACCA3′	5′GTAGCAAAATTGGAAGAGAAGG3′
Heat shock cognate 71 kDa protein	5′CCATACGCTCGATCTCTTCC3′	5′GTGAGCGTGCTATGACCAAA3′
60S ribosomal protein L3	5′CAGCCTCCACGATCTCTTTC3′	5′TCGTCATCGTGGTAAGGTCA3′
40S ribosomal protein S14	5′AATGCGGCCAATCTTCATAC3′	5′GGCACAGGATGTAGCAGAGA3′
Ubiquitin c-terminal hydrolase	5′CGGAGCTATTTCAGAGCACA3′	5′TGGAGCTGTTGTGAAATTCG3′
Protein coding gene	5′TGATCTAGCAGTAGAGGACCAA3′	5′GGTCATGAACTAGAGTCCACAGG3′
Chitinase A precursor	5′TGGCAGAGGTCAACTCGTAA3′	5′CTTCGATGGTGTCGACATTG3′
Negative regulation of transcription	5′GCGATAAGAAGGCCACAGTT3′	5′ACATACAGGCTTCCCGATTT3′
Eukaryotic translation elongation factor	5′ACGTTGTAGGGCTTGCTCTG3′	5′TGAAGGCCTACCTACCTGTCA3′

homology and a total of 29 genes were identified from the diapause and non- diapause SSH libraries of which 17 were non-diapause specific and 12 were diapause specific (Tables 2, 3). The subtractive genes identified were classified into six functional groups (regulatory, food utilization, stress response, metabolic, ribosomal, and transposable elements) (Table 4). Based on the above classification, specific primers were designed for 11 nonredundant genes (Table 1) and validated through qPCR.

Confirmation by QPCR Analysis

The upregulation and downregulation of the 11 putative genes identified from the diapause and non-diapause subtracted libraries were confirmed through qPCR. In most cases, qPCR analysis confirmed the upregulation or downregulation of the cDNAs that were identified through SSH. However, in one case, the qPCR results showed a different pattern where expression was observed in both diapause and non-diapause conditions.

Diapause Upregulated Genes

The following genes were upregulated during diapause: one regulatory gene (40S ribosomal protein S19), two stress responsive genes (heat shock cognate 71 kDa protein and ubiquitin c-terminal hydrolase), two metabolic genes (similar to chitin metabolic process and chitinase domain containing protein), six ribosomal protein genes (Bm ribosomal protein L41, 60S ribosomal protein L8,L18,L27a, L13, ribosomal protein L14), one transposable gene element (negative regulation of transcription), and two genes with unknown function (D-37 and D-57) (Fig. 1).

Non-diapause Upregulated Genes

The following genes were upregulated during non-diapause: 11 regulatory genes (pseudouridine synthase, remodelling, and splicing factor 1, 40S ribosomal protein S14 and S5, RNA binding protein, translationally controlled tumor protein homolog, eukaryotic translation elongation factor, nucleosome assembly protein, nascent protein, *B. mori* profilin protein, Bm acyl-coenzyme A dehydrogenase), one food utilization gene (taste receptor type 2 member 117), two stress response

Table 2 Non-diapause upregulated genes isolated by suppressive subtractive hybridization. Percent identities, E-value, matched Expressed Sequence Tag clone ID retrieved from Silkbase database (http://www.silkdb.org) and GenBank (http://www.ncbi.nih.gov/)

Sl. no.	Clone ID	Size (bp)	Putative identity	Percent identity	E-value	Matched Expressed Sequence Tag clone ID
During non diapause						
1	ND–8	257	Propanediol utilization protein	99	1e–98	ovS315D08f
2	ND–10	188	Predicted similar to CG7849-PB (pseudouridine synthase)	99	3e–69	fner29d24r
3	ND–21	166	Rsf1 (remodelling and splicing factor 1)	100	7e–61	ovS334D03f
4	ND–24	181	40S ribosomal protein S14	97	1e–68	wdS30543
5	ND–17	135	40S ribosomal protein S5	98	8e–19	wdV10964X
6	ND–54	230	Ubiquitin family protein	98	3e–84	fcaL-P08_pT_N24
7	ND–11	182	RNA binding	99	5e–66	fner3n16r
8	ND–60	193	Translationally controlled tumor protein homolog	99	2e–70	wdV30789
9	ND–50	231	Similar to ribosomal protein L23Ae	100	9e–78	fdpeP07_F_B21
10	ND–13	307	Taste receptor type 2 member 117	99	1e–87	fner3p04r
11	2ND–18	229	Polyubiquitin 4 UBQ4	99	1e–56	fner50g04r
12	2ND–6	405	Chitinase A precursor	99	e–117	FWDP32_FL5_O09
13	ND–55	169	Eef2protein (eukaryotic translation elongation factor)	100	2e–33	prgv0458
14	ND–1A	195	Nucleosome assembly protein	99	0.0	Ce-1048
15	ND–1B	463	Nascent protein	99	0.0	wdV30032
16	ND–64	133	*B. mori* profilin protein	99	0.0	wdV30234
17	ND–3	140	*B. mori* acyl-coenzyme A dehydrogenase (acade)	100	0.0	NM_001044207.1

Table 3 Diapause upregulated genes isolated by suppressive subtractive hybridization. Percent identities, E-value, matched EST clone ID retrieved from Silkbase database (http://www.silkdb.org) and GenBank (http://www.ncbi.nih.gov/)

Clone ID	Size (bp)	Putative identity	Percent identity	E-value	Matched EST clone ID
During diapause					
D–4	262	Heat shock cognate 71 kDa protein	99	1e–100	BmNP26_FL5_N17
D–12	185	*B. mori* ribosomal protein L41	99	6e–90	NM_001042449.1
D–28	107	60S ribosomal protein L8	99	3e–36	wdV40936
D–64	129	60S ribosomal protein L18	100	6e–45	wdV30295
D–68	228	60S ribosomal protein L27a	100	e–110	ovS334G05f
D–9	160	Ribosomal protein L14	100	3e–40	ovS325C03f
2D–12	98	Predicted similar to ENSANGP00000018877 (chitin metabolic process)	96	4e–12	fdpeP08_F_H14
2D–14	367	40S ribosomal protein S19	99	e–125	fprwP27_F_G11
D–20	213	ENSANGP00000001579 (negative regulation of transcription)	100	1e–59	BmNP15_T7_H13

Table 3 (continued)

Clone ID	Size (bp)	Putative identity	Percent identity	E-value	Matched EST clone ID
D–53	216	60S ribosomal protein L13	99	2e–53	wdV30635
2D–20	133	PREDICTED: similar to ubiquitin c-terminal hydrolase	98	8e–16	FWDP02_T7_O15
D–36	140	Chitinase domain-containing protein 1	96	4e–67	XP_001869617.1

genes (ubiquitin family protein, polyubiquitin 4), two metabolic genes (propanediol utilization protein, chitinase A precursor), and one ribosomal gene (ribosomal protein L23Ae) (Fig. 2).

Genes Unchanged in Diapauses

One ribosomal gene, 60S ribosomal protein L13, had its expression almost similar in both diapause and non-diapause conditions (Fig. 1).

The results presented herein provide a few initial clues about the molecular events that characterize the egg diapause *B. mori.* Twenty-nine genes have been identified by SSH whose expression patterns were confirmed by qPCR. Twelve genes were upregulated during diapause, 17 during non-diapause, two genes were with unknown function, and one gene remained unchanged during diapause and non-diapause. The above 32 genes were characterized into six functional groups. The qPCR analysis confirmed the expression of 11 genes selected for analysis.

Regulatory Genes

Identification of genes regulating the molecular mechanism may prove useful in understanding how *B. mori* can survive in a prolonged inactive state. A few ribosomal proteins have functions in regulating cell growth and death in addition to their roles in translation (Naora and Naora 1999; Horino et al. 1998). The appearance of ribosomal proteins in the SSH libraries suggests a possible contribution of these proteins in regulating the embryonic diapause of *B. mori.* In an earlier study on adult diapause of *Culex pipiens*, it was observed that two ribosomal proteins have been downregulated before diapause termination (Robich et al. 2007). Several studies have documented the presence of diapause-associated proteins from the fat body and haemolymph of diapausing insects (Brown and Chippendale 1978; Brown 1980; Osir et al. 1989; Salama and Miller 1992; Palli et al. 1998; Levenbook 1985). All these have proven to be storage proteins that are synthesized before the onset of diapause and are then utilized when development resumes at the termination of diapause. In this study, three 40S ribosomal proteins involved in regulatory role were found, one from diapause and two from non-diapause. The qPCR analysis of 40S ribosomal protein S14 showed an upregulation during non-diapause supporting the earlier reported data.

When diapause is terminated, one would expect to see a major shift in the patterns of gene expression. The insect rapidly increases its metabolic rate and promptly initiates development. Thus, one would predict that the genes involved in the mechanisms that suppress development would be switched off and new sets of genes involved in initiating development would be switched on (Denlinger 2000). From this perspective, 11 regulatory genes have been identified during the non-diapause period.

Pseudouridine synthases catalyze the isomerization of uridine to pseudouridine (Psi) in a variety of RNA molecules (Ramamurthy et al. 1999). Depletion of the minifly (mfl) encoded pseudouridine synthase gene of *Drosophila* causes severe reduction in size by decreasing both the number and size of wing cells suggesting that a component of the pseudouridine synthase loss of function phenotype causes defects in notch signalling (Tortoriello et al. 2010). Through qPCR analysis,

Table 4 Functional classification of genes isolated through suppressive subtractive hybridization

	Diapause	Non-diapause
Regulatory	40S ribosomal protein S19	1. Pseudouridine synthase
		2. Remodelling and splicing factor 1
		3. 40S ribosomal protein S14
		4. 40S ribosomal protein S5
		5. RNA binding
		6. Translationally controlled tumor protein homolog
		7. Eukaryotic translation elongation factor
		8. Nucleosome assembly protein
		9. Nascent protein
		10. *B. mori* profilin protein
		11. *B. mori* acyl-coenzyme A dehydrogenase
Food utilization		Taste receptor type 2 member 117
Stress response	Heat shock cognate 71 kDa protein	Ubiquitin family protein
	Similar to ubiquitin c-terminal hydrolase x4	Polyubiquitin 4 UBQ4
Metabolic genes	Chitin metabolic process	Propanediol utilization protein
	Chitinase domain-containing protein 1	Chitinase A precursor
Cytoskeletal	Nil	Nil
Ribosomal	*B. mori* ribosomal protein L41	Ribosomal protein L23Ae
	60S ribosomal protein L8	
	60S ribosomal protein L18	
	60S ribosomal protein L27a	
	Ribosomal protein L14	
	60S ribosomal protein L13	
Transposable Elements	Negative regulation of transcription	Nil

the upregulation of pseudouridine synthase gene confirms the possible role of pseudouridine synthase in cell growth and proliferation.

Remodelling and spacing factor 1 (Rsf-1) is a member of the chromatin-remodelling complex family of proteins that regulate gene expression and cell growth. Rsf-1 is a nuclear protein that acts as a histone chaperone and binds to another member of the chromatin-remodelling complex, hSNF2H. Together, this protein duplex regulates RNA transcription and DNA replication by spacing nucleosome arrays and mobilizing nucleosomes when chromatin remodelling occurs (Loyola et al. 2003).

RNA-binding proteins (RBPs) play key roles in post-transcriptional control of RNAs, which, along with transcriptional regulation, is a major way to regulate patterns of gene expression during development. Post-transcriptional control can occur at many different steps in RNA metabolism, including splicing, polyadenylation, mRNA stability, mRNA localization, and translation (Curtis et al. 1995; Wickens et al. 2000; de Moor and Richter 2001; Johnstone and Lasko 2001).

Translationally controlled tumor protein (Tctp) is known to be synthesized preferentially in cells during the early growth phase of tumors, but is also expressed in normal cells. RT-PCR analyses indicated that the BmTCTP mRNA was transcribed in all larval organs examined and was present constantly during the cell cycle of BmN4 cells (Lee et al. 2004). In this study, the qPCR analysis confirmed the upregulation of Tctp in non-diapause eggs suggesting a similar role of Tctp during embryonic development in *B. mori.*

Eukaryotic elongation factor promotes the delivery of aminoacyl-tRNA to the acceptor site of the ribosome during protein synthesis and has

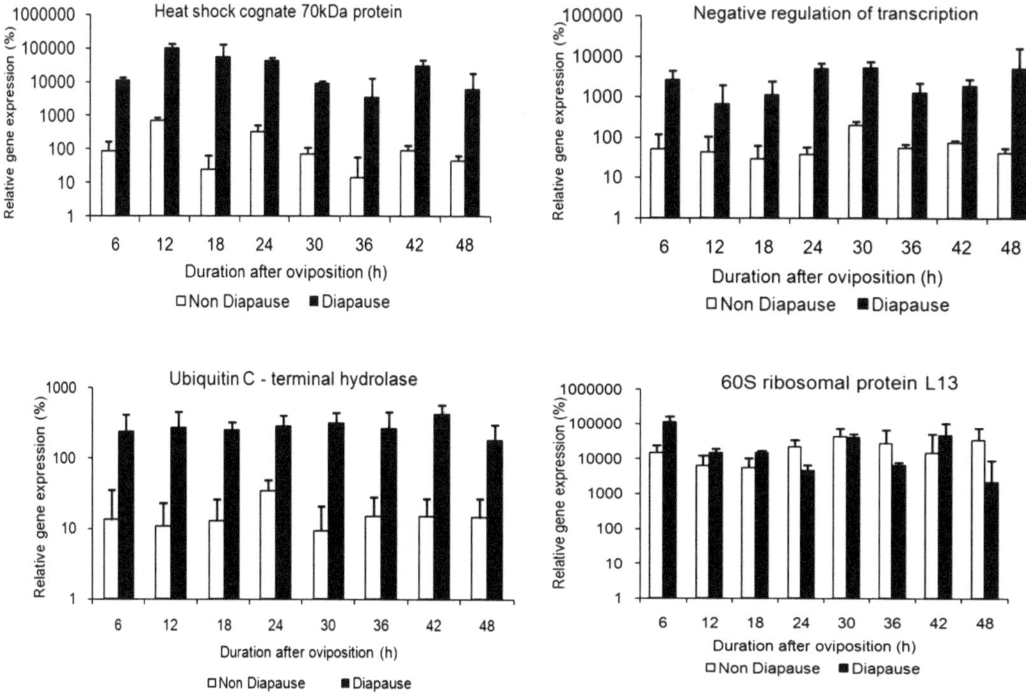

Fig. 1 Relative gene expression patterns of genes upregulated during diapause. RNA was isolated at 6 h intervals from 6 to 48 h after oviposition. The experiment was performed in triplicate and results were standardized to the expression level of the constitutive β actin gene. A non-template control (NTC) sample was also run to detect contamination if any. □-Non-diapause ■- Diapause

multiple and divergent roles in cell physiology affecting the cytoskeleton, peptide synthesis, and protein degradation (Gonen et al. 1996). qPCR analysis carried out in the study revealed the up-regulation of this gene during non-diapause demonstrating a role in cell proliferation.

Nucleosome assembly protein is an integral component in the establishment, maintenance, and dynamics of eukaryotic chromatin. It shuttles histones into the nucleus, assembles nucleosomes and promotes chromatin fluidity, thereby affecting the transcription of many genes (Park et al. 2005). The presence of nucleosome assembly protein transcripts in non-diapause eggs suggests its function in the embryonic development of *B. mori.*

Nascent proteins, a newly synthesized protein is not ready to start working in a cell immediately after it is made by the ribosome. Instead, these nascent proteins must first fold into proper 3-D shape, and possibly also receive modifications such as the addition of sugars or lipids. The site where these maturation steps take place depends upon where the final protein will be used in the cell (Sedwick 2011).

Profilin is a ubiquitous eukaryotic protein that regulates the actin cytoskeleton. In mammalian cells, profilin appears to act at the intersection of signal transduction and cytoskeletal organization systems (Machesky et al. 1990) suggesting a possible role in *B. mori* cell growth and proliferation. Acyl-CoA dehydrogenases are a class of enzymes that function to catalyze the initial step in each cycle of fatty acid β-oxidation in the mitochondria of cells (Thorpe and Kim 1995). Acyl-CoA dehydrogenases are an important class of enzymes in mammalian cells because of their role in metabolizing fatty acids present in ingested food materials. The occurrence of Acyl-CoA dehydrogenase in non-diapause eggs suggests

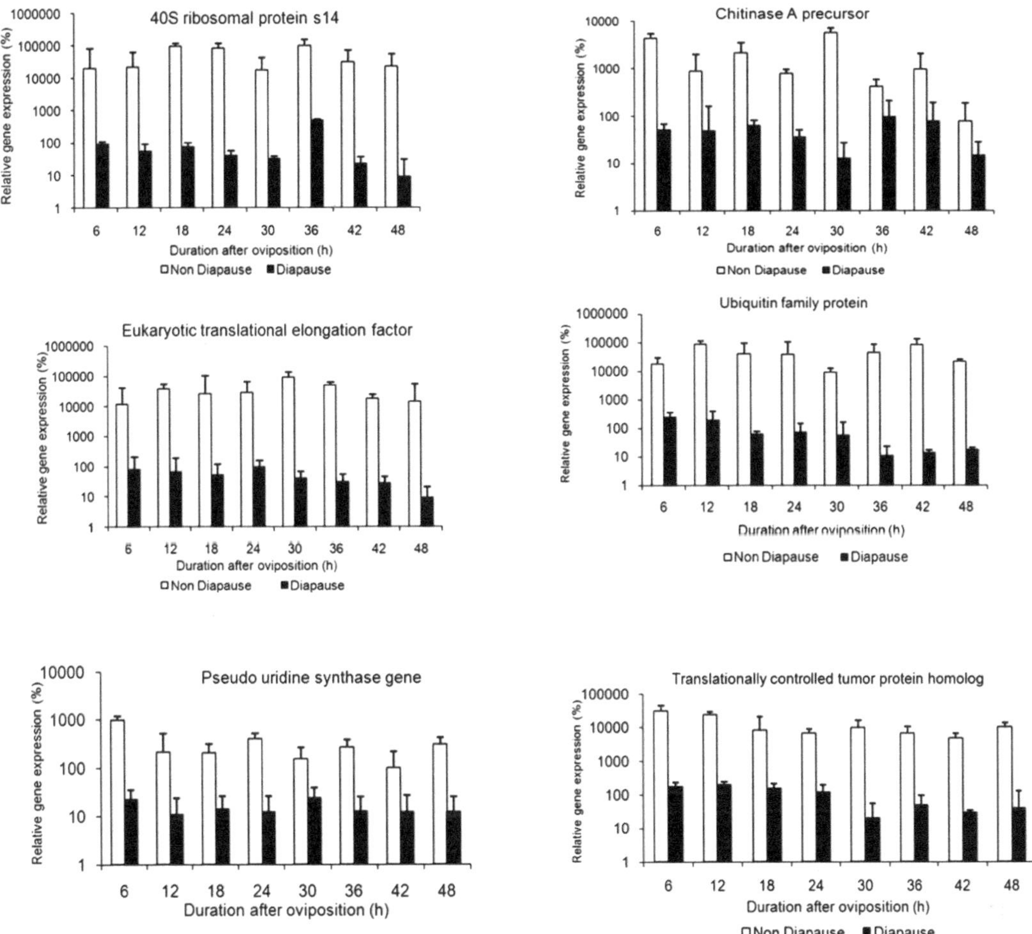

Fig. 2 Relative gene expression patterns of genes upregulated during non-diapause. RNA was isolated at 6 h intervals from 6 to 48 h after oviposition. The experiment was performed in triplicate and results were standardized to the expression level of the constitutive β actin gene. A non-template control (NTC) sample was also run to detect contamination if any. □-Non-diapause ■- Diapause

that this enzyme might be utilized in fatty acid β-oxidation as an energy source during embryonic development in *B. mori.*

Food Utilization

Only one gene related to food utilization was isolated by SSH, i.e., taste receptor type 2 member. The receptors of type II cells bind with sweet, bitter, or umami compounds (Adler et al. 2000). The actual function of the taste receptor type 2 member during embryonic development in *B. mori* is not clear.

Stress Response

In several insect species, heat shock proteins are highly upregulated upon entry into diapause (Denlinger et al. 2001). These proteins act as molecular chaperons by preventing abnormal protein folding during environmental stresses and have also been implicated in playing a role in cell cycle arrest (Feder et al. 1992). Hsp23 and hsp70 are upregulated during diapause in flesh fly *Sarcophagi crassipalpis* (Rinehart and Denlinger 2000). A small heat shock protein is observed to be slightly upregulated during adult diapause in *Cx. pipiens* (Robich et al. 2007). A heat shock cognate 71 kDa protein is found upregulated in the diapausing eggs of *B. mori* supporting the earlier data that heat shock protein might play a cryoprotective role during diapause and also as molecular chaperones in maintaining the integrity of key metabolic enzymes or structural proteins (Denlinger 2002).

Modification of proteins by ubiquitination is a fundamental mechanism for regulating numerous cellular processes, including DNA repair, cell cycle regulation, antigen presentation, cell–cell communication, cell differentiation, and apoptosis (Bheda et al. 2009). Many biological processes rely on targeted protein degradation. Ubiquitin plays a well-established role in this process, in which the covalent attachments of polyubiquitin chains to protein substrates culminate in their degradation via the proteasome. Some of these so-called "ubiquitin family proteins" have recently been shown to bind components of the 26S proteasome via their ubiquitin-like domains, thus implicating proteasome activity in pathways other than protein degradation (Walters et al. 2004). These ubiquitin family proteins might help in the degradation of proteins which misfold under stress and which can be related to the upregulation of ubiquitin family proteins in the non-diapause eggs.

Metabolic Genes

The metabolic rates in insects are typically suppressed during diapause, the metabolic suppression in egg or pupa being very extensive compared to adult diapause (Danks 1987). Two proteins of metabolic function were identified in diapausing eggs involved in chitin metabolic process and chitinase domain containing protein and one gene encoding chitinase A precursor in non-diapause eggs. Even though there is no obvious role for chitin proteins at this stage of embryonic development, this protein might have been obtained from the shell of which it is a major component. Earlier mRNAs for chitin-binding proteins that are similar to those in peritrophic membrane are upregulated in diapause-destined embryos in *Artemia.* The mRNAs especially coding for chitin reactive domains are of embryonic origin and they might show an unrecognized contribution to cyst wall assembly by the embryo (Qiu and MacRae 2007). However, chitin-degrading enzymes play a crucial role in post-embryonic development, especially during larval molt and pupation. During the molt, proteases and chitinases are synthesized by epidermal cells and accumulate in the molting fluid between the epidermis and the old cuticle (Merzendorfer and Zimoch 2003). The qPCR analysis of chitinase A precursor showed an upregulation of this gene during non-diapause suggesting its role in post-embryonic development in *B. mori.*

The propanediol utilization (pdu) operon of *Salmonella enterica* serovar *typhimurium* LT2 contains genes needed for the coenzyme B12-dependent catabolism of 1,2-propanediol (Bobik et al. 1999), but the actual pathway in insects is not yet known. However, the qPCR analysis revealed an upregulation of this gene in non-diapause eggs suggesting a possible role of this enzyme in the utilization of polyols at the termination of diapauses

Ribosomal Genes

In addition to the three ribosomal genes (two in diapause and one in non-diapause) thought to have regulatory function, six additional ribosomal genes have been identified during diapause and one gene during non-diapause. Even though these ribosomal proteins play a role as translation elongation factors, their role during diapause is not known. The ribosomal protein L23 found upregulated by qPCR during non-diapause is found to play a crucial role in cell proliferation (Jin et al. 2004).

Transposable Elements

One gene, with negative regulation of transcription, was identified to have a transposable elemental role. Transcriptional repression is required for the establishment of the temporally and spatially complex patterns of gene expression which are characteristic of eukaryotes, and is also frequently involved in the modulation of gene expression in response to changes in the microenvironment of the cell.

Microarray Analysis

The gene expression profiles of diapause-induced and non-diapause eggs of multivoltine silkworm *B. mori* were investigated at 18 and 30 h after oviposition using oligonucleotide microarrays containing 24,924 probes. Data analysis revealed upregulation of 638 genes and downregulation of 1136 genes at 18 h after oviposition. At 30 h after oviposition, 675 genes were upregulated and 595 downregulated. Further, at 18 as well as 30 h after oviposition, 115 genes were stably upregulated, while 117 were stably downregulated. The annotated genes showing fold change of one and above were classified into seven functional groups, viz., immune, metabolism, stress, signal transduction, cell cycle, transcription, and apoptosis. Hierarchical clustering based on the Pearson coefficient correlation algorithm revealed that most of the identified genes fall under the transcriptional mechanism with an increase in the number of genes expressed in diapause eggs at 30 h compared to 18 h, but a decrease in the case of non-diapause eggs. The overall genes expressed at 18 h was higher than that at 30 h under both diapause and non-diapause conditions indicating that crucial biological processes for initiation or termination of the diapause mechanism occur during 18 h time period.

This work represents the first attempt at investigating the molecular aspects of embryonic diapause in multivoltine silkworm eggs. Through SSH, the differential regulation of diapause-related genes in early embryonic diapause has been demonstrated.

Acknowledgements This research work was supported by grants from Department of Biotechnology, Ministry of Science and Technology, Government of India (BT/PR11397/PBD/19/192/2008 dated 30 June 2009).

References

Adler E, Hoon MA, Mueller KL, Chandrashekar J, Ryba NJ, Zuker CS (2000) A novel family of mammalian taste receptors. Cell 100:693–702

Bheda A, Shackelford J, Pagano JS (2009) Expression and functional studies of ubiquitin C-terminal hydrolase L1 regulated genes. PLoS One 4(8): e6764

Bobik TA, Havemann GD, Busch RJ, Williams DS, Aldrich HC (1999) The propanediol utilization (pdu) operon of *Salmonella enterica* serovar Typhimurium LT2 includes genes necessary for formation of polyhedral organelles involved in coenzyme B (12)-dependent 1, 2-propanediol degradation. J Bacteriol 181:5967–5975

Brown JJ (1980) Haemolymph protein reserves of diapausing and nondiapausing codling moth larvae, *Cydia pomonella* (L) (Lepidoptera: *Tortricidae*). J Insect Physiol 26:487–495

Brown JJ, Chippendale GM (1978) Juvenile hormone and a protein associated with the larval diapause of the Southwestern Corn Borer, *Diatraea grandiosella*. Insect Biochem 8:359–397

Clark AR, Docherty K (1993) Negative regulation of transcription in eukaryotes. Biochem J 296:521–541

Craig TL, Denlinger DL (2000) Sequence and transcription patterns of 60S ribosomal protein P0, a diapause-regulated AP endonuclease in the flesh fly, *Sarcophaga crassipalpis*. Gene 255:381–388

Curtis D, Lehmann R, Zamore PD (1995) Translational regulation in development. Cell 81:171–178

Danks HV (1987) Insect dormancy. An ecological perspective. Ottawa: Biological Survey of Canada (Terrestrial Arthropods), pp 439

Denlinger DL (2001) Time for a rest: programmed diapause in insects. In: Storey KB (ed) Molecular mechanisms of metabolic arrest: life in limbo. BIOS Scientific Publishers Ltd, Oxford, pp 155–167

Denlinger DL (2002) Regulation of diapause. Annu Rev Entomol 47:93–122

De Moor CT Richter JD (2001) Translational control in vertebrate development. In: Etkin LD. Jeon KW (ed) Cell lineage specification and patterning of the embryo. Academic, San Diego

Feder JH, Rossi JM, Solomon J, Solomon N, Lindquist S (1992) The consequences of expressing hsp70 in *Drosophila* cells at normal temperatures. Genes Dev 6:1402–1413

Flannagan RD, Tammariello SP, Joplin KH, Cikra-Irland RA, Yocum GD, Denlinger DL (1998) Diapause specific gene expression in pupae of the flesh fly *Sarcophaga crassipalpis*. Proc Natl Acad Sci U S A 95:5616–5620

Gonen H, Dickman D, Schwartz AL, Ciechanover A (1996) Protein synthesis elongation factor EF-1a is an isopeptidase essential for ubiquitin-dependent degradation of certain proteolytic substrates. Adv Exp Med Biol 389:209–219

Horino K, Nishiura H, Ohsako T, Shibuya Y, Hiraoka T, Kitamura N, Yamamoto TA (1998) Monocyte chemotactic factor, S19 ribosomal protein dimer, in phagocytic clearance of apoptotic cells. Lab Invest 78:603–617

Hwang JS, Go HJ, Goo TW, Yun EY, Choi KH, Seong SI, Lee SM, Lee BH, Kim I, Chun T., Kang SW (2055) The analysis of differentially expressed novel transcripts in diapausing and diapause-activated eggs of *Bombyx mori*. Arch Insect Biochem Physiol 59:197–201

Jin A, Itahana K, O'Keefe K, Zhang Y (2004) Inhibition of HDM2 and activation of p53 by ribosomal protein L23. Mol Cell Biol 24:7669–7680

Johnstone O, Lasko P (2001) Translational regulation and RNA localization in *Drosophila* oocytes and embryos. Annu Rev Genet 35:365–406

Kihara F, Niimi T, Yamashita O, Yaginuma T (2011) Heat shock factor binds to heat shock elements upstream of heat shock protein 70a and Samui genes to confer transcriptional activity in *Bombyx mori* diapause eggs exposed to 5 °C. Insect Biochem Mol Biol 24:4456

Krishnaswami S (1978) New technology of silkworm rearing In: bulletin No. 2 Central Sericultural Research and Training Institute, Mysore, Central Silk Board, Govt. of India, pp 1–23

Lee JM, Kusakabe T, Kawaguchi Y, Miyagawa Y, Takahashi M, Mon H, Nho SK, Koga K (2004) Molecular cloning and characterization of the translationally controlled tumor protein gene in *Bombyx mori*. Comp Biochem Physiol B 139:35–43

Levenbook L (1985) Insect storage proteins. In: Kerkut GA, Gilbert LI (eds) Comprehensive insect physiology, biochemistry and pharmacology, vol 10. Pergamon, New York, pp 307–346

Loyola A, Huang JY, LeRoy G, Hu S, Wang YH, Donnelly RJ, Lane WS, Lee SC, Reinberg D (2003) Functional analysis of the subunits of the chromatin assembly factor RSF. Mol Cell Biol 23:6759–6768

Machesky LM, Goldschmidt-Clermont PJ, Pollard TD (1990) The affinities of human platelet and *Acanthamoeba* profilin isoforms for polyphosphoinositides account for their relative abilities to inhibit phospholipase C Cell Reg 1:937–950

Merzendorfer H, Zimoch L (2003) Chitin metabolism in insects: structure, function and regulation of chitin synthases and chitinases. J Exp Biol 206:4393–412

Naora H, Naora H (1999) Involvement of ribosomal proteins in regulating cell growth and apoptosis: translational modulation or recruitment for extraribosomal activity? Immunol Cell Biol 77:197–205

Osir EO, Labongo LV, Unnithan GC (1989) A high molecular weight diapause-associated protein from the stem-borer *Busseola fusca*: purification and properties. Arch Insect Biochem 11:173–187

Palli SR, Ladd TR, Ricci AR, Primavera M, Mungrue IN, Pang ASD, Retnakaran A (1998) Synthesis of the same two proteins prior to larval diapause and pupation in the spruce budworm, *Choristoneura fumiferana*. J Insect Physiol 44:509–524

Park YJ, Chodaparambil JV, Bao Y, McBryant SJ, Luger K (2005) Nucleosome assembly protein 1 exchanges histone H2A-H2B dimers and assists nucleosome sliding. J Biol Chem 280:1817–1825

Qiu Z, MacRae TH (2007) Developmentally regulated synthesis of p8, a stress associated transcription cofactor, in diapause-destined embryos of *Artemia franciscana*. Cell Stress Chaperon 12:255–264

Ramamurthy V, Swann SL, Paulson JL, Spedaliere CJ, Mueller EG (1999) Critical aspartic acid residues in pseudouridines synthase. J Biol Chem 274:22225–22230

Rinehart JP, Denlinger DL (2000) Heat-shock protein 90 is down-regulated during pupal diapause in the flesh fly, *Sarcophaga crassipalpis*, but remains responsive to thermal stress. Insect Mol Biol 9:641–645

Robich RM, Rinehart JP, Kitchen LJ, Denlinger DL (2007) Diapause-specific gene expression in the northern house mosquito, *Culex pipiens L.*, identified by suppressive subtractive hybridization. J Insect Physiol 53:235–245

Salama MS, Miller TA (1992) A diapause associated protein of the pink bollworm *Pectinophora gossypiella* Saunders. Arch Insect Biochem Physiol 21:1–11

Sato Y, Oguchi M, Menjo N, Imai K, Saito H, Ikeda M, Isobe M, Yamashita O (1993) Precursor polyprotein for multiple neuropeptides secreted from the suboesophageal ganglion of the silkworm *Bombyx mori*: Characterization of the cDNA encoding the diapause hormone precursor and identification of additional peptides. Proc Natl Acad Sci U S A 90:3251–3255

Sedwick C (2011) SRP and NAC: guiding a nascent protein's first steps. PLoS Biol 9(7):e1001103. doi:10.1371/journal.pbio.1001103

Thorpe C, Kim JJ (1995) Structure and mechanism of action of the acyl-CoA dehydrogenases. FASEB J 9:718–725

Tortoriello G, de Celis JF, Furia M (2010) Linking pseudouridine synthases to growth, development and cell competition. FEBS J 277:3249–3263

Walters KJ, Goh AM, Wang Q, Wagner G, Howley PM (2004) Ubiquitin family proteins and their relationship to the proteasome: a structural perspective. Biochim Biophys Acta 1695(1–3):73–87

Wickens M, Goodwin EB, Kimble J, Strickland S, Hentze MW (2000) Translational control of developmental decisions. In: Sorenberg N, Hershey JWB, Mathews MB (eds) Translational control of gene expression. Cold Spring Harbor Laboratory Press, Cold Spring Harbor

Xu WH, Sato Y, Ikeda M, Yamashita O (1995) Stage-dependent and temperature-controlled expression of the gene encoding the precursor protein of diapause hormone and pheromone biosynthesis activating neuropeptide in the silkworm, *Bombyx mori*. J Biol Chem 270:3804–3808

Role of RNA Interference in Pest Management

R. Asokan, Prakash M. Navale, N. K. Krishna Kumar and M. Manamohan

Abstract

The recent demonstration of the potential of RNAi in pest management has opened a new avenue which will fuel a futuristic approach where application of chemical insecticides will no longer be needed. Identification of suitable target genes and delivery method will usher a new ecofriendly approach that is safer to nontarget organisms including humans. In addition to its role in field level pest management RNAi is also promising in other investigations such as validation of gene functions, control of insect vector transmitted plant viruses, and management of insecticide resistance.

Keywords

Genes · Gene silencing · Pest management · RNA interference

Introduction

Meeting the food security for billions of people is a Herculean task faced by the Indian agriculture today and the population is slated to increase by leaps and bounds in the ensuing years. This calls for more land to be brought under cultivation and also increasing productivity of crops. While there is no scope to further increase the area under cultivation, increasing the productivity is the only plausible and viable option to meet the demand. It has been estimated that India's population will become roughly 2,000,000,000 by 2040 and we have to go a long way from the current production level of 800 million t of major crops. Among the various biotic factors, insect pests limit the crop productivity to a larger extent and the agrochemical policy group has reported that the crop loss during the year 2007 was ₹ 1.40 lakh crores due to pests. Management of insect pests by insecticides account for 61 % of the total pesticides usage and has already resulted in many control failures due to accelerated development of resistance. Further it has also resulted in contamination of soil, water, and affected nontarget pests including human beings.

R. Asokan (✉) · P. M. Navale · M. Manamohan
Division of Biotechnology, Indian Institute of Horticultural Research, Hessaraghatta Lake (PO), Bangalore 560089, India
e-mail: asokaniihr@gmail.com

N. K. Kumar
ICAR, New Delhi 110012, India

A. K. Chakravarthy (ed.), *New Horizons in Insect Science: Towards Sustainable Pest Management*,
DOI 10.1007/978-81-322-2089-3_19, © Springer India 2015

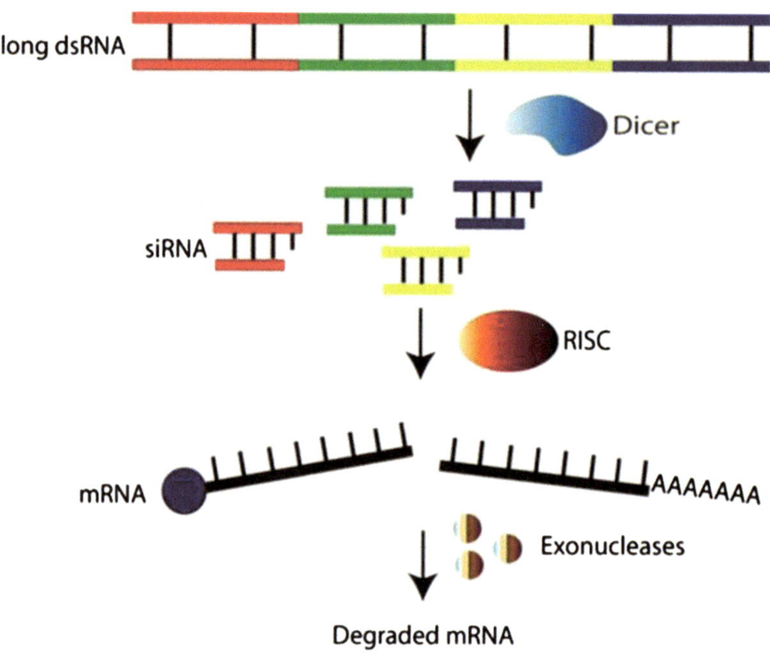

Fig. 1 General mechanism of RNAi

Therefore, there is an urgent need to look for an effective, ecofriendly alternative for pest management, where a new and novel approach called RNA interference (RNAi), popularly called Gene Silencing is poised to play a vital role. Employing this RNAi approach, it is now possible to control insect pests by silencing some of the vital genes that play an important role in insect-host plant interaction, growth and development, flight, reproduction, etc. by delivering cognate double-stranded RNA (dsRNA) either as spray or through transgenic plant. But RNAi approach has to be tailor made and cannot be universal, as the mechanism of RNAi is highly species specific. Hence the basic mechanisms underlying RNAi for selected species of insect pests that belong to three important orders of insects has to be studied thoroughly before venturing into field level application.

General Mechanism of RNAi

RNAi is an innate immune response that retard the gene expression by degrading specific mRNA molecules. RNAi mechanism is generally employed for maintaining genome integrity against the invasion by transposons and viruses. The RNAi pathway is initiated by the cellular enzyme called Dicer (RNase III), which cleaves the long double-stranded RNA (dsRNA) into short fragments of 20–25 base pairs. One of the two strands of each fragment known as guide strand is incorporated into the RNA induced silencing complex (RISC) and base pair with complementary mRNA sequences and cleaves the cognate mRNA thus arresting the target gene expression (Fig. 1). This mechanism is also known as posttranscriptional gene silencing (PTGS) in plants. In some organisms like *Caenorhabditis elegans* and ticks the silencing signal generated in a cell is communicated to other cells which are exposed to the dsRNA which is known as *systemic silencing*. Two components which are needed to elicit systemic silencing are RdRP (RNA dependent RNA polymerase) and *sid* 1 and 2 genes (systemic interference defective). Presence of at least one of them has been documented in many insects except *Drosophila melanogaster* (Fig. 2).

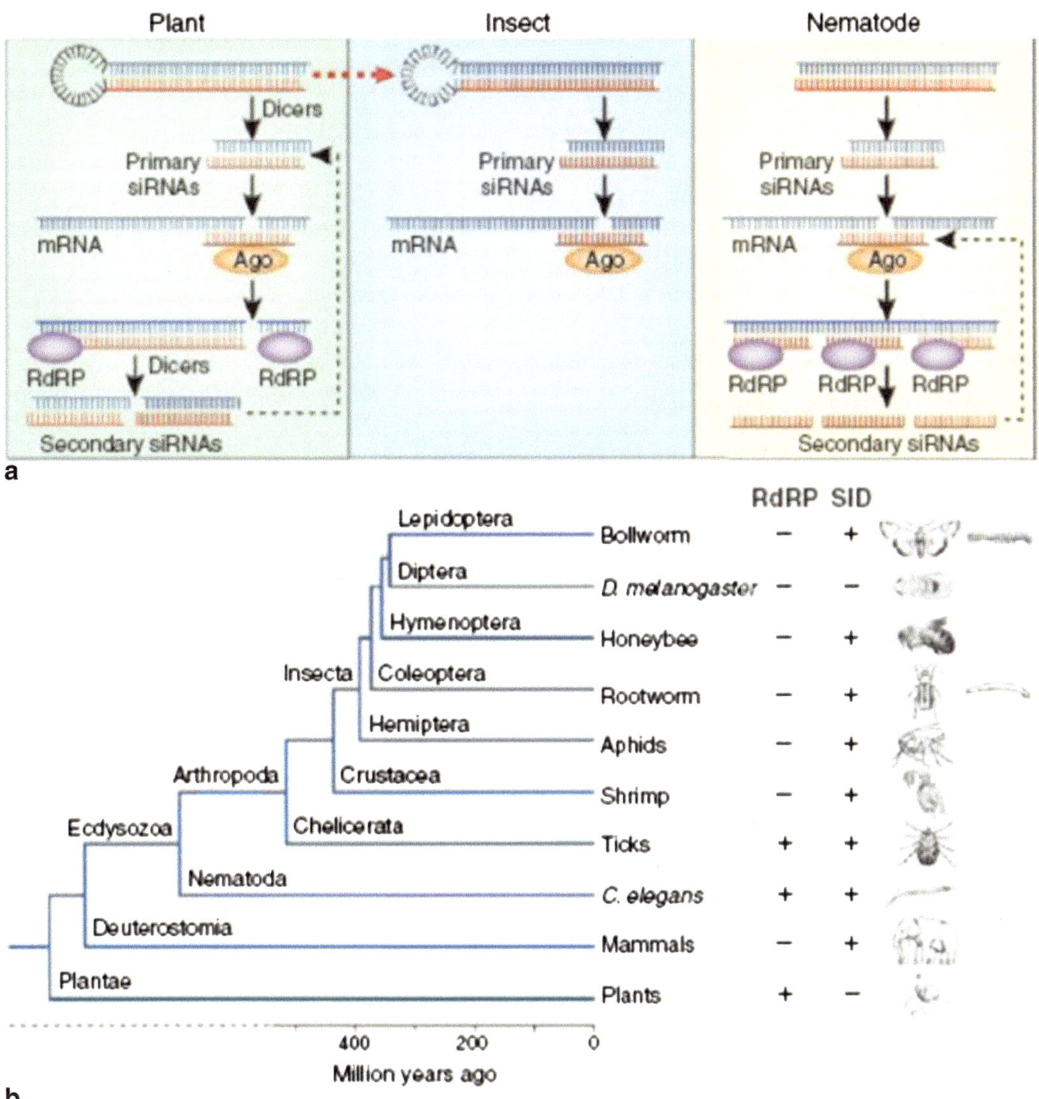

Fig. 2 Variation in theme of RNAi mechanism in various organisms. (Gordon and Waterhouse 2007)

RNAi in Insect Pest Management

Most of the studies on RNAi in entomology were on establishing the function of genes that are involved in various metabolic processes (Kramer & Bentley 2003; Fabrick et al. 2004; Dong and Friedrich 2005; Amico & Nijhout 2006; Arakane et al. 2006; Collinge et al. 2006; Lagos et al. 2006; Mutti et al. 2006). But many researchers envisaged the potential of RNAi for field level insect pest management, but success was not forthcoming due to many reasons like poor understanding on the mechanism of RNAi in different orders of insects, amplification and systemic spread of the silencing signal in the treated insects, amenability of different orders of insects for the RNAi approach, etc. In the year 1998 an experimental proof that introduction of dsRNA into the cell evoked RNAi mechanism in the free living nematode, *C. elegans* Maupas (Fire et al. 1998) opened a new exciting area of RNAi in various spheres of research including entomology. In the year 2007 two research groups from China and USA experimentally validated the utility of RNAi in insect pest management. The choice of the gene for RNAi mediated silencing could be many and generally falls into two broad categories, viz. silencing of genes that results in quick control, for example, genes involved in insect–plant interactions, digestion, moulting, etc. and silencing of genes that results in long-term management, for example, genes involved in pheromone biosynthesis, pheromone reception, migration, flight, diapause, etc. Mao et al. (2007) demonstrated that it is possible for a no chemical management of the cotton boll worm, *Helicovera armigera* Hub. by transgenic cotton mediated silencing of cytochrome P 450. Similarly Baum et al. (2007) proved that RNAi's approach is feasible in the management of the coleopteran pest the corn root worm (*Diabrotica virgifera virgifera* LeConte).

Plant Mediated-RNAi Against Insect Pests

The discovery of RNA interference was first studied in *Caenorhabditis elegans* where soaking of nematode in dsRNA solution resulted in the degradation of the target mRNA and thus affected the protein synthesis. So far more than 32 target genes have been screened for their potential for gene silencing in insects (Zhang et al. 2013). These observations showed that RNAi response in several less derived species is robust and inheritable (Ronco et al. 2008), whereas, it is otherwise in more derived species, for example, Lepidoptera exhibits variable RNAi response when administered with exogenous dsRNA (Terenius et al. 2011). The tremendous success of RNAi facilitated plant biotechnologists to utilize the various physiologically important genes from insects as ingestible insecticides through plant mediated expression of cognate dsRNA. The era of plant mediated RNAi (PM-RNAi) is a new line of defense against insects and nematodes (Huang et al. 2006; Baum et al. 2007; Mao et al. (2007).

Potential of PMRNAi in the management of root knot nematode (RKN) was successfully demonstrated by Huang et al. (2006), in *Meloidogyne incognita* L. *In vitro* studies revealed that a full length dsRNA16D10 reduced 93–97 % of transcript and 16D10 peptide by 65–69 % and reduced the extent of gall formation. Similarly Baum et al. (2007), studied 290 genes from Western corn root worm, *Diabrotica virgifera virgefer* and 14 genes like V-ATPase A, D, E and α-tubulin showed immediate response within 24 h of application of dsRNA. Similarly, susceptibility of other coleopterans such as Southern corn rootworm, *Diabrotica undecimpunctata*; Colorado potato beetle, *Leptinotarsa decemlineata* Say and cotton boll weevil, and *Anthonomus grandis* Boheman were studied for their response to RNAi. Recently dsRNA for *Mi-Rpn7* of *M, incognita* was introduced into the soyabean through hairy root culture for managing plant parasitic nematodes (Niu et al. 2012). The PMRNAi has been efficiently utilized in the control of green peach aphid *Myzus persicae* Sulzer which transmits more than 100 type of plant viruses by silencing *Rack*1 (*Receptor of Activated Kinase C*) and *C002* genes (Pitino et al. 2011). *Rack*1 based on the earlier gene silencing results as carried in the *C. elegans* (Simmer et al. 2003; Kamath et al. 2003) and *Heterorhabditis bacteriophora* Poinar (Ciche et al. 2007). Where knock down of this gene in *M. persicae* resulted in lethality in early developmental stage, stunted growth, reduced egg laying etc. *Rack*1 is the multifunctional receptor protein and one of the internal components of the circadian clock binds to the various proteins and initiates signal transduction cascades, it also functions in the actin organization. Another target gene of *M. persicae* was Mp*C002* a homologue of *COO*2 that plays an important role in insect–plant interaction and is expressed predominantly in salivary glands. Knockdown *COO*2 gene resulted in mortality and improved tolerance to peach aphid in transgenic tobacco plants.

Similarly, Mao et al. (2007) successfully silenced the allelochemical, *gossypol* detoxifying gene of *H. armigea* through artificial diet containing dsRNA for CYP450 monooxygenease gene and developing dsRNA expressing transgenic plants where there was two-fold weight loss in the treated larvae as compared to control. The above authors in the year 2013 further extended their studies and developed a new cotton expressing dsRNA for both membrane permeability enhancer gene *35S:GhCP1* and Cytochrome 450 monoxygenease downregulationg gene (*35S:dsCYP6AE14*). *GhCP1* is the serine protease of cotton that plays an important role in plant-insect interactions. The entry of *GhCP1* increases easier food absorption in the midgut. To employ RNAi mediated protection in tobbaco against *H. armigera*, Zhu et al. (2012) used 20-Hydroxyecdysone gene (*Ha*EcR), a steroid hormone required for the growth and development. In yet another study the ecdysone receptor, EcR has been tested by engineering into the tobacco which showed improved resistance to *H. armiger* and *Spodoptera exigua* Hub.

In yet another study Zha et al. (2012) demonstrated utility of RNAi in the management of *Nilaparvata lugens* Stal by expressing dsRNA for hexose transporter, trypsin like serine protease and carboxypeptidase. In this regard, the above researchers identified RNAi core machinery in *N. lugens* such as *Nlsid-1* and *Nlaub* gene, where *Nlsid -1* is needed for the dsRNA uptake and spread silencing signals among tissues. By using virus-based expression of dsRNA in plants, a new line of approach was developed in RNAi mediated management of *Manduca sexta* L. (Kumar et al. 2012). They called it as 'plant virus based dsRNA producing system'(VDPS). It was originally described and used against nematodes in the demonstration of the transient RNAi response in tobacco plants using tobacco rattle virus (Meyering-Vos and Muller A 2007; Dubreuil et al. 2009). Kumar et al. (2012) employed nicotine detoxifying cyp genes which were upregulated during nicotine detoxification, particularly, CYP4B46, CYP4M1 and CYP4M3. They selected the above three cyp targets based on the sequence homology where CYP4B46 showed 85 % homology to CYP4B45, the other two CYP4M1 and CYP4M3 showed 53 % similarity. They transformed the entire fragment individually into tobacco plants using VDPS method and CYP4B46 fragment alone transformed into the plants using *Agrobacterium* mediated transformation method. Interestingly this approach of silencing CYP4B46 did not affect larval weight but the level of transcript, whereas, the VDPS-CYP4M3 ingested larvae had gained less weight as compared to control. In another successful example, controlling *H. armigera* was achieved by the knockdown of the hormone regulating transcription factor *HR*-3 (Xiong et al. 2013). The above authors studied the effect of dsRNA on the larvae by providing different size fragments of dsRNAha*HR*-3 delivered through the artificial diet where the bacterially expressed dsRNA could cause significant mortality in 3–7 days after treatment. Molecular validation of gene silencing showed that *HaHR*-3-1 and *HaHR*-3-3 were dominant in silencing effect than *HaHR*-3-2 and *HaHR*-3-4, the repression of this protein negatively affected the development of the *H. armigera*.

Imparting Plant Resistance to Viruses

Resistance to RNA viruses is brought about by the enzyme RNA dependent RNA polymerase sequence-specific degradation of the targeted viral mRNA. Studies have indicated RNAi could be brought about by siRNA or dsRNA molecules that are cognate to the viral coat protein genes. This has been demonstrated in tobacco, squash, and papaya. But RNAi could not protect against the infection of single-stranded DNA Gemini viruses that causes severe damage to cassava and tomato (Auer and Frederick 2009).

Bacterial Blight and Fungal Resistance

As compared to the other research areas application of RNAi in offering protection against fungal and bacterial pathogens is very scanty. Some studies have indicated that small RNAs have changed their expression during pathogen attack and also regulate gene expression during the pathogen attack. Take also participate in regulation of gene expression involved in disease resistance pathways. The small RNAs silence the negative regulator molecules in the plant cell under normal circumstances, but permit quick upregulation of genes when pathogens attack. Studies have shown that silencing two bacterial genes, *iaaM* and *ipt* reduced the formation of crown gall tumors by *Agrobacterium*. This opens a new avenue for engineering trees and woody ornamental plants against crown gall formation (Auer and Frederick 2009).

MicroRNAs in Imparting Plant Resistance to Diseases

The first identified miRNA to play an active role in pathogen associated molecular pattern triggered immunity Pattern triggered immunity (PTI) was in Arabidopsis. The bacterial infection induces expression of miR393 which negatively regulates the mRNAs for the auxin receptor, including the transport inhibitor response1 for the degradation. It has been observed that over-expression of miR393 suppressed auxin signaling and restricted the growth of the bacterium, *Pseudomonas syringae* pv. Tomato. In addition to the above, miR167 and miR160 that target auxin response factors were also found to be induced by the Pseudomonas syringae pv. tomato (Pst). Hence several miRNAs that repress auxin signaling are induded by bacterial pathogens and contribute to the PTI (Padmanaban et al. 2009).

References

Amico LJ, Nijhout HF (2006) Isolation and characterization of a bombyxin-like cDNA from *Manduca sexta*. J Insect Sci 6:14

Arakane Y, Lomakin J, Muthukishnan S, Beeman RW, Gehrke SH, Kanost MR, Kramer KJ (2006) An RNAi study of the metabolic pathways for cuticle sclerotization and pigmentation in *Tribolium castaneum*. J Insect Sci 6:4

Asokan R (2007) Genetic engineering of insects. Resonance 12:47–56

Auer C, Frederick (2009) Crop improvement using small RNAs: applications and predictive ecological risk assessments. Trends Biotechnol 27:644–651

Baum JA, Bogaert T, Clinton W, Heck GR, Feldman P, Ilagan O, Johnson S, Plaetinck G, Munyikwa T, Pleau M, Vaughn T, Roberts J (2007) Control of coleopteran insect pests through RNA interference. Nat Biotechnol 25:1–5

Ciche TA, Sternberg PW (2007) Postembryonic RNAi in *Heterorhabditis bacteriophora*: a nematode insect parasite and host for insect pathogenic symbionts. BMC Dev Bio 7:101

Collinge D, Gordon K, Behm C, Whyard S (2006) Stable transformation and RNAi in *Helicoverpa armigera*. J Insect Sci 6:10

Dong Y, Fredrich M (2005) Nymphal RNAi: systemic RNAi mediated gene knockdown in juvenile grasshopper. BMC Biotechnol 5:5–13

Dubreuil G, Magliano M, Dubrana MP, Lozano J, Lecomte P, Favery B, Abad P, Rosso MN (2009) Tobacco rattle virus mediates gene silencing in a plant parasitic root-knot nematode. J Exp Bot 60(14):4041–4050

Fabrick JA, Kanost MR, Baker JE (2004) RNAi-induced silencing of embryonic tryptophan oxygenase in the pyralid moth, *Plodia interpunctella*. J Insect Sci 4:1–9

Fire A, Xu S, Montgomery MK, Kostas SA, Driver SE, Mello CC (1998) Potent and specific genetic interference by double-stranded RNA in *Caenorhabditis elegans*. Nature 391:806–811

Gordon KHJ, Waterhouse PM (2007) RNAi for insect proof plants. Nat Biotechnol 25:1231–1232

Huang G, Allen R, Davis EL, Baum TJ, Hussey RS (2006) Engineering broad root-knot resistance in transgenic plants by RNAi silencing of a conserved and essential root-knot nematode parasitism gene. Pro Nat Acad Sci 103(39):14302–14306

Kamath RS, Fraser AG, Dong Y, Poulin G, Durbin R et al (2003) Systematic functional analysis of the *Caenorhabditis elegans* genome using RNAi. Nature 421:231–237

Kramer SK, Bentley WE (2003) RNA interference as metabolic engineering tool: potential for in vivo control of protein expression in an insect larval model. Metabo Eng 5:183–190

Kumar P, Pandit SS, Baldwin IT (2012) Tobacco rattle virus vector: a rapid and transient means of silencing *Manduca sexta* genes by plant mediated RNA interference. PLoS One 7:e31347

Mao Y-B, Cai WJ, Wang J-W, Hong G-J, Tao X-Y, Wang, L-J, Huang Y-P, Chen x-Y (2007) Silencing of a cotton boll worm P450 monooxygenase gene by plant mediated RNAi impairs larval tolerance of gossypol. Nat Biotechnol 25:1307–1313

Meyering-Vos M, Muller A (2007) RNA interference suggests sulfakinins as satiety effectors in the cricket *Gryllus bimaculatus*. J Insect Physiol 53:840–848

Mutti S, Park Y, Reese JC, Reeck GR (2006) RNAi knockdown of a salivary transcript leading to lethality in pea aphid, *Acyrthosiphon pisum*. J Insect Sci 6:1–7

Niu JH, Jian H, Xu J, Chen C, Guo Q, Liu Q, Guo Y (2012) RNAi silencing of the Meloidogyne incognita Rpn7 gene reduces nematode parasitic success. European J Plant Pathol 134:131–144

Padmanaban C, Zhang X, Jin H (2009) Host small RNAs are big contributors to plant immunity. Curr Opin Plant Biol 12:465–472

Pitino M, Coleman AD, Maffei ME, Ridout CJ, Hogenhout SA (2011) Silencing of aphid genes by dsRNA feeding from plants. PLoS One 6(10):e25709

Ronco M, Uda T, Mito T, Minelli A, Noji S, Klingler M (2008) Antenna and all gnathal appendages are similarly transformed by homothorax knock-down in the cricket Gryllus bimaculatus. Dev Biol 313(1):80–92

Simmer F, Moorman C, van der Linden AM, Kuijk E, van den Berghe PV, Kamath RS, Fraser AG, Ahringer J, Plas-terk RH (2003) Genome-wide RNAi of *C. elegans* using the hypersensitive rrf-3 strain reveals novel gene functions. PLoS Biol 1(1):77–84

Terenius O, Papanicolaou A, Garbutt JS et al (2011) RNA interference in Lepidoptera: an overview of successful and unsuccessful studies and implications for experimental design. J Insect Physiol 57:231–245

Xiong Y, Zeng H, Zhang Y, Xu D, Qiu D (2013) Silencing the HaHR3 Gene by Transgenic Plant-mediated RNAi to Disrupt Helicoverpa armigera development. Int J Biol Sci 9:370–381

Zhang H, Li HC, Miao XX (2013) Feasibility, limitation and possible solutions of RNAi-based technology for insect pest control. Insect Sci 20(1):15–30

Zhu JQ, Liu S, Ma Y, Zhang JQ, Qi HS, Wei ZJ, Yao Q, Zhang WQ, Li S (2012) Improvement of pest resistance in transgenic tobacco plants expressing dsRNA of an insect-associated gene EcR. PLoS One 7(6):e38572

Molecular Characterization and Management of Shoot and Fruit Borer *Conogethes punctiferalis* Guenee (Crambidae: Lepidoptera) Populations Infesting Cardamom, Castor and Other Hosts

P. R. Shashank, B. Doddabasappa, Vasudev Kammar, A. K. Chakravarthy and Hiroshi Honda

Abstract

The genus *Conogethes* is a large, taxonomically complex cosmopolitan moth taxon. To date, 24 species have been identified by DNA bar codes and deposited in the Barcode of Life Data Systems (BOLD) (http://www.boldsystems.org). Type locality of *Conogethes punctiferalis* is India, so many closely allied species may be included but taxonomic revision of them has been neglected for a long time. Integration of several approaches like conventional taxonomy, DNA bar code, electrophysiological and behavioral responses of adults and larvae, host–plant relationship patterns, hybridization experiments, biochemical analysis, and analysis of and responses to pheromone components of *Conogethes* reared on castor and cardamom confirmed that the populations belong to two different species. Bioecology and management of *Conogethes* have been well studied. More than 31 alternate host plants for the borer have been recorded. Therefore, it has better adaptation in the cultivated ecosystems. An integrated approach involving nontoxic chemicals is desired as currently mostly insecticides are applied to suppress this pest.

Keywords

Cardamom · Castor · *Conogethes punctiferalis* · Management

A. K. Chakravarthy (✉) · V. Kammar
Department of Agricultural Entomology, Gandhi Krishi Vignan Kendra (GKVK), University of Agricultural Sciences, Bangalore, Karnataka 560065 India
e-mail: chakravarthyakshay@gmail.com

P. R. Shashank
Division of Entomology, IARI, New Delhi, India

B. Doddabasappa
College of Horticulture, Kolar, Karnataka, India

H. Honda
Faculty of Life and Environmental Sciences, University of Tsukuba, Tsukuba, Ibaraki, 305-8572 Japan

Introduction

The fruit and shoot borer *Conogethes* (= *Dichocrocis*) *punctiferalis* mainly occurs in tropical and subtropical countries (Pena et al. 2002) and is distributed in Asia and Australia. Larvae of this crambid moth are typically polyphagous pests attacking more than 120 wild and cultivated diversified plants, namely, peach, chestnut,

A. K. Chakravarthy (ed.), *New Horizons in Insect Science: Towards Sustainable Pest Management*,
DOI 10.1007/978-81-322-2089-3_20, © Springer India 2015

durian, citrus, papaya, castor, cardamom, ginger, eggplant, and maize (Sekiguchi 1974). Management of the pest by insecticides is not only undesirable, but also ineffective and expensive (Chakravarthy et al. 2012). Further, this insect has more than 30 alternate host plants (Hussain et al. 1995). Until now ten named species of genus *Conogethes* have been known from eastern Palaearctic and Indo-Australian regions (Meyrick 1884). The type locality of *C. punctiferalis* is India, so many closely allied species may be included; however, taxonomic revision of them has been neglected for a long time. *C. punctiferalis* is in focus because of the expanding host range, geographical occupancy, and complexity involved in species identifications. As such the pest, an internal tissue borer is difficult to manage in fruit orchards and plantations. This insect group is undergoing speciation, genomic changes, or evolving into new taxon. So, molecular characterization and management aspects of *Conogethes* spp. are discussed in this chapter.

Distribution and Host Range

Conogethes punctiferalis is localized to Asia, Australia, and Papua New Guinea. In Asia, it is found in, for instance, China (AQSIQ 2007), India, Indonesia, Japan, Korea, Malaysia, Taiwan, Thailand, Vietnam (Gour and Sriramulu 1992; Hang et al. 2000; Kang et al. 2004). *C. punctiferalis* is a polyphagous crop and fruit pest found to infest 30 crop plants belonging to 23 different families in India. Cardamom is the most preferred plant followed by *Hedycium* spp., *Alpinia* spp., and *Ammomum* spp. (Thyagaraj 2003). *Gossypium hirsutum* L., *Zea mays* L, *Sorghum bicolor* (L.)*Moench* spp., *Psidium guajava* L., *Curcuma longa* L., *Zingiber officinale* Roscoe, *Citrus* L., *Mangifera indica* L., and *Punica granatum* L. served as the host plants for this pest (Gurmey 1918; Tryon 1920; Flether 1922; Ballard 1924; Clausen 1927; Veitch 1931; Hutson 1937; Narasimhaswamy 1937; Sloan 1945; Twine 1971; Bilapate and Talati 1977; Cai and Mu 1993; Lu et al. 1995; Wu 1995; Konno and Shishido 1996; Peter 1996; Park et al. 1998). *Theobroma cacao* L. and *Vitis vinifera* L. were also recorded as host plants (Mohanan and Kumar 1976; Gour and Sriramulu 1992; Ram et al. 1997), and *Ricinus communis* L. served as another major host plant for this pest (Flether 1922; Ballard 1924; Pruthi 1944; Issac 1948; Bilapate and Talati 1978; Hossain et al. 1995; Sharma et al. 1995). *Diospyros virginiana* L. is also attacked by this pest (Ono 1937; Tomomatsu 1995). *C. punctiferalis* has also been recorded on yellow peach as a major pest (Konno et al. 1980; Konno et al. 1981; Kadoi and Kaneda 1990; Abe and Sanari 1992; He 1997; Kimura and Honda 1999) in Japan and other countries.

On *Musa* L., *C. punctiferalis* was recorded as a pest (Jarvis 1914). In China, *Castania mollissimia* is attacked by the pest (Ni 1998), and this pest has also been recorded to attack *Malus* Mill (Kadoi and Kaneda 1990). Conogethes has been recorded on Soybean (Anonymous 1944), *Glycine max* (L.) Merr., *Tectona grandis* L., and *Ceiba pentandra* (L.) Gaertn in Java (Tryon 1920; Kalshoven 1922, 1928), *Macadamia* F. Muell. In Queensland (Ironside and Davis 1969), *Pinus sp.* (Shiukaji 1969), *Crataegus* L. (Sun et al. 1992), and *Citrus sinensis* L. (Anonymous 1913a). *Prunus domestica* and *Persea americana* (Wang and Cai 1997), *Prunus persica var. nectarina* (Tryon 1920), *Artocarpus heterophyllus* Lam. (Devasahayam et al. 1998) were also damaged by this pest. *Sapindus sp.* was attacked by this pest (Rao 1992). *Quercus virginiana* is known to be attacked by this pest (Park et al. 1998) in temperate region.

Biosystematics

Earlier workers have put *Conogethes* under Pyralidae. Maes (1998) demonstrated that the differences in structure of tympanal organ or ears that are present at the base of abdomen called the *praecinctorium*, which joins two tympanic membranes in the Crambidae, are absent in the Pyralidae. The latest review by Munroe and Solis, in Kristensen (1999) retains the Crambidae family. Genus *Conogethes* is a large, taxonomically complex taxon belonging to superfamily Pyraloidea, family Crambidae, subfamily Pyraustinae.

The genus for a long time was under confusion because the taxonomy was based on wing venation. Since the publication of Sir George Hampson's Fauna volumes in 1896 in the "Fauna of British India" series, considerable changes in the taxonomy of this group have taken place. The species *punctiferalis* was placed in genus *Conogethes* by Meyrick (1884), although moved to *Dichocrocis* after that. It has been placed back in *Conogethes* by Munroe (Munroe 1989). Shaffer et al. (1996) placed it in *Conogethes* as a revised combination. The Natural History Museum (London) card index is yet to be updated (10/10/2012), so we considered GlobIZ (Global Information System on Pyralidae) as a better source. Hence, retention of *Conogethes* for the crambid moth is now followed by entomologists all over the world, excepting few in China. Hampson in his "Fauna of British India" reported 20 species of *Dichocrocis* mainly based on wing venation and arrangements of black spots on wings. Subsequently, few preliminary studies have been done by lepidopterists in India and abroad on this genus. Five species of *Dichocrocis* were identified, namely, *D. evaxalis* Walker, *D. punctiferalis* Walker, *D. nigrilnealis* Walker, *D. plutusalis* Walker, and *D. surusalis* Walker from light trap collections in Kerala, India (Mathew and Menon 1984).

Azam and Ali (1965) studied the morphology of larva of *D. punctiferalis* with special reference to chaetotaxy collected from castor bean (*Ricinus communis* L.). The genitalial morphology of *D. punctiferalis* and *D. plutusalis* was studied and main modification in the morphology of valvae, uncus, saccus, phallus, bursae, and ductus in these species were elucidated (Mathew and Menon 1989). Until now, ten named species of genus *Conogethes* Meyrick (1884) have been known from eastern Palaearctic and Indo-Australian regions (Shaffer et al. 1996). Inoue and Yamanaka (2006) redescribed *C. punctiferalis* along with two new (*C. parvipunctalis and C. pinicolalis*) closely allied species from eastern Palaearctic and Oriental regions. In mid-1980s, Chakravarthy (unpublished observations) found differences in morphology of *Conogethes* moths reared on castor and cardamom (*Elettaria cardamomum* Maton) in the Western Ghats of Karnataka, South India. The male genitalia also differed between the two. The *Conogethes* larvae reared on castor bean and cardamom required two different mass-rearing techniques (Chakravarthy et al. 1991). Nowadays DNA bar coding is a major tool for identification of species. Molecular taxonomy is giving an additive support for species identification with traditional taxonomy. Seventeen species (Table 1) from six countries, namely, Australia, Papua New Guinea, Cambodia, China, Indonesia, and Nepal are bar coded for genus *Conogethes* and deposited in the Barcode of Life Data Systems (BOLD) till today.

Unfortunately, there is no single species from India that is bar coded. There is a need for taxonomic revision of this genus with the help of molecular taxonomy. The species listed in Hampson's fauna differ from those that are bar coded. Recently, Shashank (2012) bar coded *Conogethes* moths reared on castor and cardamom from different geographical locations of India. In many classifications, the family Crambidae has been treated as a subfamily of the Pyralidae or snout-moths. Currently, Crambidae is treated as a full-fledged family.

In Japan

In Japan biosystematics of Japanese *Conogethes* spp. with special reference to host plant preference and reproductive isolation was done (Honda 2013). In genus *Conogethes*, *C. punctiferalis* (CPU) and *C. pinicolalis* (CPI) are the most well-known pest species of agricultural and forest plants in Japan. The *C. punctiferalis* was called as the fruit-feeding type (FFT) of the yellow peach moth in order to distinguish from other feeding group, the Pinaceae-feeding type of *C. punctiferalis*, which was registered as *C. pinicolalis* in 2006.

Morphology of male and female adults between these species is very slightly different, but some morphometric comparisons were possible in order to segregate both species. Although the *C.*

Table 1 *Conogethes* species with records on BOLD. (Source: http://www.boldsystems.org/views/speciessummary.php)

No.	Species	Specimens	Sequences	Bar codes (COI of > 500 bp)
1	*Conogethes diminutiva*	2	2	1
2	*Conogethes ersealis*	3	3	3
3	*Conogethes evaxalis*	9	8	8
4	*Conogethes haemactalis*	3	3	3
5	*Conogethes nr. diminutiva*	1	1	1
6	*Conogethes nr. haemactalis*	3	3	3
7	*Conogethes parvipunctalis*	1	1	1
8	*Conogethes pluto*	12	12	12
9	*Conogethes punctiferalis*	67	26	26
10	*Conogethes punctiferalis PS2*	1	0	0
11	*Conogethes semifascialis*	11	8	8
12	*Conogethes sp.*	1	0	0
13	*Conogethes sp.* ANIC 1	2	1	1
14	*Conogethes sp.* ANIC 2	3	3	3
15	*Conogethes sp.* ANIC 3	1	0	0
16	*Conogethes sp. complex*	4	3	3
17	*Conogethes tharsalea*	5	4	4

punctiferalis larvae were typically polyphagous, their development delayed on *C. pinicolalis* host, while *C. pinicolalis* lay eggs and feed exclusively on conifer needles including *Pinus parviflora* or *Abies homolepis*. Such host plant preference may also cause reproductive isolation between both species. In electroantennogram (EAG) (*C. punctiferalis* females responded of host plants to 17 more compounds more than males, whereas there was no sexual difference in *C. pinicolalis*. A cluster analysis of the EAG responses of each species showed a definite difference in the antennal olfactory spectra between *C. punctiferalis* and *C. pinicolalis*. Although the male moths of both species were cross-attracted to calling females and their pheromone gland extracts, a strong homogamic mating preference in laboratory test and postmating reproductive isolation between both species was also confirmed by laboratory cross tests. Female sex pheromone system of *C. punctiferalis* and *C. pinicolalis* are quite similar, which allows cross-attraction by males, and consisted of E-10-hexadecenaal (E10-16:Ald) and Z-10- hexadecenal (Z10-16:Ald) at a ratio of 95.4:4.5. The final conspecific sexual recognition in each species is accomplished with a male pheromone. E-2-methyl-2-butenoic acid was identified from hairpencil organs of *C. punctiferalis* but no pheromonal volatiles from *C. pinicolalis* hairpencils were identified. Recently, two hydrocarbons were found as pheromonal synergists in female pheromone system of *C. punctiferalis*, which functioned in a short distance from pheromone source, calling females. A similar system also was prospected in *C. pinicolalis*, but these new hydrocarbon synergists may have no contribution to reproductive isolation between *C. punctiferalis* and *C. pinicolalis* (Honda 2013).

Bioecology

Krishnamurthy et al. (1989) reported that *Conogethes* completed one life cycle on cardamom in 25–40 days with five generations a year at Mudigere, Chikmangalur, Karnataka at 26 °C and 70 % relative humidity (RH). Eggs of *C. punctiferalis* are round and light yellow in color, and

0.63 × 0.41 mm in size. After incubation of 6–7 days, the eggs turn dark brown with a dark head (Bilapate 1978; Jarvis 1914; Thyagaraj 2003). Then phenology of host plant influences the size, growth, and development of eggs (Bilapate 1978; Jacob 1981; Twine 1971). There was a significant difference in percent egg hatching from 65.0 ± 0.76 to 90.5 ± 1.38 and incubation period from 4.19 ± 0.80 to 9.35 ± 1.05 days under varied temperature and relative humidity (Thyagaraj 2003; Wang and Cai 1997). Temperature and relative humidity play an important role in egg hatching (Rajan 1965). The egg characteristics under various temperature ranges and relative humidity conditions have been extensively studied in laboratory (Kalshoven 1929; Thyagaraj 2003).

First instar (neonate) larva bored the pseudostem or capsule. On pseudostem, the larva bored at the base of leaf axis and entered inside the cardamom shoot tissue (Jacob 1981). The excreta plugged at the entry hole on the shoot indicated larval boring (Thyagaraj 2003). Larva fed on the shoot was light greenish, while those on capsules were dull yellow (Bilapate and Talati 1977; Thyagaraj 2003). The larva remains inside the pseudostem till pupation. Therefore, it is difficult to study the larval instars directly. Dyar's law was applied to record the number of larval instars, and there were five larval instars (Asante 1991; Dyar 1890; Thyagaraj 2003). Observations revealed that there was no major difference among the instars except for head and body size. All larval instars were active and when disturbed, tried to fall down with a fine silken thread (Bilapate and Talati 1978; Kondo and Miyahara 1930; Kodoi 1990; Thyagaraj 2003; Twine 1971; Young and Shaw 1962).

The growth and development of different larval instars varied with varying temperature and relative humidity. Each larval instar lasted for 3–4 days. Sloan (1945) reported that, larval period of *C. punctiferalis* lasted 3 weeks under normal conditions and 2–3 weeks in winters. Yang and Shaw (1962) studied the peach borer biology in China and reported 4–5 generations a year, the larvae overwintering in the flowers, stem, and fallen leaves. The duration of the larval stage varied from 20–23 days in August–September at 21–35 °C to 22–26 days in October–January at 14–28 °C and larvae were found in the field until February, and also the total larval period extended up to 12–14 days (Bilapate 1977; Jacob 1981; Kondo and Miyahara 1930; Twine 1971; Wang and Cai 1997; Xi et al. 1996). The larval period varied from 12.55 ± 2.00 to 19.59 ± 5.50 days and the percent survival varied from 49.6 ± 0.18 to 92.8 ± 1.39. 28.0 ± 1.0 °C and 80.0 ± 5.0 % RH were most favorable for larval development (Thyagaraj 2003).

Pupation of *Conogethes* took place in cocoons inside or between the capsules, and the pupal stage lasted for 7–10 days (Patel and Gangrade 1971; Bilapate and Talati 1978; Gour and Sriramulu 1992).The duration of development varied from 27 days at 30 °C to 48–51 days at 20 °C. Pupal period took over 8 weeks and more in winter on sorghum in Queensland (Sloan 1945).According to Wu (1995) 86.5 % larvae of *C. punctiferalis* pupated in leaf axils and in ears of maize and 13.5 % pupated in stalks. There is a clear difference in size, shape, and weight of male and female pupae. Female pupae were larger (17.81 × 6.29 mm with 0.127 gm in weight). Male pupa measured 14.30 × 4.26 mm with 0.108 gm in weight, shorter, slightly narrower, and the genital opening was located in the posterior region of the ninth abdominal segment and flanked by a pair of pads. Significant difference was also noticed between the sexes in terms of length and diameter (Thyagaraj 2003). Pupal period extended up to 7.90 ± 2.80 mean days in laboratory. There was no significant difference in the pupal period under different temperature regimes (20.0–38.0 ± 1.00 °C (Bilapate and Talati 1977; Jacob 1981; Mishra and Teotia 1965; Wang and Cai 1997). There is a morphological dimorphism in the *Conogethes* pupa that helps in sex determination.

Development and reproduction of *C. punctiferalis* were investigated at five different temperatures (15, 19, 23, 27, and 31 °C) with chestnut as food in China. The results showed that temperature had significant effects on the developmental duration, survival rate, pupal weight, and reproduction. The developmental duration at every stage reduced with increasing temperatures from

15 to 27 °C and there was a positive relationship between the developmental rate and temperature. The results provide the basis for forecasting the occurrence of the yellow peach moth in agroecosystems. Bilapate and Talati (1978) revealed that the males survived for 14.00±3.80 days as compared with 15.80±2.50 days for females. The ratio of males to females of the progeny ranged between 1:1 and 1:2. The rate of occurrence of the banded adult form, *C. punctiferalis* varied between 1 and 5 % during different months. Kaneko (1978) observed a clear abdominal constriction in female *C. punctiferalis* that had already paired. This proved a reliable indication of pairing and greatly facilitated the separation of virgin and paired females. The constriction was apparent 45–60 min after pairing. Kang et al. (2004) reported that moths of the overwintered *C. punctiferalis* generation emerged from May 20 to June 28 and reached to 50 % emergence on June 8–9 under emergence cage. Attraction of the male moths of overwintered generation to sex pheromone traps showed that the date of 50 % catches to the traps was June 17. Kuang et al. (2009) reported that, in China, adults emerged mainly between 22.00–08:00 h. The emergence rate was 92.22 %. The dates corresponding to the emergence of 16, 50, and 84 % of the total number of moths were in May 8, 11, and 15, respectively. The longevity of adults was significantly affected by adult nutritional conditions. Adults fed on water, 10 % honey solution, or 10 % sugar and vinegar liquid survived significantly longer than when they remain unfed. *C. punctiferalis* males have hair pencils at the abdominal end and thus can be differentiated from the other sex. Mating occurs only in the dark after 7.30 PM (Stanley et al. 2009).

Hybridization experiments with *Conogethes* moths reared on castor and cardamom were conducted under laboratory conditions. The mating experiments were conducted in single pair/cage and multiple pairs (4 pairs/cage). In cages where multiple pairs were enclosed, higher mating success was achieved. When the moths were reared on "shifted" plants, no mating occurred although the moths attempted to mate several times (Shashank 2012), the sexes were reproductively isolated.

Sex Ratio and Adult Longevity

Mean sex ratio worked out to be 1:1.095 (1:1.2) male to female in *C. punctiferalis* (Sithanantham and Subramaniam 1975; Bilapate and Talati 1977; Thyagaraj 2003). Life table studies pin point the key factors in the management of the pest (Moralesranous and Cote 1992). Adult longevity studies clearly indicated no significant difference between male and female moths. But laboratory studies with artificial diet showed differences in longevity between male and female, i.e., female moths survived 2–3 days more than the male (Sithanantham and Subramaniam 1975; Shanuowr et al. 1993).

Life Table

Life cycle of an insect varies with changing environmental conditions (Abraham 1965; Bilapate and Talati 1978). Insect populations are primarily controlled by weather conditions. Under laboratory conditions (28.0±1.0 °C and 80.0±5.0 % RH), the incubation period lasted for a mean of 5.3±0.49 days. Larval period 17.62±4.88 days, pupal period 8.81±0.69 days and the mean male and female longevity worked out to be 14.26±3.29 and 15.29±3.39 days, respectively, in *Conogethes* species. Sex ratio was 1:1.3. The total life cycle from egg to adult emergence required a mean of 31.75±10.16 days (Pruthi 1944; Young and Shaw 1962). Under field conditions in cardamom plantations the incubation lasts 8.51±0.65 days, larval period 25.49±4.76 days, pupal period 9.55±1.12 days, and the total life cycle from egg to adult emergence was 43.63±11.23 days. There was 8–10 days difference in the total number of days in a particular generation. However, there will be still differences in the life cycle between generation to generation due to changed environmental conditions.

Seasonal Occurrence

Shoot and fruit borer, *C. punctiferalis* occurred throughout the year on Cardamom in Western

Ghats in South India. Two peaks in the population were noticed in a year, i.e., one during April–May and the other during November–December (Thyagaraj 2003). The population coincided with the period of less or no rainfall, i.e., during premonsoon and postmonsoon periods (Ballard 1927; Ono 1937; Thyagaraj 2003). Temperature and rainfall influences greatly the growth and development of life stages (Moralesranous and Cote 1992; Rao 1992; Shanuowr et al. 1993). The overwintering of the needle feeding type/Pinaceae feeding type (PFT) of the yellow peach moth, *C. punctiferalis* were studied in China and Korea under laboratory and field conditions. Authors observed that third or fourth instar larvae rolled the needles in twigs into a bag with silk, in which they overwintered (Kang et al. 2004; Kuang et al. 2009).

Feeding Behavior

Larvae of *C. punctiferalis* exhibit diverse feeding habit; feeding on underground rhizome/root to cones/needles in pine to fruiting bodies from ground to fruit trees. On cardamom, moths lay eggs singly on the top of the leaf axis of young pseudostems; rarely two larvae are found in a pseudostem. This kind of egg laying habit probably is to avoid larval competition for food within the same pseudostem (Thyagaraj 2003). The young caterpillar bores at the base of the pedicel and later at the base (0.3 m above ground level) of the seedlings in the nursery and at the nodal region of the grown-up suckers, feeding the central tissue, and tunneled the shoot causing a dead heart. The live caterpillar indicates its presence by faecal matter (excreta) from the entry hole (Anonymous 1913a; Kalshoven 1922; Clausen 1927; Anonymous 1944; Thyagaraj 2003). Larvae feed on capsules from July to November (Anonymous 1918; Thyagaraj 2003). Second instar larvae bored into the capsule, fed on the immature seeds leaving empty capsule and then moved to another capsule or psuedostem to soil for pupation (Kalshoven 1929; Smith 1937; Sloan 1945; Sen Gupta and Behera 1955; Ironside and Davis 1969; Ram et al. 1997; Wang and Cai 1997; Ni 1998). Occasionally larvae bored into panicles also. Such panicles dry up without bearing capsules. Feeding preference test showed that among the cardamom plant parts, tender shoots were preferred the most, following young capsules (Autson 1923; Thyagaraj 2003).

Attempts have been made to mass rear *Conogethes* on artificial diet. Most insect studies deal with laboratory-reared insects, often on artificial diets. In fact, the rearing conditions and the diet composition are critical parameters for insect quality and yield. So far, artificial diet has been developed and proposed for the maintenance and continuous rearing of *C. punctiferalis* FFT using "soybean meal powder" and corn seeds (Honda et al. 1979; Utsume et al. 1990). Although there has been some success in efforts to rear successive generations by these diets, still there are lacunae.

Crop Loss

The shoot damage in cardamom due to borer *Conogethes* infestation varied from 5 to 10%. In terms of capsule yield loss (dry weight basis), it varied from 6.79 to 9.18%. Therefore, combining together, the loss was estimated to be more than 20% every year (Kapadia 1996;Thyagaraj 2003). However, the crop loss due to this pest was worked out in cardamom and economic threshold level was fixed at 10% (Anonymous 1954; Krishnamurthy et al. 1989; Ram et al. 1997). Studies on castor in Salem, Tamil Nadu, South India recorded 10.80–26.70% capsule damage (Suganthy 2011). Kapadia (1996) estimated 42.30% crop loss in castor in Indis. Fifty percent yield reduction was recorded on grapes due to *C. punctiferalis* attack (Ram et al. 1997).

Mitochondrial DNA Barcoding

The protein coding genes are the most frequently sequenced mitochondrial genes for evolutionary studies and phylogenetic analysis. Shashank (2012) did molecular characterization in *Conogethes*. Six legs of moths were taken in a 1.5 ml eppendorf tube and powdered well using liquid nitrogen. The DNA extraction was done using Cetyltrim ethyl ammonium bromide (CTAB) procedure and DNA amplification was done with procedure described by Doyle and Doyle (1990). Extracted DNA pellet was dissolved in 500 µl TE and quantified using Nanodrop DNA quantifier and electrophoresis method (0.8 % agarose gel). Two µl of isolated DNA were diluted to 1 ml with TE buffer and the absorbance at 260 and 280 nm were recorded against a buffer blank for assessment of DNA purity. A 260/280 nm ratio for all the samples was calculated to check the purity. DNA was quantified using: µg ds DNA/µl = (A260 * 40)/2. Further all samples were diluted to a final concentration of 10 ng/µl.

DNA Bar coding

The CBOL established the "All-Leps Barcodes of Life" project because the Lepidoptera is the second most diverse order of insects. There are about 180,000 known species, and it is likely that there are another 300,000 species awaiting description. The initiative involves campaigns on three geographic scales: Global (Geometridae, Saturniidae and Sphingidae), Continental (North America and Australia), and Regional (Great Smoky Mountains National Park (USA) and Area de Conservation Guanacaste) (Bravo et al. 2008). Until now, 632,006 lepidopteran species are with bar codes; of them 4443 are Crambidae (International Barcode of Life 2012). Hebert et al. (2004) studied the morphological and DNA bar coding of *Astraptes fulgerator* Walch widely distributed neotropical skipper butterfly (Lepidoptera: Hesperiidae) in north western Costa Rica with museum specimens. They showed that *A. fulgerator* is a complex of at least ten species in Costa Rica. Largely sympatric, these taxa have mostly different caterpillar food plants, mostly distinctive caterpillars, and different ecosystem preferences but only subtly differing adults with no genitalic divergence.

Conogethes Bar Code

Bar codes were generated on *Conogethes* from specimens collected on castor and cardamom from six and eight locations, respectively. The locations were chosen including all the spots in a cultivated zone of the crops. The DNA bar codes were appended to an existing dataset from BOLD for *Conogethes* species. The entire data set on *Conogethes* included 115 DNA signatures categorized into four clades and these further have been classified into more than 50 clusters. The *Conogethes* on castor and cardamom belonged to two distinct clades, moths were distinguished based on the host plants. With castor specimen matching 91 % of the standard signature in BOLD, the signature corresponded to *C. punctiferalis*. The *Conogethes* specimen on cardamom is a new signature, this DNA signature did not match with any signatures in BOLD deposited earlier. So this indicated that the moths reared on cardamom possibly belong to a new species. This needs to be further confirmed. The mean within species divergence for *C. punctiferalis* was 2.102 %. However, the pairwise divergence between *C. punctiferalis* and the moths reared on cardamom was > 5 % (Shashank 2012) (Fig. 2).

The family Crambidae, subfamily Pyraustinae, has 4585 species with bar codes. The genus *Conogethes* has 138 bar code sequences and 19 species (Table 2). Armstrong (2010) compared DNA bar coding of different populations of *Conogethes* and revealed that Australian and Asian specimens form separate clades divergent by ~6 %. The bar code data successfully distinguished *C. punctiferalis* and *C. pluto*, but unexpectedly revealed divergence between the Asian and Australian populations. Morphologically these were determined to be the same species, and distinct from other closely related species found on the east coast of Australia such as *C. haemactalis* Walker, *C. semifascialis* Walker, or *C. tharsalea* Walker.

The neighbour joining tree (NJ tree) was constructed based on all the 33 DNA bar codes using BOLD analysis tool. Based on NJ tree, two clades were recognized and those clades which includes all the *Conogethes* individuals breeding

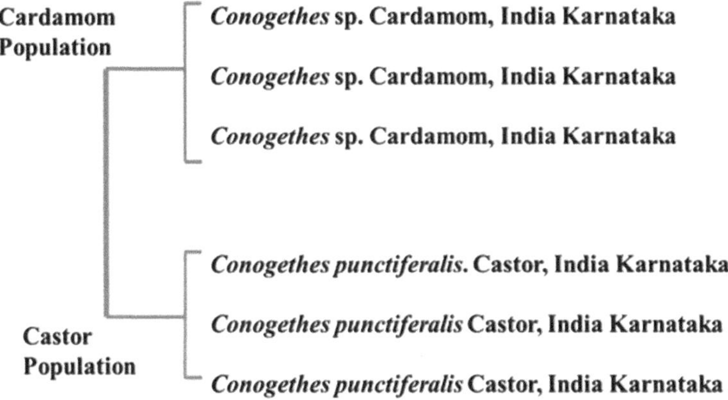

Fig. 1 Neighbor joining analysis of COI DNA bar code sequences for *Conogethes* species breeding on castor and cardamom

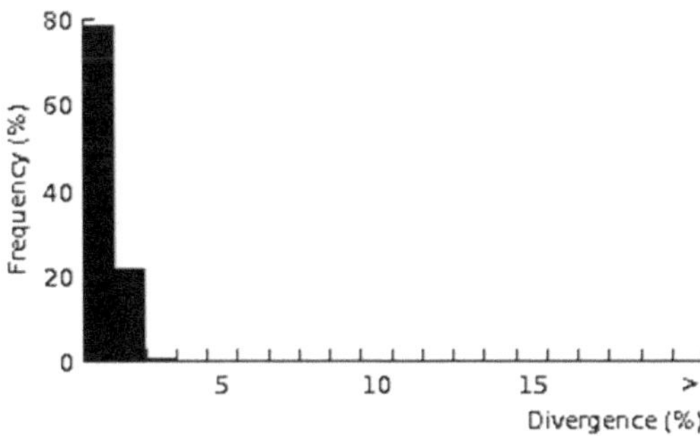

Fig. 2 Divergence in bar codes within *Conogethes* species. (Shashank 2012)

on cardamom named as *Conogethes sp.* and which include *Conogethes* from castor named as *Conogethes punctiferalis*, as shown in Fig. 1. The sequence details are analyzed carefully, submitted to NCBI for Gen Bank Accessions and subsequently uploaded to COBOL bar coding facility for formally obtaining the bar codes. All generated sequences, together with photographs and collection details, have been deposited at the BOLD (www.boldsystems.org) under the project code. DNA bar coding analyses were done through the online interface of the BOLD website. The taxon identification tree was based on the Kimura 2-parameter distance model (Kimura 1980), with the filter set to sequences with length >100 base pairs, and all codon positions included. A sequence distance summary was provided through the online interface of BOLD, with only sequences longer than 425 base pairs included in order to ensure reliable sequence distance estimates (Shashank 2012). Preliminary morphological, biological, and bar code analysis of *Conogethes* moths reared on castor, cardamom, turmeric, ginger, guava, and cocoa revealed two distinct clusters: (1) Moths on castor, guava, and cocoa fall in one cluster and (2) Moths on cardamom, ginger, and turmeric into another.

Conogethes pinicolalis Inoue and Yamanaka 2006: An Evolved Species

In Japan, workers for convenience identified two different types, in absence of precise identification the fruit tree type and conifer type from angiosperms and Pinaceae gymnosperms, respectively (Koizumi 1960). Later it was reported that the adults from coniferous plant, Japanese Cedar, *Cryptomeria japonica* (L. f.) D. Don was of the fruit tree type (Sekiguchi 1974). Then, the fruit tree type was renamed as FFT and the Conifer type, the PFT (Konno et al. 1981). Subsequently,

Table 2 DNA sequencing of *Conogethes* species. (Source: http://www.boldsystem.org/views/speciessummary.php)

Sl. No.	Species	Specimens	Sequences	Bar codes (COI of >500 bp)
1	*Conogethes diminutiva*	2	2	1
2	*Conogethes ersealis*	3	3	3
3	*Conogethes evaxalis*	9	8	8
4	*Conogethes haemactalis*	3	3	3
5	*Conogethes nr. diminutiva*	1	1	1
6	*Conogethes nr. haemactalis*	3	3	3
7	*Conogethes parvipunctalis*	1	1	1
8	*Conogethes pluto*	12	12	12
9	*Conogethes punctiferalis*	67	26	26
10	*Conogethes punctiferalis PS2*	1	0	0
11	*Conogethes semifascialis*	11	8	8
12	*Conogethes sp.*	1	0	0
13	*Conogethes* sp.ANIC 1	2	1	1
14	*Conogethes sp.* ANIC 2	3	3	3
15	*Conogethes* sp.ANIC 3	1	0	0
16	*Conogethes sp. complex*	4	3	3
17	*Conogethes tharsalea*	5	4	4

Honda and Mitsuhashi (1989) studied the morphological comparison between the FFT and PFT of *C. punctiferalis*. Nine quantitative characters were morphometrically evaluated on the male genitalia and female ovipositor. Male moths of the two types were easily distinguishable by the angular and linear characters on the valve and tegument. Differences were also observed between mandible, labrum, epipharynx, and pinacular of larvae, and cremaster of pupae. Based on these morphological differences in adults, larvae, and pupae, Honda and Mitsuhashi (1989) determined PFT of the yellow peach moth as a segregated species of the genus *Conogethes*. Until 2006 *C. pinicolalis* was deposited under *C. punctiferalis.* Now *C. pinicolalis* represents a new species of *Conogethes* and shares similar but not the same wing patterns with the other species. The present species could be identified by the following characters: the second segment of labial palp almost black; hind tibia and hind tarsus with large tufts of fuscous scales (Inoue and Yamanaka 2006).

In 2013 Vasudev Kammar studied the genetic diversity of *Conogethes* species infesting castor (CBR) and cardamom (CBE) based on COI, ITSI, and ITST gene. The multiple alignment was obtained on COI gene and most number of substitutions were (up to 35) found on turmeric population with castor population (up to 3). Pairwise genetic distance analysis indicated that high genetic divergence and phylogenetic analysis of aligned Mt COI gene sequence of *Conogethes* species breeding on castor and cardamom. The NJ tree was conducted based on all the 15 DNA bar code using BOLD analysis tool. The pair wise genetic distance analysis between the individuals varied from 0.000 to 0.074, indicating high genetic divergence. The maximum intraspecific pairwise distance in CBR was 0.022 compared to maximum intraspecific distance of CBE 0.003. The nearest neighbor distance between CBR and CBE was 5.02 % indicating wide genetic variability between two populations. This suggests that both populations differed from each other, as evident from NJ tree where they form two different clades. This may due to two species preferring to utilize host plant in two diverse habitats (Fig. 3). The NJ phylogenetic tree was constructed based on ITS1 and ITS2 gene sequence, obtaining different locations and different host plant showing sub branching and felt two clades that also indicated the divergence of the two population breed on castor and cardamom (Figs. 4 and 5).

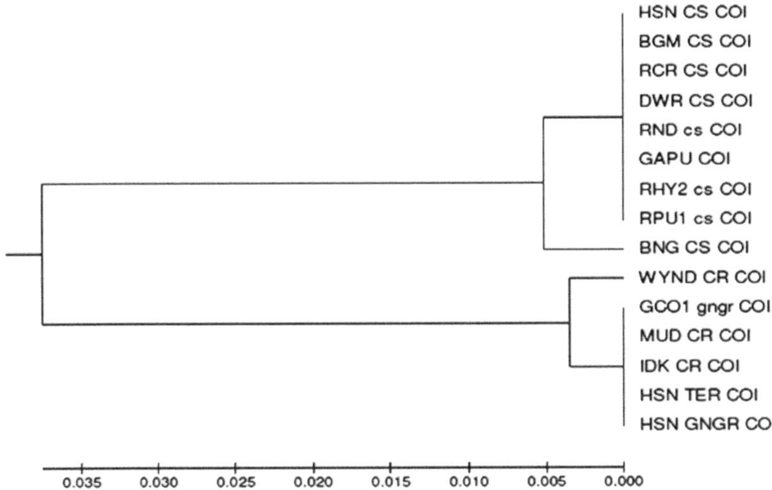

Fig. 3 Neighbor joining (NJ) of COI for *Conogethes* spp. Breeding on castor, cardamom, turmeric, and ginger. (Vasudev 2013)

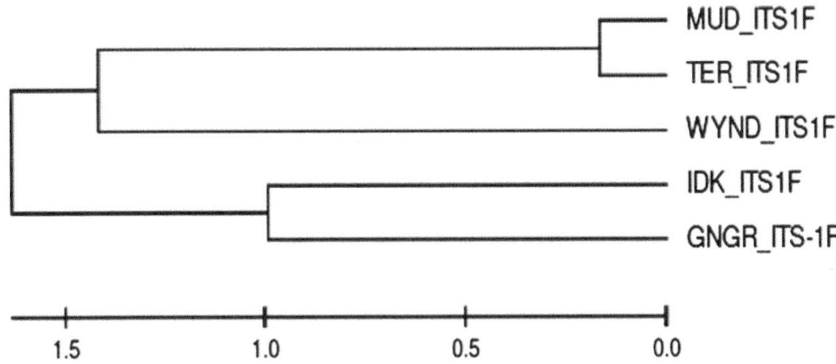

Fig. 4 Neighbor joining (NJ) phylogenetic tree of *Conogethes* spp. Population from six localities collected on various hosts for ITS1 primers. (Vasudev 2013)

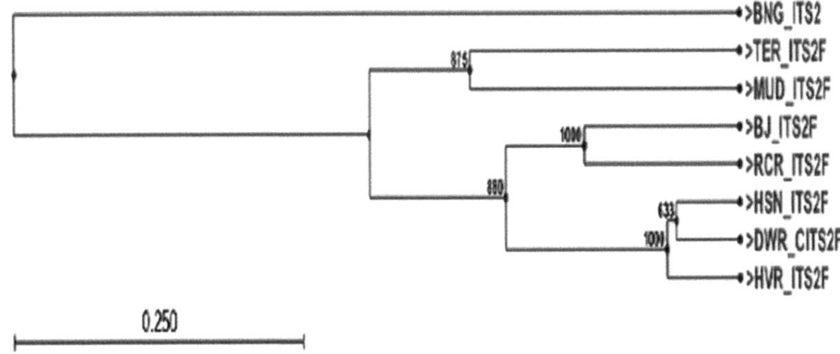

Fig. 5 Neighbor joining (NJ) phylogenetic tree of *Conogethes* spp. Population fromeight localities collected on various hosts for ITS2 primers. (Vasudev 2013)

Pheromones

Sex pheromone was detected in Japan in extracts of the abdominal tips of females of the FFT of *C. punctiferalis*. It was identified by gas liquid chromoatography, mass spectrometry, ozonolysis, and electro antennography as (E)-10-hexadecenal. The synthetic compound was attractive to males (10 ng being equivalent to 1 live virgin female) in the laboratory, but in the field, traps baited with this compound trapped only a few males. However, four times as many males were trapped with a mixture of (E)—and (Z)-10 hexadecenal in 9:1. It is therefore concluded that (Z)-10-hexadecenal is a minor component of the sex pheromone of *C. punctiferalis*. Males of the type *C. punctiferalis* feeding on the leaves of Pinaceae were also caught in traps baited with the mixture at 9:1 and 8:2 (Konno et al. 1982).

((E)—and (Z)-Tetradecenyl formate) for the polyphagous pest *C. punctiferalis*. A 10:1 mixture of (E) -and (Z)-tetradecenyl formate was found as attractive as the natural pheromone for adults of the crambid.

Chakravarthy and Thyagaraj (1997) studied activity of 7 pheromone compounds ((Z)-9-hexadecenyl acetate, (Z)-7-tetradecenal,(E)-11 -tetradecenal, (Z)-11-tetradecenal, (Z)-11-hexadecenal, (E)-10-hexadecenal, and (Z)-10-hexadecenal) against *C. punctiferalis.* Tests were carried out in laboratory and cardamom fields in Mudigere and Sakleshpur, Karnataka, India during September–December between 1985 and 1993. In laboratory, males responded to (E)-10-hexadecenal, (Z)-10-hexadecenaland (Z)-11-hexadecenal at 1000 ng. A blend of (E)-10-hexadecenal and (Z)-10-hexadecenal at 9:1 had maximum attractancy. Field trials with this blend gave positive response from moths. Laboratory studies on the mating behavior of *C. punctiferalis* showed that an airborne sex pheromone is released from the calling female. Males were attracted to virgin female extract in laboratory tests suggesting that the extract contained pheromone components for attraction. (E)-10-hexadecenal and (Z)-10-hexadecenal at 1000 ng concentration when tested separately, and a blend of these two compounds at 9:1 attracted maximum number of male moths in the laboratory. In field trials at Sakaleshpur and Mudigere, male moths were attracted to traps baited with (E)-10-hexadecenal and (Z)-10-hexadecenal at 9:1 in cardamom plantations (Chakravarthy and Thyagaraj 1998).

CH_3 CH_3

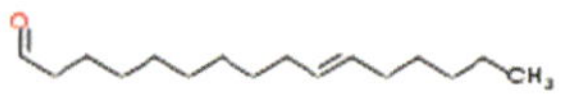

(E)-10-hexadecenal 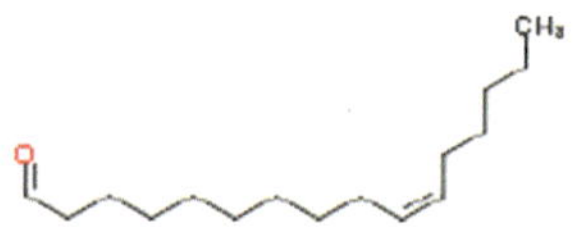 (Z)-10-hexadecenal

Source: http://www.pherobase.com/database/kovats/kovats-detail-Z10-16Ald.php

Konno (1986) investigated the relationship between daily changes in the sex pheromone quantity and calling behavior in females of *C. punctiferalis* in the laboratory at 23 °C, 70–80 % RH, and LD15:9. He recorded that quantity of the sex pheromone, (E)-10-hexadecenal, in the pheromone gland increased when the lights were turned off, reached a maximum after 5 h and then decreased. On the other hand, calling started from 5 h after light-off and reached to a maximum of 7.5 after light-off and then decreased. Similar results were obtained by Rajabaskar and Regupathy (2012) on cardamom in India. Nakano et al. (2012) detected ultrasound produced by male *C. punctiferalis* for mating. Males developed mesothoracic tymbal organs for generarating ultrasonic clicks in mating. Mori et al. (1990) studied and described the synthesis and biological evaluation of 2 sex pheromone mimics

The sex pheromone components of *Conogethes* in Japan, Korea, China, and India show variations (Table 3) (http://www.pherobase.com). The host plant interactions also vary. *Conogethes*

Table 3 Pheromone components of *C. punctiferalis* reported by workers. (Source of the chemical signal: *F* Female, *L* Lure, *M* Male. Category of the chemical signal *A* Attractant, *P* Pheromone (*) indicates that compound is active. Source: http://www.pherobase.com/database/genus/genus-Conogethes.php)

Author	Year	Journal	Binary mixture	Relative ratio of the component	Category & Source of Chemical	Country
Jung et al.	2000	*Korean Journal of Applied Entomology* 39:105	E10–16Ald Z10–16Ald	80 20	P L	Korea
Kimura T.	1999	*Applied Entomology and Zoology* 34:147	Tiglic acid		P M	Japan
Boo K. S.	1998	*Journal of Asia Pacific Entomology* 1:17	E10–16Ald Z10–16Ald	100 8		Japan
Boo K. S.	1998	*Journal of Asia Pacific Entomology 1*:17	E10–16Ald Z10–16Ald	100 11	P L	China
Boo K. S.	1998	*Journal of Asia Pacific Entomology* 1:17	E10–16Ald Z10–16Ald	100 43	P L	Korea
Chakravarthy A. K. & Thagaraj	1998	*Pest Management in Horticultural Ecosystems* 4:78	E10–16Ald Z10–16Ald	9 1	P L	India
Liu et al.	1994	*Entomologia Sinica* 1:*150*	E10–16Ald Z10–16Ald 16Ald	80.4* 6.6 13	P F	China
Mori et al.	1990	*Liebig's Annalender Chemie* 12:1257	E8–14-formate Z8–14-formate	10 1	A L	Japan
Konno et al.	1982	*Applied Entomology and Zoology* 17:207	E10–16Ald Z10–16Ald	9 1	P L	Japan

represents a cryptic species complex (Konno et al. 1980). The nonpolar components of female body wax and pheromone gland extracts of yellow peach moth synergistically enhanced male behavioral responses from close to pheromone resources in wind tunnel experiments in Japan (Xiao and Matsuyana 2011). For efficient management of this serious pest, studies on molecular genetic aspects are urgently required. Studies are needed on sex pheromone components, morphology and anatomy, and host plant interactions for the effective management of the pest on a variety of crops and cultivated ecosystems. So, collaborative and multidisciplinary research is essential. Literature concerning the status of shoot and fruit borer, *Conogethes* spp. in the Orient has recently been reviewed (Chakravarthy et al. 2012).

Gas Chromatography and Gas Chromatography Linked Electroantennogram (GC-EAG)

The pheromone gland extracts were analyzed using gas chromatography (GC) and are presented in Figs. 6 and 7. Before analyzing the abdominal gland extracts the previously identified synthetic pheromone E10-16: Ald was analysed. Sixty female abdominal gland extracts of *Conogethes* bred on castor was analyzed in GC (Fig. 6) and the observations revealed that there was a small peak at 31.79 retention time. This peak indicates the presence of E10-16: Ald when compared with standard peak. Fifty-one female abdominal glands extracts of *Conogethes* bred on cardamom was analyzed in GC and the observations revealed there was a small peak at

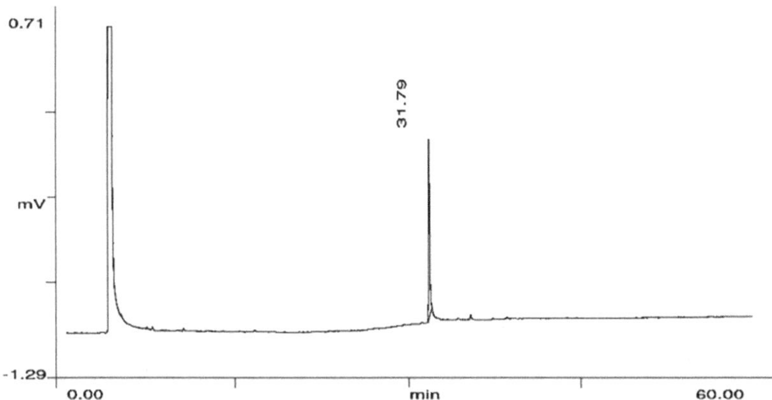

Fig. 6 Gas chromatogram (GC) of synthetic E10:16Ald (Retention time indicated). (Shashank 2012)

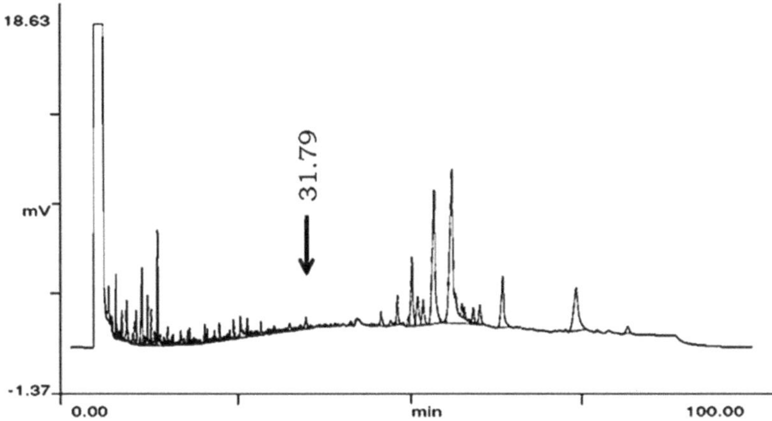

Fig. 7 Gas chromatogram (GC) of female abdominal extracts of *Conogethes* breeding on castor (60 FE; *arrow*, peak at which E10: 16Ald activity was detected). (Shashank 2012)

31.86 retention time, which was almost similar to E10-16:Ald. In this case, there were many small peaks which overlapped on specific peak. Thus, GC analysis indicated the presence of major pheromone compound in both populations.

Host—Plant Interactions

Egg incubation period (in days) was the least (2.67 days) on spineless spike castor as also the larval period (26.68 days) compared to compact spiny and spiny loose castor types. The trend in the egg incubation, larval and pupal periods was consistent across three castor types. *Conogethes* larvae preferred compact spiny castor type for feeding over other two types. *Conogethes* larvae and pupae gained more weight on compact spiny castor type than on the spineless and spiny loose castor types. This is obvious because compact spiny type was more suitable to *Conogethes* than the other two castor types. There were statistically significant differences between the spineless spike and spiny loose spike types and the compact spiny types at 1 % ($p<0.001$) level of significance. Female pupae weighed more than the male. Studies on the biology of *Conogethes* on castor and cardamom revealed that the insect could complete the life cycle on the three types of castor and cardamom. The insect required longer period (in days) of time to complete the life cycle on cardamom than on castor, i.e., on an average 34.5 days on castor compared to 39 days on cardamom (Doddabasappa 2012).

When *Conogethes* reared on cardamom were implanted on castor, larvae suffered cent percent mortality. Besides, larvae completed the period early probably because of physiological stress throughout by host plant shift (11–12 days). When castor larvae were on castor itself, larvae took 14–16 days. This is the normal larval period and larvae did not suffer from any mortality.

When *Conogethes* reared on castor were implanted on cardamom, larvae suffered cent percent mortality. Larvae completed the period early probably because of physiological stress brought out by host plant shift (13–14 days). When cardamom larvae were implanted on cardamom, larvae took 20–22 days. This is the normal larval period and larvae did not suffer from mortality. The drastic reduction in survival of the larvae, pupae, and insect as a whole may be attributed mainly to antibiotic factors. Antixenosis in combination with antibiosis might play a role in reduced pest survival. The antibiotic factors in terms of antimetabolites/antidigestible chemical components and antixenosis in terms of presence of trichome on castor types may impede larval movement on the host, the spines inhibited consumption of the host material. It is interesting to record here that the mortality of neonate larvae recorded was much higher compared with the later instar larvae (Doddabasappa 2012). Davis et al. (1989) too found reduced larval survival and larval weight of southwestern corn borer, *Diatraea grandisella* Dyar with increased development time when larvae were fed on whorl leaf tissue from resistant genotype. Resistance was identified as a combination of larval nonpreference and antibiosis.

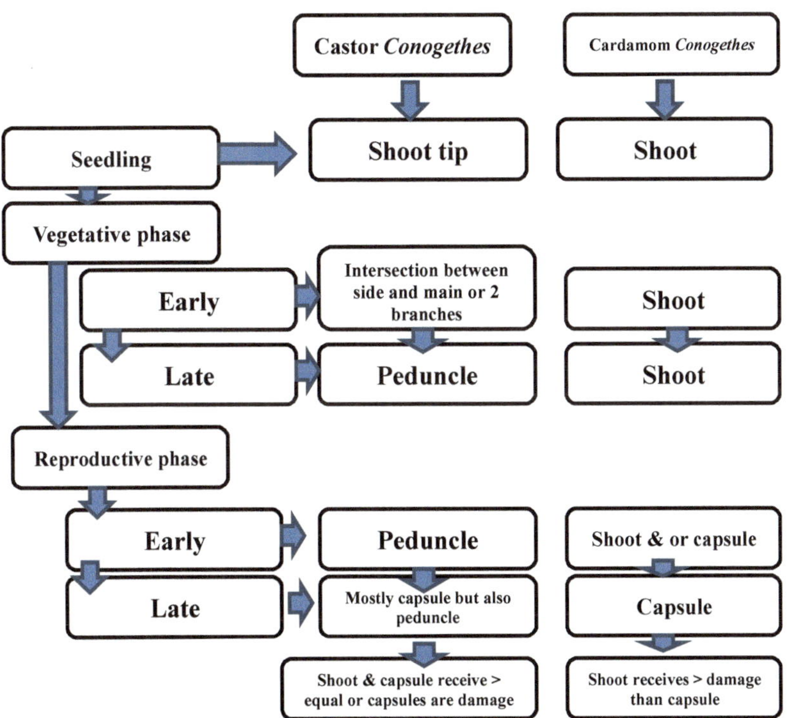

A flow chart showing pattern of Conogethes infestation on castor and cardamom (Doddabasappa 2012)

Table 4 Behavioural responses of *Conogethes* to host and shifted host plants. (Doddabasappa 2012)

Attribute	Host plant	Shifted plant
Movement	Direct (taxis)	Random (Kinesis)
Time budgeting	Less for movements, more for feeding	Negligible or Very less for feeding
Energetics	Normal growth and development	Abnormal, highly reduced
Establishment on plant	Occured	Does not occur
As food source	Suitable	Unsuitable
Feeding behavior of larvae	Normal	Deviate from the normal
Ovipositional behavior of moths	Normal	Deviate from the normal

Consideration of the sequential behavioral steps in host selection raises a number of issues that have consequences for host specificity testing. Much of the progress in applying the concepts of insect behavior to host specificity testing has been made by examining this process (Wapshere 1989; Marohasy 1998). Possibly, the most important consequences are those that stem from the absence of early steps in the host selection sequence in experimental arenas. To find whether *Conogethes* populations sustaining on castor and cardamom are genetically homogenous, studies on the host shift involving two aforesaid plants were conducted. Results showed that the *Conogethes* originating from castor could not survive on cardamom and vice versa. Larvae suffered 100% mortality when fed on castor and cardamom after host shift. The reduced plant consumption and subsequent reduced larval survival may also be due to Hopkins' host-selection principle (HHSP). HHSP refers to the observation that many adult insects demonstrate a preference for the host species on which they themselves developed as larvae. Although the practicality of HHSP has been debated significantly since its first proposal in 1916, in the case of *Conogethes* a cent percent mortality of insects was recorded and not even a single insect completed the life cycle normally (Hopkins 1917). This suggest that, although there is an effect of host shift, but the entire impact on the life history of insect is not due to the HHSP but in a major way due to the unsuitability of the plant as a food source for the candidate species (Doddabasappa 2012).

When nonhost plant like *Ocimum* was offered to these *Conogethes* larvae, none of the larvae orientated and none of them showed initial biting responses. This observation suggested that on the same host higher number of larvae oriented toward and moderate number of larvae orientated toward plants related to *Conogethes* species to the host plant and none of the larvae showed orientation and biting responses on the nonhost plant (Table 4).Host selection is influenced primarily by moth oviposition, but neonate ballooning and larval movement also are important (Ross and Ostlie 1990). Learning about feeding preferences of the neonates for different plant hosts will help decipher complex plant–insect interactions in different cultivated ecosystems and may help predict the degree of *Conogethes* infestations in natural and cultivated ecosystems.

Quarantine importance

The distribution of *C. punctiferalis* extend from Asia to Australia (CABI 2011), representing a complex species (Robinson et al. 1994). According to the unpublished reports of the Food and Environment Research Agency (FERA, UK) larvae of *C. punctiferalis* have been detected inside tropical fruits 18 times in the last 5 years at three points (in England and Wales) between 2007 and 2011. Due to report of damage to apples in North China (CABI 2011), interceptions of the pest in fruits from Pakistan, and a pest concern to many countries like New Zealand, South Africa, Canada, and the USA, a Rapid Pest Risk Analysis on *Conogethes* was conducted.

Since the members of species within the complex are unknown and their biology cannot be distinguished, *Conogethes* represent all the more important species from quarantine point of view. So the Pest Risk Analysis (PRA) assessment has

Table 5 Interceptions of *Conogethes punctiferalis* by the PHSI in England and Wales between 2007 and 2011. (Anastariq 2012)

Host	Origin	Year(s)
Annona squamosa (sugar apple)	India	2011
Mangifera indica (mango)	India	2011
Psidium guajava (guava)	India	2011
Psidium guajava	Pakistan	2008 (twice), 2009 (twice), 2011 (eight times)
Psidium guajava	Sri Lanka	2011
Psidium sp.	Thailand	2011
Psidium sp.	Unknown	2007

been made on all putative species within *C. punctiferalis* species complex. Although *C. punctiferalis* is not listed in the European Commission (EC) plant Health Directive nor in any of the European and Mediterranean plant protection organization (EPPO) lists, it is an important quarantine pest. Being a highly polyphagous pest, larvae of *C. punctiferalis* attack fruits, seeds, and stems of diverse plants. While much of the species distribution is in the subtropics, *C. punctiferalis* has also been recorded from Hokkaido prefecture, North Japan (Inoue and Yamanaka 2006), and north China (CABI 2011). Currently *C. punctiferalis* is a serious pest for several plants of economic importance for more than two centuries worldwide.

Management

Since many plants serve as alternate hosts, the pest has adapted well in agroecosystems. So, tactfully one has to tackle this pest. Individual methods like cultural, mechanical, biological, and chemical are partially effective in keeping this pest below the economic threshold level. Hence, an integrated approach is a must to manage this pest. Some of the cultural practices like increased planting distance reduced the pest damage (Sharma et al.1992). Reducing the nitrogenous fertilizer levels also minimizes the pest damage to a greater extent (Chakravarthy and Thyagaraj 1999). Clean cultivation in cardamom plantations can reduce the borer damage (Sharma 1992). Mechanical methods also checked the pest to the extent of 5–10 % (Butani 1978; Tomomatsu et al. 1995; Chakravarthy et al. 1997; Boo 1998) in crops like castor and pomegranate.

Some predators and parasites were also identified and can be exploited in the management of this pest (Clausen 1927; Rodrigo 1940; Abe and Sanari 1992; Anandraj and Peter 1996). Intercropping is an important tool for *C. punctiferalis* management. Intercropping is based on the principle of reducing pests by increasing diversity in an agroecosystem. In semiarid tropics in India introducing cluster bean, cowpea, blackgram, and/or groundnut as intercrops in castor (1:2 proportions) reduced *C. punctiferalis* infestation and build up of natural enemies of crop pests has been recorded (Rao et al. 2012). There are proven insecticides that can be used (Rama 1980; He 1997; Chakravarthy et al. 1997; Thyagaraj 2003). Individual management methods were evaluated and compared with the integrated method (Thyagaraj 2003) in cardamom plantations in Western Ghats of Karnataka. Cardamom is an export oriented crop, residual effects of toxic compounds are to be avoided or carefully used (Devasahayam 1998; Thyagaraj 2003).

Therefore, an integrated approach embracing thrashing (removing dried leaves, suckers, panicle, etc., during February and March), and the timely spraying of monocrotophos 36 % SL @ 15 ml/10 L of water (March and April) followed by spraying phosalone 30 EC. @ 20 ml/10 L of water twice at an interval of 25–30 days after each spray could be adopted for better, economical, and ecofriendly management of the pest (Krishnamurthy et al.1989; Devasahayam 1998; Thyagaraj 2003) in cardamom plantations. The pesticide effect to *C. punctiferalis* of 2 % thiacloprid DP was tested in Chestnut Orchard in July, 2010 in China. Results showed that orchards sprayed with 2 %Thiacloprid DP achieved significant results to kill the borer. The killing efficiency became stronger with the dose increased (Yuan Yuan et al. 2011). The castor ecosystem is entirely different from cardamom plantations. Most of the work on management of *Conogethes* on castor is by insecticides. Insecticides such as fenitrothion (0.05 %), dimethoate 30 EC (0.05 %), endosulfan (0.07 %), carbaryl (0.15 %), monocrotophos (0.05 %), and quinalphos (0.05 %) have proved effective (Saroja et al. 1973; Patel et al.1988).

In Japan, entomopathogenic fungi detrimental to *Conogethes* eggs have been identified. Application of microbial suspensions on emerging fruiting bodies can avert the need for application of chemical insecticides. Intercropping, mixed cropping, use of pheromone traps and timely harvests, and destruction of affected plant parts can substantially contribute to borer suppression.

References

Abe Y, Sanari G (1992) Notes on the peach moth *Conogethes punctiferalis* attacking cynipid and aphid galls. Japanese J Entomol 60(1):108

Abraham P (1965) The cardamom in India. ICAR Farm B 37:6–8

Abraham EV, Padmanabhan MD (1967) Pests that damage cocoa in Madras. Indian Hortic 11(3):11–12

Anandraj M, Peter KV (1996) Biological control in spices. Indian Institute of Spice Research, Calicut, p 52

Anonymous (1913a) Insect injurious to Papaw apple. Queens Land Agric J 3:33–35

Anonymous (1913b) The cotton bollworm. Queens Land Agric J 3:142

Anonymous (1918) Cultivation of castor. Tropic Agric 5:297–301

Anonymous (1944) Annual report of dept. of agriculture and stock for the year, Brisbane. Queens Land Agric J 3:88

Anonymous (1954) Pests of cardamom. Memories of the Department of Agriculture, Madras, pp 998–1000

AQSIQ (2007) Chinese *Table Grape Export Technical Information*. Animal and Plant Quarantine Service of the People's Republic of China. pp 99

Armstrong K (2010) DNA barcoding: a new module in New Zealand's plant biosecurity diagnostic toolbox. B OEPP/EPPO 40:91–100

Asante SK, Danthanarayana W, Heatwol H (1991) Bionomics and population growth static of aptirousvirginoparae of woody aphid, *Eriosomal anigereum*, at constant temperature. Entomol Exp App 60:261–270

Autson JC (1923) Report of entomologist, cyclone department of agriculture report, 1921, pp C-21–26

Azam KM, Ali MH (1965) Morphology of larva of *Dichocrocis punctiferalis* Guen., the castor shoot and capsule borer with special reference to chaetotaxy (Lepidoptera: Pyralvdae). Indian J Entomol 27(4):423–431

Ballard E (1924) Entomological Note. Queens Land Agric J 11:31

Ballard E (1927) Some insects associated with cotton in Papua and in the mandated territory of New Guinea. Bull Entomol Res 27:295–300

Bilapate GG, Talati GM (1978) Some studies on bionomics of castor short and fruit borer *Dichocrocis punctiferalis* from Gujarat. J Maharashtra Agric Univ 3(1):75–76

Bilapate G, Talati GM (1977) Effect of different food plants on the larval and papal development of *Dichocrocis punctiferalis*. Madras Agric J 64(1):53–54

Boo KS (1998) Variation in sex pheromone composition of a few selected lepidopteron. J Asia Pacific Entomol 1(1):17–23

Bravo JP, Silva JLC, Munhoz REF, Fernandez MA (2008) DNA barcode information for the sugar cane moth borer *Diatraea saccharalis*. Gen Mol Res 7(3):741–748

Butani DK (1978) Insect pests of fruit crops and their control. Pesticides 12(8):53–59

Cai RX, Mu ZL (1993) Trap trials for *Conogethes punctiferalis* with sex pheromones in citrus orchards. China Citrus 22(1):33

Chakravarthy AK, Thyagaraj NE (1997) Response of cardamom shoot and fruit borer to different pheromone compounds. Insect Environ 2(4):127–128

Chakravarthy AK, Thyagaraj NE (1998) Evaluation of selected synthetic sex pheromones of the cardamom shoot and fruit borer, *C. punctiferalis* in Karnataka. Pest Manag Hort Eco 4(2):78–82

Chakravarthy AK, Thyagaraj NE (1999) Effect of different NPK fertilizer levels on the incidence of cardamom pests. Insect Environ 4(4):139–140

Chakravarthy AK, Honda H, Thyagaraj NE (1991) Comparison of containers for larval rearing in stalk and fruit feeding type of *Conogethes punctiferalis* (Guen.) Lepidoptera: Pyralidae. Placrosym IX:127–131

Chakravarthy AK, Shivanna S, Chandrappa L, Naik L, (1997) Insect pests on castor in Karnataka. Insect Environ 2(4):135

Chakravarthy AK, Shashank PR, Doddabasappa B, Thyagaraj NE (2012) Status of shoot and fruit borer, *Conogethes* spp (Lepidoptera: Crambidae) in the Orient: biosystematics, biology and management. J Insect Sci 25(2):107–117

Clausen CP (1927) The citrus insects in Japan. Technol Bull USDA 15:15

Devasahayam S, Koya MA, Reddy PP, Kumar WKK, Varghese A (1998) IPM in spices, challenges for the future. Advances in IPM for horticultural crops: environmental complications and thrust, Bangalore. India Oct 15-17(1997):157–164

Doddabasappa B (2012) Factors associated with resistance mechanism with shoot and fruit borer *Conogethes punctiferalis* Guenee (Lepidoptera, Crambidae) population infesting caster and cardamom. PhD Dissertation, UAS, GKVK, Bangalore

Doyle JJ, Doyle JL (1990) A rapid total DNA isolation preparation procedure for fresh plant tissue. Focus 12:13–15

Du YL, Guo HM, Sun SL, Zhang MZ, Ai H, Wang JB, Qin L (2012) Effect of temperature on the development and reproduction of the yellow peach moth, *Conogethes punctiferalis* (Lepidoptera: Pyralidae). Acta Entomol Sinica 55(5):561–569

Dyar HG (1890) The number of molts of Lepidopterus larvae. Psyche 5:420–422

Flether TB (1922) Report of the Imperial Entomologist. Scientific reports of the Imperial Institute of Agricultural Research. Pusa. pp 51–67

Gour TB, Sriramulu M (1992) Grapevine, *Vitis vinifera*—a new host of castor shoot and capsule borer *Conogethes punctiferalis*. Trop Pest Manag 38(4):459

Gurmey WB (1918) The insect pests of maize. Agr Gazetteer Bydney 9:641–650

Hampson (1863) The Fauna of British India. Cyclone Moths IV:305–312

Hang HL, Yan KF, Sun XH, MA JX (2000) Investigation on the kinds of fruit moth in the western part of Henan province and their control. China Fruits 2:44–45

He RQ (1997) Study on the control of yellow peach moth in china. South China Fruits 26(4):33

Hebert PDN, Penton EH, Burns JM, Janzen DH, Hallwachs W (2004) Ten species in one: DNA barcoding reveals cryptic species in the neotropical skipper butterfly *Astraptes fulgerator*. Proc Natl Acad Sci U S A 101(41):14812–14817

Honda H (2013) Bio systematics of Japanese *Conogethes* spp. With special reference to host plant preference and reproductive isolation. New Horizons in Insect Science. Abstracts ICIS-2013 (eds) Chakravarthy AK, Ashok Kumar CT, Varghese A, Thyagaraj NE, Bangalore, pp 136

Honda H, Mitsuhashi W (1989) Morphological and morphometrical differences between the fruit and Pinaceae Feeding type of yellow peach moth, *Conogethes punctiferalis* (Guenee) (Lepidoptera:Pyralidae). J Appl Entomol Zool 24(1):1–10

Honda H, Kaneka JI, Konno Y, Matsumoto Y (1979) A simple method for most rearing of the yellow peach moth *Dichocrocis punctiferalis* on an artificial diet. J Appl Entomol Zool 14(4):464–468

Hopkins AD (1917) A discussion of C. G. Hewitt's paper on "Insect Behavior". J Econ Entomol 10:92–93. http://www.boldsystems.org/views/speciessummary.php

Hossain MM, Ali SH, Rahim A (1995) Some lepidopteron and homopterous and pests and parasites of castor. Bangladesh J Sci Ind Res 30(23):265–67

Hussain MM, Ali SH, Rahim A (1995) Some lepidopteron and homopterous and pests and parasites of castor. Bangladesh J Sci Indu Res 30(23):265–267

Hutson JC (1937) Report on the work of the Entomological Division. Report of The Director of Agriculture Ceylon, 1936. pp 22–28

Inoue H, Yamanaka H (2006) Redescription of *Conogethes punctiferalis* (Guenee) and description of two new closely allied species from Eastern Palaearctic and Oriental regions (Pyralidae, Pyraustinae). Tinea 19(2):80–91

Ironside DA, Davis JJ (1969) Major Macadamia pests. Queens Land Agric J 95(7):468–471

Issac PV (1948) Imperial Entomologist.Scientific reports of the Agricultural Research Institute. Pusa. pp 73–79

Jacob SA (1981) Biology of *Conogethes punctiferalis* on turmeric. J Plantation Crops 9(2):119–123

Jarvis E (1914) A new fruit boring caterpillar of Banana occurring at Tweed keeds. Queens Land Agric J 5:280–284

Jung JK, Han KS, Choi KS, Boo KS (2000) Sex pheromone composition for field-trapping of *Dichocrocis punctiferalis* (Lepidoptera: Pyralidae) males. Korean J Appl Entomol 39:105–110

Kalshoven LGE (1922) Notes on forest zoology for the Dutch East India. Tectona 15:944–959

Kalshoven LGE (1929) Injuries, diseases and pests of Teak forests in Java. Tectona 21:593–523

Kang CH, Lee SM, Chung YJ, Choi KS, Park CG (2004) Overwintering ecology of the peach pyralid moth, *Dichocrosis punctiferalis* in southern regions of Korea. Korean J Appl Entomol 43(3):201–209

Kapadia MN (1996) Estimation of loses due to pod borer in oil seed crops. J Oil Seeds Res 13(1):139–140

Kimura M (1980) A simple method for estimating evolutionary rate of base substitutions through comparative studies of nucleotide sequences. J Mol Evol 16:111–120

Kimura T, Honda N (1999) Identification and possible functions of the hair pencil scent of the yellow peach moth *Conogethes punctiferalis* in Japan. Appl Entomol Zool 34(1):147–153

Kodoi M, Kaneda M (1990) Development of yellow peach mouth *Conogethes punctiferalis*on apple fruit. Res Bull Plant Protec Serv Japan 26:61–63

Koizumi K (1960) Two forms of *Dichocrocis punctiferalis* (Guenee) presumably representing separate species. The main purport of a lecture the 20th annual meeting of the Entomological Society of Japan, pp 8–9

Kondo T, Miyahara T (1930) Fruit borer and a summary of their life histories in Knangtuny. China J Plant Protec 17:85–94

Konno Y (1986) Time-lag between sex pheromone content and the calling behavior in the yellow peach moth, *Conogethes punctiferalis* (Guenee) (Lepidoptera: Pyralidae). Appl Entomol Zool 21:622–624

Konno Y, Shishido T (1996) Glutathione—dependent o-alkyl and oral conjunctions for dicapthon and fertikothine in several insects. J Pesticide Sci 21(4):430–433

Konno Y, Honda N, Matsumoto Y (1980) Observation on the mating behaviour and biology for the six pheromone of yellow peach moth, *Dichocrocis punctiferalis*. Appl Entomol Zool 15(3):321–322

Konno Y, Honda N, Matsumoto Y (1981) Mechanism of reproductive isolation between the fruit feeding and pin ace feeding type of yellow peach moth, *Dicho-*

crocis punctiferalis. Japanese J Appl Entomol Zool 25(4):253–258

Konno Y, Arai K, Sekiguchi K, Matsumoto Y (1982) (E)-10-hexadecenal, a sex pheromone component of the yellow peach moth, *Dichocrocis punctiferalis* Guenée (Lepidoptera: Pyralidae). Appl Entomol Zool 17:207–217

koryeinska A (2012) Food and environment research agency (FERA) Rapid arresment of the need for detail pest risk analysis of *Conogethes punctiferalis* FERA 4th version

Krishnamurthy K, Khan MM, Avadhani KK, Venkatesh J, Siddaramaiah L, Chakravarthy AK, Gurumurthy BR (1989) Three decades of cardamom Research, Regional Research Station (1958–1988). Station Technol Bull 2:44–68

Kristensen NP (1999) Lepidoptera, Moths and Butterflies. Volume 1: evolution, systematics, and biogeography (Handbuch der Zoologie [founded by Willy Kukenthal; ed.: Fischer, M.]. Band 4, Teilband 35). Walter de Gruyter, Berlin, pp 217–232

Kuang MH, Liu SW, Ji BZ, Gao JY, Gao YG, Wang GX (2009) Investigation on the overwintering status and bionomics of needle feeding type of *Conogethes punctiferalis*. Chinese Bull Entomol 46(4):569–573

Lall BS, Zafer JM, Yazdani SS (1980) Preliminary studies on comparative susceptibility of different castor cultivars against *Dichocrocis punctiferalis*. Indian J Entomol 42(2):299–301

Liu MY, Tian Y, Li YX (1994) Identification of minor components of the sex pheromone of yellow peach moth, *Dichocrocis punctiferalis* Guenee, and field trials. Entomol Sinica 1:150–155

Lu Z, Yang ZF, Wang GY, Bo WL (1995) The niche of the corn and peach borers and interspecific competition between them in corn. Acta Agric Zhajiangensis 7(1):31–34

Maes KVN (1998) Pyraloidea: Crambidae, Pyralidae pp. 87–98, 426–435. In: Polaszek A (Ed) African cereal stem borers: economic importance, taxonomy, natural enemies and control. CAB International, Oxon, pp 74

Marohasy J (1998) The design and integration of host specificity tests for weed biological control with particular reference to insect behaviour. Biocontrol News Inform 19:13–20

Mathew G, Menon MGR (1984) The pyralid fauna (Lepidoptera: Pyralidae) of Kerala (India). J Entomol Res 8(1):5–13

Mathew G, Menon MGR (1989) Identification of some Indian Pyraustinae (Lepidoptera: Pyraustidae). J Entomol Res 13(1):76–89

Meyrick E (1884) On the classification of the Australian Pyralidina. Trans Ento Soc Lond 32(3):277–350

Mishra SC, Teotia TPS (1965) Studies on the biology of the castor shoot and capsule borer. Indian oil seeds J 9(3):184–187

Mohanan RC, Kumar RH (1976) The castor capsule borer *Dichocrocis punctiferalis* as a pest of cocoa is India. Agr Res J Kerala 14(1):79–80

Moralesranous JA, Cote JR (1992) Rate of increase and adult longevity of *Catolocus grandis* in the laboratory at four temperature. Environ Entomol 21:620–627

Mori K, Watanabe H, Fujiwhara M, Kuwachau (1990) (E)—and (Z). Tetra decenyl format, potent sex pheromone mimics against the yellow peach moth. Liebigs Annalen der Chemie 12:1257–1259

Munroe E (1989) Changes in classification and names of Hawaiian Pyralidae since the publication of Insects of Hawaii, vol 8, by Zimmerman EC (1958). Bishop Museum Occasional Papers 29, pp 199–212

Nakano R, Takanashi T, Ihara F, Mishiro K, Toyama M, Ishikawa Y (2012) Ultrasonic courtship song of the yellow peach moth, *Conogethes punctiferalis* (Lepidoptera: Crambidae). Appl Entomol Zool 47(2):87–93

Narasimha Swamy RL (1937) Staminodes in *Elettaria cardamomum*. Maton Current Sci 5(8):427

Ni DW (1998) Main pests and diseases of *Castame amollissinia* in Zhejiang and their control technique. J Zhejany Forestry Sci Tech 6:68–71

Ono K (1937) On the insect pests of persimmon. Insect world 41(5):195–199

Park JD, Lec SG, Kim CS, Byun RK (1998) The bionomics of the oak nut weevil, *Mehorisur sulus* and the insect pests of the corer in Korea. J Forest Sci 5:151–156

Patel RK, Gangrade GA (1971) Note on the biology of castor capsule borer, *Dichocrocis punctiferalis* Gn. (Lepidoptera: Pyralidae). Indian J Agril Sci 41(5):443–444

Patel MM, Naik MM, Vyas HN (1988) Field evaluation of certain insecticides against castor capsule borer. Agr Sci Digest 8(3):122–124

Pena JE, Nadel H, Pereira MB, Smith D (2002) Pollinators and pests of *Annona* Sp. In: Pena J, Sharp J, Wysoki (eds) Tropical fruit pests and pollinators: biology, economic importance, natural enemies and control. CABI Publication, UK, p 208

Peter KV (1996) Major spices, 50 years of crop science research in India. Indian Council of Agricultural Research, New Delhi, pp 771–796 (CAR)

Peterson D (1967) A quick method for sex determination of codling moth. J Eco Entomol 58:576

Pruthi HS (1944) Report of the Imperial Entomologist. Abriay. Scientific reports of the Agricultural Research Institute. Pusa, pp 64–71

Rajan SV (1965) Protect your cardamom against insect pests. Indian Farm 14(11):21–24

Ram KT, Hugar PS, Somasekhar Patil BV (1997) Record of *Conogethes punctiferalis* (Guenee) on grapes. Pest Manag Hortic Ecosyst 3:37–39, 137

Rama CSV (1980) Plant protection in plantation crops through the soil. J Plant Crops 8(1):1–12

Rao AS (1992) Preliminary studies on the seasonal occurrence of insect pests a soap nut (*Sapindus* sp). Indian Forester 118(6):432–437

Rao MS, Rao CAR, Srinivas K, Pratibha G, Sekhar SMV, Vani GS, Venkateswarlu B (2012) Intercropping for management of insect pests of castor in the semi-arid tropics of India. J Insect Sci U S A 12:14

Robinson DN, Cant K, Cooley L (1994) Morphogenesis of Drosophila ovarian ring canals Development, 120(7):2015–2025

Rodrigo E (1940) Administration report of the Acting Director of Agriculture for 1940. RAE 30:338

Ross SE, Ostlie KR (1990) Dispersal and survival of early instars of European corn borer (Lepidoptera: Pyralidae) in field corn. J Econ Entomol 83:831–836

Saroja R, Lewin HD, Padmanabhan MD (1973) Control of the pests of castor (*Ricinus communis* L.) with insecticides. Madras Agric J 60:484–486

Sekiguchi K (1974) Morphology, biology and control of the yellow peach moth, 402 *Dichocrocis punctiferalis* Guenée (Lepidoptera: Pyralidae). Bull Ibaraki Hort 403 Expt Stn Special Issue, pp 89

Senguptha GC, Behera BK (1955) Some new records of crop pests from India. Indian J Entomol 17:283–285

Shaffer M, Nielsen ES, Horak M (1996) Pyralidae. In: Nielsen ES, Edwards ED, Rangsi TV (eds) Checklist of the lepidoptera of Australia. Monograph Australia lepidoptera, vol 4, pp 164–199

Shanuowr TG, Guttierray AP, Wightman A (1993) Effect of temperature on development rate, fecundity and longevity of ground nut leaf minor, *Aproaerema modicella* (Lepidoptera: Gelechiidae) in India. Bull Entomol Res 83:413–419

Sharma YR, Devasahayam S, Anandraj M (1992) Black Pepper and cardamom: problems and prospects. Proceedings of National seminar on black pepper and cardamom. May, 17–18 1992, Calicut, Kerala, India

Sharma ML, Shukla A, Reddy RK (1995) Field testing of caster cultivar for yield potential and resistance against shoot and capsule borer *Conogethes punctiferalis*. Crop Res 10(1):54–58

Shashank PR (2012) Biosystematics and pheromone components of *Conogethes punctiferalis* Guenee (Lepidoptera: Crambidae) with special reference to populations infesting castor (*Ricinus communis* L) and cardamom (*Elettaria cardamomum* Moton). Ph D thesis submitted to UAS, GKVK, Bangalore

Shiukaji N (1969) Notes on the parasites attacks the fruit type of and confer type of *Dichocrocis punctiferalis*. Japanese J Appl Entomol Zool 13(2):91–93

Sithananthan S, Subramaniam TR (1975) Sex determination of pupae of Sorghum stalk borer, *Chilo zonallus*. Madras Agric J 62(2):62–63

Sloan WJS (1945) Insect pests of grain sorghum. Queensland Agric J 61(4):221–239

Smith JH (1937) Insects and allied pests of the Papua. Queensland Agric J 48:553–557

Stanley J, Chandrasekaran S, Preetha G (2009) *Conogethes punctiferalis* (Lepidoptera: Pyralidae) its biology and field parasitization. Indian J Agric Sci 79(11):906–909

Suganthy M (2011) Field screening of promising castor cultures against capsule borer, *Conogethes punctiferalis* and leafhopper, *Empoasca flavescens*. Madras Agric J 98(4/6):178–179

Sun ZP, Wang W, Mung FY (1992) Survey of pests and diseases of Chinese hawthorn. Bull Agric Sci Techn 44:25–26

Thyagaraj NE (2003) Integrated Management of some important cardamom pests of hill region of Karnataka, South India. Ph D Thesis, Dr BR Ambedkar University, Uttar Pradesh, India

Tomomatsu S, Sakaguchi T, Ogino T, Hiramatsu T, Kaukami F (1995) Methyl bromide fumigator for quarantine control of persimmon fruit moth and yellow Peach moth on Japanese persimmon. Res Bull Plant Prot Serv 31:67–73

Tryon H (1920) Report of the Entomologist, Vegetable pathologist. Queensland annual report of department of agriculture for 1916–1917, pp 49–63

Twine PH (1971) Insect pests of maize. Queens Land Agr J 97(10):523–531

Utsume K, Aihara E, Watanabe N (1990) Artificial diet of yellow peach moth *Conogethes punctiferalis* eating corn seeds. Res Bull Plant Prot Serv Japan 26:65–67

Vasudev K (2013) Molecular diversity of Conogethes spp. Guenèe,(Lepidoptera: Crambidae) infesting castor andcardamom, M Sc Dissertation, University of Agricultural sciences, GKVK, Bangalore, pp 55–66

Veitch R (1931) Report of the chief entomologist, Annual report of department of agriculture. Queens Land Agric J 1930–1931:43–46

Wang YX, Cai RT (1997) The emergence and control of *Dichocrosis punctiferalis* for Younai plum variety. South China Fruits 26(3):45

Wapshere AJ (1989) A testing sequence for reducing rejection of potential biological control agents for weeds. Ann Appl Biol 114:515–526

Wu LM (1995) A study on the pupation pattern of three pyralid caterpillars helps is maize field. Plant Prot 21(3):24–26

Xi K, Xu FY, Dai YS (1996) Studies on the Chinese chestnut cone life table. J Joangsu For Sci Technol 23(4):6–10

Xiao WH, Matsuyama S (2011) Monoenyl Hydrocarbons in female body wax of the yellow peach moth as synergists of aldehyde pheromone components. Appl Entomol Zool 46(2):239–246

Young HY, Shaw KY (1962) Preliminary studies on the peach borer *Dichocrocis punctiferalis*. Entomol Sinica 1(2):109–134

Yuan Zuo, Wang Mingyue, Lu Junmei, Li Qiliang (2011) Efficacy of 2% Thiacloprid DP to control *Dichocrocis punctiferalis* in chestnut orchard. J Jiangsu For Sci Technol 2011(4):21–23

Part VI

Insect Semiochemicals

Use of Insect Pheromones in Vegetable Pest Management: Successes and Struggles

R. Srinivasan, M-Y Lin, F-C Su, Sopana Yule, Chuanpit Khumsuwan, Thanh Hien, Vu Manh Hai, Le Duc Khanh and K. R. M. Bhanu

Abstract

Insect pheromones can be used to monitor, mass-trap, and/or disrupt the mating process of insect pests. Sex pheromones of major lepidopteran pests such as tomato fruit worm (*Helicoverpa armigera*) and common armyworm (*Spodoptera litura*) are widely used as monitoring lures in tropical vegetable production systems. The use of sex pheromone traps as a mass-trapping tool against polyphagous insects is limited. However, such traps are highly effective in reducing the damage and yield losses caused by monophagous insects such as eggplant fruit and shoot borer (*Leucinodes orbonalis*), as demonstrated in the Indo-Gangetic Plains of South Asia by AVRDC—The World Vegetable Center. By definition, pheromones are species-specific, yet "cross-talks" were observed when we attempted to refine sex pheromones for monitoring and mass-trapping legume pod borer (*Maruca vitrata*) and cucumber moth (*Diaphania indica*). In addition to sex pheromones, the possible use of aggregation pheromone as a pest management tool has been attempted in striped flea beetle (*Phyllotreta striolata*) on vegetable brassicas. Recent research indicated that pheromones act synergistically when combined with host plant volatiles, a process validated against *P. striolata* at AVRDC.

R. Srinivasan (✉) · M.-Y. Lin · F.-C. Su
AVRDC—The World Vegetable Center, Shanhua, Tainan 74199, Taiwan
e-mail: srini.ramasamy@worldveg.org

S. Yule · C. Khumsuwan
AVRDC—The World Vegetable Center, East and Southeast Asia, Research and Training Station, Kasetsart University, Kamphaeng Saen, Nakhon Pathom 73140, Thailand

T. Hien · Vu M. Hai · Le. D. Khanh
Vietnam Academy of Agricultural Sciences, Thanh Tri, Hanoi, Vietnam

K. R. M. Bhanu
Biocontrol Research Laboratories, Pest Control (India) Pvt. Ltd., Bangalore, Karnataka 561203, India

A. K. Chakravarthy (ed.), *New Horizons in Insect Science: Towards Sustainable Pest Management*,
DOI 10.1007/978-81-322-2089-3_21, © Springer India 2015

Keywords

Pheromones · *Helicoverpa armigera* · Spodoptera litura · Maruca vitrata · Diaphania indica · Phyllotreta striolata

Introduction

By definition, pheromones are substances secreted by an individual and received by a conspecific in whom it elicits a specific reaction (*e.g.*, behavior or developmental process) (Karlson and Lüscher 1959). Depending on the modes of action, the pheromones give a "releaser" effect (an immediate and reversible behavior change in the receiver) or a "primer" effect (eliciting a neuroendocrine or developmental change). The pheromone compounds that give a releaser effect are quite important, since they are responsible for a change in the behavior such as alarm, aggregation, trail, or sex, which could be exploited in pest management. Although sex pheromones have been reported in several insect orders, use of long-distance sex pheromones for mate recognition has been widely studied in Lepidoptera (Löfstedt and Kozlov 1997). After the identification of first moth sex pheromone in silkworm (*Bombyx mori*), several hundred pheromone compounds have been identified in Lepidoptera.

The pheromones are used to monitor, mass-trap, and/or disrupt the mating process of selected insect pests in vegetable production systems in the tropics. For instance, placing a high concentration of sex pheromone in a slow-release formulation at 5- and 10-m grid in the field resulted in drastic reduction of tomato fruit worm (*Helicoverpa armigera*) male moths being attracted to virgin females, which adversely affects mating in *H. armigera* (AVRDC 1988). An integrated pest management (IPM) strategy based on sex pheromones for managing eggplant fruit and shoot borer (*Leucinodes orbonalis*) reduced the pesticide use significantly in the Indo-Gangetic plains of South Asia (Alam et al. 2006).

Besides sex pheromones, aggregation pheromones are the other group that is being used in pest management. Aggregation pheromones are produced by one or both sexes of an insect species to increase the density of conspecifics for feeding, mating, and protection. Several coleopterans such as Bostrichids, Cerambycids, and Chrysomelids are known to produce aggregation pheromones (Edde 2005; Soroka et al. 2005; Teale et al. 2011). For instance, the active male-derived aggregation pheromone compound of striped flea beetle (*Phyllotreta striolata*) attracted significantly high numbers of *P. striolata* either alone or in combination with the host plant volatiles (Beran et al. 2011). Thus, the sex and aggregation pheromones have become an important component in the IPM strategies. This chapter aims to compile few case studies on the attempts in developing and using selected insect pheromones as pest-management tools at AVRDC—The World Vegetable Center.

Sex Pheromone of Common Armyworm, *Spodoptera litura* F. (Lepidoptera: Noctuidae)

After the isolation, identification, and synthesis of sex pheromone of *S. litura* by Tamaki et al. (1973), significant progress has been made in the use of this pheromone. Researches from different parts of the world have confirmed that the adoption of sex pheromone lures can effectively monitor and/or suppress adult population, largely decrease larvae or egg mass density and thus reducing the damage rate of *S. litura* on various crops (Singh and Sachan 1993; Arida et al. 2002; Yang et al. 2009). An experiment was set up at AVRDC—The World Vegetable Center, during 2005–2007 to continuously monitor the *S. litura* population in vegetable fields. The synthetic lures were obtained from the Taiwan Agricultural Research Institute (TARI), Taichung, and plastic funnel traps with two windows (2 × 2 cm) on two opposite sides were used in the experiment. The results have indicated that *S. litura* has only one peak in Taiwan during November, with population exceeding 5000 moths per trap per month

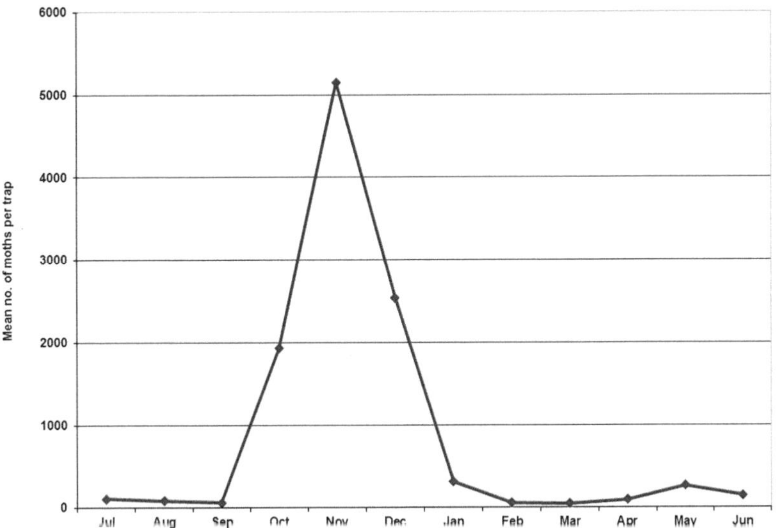

Fig. 1 Monitoring of *Spodoptera litura* population using sex pheromone lure during 2005–2007

(Fig. 1). The population usually starts increasing in October, reaches a peak in November, and then gradually declines in December. Thus, *S. litura* pheromone traps could be used to predict the population build-up in the field, and to initiate the IPM interventions. However, it cannot be solely used as a pest-management tool because of its polyphagy (several cultivated and wild plant species as host in at least 40 families) (Brown and Dewhurst 1975), and great potential to undertake long-distance migratory flights (>83 km for a male *S. litura* moth in a day) (Tu et al. 2010).

Sex Pheromone of Eggplant Fruit and Shoot Borer, *Leucinodes orbonalis* (Guenée) (Lepidoptera: Crambidae)

(E)-11-hexadecenyl acetate (E11-16: Ac) was identified as the major component of *L. orbonalis* sex pheromone in China (Zhu et al. 1987). Attygalle et al. (1988) and Gunawardena et al. (1989) also identified the presence of this compound from the sex pheromone glands of *L. orbonalis* in Sri Lanka. In addition, they also identified (E)-11-hexadecen-1-ol (E11-16: OH) as a minor component. Cork et al. (2001) also confirmed the presence of E11-16: Ac as a major component, and E11-16: OH as a minor component in the pheromone gland extracts of *L. orbonalis* from India and Taiwan, and found that E11-16: Ac and E11-16: OH (100:1 ratio) attracted significantly more numbers of male moths. Hence, the *L. orbonalis* sex pheromone was included as a potential component in the *L. orbonalis* IPM program that was implemented by AVRDC—The World Vegetable Center in the Indo-Gangetic Plains of South Asia (Alam et al. 2003). Because of the monophagous nature of *L. orbonalis*, the pheromone traps are highly effective in mass-trapping the male moths, when the traps are deployed in the entire community in a region. Thus, the *L. orbonalis* sex pheromone trap as a component of IPM significantly reduced the fruit damage, and increased the yield (Alam et al. 2003; Cork et al. 2003). Due to its high success and demand, about nine small and medium sized enterprises in India are currently selling sex pheromone lures and traps of *L. orbonalis* throughout the country (Alam et al. 2006). With the recent opening of the registration system for biopesticides in Bangladesh, several companies have got registration for the *L. orbonalis* sex pheromones. For instance, the sales of *L. orbonalis* pheromone lures and the eggplant acreage under pheromone-based IPM have almost doubled in 2012 compared to the previous year in Bangladesh (Mr. Kbd. Md. Ibrahim Khalil, Ispahani Agro Ltd., Personal Communication). Hence, *L. orbonalis* pheromone-

Table 1 Total catches of *Maruca vitrata* and *Spodoptera litura* per trap in yard long bean fields using delta traps at two different trap heights in Kamphaeng Saen (Thailand) and Hanoi (Vietnam)

Lure	Height (cm)	Thailand		Vietnam	
		M. vitrata	*S. litura*	*M. vitrata*	*S. litura*
Ratio 100:5:5	120	0.00	7.75 c	0.00	1.33 c
	170	0.00	15.25 bc	0.00	3.67 bc
*EE*10,12–16:Ald	120	0.25	19.25 ab	0.00	6.33 b
	170	0.25	21.75 a	0.00	10.67 a
Check	120	0.25	8.75 bc	0.00	3.00 bc
	170	0.75	19.00 bc	0.00	1.00 c

Means within a column followed by different letters indicate significant differences (LSD following ANOVA, $p \leq 0.05$)

Table 2 Mean trap catches of *Maruca vitrata* and *Spodoptera litura* in *Sesbania grandiflora* fields at AVRDC, Taiwan

Pheromone lure	Mean number of moths per trap	
	M. vitrata	*S. litura*
A	2.33	17.00 a
B	1.33	8.67 ab
C	1.00	21.00 a
D	0.67	15.00 ab
E	1.33	16.67 a
Check	2.67	2.00 b

Means within a column followed by different letters indicate significant differences (LSD following ANOVA, $p \leq 0.05$)

based IPM strategy has high potential in reducing the pesticide misuse in the eggplant production systems of South and Southeast Asia.

Sex Pheromone of Legume Pod Borer, *Maruca vitrata* F. (Lepidoptera: Crambidae)

Sex pheromone components of *M. vitrata* have been already identified. The major compound is (*E*, *E*)-10,12-hexadecadienal (*EE*10,12-16:Ald) (Adati and Tatsuki 1999), whereas the minor components are (*E*, *E*)-10,12-hexadecadienol (*EE*10,12-16:OH) and (*E*)-10-hexadecenal (*E*10-16:Ald) (Downham et al. 2003). A synthetic pheromone lure for *M. vitrata* consisting of *EE*10,12-16:Ald, *EE*10,12-16:OH, and *E*10-16:Ald in the ratio of 100:5:5 was attractive to male moths in Benin and Ghana, while *EE*10,12-16:Ald alone was most effective in Burkina Faso (Downham et al. 2004). Neither pheromone was effective against *M. vitrata* in Taiwan, although *EE*10,12-16:Ald alone attracted significantly higher male moths of *S. litura* (Schläger et al. 2012). Hence, we evaluated these pheromone blends against *M. vitrata* in Thailand and Vietnam during 2012. The lures were tested using sticky delta traps in two heights (120 and 170 cm above ground) in the yard-long bean fields. In both Thailand and Vietnam, none of the pheromone blend was attractive to the male moths of *M. vitrata* (Table 1). However, traps containing *EE*10,12-16:Ald alone lures attracted significantly higher male moths of *S. litura*, especially at a height of 170 cm.

Five different pheromone blends were subsequently developed at Bio-Control Research Laboratories (BCRL), India. They were—(A) *EE*10,12-16:Ald + *EE*10,12-16:OH in 100:5 ratio, (B) *EE*10,12-16:Ald + *EE*10,12-16:OH in 100:10 ratio, (C) *EE*10,12-16:Ald + *E*10-16:Ald in 100:5 ratio, (D) *EE*10,12-16:Ald + *E*10-16:Ald in 100:10 ratio, and (E) *EE*10,12-16:Ald + *EE*10,12-16:OH + *E*10-16:Ald in 100:10:10 ratio. These lures were tested using sticky delta traps in the *Sesbania grandiflora* field at AVRDC—The World Vegetable Center during 2012–2013. None of these pheromone blends was attractive to the male moths of *M. vitrata*. However, the lures were significantly attractive to the male moths of *S. litura* (Table 2). Although *M. vitrata* and *S. litura* do not share any pheromone compounds in common, the attraction of *S. litura* to *EE*10,12-16:Ald, the major component in *M. vitrata* pheromone has been confirmed in this study in both Thailand and Vietnam. Similar results were already obtained in Taiwan (Schläger et al. 2012).

This is a rare example for cross-attraction between species sharing the same habitat in the

Table 3 Mean trap catches of *Diaphania indica* and *Spodoptera exigua* in cucurbit fields at AVRDC, Taiwan

Pheromone lure	Mean number of moths per trap per week (cucumber field, 2010)		Mean number of moths per trap per week (bitter-gourd field, 2012)	
	D. indica	*S. exigua*	*D. indica*	*S. exigua*
Qlure-DII	0.75	250.25 a	1.00	13.43
Check	1.50	11.00 b	0.57	4.43

Means within a column followed by different letters indicate significant differences (LSD following ANOVA, $p \leq 0.05$)

field. However, none of the pheromone blends was found to be attractive to the target insect pest, *M. vitrata*, which implied that additional components may be present in its pheromone, besides the known compounds. Hence, future research activities should focus on the identification of these missing links in the *M. vitrata* pheromone chain to develop the most effective lures.

Sex Pheromone of Cucumber Moth, *Diaphania indica* (Saunders 1851) (Lepidoptera: Crambidae)

Three sex pheromone components of *D. indica, viz., E*11-16:Ald, *EE*10,12-16:Ald, and hexadecanal (16:Ald) were already identified (Wakamura et al. 1998). *E*11-16:Ald and *EE*10,12-16:Ald are the two major components. Synthetic mixture of *E*11-16:Ald and *EE*10,12-16:Ald attracted male moths of *D. indica* in the field (Wakamura et al. 1998). It is interesting to note that *EE*10,12-16:Ald, which is a major component in the pheromone of *M. vitrata* is one of the major components in *D. indica* as well. The commercially available *D. indica* pheromone lures (Qlure-DII) were purchased from Russell IPM Company, UK. The lures were tested using sticky delta traps in the cucumber field during 2010 at AVRDC—The World Vegetable Center. However, the season-long trial confirmed that the lure was not attractive to the male moths of *D. indica* (Table 3). However, traps containing *D. indica* lures overwhelmingly attracted male moths of *S. exigua*. Similar trial was conducted during 2012 in bitter-gourd field. In this trial also, no *D. indica* male moths was attracted by the pheromone lures. Although *S. exigua* male moths were attracted by the Qlure-DII, it was not significantly different from the untreated check (traps without pheromone lures). Hence, it has become imperative to refine the sex pheromone lures of *D. indica* to use it in the IPM strategies.

Aggregation Pheromone of Striped Flea Beetle, *Phyllotreta striolata* F. (Coleoptera: Chrysomelidae)

Flea beetles (*Phyllotreta* spp.) are known to produce aggregation pheromones (Peng et al. 1999), and mostly they are the male-specific compounds. Synthetic compounds of these aggregation pheromones attracted both sexes of the flea beetles (Soroka et al. 2005). Since information on the aggregation pheromone of *P. striolata* was scanty, efforts were made to identify the aggregation pheromones in *P. striolata*. An active male-specific compound from *P. striolata* was identified as (+)-(6R,7S)-himachala-9,11-diene. Under laboratory conditions, the activity of this synthetic pheromone either alone or in combination with the host plant volatile (allyl isothiocyanate, AITC) attracted significantly high numbers of *P. striolata* (Beran et al. 2011). Subsequently, the synthetic aggregation pheromone was evaluated either alone or in combination with AITC against *P. striolata* under field conditions on radish at AVRDC—The World Vegetable Center during 2011. The results indicated that the pheromone alone was ineffective in attracting the *P. striolata* beetles (Table 4). However, the pheromone compound, when combined with AITC, enhanced its attraction. The result is also consistent with our earlier findings (Beran et al. 2011). This may be due to the fact that *P. striolata* aggregation pheromone might

Table 4 Mean trap catches of *Phyllotreta striolata* in Radish fields at AVRDC, Taiwan

Lure	Mean number of beetles per trap per week
Control	5.00 b
AITC (1X)	30.67 a
Aggregation pheromone (1X)	6.67 b
AITC (0.5X)+Aggregation pheromone (0.5X)	54.67 a
AITC (1X)+Aggregation pheromone (1X)	59.33 a

Means within a column followed by different letters indicate significant differences (LSD following ANOVA, $p \leq 0.05$)

have more than one active compound, which should be identified to improve its efficiency.

Conclusions

Insect pheromones are useful as a monitoring (e.g., *S. litura*) or a mass-trapping (e.g., *L. orbonalis*) tool in IPM strategies in tropical vegetable production systems. However, a thorough understanding on various factors, such as insights on the complete list of pheromone components and their ratio in an insect species, genetic variations among the populations of a pest species, variations in local weather factors in a region, etc., is highly imperative before designing a pheromone-based pest management strategy. For instance, the pheromone-based monitoring is still a challenge for pest species such as *M. vitrata*, *D. indica*, and *P. striolata*, although the major constituents of their pheromone compounds are already known. Hence, additional research is required to refine and develop the most effective pheromone blends for these key pest organisms in tropical Asia and Africa.

References

Adati T, Tatsuki S (1999) Identification of female sex pheromone of the legume pod borer, *Maruca vitrata* and antagonistic effects of geometrical isomers. J Chem Ecol 25:105–116

Alam SN, Rashid MA, Rouf FMA, Jhala RC, Patel JR, Satpathy S, Shivalingaswamy TM, Rai S, Wahundeniya I, Cork A, Ammaranan C, Talekar NS (2003) Development of an integrated pest management strategy for eggplant fruit and shoot borer in South Asia, Technical Bulletin TB28, AVRDC—The World Vegetable Center, Shanhua, Taiwan, p 66

Alam SN, Hossain MI, Rouf FMA, Jhala RC, Patel MG, Rath LK, Sengupta A, Baral K, Shylesha AN, Satpathy S, Shivalingaswamy TM, Cork A, Talekar NS (2006) Implementation and promotion of an IPM strategy for control of eggplant fruit and shoot borer in South Asia. Technical Bulletin No. 36. AVRDC publication number 06-672. AVRDC—The World Vegetable Center, Shanhua, Taiwan. p 74

Arida GS, Punzal BS, Ravina CC Jr, Rajotte EG, Talekar NS (2002) Insecticide application against *Spodoptera litura* (F.) and *S. exigua* (Hubner) based on trend in sex pheromone trap catches. IPM CRSP annual report. http://www.oired.vt.edu/ipmcrsp/Publications/AnnualReports/2002/Philippines/phil_topic6.pdf. Accessed 27 Dec 2012

Attygalle AB, Schwarz J, Gunawardena NE (1988) Sex pheromone of brinjal shoot and pod borer *Leucinodes orbonalis* Guénee (Lepidoptera: Pyralidae: Pyraustinae). Z Naturforschung 43c:790–792

AVRDC (1988) Progress report. Asian Vegetable Research and Development Center. Shanhua, Taiwan

Beran F, Mewis I, Srinivasan R, Svoboda J, Vial C, Mosimann H, Boland W, Büttner C, Ulrichs C, Hansson BS, Reinecke A (2011) Male *Phyllotreta striolata* (F.) produce an aggregation pheromone: identification of male-specific compounds and interaction with host plant volatiles. J Chem Ecol 37:85–97

Brown ES, Dewhurst CF (1975) The genus *Spodoptera* in Africa and the Near East. B Entomol Res 65:221–262

Cork A, Alam SN, Das A, Das CS, Ghosh GC, Farman DI, Hall DR, Maslen NR, Vedham K, Phythiam SJ, Rouf FMA, Srinivasan K (2001) Female sex pheromone of brinjal fruit and shoot borer, *Leucinodes orbonalis* blend optimization. J Chem Ecol 27(9):1867–1877

Cork A, Alam SN, Rouf FMA, Talekar NS (2003) Female pheromone of brinjal fruit and shoot borer, *Leucinodes orbonalis*: trap optimization and preliminary mass trapping trials. B Entomol Res 93:107–113

Downham MCA, Hall DR, Chamberlain DJ, Cork A, Farman DI, Tamo M, Dahounto D, Datinon B, Adetonah S (2003) Minor components in the sex pheromone of legume pod-borer: *Maruca vitrata* development of an attractive blend. J Chem Ecol 29:989–1012

Downham MCA, Tamo M, Hall DR, Datinon B, Adetonah S, Farman DI (2004) Developing pheromone traps and lures for *Maruca vitrata* in Benin, West Africa. Entom Exp App 110:151–158

Edde PA (2005) Studies with the aggregation pheromones of *Ryhzopertha dominica* (Coleoptera: Bostrichidae): habitat affinities, seasonal flight activity, and pheromone-mediated host selection behavior. Doctoral dissertation, Oklahoma State University, USA. p 207. http://digital.library.okstate.edu/etd/umi-okstate-1558.pdf. Accessed 25 Sept 2013

Gunawardena NE, Attygalle AB, Herath HMWKB (1989) The sex pheromone of the brinjal pest, *Leucinodes orbonalis* Guenee (Lepidoptera): problems and perspectives. J Natl Sci Cou Sri Lanka 17(2):161–171

Karlson P, Lüscher M (1959) "Pheromones": a new term for a class of biologically active substances. Nature 183:55–56

Löfstedt C, Kozlov M (1997) A phylogenetic analysis of pheromone communication in primitive moths. In: Cardé RT, Minks AK (eds) Insect pheromone research, new directions. Chapman & Hall, New York, pp 473–489

Peng C, Bartelt RJ, Weiss MJ (1999) Male crucifer flea beetles produce an aggregation pheromone. Physiol Entomol 24:98–99

Schläger S, Ulrichs C, Srinivasan R, Beran F, Bhanu KRM, Mewis I, Schreiner M (2012) Developing pheromone traps and lures for *Maruca vitrata* in Taiwan. Gesunde Pflanzen 64(4):183–186

Singh KN, Sachan GC (1993) Asessment of the use of sex pheromone traps in the management of *Spodoptera litura* F. Indian J Plant Protect 21(1):7–13

Soroka JJ, Bartelt RJ, Zilkowski BW, Cosse AA (2005) Responses of flea beetle *Phyllotreta cruciferae* to synthetic aggregation pheromone components and host plant volatiles in field trials. J Chem Ecol 31:1829–1843

Teale SA, Wickham JD, Zhang F, Su J, Chen Y, Xiao W, Hanks LM, Millar JG (2011) A male-produced aggregation pheromone of Monochamus alternatus (Coleoptera: Cerambycidae), a major vector of pine wood nematode. J Econ Entomol 104(5):1592–1598

Tu YG, Wu KM, Xue FS, Lu YH (2010) Laboratory evaluation of flight activity of the common cutworm, *Spodoptera litura* (Lepidoptera: Noctuidae). Insect Sci 17(1):53–59

Wakamura S, Norio A, Kinue K, Yasuda T (1998) Sex pheromone of the cotton caterpillar, *Diaphania indica* (Saunders) (Lepidoptera: Pyralidae): identification and field attraction. Appl Entomol Zool 33:429–434

Yang S, Yang S, Sun W, Lv J, Kuang R (2009) Use of sex pheromone for control of Spodoptera litura (Lepidoptera: Noctuidae). J Entomol ResSoc 11(1):27–36

Zhu P, Kong F, Yu S, Yu Y, Jin S, Hu X, Xu J (1987) Identification of the sex pheromone of eggplant borer, *Leucinodes orbonalis* Guénee (Lepidoptera: Pyralidae). Z Naturforschung 42c:1347–1348

Glimpses of Semiochemical Research Applications in Indian Horticulture: Present Status and Future Perspectives

Kamala Jayanthi PD, Ravindra Mahadappa Aurade, Vivek Kempraj, A. K. Chakravarthy and Abraham Verghese

Abstract

Pests of horticulture importance are evolving rapidly with changing climatic conditions, intensive farming practices, and constant selection pressures exerted through insecticides. With worldwide interest in environmental protection, chemical insecticides have become objects of scientific and popular protest. Critics charge chemical insecticides of their danger in provoking the development of resistant strains of pests, sabotaging ecological systems, and poisoning the environment. These liabilities of chemical insecticides have paved way to nonchemical methods, which use natural processes and mechanisms against insect pests. Of several natural processes available for exploitation of management of insect pests, semiochemicals are less exploited inspite of their ability in making integrated pest management (IPM) programs sustainable in the long run. The past, present, and future of integrating these viable alternatives with IPM programs against horticultural insect pests in India is discussed.

Keywords

Horticulture · Intensive farming · IPM · Semiochemicals

Kamala Jayanthi PD (✉) · R. Mahadappa Aurade · V. Kempraj
Division of Entomology and Nematology, Indian Institute of Horticultural Research, Hesseraghatta Lake P O, Bangalore 560089, Karnataka, India
e-mail: jaiinsect@gmail.com

A. K. Chakravarthy
Department of Agricultural Entomology, Gandhi Krishi Vignan Kendra (GKVK), University of Agricultural Sciences, Bangalore 560065, Karnataka, India

A. Verghese
National Bureau of Agriculturally Important Insects, P. Bag No. 2491, H A Farm Post, Bellary Road, Bangalore 560024, Karnataka, India

Introduction

"Certainly insects cannot think, but they can react"—Chemical cues (= semiochemicals/infochemicals) are used by insects to interact with their environment for survival and reproduction. This reliance of insects on chemical cues offers a number of opportunities for their control (Bruce 2010). Recently, semiochemicals are being increasingly used as important components

A. K. Chakravarthy (ed.), *New Horizons in Insect Science: Towards Sustainable Pest Management*,
DOI 10.1007/978-81-322-2089-3_22, © Springer India 2015

of integrated pest management (IPM) for a number of insect pests worldwide. Push–pull strategies or stimulo-deterrent diversionary strategies (SDDS) (= behavioral manipulation methods) uses repellent/deterrent (push) and attractant/stimulant (pull) stimuli to direct the movement of pest or beneficial insects for sustainable pest management. However, their potential is less exploited in horticulture. This may be mainly due to lack of thorough understanding of chemical mediated processes or chemical ecology of the intended pests. Therefore, development of reliable, robust, and sustainable push–pull strategies requires a clear scientific understanding of behavioral and chemical ecology of pest, its interactions with hosts, conspecifics and natural enemies in order to underpin key processes that can be exploited as weak links. Moreover, to understand/manipulate various semiochemicals (pheromones, allelochemicals-kairomones, allomones, and synnomones) and maximize their usefulness in IPM, collaborations between entomologists and chemists is essential. Presently, semiochemicals that induce behavioral responses to over 1500 insect species are described (Arn et al. 1992). Apart, there is a growing awareness about the complex interaction between host plants and insects (Landolt and Phillips 1997, Bruce and Pickett 2011). Despite the importance of semiochemicals, there is no information available on potential behavior modifying cues involved in tritrophic interactions and application of this technology for horticultural IPM in India. An adaptive approach to take advantage of collaborative work between biology, chemical ecology, physiology, analytical chemistry, and molecular biology in order to elucidate, develop, and implement the field application of semiochemicals for the sustainable management of horticultural insect pests is paramount for India.

Fruit Crop Pests

The major fruit crops grown in India and their insect pests along with the status of semiochemicals in their management are outlined in Table 1. Overview of major potential fruit crop pests in India reveals significant losses due to lepidopterans, coleopterans, dipterans, and homopterans where the practical feasibility of isolation, identification, and application of semiochemicals is quite possible. At present, in India the focal pest species where semiochemicals are playing a major role in integrated management programs are very limited. It is largely confined to tephritid fruit flies through male annihilation technique (MAT) in mango as well as in other crops viz., guava, citrus (Verghese and Jayanthi 2001). No serious efforts have been made about chemo-behavioral strategies of fruit flies in India involving host kairomones and male based sex pheromones.

Studies have revealed that visual and chemical cues play an important role in the host-finding behavior of fruit flies (Kamala Jayanthi and Verghese 2011; Kamala Jayanthi et al. 2012). Since, the host plant is the focal point for the ecological behavior of fruit flies involving host seeking, adult feeding, mating behavior, oviposition, and egg development, these functions are strongly modulated by chemical cues (Drew 1989). Volatile fruit odors have been used successfully as attractants for the apple maggot fly, *Rhagoletis pomonella* (Walsh; Reissig et al. 1985; Jones 1988; Jones and Davis 1989; Agnello et al. 1990). Volatile fruit odors have also been investigated as potential attractants for the Mediterranean fruit fly, *Ceratitis capitata* (Wiedemann; Prokopy and Vargas 1996; Warthen et al. 1997; Prokopy et al. 1998), the Mexican fruit fly, *Anastrepha ludens* Loew (Robacker 1992; Robacker and Andheath 1996), and Caribbean fruit fly, *Anastrepha suspensa* (Loew) (Nigg et al. 1994). However, such studies are limited in case of *Bactrocera dorsalis* complex in India (Kamala Jayanthi et al. 2012, 2013, 2013a).

Apart from tephritid fruit flies, the other important pests of national importance in fruit crops where semiochemicals can play a key role in strengthening the existing IPM strategies are mango stone weevil (MSW), *Sternochetus mangiferae* (Fabricius), inflorescence hoppers, *Idioscopus* spp, red banded caterpillar, *Deonalis albizonalis* (Hampson), fruit borer, *Citripestis eutraphera* (Meyrick), leaf gall midge, *Procontarinia matteiana* Kieffer and Cecconi, inflorescence midge, *Erosomyia indica* Grover, shoot

Table 1 Status of semiochemical technologies in important pests of horticultural crops

Crop	Major economically significant pests	Status of semiochemicals in pest management		
		Global	India	Future scope
Fruits				
Mango	Fruit flies, *Bactrocera* spp., (Tephritidae: Diptera)	MA	MA	K, SP,+
	Stone weevil, *Sternochetus mangiferae* (Fab.) (Curculionidae: Coleoptera)	–	–	K, SP, AgP,+
	Hoppers, *Ideoscopus* spp, (Cicadellidae: Hemiptera),	–	–	K, SP, AP
	Gall midges, *Procontarinia* spp, *Erosymia indica* (Grover) (Cecidomyiidae: Diptera)	–	–	K, SP
	Psyllids, *Apsylla cistellata* (Buckton), (Psyllidae: Hemiptera)	–	–	K, SP
	Leaf webbers, *Orthaga* spp., (Pyralidae: Lepidoptera)	–	–	K, SP
	Leaf miner, *Acrocercops syngramma* (Meyrick) (Gracillaridae:Lepidoptera)	–	–	K, SP
	Leaf eating weevils, *Deporaus marginatus* (Pascal) (Curculionidae: Coleoptera)	–	–	K, SP
	Shoot borers, *Chlumetia transversa* (Walker) (Noctuidae: Lepidoptera)	–	–	K, SP
	Stem borer, *Batocera rufomaculata* (DeGeer) (Cerambycidae: Coleptera)	–	–	K, SP
Banana	Rhizome weevil, *Cosmopolites sordidus* (Germar) (Curculionidae: Coleoptera)	SP	SP	K, SP,+
	Stem borer, *Odoiporus longicollis* (Olivier), (Curculionidae: Coleoptera)	–	–	K.,SP, AgP,+
	Fruit scarring beetle, *Colaspis hypochlore* Lef., (Chrysomelidae: Coleoptera)	–	–	K, SP
Citrus	Citrus leaf miner *Phyllocnistis citrella* (Stainton) (Gracillaridae: Lepidoptera)	SP	–	K, SP,+
	Citrus psylla, *Diaphorina cirti* (Kuwayama), (Psyllidae: Hemipteara)	–	–	K, SP,+
	Whiteflies, *Aleurocanthus citriperdus* (Quaintance and Baker)	–	–	K, SP,+
	Lemon butterflies, *Papilio spp* (Linn.), (Papilionidae: Lepidoptera)	–	–	K, SP,+(K, P reported for *P. polytes*)
	Black aphids, *Toxoptera* spp. (Aphididae: Hemiptera)	–	–	K, SP, AlP
	Blackfly, *Aleurocanthus woglumi* Ashby, (Aleyrodidae: Hemiptera)	–	–	K, SP, AlP
	Bark eating caterpillars, *Indarbela quadrinotata* (Walker) (Indarbelidae: Lepidoptera)	–	–	SP
	Stem boring beetles, *Chelidonium* spp, (Cerambycidae: Coleoptera)	–	–	K, SP,+
	Fruit sucking moths, *Eudocima* spp. (Noctuidae: Lepidoptera)	–	–	K, SP,+
Guava	Fruit flies, *Bactrocera* spp., (Tephritidae: Diptera)	MA	MA	K, SP,+
Grapes	Flea beetles, *Scelodonta strigicollis* (Mostschulsky) (Chrysomelidae: Coleoptera)	–	–	K
	Berry webber, *Adoxophyes privatana* (Walker) (Tortricidae: Lepidoptera)	SP	–	SP
	Shot hole borer, *Xyleborus* spp, (Scolytidae: Coleoptera)	–	–	K, AgP,+

Table 1 (continued)

Crop	Major economically significant pests	Status of semiochemicals in pest management		
		Global	India	Future scope
Sapota	Chikoo moth, *Nephopteryx eugraphella* (Rag.) (Pyralidae: Lepidoptera)	–	–	SP
	Leaf miner, *Acrocercops gemoniella* (Stainton) (Gracillaridae: Lepidoptera)	–	–	SP
	Bud borers, *Anarsia species* (Gelichiidae: Lepidotera)	–	–	SP
	Seed borer, *Trymalitis margarias* Meyrick (Tortricidae: Lepidoptera)	–	–	SP
Pomegranate	Pomegranate butterfly, *Deudorix isocrates* (Fab.) (Lycaenidae: Lepidoptera)	–	–	K, SP,+
Ber	Fruit flies, *Carpomyia vesuviana* (Costa), (Tephritidae: Diptera)	–	–	K, SP, +
	Fruit borers, *Meridarchis scyrodes* (Mey.), (Carposinidae: Lepidoptera)	–	–	SP
Litchi	Litchi stink bug, *Tessaratoma papillosa* Drury (Tessarotomidae: Hemiptera)	–	–	K, SP, +
Temperate fruits	Codling moth, *Cydia pomonella* (L.), (Tortricidae: Lepidoptera)	SP	SP	K, SP,+
	Stem borer, *Apriona cinera* (Chevrolet), (Cerambycidae: Coleoptera)	–	–	K, AgP, SP+
	Tent caterpillar, *Malacosoma indica* (Walk.) (Lasiocampidae: Lepidoptera)	–	–	SP
	Leopard moths, *Zeuzera spp* (Cossidae: Lepidoptera)	–	–	SP(A, SP are available for *Z. pyrina*)
	Leaf roller, *Archips termias* (Meyrick), (Tortricidae: Lepidoptera)	P	–	SP (SP are available for other *Archips* spp)
	Peach twig borer, *Anarsia lineatella* Zeller (Tortricidae: Lepidoptera)	P	–	SP
	Walnut weevil, *Alcidodes porrectirostris* Marshall (Curculionidae: Coleoptera)	–	–	K, SP,+
Vegetables				
Tomato	Fruit borer, *Helicoverpa armigera* Hub. (Lepidoptera: Noctuidae)	SP	SP	K,+
	Tobacco caterpillar, *Spodoptera litura* Fab. (Lepidoptera: Noctuidae)	SP	SP	K,+
	Serpentine leaf miner, *Liriomyza trifolii* (Burgess) (Agromyzidae: Diptera)	–	–	SP
	Whiteflies, *Bemisia tabaci* (Genn.), (Aleyrodidae: Hemiptera)	–	–	SP, HD
Brinjal	Fruit and shoot borer, *Leucinodes orbonalis* Guen. (Pyralidae: Lepidoptera)	SP	SP	K, SP,+
	Ash weevil, *Myllocerus* spp., (Curculionidae: Coleoptera)	–	–	SP, K,+
	Epilachna beetle, *Epilachna viginctioctopunctata* Fab. (Coccinellidae: Coleoptera)	–	–	K, SP,+
	Mealy bug, *Coccidohystrix insolita* (Green) (Pseudococcidae: Hemiptera)	–	–	K, SP,+
	Gall midge, *Asphondylia* sp. (Cecidomyiidae: Diptera)	–	–	K, SP,+

Table 1 (continued)

Crop	Major economically significant pests	Status of semiochemicals in pest management		
		Global	India	Future scope
Chilli/ Capsicum	Thrips, *Scirtothrips dorsalis* Hood (Thripidae: Thysanoptera)	–	–	A, HD
	Green peach aphid, *Myzus persicae* Sulzer, (Aphididae: Hemiptera)	A, AlP	–	A, AlP,+
	Gall midge, *Asphondylia capsici* Barnes (Cecidomyiidae: Diptera)	–	–	K, SP,+
Okra	Leafhoppers, *Amrasca biguttula* Ishida, (Cicadellidae: Hemiptera)	–	–	K, Agp
	Shoot and fruit borer, *Earias spp* (Noctuidae:Lepidoptera)	SP	–	SP
	Aphids, *Aphis gossypii* Glover, (Aphididae: Hemiptera)	SP	–	SP, HD
	Red cotton bug, *Dysdercus cingulatus* Fab., (Pyrrhocoridae:Hemiptera)	–	–	K, SP,+
Cruciferous vegetables	Diamondback moth, *Plutella xylostella*,(L), (Plutellidae: Lepidoptera)	SP, K	SP	SP, K,+
	Leaf webber, *Crocidolomia binotalis* (Zeller), (Pyralidae: Lepidoptera)	SP	–	SP, K,+
	Stem borer, *Hellula undalis* (Fab.), (Crambidae: Lepidoptera)	SP	–	SP, K,+
	Cabbage butterfly, *Pieris brassicae* (L), (Pieridae: Lepidoptera)	A	–	K, SP,+
	Aphids, *Brevicoryne brassicae* (Linn), *Lipaphis erysimi* (Kalt.) (Aphididae: Hemiptera)	–	–	K, AlP, HD+
Leguminous vegetables	Stem fly, *Ophiomiia phaseoli* (Tryon), (Agromyzidae: Diptera)	–	–	K, SP.+
	Pod borers, *Lampedis boeticus* (L.), (Lycaenidae: Lepidoptera)	–	–	K, SP,+
	Aphids *Aphis craccivora* (Koch), *Acrythosiphon pisum* (Harris) (Aphididae: Hemiptera)	–	–	K, SP, HD,+(P available for *A. fabae*, *A. gossypii)*
	Bruchids, *Callosobruchus chinensis* (L.), (Bruchidae: Coleoptera)	A, SP	–	K, SP,+
Cucurbits	Fruit flies *Bactrocera cuccurbitae* (Coq.), (Tephritidae: Diptera)	MA	MA	K, SP,+
	Red pumpkin beetle, *Aulacophora fovicollis* (Lucas) (Chrysomelidae: Coleoptera)	–	–	K, SP,+ (A available for *A. fermoralis*)
	Leaf eating caterpillar, *Diaphania indica* (Saund) (Crambidae: Lepidoptera)	SP		SP
Onion/ Garlic	Thrips, *Thrips tabaci Lindeman*, (Thripidae: Thysanoptera)	K, A	–	K, HD,+
Tuber crops				
Potato	Tuber moth, *Phthorimaea operculella* (Zeller) (Gelechiidae: Lepidoptera)	P	–	SP, K,+
	White grubs, *Holotrichia* sp., (Scarabaeidae: Coleoptera)	–	–	K, SP,+
	Green leaf hopper, *Empoasca kerri* Pruthi (Cicadellidae: Hemiptera)	–	–	K, HD,+(A, PgS available for *E. vitis*, *E. fabae*)
	Green peach aphid, *Myzus persicae* Sulzer, (Aphididae:Hemiptera)	A, SP	–	A, SP, HD,+
Sweet potato	Sweet potato weevil, *Cylas formicarius* Fab., (Curculionidae: Coleoptera)	K, SP	K, SP	K, SP,+

Table 1 (continued)

Crop	Major economically significant pests	Status of semiochemicals in pest management		
		Global	India	Future scope
Flowers	sucking pests	–	–	HD
Plantation crops				
Coconut	Rhinoceros beetle, *Oryctes rhinoceros* (L.) (Scarabaeidae: Coleoptera)	AgP, A	AgP	K, SP,+
	Red palm weevil, *Rhynchophorus ferrugineus* (Olivier) (Curculionidae: Coleoptera)	AgP, A	AgP	K, AgP,+
	Black headed caterpillar, *Opisina arenosella* Walker (Crystophasidae: Lepidoptera)	SP	SP	K,+
Cashew	Stem/root borer, *Plocaederus ferrugineus* L. (Cerambycidae: Coleoptera)	–	–	K, SP,+
	Tea mosquito bug, *Helopeltis antonii* Sign., (Miridae: Heteroptera)	–	–	K, SP,+
Coffee	White stem borer, *Xylotrechus quadripes* (Chevrolat) (Cerambycidae: Coleoptera)	A, AgP, CP, CuH	AgP	K, AgP, SP,+
	Coffee berry borer, *Hypothenemus hampei* (Ferrari) (Curculionidae: Coleoptera)	A	A	K, SP,+
	Shot hole borer, *Xylosandrus compactus* (Eichhoff) (Scolytidae: Coleoptera)	A	–	K, SP, AgP,+
Spice crops				
Arecanut	Root grub, *Leucopholis burmeisteri* Brenske Scarabaeidae: Coleoptera)	SP	–	K, AgP, SP,+
	Inflorescence caterpillar, *Tirathaba mundella*Walker (Pyralidae: Lepidoptera)	–	–	K, P,+
Ginger/ Turmeric	Shoot borer, *Conogethes punctiferalis* Guen. (Crambidae: Lepidoptera)	SP	SP	K, SP,+
	Leaf roller, *Udaspes folus* Cram., (Hesperiidae: Lepidoptera)	–	–	SP
	Thrips, *Panchaetothrips indicus* Bagnall (Thripidae: Thysanoptera)	–	–	A, HD
	Cigarette beetle, *Lasioderma serricorne* (F.) (Anobiidae: Coleoptera)	Al, MaP	–	Al, MaP, SP,+
Cardmom	Shoot borer, *Conogethes (= Dichocrocis) punctiferalis* Guen. (Crambidae: Lepidoptera)	SP	SP	K, SP,+
	White flies, *Dialeurodes cardamomi* (David and Subr.) (Dialeurodidae: Hemiptera)	–	–	A, Al, HD,+
	Thrips, *Sciothrips cardamomi,* (Thripidae: Thysanopotera)	–	–	A, Al, HD,+
Pepper	Pollu beetle, *Longitarsus nigripennis,* (Chrysomelidae: Coleoptera)	SP	–	SP, K,+ (SP are available for other *Longitarsus* spp)

A attractant, *AgP* aggregation pheromone, *Al* Allomone, *AlP* alarm pheromone, *CP* contact pheromone, CuH cuticular hydrocarbons, *HD* host plant defense, *K* kairomone, *MaP* Marking pheromone, *PgS* phagostimulant, *SP* sex pheromone, + mixtures

gall psylla, *Apsylla cistellata* (Buckton), early shoot borer, *Chlumetia transvera* (Walker), webber, *Othaga exvinaceae* (Hampson), stem borer, *Batocera rufomaculata* (DeGeer), etc. in mango; asian citrus psylla, *Diaphorina citri* Kuwayama, leaf miner, *Phyllocnistis citrella* Stainton, blackfly, *Aleurocanthus woglumi* Ashby, black aphid, *Toxoptera aurantii* (Boyer de Fonscolombe), brown aphid *Toxoptera citricida* (Kirkaldy), fruit sucking moth, *Eudocima* spp. in case of citrus; rhizome weevil, *Cosmopolites sordidus*

(Germar), pseudostem weevil, *Odoiporus longicollis* Olivier in banana; anar butterfly, *Deudorix isocrates* (Fab.) and fruit sucking moth, *Eudocima spp* in pomegranate; seed borer, *Trymalitis margarias* Meyrick, chiku bud borer, *Anarsia achrasella* Bradley, chiku moth, *Nephopteryx eugraphella* Ragonot in sapota; codling moth, *Cydia pomonella* in apple; Ber fruit borer, *Meridarchis scyrodes* Meyrick and ber fruit fly, *Carpomyia vesuviana* Costa in ber; fruit borer complex viz., *Conopomorpha cramerella* (Snellen), *Platepeplus aprobola* Meyer, *Dichocrosis* sp., litchi bug, *Tessarotoma javanica* Thumb in Litchi.

The MSW, *S. mangiferae* is monophagous to mango and carries quarantine regulations. The IPM strategies involve chemical as well as cultural measures only to manage this pest so far. Infochemical research revealed that volatile odors surrounding weevils, frass, and mango leaves were chemically distinguishable through Gas chromatography mass spectrometry (GC-MS) (Andrew 2011). The compounds identified in weevil odors were found to be of mango origin with varied levels of relative abundance suggesting a sequestration by the weevils and a possible role in chemical communication. The behavioral experiments further demonstrated that weevils are attracted to conspecifics possibly through an actively released aggregation pheromone. Nevertheless, role of phyto-semiochemicals have not been explored in this monophagous pest. Studies on the semiochemistry of closely related species *S. frigidus* (mango pulp weevil) that is restricted to North east (NE) India revealed that floral volatiles of *Mangifera indica* L. (cv. Carabao) are attractive to adult weevils (De Jesus et al. 2003). Further, a six component blend of floral volatiles containing acetic acid (0.70 %), decane (0.3 %), acetone (4.4 %), linalool (82.2 %), ethyl benzoate (11.4 %), and 2-methyl heptenone (1.0 %) elicited 70 % attraction response in *S. frigidus*. This blend was better than each individual component. The hexane extract of 70-day-old green "carabao" mango fruit yielded two active fractions (fractions 5 and 6 at 4.6 mg and 11.7 mg per kg. fruit, respectively) eliciting oviposition by gravid female weevils. The fatty acids found in the active fractions viz., myristic and oleic acids elicited significantly higher oviposition response (Averages of 5.7 and 4.0 eggs, respectively) compared to all other treatments including the control (average of 1.5 eggs) representing an important step in the development of bait traps (de Jesus et al. 2004). Similarly, in MSW, semiochemicals involving either aggregation/sex pheromones or plant kairomones appear to offer the best chance of developing practical systems for orchard monitoring and may provide additional opportunities for controlling this quarantine pest through mating disruption or mass-trapping.

The stem borer, *Batocera rufomaculata* (Sub family: Lamiinae, Family: Cerambycidae, Order: Coleoptera), a longhorned beetle is increasingly becoming a menace in older (>10 years) mango orchards across the country. It is highly polyphagous and about 50 host plant species of 18 different plant families are attacked (CABI 2007). Several thousands of trees have been lost in the last decade. The borer, a beetle of almost 4–5 cm length, thought to be univoltine, lays eggs on the main trunk of relatively older mango trees mainly during monsoon season. The grubs bore and feed on vascular tissues thereby interrupting nutrient and water transport. The symptoms of damage include active frass that shove out from infested holes and sometimes sap oozing. Our observation revealed existence of overlapping generations as against previously thought and reinfestation of already damaged trees (with fresh frass) are quite common. There is a growing body of evidence that hydrocarbons within the epicuticular wax layer of females serve as contact pheromones, and play important roles in the mating systems of longhorned beetles (Ginzel and Hanks 2003; Ginzel et al. 2006; Ginzel 2010; Spikes et al. 2010). Males from diverse subfamilies of the Cerambycidae orient to females only after contacting them with their antennae, suggesting that males recognize potential mates by contact chemoreception (Ginzel 2010). Nevertheless, these signals have been identified already for several species in the phylogenetically advanced subfamilies like Lamiinae (Ginzel 2010).

Further, long range mate location mediated by pheromones has been documented in several sub families of Cerambycidae including Lamiinae, through both male-produced (Lacey et al. 2004;

Reddy et al. 2005a) as well as female-produced sex pheromones (Rodstein et al. 2011; Ray et al. 2011). Further, evidences suggest that attraction in longhorned beetles is likely to be a combination of host kairomones and pheromones (Nehme et al. 2010; Saint- Germaine et al. 2007; Smith et al. 2007, 2008) indicating at least two strategies of long-range attraction that depend largely on the condition of the larval host (Hanks 1999; Allison et al. 2004; Millar et al. 2009) have emerged. For species utilizing stressed hosts as noticed in the case of *B. rufomaculata*, which tend to be ephemeral resources, both sexes are reported to get attracted to host kairomones (Ginzel and Hanks 2005), male-produced pheromones (Hanks et al. 2007; Lacey et al. 2007a, b, 2008, 2009; Ray et al. 2009a, b), host kairomones + male produced aggregation pheromones (Silk et al. 2007), or host/bark kairomones + male produced aggregation pheromones (Pajares et al. 2010). In a few species, females are attracted to male-produced sex pheromones (Lacey et al. 2004; Hall et al. 2006; Hanks et al. 2007) or a combination of host kairomones + male produced sex pheromones (Fonseca and Zarbin 2009).

The role of semiochemicals in mate location in majority of cerambycids apparently appears quite complex involving possible role of male and female pheromones and kairomones (Wickham et al. 2012). A detailed multi-faceted behavioral approach to understand the mate/host location perhaps will provide a needed tool for pest management, survey and detection, and control for this important polyphagous cerambycid.

Similarly, there are a couple of serious curculionid weevil pests in banana viz., rhizome weevil, *C. sordidus*; pseudostem weevil, *O. longicollis* that are highly monophagous and specific to banana. Of these, for rhizome borer, *C. sordidus*, a commercial male aggregation pheromone—Sordidin was identified by Budenberg et al. (1993) and is popular among banana farmers. In case of pseudostem borer, *O. longicollis*, the weevils were already known to get attracted to cut stems of host plant (Sahayaraj and Kombaiah 2009; Palanichamy and Ya-Ping 2011; Kamala Jayanthi et al. 2012) and conspecifics (Prasuna et al. 2008). Baiting with male aggregation pheromone (2-methyl-4-heptanol) of *O. longicollis* in conjunction with host plant extract in funnel traps attracted significantly more weevils than traps baited with either pheromone or host plant extract alone (Gunawardena et al. 1997, 1999; Palanichamy and Ya-Ping 2011). Further, evidence for a female-produced sex pheromone which is attractive to male weevils was also reported (Ravi and Palaniswami 2002). Nevertheless, all these studies are limited laboratory/field trials and there is no commercial availability of phyto-semiochemicals that will attract weevils to strengthen the management programs to date. Semiochemical based trapping in banana weevil management has potential either in mass trapping or as part of IPM programs. Accurate identification, isolation, characterization, synthesis and formulation of these semiochemicals through consistent behavioral bioassays will be a boon to farmers.

The chemical ecology of homopterans from host plants to consepcific interactions were compiled in detail by Kristoffersen (2003) where research concerning homopteran behaviors and semiochemicals (= sex pheromones, alarm pheromones, aggregation pheromones, spacing pheromones, host plant volatiles and their synergistic interactions) though quite scarce. The positive results of aphid and psylloid studies certainly encourage similar attempts in other major pestriferous homoptera in mango and citrus.

There are several species of hoppers damaging mango crop all over India. Of these *Ideoscopus* spp are very important mainly attacking inflorescence and causing economic damage. The *Ideoscopus* spp are highly host specific; *I. clypealis* breeds only on inflorescence stalks of mango; where as *I. nitidulus, Amrasca splendens*, and *Amritiodus atkinsoni* though specific to mango can breed both on new flushes of leaves as well as inflorescence stalk. *Amrasca biguttula biguttula* can however feed and breed on a number of hosts of several families such as solanaceae, cucurbitaceae, malvaceae, asteraceae etc. Apart from cultural and biological methods, chemical option is still the major weapon for controlling this pest. There are no studies on exploring the role of host plant cues involved as stimulants for oviposition and also for breeding. Further, any insect borne semiochemicals like aggregation pheromones helping the hoppers to stay as broods is less understood. Studies in other cicadellids indicated the role of olfactory cues in host plant detection of American grapevine leafhopper nymphs, *Scaphoideus titanus* Ball (Mazzoni et al. 2009; Hegde et al. 2012).

The regulated pest, mango fruit borer or Red banded mango caterpillar (RBMC), *Deanolis albizonalis* Hampson is causing alarming damage to all stages of mango fruits (Sujatha and Zaheruddeen 2002). An effective pheromone lure has been identified by Hortresearch in New Zealand that trapped hundreds of male moths and was significantly more attractive than virgin female moths, catching six times more moths than caged virgin females suggesting the possibility of routine trapping of male moths for monitoring as well as for mating disruption (Gibb et al. 2006, 2007). Initial testing suggested the lure was effective for at least four weeks in the field under tropical conditions and further work needs to be done on trap type, lure matrices (instead of rubber septa), mating behavior, dispersal, and lure attractancy range (Gibb et al. 2006, 2007). Further, Yarrow and Chandler (2007) suggested that trap clearance is needed to be done weekly to fortnightly in tropical areas to minimize degradation of trap catches. Further, the lure impregnated septum used with a delta trap/sticky mat can also be used as a supplementary early warning to growers. In India, attempts were made to use the reported sex pheromone fractions to Indian RBMC populations, but did not generate any moth response under field conditions indicating the need to study the behavioral ecology and strain differences in Indian RBMC populations. Limited studies on intermittent calling behavior of *D. albizonalis* and female sex pheromone were carried out in India (Sujatha et al. 2002).

Until recently, the leaf webber, *O. exvinaceae* is a major problem in neglected orchards where infestation starts from April and continues up to December. Literature states that after December, the webber moth undergoes hibernation in pupal form in soil until April. During the summer showers the moths from hibernating pupae emerges and fresh infestation starts from April onwards. However, the change in precipitation pattern resulting in intermittent showers after December lead to continuous webber infestation even in well maintained orchards. Recent studies clearly proved that *O. exvinaceae* tags along with conspecifics through multiple ovipositions by several conspecific gravid females into the same web and/or nearby web (Kamala Jayanthi et al. 2013b). Usually, the mother moths depend on several cues viz., secondary compounds, visual signals (plant and leaf shape), presence of natural enemies or mutualists, presence of conspecific immatures, microclimate (Rausher 1978; Williams and Gilbert 1981; Freitas and Oliveira 1992) to choose the suitable oviposition site for its progeny survival. Further experiments to find out the cues aiding the gravid female to oviposit within the same web or near the already existing web and female based sex pheromone (In the Lepidoptera, with the exception of butterflies, mate finding is primarily mediated by female-emitted sex pheromones and finely tuned male responses, Wyatt 2003) will definitely help in luring both sexes into traps. Similar attempts to locate female-based sex pheromones may yield positive results in another lepidopteran, early shoot borer, *C. transversa*. Successful synthesis of codlemone, the codling moth female sex pheromone blend has led to behavior based monitoring and management of codling moth infestations in apple (Lo et al. 2013).

Pomegranate fruit borer, *Deudorix isocrates* (Fab.), a polyphagous pest with a wide range of host plants such as plums, peaches, mulberry, litchi, sapota, guava, tamarind, pears, citrus, litchi, ber, anola, and apple. Pomegranate is the most preferred host in which the pest may destroy up to 50 % of the fruits. The management includes several cultural, mechanical, chemical, botanical and biocontrol agents, but still leaves scope for strengthening of IPM through semiochemicals. Existence of strong female based sex pheromone communication system in pomegranate fruit borer was identified by Indian Institute of Chemical Technology (IICT) in collaboration with Mahatma Phule Krishi Vidyapeeth (MPKV), Rahuri. Three sex pheromone components have been identified and synthesized successfully in the lab and confirmed their bioactivity by GC-EAD (Seema 2009).

The role of semiochemicals was less explored for the notorious fruit piercing moths, *Eudocima* spp that are serious pests on commercially important crops including pomegranate, citrus, guava, mango, papaya, litchi, carambola, grapes, eggplant tomato, etc. This pest occurs not only in the tropics all around the world including Indo-Australian-Pacific region but also in Africa. Moths feed at night by penetrating the rind of ripe fruits with their strong proboscis and suck the juice. Internal injury consists of a bruised dry area beneath the skin. Secondary rots develop at the puncture site. Fermenting fruits are often visited and fed on by secondary-moth feeders taking advantage of the access hole drilled by this fruit piercing moth. The management for this moth includes insecticidal control, night watching, hand collection of moths, moth destruction using light traps, bonfires, and altering cropping period that have limited impact. No attempts to explore the possibility of phytosemiochemicals were made. However, semiochemicals may have tremendous impact on the moth, which breeds on a specific creeper host, *Tenospora cordifolia* L.

Usually cecidomyids select hosts and mates by means of olfactory signals. Semiochemical trap can be a sensitive monitoring tool for detection, timing treatments, monitoring population trend, etc. in gall midges owing to their minute, inconspicuous body size and sudden outbreaks. In the recent past, numbers of chemical identifications of gall midge sex pheromones have been accumulated highlighting the feasibility of practical pheromone based field monitoring in these difficult to detect insects (Bruce and Smart 2009). In mango, leaf gall midges viz., *Procontarinia matteiana* Kieffer and Cecconi, inflorescence midge, *E. indica*, are attaining serious status causing severe damage. Similarly, in case of shoot gall psylla, *A. cistallata*, an unique univoltine species that causes huge direct damage through transformation of reproductive and vegetative buds in to galls (Jha et al. 2013); optimizing a *sex* pheromone-based method of monitoring may help to attract psyllids to traps as being practiced in several other crops (Pear psylla, *Cacopsylla pyricola* (Forster), Guédot et al. 2009). Similarly, a psyllid repellent that was discovered by scientists exploring why citrus trees planted near guava trees had fewer citrus psyllids revealed that the compound dimethyl disulphide (DMSD), identified in volatiles emitted by the guava trees was found in laboratory tests to be highly repellent to citrus psyllid. Recent trials have shown that the potato psyllid is also repelled by the compound.

Vegetable Crop Pests

At present the prospects of using semiochemical technologies for controlling various insect pests of vegetable crops are mainly limited to sex pheromones of lepidopterans and tephritid fruit flies. Among the lepidopterans, (Z) 11 hexadecanal + (Z) hexadecanal (97:3) (popular as Helilure- for *Helicoverpa armigera*); (Z, E), 9,11 tetradecanyl acetate + (Z, E) 9,12-dienyl acetate (19:1) (popular as Spodolure for *Spodoptera litura*); (E)-11 hexadecenyl acetate + (E)-11-hexadecen-1-ol (100:1) (popular as Leucilure for brinjal fruit and shoot borer, *Leucinodes orbonalis*); (Z)-hexadecanal-11-enal + (Z)-exzadec-11-enyl acetate (popular as Nomate-DBM, Checkmate for diamondback moth-DBM, *Plutella xylostella*) and tephritid fruit flies 4-(4-hydroxyphenyl)-2-butanone acetate (as Cuelure for cucurbit fruit fly, *B. cucurbitae*) are commercially available and fitting well in to the current IPM programs. The possibility of exploring and integrating potent viable semiochemical approaches of either insect or plant derived chemical cues for several priority target pests across vegetable crops viz., solanaceae, cruciferaceae, cucurbitaceae, etc where semiochemical scientific input is lacking will pave the way for new interventions to make current IPM programs more robust (Table 1).

Among vegetable crops, except for major lepidopteran pests (*H. armigera, S. litura* and *L. orbonalis, P. xylostella*) where sex pheromones are being used predominantly for monitoring and to an extent in the management of this pest (Tamhankar et al. 2003; Anju et al. 2004; Sun et al. 2002, 2003; Gedia et al. 2007; Lognathan 2000; Zhu et al. 1987; Attygalle et al. 1988; Kong et al. 1990; Srinivasan and Babu 2000; Cork et al. 2001; Andagopal et al. 2010), efforts were limited to explore the semiochemical possibilities for other sucking pests viz., whiteflies, aphids, hoppers where a clear scientific understanding of chemical ecology interactions between host plant-pest-natural enemy will help to design reliable, robust, and sustainable IPM components. Globally, much of success in utilizing the host plant defense was already realized against several sucking pests viz., grain aphid, *Sitobion avenae*; oat aphid, *Rhopalosiphum padi*, etc. Field plots of wheat sprayed hydraulically with volatile plant activator, *cis*-jasmone at 50 g ha^{-1} in 200 l ha^{-1}, in mid May and early June reduced aphid infestations consistently (Bruce et al. 2003). Application of methyl salicylate, a plant signal associated with oat aphid, *R. padi* either as an aqueous emulsion or from slow release vials significantly reduced settling of aphid spring migrants and served as an attractant for beneficial insects (James and Price 2004).

Further, role of host plant volatiles and host plant cues were less explored even for major lepidopteran pests. Application of aqueous jasmonic acid (JA-signalling pathway best known to regulate plant defence against herbivore attack) to tomato plants with backpack sprayer received 60% less leaf damage from herbivory than control plants with enhanced parasitism of lepidopteran larvae (Thaler 1999a, b). The push–pull system of IPM (where companion plants are used instead of synthetic chemicals to deliver semiochemicals in the field) developed for small holder agriculture has been used with much success in maize and sorghum in eastern Africa (Hassanali et al. 2008; Khan et al. 2008).

A study to explore the oviposition preferences of the gravid female moths of *H.armigera* to identify the strong kairomone source(s) from its favored hosts viz., marigold, maize, sunflower, and pigeonpea revealed that young pods of pigeonpea are the best kairomone source that could be used for the management of *H.armigera* in a cotton ecosystem (Anitha and Peter 2011). Funnel traps baited with synthetic floral odors of African marigold, *Tagetes erecta* and sweet pea, *Lathyrus odoratus* caught significantly more *H. armigera* (Bruce and Cork 2001; Bruce et al. 2002). The marigold blend contained benzaldehyde, (+)-linalool, phenylacetaldehyde and (S)-(-)-limonene, and the sweet pea blend (-)-linalool, phenylacetaldehyde, benzyl alcohol and diacetone (4-hydroxy-4-methyl-2-pentanone) in natural ratio. Although the target specificity and level of attraction obtained with floral traps was too low for mass trapping, the floral cues could possibly be used for monitoring female *H. armigera* populations in their integrated control. The electrophysiological responses of *H. armigera* to a range of putative kairomonal compounds showed that of nine host plant-produced terpenoids tested, ocimene and beta-phellandrene elicited the highest responses and of the six aromatic compounds tested phenylacetaldehyde and benzaldehyde elicited the largest responses (Burquiere et al. 2001). Several studies to understand the role of kairomones from lepidopteron pests in eliciting a host searching behavior of their natural enemies (Ballal and Singh 1999; Bakthavatsalam et al. 2000, 2007; Sahayaraj and Paulraj 2001; Singh et al. 2002; Sahayaraj 2008; Srivastava et al. 2008; Maruthadurai et al. 2011) with a aim to manipulate role of entomophagous insects in biological control programs has already been attempted.

Potential of trap crops for integrated management of several vegetable pests viz., *H. armigera*, *S. litura* (Zhong-Shi et al. 2010, 2012), major cabbage pests (Srinivasan and Krishnamoorthy 1991; Muniappan et al. 2001) was well established. Nevertheless, elucidation and identification of specific chemical compounds responsible for attractiveness of trap crops to main crop will further ease the crop-phenology related manipulations and pave way for push–pull strategies. Further, enhancing the efficacy of mass trapping programs through combinations of pheromones with attractive host-plant kairomones have been attempted in several pests (Stelinski et al. 2013; Ryall et al. 2013). This demand in depth studies to determine the optimal combinations of attractants and trap designs to maximize target pest capture. In field experiments, significantly more diamondback moths were captured in traps baited with synthetic sex pheromone with either (*Z*)-3-hexenyl acetate alone or a blend of (*Z*)-3-hexenyl acetate, (*Z*)-3-hexen-1-ol, and (*E*)-2-hexenal compared with sex pheromone alone and other blend mixtures demonstrating that green leaf volatiles (GLVs) could be used to enhance the attraction of *P. xylostella* males to sex pheromone-baited traps (Li et al. 2012). A similar phytochemical associations were reported in the family Cucurbitaceae, which include squash, melons and cucumbers against chrysomelid leaf beetles (*Diabrotica* spp/*Aulacophora* spp) and tephritid fruit flies (*B. cucurbitae*).

Cucurbitacins, the compounds responsible for bitterness in cucurbits were reported as feeding stimulants for red pumpkin beetles, *Aulacophora foveicollis* (Mehta and Sandhu 1990) and as attractants for *B. cucurbitae* (Siderhurst and Jang 2010). Thus recent research suggests that exploitation of kairomone based monitoring will result in more robust semiochemical based control programs that use concept of lure and kill, consisting of an attractant and a killing agent.

Ornamental Crops

Despite considerable economic importance, there are no systematic studies on flower and medicinal crop pests to explore the feasibility of semiochemical based management techniques in India. Keeping this in view, the important pests where semiochemicals have scope to play role are listed in Table 1. Major damage in crops like rose, gladiolus, chrysanthemum, jasmine, tuberose, aster, marigold, etc is mainly reported to be due to sucking insect pests viz., aphids, thrips, whiteflies, and mites. The thrust till recently had been on standardization of plant protection management using combination of biological and chemical methods. In view of the fact that most of the sucking pests do respond to semiochemical mediated communication, efforts in this direction to explore potent phyto-semiochemicals will definitely add up to their existing management strategies.

Tuber Crop Pests

The major pests distributed across the important cultivated tuber crops are listed in Table 1. The existing tuber crop pest management strategies that include semiochemical interventions are reported in sweet potato weevil (SPW), *Cylas formicarius* (Fab.) (Pillai et al. 1993). Management of SPW took a tremendous path after the discovery of the sex pheromone, (Z)-3-dodecen-1-ol (E)-2-butenoate by Heath et al. (1986) and thus become an important precise component in the monitoring, control, and subsequent eradication programs in different parts of sweet potato growing countries globally (Rajasekhara Rao et al. 2010). Isolation of boehmeryl acetate (a pentacyclic triterpenoid) that serves as a potent kairomone for attracting both sexes lead to pheromone–kairomone combinations that contributed to significant reduction in weevil populations and subsequent tuber damage (Palaniswami et al. 2000). A new SPW pheromone formulation, a combination of visual stimulation, pheromone and insecticide exhibited a synergistic response among weevils thereby lowering the cost of application (Yasuda et al. 2004).

In another important gelichiid pest, *Phthorimaea operculella* that causes huge damage to potato both in field and storage, the existing sex pheromones (mixtures of trans-4, cis 7-tridecadienil-1-ol-acetate and trans-4, cis7, cis 10 tridecatrienil-1-ol-acetate compounds in a ratio of 1:1.5) are being used as an ideal tools for monitoring moth flight activity than means of control (Raman 1982, 1988). A comprehensive analysis of olfactory sensitivities of *P. operculella* moths to broad range of host volatiles revealed fatty acid derivatives as important link in host location process of this oligophagous pest (Das et al. 2007). A new strategy involving attract-and-kill through optimizing pheromone and insecticide was also worked out (Kroschel and Zegarra 2010).

Plantation Crop Pests

Developing semiochemical based management technologies is highly relevant for pests of plantation crops where application of pesticides is not practical. Several groups have been working in this direction on several pests viz., rhinoceros beetle (RB), *Oryctes rhinoceros* (= rhinolure), red palm weevil (RPW), *Rhynchophorus ferrugineus* (= ferrugincol), black headed caterpillar, *Opisina arenosella,* coffee white stem borer (CWSB), *Xylotrechus quadripes*, coffee berry borer, *Hypothenemus hamper*, etc. and integration of semiochemicals with IPM is already in place (Table 1). Of three sex-specific compounds, ethyl 4-methyloctanoate (E4-MO), ethyl 4-methylheptanoate, and 4-methyloctanoic acid produced by male *O. rhinoceros* beetles, the first is an aggregation pheromone. In field trapping experiments, (4S)-ethyl 4-methyloctanoate and the racemic mixture were equally attractive and ten times more effective in attracting beetles than ethyl chrysanthemumate (EC), a previously recommended attractant indicating the potential of using ethyl 4-methyloctanoate in operational programs to control *O. rhinoceros* in plantation crops (Hallett et al. 1995). Thus the discovery of E4-MO as the male-produced aggregation pheromone and its commercial synthesis led to E4-MO rapidly superseding the earlier synthetic attractant, EC. Several trap designs, lure sources, dispensers have been tested (Bhanu et al. 2011) for use on a commercial scale.

Usually, the aggregation pheromones of RPW and RB are used for monitoring and mass trapping to manage the pest under economic threshold level and these lures predominantly attracted virgin and gravid females of RPW and RB. Female sex pheromone of black headed caterpillar, *O. arenosella* was also identified and developed in India by Pest Control India (PCI) with detailed studies on dispensers, dosage and traps to suit field conditions. Thus, it is already a quite established fact that pheromone lures can be used as a component of IPM against major coconut pests.

In case of CWSB, male produced sex pheromone, (S)-2-hydroxy-3-decanone was identified as the major and potent component for female attraction (Hall et al. 2006). Field trials in India using different trap designs viz., sticky, cross-vane traps, etc., along with different racemic mixtures have been tried for trapping females (Venkatesha et al. 2001; Hall et al. 2006). However, the pheromone-based trapping of females does not seem to be a viable strategy in managing the CWSB infestation in China as female beetles are not directly attracted to the pheromone source because of the complex mating behavior (i.e., attraction of potential mates by both sexes, repeated landing of a female next to a male, males dashing to a nearby female, rejection of mating attempts by females, post-mating female guarding by males, and size-dependent mating success of males). However, in India, pheromone traps can successfully trap the female beetles (Hall et al. 2006). Although the behavior of the beetles from both China and India is similar, larger plantation area and greater CWSB population density in India may have contributed to a higher pheromone trapping of females, compared to China (Rhainds et al. 2001). Response of coffee berry borer, *Hypothenemus hampei* Ferrari to host volatiles and their role in monitoring and management has been studied in detail (Mendesil et al. 2009; Da Silva et al. 2006; Saravanan and Chozhan 2003).

In case of oil palm bunch moth *Tirathaba mundella* Wlk., identification and testing of male sex pheromone components viz., (3S, 6S)-2, 2,6-trimethyl -6-vinyl-tetrahydro-pyran-3-ol, 4-hydroxy-3-methoxy-benzaldehyde (= vanillin), 6,10,14-trimethyl-2-pentadecanone and 6,10, 14-trimethyl-2-pentadecanol elicited antennal responses in female antenna (Yorianta Sasaerila et al. 2003).

Spice Crop Pests

At present, the reality of using semiochemicals for insect pest management in these crops are limited by both in terms of quantum of original research and standardization of IPM programs which are certainly the cases. Of various insect pests infesting spices, the scope of semiochemical application technology do exists for several lepidopteran and sucking pests that are most destructive and persistent in spice crop ecosystems (Table 1). Semiochemical work on cardamom shoot and capsule borer *Conogethes punctiferalis* Guenee (Lepidoptera: Pyralidae) is directed extensively on the sex pheromones in order to monitor moth populations. The calling behavior and attractive responses of male *C. puntiferalis* to the female crude extract and synthetic blend was studied in detail by Rajabaskar and Regupathy (2012) who found that attraction of male moths to synthetic blend (E-10-hexadecenal and Z-10-hexadecenal) was maximum at 90:10 and followed by 80:20 ratio. However, the field trapping studies by JinKyo et al. (2000) revealed the best attraction of males to blend at 70:30 ratio of E-10-hexadecenal and Z-10-hexadecenal for hair pencil extrusion and at 80:20 ratio for the flying upwind response with the highest attractiveness in fields between 70:30 and 80:20. Apart from field pests, the stored-product pests are also responsible for tremendous damage and economic losses during post-harvest phase of spices. Of which in the cigarette beetle, *Lasioderma serricorne* (F.), also known as the tobacco beetle, synthetic serricornin (4,6-dimethyl-7-hydroxy-nonan-3-one), a female sex pheromone is commercially available (Chuman et al. 1985) with noted variation in the chemical and isomeric purity of synthetic compound based on manufacturer.

Scope

Future applications of semiochemicals depend on the availability of the potential cues that enable efficient manipulation of mate- and host-finding behavior in horticultural pests. It is now within our reach to facilitate the discovery of relevant chemical cues with emerging molecular/sensitive biochemical-behavioral equipments. Nevertheless, detection and identification of potential semiochemicals for several Indian horticultural crop pests are still rudimentary and particularly needed. Our thorough understanding of insect–insect/insect–plant interactions via semiochemicals, that are involved in tritrophic interactions, could form an integral component for updating the current IPM programs. To understand the role of semiochemicals in insect–plant interactions, in addition to studying the behavioral responses of host plant-pest-natural enemies, we should be able to elucidate the origin of these chemical cues and also locate the trophic level at which they are potentially active. Of all these, the most promising will be studying the combination of host plant derived kairomones and sex/aggregation pheromones for developing sound behavior modifying pest management tactics. This will help to formulate end-to-end programs with viable behavioral cues that can be incorporated into current IPM programs. Many of the successful early studies in this field are the result of collaborations between biologists and chemists that helped to elucidate the structures of the chemical cues. Thus, customized work plan to take advantage of collaborative work on the biology, chemical ecology, physiology, analytical chemistry, and molecular biology is the need of the hour in order to elucidate, develop, and standardize the field application of sustainable semiochemical based IPM.

References

Agnello AM, Spangler SM, Ressig WH (1990) Development and evaluation of a more efficient monitoring system for apple maggot (Diptera: Tephritidae). J Econ Entomol 83:539–546

Allison JD, Borden JH, Seybold SJ (2004) A review of the chemical ecology of the Cerambycidae (Coleoptera). Chemoecology 14:123–150

Andagopal VN, Prasad TV, Jethwa D, Gedia MV, Koradia VG, Patel HV (2010) Evaluation of pheromone traps against brinjal fruit and shoot borer, *Leucinodes orbonalis* Guenee. Indian J Entomol 72:7–10

Andrew RH (2011) The role of host odours in the biology of a quarantine pest, the mango weevil. The 6th Asia-Pacific conference on chemical ecology, Chemical ecology future: green chemoecology in twenty first century during October 11–15, 2011 Beijing, China. P-33-H. pp 150

Anitha G, Peter AJ (2011) Studies on host plant kairomones for trapping *Helicoverpa armigera* in cotton ecosystem. Indian J Plant Prot 39:271–277

Anju D, Kanaujia KR, Sudha K (2004) Efficiency of different pheromonal blends to monitor *Helicoverpa armigera* (Hübner) moth catches. Indian J Plant Prot 32:147–148

Arn H, Toth M, Priesner E (1992) List of sex pheromones of lepidoptera and related attractants, 2nd eds. International Organization of Biological Control Montfavet, France

Attygalle AB, Schwarz J, Gunawardena NE (1988) Sex pheromone of brinjal shoot and pod borer *Leucinodis orbonalis* Guenee (Lepidoptera: Pyralidae: Pyraustinae). Z Naturforsch C 43:790–792

Bakthavatsalam N, Singh SP, Tondan PL, Chaudhary M, Preethi S (2000) Electrophysiological response of *Chrysoperla carnea* (Stephens) (Neuroptera: Chrysopidae) to some potential kairomonal substances. J Entomol Res 24:109–114

Bakthavatsalam N, Tandon PL, Ballal CR (2007) Behavioural responses of *Campoletis chlorideae* Uchida (Hymenoptera: Ichneumonidae) to kairomonal substances. J Entomol Res 31:217–223

Ballal CR, Singh SP (1999) Host plant-mediated orientational and ovipositional behavior of three species of chrysopids (Neuroptera: Chrysopidae). Biol Control 16:47–53

Bhanu KRM, Hall DR, Mathew T, Malvika C, Prabhakara MS, Awalekar RV, Jayanth KP (2011) Monitoring of *Opisina arenosella* by using female sex pheromones. APACE 39

Bruce TJA (2010) Exploiting plant signals in sustainable agriculture. In: Ninkovic V, and Baluška F (eds) Plant communication from ecological perspective ch 12. Springer, Berlin, pp 215–227

Bruce TJ, Cork A (2001) Electrophysiological and behavioral responses of female *Helicoverpa armigera* to compounds identified in flowers of African marigold, *Tagetes erecta*. J Chem Ecol 27:1119–1131

Bruce TJA, Pickett JA (2011) Perception of plant volatile blends by herbivorous insects-finding the right mix. Phytochemistry 72:1605–1611

Bruce TJA, Smart LE (2009) Orange wheat blossom midge, *Sitodiplosis mosellana*, management. Outlooks Pest Manage 20:89–92

Bruce WB, Edmeades GO, Barker TC (2002) Molecular and physiological approaches to maize improvement for drought tolerance. J Exp Bot 53:13–25

Bruce CR, Anderson MJ, Carey AL, Newman DG, Bonen A, Kriketos AD, Cooney GJ, Hawley JA (2003) Muscle oxidative capacity is a better predictor of insulin sensitivity than lipid status. J Clin Endocrinol Metab 88:5444–5451

Budenberg WJ, Ndiege IO, Karago FW, Hansson BS (1993a) Behavior and electrophysiological response of the banana weevil, *Cosmopolites sordidus* to host plant. J Chem Ecol 19:267–278

Burquiere L, Marion-Poll F, Cork A (2001) Electrophysiological responses of female *Helicoverpa armigera* (Hübner) (Lepidoptera;Noctuidae) to synthetic host odours. J Insect Physiol 47:509–514

CABI (2007) CABI Crop protection compendium

Chuman T, Mochizuki K, Mori M, Kohno M, Kato K, Noguchi M (1985) *Lasioderma* chemistry, sex pheromone of cigarette beetle (*Lasioderma serricorne* F.). J Chem Ecol 11:417–434

Cork A, Alam SN, Das A, Das CS, Ghosh GC, Farman DI, Hall DR, Maslen NR, Vedham K, Phythian SJ, Rouf FMA, Srinivasan K (2001) Female sex pheromone of brinjal fruit and shoot borer, *Leucinodes orbonalis* blend optimization. J Chem Ecol 27:1867–1877

Da Silva F, Ventura MU, Morales L (2006) Capture of *Hypothenemus hampei* Ferrari (Coleoptera, Scolytidae) in response to trap characteristics. Sci Agric 63:567–571

Das PD, Raina R, Prasad AR, Sen A (2007) Electroantenogram responses of the potato tuber moth, *Phthorimaea operculella* (Lepidoptera; Gelichiidae) to plant volatiles. J Biosci 32:339–349

De Jesus LRA, Calumpang SMF, Medina JR, Anju OK (2003) Feeding and reproductive behavior of the mango pulp weevil, *Sternochetus frigidus* (Fabr.) at different phenological stages of mango *Mangifera indica* L. Philipp Agric Sci 86:282–289

De Jesus LRA, Calumpang SMF, Medina JR, Kanju ohsawa (2004) Floral volatiles of *Mangifera indica* L. (cv. Carabao) attractive to *Sternochetus frigidus* (Fabr) (Coleoptera: Curculionidae). Philipp Agric Sci 87:23–35

Drew RAI (1989) The tropical fruit flies (Diptera: Tephritidae: Dacinae) of the Australasian and Oceanian regions. Mem Queensland Mus 26:1–521

Fonseca MG, Zarbin PHG (2009) Mating behavior and evidence for sex-specific pheromone in *Hedypathes betulinus* (Coleoptera: Cerambycidae: Lamiinae). J Appl Entomol 133:695–701

Freitas AVL, Oliveira PS (1992) Biology and behavior of the Neotropical butterfly *Eunica bechina* (Nymphalidae) with special reference to larval defense against ant predation. J Res Lepid 31:1–11

Gedia MV, Vyas HJ, Acharya MF (2007) Predictive relationship between spodoptera *litura* male moths in pheromone trap and their oviposition on cotton. Indian J Plant Prot 35:344–345

Gibb AR, Bunn B, Pinese B, Tenakani D, Kawi AP, Ramankutty P, Suckling DM (2006) Identification of the Sex Pheromone of the Red Banded Mango Caterpillar *Deanolis sublimbalis* Snellen. Hortresearch, Auckland, pp 20

Gibb AR, Pinese B, Tenakanai D, Kawi AP, Bunn B, Ramankutty P, Suckling DM (2007) (Z)-11-Hexadecenal and (3Z,6Z,9Z)-Tricosatriene: sex pheromone components of the red banded mango caterpillar *Deanolis sublimbalis*. J Chem Ecol 33:579–589

Ginzel MD, Hanks LM (2003) Contact pheromones as mate recognition cues of four species of longhorned beetles (Coleoptera: Cerambycidae). J Insect Behav 16:181–187

Ginzel MD, Hanks LM (2005) Role of host plant volatiles in mate location for three species of longhorned beetles. J Chem Ecol 31:213–217

Ginzel MD, Moreira JA, Ray AM, Millar JG, Hanks LM (2006) (Z)-9-nonacosene-major component of the con-tact sex pheromone of the beetle *Megacyllene caryae*. J Chem Ecol 32:435–451

Guédot C, Horton DR, Landolt PJ (2009) Attraction of male winter form pear psylla to female-produced volatiles and to female extracts and evidence of male-male repellency. Entomol Exp Appl 130:191–197

Gunawardena NE, Dissanayake S, Herath KB, Attygalle AB (1997) 2-Methyl-4-heptanol and banana stem tissue baited trap to reduce the adult population of the banana stem borer, *Odoiporus longicollis*. Patent SriLanka No 11337

Gunawardena NE, Dissanayake S, Herath KB, Attygalle AB (1999) Isolation, identification and synthesis of the aggregation pheromone of the banana stem borer, *Odoiporus longicollis*. Chem in Srilanka 16:38

Hall DR, Cork A, Phythian SJ, Chittamuru S, Jayarama, Venkatesha MG, Sreedharan K, Seetharama HG, Vinod Kumar PK, Naidu R (2006) Identification of components of male-produced pheromone of coffee white stemborer, *Xylotrechus quadripes* Chevrolat (Coleoptera: Cerambycidae). J Chem Ecol, 32(1): 213-237.

Hallett HR, Perez AL, Gries G, Gries R, Pierce HD, Yue J, Oehlschlager AC, Gonzalez LM, Boreden JH (1995) Aggregation pheromone of coconut rhinoceros beetle, *Oryctes rhinoceros* (L.) (coleoptera: Scarabaeidae). J Chem Ecol 21:1549–1570

Hanks LM (1999) Influence of the larval host plant on reproductive strategies of cerambycid beetles. Annu Rev Entomol 44:483–505

Hanks LM, Millar JG, Moreira JA, Barbour JD, Lacey ES, McElfresh JS, Reuter FR, Ray AM (2007) Using generic pheromone lures to expedite identification of aggregation pheromones for the cerambycid beetles *Xylotrechus nauticus, Phymatodes lecontei*, and *Neoclytus modestus modestus*. J Chem Ecol 33:889–907

Hassanali A, Herren H, Khan ZR, Pickett JA, Woodcock CM (2008) Integrated pest management: the push-pull approach for controlling insect pests and weeds of cereals, and its potential for other agricultural systems including animal husbandry. Phil Trans R Soc B 363:611–621

Heath RR, Coffelt JA, Sonnet PE, Proshold FI, Dueben B, Tumlinson JH (1986) Identification of a sex pheromone producted by female sweet potato weevil, *Cylas formicarius elegantulus* (Summers). J Chem Ecol 12:1489–1503

Hegde M, Oliveira JN, da Costa JG, Loza-Reyes E, Bleicher E, Santana AEG, Caulfield JC, Mayon P, Dewhirst SY, Bruce TJA, Pickett JA, Birkett MA (2012) Aphid antixenosis in cotton is activated by the natural plant defence elicitor *cis*-jasmone. Phytochemistry 78:81–88

James DG, Price ST (2004) Field-testing of methyl salicylate for recruitment and retention of beneficial insects in grapes and hops. J Chem Ecol 30:1613–1628

Jha S, Samui G, Mondal S, Hasan A (2013) Studies on bio-ecology of mango shoot gall psyllid (*Apsylla cistellata* buckton) and its management. Acta Hort (Ishs) 992:467–474

JinKyo J, Han KS, Choi KS, Boo KS (2000) Sex pheromone composition for field-trapping of *Dichocrocis punctiferalis* (Lepidoptera: Pyralidae) males. Korean J Appl Entomol 39:105–110.

Jones VP (1988) Longevity of apple Maggot (Diptera:Tephritidae) lures under laboratory and field conditions in Utah. Environ Entomol 17:704–708

Jones VP, and Davis DW (1989) Evaluation of traps for apple maggot populations associated with cherry and hawthorn in Utah. Environ Entomol 18:521–525.

Kamala Jayanthi PD, Christine MW, John C, Michael AB, Toby JAB (2012) Isolation and identification of host cues from mango, *Mangifera indica*, that attract gravid female oriental fruit fly, *Bactrocera dorsalis*. J Chem Ecol 38(4):361–369

Kamala Jayanthi, P.D., Verghese Abraham (2011). Host-plant phenology and weather based forecasting models for population prediction of the oriental fruit fly, *Bactrocera dorsalis (Hendel), Crop Protection* 30 (12):1557–1562.

Kamala Jayanthi PD, Christine MW, John C, Michael AB, Toby JAB (2012) Potential of host-specific kairomone blends for the management of mango fruit fly, *Bactrocera dorsalis* (Diptera:Tephritidae). Presented in International symposium on "New Processes and applications for plant and microbial products" ISNPMP-2012

Kamala Jayanthi PD, Vivek K, Ravindra MA, Abraham V, Bakthavatsalam N, Ravindra KV (2012) Sex linked differential response of adult banana weevil, *Odoiporus longicollis* Olivier (Coleoptera: Curculionidae) to host cues: In IV National Symposium on Plant Protection in Horticultural Crops: Emerging challenges and sustainable pest management, pp 82

Kamala Jayanthi PD, Ravindra MA, Vivek K, Bakthavatsalam N, Abraham V (2013) Specific chemical cues trigger the ovipositional decision-making behaviour of oriental fruit fly, *Bactrocera dorsalis*. Presented at international insect science congress held in Bangalore

Kamala Jayanthi PD, Vivek K, Ravindra MA, Abraham V, Toby JAB (2013a). A mango volatile cue γ-octalactone triggers a congenital recognition template for oviposition in *Bactrocera dorsalis*. Submitted to international chemical ecology conference, 19–23 Aug, 2013 Melbourne, Australia

Kamala Jayanthi PD, Ravindra MA, Vivek K, Abraham V (2013b) Mango webber, *Orthaga exvinaceae* keeps tagging along!. Presented at international insect science congress held in Bangalore

Khan ZR, Midega CAO, Amudavi DM, Hassanali A, Pickett JA (2008) On-farm evaluation of the 'push–pull' technology for the control of stemborers and Striga weed on maize in western Kenya. Field Crops Res 106: 224–233

Kong FL, Yu SD, Yu YQ, Hu XH, Xu JW, Zhu PC (1990) Sex pheromone of eggplant borer, *Leucinodes orbonalis* Guenée. Youji Huaxue 10:258–260

Kristoffersen L (2003) (E) The chemical ecology of Homoptera—from host plants to conspecific interactions. Deparment of Ecology. Lund University, Lund, pp 38

Kroschel J, Zegarra O (2010) Attract-and-kill: a new strategy for the management of the potato tuber moths *Phthorimaea operculella* (Zeller) and *Symmetrisch-*

ema tangolias (Gyen) in potato: laboratory experiments towards optimising pheromone and insecticide concentration. Pest Manag Sci 66:490–496

Lacey ES, Ginzel MD, Millar JG, Hanks LM (2004) Male-produced aggregation pheromone of the cerambycid beetle *Neoclytus acuminatus acuminatus*. J Chem Ecol 30:1493–1507

Lacey ES, Moreira JA, Millar JG, Ray AM, Hanks LM (2007a) Male produced aggregation pheromone of the cerambycid beetle *Neoclytus mucronatus mucronatus*. Entomol Exp Appl 122:171–179

Lacey ES, Ray AM, Hanks LM (2007b) Calling behavior of the cerambycid beetle *Neoclytus acuminatus acuminatus* (F.). J Insect Behav 20:117–128

Lacey ES, Moreira JA, Millar JG, Hanks LM (2008) A male-produced aggregation pheromone blend consisting of alkanediols, terpenoids, and an aromatic alcohol from the cerambycid beetle *Megacyllene caryae*. J Chem Ecol 34:408–417

Lacey ES, Millar JG, Moreira JA, Hanks LM (2009) Male-produced aggregation pheromones of the cerambycid beetles *Xylotrechus colonus* and *Sarosesthes fulminans*. J Chem Ecol 35:733–740

Landolt PJ, Philips TW (1997) Host plant influences on sex pheromone behaviour of phytophagous insects. Annu Rev Entomol 42:371–391

Li P, Zhu J, Qin Y (2012) Enhanced attraction of *Plutella xylostella* (Lepidoptera: Plutellidae) to pheromone baited traps with the addition of green leaf volatiles. J Econ Entomol 105:1149–1156

Lo PL, Walker JTS, Horner RM, Hedderley DI (2013) Development of multiple species mating disruption to control codling moth and leafrollers (Lepidoptera: Tortricidae). New Zealand Plant Protection 66:264–269

Lognathan M (2000) Monitoring of *Spodoptera litura* (F.) in tobacco using pheromone trap. J Exper Zool India 3:141–142

Maruthadurai R, Gautam RD, Archna (2011) Behavioural response of *Trichogramma chilonis* Ishii (Hymenoptera: Trichogrammatidae) to kairomones. Indian J Entom 73:247–252

Mazzoni V, Ioriatti C, Trona F, Lucchi A, De Cristofaro A, Anfora G (2009) Study on the role of olfaction in host plant detection of *Scaphoideus titanus* Hemiptera: Cicadellidae) Nymphs. J Econ Entomol 102:974–980

Mehta PK, Sandhu GS (1990) Monitoring of red pumpkin beetle (*Aulacophora foveicollis*) by cucurbitacins as kairomones through poison baits. Indian J Agr Sci 60:225–226

Mendesil E, Bruce TJA, Woodcock CM, Caulfield JC, Seyoum E, Pickett JA (2009) Semiochemicals used in host location by the coffee berry borer, *Hypothenemus hampei*. J Chem Ecol 35:944–950

Millar JG, Hanks LM, Moreira JA, Barbour JD, Lacey ES (2009) Pheromone chemistry of cerambycid beetles. In: Nakamuta K, Millar JG (eds) Chem Eco Woodboring Ins pp 52–79

Muniappan R, Junard C, Jesse B (2001) Trap crops for diamondback moth and other crucifer pests in Guam. Presented in the management of diamondback moth and other crucifer pests. Proceedings of the 4th International Workshop, Nov. 2001, Melbourne, Australia 141–146

Nehme ME, Keena MA, Zhang A, Baker TC, Xu Z, Hoover K (2010) Evaluating the use of male-produced pheromone components and plant volatiles in two trap designs to monitor *Anoplophora glabripennis*. Environ Entomol 39:169–176

Nigg HN, Mallory LL, Simpson SE, Callaham SB, Toth JP, Fraser S, Klim M, Nagy S, Nation JL, Attaway JA (1994) Caribbean fruit fly, *Anastrepha suspensa* (Loew), attraction to host fruit and host kairomones. J Chem Ecol 20:727–743

Pajares JA, Alvarez G, Ibeas F, Gallego D, Hall DR, Farman DI (2010) Identification and field activity of a male-produced aggregation pheromone in the pine sawyer beetle, *Monochamus galloprovincialis*. J Chem Ecol 36:570–583

Palanichamy MG, Ya-Ping Z (2011) Identifying potential pitfalls in interpreting mitochondrial DNA mutations of male infertility cases. Indian J Med Res 134:447–451

Palaniswami MS, Sreeja VG, Roshan CJ (2000) A potent kairomone for the management of sweet potato weevil *Cylas formicarius* F. Entomon 25:269–280

Pillai KS, Palaniswami MS, Rajamma P, Mohandas C, Jayaprakas CA (1993) Pest management in tuber crops. Indian Hortic 38:20–23

Prasuna AL, Jyothi KN, Prasad AR, Yadav JS, Padmanaban B (2008) Olfactory responses of banana pseudostem weevil, *Odoiporus longicollis* Olivier (Coleoptera: Curculionidae) to semiochemicals from conspecifics and host plant. Current Sci 94:896–900

Prokopy RJ, Andvargas RI (1996) Attraction of *Ceratitis capitata* (Diptera: Tephritidae) flies to odor of coffee fruit. J Chem Ecol 22:807–820

Prokopy RJ, Hu X, Jang EB, Vargas RI, Warthen JD (1998) Attraction of mature *Ceratitis capitata* females to 2-heptanone, a component of coffee fruit odor. J Chem Ecol 24(8):1293–1304

Rajabaskar D, Regupathy A (2012) Calling behavior and attractive response of cardamom shoot and capsule borer *Conogethes punctiferalis* Guenee (Lepidoptera: Pyralidae) to the female crude extract and synthetic blend. J Entom 9:296–301

Rajasekhara Rao K, Naskar SK, Palaniswami MS, Ray RC (2010) Management of sweet potato weevil, *Cyclas formicarius* (Fab.): An overview. J Root Crops 36:14–26

Raman KV (1982) Field trials with the sex pheromone of the potato tuber worm (Lepidoptera: Gelechiidae). Environ Entomol 11:367–370

Raman KV (1988) Control of potato tuber moth *Phthorimaea operculella* with sex pheromones in Peru. Agric Ecosyst Environ 21:85–89

Ravi G, Palaniswami MS (2002) Evidence for a female-produced sex pheromone in the banana pseudostem weevil, *Odoiporus longicollis* Olivier. Current Sci 83:893–898

Ray AM, Millar JG, McElfresh JS, Swift IP, Barbour JD, Hanks LM (2009a) Male-produced aggregation pheromone of the cerambycid beetle *Rosalia funebris*. J Chem Ecol 35:96–103

Ray AM, Swift IP, Moreira JA, Hanks LM (2009b) (R)-3-hydroxyhexan-2-one is a major pheromone component of *Anelaphus inflaticollis* (Coleoptera: Cerambycidae). Environ Entomol 38:1462–1466

Ray AM, Zunic A, Alten RL, McElfresh JS, Hanks LM, Millar JG (2011) Cis-vaccenyl acetate, a female-produced sex pheromone component of *Ortholeptura valida* a long horned beetle in the subfamily Lepturinae. J Chem Ecol 37:172–178

Reddy GV, Fettkother R, Noldt U, Dettner K (2005a) Enhancement of attraction and trap catches of the old-house borer, *Hylotrupes bajulus* (Coleoptera: Cerambycidae), by combination of male sex pheromone and monoterpenes. Pest Manag Sci 61:699–670

Reissig WH, Stanley BH, Roelofs WL, Schwartz MR (1985) Tests of synthetic apple volatiles in traps as attractants for apple maggots flies (Diptera: Tephritidae)in commercial apple orchards. Environ Entomol 14:55–59

Robacker DC (1992) Effects of shape and size of colored traps on attractiveness to irradiated, laboratory strain mexican fruit flies (Diptera: Tephritidae). Fla Entomol 75:230–241

Rhainds M, Lan CC, King S, Gries R, Mo LZ, Gries G (2001) Pheromone com-munication and mating behaviour of coffee white stem borer, *Xylotrechus quadripes* Chevrolat (Coleoptera: Cerambycidae). Appl Entomol Zool 36:299–309.

Robacker DC, Andheath RR (1996) Attraction of mexican fruit flies (Diptera: Tephritidae) to lures emitting host-fruit volatiles in a citrus orchard. Fla Entomol 79:600–602

Rodstein J, Millar JG, Barbour JD, McElfresh JS, Wright IM, Barbour KS, Ray AM, Hanks LM (2011) Determination of the relative and absolute configurations of the female-produced sex pheromone of the Cerambycid beetle *Prionus californicus*. J Chem Ecol 37:114–124

Ryall KL, Fidgen JG, Silk PJ, Scarr TA (2013) Efficacy of the pheromone, (3Z)-lactone, and the host kairomone, (3Z)-hexenol, at detecting early infestation of the emerald ash borer, *Agrilus planipennis*. Entomol Exp Appl 147:126–131

Sahayaraj K, Paulraj MG (2001) Behaviour of *Rhynocorism arginatus* (Fabricius) (Heteroptera: Reduviidae) to chemical cues from three lepidopteran pests. J Biol Control 15:1–4

Saint-Germain M, Buddle C, Drapeau P (2007) Primary attraction and random landing in host-selection by wood-feeding insects: a matter of scale? Agric For Entomol 9:227–235

Saravanan PA, Chozhan K (2003) Monitoring and management of coffee berry borer, *Hypothenemus hampei* Ferrari (Scolytidae: Coleoptera). Crop Res 26:154–158

Sasaerila Y, Gries R, Gries G, Khaskin G, King S, Takacs S, Hardi (2003) Sex pheromone component of male *Tirthaba mundella* (Lepidoptera: Pyralidae). Chemoecol 13:89–93

Seema W (2009) Biotechnological approaches in the management of plant pests, diseases and weeds for sustainable agriculture. J Biopesticides 2:115–134

Siderhurst M, Jang E (2010) Cucumber volatile blend attractive to female melon fly, *Bactrocera cucurbitae* (Coquillett). J Chem Ecol 36(7):699–708

Silk PJ, Sweeney J, Wu J, Price J, Gutowski JM, Ket-tela EG (2007) Evidence for a male-produced pheromone in *Tetropium fuscum* (F.) and *Tetropium cinnamopterum* (Kirby) (Coleoptera: Cerambycidae). Naturwissenschaften 94:697–701

Singh PK, Parsek MR, Greenberg EP, Welsh MJ (2002) A component of innate immunity prevents bacterial biofilm development. Nature 417:552–555

Smith M, Tobin P, Wu J, He W, Xu X, Gries G, Gries R, Borden J, Turgeon J, De Groot P (2007) Detection and monitoring of the Asian longhorned beetle: update on sentinal trees, attract and kill, and artificial lure studies. In: Mastro V, Lance D, Reardon R Parra G (eds) Emerald Ash Borer Research Technology Development Meeting

Smith M, Tobin P, Wu J, He W, Xu X, Gries G, Gries R, Borden J, Turgeon J, De Groot P (2008) Behavioral ecology of host selection in the Asian longhorned beetle: implications for surveying, detecting, and monitoring adult beetles. In: Gottschalk KW (ed) Proceedings of the 18th U.S. Department of Agriculture Interagency Research Forum on Gypsy Moth and Other Invasive Species. Annapolis, Mary-land

Spikes AE, Matthew AP, Jocelyn GM, Jardel AM, Paul BH, Nathan MS, Matthew DG (2010) First Contact Pheromone Identified for a Longhorned Beetle (Coleoptera:Cerambycidae) in the Subfamily Prioninae J Chem Ecol 36:943–954

Srinivasan G, Babu PCS (2000) Sex pheromones for brinjal shoot and fruit borer, *Leucinodes orbonalis*. Indian J Entomol 62:225–226

Srinivasan K, Krishna Moorthy PN (1991) Indian mustard as a trap crop for management of major lepidopterous pests on cabbage. Tropical Pest Manage 37:26–32

Srivastava M, Paul AVN, Dureja P, Singh AK (2008) Response of the egg parasitoid *Trichogramma chilonis* Ishii (Hymenoptera: Trichogrammatidae) to kairomones from three host insects. J Biol Control 22:333–340

Stelinski LL, Gut LJ, Miller JR (2013) An attempt to increase efficacy of moth mating disruption by co-releasing pheromones with kairomones and to understand possible underlying mechanisms of this technique. Environ Entomol 42:158–166

Sujatha A, Zaheruddeen SM (2002) Biology of pyralid fruit borer, *Deanolis albizonalis* Hampson: a new pest of mango. J App Zool Res 13:1–5

Sujatha A, Zaheruddeen SM, ArjunaRao A, Nagalingam B (2002) Electronantennogram and wind tunnel studies on olfactory responses of mango fruit borer *Deanolis albizonalis* (Hampson) to female sex pheromone. Pest Manag Hortic Ecosys 8:83–89

Sun F, Hu YY, Du JW (2002) The sex pheromone communication system of *Spodoptera litura* (Fabricius). Acta Entomol Sin 45:404–407

Sun F, Du JW, Chen TH (2003) The behavioral responses of *Spodoptera litura* (F.) males to the female sex pheromone in wind tunnel and field trapping tests. Acta Entomol Sin 46:126–130

Tamhankar AJ, Rajendran TP, Rao NH, Lavekar RC, Jeyakumar P, Monga D, Bambawale OM (2003) Variability in response of *Helicoverpa armigera* males from different locations in India to varying blends of female sex pheromone suggests male sex pheromone response polymorphism. Current Sci 84:448–450

Thaler JS (1999a) Induced resistance in agricultural crops: effects of jasmonic acid on herbivory and yield in tomato plants. Environ Entomol 28:30–37

Thaler JS (1999b) Jasmonate-inducible plant defenses cause increased parasitism of herbivores. Nature 399:686–688

Verghese, A. and Kamala Jayanthi, P. D. (2001) Integrated pest management in fruits, In *Pest Management in Horticultural Ecosystems,* (Eds. P. Parvatha Reddy, Abraham Verghese and N. K. Krishna Kumar), Published by Capital Publishing Company, New Delhi. pp. 1–23

Venkatesha MG, Jayarama DR, Hall AC, Seetharama HG, D'Souza MV, Srinivasan CS, Naidu R (2001) Behavioural responses of coffee white stem borer, *Xylotrechus quadripes* Chevr. (Coleoptera: Cerambycidae) to synthetic male sex pheromone. Entomon (Spl. Issue):293–295

Warthen JD, Cunningham RT, Leonhardt BA, Cook JM, Avery JW, Harte EM (1997) Improved controlled-release formulations for a new trap design for male mediterranean fruit flies: the C & C trap. J Chem Ecol 23:1471–1486

Wickhan JD, Xu Z, Teale SA (2012) Evidence for a female-produced, long range pheromone of *Anoplophora glabripennis* (Coleoptera: Cerambycidae). Insect Sci 19:355–371

Williams K, Gilbert LE (1981) Insects as agents of visual selection on plant vegetation morphology: egg mimics of Passiflora reduce egg-laying by *Heliconius* butterflies. Science 212:467–469

Wyatt TD (2003) Pheromones and Animal Behaviour. Communication by Smell and Taste. Cambridge University Press, University of Oxford

Yarrow R, Chandler L (2007) Informal testing of RBMC pheromone over the early mango fruiting season at Lockerbie Station, via Bamaga, Cape York Peninsula. The Amistar sixth Australian mango conference. Conference Proceedings. Gold Coast, Queensland, pp 24–25

Yasuda K, Hongo T, Etoo N (2004) A new type of pheromone formulation with visual stimulation for control of sweet potato weevil, *Cylas formicarius* (Fab.). Jap J Appl Entomol Zool 48:87–93

Zhong-Shi Z, Zepeng C, Zaifu X (2010) Potential of trap crops for integrated management of the tropical armyworm, *Spodoptera litura* in tobacco. J Insect Sci 10:117

Zhong-Shi Z, Ze-Peng C, Zai-Fu X (2012) Effects of three *Spodoptera litura* control strategies on arthropod diversity and abundance in tobacco agro ecosystems in South China. Pakistan J Zool 44:151–157

Zhu P-C, Kong F-L, Yu S-D, Yu Y-Q, Jin S-P, Hu X-H, Xu. J-W (1987) Identification of the sex pheromone of eggplant borer *Leucinodes orbonalis* Guenée (Lepidoptera: Pyralidae). Z Naturforsch C 42:1347–1348

Part VII

Insect Pest Management: Using Host Plant Resistance, Biological Control

Impact of Gall Midge, *Orseolia Oryzae* (Wood-Mason) Infestation on Total Phenols, Proline and Indole Acetic Acid in Paddy (*Oryza Sativa* Linn.) Genotypes

Vijay Kumar Lingaraj, A. K. Chakravarthy and Siddanagowda Ujjanagowda Patil

Abstract

The impact of gall midge infestation on total phenols, proline and indole acetic acid (IAA) contents in rice genotypes was investigated under laboratory conditions. The resistant rice genotypes showed an increased phenol and decreased proline and IAA contents in the growing apical meristem due to infestation. Significant and rapid accumulation of total phenols (0.24–0.59 mg/g) was observed at the third, seventh, and the ninth day after midge infestation in resistant genotypes compared to susceptible ones. The rapid accumulation of phenols in resistant genotypes following gall midge infestation highlights the inducible biochemical pathways involving synthesis of phenolic precursors and further oxidation into toxic quinones. Even in the susceptible genotypes, a slight increase (0.22–0.31 mg/g) was observed initially. It may be associated with gall initiation process because phenols act as IAA-oxidase inhibitors resulting in hyper-auxinity in gall tissue that leads to formation of nutritive tissue on which the gall midge feed. Significant and rapid accumulation of proline (2.96−13.50 µg/g) from the third, seventh and ninth day after infestation was recorded only in susceptible genotypes indicating their role under stress conditions. Likewise, the higher accumulation of IAA (3.16−8.18 µg/g) was observed in all the susceptible genotypes. Thus, rapid accumulation of IAA in relation to insect infestation in susceptible genotypes clearly indicated their role in gall formation because IAA acts as a growth regulator.

Keywords

Asian rice gall midge · Rice genotypes · Resistance

A. K. Chakravarthy (ed.), *New Horizons in Insect Science: Towards Sustainable Pest Management*,
DOI 10.1007/978-81-322-2089-3_23,

V. K. Lingaraj (✉)
Department of Agricultural Entomology, College of Agriculture, UAS (B), Visveswarayya Canal Farm, Mandya, Karnataka, 571405 India
e-mail: vkumaruasb@gmail.com

A. K. Chakravarthy
Department of Agricultural Entomology, Gandhi Krishi Vignan Kendra (GKVK), University of Agricultural Sciences, Bangalore Karnataka 560065 India

S. U. Patil
Zonal Agricultural Research Station (ZARS), Bramhavar, Karnataka, 576213 India

Introduction

The Asian rice gall midge, *Orseolia oryzae* (Wood-Mason) (Diptera: Ceceidomyiidae), is a major insect pest of rice in Asia (Bentur et al. 2003). In India, gall midge has been reported from almost all the rice growing states except Western Uttar Pradesh, Uttarakhand, Punjab, Haryana and the hill states of Himachal Pradesh and Jammu and Kashmir (Bentur et al. 1992). The insect being endoparasitic, use of resistant varieties is the most economical and feasible tool for its suppression (Heinrichs and Pathak 1981; Khush 1997; Mathur et al. 1999). But the emergence of new virulent biotypes of gall midge in popular rice varieties is capable of overcoming resistance and this is a cause for concern. To date six biotypes of gall midge were identified and characterized in India (Bentur et al. 2003). Widespread cultivation of high-yielding varieties made a radical change in the pest status of rice gall midge in coastal Karnataka. In the past few years, rice growers faced a substantial loss in yield because of this pest.

A wide range of allelochemical compounds present in the plants play an important defensive role against insects and other herbivores. Several instances of associations have been reported between phenolics and the resistance of plants to insect damage (Panda and Khush 1995). Proline is a basic amino acid found in high percentage in basic protein. Free proline is said to play a role in plants under stress conditions (Singh et al. 1972; Blum and Ebercon 1976). Though the molecular mechanism has not yet been established for the increased level of proline, one of the hypotheses refers to the breakdown of proteins into amino acids and conversion to proline for storage. Many workers have been reported a severalfold increase in the proline content under entomological, pathological and other physiological stress conditions (Mohanty and Sridhar 1982; Roy et al. 1988). Indole acetic acid (IAA) is an important hormone involved in plant growth and development. As this phytohormone is intimately involved in the biochemical, physiological and genetic functions of plants under stress conditions (Balasubramanian and Purushothaman 1971), information on the exact amount of IAA is essential. Since the gall was developed upon infestation by the gall midge, the present investigation was made to know the changes in phenols, free proline and IAA profile due to infestation of gall midge.

Material and Methods

Extraction

Un-infested vegetative shoot epics of 0.5 cm (approx.) from 30-day-old plants of test entries were collected after stripling leaves and leaf sheaths. Five replications were maintained for each genotype. The collected plant samples were thoroughly washed with distilled water and dried under shade. One gram plant sample piece of all genotypes were taken in separate conical flasks and 15 ml, 80 % ethanol, was added. It was refluxed for 30 min on hot water bath. After boiling, the extract was cooled and the pieces of tissues were ground thoroughly in a mortar with pestle in slight ethanol. The supernatant was decanted into another flask and residue was re-extracted with small quantity of hot ethanol and decanted. The extract was filtered through Whatman No.1 filter paper and made up to a known volume with 80 % ethanol. The ethanol part of (alcoholic) extract was stored in refrigerator at 4 °C and was used for the estimation.

Total Phenols, Proline and IAA Profile

Four resistant viz., JGL 13595, RP 4639-233, RDR 987, Abhaya and three highly susceptible viz., Jaya, IR 20 and TN1 rice genotypes were selected based on the field and laboratory evaluation against local gall midge populations from wet and winter seasons of 2005 and 2006. The genotypes were grown in plastic trays as a row of 25–30 seedlings and exposed to gall midge populations (50 females and 10 males) 8 to 10 days after sowing. The next day the trays were shifted to shallow trays containing water, and plants were frequently (once in 2 h) sprayed with water using hand atomizer to create high relative humidity for egg incubation and hatching. Eggs hatched on the third day were treated as day 0 of infestation. Stem bases, 3–5 cm in length, were cut from five seedlings/replication; pooled and fresh weights were recorded before phenol extraction in methanol at 60 °C for 20 min. Five replications variety were maintained. The plants were sampled on 0, 1, 3, 5 and 7 days after infestation. The total phenols was estimated following the Folin–Ciocalteau method (Bray and Thorpe 1954), proline by the acid ninhydrin method (Sadasivam and Manickam 1996) and IAA by the fluorimetric method of Knegt and Bruinsma (1973). The biochemical profiles in gall midge resistant and susceptible genotypes were compared with a noninfested, susceptible TN1. The data on biochemical contents were subjected to analysis of variance (ANOVA) and means were separated by LSD at $P<0.001$ (SPSS Inc 1999; SAS 2007).

Results and Discussion

Total Phenols

On the day of infestation each resistant and susceptible genotypes differed significantly with respect to total phenols. Further, the day of infestation had a significant effect on the total phenol content in all the genotypes. All the resistant genotypes showed increased level of phenols from 3–7 days after infestation compared to a susceptible and noninfested TN 1.

On the third day subsequent to infestation all the resistant genotypes viz., JGL 13595, RP 4639-233 RDR 987, OR 1941-8, NDR 2063, JGL 11605 and Abhaya recorded significantly higher total phenols compared to a susceptible and noninfested TN 1. Similarly, on the fifth and seventh day after infestation, in all the resistant genotypes there was a rapid increase in accumulation of total phenols compared to susceptible genotypes viz., Jaya, IR 20 and TN 1 and the day effect on the level of phenol contents in resistant genotypes was found significant. The total phenol content in resistant genotypes viz., JGL 13595, RP 4639-233, RDR 987, OR 1941-8, NDR 2063, JGL 11605 and Abhaya on the seventh day after infestation was 0.576, 0.550, 0.547, 0.510, 0.520, 0.549 and 0.596 mg/g, respectively, as against the day of infestation (0.240, 0.242, 0.290, 0.216, 0.180, 0.230 and 0.242 mg/g, respectively) in Fig. 1.

Thus, there was a higher accumulation of total phenols in resistant genotypes compared to susceptible genotype in relation to gall midge infestation. Thus, the rapid accumulation of total phenols in all the resistant genotypes following gall midge infestation highlights the inducible biochemical pathways, probably involving synthesis of phenolic precursors and their further oxidation into toxic quinones in the insect.

The present investigation was in close agreement with the study made by Amudhan et al. (1999). Similarly, higher concentration of polyphenols has been reported in shoot apices of resistant rice cultivars such as Shakti, Leauang 152 (Vidyachandra et al. 1981), Ptb 18 (Rajamani 1982), IET 7009, IET 7008 and Siam 29 (Joshi and Venugopal 1984). Increased phenolic content in the growing point of resistant rice cultivars during early infestation by gall midge has been reported (IRRI 1977). In the present investigation, the amount of total phenols in relation to gall midge infestation could be related to resistance. Even in susceptible genotypes such as Jaya, IR 20 and TN1 in the present investigation, slight increase in phenol content was observed, and it may be associated with the gall midge initi-

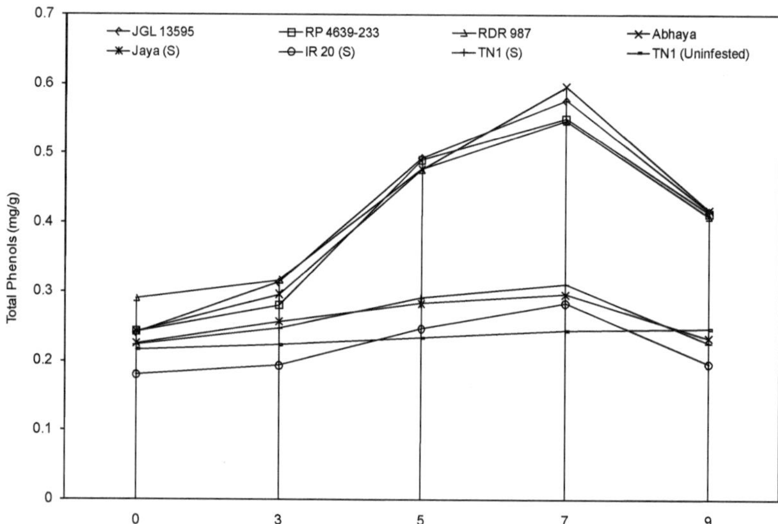

Fig. 1 Effect of gall midge infestation on total phenols (days after infestation)

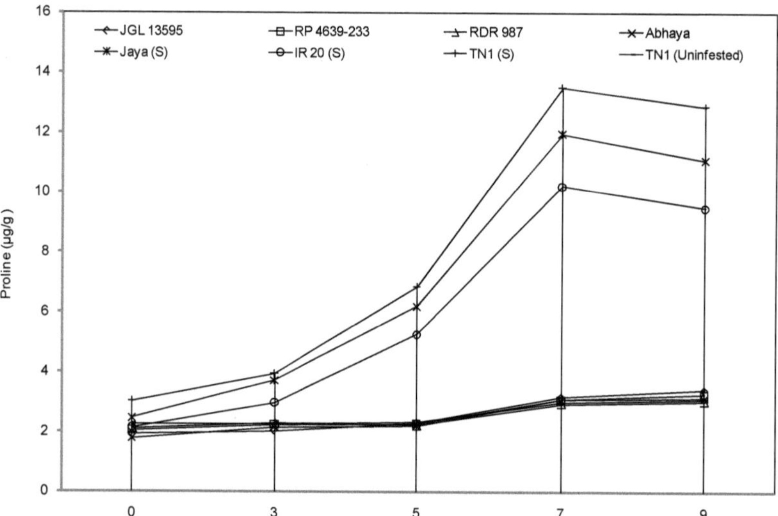

Fig. 2 Effect of gall midge infestation on total proline (days after infestation)

ation process because phenols act as IAA oxidase inhibitors (Amudhan et al. 1999), resulting in hyperauxinity in gall tissue that leads to formation of nutritive tissue on which the gall formers feed.

Proline

On the day of infestation significant differences were observed between the resistant and susceptible genotypes (Fig. 2). Further, the day of infestation had a significant effect on the proline content in all the genotypes observed. Among the genotypes, significantly higher, rapid accumulation of proline on the third, fifth and seventh day after infestation was observed in the infested susceptible genotypes. In resistant genotypes this phenomenon was not evident. On the third day after infestation, the proline content in susceptible genotypes viz., Jaya IR 20 and TN 1 was 3.70,

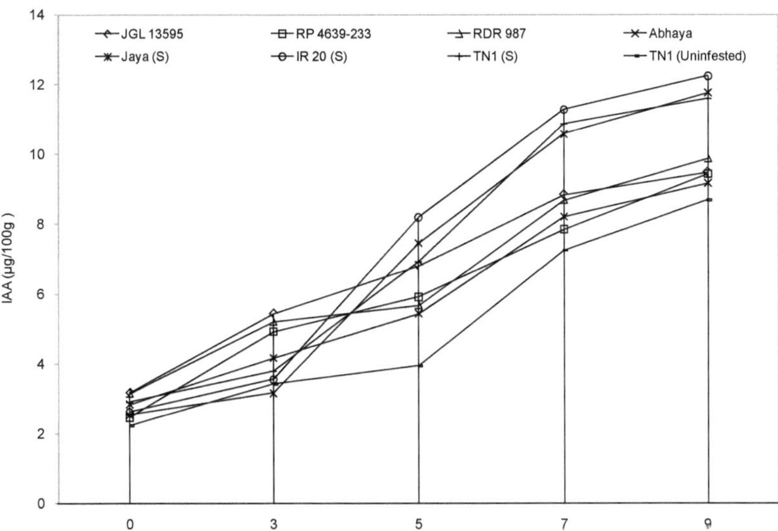

Fig. 3 Effect of gall midge infestation on indole acetic acid (days after infestation)

2.96 and 3.92 µg/g, respectively, and it increased to 6.19, 5.26 and 6.86 µg/g, respectively, on the fifth day and 11.96, 10.21 and 13.50 µg/g on the seventh day after infestation, respectively. Thus, the rapid accumulation of proline in relation to insect infestation in susceptible genotypes indicated clearly their role under stress conditions.

Among the genotypes, studies for proline profile in relation to gall midge infestation were significantly higher with rapid accumulation of proline on the third, fifth and seventh day after infestation in susceptible genotypes. In resistant genotypes this phenomenon was not evident. On the third day after infestation, the proline content in susceptible genotypes viz., Jaya, IR20 and TN1 was 3.70, 2.96 and 3.92 µg/g and it increased to 6.19, 5.26 and 6.86 µg/g on the fifth and 11.96, 10.21 and 13.50 µg/g on the seventh day after infestation, respectively.

Thus, the rapid accumulation of proline in relation to insect infestation in susceptible genotypes indicated clearly their role under stress conditions. Silver shoots caused by gall midge infestation contained higher levels of proline than the healthy leaves. Even the healthy leaves from the damaged plant contained more amount of proline than the healthy leaves of the noninfested plant as reported by Roy et al. (1988). In the present study also higher accumulation of proline in susceptible (infested) genotypes was evident.

These observations clearly showed that insect damage stimulates increase in free proline level in rice and the concentration increases in the insect damaged plants compared to the noninfested plants (resistant). A similar trend of proline accumulation was reported in brown plant hopper infestation, where the stems of the affected plants (hopper burnt hills) possessed higher concentration of proline than the stems of noninfested plants (Roy et al. 1988). Thus, in physiological stress, biotic stress caused by diseases (Mohanty and Sridhar 1982) in the present study also makes it evident that the gall midge infestation also induces accumulation of free proline in rice plants.

Indole Acetic Acid (IAA)

The data on IAA profile in resistant and susceptible genotypes in relation to gall midge infestation are presented in Fig. 3. On the day of infestation, significant differences among the genotypes were observed. Further, the day of infestation had significant effect on the IAA. Among the test genotypes, significantly higher accumulation of IAA on the third, fifth and seventh day after infestation was noticed in the infested susceptible genotypes. In resistant genotypes this phenomenon was not evident. On the third day after infestation, the IAA content in susceptible genotypes viz., Jaya, IR20 and TN 1 was 3.16, 3.56 and 3.78 µg/100 g, respectively, and it increased to 7.46, 8.18 and 6.91 µg/100 g, respectively, on the fifth day and 10.58, 11.26 and 10.86 µg/100 g on the seventh day after infestation, respectively. Thus, the rapid accumulation of IAA in relation to insect infestation in susceptible genotypes clearly indicated their role in gall formation because IAA acts as a growth regulator. In the phenol profile studies there was a slight increase in the level of phenols even in the susceptible genotypes viz., Jaya, IR 20 and TN 1. This slight increase in phenols acts as IAA-oxidase inhibitors resulting in hyperauxinity in gall tissue on which the gall midge feed.

As observed in the accumulation of proline in susceptible genotypes, significantly higher accumulation of IAA levels on the third, fifth and seventh day after infestation was observed in all susceptible genotypes, while in the case of resistant entries this phenomenon was not observed. On the third day after infestation, the IAA content in susceptible genotypes such as Jaya, IR 20 and TN1 was 3.16, 3.56 and 3.78 µg/100 g, respectively, and it was increased to 7.46, 8.18 and 6.91 µg/100 g on the seventh day after infestation, respectively. Thus, the rapid accumulation of IAA in relation to insect infestation in susceptible genotypes in the present study clearly indicated their role in gall formation because IAA acts as a growth regulator.

The present investigation corroborates the study by Balasubramanian and Purushothaman (1971) who also reported higher tryptophan and IAA level in galled shoot than healthy tissues. In the present study the increase in IAA level in susceptible genotypes of rice might be attributed to the release of IAA from IAA–protein complex. The IAA might also have originated from structural proteins by the action of proteolytic enzymes releasing a number of amino acids, including proline. Similarly, the present phenol profile study indicated the slight increase of phenols even in susceptible genotypes due to gall midge infestation. This slight increase in phenols acts as IAA-oxidase inhebitors (Amudhan et al. 1999) which might have resulted in hyperauxinity due to changes in the auxin level due to host parasite interactions in the gall tissue on which the gall midge feed.

Orseolia oryzae continues to be one of the most important endemic pests of rice in Coastal Karnataka. All cultivated rice varieties, either local or released as resistant, have become susceptible during the past decade. Concomitantly, the *O. oryzae* population has evolved into a complex, embracing more than one biotype, i.e. the genetically homogenous population has become heterogeneous, rendering pest management difficult. This study has clearly revealed the biochemical profile of selected rice genotypes and induced resistance in these rice genotypes against the rice gall midge damage.

Acknowledgements We thank J.S. Bentur (AICRIP) and I.C. Pasalu (DRR, Hyderabad). VK Lingaraj also acknowledges a fellowship of the University Grants Commission (UGC) sponsored by the Ministry of Social Justice and Empowerment, Government of India (UGC F.16-93/SA-II/RGNF/KAR-6/2005–06). Thanks are also due to director of research, University of Agricultural Sciences, Bangalore and to the farmers of southern and coastal regions of Karnataka.

References

Amudhan S, Rao PU, Bentur JS (1999) Total phenol profile in some rice varieties in relation to infestation by Asian rice gall midge *Orseolia oryzae* (Wood-Mason). Curr Sci 76:1577–1580

Balasubramanian M, Purushothaman D (1971) Effect of gall midge (*Pachydiplosis oryzae* Wood-Mason) on indole acetic acid and tryptophan contents in rice (*Oryza sativa* L.). Indian J Exp Biol 9:526–527

Bentur JS, Pasalu IC, Kalode MB (1992) Inheritance of virulence in rice-gall midge (*Orseolia oryzae*). Indian J Agric Sci 62:492–493

Bentur JS, Pasalu IC, Sharma NP, Rao PU, Mishra B (2003) Gall midge resistance in rice: current status in India and future strategies 2003. DRR Research Paper Series 01/2003. Directorate of Rice Research, Rajendranagar, Hyderabad

Blum A, Ebercon A (1976) Genetic responses in sorghum to drought stress: free proline accumulation and drought resistance. Crop Sci 16:428–431

Bray HG, Thorpe WV (1954) Analysis of phenolic compounds of interest in metabolism. Methods Biochem Anal 1:27–52

Heinrichs EA, Pathak PK (1981) Resistance to the gall midge, *Orseolia oryzae* (Wood-Mason) in rice. Insect Sci Appl 1:123–132

IRRI (1977) International Rice Research Institute. Final report of the second International Rice Gall midge Nursery, International Rice Research Institute, Los Banos, Philippines

Joshi RC, Venugopal MS (1984) Mechanism of resistance in rice to gall midge *Orseolia oryzae* (Wood-Mason): antibiosis. Indian J Entomol 46:479–481

Khush GS (1997) Breeding for resistance in rice. Ann N Y Acad Sci 287:296–308

Knegt E, Bruinsma J (1973) Estimation of indole acetic acid. Phytochemistry 12:753

Mathur KC, Reddy PR, Rajamani S, Moorthi BTS (1999) Integrated pest management in rice to improve productivity and sustainability. Oryza 36:195–207

Mohanty HK, Sridhar S (1982) Physiology of rice tungro virus and proline accumulation due to infection. Physiol Plant 56:89–93

Panda N, Khush GS (1995) Host-plant resistance to insects. CABI Publishers, International Rice Research Institute, Manila

Rajamani S (1982) Factors affecting gall midge resistance in paddy. Ph.D. thesis, Orissa University of Agriculture and Technology, Bhubaneswar, India.

Roy JK, Satapathy G, Nanda BB (1988) Proline accumulation in insect-damaged rice plants. Oryza 25:420–421

Sadasivam S, Manickam A (1996) Biochemical methods, 2nd edn. New Age International Publishers (P) Limited, New Delhi

SAS (2007) PROC user's manual, version 10th ed., SAS Institute, Cary, North Carolina.

Singh TN, Aspinall D, Paleg LG (1972) Proline accumulation and varietal adaptability to drought in barley a potential metabolic measure of drought resistance. Nat Biol 236:188–190

SPSS Inc (1999) SPSS syntax reference guide, version 10.0, SPSS Inc., 233 South Wacker Drive, Chicago

Vidyachandra B, Roy JK, Das B (1981) Chemical differences in rice varieties susceptible or resistance to gall midge and stem borer. Int Rice Res Newsl 6:7

Antixenosis and Antibiosis Component of Rice Resistance to Asian Rice Gall Midge, *Orseolia oryzae* (Wood-Mason)

Vijay Kumar Lingaraj, A. K. Chakravarthy and Siddanagowda Ujjanagowda Patil

Abstract

Antixenosis and antibiosis component of resistance to Asian rice gall midge, *Orseolia oryzae* was investigated in 23 selected rice genotypes under greenhouse conditions in a free-choice test. The higher number of adults was found settled on TN1 and Jaya (0.35 and 0.32 adults/plant) compared to resistant genotypes. However, the resistant genotypes, viz., RP 4647-1073, MTU 1075, and RP 4644-1183 were also recorded higher number of adults per plant 6 h after adult release besides showing resistance. So no distinct antixenosis in terms of numbers of adults settled was evident between resistant and susceptible genotypes. But significantly higher number of eggs were recorded on susceptible genotypes TN1 (14.44 eggs/plant) and Jaya (13.10 egg/plant) compared to resistant. Nevertheless, these differences could not be linked to resistance against gall midge, in view of the short adult life span of one day and total inactivity of adults during the day time. It is not surprising if antixenosis component is not clearly expressed in case of rice gall midge. Similarly, the maggots survived on all the resistant genotypes remained in the first instar while in susceptible genotypes they reached second instar at 7 days, third instar at 14 days, and pupal stage at 18 days after adult release. However, in resistant genotypes despite supporting the maggot development up to second instar caused mortality without manifestation of silver shoot. In this study, genotypes such as NDR 2063, JGL 11459, and JGL 13376, despite recording comparable proportion of eggs to the susceptible TN1 and Jaya, continued to be resistant.

Keywords

Asian rice gall midge · Resistance · Antixenosis · Antibiosis · Rice genotypes

V. K. Lingaraj (✉) · A. K. Chakravarthy · S. U. Patil
Department of Agricultural Entomology, College of Agriculture, UAS (B), Visveswarayya Canal Farm, Mandya, Karnataka 571405, India
e-mail: vkumaruasb@gmail.com

A. K. Chakravarthy (ed.), *New Horizons in Insect Science: Towards Sustainable Pest Management*, DOI 10.1007/978-81-322-2089-3_24, © Springer India 2015

Introduction

The Asian rice gall midge, *Orseolia oryzae* (Wood-Mason) (Diptera:Ceceidomyiidae) is a major insect pest of rice in several Asian countries (Bentur et al. 2003). In India, gall midge has been reported from almost all the rice growing states except the Western Uttar Pradesh, Uttaranchal, Punjab, Haryana, Hill states of Himachal Pradesh, and Jammu and Kashmir (Bentur et al. 1992). The insect being endoparasitic, the use of resistant varieties is the most economical and feasible tool to its supression (Heinrichs and Pathak 1981; Khush 1997; Mathur et al. 1999). But the emergence of new virulent biotypes of gall midge in popular rice varieties capable of overcoming resistance and this is a cause for concern. So far six biotypes of gall midge were identified and characterized in India (Bentur et al. 2003). Widespread cultivation of high yielding varieties made a radical change in the pest status of rice gall midge in coastal Karnataka. Recently, the emergence of resistance-breaking biotypes of rice gall midge has also contributed to its spread in new regions and a change in its status by evolving different resistance mechanism in resistant rice cultivars and donors.

Early studies to correlated morphological differences in attributes such as color and hairiness of the leaf or compactness of the leaf sheath with resistance (Rao et al. 1971). Similarly, the density and length of trichomes were negatively correlated with gall midge incidence (Joshi 1982; Devaiah 1984) while tillering pattern, leaf sheath compactness and interspace had no influence. Soon it was noted that resistant varieties offered no mechanical barrier since maggots could reach the apical meristem in all varieties (Shastry et al. 1972; Sain 1988). No distinct oviposition preferences were noted among resistant and susceptible varieties (Hidaka 1974; Kalode 1980; Kalode et al. 1983; Sain and Kalode 1994). The primary component of varietal resistance against gall midge is antibiosis (Modder and Alagoda 1971; Hidaka 1974; Kalode 1980). Understanding rice gall midge interaction and the genetics of resistance is important for breeders and entomologists in crop improvement (Bentur et al. 2004).

Material and Methods

Gall Midge Culture

Fertile soil was collected from the field and fertilizers were mixed thoroughly. Twice in a week seeds of susceptible variety TN 1 were soaked for germination and sown in the plastic pots of 8 cm diameter and 10 cm height, 2 days later at the rate of 50–75 seeds per pot. The potted plants were kept in the greenhouse with adequate light for 10 to 15 days after sowing. Eight potted plants were kept inside oviposition cage covered by polythene cover. During evening 19.00–22.00 h, the adults of rice gall midge were collected near light source using aspirator developed by the Directorate of Rice Research (DRR), Hyderabad, and were released inside oviposition cage for infestation. Twenty-five females and twenty-five males were released inside oviposition cage containing 8–10 potted plants of 15 to 20 days old seedlings. Two cages were daily infested for routing rearing during the study period. Adults were provided with fresh 15–20 days old potted plants daily for oviposition. Two days after adult release, the potted plants were sprayed with water periodically at 2–3 h intervals to moisten the plants for egg hatching and for better movement of newly hatched maggot to reach the apical meristem region for better establishment and development. The potted plants were transferred to shallow water tray and water level of 2–3 cm above the basal part of the plant was maintained to create optimum humidity and to prevent natural pasasitization and predation of maggot. On gall formation, the potted plants were shifted from water tray to the adult emergence cage. The adults were collected every morning between 18.00 and 09.00 h carefully with an aspirator for studies.

Four hundred and sixteen rice genotypes under All India Coordinated Rice Improvement Programme were evaluated under field conditions during wet 2005 and 2006 at Agricultural Research Station, Kankanady, Mangalore (12°54′N, 74°51′E; 30 m), Karnataka, South India. Each entry was rated either resistant with <10% seedling damage or susceptible with higher damage (Kalode and Bentur 1989). The genotypes recording $<10\%$ seedling damage in preliminary evaluation were selected for replicated tests. Only the genotypes recording no damage in replicated tests were rated as resistant and were used for studying mechanism of resistance. After screening, the selected rice genotypes were used for resistance studies.

Nature of Resistance

Antixenosis and antibiosis were studied together. The test resistant and susceptible entries from the trials, viz., gall midge screening trial (GMS) and gall midge biotype trial (GMBT), National screening trial 1 (NSN 1) were sown as row in plastic trays ($42\times30\times8$ cm) with 2–3 cm from plant to plant and 4 cm between rows. Each test entry was represented by ten seedlings of 20 days age in a tray, as replicate. Such four trays were maintained as four replications. A day before infestation, the trays were covered with cage made out of plastic sheets. From the stock culture, 20 female and 10 male adult gall midges were released on the test seedlings in each replication. Thus, in the free choice test gall midges were allowed to settle on test seedlings of the entries. The number of adults settled on each entry after 6 h and the number of eggs laid on seedling of the test entry 48 h after infestation were counted using hand lens. Antibiosis was studied on the same plants by recording maggot survival through periodic dissection and recording the number of surviving and dead insects at 7, 14, and 18 days after release. The data obtained were subjected to analysis of variance (ANOVA) and means were separated by Duncan's multiple range test (DMRT) (Duncan 1955).

Results and Discussion

Nature of Resistance

Antixenosis

The results revealed that, significantly higher number of adults ($P<0.05$) settled on susceptible checks TN 1 (0.35 per plant) and Jaya (0.32 per plant) compared to resistant genotypes. For instance, the number of adults that settled on JGL 13595, NDR 2063, Abhaya, and RDR 987 was 0.11, 0.11, 0.12, and 0.12 per plant, respectively, and was on par with each other. However, on the resistant genotypes, viz., RP 4647-1073, MTU 1075, and RP 4644-1183 recorded higher number of adults after release besides showing resistance. So no distinct antixenosis in terms of number of adults settled was evident between resistant and susceptible genotypes. But, significantly higher number of eggs ($P<0.05$) were noticed on TN 1 (14.44 eggs per plant) and Jaya (13.10 eggs per plant) compared to resistant genotypes. Among the resistant genotypes, significantly lower eggs were observed on RP 4613–260 (3.88 eggs per plant) followed by OR 2093-4 (6.77 eggs per plant). Likewise, the number of eggs on RP 4647-1073, Abhaya, RDR 987, NDR 3110, RP 4644-1183, MTU 1075, and OR 1914-8 was 8.33, 8.44, 8.77, 9.33, 9.44, 9.77, and 9.88, respectively, and was on par with each other (Table 1).

Similarly, in 2006 wet, significantly higher number of adults ($P<0.05$) settled on susceptible check TN 1 and Jaya, which recorded 0.54 and 0.34 adults per plant, respectively, compared to resistant genotypes. Among the resistant genotypes, few adults were found on JGL 13376, JGL 13418, JGL 11605, RP 4643-713, and JGL 11459 which recorded 0.05, 0.15, 0.21, 0.21, and 0.23 adults per plant, respectively, and differed significantly ($P<0.05$) from each other. Thus, distinct antixenosis in terms of number of adults settled was noticed between resistant and susceptible genotypes. Maximum (23.21 and 16.33 eggs per plant) numbers of eggs were noticed on susceptible checks TN1 and Jaya, and were significantly higher ($P<0.05$) compared to resistant genotypes. Likewise, the number of eggs laid by the females on JGL 13418, RP 4643-713, JGL 11605, and OR 1967-15 was 4.33, 5.11, 5.44, and 5.44 eggs per plant, respectively, and they were at par with each other. Thus, distinct ovipositional preference was observed between the resistant and susceptible genotypes (Table 2).

Earlier studies on the mechanism of resistance to gall midge suggested the involvement of biophysical characters such as hairiness of leaf blade or compactness of leaf sheath as a factor conferring resistance (CRRI 1952; Israel et al. 1961;

Table 1 Number of gall midge adults settled, eggs laid, and maggot survival on test rice genotypes, 2005

Genotypes	Antixenosis		Antibiosis						Plants damaged (%) after 35 days
	Number of adults settled/plant	Number of eggs laid/plant	7 days after adult release		14 days after adult release		18 days after adult release		
			Number of live maggots	Instar	Number of live maggots	Instar	Number of live maggots	Instar	
JGL 13595	0.11 (0.78)^{a}	10.22^{cde}	3.44 (1.98)^{cd}	I	1.00	I	0.00	–	0.00
RDR 897	0.12 (0.78)^{a}	8.77^{cd}	1.22 (1.31)^{ab}	I	0.00	–	0.00	–	0.00
RP 4613-260	0.17 (0.82)^{b}	3.88^{a}	2.66 (1.76)^{d}	I	0.00	–	0.00	–	0.00
RP 4644-1183	0.28 (0.88)^{cd}	9.44^{de}	2.10 (1.60)^{bcd}	I	0.00	–	0.00	–	0.00
RP 4647-1073	0.32 (0.90)^{de}	8.33^{bc}	3.44 (1.98)^{cd}	I	1.00	II	0.00	–	0.00
OR 1914–8	0.25 (0.86)^{c}	9.88^{cde}	1.33 (1.34)^{ab}	I	0.00	–	0.00	–	0.00
OR 2093-4	0.32 (0.90)^{de}	6.77^{b}	1.11 (1.26)^{a}	I	0.00	–	0.00	–	0.00
NDR 2063	0.11 (0.78)^{a}	11.11^{e}	1.55 (1.42)^{abc}	I	0.00	–	0.00	–	0.00
MTU 1075	0.32 (0.90)^{de}	9.77^{cde}	1.88 (1.52)^{abcd}	I	0.00	–	0.00	–	0.00
NDR 3110	0.25 (0.86)^{c}	9.33^{cde}	1.44 (1.39)^{abc}	I	0.00	–	0.00	–	0.00
Abhaya	0.12 (0.78)^{a}	8.44^{bcd}	1.88 (1.54)^{abcd}	I	0.00	–	0.00	–	0.00
Jaya (S)	0.32 (0.90)^{de}	13.10^{f}	4.22 (2.16)^{e}	II	1.44	II & III	1.00	Pupa	100.0
TN 1 (S)	0.35 (0.92)^{e}	14.44^{f}	5.77 (2.50)^{f}	II	1.22	II & III	1.00	Pupa	100.0

Figures in parentheses are square root transformation values; values in the column followed by common letters are nonsignificant at $p=0.05$ as per DMRT (Duncan's multiple range test)

Table 2 Number of gall midge adults settled, eggs laid, and maggot survival on test rice genotypes, 2006

Genotype	Antixenosis		Antibiosis						Plants damaged (%) after 35 days
	Number of adults settled/plant	Number of eggs laid/plant	7 days after adult release		14 days after adult release		18 days after adult release		
			Number of live maggots	Instar	Number of live maggots	Instar	Number of live maggots	Instar	
JGL 11605	0.21 (0.84)^{c}	5.44^{c}	2.33 (1.68)^{de}	I	0.00	–	0.00	–	0.00
JGL 11459	0.23 (0.85)^{c}	13.55^{f}	1.66 (1.46)^{bc}	I	0.00	–	0.00	–	0.00
JGL 13376	0.05 (0.74)^{a}	14.55^{g}	3.10 (1.89)^{e}	I	1.00	II	0.00	–	0.00
JGL 13418	0.15 (0.80)^{b}	4.33^{ab}	1.33 (1.34)^{b}	I	0.00	–	0.00	–	0.00
RP 4643-713	0.21 (0.84)^{c}	5.11^{bc}	1.99 (1.57)^{cd}	I	0.00	–	0.00	–	0.00
OR 1967-15	0.31 (0.89)^{f}	5.44^{c}	0.88 (1.16)^{a}	I	0.00	–	0.00	–	0.00
NDR 9930095	0.25 (0.86)^{cd}	10.77^{de}	3.33 (1.95)^{e}	I	0.05	II	0.00	–	0.00
R-1249-1196-2-1	0.28 (0.88)^{ef}	10.22^{d}	2.77 (1.80)^{ef}	I	0.00	–	0.00	–	0.00
WGL 157	0.34 (0.91)^{g}	3.66^{a}	0.77 (1.12)^{a}	I	0.00	–	0.00	–	0.00
RDR 918	0.27 (0.87)^{de}	11.33^{e}	2.88 (1.83)^{ef}	I	0.00	–	0.00	–	0.00
Jaya (S)	0.34 (0.91)^{g}	16.33^{h}	5.33 (2.41)^{g}	II	1.44	II & III	1.00	Pupa	100.0
TN 1 (S)	0.54 (1.01)^{h}	23.21^{i}	6.33 (2.61)^{g}	II	1.55	II & III	1.00	Pupa	100.0

Figures in parentheses are square root transformation values; values in the column followed by common letters are nonsignificant at $p=0.05$ as per DMRT

Venkataswamy 1966; Roy et al. 1971). Subsequent studies did not confirm this view (Shastry et al. 1972; Rao 1972; Kalode 1973). Antixenosis studies have indicated greater attraction of adult midges toward the susceptible genotypes than the resistant genotypes (Kalode et al. 1977). In this study, significant differences were recorded on the number of adults settled 6 h after adult release further, Kalode et al. (1983) also observed the similar results. But this study is contradicted by the reports of Sain and Kalode (1994) who observed nonsignificant differences with respect to the number of adults settled between resistant and susceptible cultivars. Studies in India (AICRIP 1969), Sri Lanka (Modder and Alagoda 1971), and Thailand (Hidaka et al. 1974) indicated ovipositional antixenosis by the adult female gall midge. Kalode (1980) and Sain and Kalode (1994) also found significant differences in the number of eggs laid on resistant and susceptible rice varieties. This study also revealed significant differences in the number of eggs oviposited on the resistant and susceptible cultivars.

Nevertheless, these differences could not be linked to resistance against gall midge. In view of the short adult life span of one day and total inactivity of adults during the day time, it is not surprising if antixenosis component is not clearly expressed in rice gall midge. In this study, NDR 2063, JGL 11459, and JGL 13376 despite recording comparable proportion of eggs to the susceptible TN1 and Jaya, continued to be resistant.

Antibiosis

Among the genotypes, maximum number of live maggots after 7 days of release was observed on TN 1 (5.77 per plant) and Jaya (4.22 per plant), and was significantly superior compared to the rest. The number of live maggots at 7 days after release on OR 2093-4, RDR 987, OR 1914-8, NDR 3110, NDR 2063, MTU 1075, RP 4644-1183, RP 4613-260, RP 4647-1073, and JGL 13595 was 1.11, 1.22, 1.33, 1.44, 1.55, 1.88, 2.10, 2.66, 3.44, and 3.44 maggots per plant, respectively. But, the maggots in all the resistant genotypes tested remained in the first instar while in susceptible genotypes the maggots reached the second instar. At 14th day after adult release, in all the resistant genotypes the mortality of maggots was evident but in JGL 13595 and RP 4647-1073, live maggots were observed. But still the maggots remained in the first instar only, while in susceptible genotypes they moulted to second and third instars (Table 1).

Observations on 18th day revealed high maggot mortality in all the resistant donors including JGL 13595 and RP 4647-1073, while most of the surviving insects on TN 1 and Jaya were in pupal stage. Not only survivals of maggots were adversely affected on resistant varieties, even the development of surviving insects was retarded. In 2006 wet, 7 days after adult release, significantly higher numbers of live maggots ($P<0.05$) were noticed in TN 1 and Jaya which recorded 6.33 and 5.33 maggots/plant and were on par with each other. Among the resistant genotypes, lower (0.77 maggot per plant) number of maggots was observed in WGL 157 and significantly differed other genotypes. Likewise, the number of live maggots on OR 1967-15, JGL 13418, JGL 11459, RP 4643-713, and JGL 11605 was 0.88, 1.33, 1.66, 1.99, and 2.33 maggots/plant, respectively. Though the maggot survival was noticed on resistant genotypes at 7 days of adult release, the maggots remained in the first instar only, while in the susceptible genotypes, TN 1 and Jaya, the maggots reached the second instars (Table 2).

After 14 days of adult release, on resistant genotypes maggot mortality was observed except in JGL 13376 and NDR 9930095, where the maggots remained in first instar only. But, on susceptible genotypes, TN 1 and Jaya second- and third-instar maggots were observed. Further, subsequent observations on 18th day of adult release indicated higher maggot mortality on the resistant genotypes including JGL 13376 and NDR 9930095. Most of the surviving maggots in TN 1 and Jaya were already in pupal stage on 18th day. Thus, antibiosis is evident; not only in survival of maggots were adversely affected on resistant genotypes, but even the development of surviving maggots was retarded.

So, in the present investigation distinct antibiosis effects leading to mortality of maggots on the resistant cultivars were evident. This study was in close agreement with studies conducted by several workers such as Modder and Alagoda (1971), Hidaka and Vungsilabutr (1971), Hidaka (1974), and Kalode (1980). The present investigation has also confirmed antibiosis effects on resistant rice cultivars studied by Sain and Kalode (1994). Presence of live maggots in all the test entries on 7th day after adult release further supports the view that mechanical or biophysical barrier preventing entry of maggots may not be playing a critical role in resistance.

Hidaka and Vungsilabutr (1971) observed failure in moulting of first-instar maggot on W1263. A predominant antibiosis component leading to mortality of first-instar larvae has been also observed by many workers (Pathak and Heinrichs 1982; Mathur and Rajamani 1984). Mounting inhibition was presumed to be a cause of resistance, but was not attributed to nutritional inadequacy. In the present investigation, maggots remained in the first instar in all the resistant genotypes. However, some of the test cultivars despite supporting the maggot development up to second instar caused mortality without manifestation of silver shoots. Thus, failure to moult cannot be the sole case of maggot mortality, but other factors like nutritional deficiency, or the presence of folic chemicals also might have influenced survival and development. Joshi and Venugopal (1984) also reported reduced maggot development and lower weight of adult females on the resistant cultivar IET 7008 than on susceptible TN1 and Co 42.

Acknowledgments The authors would like to thank J. S. Bentur (AICRIP) and I. C. Pasalu (DRR, Hyderabad) for technical advice. V. K. Lingaraj acknowledges a fellowship by the University Grants Commission (UGC) sponsored by the Ministry of Social Justice and Empowerment, Government of India (UGC F.16-93/SA-II/RGNF/KAR-6/2005-06). Thanks are also due to the Director of Research, University of Agricultural Sciences, Bangalore and to farmers of southern and coastal regions of Karnataka.

References

AICRIP (1969) All India Coordinated Rice Improvement Project. Progress Report, kharif. Indian Council of Agricultural Research, New Delhi, India

Bentur JS, Pasalu IC, Kalode MB (1992) Inheritance of virulence in rice-gall midge (*Orseolia oryzae*). Indian J Agric Sci 62:492–493

Bentur JS, Pasalu IC, Sharma NP, Rao PU, Mishra B (2003) Gall midge resistance in rice: current status in India and future strategies 2003. DRR Research Paper Series 01/2003. Directorate of Rice Research, Rajendranagar, Hyderabad, India

Bentur JS, Amudhan S, Pasalu IC, Sharma NP, Rao UP (2004) Insect-host plant interactions in gall midge-rice. In: Bennet J, Bentur JS, Pasalu IC, Krishnaiah K (ed) New approaches to rice gall midge resistance in rice. Proceedings of the International Workshop, 22–24 November 1998, Hyderabad, India. International Rice Research Institute, Los Banos, Philippines

CRRI (1952) Annual Report for 1952. Central Rice Research Institute, Cuttack, India.

Devaiah MM (1984) Further studies on the gall midge, Orseolia oryzae (Wood-Mason) (Cecidomyiidae: Diptera). M.S. thesis, Tamil Nadu Agricultural University, Coimbatore, TN, India

Duncan DB (1955) Multiple range and multiple F-test. Biometrics 11:1–14

Heinrichs EA, Pathak PK (1981) Resistance to the gall midge, *Orseolia oryzae* (Wood-Mason) in rice. Insect Sci Appl 1:123–132

Hidaka T (1974) Recent studies on the rice gall midge, *Orseolia oryzae* (Wood-Mason) (Cecidomyiidae: Diptera). Rev Plant Prot Res 7:99–143

Hidaka T, Vungsilabutr P (1971) Study on larval development of rice gall midge on susceptible and resistant rice varieties. Annual Report, Rice Department, Ministry of Agriculture, Bangkok, Thailand

Hidaka T, Vungsilabutor P, Kadkao S (1974) Studies on ecology and control of rice gall midge in Thailand. Trop Agric Res Cent Tech Bull 6:113

Israel P, Vedamurthy G, Rao YS (1961) Varietal resistance to the gall midge, Pachydiplosis oryzae (Wood-Mason) and other insect pests of paddy. Working paper No. 21, 9th meeting working party of rice production and protection. Food and Agriculture Organization, International Rice Commission, New Delhi, India

Joshi RC (1982) Recent studies on rice gall midge, Orseolia oryzae (Wood-Mason) (Cecidomyiidae: Diptera). M.S. thesis, Tamil Nadu Agricultural University, Coimbatore, TN, India

Joshi RC, Venugopal MS (1984) Mechanism of resistance in rice to gall midge *Orseolia oryzae* (Wood-Mason): antibiosis. Indian J Entomol 46:479–481

Kalode MB (1973) Hypersensitive reaction to insect pests in rice varieties Summer Instt. on insect-plant relationship, Indian Council of Agricultural Research, NewDelhi, India

Kalode MB (1980) The rice gall midge: varietal resistance and chemical control. In: Rice improvement in China and other Asian countries. International Rice Research Institute, Manila, Philippines.

Kalode MB, Bentur JS (1989) Characterization of Indian biotypes of the rice gall midge *Orseolia oryzae* (Wood-Mason) (Diptera: Cecidomyiidae). Insect Sci Appl 10:219–224

Kalode MB, Pophaly DJ, Kasiviswanathan PR, Sreeramulu M (1977) Studies on resistance and mass rearing of rice gall midge, *Orseolia oryzae* (Wood-Mason). Madras Agric J 64:733–739

Kalode MB, Sain M, Bentur JS, Pophaly DJ, Raghavaiah G (1983) New donors and mechanism of resistance to gall midge in rice grown in greenhouse. Indian J. Agric. Sci 53:483–485

Khush GS (1997) Breeding for resistance in rice. Ann N Y Acad Sci 287:296–308

Mathur KC, Rajamani S (1984) *Orseolia* and rice: cecidogenous interactions. Proc. Indian Acad Sci 93:283–292

Mathur KC, Reddy PR, Rajamani S, Moorthi BTS (1999) Integrated pest management in rice to improve productivity and sustainability. Oryza 36:195–207

Modder WWD, Alagoda A (1971) A comparison of the susceptibility of rice IR 8 and Warangal 1263, to attack by the gall midge, Pachydiplosis oryzae (Wood-Mason). Bull Entomol Res 61:745–753

Pathak PK, Heinrichs EA (1982) A bibliography of varietal resistance to the rice gall midge, Orseolia oryzae (Wood-Mason). Commonwealth Agricultural Bureaux, London, United Kingdom.

Rao PPS (1972) Ecology and control of *Tryporyza incertulus* (Walker) and *Pachydiplosis oryzae* (Wood-Mason) in rice. Ph.D. dissertation, Utkal University, Bhubaneswar, India.

Rao YS, Israel P, Yadava CP, Roy JK (1971) Nature of gall midge resistance in rice. Curr Sci 40:497–498

Roy JK, Israel P, Panwar MS (1971) Breeding for insect resistance in rice. Oryza 8:129–134

Sain M (1988) Studies on varietal resistance and bionomics of rice gall midge, *Orseolia oryzae* (Wood-Mason). Ph.D. dissertation, Osmania University, Hyderabad, India

Sain M, Kalode MB (1994) Greenhouse evaluation of rice cultivars for resistance to gall midge, *Orseolia oryzae* (Wood-Mason) and studies on mechanism of resistance. Insect Sci Appl 15:67–74

Shastry SVS, Freeman WM, Seshu DV, Israel P, Roy JK (1972) Host plant resistance to rice gall midge. In Rice Breeding. International Rice Research Institute, Los Banos, Philippines

Venkataswamy T (1966) Association between hairiness and resistance to gall midge (*Pachydiplosis oryzae* Wood-Mason) in rice. Andhra Agric J 13:149

Integrated Management of Insect-Induced Reddening in Bt Cotton Hybrids

Vijaykumar N. Ghante, Rajesh Chowdary, M. Bheemanna, Hosamani Arunkumar and Ranjith Kumar

Abstract

Studies on the factors responsible for reddening in Bt cotton are due to increased incidence of sucking pests, especially leafhoppers, use of susceptible hybrids and improper nutrient management. Validation of integrated module comprised use of sucking pest-tolerant cotton hybrid; balanced and timely soil application of macro- and micro-fertilizers as per recommendation based on site-specific soil test results; threshold-based management of sucking pests, especially leafhoppers, using systemic insecticides; foliar applications of Mancozeb 75WP at 2 g/L against leaf spot disease; three foliar applications of water-soluble multinutrient mixtures at 4 ml/L during 60, 75 and 90 DAS and weeding during first 40 DAS proved best for the management of Bt cotton reddening. Differential response to reddening was found among the cultivated commercial Bt cotton hybrids. Among the hybrids tested, Jackpot, MRC-7347, Marvel and MRC-7351 were found tolerant to sucking pests and, in turn, the manifestation of reddening symptoms. Bt hybrids such as RCH 2, RCH 530, Brahma, Arya and Bunny were highly sensitive to insect-induced reddening.

Keywords

Bt cotton · Integrated management · Leafhoppers · Reddening

Introduction

Reddening in Bt cotton is one of the most complex problems in Bt cotton-growing areas of India. Thousands of acres of cotton were found to be affected by this problem in major cotton-growing belts of many states. Worldwide studies revealed multifaceted factors responsible for reddening in cotton (Praharaj and Sankaranarayanan 2010). Both biotic and abiotic factors have been

V. N. Ghante (✉) · R. Chowdary · M. Bheemanna · H. Arunkumar · R. Kumar
Main Agricultural Research Station, UAS, Raichur, Karnataka, 584102, India
e-mail: vijayent1@rediffmail.com

A. K. Chakravarthy (ed.), *New Horizons in Insect Science: Towards Sustainable Pest Management*, DOI 10.1007/978-81-322-2089-3_25, © Springer India 2015

considered as possible causative agents for reddening. As many as 40 reasons were documented as factors influencing reddening in Bt cotton (Ghante et al. 2010). Thus, a concerted effort involving physiological, biochemical, nutritional and biotic factors is required for diagnosing, confirmation and management of red leaf symptoms in Bt cotton (Praharaj and Sankaranarayanan 2010). Survey work and field experiments for reddening management on Bt cotton were undertaken with the aim to understand (a) sucking pests (b) multinutrient management for Bt cotton reddening and (c) identification of tolerant Bt hybrids for reddening.

Material and Methods

Surveys were carried out across different Bt cotton-growing belts of North Karnataka to know the causes for reddening in Bt cotton through a set of questionnaire. Thereafter, experiments were designed and executed to validate the biotic and abiotic factors known to cause the problem. Pot culture experiments were conducted to know the role of biotic factors such as insects and disease-causing pathogens in symptom manifestation. Field experiments with Bt hybrids, susceptible to sucking pests were conducted with and without sucking pest management and red leaf index (RLI) at 60 and 100 DAS to know the role of sucking insects, especially green leafhopper. RLI is the manifestation of all the factors responsible for reddening in leaves in cotton. RLI indicates the degree of leaf reddening quantitatively (Dastur et al. 1952). As the performance of Bt cotton hybrids depends on the interaction of hybrids and environment, a total of 14 Bt cotton hybrids with one non-Bt check (DCH-32) (most preferred by farmers) was screened for reaction to reddening under recommended package of practices and unprotected conditions.

Integrated module includes (1) use of leaf reddening–tolerant cotton hybrid, namely Jackpot, MRC 7351, MRC 7347 and Marvel; (2) balanced and timely soil application of macro- and micro-fertilizers; (3) threshold-based management of sucking pests, especially leafhopper, using effective systemic insecticides like Acephate 75 SP at 1 g/L, Clothianidin 50WG at 0.07 g/L and Thiamethoxam at 0.25 g/L; (4) seed treatment with Trichoderma 4 g/kg seeds against wilt and foliar application of Mancozeb 75WP at 2 g/L against leaf spot diseases; (5) three foliar applications of water-soluble multinutrient mixtures at 4 ml/L during 60, 75 and 90 DAS; (6) recommended plant density was adopted for variety/hybrid to avoid competition for light, moisture and nutrients; and (7) weeding during first 40 DAS.

Results and Discussion

Comprehensive studies conducted for 3 years on reddening in Bt cotton revealed that it is a complex phenomenon caused due to multiple factors. Reddening is of two types: in the first, the change in colour is from green to yellow and then to red; in the second type, the change is directly from green to red (Fig. 1). Reddening was seen irrespective of the crop stage (Fig. 2 and 3), and in most of the cases, the reddening at the early phase of the crop was attributed to severe incidences of cotton leafhopper. The reddening noticed during the later phase of the crop was specifically due to nutrient deficiencies.

The role of sucking pests in Bt cotton leaf reddening was confirmed by an experiment of sucking pest management in leafhoppers susceptible to non-hairy cotton hybrid RCH-2 Bt. The crop treated for effective control of leafhoppers was free from reddening compared to the untreated one (Table 1). Reddening with change in the shape of the leaf (curling/cupping) was observed under field conditions due to desapping by leafhopper, while reddening caused by nutrient deficiencies usually have normal leaves without any change in the shape of the leaf (Fig. 4). Leafhopper plays a major role in the manifestation of reddening symptoms under field conditions. Yield losses of 60–75 % were seen due to occurrence of reddening in the early stages of squaring to flowering, and losses of about 10–25 % were recorded in the fields when reddening was observed at post-

Fig. 1 Reddening type in Bt cotton

Fig. 2 Reddening at different stage of Bt cotton. **a** Seedling, **b** vegetative, **c** boll formation, **d** boll bursting

boll formation. From the study, it is clear that the onset of reddening is not harmful if it commences at late reproductive stage (Dhopte 2001).

Pot culture studies involving artificial infestation with sucking insects confirmed the finding of insect-induced reddening by cotton leafhopper. Other sucking pests such as red spider mites, whiteflies, aphids and thrips do induce reddening due to feeding, but results were not consistent.

Infestation of leafhopper on Bt cotton has been widespread since 2002. Every Bt cotton seed has been treated with the highly effective insecticide, imidacloprid. Farmers have also been spraying the chemical on cotton crops to control leafhoppers. Recently, leafhoppers were found to develop resistance to imidacloprid (neonicotinoid) and development of resistance to the widely used neonicotinoid insecticides to control leafhopper incidence is aggravating the reddening in cotton. Bt cotton is non-toxic to any of the sucking pests of cotton. Since the donor parent, Coker 312, is known to be highly susceptible to sucking pests such as leafhoppers and thrips, the hybrids showed slightly enhanced susceptibility to these pests, especially if the recurrent parent did not possess inherent resistance (Khadi 2007).

Fig. 3 Reddening symptoms on various parts of the Bt cotton. **a** Leaves, **b** square, **c** boll, **d** bursting boll

Yellowing and reddening in leaves were considered as hopper burn, cupping (an inward curl of the leaf) and leaf fall, and in severe cases the vigour of the plant is diminished and it does not grow, were due to the leafhoppers attack which starts early in the season. Differential response to reddening was found among the cultivated commercial Bt cotton hybrids. Among the hybrids tested, Jackpot, MRC-7347, Marvel and MRC-7351 were found more tolerant to the manifestation of reddening symptoms. Bt hybrids such as RCH 2, RCH 530, Brahma, Arya and Bunny were found to be highly sensitive to reddening whereas

Table 1 Severity of reddening in Bt cotton with and without sucking pest management

Treatment	Reddening severity (RLI)		Average LH/plant	Yield/ha (qtl)
	60 DAS	100 DAS		
RCH 2 with sucking pest management	1.21	1.52	1.87	26.52
RCH 2 without sucking pest management	2.89	3.32	11.28	15.92

Fig. 4 **a** Differentiation of reddening due to cotton leafhoppers, **b** nutrient deficiency

Ankur Bt, Tulsi-144, VICH-32 and Pratika were found to be moderately tolerant (Table 2).

Bt cotton hybrids are more prone to red leaf symptoms. More than 500 Bt cotton hybrids have been released for cultivation in India. The expression of reddening depends on the interaction of the hybrid and the environment. Cotton varieties with hairy leaves are less susceptible to leafhopper attacks than varieties with glabrous leaves (Praharaj and Sankaranarayanan 2010). The phenomenon of leaf reddening is genetically controlled and some genome is responsible for anthocyanin synthesis. A differential response of different hybrids was observed with respect to specific activity of phenylalanine ammonia lyase, tyrosine ammonia lyase and phenol content of leaves, which play a major role in flavonoid biosynthesis (Koukol and Conn 1961). Bt hybrids should have a strong photosynthetic system and stay green so that the hybrid is capable to withstand enhanced Bt gene-induced reproductive load (Patil et al. 2010).

Among the different nutrients tested through foliar application during reproductive stage of Bt cotton crop, multinutrient mixture (both major and micro nutrients) proved best in managing the reddening due to nutrient deficiencies and recorded very less values for RLI (Table 3). The mismatch in nutrient application and crop removal resulted in multiple nutrient deficiencies (Naidu et al. 2011). Foliar application/feeding is one of the most efficient ways of supplying essential nutrients to the crop at the appropriate stage. Through foliar nutrition, the nutrients are taken into the foliage and distributed (transported) to all parts of the plant within a short period to get the needed effect. It is also effective in correcting the mid-season discrepancies in crop growth that may be due to either intensive growth or inappropriate supply of nutrients from the soil under abiotic stress conditions (Rathinavel et al. 1999).

Based on the confirmation of the major factors influencing reddening in Bt cotton (i.e., insect pests, nutrients and Bt hybrid), multidisciplinary crop management practices (MDCMP) involving balanced and timely applications of macro- and micro-fertilizers based on site-specific soil test results, use of leaf reddening-tolerant cotton hybrid, effective management of sucking pests specially leafhoppers, effective management of diseases and 2–3 foliar applications of water-soluble multinutrient mixtures during reproductive stage of the crop were evaluated for the management of leaf reddening in Bt cotton and compared with farmers' practice. An integrated approach was found (Table 4).

Table 2 Screening of Bt cotton hybrids against leaf reddening

Bt hybrids	Reddening severity	RLI		Leafhoppers/plant
		90 DAS	100 DAS	
RCH-530	High	2.78	3.23	6.74
RCH-2	High	3.24	3.54	7.23
MRC 7351	Low	0.98	1.02	1.86
MRC 7347	Low	0.97	1.02	1.82
Marvel	Low	0.87	0.99	1.97
Brahma	High	2.87	3.23	5.23
Arya	High	3.24	3.89	4.85
Bunny	High	3.21	3.78	5.64
Jackpot	Very low	0.65	0.87	1.12
Ankur	Medium	1.98	2.24	2.25
Tulsi-144	Medium	1.65	2.10	2.46
VICH-303	Medium	1.58	2.12	2.38
DCH-32	High	2.67	3.46	7.88
Pratika	Medium	1.34	2.34	2.57

Table 3 Evaluation of different nutrient products on the incidence of reddening in Bt cotton

Treatment	Reddening severity (RLI)		Average LH/plant	Yield/ha (qtl)
	90 DAS	100 DAS		
Multinutrient mix	0.87	1.00	3.29	24.67
MgSO4	1.17	1.63	3.45	23.66
19:19:19	1.20	2.02	2.98	23.72
KNO3	1.19	1.66	3.49	23.69
Control	2.96	3.83	13.27	14.54
Sem	0.07	0.16	0.05	0.12
CD	0.20	0.50	0.16	0.35

Table 4 Red leaf index (RLI) and economics of different modules for management of reddening in Bt cotton

Modules	RLI		Average Yield (qtl/ha)	Net Return
	60 DAS	100 DAS		
Integrated module	0.67	1.12	26.50	76450
FP	1.78	2.89	21.40	56717
SEM	0.17	0.40	0.94	3906.6
CD	0.48	1.18	2.84	11720

References

Dastur RH, Singh K, Kanwar SR (1952) Investigations on the red leaf in American cottons in Malwa and Bombay Karnataka. Indian Cotton Grow Rev 6:193–204

Dhopte AM (2001) Leaf reddening in cotton: impact, causes and remedies. Kalyani Publishers, New Delhi, p 217

Ghante VN, Ranjithkumar L, Chowdary Rajesh (2010) Management of leaf reddening in Indian Bt cotton needs a multidisciplinary approach. Green Farming 1(6):668

Khadi BM (2007) Impact of Bt-cotton on agriculture in India. www.gmo-safety.eu/pdf/biosafenet/Khadi.pdf. Accessed 27 Oct 2013

Koukal J, Conn BE (1961) Metabolism of aromatic compounds in higher plants. J Biol Chem 236:2692–2698

Naidu TCM, Mahalakshmi BK (2011) Reddening of Bt cotton under waterlogged conditions. J Cotton Res Dev 25(1):71–73

Patil SS, Manjula M, Alkuddsi YA, Patil BC, Rajeev S, Swathi P, Ranganath HM(2010) Mismatch in source and sink as a reason determining realized potentiality of Bt cotton hybrids. 2nd annual biotechnology conference for students (Abcs-2010), held at School Of Biotechnology International Institute of Information 2 Technology (IIT, Pune) during 13 & 14 Nov, 2010, p 14

Praharaj CS, Sankaranarayanan K (2010) Red leaf symptoms in cotton. Indian farming. September, pp 12–14

Rathinavel K, Dharmalingam C, Paneersewam S (1999) Effect of micronutrient on the productivity and quality of cotton seed cv. TCB 209 (*Gossypium barbadense* L.). Madras Agric J 86:313–316

Inducing Resistance in Sesamum Accessions Against Shoot Webber and Capsule Borer, *Antigastra catalaunalis* Duponchel Through Mutation Breeding

M. Saravanaraman, K. Balaji and V. Selvanarayanan

Abstract

Sesame (*Sesamum indicum* L.) is one of the ancient oil seed crops cultivated extensively in several countries of Asia and Africa. Incidence of insect pests and diseases cause severe yield losses in sesame. Among the insect pests, shoot webber, and capsule borer, *Antigastra catalaunalis* Duponchel (Pyraustidae: Lepidoptera) is predominant throughout India. To identify insect tolerant sesame, 140 sesame accessions were evaluated under glasshouse and field conditions for resistance against *A. catalaunalis* at the Faculty of Agriculture, Annamalai University during 2005–2007. Among the 140 accessions, 14 accessions were found resistant to *A. catalaunalis*. Based on this earlier work and reports from other sources, in the present work, accessions namely IVTS-2001–7(TKG-22), NIC-7875, NIC-16278, NIC-17345, NIC-7908, KMR-102, KMR-63, KMR-56, TMV-3 and SVPRI-1 were selected for inducing mutation. These accessions were evaluated for the LD50 with regard to the physical mutagen namely Gamma rays and chemical mutagens namely EMS (Ethyl methane sulfonate) and DES (Di ethyl sulfate). For gamma rays, the LD_{50} was at 50 krad while EMS and DES registered 0.06 and 0.5% respectively as LD_{50}. Upon evaluating the first mutant (M_1) and second mutant (M_2) generations under screenhouse conditions, based on leaf, flower, and capsule damage, mutants of four accessions namely IVTS 2001-7, NIC-7875, NIC-16278, and TMV-3 were rated resistant. To ascertain the segregation of such desirable traits in the mutant generations, concerted field evaluation needs to be conducted till stabilization of phenotypic variability in the offsprings.

Keywords

Capsule borer · Resistant · Sesame · Shoot webber

M. Saravanaraman (✉) · K. Balaji · V. Selvanarayanan
Department of Entomology, Faculty of Agriculture, Annamalai University, Annamalainagar 608002 Tamil Nadu, India
e-mail: selvaento@gmail.com

Introduction

Sesame (*Sesamum indicum* Linn.) (Pedaliaceae) is an important oilseed crop grown in India and is regarded as the "Queen of Oilseeds". Sesame

A. K. Chakravarthy (ed.), *New Horizons in Insect Science: Towards Sustainable Pest Management*,
DOI 10.1007/978-81-322-2089-3_26, © Springer India 2015

seeds contain 45–55% of oil, which is high in quality because of the presence of antioxidant sesamol. In most of the countries including India, sesame is an underutilized crop of local importance, which warrants improved use and conservation. In India, it is being cultivated under both rainfed and irrigated conditions. Among the sesame cultivating countries, though India ranks first in the production, the productivity is comparatively less (413 kg/ha). This shortfall in the productivity is attributed to the incidence of insect pests.

Among the key insect pests, shoot webber, and capsule borer, *Antigastra catalaunalis* (Duponchel) (Pyraustidae: Lepidoptera) is the most serious in India. It occurs regularly and infests the crop during seedling, flowering, and maturity stages of crop growth and causes up to 90% yield losses (Cheema and Singh 1987). But the attack is more severe during dry seasons and after initiation of flowering. *A. catalaunalis* feeds on tender foliage by webbing the top leaves, bores into the pods and shoots (Narayanan and Nadarajan 2005). Keeping in view the ill effects of chemical control of this insect pest, exploiting host plant resistance is considered economically viable, technically feasible, socially acceptable, and environmentally compassionate. But, sesame varieties currently under use are susceptible to pests and diseases and if tolerant, they have either limited yield potential (Rohilla et al. 2003) or limited adaptability to wider geographic area.

Hence, in India, increasing research efforts are being initiated to identify, exploit, and utilize sesame varieties possessing pest and disease resistance in addition to higher yield potential and better adaptability to specific locations. Realizing this, the present study was carried out to explore and exploit host plant resistance in sesame against *A. catalaunalis*.

Gene Pool of Sesame

The genus *Sesamum*, *(Pedaliaceae)* includes many species. Of these species, few are domesticated like *Sesamum indicum* Linn. while many others are wild relatives like *Sesamum prostratum* Retz. This wider gene pool offers diverse sources of pest resistance well exploited for development of insect tolerant/resistant varieties.

Resistance in Wild Species of Sesame Against *A. catalaunalis*

Sesamum prostratum, a wild species was found resistant to *A. catalaunalis* (Mukherjee 1947). Crosses between *Sesamum indicum* Linn. and *Sesamum prostratum* (Ramanujam 1942) led to the hybrid *Sesamum indicatum* Linn., which was found susceptible to *Antigastra catalaunalis*. The wild species *Sesamum prostratum, Sesamum malabaricum, Sesamum alatum* Thonn., *Sesamum lacinatum*, and *Sesamum radiatum* were found to possess resistance against *A. catalaunalis* (Nath and Agarwal 1982; Thangavelu et al. 1989; Muralibaskaran et al. 1990; Ahuja and Kalyan 2001). Philip Sridhar and Gopalan (2002) reported that the wild species *S. alatum*, *S. laciniatum* Klein., and *S. prostratum* exhibited high resistance under field conditions.

Resistance in Domesticated Sesame Against *A. catalaunalis*

In India, because of the wider diversity in the available *Sesamum* gene pool, though many sesame lines are being maintained under All India Coordinated Research Project on Oilseeds (Sesame and Niger) at Jabalpur, Madhya Pradesh, and at other Regional Research Stations, only few lines possess insect and/or disease resistance traits, but may be either wild lines or lines with lesser yield potential or otherwise may be suited to specific areas (DOR 1999; Gupta 2004).

Attempts were made in Tamil Nadu to identify and exploit host plant resistance in sesame gene pool (Muralibaskaran et al. 1990; Selvanarayanan and Baskaran 1996b; Manisegaran et al. 2001; Philip Sridhar and Gopalan 2002; Vijai Anandh and Selvanarayanan 2005). But, success had not been attained in the development of sesame varieties with higher yield potential and insect and/or disease resistance as well as adaptability to

Table 1 Reports from Tamil Nadu on resistance in sesame against *A. catalaunalis*

Name of tolerant/resistant varieties/cultivars/accessions	Reference
ES 22, Si 250, ES 12	Mahadevan 1988
SI-53, SI-75, SI-810, SI-882, SI-935, SI-968, SI-970, SI-1002, SI-1004, SI-1029, SI-1671, SI-3315/6, SI-3315/11, PDK-31, PDK-20-3-1, PDK-59-1	Mahadevan et al. 1989
ES 22, SI-250, TMV 3, *S. alatum*	Muralibaskaran and Mahadevan 1989
SI-250, ES 22, PDK 31, SI-810, B 67, C 1036	Thangavelu et al. 1989
ES 22, SI 250	Muralibaskaran et al. 1990
SVPR-1	Jebaraj et al. 1993
ES 22, SI-250, *S. alatum*	Muralibaskaran et al. 1994
CO-1, TMV-3	Selvanarayanan and Baskaran 1996a
IVTS 2001-20, IVTS 2001-23, IVTS 2001-24, IVTS 2001-25, IVTS 2001-24	Vijai Anandh 2003

wider areas. In spite of this, many cultivars were reported promising at specific areas by workers (Tables 1 and 2).

In addition to the above, under the All India Coordinated Research Project on Oilseeds program being conducted by Indian Council of Agricultural Research many varieties/cultivars have been identified (Table 3).

Screening Sesame Accessions for Resistance Against *A. catalaunalis*

An attempt was made to study the biology, varietal resistance, and management of *A. catalaunalis* during 1989–1991 at Annamalai University, Tamil Nadu, India (Selvanarayanan 1991). Later, 40 sesame lines were gathered from various sources and screened for field resistance at Annamalai University Experimental Farm as well as at a popular sesame tract namely Vridhachalam, in Cuddalore district, Tami Nadu, India (Vijai Anandh 2003). Subsequently, 140 sesame accessions were evaluated for resistance against *A. catalaunalis* under field conditions during two seasons at the experimental Farm of Faculty of Agriculture, Annamalai University, Tamil Nadu, India (Balaji 2006; Balaji and Selvanarayanan 2009).

The sesame accessions were sown on the ridges of 2-m length with a spacing of 30 cm between the rows and 30 cm between plants. Ten plants were maintained per replication and three replications were maintained per accession. A known susceptible check, namely, SVPR-1 (Vijai Anandh 2003) was maintained at one row for every five rows of the test accessions as well as two rows around the experimental field as augumentor rows. Recommended agronomic practices were followed and no pesticide was applied. The per cent leaf, flower, and capsule damage caused by *A. catalaunalis* was recorded respectively from 15, 30, and 50 days after sowing onwards till harvest at weekly interval by observing five plants selected randomly/replication and the mean percentage damage was computed.

Based on the intensity of damage assessed on different plant parts, the accessions were categorized adopting the score chart formulated by Philip Sridhar and Gopalan (2002), but with the modification in fixing the rating scales at a lower range based on the maximum and minimum damage by *A. catalaunalis* (Balaji 2006; Table 4). As the damage on reproductive parts such as flowers and capsules influence yield more than the leaf damage, lesser flower and capsule damage and more leaf damage were equated to a particular score (Table 5).

Based on per cent damage on leaves, flowers, and capsules for an accession, individual scoring was given and subsequently were summed up to calculate the cumulative score and accordingly the scoring grade (1–9) was allotted by referring the grade chart and resistance rating was made.

In the first field evaluation, based on leaf and/or flower damage, only few accessions were rated highly resistant or resistant or moderately resistant. Whereas several accessions were ei-

Table 2 Reports from other parts of India on resistance in sesame against *A. catalaunalis*

Name of tolerant/resistant varieties/cultivars/accessions	Reference
NP 29	Bhattacharjee and Lal 1962
C 1036, Chanda-3	Prasad 1970
M 3-2	Jakhmola and Yadav 1974
N66-276, N66-250, N66-4	Yadav and Lal 1976
IS-2, IS-22, IS-36	Singh and Ghewande 1981
TC 289, Pb. Til No.1	Cheema et al. 1982
TMV 2, B 14, T4	Jakhmola 1983
B 67, RAUSS 17-6	Yadav 1985
C1036, JT, CT, Anand 74	Singh and Dhamdhere 1986
B 67, RAUSS-17, RAUSS-17-4	Singh 1987
Pb. Til-1, HT-111, HT 12, TLC 75, TLC 77	Singh 1988
C-7	Tiwari and Shaw 1988
RAUSS 17-4	Surendarkumar and Rohilla 1989
TLC 83,TLC 77	Singh et al. 1990
IVTS-17	Gurs and Hussain 1998
OMT-30, OMT-32, B67, Bolangir local	Rath et al. 1998
ES 12, ES 22, SI-3317/11, SI-250, IS-210-1, Zodge-3	Ahuja and Kalyan 2001
JCS-9426, OS-15, OS-5, JTS-104, TC-22, RT-28, RT-238	Patil et al. 2001
Bolangir Local, Kalika, Uma, OTM 10	Patra 2001
JT-7, CT-785, JLT-26, TK-6-21, TKG-22	Gupta et al. 2002
SI-73, SI-1406, SI-1729, SI-3239, IC-132246, IC-204137, IC-205071, IC-205082, IC-205304	Shrivastava et al. 2002
ES-22, SI-250, IS-23-1, KIS-305, ES-12	Singh 2002
PR 19/9, S17	Talpur et al. 2002
RAUSS-17-4, B-67	Rohilla et al. 2003
SI-232–2, SI-911, SI-928, Si-934, SI-1400, SI-1496, SI-1556, SI-1843, SI-2174-2, SI-2253, SI-2584, SI-3175, TZA/91-640, TC-204788, IC-205595	Gupta 2004
SI 250, ES 22, Uma	Karuppaiah and Nadarajan 2011

Table 3 Reports from all India coordinated research project on oilseeds on resistance in sesame against *A. catalaunalis*

Name of tolerant/resistant varieties/cultivars/accessions	Reference
HCH, RUSS, DDI-1484, Hogular, ES-12,ESS-22, RCR-5, Jodge 10	AICORPO 1987
RTS-152 (Immune), SI 32-6, ISI 178-2, IS-42, SI-79, SI-197, SI-2225, RTS-165, GGK-7, IS-423, IS-79-2, 84-B, ES 98-3-84-B, ES 102, OLT-12, SI 837, TLC 88B, TLC 88A, TLC 74	AICORPO 1989
K-382-2, JS-80, KRR-1, RT-4, RT-106, K-302	DOR 1995
SI-898, 1345, 172, ES-31–2, ES-93-3-84, IS 194, 245, 285, 347, BS 10, EC 303430, ES 22, SI-1030, SI-3257, SI-3315, IS-733, SI-250	DOR 1997
VDV-8/145, NKD-1089, SI-116, SI-810, SI-926, SI-1887, SI-2584, SI-107, SI-1250, SI-1258, SI-1887	DOR 1999
IC-132089, IC-132378, IC-204670, IC-431102, IC-132176, IC-204079, IC-204013, IC-204045	DOR 2002

Table 4 Scoring sesame accessions for resistance to *A. catalaunalis*

Per cent damage				Resistance rating
Leaf	Flower	Capsule	Score	
0–2	0–1	0–1	1	Highly resistant (HR)
>2–4	>1–2	>1–2	3	Resistant (R)
>4–6	>2–3	>2–3	5	Moderately resistant (MR)
>6–8	>3–4	>3–4	7	Susceptible (S)
>8	>4	>4	9	Highly susceptible (HS)

Table 5 Cumulative scoring of sesame accessions for resistance to *A. catalaunalis*

Cumulative score	Grade	Resistance rating
0–1	1	HR
>1–3	3	R
>3–5	5	MR
>5–7	7	S
>7	9	HS

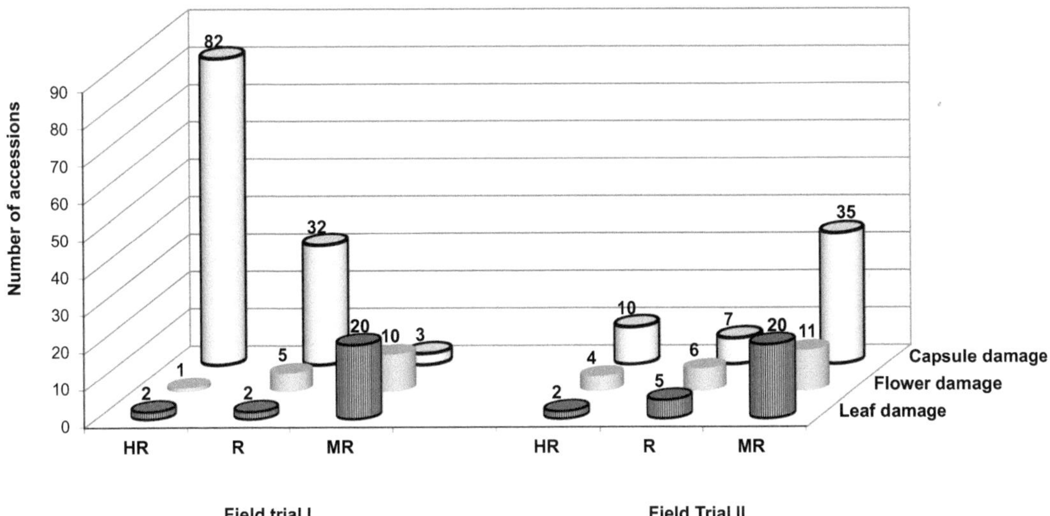

Fig. 1 Resistance reaction of sesame accessions to *A. catalaunalis*

ther susceptible or highly susceptible (Fig. 1). In contrast, based on capsule damage, 58.57 % of the accessions was rated highly resistant, while 22.85 % was rated as resistant. In the second field evaluation, many accessions were rated susceptible or highly susceptible, where as only lesser proportion was rated "HR" or "R" or "MR". A significant difference was observed between two seasons with regard to total number of accessions observed under each category for the respective plant part. Based on the above, five accessions tolerant to *A. catalaunalis* were selected (Balaji 2006). Culminating from these reports and collections from All India Coordinated Research Project on Niger and Sesame at Jabalpur, Madhya Pradesh and Regional Research Station, Vridachalam, Tamilnadu, accessions namely NIC-7875, NIC-78, IVTS-2001-7, NIC-16278, NIC-17345, KMR 63, KMR-102, KMR 56, TMV-3, and SVPR-1 were selected for inducing mutagenesis.

Mutation Breeding for Sesame Varietal Development

Crop improvement in sesame for such desirable attributes is being attempted through conventional breeding methods, by exploiting the natural variability available in the germplasm. However, for changing the plant type, if adequate variability is not available in the existing germplasm, under such circumstances, mutation breeding can be effectively employed as an alternative or supplemental source (Anitha Vaseline et al. 2000). In contrast to conventional breeding which may require more time to develop a promising variety, mutation breeding yields desirable crop varieties in a shorter time. Considering the above, attempts were made at Annamalai University, India to develop high yielding, shoot webber resistant sesame lines from already selected tolerant lines.

Mutagenesis

The parent materials chosen from earlier works were subjected to mutation breeding using physical and chemical mutagens. For physical mutagenesis, gamma rays were used. Freshly collected sesame seeds were irradiated with 40, 50, and 60 krad, respectively. Irradiation was conducted at Centre for Application of Radioisotopes and Radiation Technology (CARRT), Mangalore University, Mangalore, India. Irradiated seeds were sown in cement pots and maintained individually.

For chemical mutagenesis, two chemical mutagens namely Ethyl Methane Sulfonate (EMS) and Diethyl Sulfate (DES) were used. To determine the LD_{50} concentration for EMS, freshly collected parent seeds were treated with five concentrations of EMS namely 0.04, 0.05, 0.06, 0.07, and 0.08%. The treated seeds were placed on moist filter paper inside a petri plate and observed for germination. Based on the germination percentage, LD_{50} was ascertained. For every treatment, five replications were maintained. Similarly, to ascertain the LD_{50} for DES, the solution was freshly prepared once in every half an hour because of the shorter life period of the chemical. Five concentrations namely 0.3, 0.4, 0.5, 0.6, and 0.7% were evaluated. Based on the LD_{50} values for each chemical, the test dosages were fixed. The treated seeds were washed thoroughly several times with distilled water and sown in cement pots under screenhouse conditions and maintained as individual plants in each pot. Untreated seeds were used as control. The plants thus raised were evaluated for resistance against *A. catalaunalis*.

Mass Culturing *A. catalaunalis*

Sesame plants raised in earthen pots (30 cm diameter and 30 cm high) were used for mass culturing of *A. catalaunalis*. Seeds of sesame were sown in the potting mixture comprising soil, farmyard manure, and sand in the ratio of 2:1:1 which was further amended with urea @ 2–3 g/pot. Sesame plants were raised at regular interval so as to maintain a continuous stock of young plants.

Mass culturing of *A. catalaunalis* was done by collecting larvae from infested sesame fields. The larvae were released on potted plants. The potted plants were placed inside a cage consisting of nylon net cloth affixed to an iron frame (2.5 × 1.5 × 1.5 m). The cage was permanently placed in a sunny area and 20 plants were enclosed per cage. Fresh plants were provided as and when needed for larval infestation. Thus the larvae were allowed to grow without any disturbance until pupation. The pupae along with silken cocoons were collected in small plastic cups having a layer of cotton and kept inside a specially designed adult emergence and oviposition cage. The cage consisted of a cylindrical glass jar (15 cm high and 10 cm diameter) above which mylar film sheet rolled cylindrically (30 cm high and 10 cm diameter) was affixed. Fifteen days old sesame seedlings were raised in the nursery bag (12 × 8 cm) and kept inside the cage. Strips of cloth (15 × 5 cm) were hung from the top of the cage, so as to enable emerging moths for stretching their wings and cuticle hardening. The adults were fed with 10% sucrose solution through soaked cotton wigs placed inside. Adults were

Table 6 Resistance rating of M_1 generation of sesame accessions against *A. catalaunalis* (Physical mutation)

Name of the accession	Leaf damage (%)	Resistance rating	Flower damage (%)	Resistance rating	Capsule damage (%)	Resistance rating
NIC 7875	3.1 (10.14)	R	1.5 (7.03)	R	1.6 (7.26)	R
NIC 7908	2.5 (9.09)	MR	3.0 (9.97)	MR	2.5 (9.09)	MR
IVTS 2001-7	1.7 (7.49)	R	1.5 (7.03)	R	1.0 (5.73)	HR
NIC 16278	3.0 (9.97)	R	2.0 (8.13)	R	1.2 (6.28)	R
NIC 17345	5.0 (12.92)	MR	3.0 (9.97)	MR	2.0 (8.13)	MR
KMR-56	4.5 (12.24)	MR	2.4 (8.91)	MR	2.0 (8.13)	MR
KMR-63	4.0 (11.53)	MR	2.8 (9.63)	MR	2.5 (9.09)	MR
KMR-102	6.0 (14.17)	MR	3.0 (9.97)	MR	2.7 (9.45)	MR
TMV-3	2.4 (8.91)	R	2.0 (8.13)	R	2.2 (8.52)	R
SVPRI-1 ('S' check)	15.2 (22.46)	HS	6.2 (14.41)	HS	6.2 (14.41)	HS
C.D.	1.05		0.92		0.86	

Table 7 Resistance rating of M_1 generation of sesame accessions against *A. catalaunalis* (Chemical mutation)

Name of the accTession	Leaf damage (%)	Resistance rating	Flower damage (%)	Resistance rating	Capsule damage (%)	Resistance rating
NIC 7875	3.4 (10.62)	R	2.2 (8.53)	R	3.7 (11.09)	MR
NIC 7908	1.9 (7.92)	R	3.6 (10.93)	MR	2.0 (8.13)	MR
IVTS 2001-7	1.4 (6.79)	HR	1.8 (7.71)	R	0.9 (5.44)	HR
NIC 16278	3.6 (10.93)	R	2.9 (9.80)	MR	1.8 (7.71)	R
NIC 17345	1.9 (7.92)	R	3.2 (10.30)	S	2.9 (9.80)	MR
KMR-56	4.9 (12.79)	MR	2.7 (9.46)	R	3.1 (10.14)	MR
KMR-63	2.1 (8.33)	R	4.1 (11.68)	R	1.8 (7.71)	R
KMR-102	6.4 (14.65)	S	2.9 (9.80)	MR	2.5 (9.10)	MR
TMV-3	2.1 (8.33)	R	0.8 (5.13)	HR	1.3 (6.55)	R
SVPRI-1 ('S' check)	13.1 (21.22)	HS	9.6 (18.05)	HS	7.3 (15.68)	HS
C.D.	0.98		0.67		0.38	

allowed to mate and lay eggs on the seedlings. Neonate larvae were transferred to potted plants using a fine camel hairbrush.

Screening M_1 Generation

Based on LD50 evaluation, a dose of 50 krad was identified as the LD_{50} value for physical mutant while for the chemical mutagens EMS and DES, 0.5 and 0.06 % were identified as $LD_{50,}$ respectively. Upon evaluating the first mutant generation for resistance against *A. catalaunalis*, several leaf and growth habit mutants were recovered. Chlorophyll mutants were also observed wherein mostly pale-green or yellowish green leaves were seen. One mutant was found to have more vigorous growth while another mutant plant had tripod at single node, which is peculiar. Based on leaf, flower and capsule damage, mutants of four accessions namely IVTS 2001-7, NIC-7875, NIC-16278 and TMV-3 were rated resistant, while four accessions were moderately resistant and two were susceptible (Tables 6, 7).

Screening M_2 Generation

Plants of M_2 generation of the four accessions namely IVTS 2001-7, NIC-16278, KMR 63, and TMV-3 were rated as resistant, while another four accessions were moderately resistant and two accessions were susceptible (Tables 8, 9). Hence, it may be concluded that to ascertain the segregation of such desirable traits in the mutant generations, concerted field evaluation needs to

Table 8 Resistance rating of M_2 generation of sesame accessions against *A. catalaunalis* (Physical mutation)

Name of the accession	Leaf damage (%)	Resistance rating	Flower damage (%)	Resistance rating	Capsule damage (%)	Resistance rating
NIC 7875	3.4 (10.62)	R	2.2 (8.53)	R	3.7 (11.09)	MR
NIC 7908	1.9 (7.92)	R	3.6 (10.93)	MR	2.0 (8.13)	MR
IVTS 2001–7	1.4 (6.79)	HR	1.8 (7.71)	R	0.9 (5.44)	HR
NIC 16278	3.6 (10.93)	R	2.9 (9.80)	MR	1.8 (7.71)	R
NIC 17345	1.9 (7.92)	R	3.2 (10.30)	S	2.9 (9.80)	MR
KMR-56	4.9 (12.79)	MR	2.7 (9.46)	R	3.1 (10.14)	MR
KMR-63	2.1 (8.33)	R	4.1 (11.68)	R	2.5 (9.10)	MR
KMR-102	6.4 (14.65)	S	2.9 (9.80)	MR	1.8 (7.71)	R
TMV-3	2.1 (8.33)	R	0.8 (5.13)	HR	1.3 (6.55)	R
SVPRI-1 ('S' check)	13.1 (21.22)	HS	9.6 (18.05)	HS	7.3 (15.68)	HS
C.D.	0.98		0.67		0.38	

Table 9 Resistance rating of M_2 generation of sesame accessions against *A. catalaunalis* (Chemical mutation)

Name of the accession	Leaf damage (%)	Resistance rating	Flower damage (%)	Resistance rating	Capsule damage (%)	Resistance rating
NIC 7875	2.6 (9.28)	R	1.1 (6.02)	R	2.2 (8.53)	MR
NIC 7908	2.0 (8.13)	HR	2.8 (9.63)	MR	1.9 (7.92)	R
IVTS 2001-7	1.9 (7.92)	HR	2.0 (8.13)	R	1.0 (5.74)	HR
NIC 16278	2.5 (9.10)	R	1.9 (7.92)	HR	2.1 (8.33)	R
NIC 17345	2.9 (9.80)	R	4.1 (11.68)	S	2.9 (9.80)	MR
KMR-56	3.1 (10.14)	R	2.9 (9.80)	MR	3.5 (10.78)	MR
KMR-63	4.0 (11.54)	MR	3.8 (11.24)	S	0.7 (4.80)	HR
KMR-102	5.3 (13.31)	MR	2.7 (9.46)	R	2.4 (8.91)	MR
TMV-3	3.3 (10.47)	R	1.5 (7.03)	R	0.7 (4.80)	HR
SVPRI-1 ('S' check)	18.7 (25.62)	HS	7.33 (15.68)	HS	8.01 (16.44)	HS
C.D.	1.07		0.73		0.54	

be conducted till stabilization of phenotypic variability in the offsprings. Further, the actual mechanisms of resistance in such offsprings needs to be explored.

Acknowledgement The authors are thankful to Project Coordinator; All India Coordinated Research Project on Oilseeds (Sesame and Niger), Jabalpur, India and Regional Research station, Tamil Nadu Agricultural University, Vridhachalam, India for sparing sesame accessions. Thanks are due to Dr. Shomasekarappa, Centre for Application of Radioisotopes & Radiation Technology (CARRT), Mangalore University, Mangalore for his help in irradiation. Financial support by Department of Science and Technology, Government of India is gratefully acknowledged. The authors thank Annamalai University for according permission to conduct the research and for providing all the facilities.

References

Ahuja DB, Kalyan RK (2001) Field screening of genotypes of sesame against leaf webber capsule borer, *Antigastra catalaunalis* Dup., gall fly *Aspondylia sesami* Felt. and mite *Polyphagotarsonemus latus* (Banks). Pest Manag Econ Zool 29(1):5–9

AICORPO (1987) Annual Progress Report. All India Coordinated Research Project on Oilseeds Research, Rajendra Nagar, Hyderabad

AICORPO (1989) Annual Progress Report. All India Coordinated Research Project on Oilseeds Research, Rajendra Nagar, Hyderabad

Anitha Vaseline Y, Saravanan K, Ganesan J (2000) Studies on variability, heritability and genetic advance in mutant population for certain characters in sesame (*Sesamum indicum* L.). Sesamum Safflower Newsl 15:39–43

Balaji K (2006) Host plant resistance in sesame against shoot webber and capsule borer, Antigastra catalaunalis (Duponchel) and phyllody disease. Ph. D. Dissertation, Annamalai University, Annamalai Nagar, Tamilnadu, India

Balaji K, Selvanarayanan V (2009) Evaluation of resistance in sesame germplasm against shoot webber and capsule borer, *Antigastra catalaunalis*. Indian J Plant Prot 37(1&2):35–38

Bhattacharjee NS, Lal R (1962) Studies on the varietal susceptibility of 'til' (*Sesamum orientale*) to the attack of *Antigastra catalaunalis* (Duponchel). Indian J Entomol 24:56–63

Cheema JS, Singh G (1987) Biology of sesame leaf webber and capsule borer, *Antigastra catalaunalis* (Duponchel) (Pyralidae: Lepidoptera) in Punjab. J Res Punjab Agric Univ 24:65–74

Cheema JS, Singh G, Grewal GS, Labana KS (1982) Comparative incidence of sesamum shoot webber and leaf webber, *Antigastra catalaunalis* (Duponchel) in different cultivars of sesame. J Res Punjab Agric Univ 19:170–171

DOR (1995) Annual Progress Report, Sesame and Niger, Directorate of Oilseeds Research, Hyderabad, India

DOR (1997) Annual Progress Report, Sesame and Niger, Directorate of Oilseeds Research, Hyderabad, India

DOR (1999) Annual Progress Report, Sesame and Niger, Directorate of Oilseeds Research, Hyderabad, India

DOR (2002) Annual Progress Report, Sesame and Niger, Directorate of Oilseeds Research, Hyderabad, India

Gupta MP (2004) Field screening of sesame germplasm against leaf webber and capsule borer, *Antigastra catalaunalis* Dup. Indian J Plant Prot 32(1):42–44

Gupta MP, Rai HS, Chaurasia SK (2002) Incidence and avoidable losses due to leaf roller and capsule borer, *Antigastra catalaunalis* Dup. in sesame. Ann Plant Prot Sci 10(2):202–206

Gurs SD, Hussian R (1988) Evaluation of sesame germplasm against the incidence of *Antigastra catalaunalis* (Dup.). Insect Env 3(2):50–51

Jakhmola SS (1983) Natural enemies of til leaf roller and capsule borer, *Antigastra catalaunalis* Dup. Bull Entomol 24:147–148

Jakhmola SS, Yadav HS (1974) Varietal susceptibility of sesame (*Sesamum indicum* L.) to *Antigastra catalaunalis* (Dup.). JNKVV Res J 8:285

Jebaraj S, Mohamed Sheriff N, Muralibaskaran RK, Ramalingam A, Rajaratinam S (1993) SVPR-1: a new high yielding white seeded sesame (*Sesamum indicum*). Madras Agric J 80(11):606–608

Karuppaiah V, Nadarajan L (2011) Evaluation of sesame genotypes for resistance to sesamum leaf roller and capsule borer, *Antigastra catalaunalis* Duponchel. (Pyraustidae: Lepidoptera). Arch Phytopathol Plant Prot 44(9):882–887

Mahadeven NR (1988) Screening of sesamum entries for resistance to leaf webber, *Antigastra catalaunalis* (Duponchel). Madras Agric J 75(5–6):224

Mahadevan NR, Ganesh SK, Kandasamy G, Sivaraman MR (1989) Evaluation of sesame entries for resistance to leaf webber and capsule borer, *Antigastra catalaunalis* Duponchel. Madras Agric J 76(5):300

Manisegaren S, Manimegalai N, Pushpa J, Mohammed SEN (2001) Non-preference mechanism of resistance in sesame to shoot webber and capsule borer, *Antigastra catalaunalis*. Ann Plant Prot Sci 9(1):123–124

Mukherjee S (1947) On the probable role of soluble solids on the causation of resistance or susceptibility of the Indian species of sesamum to the larval attack of *Antigastra catalaunalis* (Lepidoptera: Pyralidae). Nature 160:95

Muralibaskaran RK, Mahadevan NR (1989) Control of shoot and pod borer, (*Antigastra catalaunalis*). II—Reaction of the pest to tolerant sesamum entries. Sesame Safflower Newsl 4:25–27

Muralibaskaran RK, Ganesh SK, Thangavelu S (1990) Germplasm screening against sesame leaf roller and pod borer. Sesame Safflower Newsl 5:40–45

Muralibaskaran RK, Ganesh SK, Thangavelu S (1994) Germplasm screening against sesame leaf roller and pod borer, *Antigastra catalaunalis*. Madras Agric J 81(11):618 621

Narayanan US, Nadarajan L (2005) Evidence for amale-produced sex pheromone in sesame leaf webber, *Antigastra catalaunalis* Duponchel (Pyraustidae: Lepidoptera). J Curr Sci 88(4)

Nath V, Agarwal RA (1982) Insect pests of crops and their control. Bharathi Publications, Indian Agricultural Research Institute, New Delhi. p 94

Patil SK, Dharne PK, Shambarkar DA (2001) Response of different promising genotypes of sesame to major insect pests and leaf spot disease. Sesame Safflower Neswl 16:72–74

Patra AK (2001) Response of sesame (*Sesamum indicum*) varieties to dates of sowing during rainy season. Indian J Agric Sci 71(8):523–524

Philip Sridhar R Gopalan M (2002) Studies on screening and mechanism of resistance against the shoot webber, *Antigastra catalaunalis* (Duponchel). Entomon 27:365–373

Prasad SK (1970) Varietal difference in the incidence of *Antigastra catalaunalis* (Duponchel) on *Sesamum indicum* L. Indian J Entomol 32:279–282

Ramanujam S (1942) An interspecific hybrid in sesamum: *S. orientale* × *S. prostratum*. Curr Sci 2:426–428

Rath LK, Mishra A, Mukherjee SK (1998) Varietal reaction in sesamum to leaf webber and capsule borer, *Antigastra catalaunalis* (Dup). Insect Env 3(2):50–51

Rohilla HR, Chillar BS, Harvir Singh (2003) Assessment of yield losses caused by *Antigastra catalaunalis* (Dup) in different genotypes of sesame under various agroclimatic conditions of Haryana. J Oilseeds Res 20(2):315–316

Selvanarayanan V (1991) Biology and control of sesame (Sesamum indicum L.) leaf webber, Antigastra catalaunalis (Duponchel) (Pyralidae: Lepidoptera) M. Sc. (Ag.) Dissertation, Annamalai University, Annamalai Nagar, India

Selvanarayanan V, Baskaran P (1996a) Efficacy of non-conventional insecticides against the sesame shoot webber and capsule borer, *Antigastra catalaunalis* Dup. Indian J Entomol 58(4):326–336

Selvanarayanan V, Baskaran P (1996b) Varietal response of sesame to the shoot webber and capsule borer, *Antigastra catalaunalis* Duponchel. (Lepidoptera: Pyraustidae). Int J Pest Manag 42(4):335–336

Shrivastava N, Duhoon SS, Raghuwansi KMS (2002) Testing of different germplasm lines of sesame against leaf roller and capsule borer, *Antigastra catalaunalis* Duponchel (Lepidoptera: Pyraustidae). Sesame Safflower Newsl 17:49–51

Singh H (1987) Key pests of minor oilseeds and their management. In: Mathur YK et al (eds) Recent advances in entomology, Gopal Prakeshan, Parade, pp 353–362

Singh R (1988) Evaluation of sesame germplasm/varieties against leaf roller and pod borer, Antigastra catalaunalis (Duponchel). M. Sc. (Ag.) Dissertation, Haryana Agricultural University, Hisar, India

Singh V (2002) Geographical variation in sesame shoot webber and capsule borer, *Antigastra catalaunalis* Duponchal (Lepidoptera: Pyraustidae) populations in India. Sesame Safflower Newsl 17:49–51

Singh UC, Dhamdhere SV (1986) Response of sesame varieties to *Antigastra catalaunalis* Dup. Narendra Deva J Agri Res 1(2):109–111

Singh V, Ghewande RB (1981) Plant protection in oilseeds. Pesticides Annual (1980–81), pp 30–45

Singh H, Jagan RS, Kharub SS (1990) Antibiosis in some sesame germplasm genotypes against shoot webber and pod borer, *Antigastra catalaunalis* (Duponchel). J Insect Sci 3:176–176

Surendarkumar M, Rohilla HR (1989) Studies on the susceptibility/resistance of sesame genotypes to the infestation of 'til' leaf roller and pod borer, *Antigastra catalaunalis* (Dup.). Sesame Safflower Newsl 4:86–87

Talpur MA, Khuhro RD, Rustamani MA, Nizamani IA (2002) Insect pests associated with sesame at Tando Jam, Pakistan. J Appl Sci 2(7):723–726

Thangavelu S, Kandaswamy G, SivanandamM, Muralibaskaran RK (1989) Problem and Progress of Sesamum Production in India. p 87–91. In: Omran AU (ed) Proc. Oilcrops network: Sesame and Safflower, Agricultural Research Centre, Giza, Egypt and International Development research Center, Canada

Tiwari NR, Shaw SS (1988) Assessment of loss caused by *Antigastra catalaunalis* (Dup.) in different varieties of sesamum. Agric Sci Digest 8(1):1–4

Vijai Anandh G (2003) Host plant resistance in sesame against shoot webber and capsule borer, *Antigastra catalaunalis* Dup. and phyllody disease. M. Sc. (Ag.) Dissertation, Annamalai University, Annamalai Nagar, India

Vijai Anandh G Selvanarayanan V (2005) Reaction of sesame germplasm against shoot webber *Antigastra catalaunalis* Duponchel and phyllody. Indian J Plant Prot 31(1):35–38

Yadav SK (1985) Studies on the susceptibility or resistance of sesame genotypes to the infestation of 'til' leaf roller, Antigastra catalaunalis (Dup.). M. Sc. (Ag.) Dissertation, Haryana Agricultural University, Hisar, India

Yadav RD, Lal BS (1976) Field trials and relative susceptibility of some important sesame varieties to *Antigastra catalaunalis* (Duponchel). Oilseeds J 6:34–35

Redesigning Research on Crop Resistance to Insects: Experiences with Tomato

V. Selvanarayanan

Abstract

Exploitation of host plant resistance (HPR) against insect pests has gained greater importance in intensive agriculture since the days of Reginald H. Painter. In spite of the initial euphoria mentioning few success stories, currently it warrants for introspection and redesigning of research strategies in developing insect tolerant/resistant crop cultivars. Conventional tools of resistance breeding, though successful, have lost its legacy because of temporal, manpower, and cost considerations. The declined output in terms of research publications or pest resilient crop cultivars stands testimony to the dwindling number of entomologists and breeders concentrating on such conventional tools. Many of the novel molecular approaches such as marker aided selection, mapping QTL have gained much prominence and few successful outputs are being popularized. Though many preliminary laboratory reports on molecular techniques in HPR are widely surfacing, their field success is less validated. The success of pest resilient cultivars depends on various factors including preference by growers based on better yield and economic parameters and in turn preference by consumers based on organoleptic characters. The sustained use of genetically modified crops is still suspected and debated. Further, the consumerist enigma attached to its wider adoption needs to be rusticated through ostensible safety studies. Our two decade-long experience in exploiting the resistant traits in a huge germplasm of tomato has given a far insight into the trends that were followed earlier, failures faced in developing insect tolerant cultivars and the conceptual changes visualized for the future, which are highlighted in this chapter.

Keywords

Breeding · Insect Pests · Resistance · Tomato

V. Selvanarayanan (✉)
Department of Entomology, Faculty of Agriculture,
Annamalai University, Annamalai Nagar, Cuddalore
District, Tamil Nadu 608002, India
e-mail: selvaento@gmail.com

A. K. Chakravarthy (ed.), *New Horizons in Insect Science: Towards Sustainable Pest Management*,
DOI 10.1007/978-81-322-2089-3_27, © Springer India 2015

Introduction

Since the advent of intensive agriculture, exploration for pest-resistant crop cultivars and their widespread exploitation is continually practiced. Though successful initially, those crop cultivars lost their glory later due to selection pressure by the target pest as well as other factors. Conventional breeding methods for exploitation of insect-resistant traits have earlier yielded promising cultivars. Due to practical limitations such as time and manpower requirements and cost considerations, such methods are in wane paving way for novel molecular approaches. This is evidenced by a brief perusal of the papers published in the Annual Reviews of Entomology since 2000, wherein a major proportion of the papers related to host plant resistance was focused on genetically modified crops or on molecular bases of plant resistance. Though many molecular techniques are available, the field success of outputs is less validated. Further, genetically modified food crops still faces the consumerist enigma, warranting for more detailed safety studies. The success of pest resilient cultivars depend on factors including preference by the growers based on better yield and economic parameters and preference by the consumers based on organoleptic characters. This chapter highlights insect resistance in huge germplasm of tomato, the paradigm shifts in research approaches, attempts, and the results therein.

Insect Resistance in Tomato Germplasm

Tomato (*Lycopersicon esculentum* Mill.), an important vegetable is commercially cultivated throughout the world both for fresh consumption and processing industries. In India, among the insect pests damaging tomato, the fruitworm, *Helicoverpa armigera* Hubner (Lepidoptera: Noctuidae) is a polyphagous, widely distributed insect pest inflicting heavy loss. Its young larvae feed on leaves or flowers, while matured instars bore into the fruits. Occasionally, vine boring is also witnessed (Gopalakrishnan 2006). Among the other insect pests, whitefly, *Bemisia tabaci* Gennadius (Hemiptera: Aleyrodidae) is also predominant in India. Besides desapping and reducing the vigor of plants, it also vectors leaf curl disease. In addition to these two insect pests, leaf caterpillar, *Spodoptera litura* Fabricius (Lepidoptera: Noctuidae) and serpentine leaf miner, *Liriomyza trifolii* Blan. (Diptera: Agromyzidae) are also commonly infesting tomato. Dependence on chemical insecticides for managing these insect pests is discouraged and warrants for exploration of insect tolerant or resistant tomatoes.

The gene pool of tomato (*Lycopersicon* spp.) is diverse and offers an ample scope for exploitation and utilization in pest and disease resistance. Resistance to 16 pest species has been reported in accessions of wild lines such as *Lycopersicon hirsutum* f. *typicum* and *L. hirsutum* f. *glabratum* (Dimock 1981; Dimock and Kennedy 1983; Kauffman 1987). Both physical (Gentile et al. 1969) and chemical (Williams et al. 1980) factors have been reported in these accessions where in the trichome mediated resistance playing a major role. Other physical and chemical factors associated with the leaf lamella were also reported earlier (Quiros et al. 1977; Elliger et al. 1981). Host plant resistance in tomato against *H. armigera*, *B. tabaci* and other pests has been widely explored and exploited. Reports on resistance of tomatoes against key insect pests are listed below (Table 1 to 6).

Evaluation of Insect Resistance in Tomato Germplasm

Considering the above, a huge tomato germplasm comprising 321 accessions were gathered from various sources. This germplasm included few wild species, land races, hybrids, and many cultivars (89 % cultivated species, *L. esculentum*, 10 % wild relative, *L. pimpinellifolium* and 1 % suspected cross of this two). These accessions were screened for resistance against *H. armigera*, both under field and glasshouse conditions at Annamalai Nagar, Tamil Nadu, India from 1998 to 2000. In the field screening, larval population and fruit damage was recorded while in the glass-

Table 1 Evaluation of resistance in tomato against fruitworm, *H. armigera* in the global perspective

Plant spp. (variety/ accession/line/cultivar)	Reports on resistance	Reference
1030 Accessions	cv. Tiny Tim to be resistant	Fery and Cuthbert 1974
Popular cultivars	cv. Pearson to be resistant	Schuster 1977
L. hirsutum accessions	High level of antibiosis	Juvik 1980; Elliger et al. 1981; Isman and Duffey 1982
3000 Accessions	Few cultivars and many wild varieties found resistant	AVRDC 1981, 1985
Wild accessions	High level of resistance	Juvik et al. 1982
Few cultivars	cv. Parker, Bonus and VEN−8	Lal 1985
Few cultivars	Tolerance in cv. Red Clande and Urbana	Farid 1987
800 Accessions	Wild accessions free from damage	Opena et al. 1989
8 Accessions	cv. Rashen and Yusuporskii 40 found resistant	Todzhaer and Ruzmetor 1993
12 Genotypes	cv. Bingo most promising	Santiago et al. 1998

Table 2 Evaluation of resistance in tomato against fruitworm, *H. armigera* in the Indian perspective

Plant spp. (variety/accession/line/ cultivar)	Reports on resistance	Reference
124 Genotypes	Biochemials like ascorbic acid	Kashyap and Verma 1987
Wild types and cv. Paiyur 1	Less infestation in *L. pimpinellifolium* and cv. Paiyur 1	Sivaprakasam 1988
19 Wild types and cultivars	*L. hirsutum f. glabratum*	Kashyap et al. 1990
Four cultivars	Less damage in cv. Kanchan 3	Sharma et al. 1990
250 Accessions	Less damage in Punjab Kesari, Punjab Chuhara, Sel 152 and Roma	Singh et al. 1990
70 Genotypes including 31 crosses	Resistance in No. 122775 and the cross Rick x Solan Gola	Kakar et al. 1990
69 Accessions	Less damage in LE 206 x CO3	Janarthanam et al. 1992
12 Cultivars	cv. Pusa Ruby found tolerant	Shukla and Sharma 1993
16 Varieties	cv. RT6–2, BT10, BT17, T30 and T32 were resistant	Mishra and Mishra 1993
Wild accessions and cultivars	LA 2992, LA 2449, LA 2531 found resistant	Brar et al. 1995
Hybrids and breeding lines	Less damage in Hybrids TH 802, TH2920, TH818 and breeding lines No.48, 29, 21	Kaur et al. 1996
35 Varieties	cv. Angurtala highly resistant	Mishra et al. 1996
Ten cultivars	cv. CO 3 found susceptible	Sivaprakasam 1996
28 Genotypes	Wild accessions and cv. HT 64 found resistant	Rath and Nath 1997
Two indeterminate varieties	Less fruit damage in Hybrid Naveen	Kaur and Singh 1997
10 Cultivars and 12 hybrids	Hybrids and cv. Roma tolerant	Pandey et al. 1997
Popular cultivars and hybrids	Hybrid Meenakshi susceptible than cultivars	Sankhyan and Verma 1993
Few varieties	cv. S12 resistant; cv. HS10 susceptible	Thakur et al. 1998
Popular varieties	Least damage in cv. HT 64 and HT 50	Shukla and Yadav 1998
24 Cultivars	Pusa Early Dwarf, Arka Vikas, Pusa Gaurav less susceptible	Chandrakar et al. 1998
cv. Pusa Ruby and 6 hybrids	Hybrid Arjuna found resistant	Chaudhuri et al. 2000

house screening, feeding damage was recorded to rate the resistance of the accessions (Table 7).

Based on the above-said resistance rating, a major proportion of the germplasm was found either susceptible or highly susceptible (Fig. 1) and five accessions namely Varushanadu Local, PT 4287, Seijima Jeisei, Ac 238 and Roma were identified as resistant/tolerant (Selvanarayanan

Table 3 Evaluation of insect resistance in tomato at Annamalai University

Plant spp. (variety/accession/line/cultivar)	Reports on resistance	Reference
32 accessions	Less damage in No. 1101, Hybrid F3 and No. 986	Krishnakumar 1993
321 accessions (285 *L. esculentum*; 32 *L. pimpinellifolium*; four suspected crosses)	Varushanadu Local, Seijima Jeisei, Ac 238, Roma found resistant	Selvanarayanan 2000
F1 Hybrids	Hybrid of Ac 238 × Roma tolerant	Dhakshinamoorthy 2002
Mutants	Mutants of Varushanadu Local tolerant	Gopalakrishnan 2010
Backcross generations	(PKM1 × VL) × PKM1 (BC_1) tolerant	Manikandan 2012

Table 4 Reports on resistance in tomato against whitefly, *B.tabaci*/ *B. argentifolii Trialeurodes vaporiorum* W

Plant spp. (variety/accession/line/cultivar)	Reports on resistance	Reference
L. pennellii	Significant levels of resistance	Gentile et al. 1968; de Ponti et al. 1975
L. hirsutum	High level of antibiosis	Juvik 1980
Four cultivars	Glandular trichome mediated resistance	Jennifer and Kisha 1981
L. pennellii ecotypes	Trichome mediated resistance	Shevach-Urkin 1983; Dahan 1985; Morag 1986
L. pennellii and *L. hirsutum*	Glandular trichome secretions	Berlinger et al.1984
L. pennellii accessions	Sticky exudates from type IV glandular trichomes	Goffreda et al. 1989; Steffens and Walters 1991
Ecotypes of *L. hirsutum* f. glabratum	Glandular secretions	Berlinger and Dahan 1987
Seven accessions	Varying susceptibility to TLCV	Banerjee and Kalloo 1987a
L. hirsutum f. glabratum	Biochemicals mediated resistance	Banerjee and Kalloo 1987b
Wild accessions	Acyl sugar based resistance	Liedl et al. 1995
cv. Pusa Ruby and 6 hybrids	Hybrid Abinash susceptible to leaf miner but tolerant to whitefly	Chaudhuri et al. 2000
Selected cultivars	L 27, 8484, Fiona and TY 172 found resistant and lower incidence of leaf curl	Lapidot et al. 2001
L. hirsutum var. hirsutum	Leaf glandular trichome based resistance	Freitas et al. 2002
Hybrid derivatives	Hybrid derivatives of Varushanadu Local tolerant	Muthukumaran 2004
Mutants	Mutants of Varushanadu Local tolerant	Gopalakrishnan 2010
Backcross generations	(PKM1 × VL) × PKM1 (BC_1) tolerant	Manikandan 2012

2000). As the young larvae of *H. armigera* feed on tender foliage and flowers while older larvae feed on the fruits, a promising resistant/tolerant accession should have the defense traits both in the foliage and fruits. Hence, in vitro and in situ studies were made to ascertain the mechanisms and factors of resistance in the selected accessions in comparison with the susceptible check, I 979. Feeding preference of *H. armigera* toward the foliage and fruits of the selected accessions was less compared to the susceptible check. In contrast, the selected accessions recorded higher ovipositional preference but egg hatching got impaired considerably (Selvanarayanan and Narayanasamy 2004). Larval feeding on the foliage and fruits of these accessions exerted pronounced antibiosis on *H. armigera* as evidenced by higher larval and pupal mortality; prolongation of larval and pupal tenure; decline in emergence and longevity of adults and also higher malformation in the adults (Selvanarayanan and Narayanasamy 2006; Selvanarayanan 2011) as observed earlier by Farrar and Kennedy (1990).

Table 5 Reports on resistance in tomato against serpentine leaf miner, *L. trifolii*

Plant spp. (variety/accession/line/cultivar)	Reports on resistance	Reference
L. hirsutum accessions	*L. hirsutum f. typicum*, *L. hirsutum f. glabratum*	Farrar and Kennedy 1992
Wild accessions and cultivars	LA 1663 found resistant	Eigenbrode et al. 1993
cv. Pusa Ruby and six hybrids	Hybrid Abinash susceptible to leaf miner but tolerant to whitefly	Chaudhuri et al. 2000
10 tomato accessions	cv. Varalakshmi found less susceptible	Tandon and Bakthavatsalam 2003
Hybrid derivatives	Hybrid derivatives of Varushanadu Local tolerant	Muthukumaran 2004
Backcross generations	(PKM1 × VL) × PKM1 (BC_1) tolerant	Manikandan 2012

Table 6 Reports on resistance in tomato against *S. litura/ S.littoralis*

Plant spp. (variety/accession/line/cultivar)	Reports on resistance	Reference
Wild accessions	High level of resistance	Juvik et al. 1982
Cultivars and wild accessions	Fruit skin toughness	Juvik and Stevens 1982
Wild accessions and cultivars	Antibiosis in wild lines	Eigenbrode and Trumble 1993
Five *L. pennellii* accessions and two cultivars	*L. pennellii* found resistant	Berlinger and Dahan. 1987
Hybrid derivatives	Hybrid derivatives of Varushanadu Local tolerant	Muthukumaran 2004
Backcross generations	(PKM1 × VL) × PKM1 (BC_1) tolerant	Manikandan 2012

Table 7 Criteria used to determine resistance ratings of 321 tomato accessions under glasshouse and field conditions. (Selvanarayanan 2000; Selvanarayanan and Narayanasamy 2006)

Rating	Glasshouse		Field	
	Percent foliage damage	Percent fruit damage	Number of larvae/plants	Percent fruit damage
Highly resistant (HR)	<10.0	<15.0	0.0–0.30	<15.0
Moderately resistant (MR)	10.1–25.0	15.1–30.0	0.31–0.60	15.1–30.0
Susceptible (S)	25.1–40.0	30.1–45.0	0.61–0.90	30.1–45.0
Highly susceptible (HS)	>40.0	>45.0	>0.90	>45.0

Antixenosis and/or antibiosis mechanisms of resistance operating in insect resistant/tolerant crop varieties may be attributed to the biophysical and/or biochemical factors of resistance. Various biophysical parameters of resistance namely trichome density, calyx area, thickness and toughness of the fruit rind, fruit sepal thickness, seed: pulp ratio were examined among which, trichome density was found to exert a significant negative correlation with larval feeding (Fig. 2). Trichomes or plant hairs on the foliage of *Lycopersicon* spp. have been categorized as types I–VII (Luckwill 1943). Among these, types I, IV, VI, and VII are glandular and types II, III, and V are non-glandular. In this study, the selected accessions had two non-glandular (III and V) and three glandular types (I, VI, VII) (Selvanarayanan 2011) (Fig. 3). Though the trichome density exerts a negative influence on pest species, it also negates the role of natural enemies as inferred earlier by Kennedy (2003).

Various biochemical factors of resistance namely, reducing sugars, non-reducing sugars, total sugars, amino acids in the foliage and fruits; acidity, ascorbic acid, lycopene in the fruits; total phenols, O. D. phenols and chlorogenic acid in the foliage were estimated. Among them, phenol content of the foliage and acidity of the fruits exerted a sig-

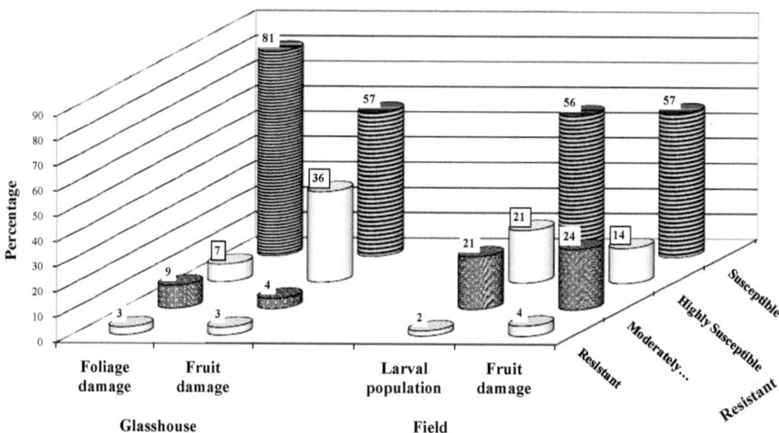

Fig. 1 Resistance rating of the tomato accessions under glasshouse and field conditions

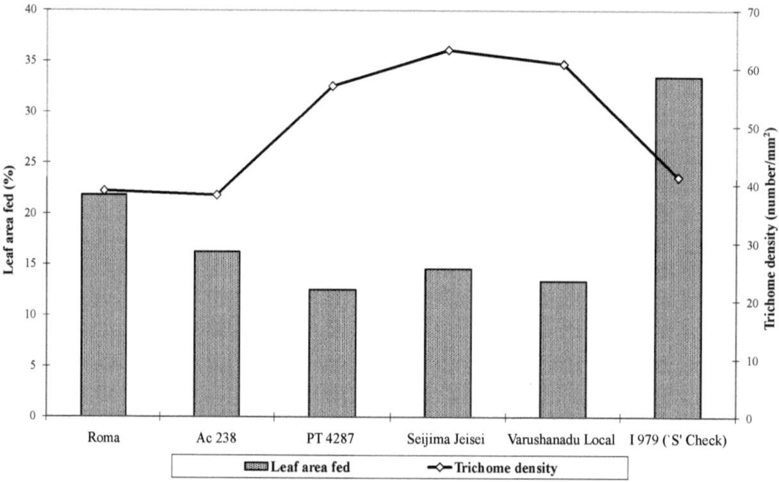

Fig. 2 Interaction between trichome density of tomato accessions and *Helicoverpa armigera* (*Hubner*) feeding

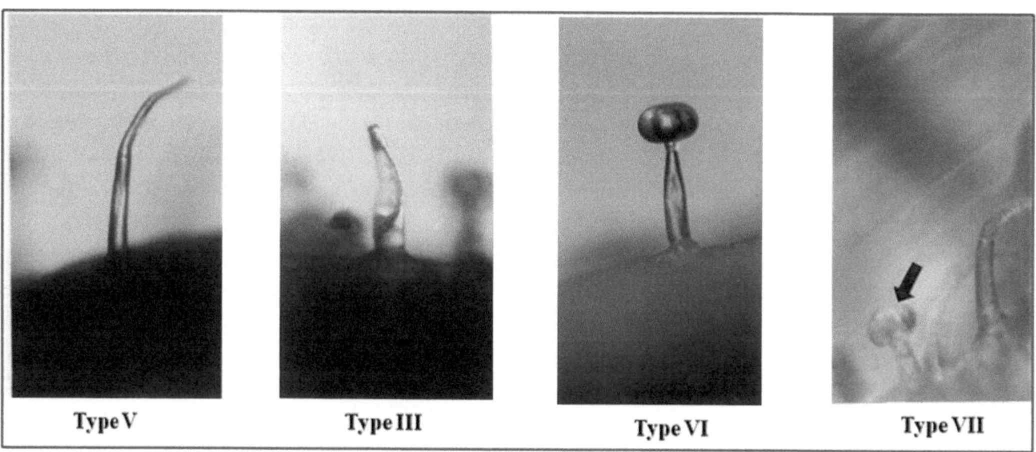

Fig. 3 Types of trichomes on the foliage of selected tomato accessions

Table 8 Biochemical constituents of the foliage of selected tomato accessions

Name of the accession	Phenol (mg/g of dry leaf)		Chlorogenic Acid (mg/g)	Acidity (% citric acid)
	O.D. Phenol	Total		
Roma	0.103	0.190	0.010	0.62
Ac 238	0.096	0.185	0.010	0.67
PT 4287	0.140	0.198	0.017	0.80
Seijima Jeisei	0.142	0.184	0.012	0.78
Varushanadu local	0.146	0.202	0.016	0.83
I 979 (Susceptible check)	0.082	0.172	0.007	0.60
C.D. (p=0.05)	0.02	0.02	0.00	0.08
Correlation coefficient (r)	−0.79**	−0.77**	−0.88**	−0.85**

Each value is a mean of five replications
Significant at 5% level

Table 9 Field screening of selected tomato accessions and their hybrid derivatives against major insect pests

Accession	*Helicoverpa armigera* (Hubner)		*Spodoptera litura* Fab. larval population/plant	Percent leaf damage by *Liriomyza trifolii* Blan	Population of *Bemisia tabaci* Genn./plant
	Larval population/plant	Fruit damage (%)			
Ac 238	0.53	17.65 (24.84)	0.38	19.17	0.56
Roma	0.77	16.81 (24.20)	0.16	19.44	0.00
Seijima Jeisei	0.42	15.50 (23.18)	0.06	19.96	0.36
Varushanadu Local	0.50	12.48 (20.69)	0.30	19.89	0.56
HY1F1	0.66	6.20 (14.42)	0.22	19.81	0.08
HY2F1	0.42	5.70 (13.81)	0.26	18.78	0.44
HY3F1	0.32	5.35 (13.37)	0.06	3.48	0.00
HY1F2	0.52	7.40 (15.79)	0.06	20.54	0.62
HY2F2	0.45	6.75 (15.06)	0.28	19.61	0.18
HY3F2	0.36	5.80 (13.94)	0.08	5.14	0.00
I 979	1.83	32.24 (34.60)	0.42	25.20	0.86
C.D. (p=0.05)	0.18	3.56	0.07	4.41	0.11

Each value is a mean of five replications
Values in parentheses are arcsine transformed

nificant negative correlation with larval feeding (Table 8). In addition to phenols, other secondary metabolites on wild tomatoes were reported to confer insect resistance. The type VI glandular trichomes of *L. hirsutum f. glabratum* were reported to possess methyl ketones such as 2-tridecanone and 2-undecanone, (Farrar et al. 1992) while that of *L. pennelli* possess acyl sugars (Juvik et al. 1994).

Exploiting Insect Resistance in Selected Tomato Accessions Through Hybridization

Realizing the value of the selected accessions, they were subjected to intercrossing by conventional hybridization (emasculation and pollination). Among the many intercrosses, Varushanadu Local x Ac 238; Varushanadu Local x Roma and Ac 238 x Roma yielded Hybrid 1, Hybrid 2 and Hybrid 3 respectively. The selected accessions and their hybrid derivatives were evaluated for insect resistant traits, both in the glasshouse and field, in comparison with a susceptible check against fruit worm, *H. armigera* (Dhakshinamoorthy 2002), serpentine leaf miner, *L. trifolii*, whitefly, *B. tabaci*, and leaf caterpillar, *S. litura* (Muthukumaran 2004). Though the hybrids recorded tolerance to many of the pests at the field level, a wider variation was observed among them with regard to population of *H. armigera*, *S. litura* and *B. tabaci*, besides fruit and leaf damage (Table 9). Similarly, under glasshouse conditions also, preference of these insects toward the

Table 10 Preference of major insect pests toward selected tomato accessions and their derivatives

Accession	*Helicoverpa armigera* (Hubner) damage		*Spodoptera litura* Fab	*Liriomyza trifolii* Blan
	Foliage (%)	Fruit (g)	Percent leaf damage	Percent infestation
Ac 238	14.54 (22.40)	3.28	99.09 (85.89)	39.90 (39.11)
Roma	14.54 (22.40)	2.53	100.00 (90.00)	31.04 (33.72)
Seijima Jeisei	12.80 (20.96)	1.82	23.63 (29.08)	39.99 (38.87)
Varushanadu Local	13.86 (21.85)	1.59	25.56 (30.35)	46.66 (43.05)
HY1F1	12.59 (20.79)	2.23	100.00 (90.00)	41.33 (39.67)
HY2F1	12.20 (20.44)	1.35	100.00 (90.00)	49.66 (44.89)
HY3F1	10.19 (18.64)	1.28	100.00 (90.00)	6.66 (14.94)
HY1F2	12.80 (20.96)	2.04	100.00 (90.00)	36.66 (36.92)
HY2F2	12.10 (20.36)	1.38	99.99 (87.87)	44.28 (41.64)
HY3F2	11.20 (19.55)	1.32	98.63 (83.45)	13.33 (13.82)
I 979	26.45 (30.96)	6.26	93.40 (75.14)	37.61 (37.46)
C.D. (p=0.05)	0.26	0.47	2.45	13.68

Each value is a mean of five replications
Values in parentheses are arcsine transformed

selected accessions and their hybrid derivatives were found varying (Table 10) (Selvanarayanan 2011). The major biophysical factor namely trichome density and the major bio-chemical factor namely phenol content was found varying in the F_1 and F_2 generations. This suggested the possibility of segregation of the defense traits among the generations and hence a promising hybrid with higher yield potential could not be developed.

Exploiting Insect Resistance in Selected Tomato Accessions Through Backcross Breeding

Upon realizing that the conventional hybridization approach is proving futile in developing insect tolerant/resistant tomato varieties, backcross breeding strategy may be adopted, provided the segregation of desirable traits is well ascertained and exploited. The most promising insect tolerant accession namely Varushanadu Local was backcrossed with the most popular cultivar namely PKM 1 (Manikandan 2012). The resultant F1 and the backcross progenies were evaluated for resistance under semi-field conditions against key insect pests. Both the selected accessions and the hybrid derivatives varied widely among themselves with regard to resistance potential. Significant variation was observed in the preliminary semi-field screening. The accession VL, (PKM1 × VL) × PKM1 (BC_1) and (VL × PKM1) × PKM1 (BC_1) recorded lesser population of *H. armigera* and *S. litura* larvae, lesser population of whitefly as well as lesser damage by leaf miner (Table 11). Further studies on the hybrid derivatives may unravel the nature of segregation and stabilization of the desirable traits.

Mutation Breeding of Selected Tomato Accessions

If the natural variability available in the germplasm could not be exploited through conventional breeding methods, mutation breeding may be attempted to increase variability in morphological and physiological characters besides inducing new plant ideotypes. Mutation breeding is relatively a quicker method for crop improvement and it has an added advantage over hybridization since the basic genotype of a variety is slightly altered (Gopalakrishnan 2010).

Keeping the above approach in view, chemical mutagens namely Ethyl Methane Sulfonate (EMS) and Di Ethyl Sulfate (DES) and physical mutagen namely gamma rays were used for inducing mutation in the selected tomato accessions (Gopalakrishnan 2010). EMS and DES at various doses were evaluated and the LD_{50} dose was fixed at the dose recording 50% seed

Table 11 Evaluation of backcross progenies of promising tomato accession against key insect pests

Accession/Hybrid	Mean larval/nymphal population per plant			Percent leaf infestation *L. trifolii*
	H. armigera	*S. litura*	*B. tabaci*	
PKM1	0.32	0.36	1.07	21.42
Varushanadu Local (VL)	0.06	0.06	0.05	8.12
PKM1 × VL (F_1)	0.26	0.30	0.45	20.48
VL × PKM1 (F_1)	0.08	0.18	0.21	18.24
(PKM1 × VL) × PKM1 (BC_1)	0.06	0.08	0.08	9.86
(VL × PKM1) × PKM1 (BC_1)	0.08	0.08	0.11	17.22
(VL X PKM1) × VL (BC_1)	0.16	0.26	0.26	18.96
(PKM1 X VL) × VL (BC_1)	0.18	0.24	0.23	18.52
C.D. (p=0.05)	0.01	0.02	0.05	1.56

Each value is a mean of five replications

Table 12 Field evaluation of tomato mutants against *H. armigera*

Mutagen	Larval population per plant					
	M1		M2		M3	
Conc./Dose	VL	I 979	VL	I 979	VL	I 979
DES						
0.09%	1.98	4.66	1.49	4.66	3.49	5.89
0.1%	0.89	3.55	1.94	4.83	3.16	4.98
0.2%	1.94	3.16	2.39	4.61	2.44	4.22
EMS						
0.2%	2.21	3.83	2.05	5.16	3.32	5.61
0.3%	2.49	5.28	1.44	3.67	3.30	5.05
0.4%	2.55	4.17	0.94	3.94	1.22	3.44
Gamma rays						
5 KR	2.43	4.38	2.44	3.50	1.98	4.77
10 KR	2.11	4.56	2.16	4.94	3.50	6.22
15 KR	2.38	5.22	3.27	4.60	3.99	5.83
Control	4.43	6.88	4.39	6.67	5.55	7.61
C.D. (p=0.05)	0.56		0.75		0.37	

germination. Based on these values, seeds, presoaked in distilled water for 6 h were treated with DES @ 0.09, 0.1, and 0.2% and EMS @ 0.2, 0.3, and 0.4% for 12 h. After the treatment, seeds were washed in sodium thiosulfate buffer and distilled water.

For physical mutation, the seeds of the selected tomato accessions were treated with gamma rays at the radiation frequency of 5, 10, and 15 kR in the Gamma Chamber 1200, supplied by Board of Radiation and Isotope Technology (BRIT), Department of Atomic Energy (DAE), Government of India, Mumbai, India and established at the Centre for Plant Breeding and Genetics, Tamil Nadu Agricultural University, Coimbatore, India.

The mutagenized seeds were sown in earthen pots and 25 days old seedlings were transplanted in the field for resistance evaluation against *H. armigera* and *B. tabaci.* Three field trials were conducted respectively for the M1, M2, and M3 generations of mutagenized tomato accessions. Population of *H. armigera* larvae and *B. tabaci* nymphs was recorded and the mutant generations were rated for resistance.

In the field trials, larval population of *H.armigera* was found the least in mutagenized tomato accession Varushanadu Local in all the three generations (Table 12). In the M1 generation, plants of this accession mutagenized with DES @ 0.1% recorded the least population. But in the M2 and M3 generation, plants mutagenized with EMS @ 0.4% recorded the least population. With regard to whitefly population, wider variation was observed among mutagens and their

Table 13 Field evaluation of tomato mutant generations against *B. tabaci*

Mutagen	Population of nymphs per plant					
	M1		M2		M3	
Conc	VL	I 979	VL	I 979	VL	I 979
DES						
0.09%	20.78	30.39	21.39	25.94	14.56	17.94
0.1%	19.39	29.28	14.94	21.88	12.22	15.00
0.2%	18.38	26.83	14.33	21.72	10.56	13.06
EMS						
0.2%	19.72	27.78	13.11	19.55	11.39	14.39
0.3%	20.72	27.05	14.89	21.28	10.98	14.83
0.4%	23.50	30.33	16.83	23.38	13.89	17.67
Gamma rays						
5 KR	19.88	29.33	16.94	23.67	13.28	17.34
10 KR	20.83	28.61	16.00	22.61	13.11	17.67
15 KR	17.94	30.22	14.98	22.27	11.78	15.22
Control	31.22	38.45	31.33	42.89	22.34	28.22
C.D. (p=0.05)	2.33		3.10		2.11	

concentrations (Table 13). In the M1 generation, plants of Varushanadu Local mutagenized with DES @ 0.2% recorded the least population of whitefly whereas in M2 and M3 generation, plants mutagenized with EMS @ 0.3% and Gamma rays @ 5 KR recorded the least population of whitefly. Though whitefly population was noticed, symptoms of leaf curl disease could not be observed in all the three generation suggesting that the population present may not be active transmitters (Gopalakrishnan 2010).

Inherent Insect Resistance Vis-A-Vis Induced Insect Resistance in Tomato

In view of the failure of conventional hybridization in exploiting the inherent resistance in the selected accessions, inducing insect resistance using external inputs was attempted. Arbuscular mycorrhizal fungi (AMF), which form symbiotic associations with root systems of most agricultural, horticultural and hardwood crop species, have been suggested as widespread potential bioprotective agents. Fungal symbiosis in various crop plants confers tolerance and/or resistance against many phytophagous insects, nematodes and pathogens. Co-inoculation of AM fungus, *Glomus* spp. with *Azospirillum brasilense* (Linn.) reduced the incidence of leaf folder, *Cnaphalocrocis medinalis* Guen. in upland paddy var. PKM–1 (Amutha et al. 2003) while individual inoculation of *G. intraradices* in pea plants conferred resistance against the adult weevils, *Sitona lineatus* (Wamberg et al. 2003). In view of the above, the influence of AM fungal inoculation in inducing resistance in the selected tomato accessions against fruit worm, *H. armigera* and leaf caterpillar, *S. litura* was analysed. The resistant tomato accession namely Varushanadu Local and the susceptible check, I 979 identified earlier were tested for their interaction with four AM fungi *viz., Glomus fasciculatum* (Thaxt.)Gerd. and Trappe, *Glomus mosseae* (Nicol. and Gerd.), *Acaulospora laevis* (Gerd. and Trappe) and *Gigaspora margarita* (Becker and Hall) under pot culture conditions and also under field conditions for two seasons (Gopalakrishnan 2006).

The length of the trichomes on tomato foliage was found increased, but their density was not enhanced considerably in AM fungal inoculated plants. On analyzing the phenol content of the leaves of the test accessions, it was found that *G. margarita* inoculated plants of the resistant accession Varushanadu Local contained higher amount (0.3303 mg/g of leaf) followed by *G. mosseae, G. fasciculatum, A. laevis* (0.3106, 0.3105, 0.3022 mg/g respectively) as against the least in control plants (0.3007 mg/g). Based on the above, it is inferred that the role of Arbuscular

Table 14 Influence of AM fungi on larval population of *H. armigera and S. litura* on tomato

AMF	Larval population/plant			
	H. armigera		*S. litura*	
	Varushanadu Local	I 979	Varushanadu Local	I 979
Glomus fasciculatum (Thaxt.)Gerd. and Trappe	1.00	1.83	0.67	1.20
Glomus mosseae (Nicol. and Gerd.)	0.67	1.73	0.93	1.53
Gigaspora. Margarita (Becker and Hall)	1.27	2.00	1.10	1.70
Acaulospora laevis (Gerd. and Trappe)	1.47	2.67	1.13	1.40
Control	2.47	2.30	0.67	1.33
C.D. (p=0.05) Acc./ AMF	0.36		0.16	
Acc. × AMF	0.80		0.36	

Each value is a mean of five replications

Table 15 Influence of AM fungi on feeding preference of *H. armigera* and *S. litura* on tomato accessions

AMF	Leaf damage (%)			
	H. armigera		*S. litura*	
	Varushanadu Local	I 979	Varushanadu Local	I 979
Glomus fasciculatum (Thaxt.)Gerd. and Trappe	10.25	23.53	8.51	70.97
Glomus mosseae (Nicol. and Gerd.)	10.23	23.55	7.84	70.61
Gigaspora. Margarita (Becker and Hall)	10.21	23.50	8.51	70.62
Acaulospora laevis (Gerd. and Trappe)	10.23	23.84	8.29	70.96
Control	11.07	24.49	8.27	70.63
C.D. (p=0.05) Acc./ AMF	0.078		0.24	
Acc. × AMF	0.17		0.54	

Each value is a mean of five replications

Mycorrhizal Fungi (AMF) inoculation in inducing insect resistance/tolerance could not be definitely explained though yield enhancement was achieved (Gopalakrishnan 2006; Selvanarayanan 2009; Selvanarayanan 2011).

In the field evaluation, larval population of *H. armigera* was found to be less in the AM fungal inoculated plants as against the uninoculated plants of both the resistant and susceptible accessions (Table 14). But higher larval population of *S. litura* was recorded in inoculated plants than that of uninoculated plants. Similarly, in the pot culture studies, feeding preference of *H. armigera* was high toward uninoculated plants of both the accessions in contrast to *S. litura* which preferred the inoculated plants (Table 15). *G. margarita* inoculated plants recorded lesser feeding preference by *H. armigera* while *G. mosseae* inoculated plants were less preferred by *S. litura* (Selvanarayanan 2011).

Problems and Prospects of Resistance Breeding of Tomatoes

In resistance breeding programs, segregation of desirable traits in the generations is a common phenomenon and for obtaining sustenance of such traits, evaluation is to be done for many generations. This needs not only concerted perseverance but sufficient funding sources.

Any new promising variety that is developed by conventional or other novel methods of resistance breeding should be subjected to molecular studies. Mapping of the genes of the generations and their parents using marker aided selection or identifying the QTL may unravel the genes responsible for such desirable traits. DNA-sequencing will therefore be extremely useful in providing the information necessary for future crop improvement in tomato. Under the International Solanaceae Genome Project (SOL), novel DNA sequencing technologies collectively referred to as Next Generation Sequencing (NGS) technologies had emerged, which may enable faster and better sequencing.

A cultivar that is developed by any of the above methods, even if less susceptible or tolerant to insect pests with desirable yield attributes may be subjected for inducing the resistance using suitable strain of bio-inoculants such as arbuscular mycorrhizal fungi. Elaborate studies and large scale evaluation of strains suitable for varied agro-climatic conditions are warranted. These efforts may culminate in developing insect tolerant/resistant tomatoes with promising yield traits that enable eco-friendly management of the major insect pests. Developing such an insect tolerant/resistant tomato is pivotal in implementing integrated management of insect pests.

References

Amutha M, Mahadevan BR, Chozhan K, Gajendran G (2003) Synergist effect of *Azospirillum brasilense* (Linn.) and vesicular arbuscular mycorrhiza (*Glomus* spp.) on management of leaf folder in rice. In: Proceedings of a national symposium on sustainable insect pest management, Entomology Research Institute, Loyola College, Chennai, abstr. p 48, 6–7 Feb

AVRDC (1981) Screening for resistance to tomato fruit worm (*Heliothis armigera*). AVRDC Progress Report-1980, AVRDC, Taiwan

AVRDC (1985) Screening hairy tomatoes for resistance to tomato fruitworm. AVRDC Progress Report, AVRDC, Taiwan

Banerjee MK, Kalloo G (1987a) Sources and inheritance of resistance to leaf curl virus in *Lycopersicon*. Theor Appl Genet 73:707–710

Banerjee MK, Kalloo G (1987b) Inheritance of tomato leaf curl virus resistance in *Lycopersicon hirsutum f. glabratum*. Euphytica 36:581–584

Berlinger MJ, Dahan R Shorach-Urkin E (1984) Resistance to the tobacco whitefly, *Bemisia tabaci* in tomato and related species: a quick screening method. Bull IOBC/WPRS 4:39–40

Berlinger MJ, Dahan R (1987) Breeding for resistance to virus transmission by whiteflies in tomatoes. Insect Sci Appl 8:783–784

Brar KS, Raman H, Dhaliwal HS, Cheema DS (1995) Field screening of different species of *Lycopersicon* against tomato fruit borer, *Helicoverpa armigera* (Hubner). J Insect Sci 8(2):196–197

Chandrakar G, Ganguli RN, Kaushik DK, Dubey VK (1998) Resistance to the fruit borer, *Heliothis armigera* in tomato, pp 73–74. In: Reddy PP, Kumar NKK, Verghese A (eds) Adv. in IPM for Hortl. Crops, Proceedings first National Symposium on Pest Management in Horticulture Crops. Environmental implications and thrusts, 15–17 Oct 1997, Bangalore

Chaudhuri N, Deb DC, Senapati SK (2000) Evaluation of commonly grown tomato variety and hybrids in Terai region of West Bengal. Environ Ecol 18(4):933–939

Dahan R (1985) *Lycoperiscon pennellii* as a source of resistance to the tobacco whitefly, *Bemisia tabaci* on tomato. M. Sc. Dissertation, Hebrew University, Jerusalem

de Ponti OMB, Pet G, Hogenboom NG (1975) Resistance to the glasshouse whitefly, *Trialeurodes vaporariorum* in tomato (*Lycopersicon esculentum*) and related species. Euphytica 24:645–649

Dhakshinamoorthy G (2002) Studies on resistance of tomato against fruit borer, *Helicoverpa armigera* (Hubner). M. Sc. Dissertation, Annamalai University, India, p 110

Dimock MB (1981) Chemical and physiological studies of antibiosis in the wild tomato, *Lycopersicon hirsutum* f. *glabratum* to the tomato fruit worm, *Heliothis zea*. M. S. Dissertation, North Carolina State University, Raleigh, p 83

Dimock MB, Kennedy GG (1983) The role of glandular trichomes in the resistance of *Lycopersicon hirsutum* f. *glabratum* to *Heliothis zea*. Entomol Exp Appl 33:263–268

Eigenbrode SD, Trumble JT (1993) Antibiosis to beet armyworm (*Spodoptera exigua*) in *Lycopersicon* accessions. Hort Sci 28:932–934

Eigenbrode SD, Trumble JT, Jones RA (1993) Resistance to beet army worm, hemipterans and *Liriomyza* spp. in *Lycopersicon* accessions. J Am Soc Hort Sci 118(4):525–530

Elliger CA, Wong Y, Chan BG, Waiss AC Jr (1981) Growth inhibitors in tomato (*Lycpersicon*) to tomato fruitworm (*Heliothis zea*). J Chem Ecol 7:753–758

Farid A (1987) Some bio-ecological features of *Heliothis armigera* Hb. on tomato in Djiroft. Entomologie et phytopathologie appliquees 54:5–6

Farrar RR Jr, Kennedy GG (1990) Growth inhibitors in host plant resistance to insects: examples from a wild tomato with *Heliothis zea* (Noctuidae: Lepidoptera). J Entomol Sci 25:46–50

Farrar RR Jr, Kennedy GG (1992) Insect and mite resistance in tomato. In: Kalloo G (ed) Monographs on theoretical and applied genetics, 14, genetic improvement of tomato. Springer, Berlin, pp 121–142

Farrar RR Jr, Kennedy GG, Kashyap RK (1992) Influence of life history differences of two tachinid parasitoids of *Helicoverpa zea* (Boddie) (Lepidoptera: Noctuidae) and their interactions with glandular trichome/methyl ketone based insect resistance in tomato. J Chem Ecol 18:499–515

Fery RL, Cuthbert FP Jr (1974) Resistance of tomato cultivars to the fruit worm, *Heliothis zea* (Boddie). Hort Sci 9:469–470

Freitas JA, Maluf WR, Cardosa M, Das G, Gomes LAA, Bearzotti E (2002) Inheritance of foliar zingiberene contents and their relationship to trichome densities and whitefly resistance in tomatoes. Euphytica 127(2):275–287

Gentile AG, Webb RE, Stoner AK (1968) Resistance in *Lycopersicon* and *Solanum* to greenhouse whiteflies. J Econ Entomol 61:1355–1357

Gentile AG, Webb RE, Stoner AK (1969) *Lycopersicon* and *Solanum* spp. resistant to the carmine and two spotted spider mite. J Econ Entomol 62:834–836

Goffreda JC, Mutschler MA, Ave DA, Tingey WM, Steffens JC (1989) Aphid deterrence by glucose esters in glandular trichome exudate of the wild tomato, *Lycopersicon pennellii*. J Chem Ecol 15:2135–2147

Gopalakrishnan R (2006) Induced resistance in tomato against key insect pests. M. Sc. Dissertation, Annamalai University, India, p 146

Gopalakrishnan R (2010) Induced mutagenesis in selected tomato accessions for resistance against fruitworm, *Helicoverpa armigera* (Hubner) and whitefly, *Bemisia tabaci* Genn. Ph. D. Dissertation, Annamalai University, India, p 214

Isman MB, Duffey SS (1982) Toxicity of tomato phenolic compounds to the fruitworm, *Heliothis zea*. Entomol Exp Appl 31:370–376

Janarthanam R, Rajendiran R, Raghumoorthi KN, Arumugam R (1992) Field evaluation of tomato germplasm types for resistance to tomato fruit borer. Ann Plant Resist Insects Newsl 18:63

Jennifer J, Kisha M (1981) Observations on the trapping of the whitefly, *Bemisia tabaci* by glandular hairs on tomato leaves. J Appl Biol 97:123–127

Juvik JA (1980) Assessment of the alkaloid, α-tomatine, as a potential source of natural insect resistance in tomato cultivars. Ph. D. Dissertation, University of California, Davis, p 150

Juvik JA, Stevens MA (1982) Physiological mechanisms of host plant resistance in the genus *Lycopersicon* to *Heliothis zea* and *Spodoptera exigua*, two insect pests of the cultivated tomato. J Am Soc Hort Sci 107:1065–1069

Juvik JA, Berlinger MJ, Ben David T, Rudich J, David T (1982) Resistance among accessions of the genera *Lycopersicon* and *Solanum* to four of the main insect pests of tomato in Israel. Phytoparasitica 10:145–156

Juvik JA, Shapiro JA, Young TE, Mutschler MA (1994) Acylglucose from wild tomatoes alter behaviour and reduce growth and survival of *Helicoverpa zea* and *Spodoptera exigua* (Lepidoptera: Noctuidae). J Econ Entomol 87:482–492

Kakar KL, Bhalla OP, Dhaliwal MS (1990) Screening of tomato germplasm and breeding for resistance against fruit borer, *Heliothis armigera* Hub. J Insect Sci 3:57–61

Kashyap RK, Verma AN (1987) Factors imparting resistance to fruit damage by *Heliothis armigera* (Hubner) in some tomato phenotypes. Insect Sci Appl 8:111–114

Kashyap RK, Banerjee MK, Kalloo G, Verma AN (1990) Survival and development of fruit borer, *Heliothis armigera* (Hubner), (Lepidoptera: Noctuidae) on *Lycopersicon* spp. Insect Sci Appl 11:877–881

Kauffman WC (1987) Influence of 2-tridecanone based resistance of a wild tomato on parasitoids and predators of the tomato fruitworm, *Heliothis zea* (Boddie). Ph. D. Dissertation, North Carolina State University, Raleigh, U.S.A.

Kaur S, Cheema DS, Singh S, Singh D (1996) An integrated approach in the management of pests and diseases in tomato. J Insect Sci 9:100–101

Kaur S, Singh D (1997) Management of *Helicoverpa armigera* (Hubner) through planting periods and staking of tomato crop. J Insect Sci 10(2):160–162

Kennedy GG (2003) Tomato, pests, parasitoids and predators: tritrophic interactions involving the genus *Lycopersicon*. Ann Rev Entomol 48:51–72

Krishnakumar J (1993) Screening tomato germplasm for resistance against *Heliothis armigera* (Hubner) (Lepidoptera: Noctuidae). M.Sc. Dissertation, Annamalai University, India, p 101

Lal OP (1985) Field resistance of some tomato cultivars against the fruit worm, *Heliothis armigera* in Tripoli. Bull Ent 26:96–97

Lapidot M, Friedmann M, Pilowsky M, Ben-Joseph R, Cohen S (2001) Effect of host plant resistance to tomato yellow leaf curl virus (TYLCV) on virus acquisition and transmission by its whitefly vector. Phytopathology 91(12):1209–1213

Liedal BE, Lawson DM, White KK, Shapiro JA, Cohen DE, Carson WG, Trumble JT, Mutschler MA (1995) Acylsugars of wild tomato, *Lycopersicon pennellii* alters settling and reduces oviposition of *Bemisia argentifolii* (Homoptera: Aleyrodidae). J Econ Entomol 88(3):742–748

Luckwill LC (1943) The genus *Lycopersicon*: a historical, biological and taxonomic survey of the wild and cultivated tomatoes, Aberdeen University Press, Aberdeen (Aberdeen University Studies number 120)

Manikandan KR (2012) Evaluation of resistance in certain *Lycopersicon* hybrids to selected insect pests, M.Sc. (Ag.) Dissertation, Annamalai University, India, p 103

Mishra NC, Mishra SN (1993) Performance of tomato varieties against wilt, *Fusarium oxysporum* f. sp. *lycopersici* and fruit borer, *Helicoverpa armigera* in the north-eastern Ghat zone of Orissa. Indian J Pl Prot 21(2):183–186

Mishra PN, Singh MP, Nautiyal MC (1996) Varietal resistance in tomato against fruit borer, *Heliothis armigera*. Indian J Ent 58:222–225

Morag Y (1986) The wild tomato, *Lycopersicon hirsutum* as a source for resistance to the tobacco whitefly, *Bemisia tabaci* on tomato. M. Sc. Dissertation, Hebrew University, Jerusalem

Muthukumaran N (2004) Resistance of *Lycopersicon* accessions and their hybrid derivatives against certain insect pests. M.Sc. (Ag.) Dissertation, Annamalai University, India, p 94

Opena RT, Green SK, Talekar NS, Chen JT (1989) Genetic improvement of tomato adaptability to the tropics. Progress and future prospects. pp 70–85. In: Proc. Int. Symp. Integ. Mgt. Practices, AVRDC, Taiwan

Pandey RR, Gurung TB, Gc YD, Gurung G (1997) Monitoring and management of tomato fruit worm (*Heliothis armigera*), its egg parasite (*Trichogramma ishii*) in the Western Hills. Working paper-Lumle Regional Agrl. Res. Center, No. 97/24, p 14

Quiros CF, Stevens MA, Rick CM, Kok-Yokomi ML (1977) Resistance in the tomato to the pink form of the potato aphid (*Macrosiphum euphorbiae* Thomas): the role of anatomy, epidermal hairs and foliage composition. J Am Soc Hort Sci 102:166–171

Rath PC, Nath P (1997) Screening of some tomato genotypes for susceptibility to the fruit borer, *Helicoverpa armigera* Hubner at Varanasi. Veg Sci 24:153–156

Sankhyan S, Verma AK (1993) Life fecundity tables for the tomato fruit borer *Heliothis armigera* (Hubner) on tomato, *Lycopersicon esculentum* Mill. Pest Manage Econ Zool 1:80–84

Santiago J, Mendoza M, Borrego F (1998) Evaluation of tomato (*Lycopersion esculentum* Mill.) in greenhouse conditions: phenological and physiological criteria. Agronomic Mesoamericana 9(1):59–65

Schuster DJ (1977) Effect of tomato cultivars on insect's damage and chemical control. Fla Entomol 60:227–232

Selvanarayanan V (2000) Host plant resistance in tomato against fruit borer, *Helicoverpa armigera* (Huebner). Ph. D. Dissertation, Annamalai University, India. p 251

Selvanarayanan V, Narayanasamy P (2004) Antixenosis resistance in tomato against fruit borer, *Helicoverpa armigera* (Hubner). Int J Trop Ins Sci 24(3):201–206

Selvanarayanan V, Narayanasamy P (2006) Factors of resistance in tomato accessions against the fruit worm, *Helicoverpa armigera* (Hubner). Crop Prot 25:1075–1079

Selvanarayanan V (2009) Reviving cultural strategies and redefining nutrient supplies for an effective insect pest management, pp 226–229. In: Ignacimuthu S, David BV (eds) Ecofriendly insect pest management. Elite, New Delhi

Selvanarayanan V (2011) Research on insect resistance in tomato: present trends and future thrusts, pp 365–371. In: Ambrose DP (ed) Insect pest management, a current scenario, entomology research unit. St. Xavier's College, Palayamkottai, Tamil Nadu p 594

Sharma ML, Sharma TR, Sharma RC, Nandkar DN (1990) Field evaluation of tomato cultivars to fruit borer (*Heliothis armigera* Hubner) in Satpura plateau region of Madhya Pradesh. Orissa J Hort 18:22–25

Shevach-Urkin E (1983) Comparison of the resistance to the tobacco whitefly, *Bemisia tabaci* (Gennadius) in the cultivated tomato, *Lycopersicon esculentum* and the wild species *L. hirsutum* var. glabratum C.H. Mull. M. Sc. Dissertation, Hebrew University, Jerusalem

Shukla A, Sharma TR (1993) Field evaluation of tomato cultivars against fruit borer (*Heliothis armigera* Hubner) in Satpura plateau region of Madhya Pradesh. Jawaharlal Nehru Krishi Vishwa Vidyalaya Res J 27:140–142

Shukla A, Yadav HS (1998) Resistance in crops against *Helicoverpa armigera* (Hubner): a review. Adv Plant Sci 11:265–270

Singh D, Gill HS, Sharma BR (1990) Review of work on screening germplasm against insect pest resistance (1979–1989) XII Vegetable workshop, 4–7 June 1990, Solan, Himachal Pradesh, India

Sivaprakasam N (1996) Ovipositional preference of *Helicoverpa armigera* to tomato cultivars. Madras Agric J 83:306–307

Sivaprakasam N (1988) Studies on the tomato fruit borer, *Heliothis armigera* Hubner. M. Sc. Dissertation, Tamilnadu Agrl. Univ., Coimbatore, India

Steffens JC, Walters DS (1991) Biochemical aspects of glandular trichome mediated insect resistance in the Solanaceae. pp 136–149. In: Hedin PA (ed) Naturally occurring pest bioregulators. ACS Symp. Ser. 449, American Chemical Society, Washington DC

Tandon PL, Bakthavatsalam N (2003) Relative susceptibility of tomato, *Lycoperiscon esculentum* (Mill.) genotypes to American serpentine leaf miner, *Liriomyza trifolii* (Burgess) (Diptera: Agromyzidae). Insect Environ 8(4):172–173

Thakur SS, Chandel KS, Kashyap NP (1998) Field evaluation of tomato varieties against *Helicoverpa armigera* (Hubner) in the higher hills of Himachal Pradesh. Insect Environ 4:51–52

Todzhaer ShT, Ruzmetor P (1993) Harmfulness of the cotton moth on tomato plants. Zashchita Rastenii (Moskva) 12:33

Wamberg C, Christensen S, Jakobsen I (2003) Interaction between foliar-feeding insects, mycorrhizal fungi, and rhizosphere protozoa on pea plants. Pedobiologia 47(3):281–287

Williams WG, Kennedy GG, Yamamoto RT, Thacker JD, Bordner J (1980) 2-Tridecanone: a naturally occurring insecticide from the wild tomato *Lycopersicon hirsutum f. glabratum*. Science 207:888–889

Role of Parasitoids and Predators in the Management of Insect Pests

Chandish R. Ballal and Abraham Verghese

Abstract

With increasing hazards due to chemical/synthetic pesticides, the only answer to mitigate these ill-effects is use of safe alternatives. Amongst them, use of natural enemies comprising of parasitoids, predators, entomopathogens, etc. as biological control agents is the most effective, environmentally sound and cost-effective pest management approach to control insect pests. It is anticipated that biological control will play an increasingly important role in integrated pest management (IPM) programs as broad-spectrum pesticide use continues to decline. Moreover, biological control is a cornerstone of organic farming, and the production of organic commodities in developed countries like the USA continues to increase at roughly 20% per year (USDA-ERS 2002). Organic farming is no longer considered a cottage industry since retail sales hit $ 14.6 billion in 2006. For a given arthropod pest or weed, a pool of natural enemies often exists which consists of vertebrates, invertebrates and microorganisms. The fundamental problem in applied biological control is to select an appropriate species or combination of species from this pool that will bring about the desired level of pest suppression with minimal impact on non-target species.

Keywords

Predators · Parasitoids · Entomopathogens

C. R. Ballal (✉) · A. Verghese
National Bureau of Agricultural Insect Resources (NBAIR) (erstwhile Project Directorate of Biological Control) (ICAR), H. A. Farm P. O., Bellary Road, Bangalore, 560024 Karnataka, India
e-mail: chandishballal.nbaii@gmail.com

Introduction

Integrated Pest Management (IPM) programs based largely on biological control are of great benefit to agriculture, the quality of rural life and the consumer. Reductions in insecticide, acaricide and herbicide applications should allow farmers

A. K. Chakravarthy (ed.), *New Horizons in Insect Science: Towards Sustainable Pest Management*,
DOI 10.1007/978-81-322-2089-3_28,

to reduce production costs and make adjustments for a more sustainable agriculture. Reduced pesticide use will enhance the quality of rural life by decreasing ground and surface water contamination, reducing effects on non-target species (including wildlife) and increasing safety of farm workers and other rural residents. These benefits also accrue at the interface of urban and agricultural environments, where there is an increasing opposition to pesticide use by stakeholders. The reduction in pesticide residues in food is also desirable, although controversy remains over the extent and public health significance of such residues. Background information and justification: Despite many advances in recent years, our practical and conceptual understanding of success and failure in applied biological control fall short of meeting certain current and future requirements. For example, in classical biological control, the rate of establishment of natural enemies is relatively low in the case of arthropod pests (ca. 34 %) (Kimberling 2004); further research into the genetics and ecology of colonization is clearly warranted. In the future, classical biological control should ideally be able to predict (1) the appropriate species (or biotype) or combination of species (and/or biotypes) to release for control of a target pest in a given situation; and (2) the environmental impact resulting from the introduction of an exotic enemy. Non-target impacts to plants or insects from bio-control agents are of great concern to conservation biologists, environmentalists and federal agencies.

More than one-and-half million insect species occur in this world, out of which only about 15,000 (1.0 %) have attained the status of pests while the others, many of which have pestilent potential, remain at low levels. One of the major reasons for the secondary status of such insects is the perpetual regulatory action exerted on them by their natural enemies. This in itself reflects the great potential of biological control, which can be exploited for management of some of our major pests, diseases and weeds by restoring the natural balance through purposeful human intervention. Such an approach could be classical biological control for invasive species, or generally by augmentation or conservation for indigenous pests.

A worldwide review reveals that there have been altogether 120 successful cases of classical biological control of insect pests of which 42 have been completely controlled, 40 substantially controlled and 30 partially controlled. These include pests, diseases and weeds. There are also a number of successful cases by augmentation of natural enemies in several countries. India is rated as one of the top 10 countries in the world in the area of biological control.

In India, innumerable attempts have been made to augment the populations of promising indigenous natural enemies like trichogrammatids, bethylids, chrysopids, ladybird beetles, nuclear polyhedrosis viruses, etc. to control pests of sugarcane, cotton, coconut, coffee, grapevine, tomato, sunflower, etc. To support such augmentative programmes, mass-production of natural enemies is a necessity. Thus, commercial production of biocontrol agents has a great potential which has already received considerable attention in recent years.

Where success has been achieved in classical biological control, the underlying ecological mechanisms are not always clear. After 100 years of effort, we still do not fully understand the mechanisms by which a successful natural enemy operates in nature, or why a particular organism is successful in one situation and unsuccessful in another. Basic research in augmentation and conservation of natural enemies is also needed. In augmentation, we urgently need a coherent theory of inundative/inoculative release as well as basic efficacy data in order to more readily incorporate commercially available predators and parasitoids of arthropod pests into IPM systems. The genetics of mass production must be evaluated experimentally so that quality control procedures can become a regular practice in the commercial production of natural enemies. Advances in the nutrition of parasitoids and predators are needed. Continued commitment to conservation of natural enemies is required, including innovative ways of integrating pesticides and cultural controls with key natural enemy species. Global warming has now been accepted as a serious threat to our natural and agroecosytems. It will be imperative that biological control scientists watch for the effects of climate change on arthropod pests that have been kept in check by natural enemies. Products of biotechnology designed for pest control must also be assessed and incorporated (where appropriate) into IPM programs. In the past five years, scientists have examined interactions between transgenic crops and biological control species, and these studies will increase as more such crops are approved. Finally, biological control scientists are providing management professionals with the sustainable and effective tools with which to manage the relentless pressure of invasive species on natural and agricultural ecosystems.

Exotic pests continue to arrive and many of these will become permanently established. For such pests, the use of classical biological control should remain a high priority. At the same time, our IPM programs must be continuously evaluated, refined and adjusted in response to changes in newer and more specific control technologies and production practices.

Transcending the coordination and cooperation on a given pest is an important shared need for advances in regulatory policy, general methodologies for release and evaluation of natural enemies, and the need to develop sound ecological theory concerning pest population dynamics, predator–prey interactions, and the genetics of colonization in biological control. The advancement in agricultural technology has brought about remarkable changes in the agricultural sector. These changes have been accompanied by excessive use of pesticides. Worldwide, there are 500 species of resistant insects, mites and ticks compared with only 25 in 1955, coupled with this has been the well-publicized environmental effects, such as toxic residues on produce, destruction of beneficial insects and other non-target organisms, and human poisoning. The World Health Organisation (WHO) estimates that worldwide over a million people are poisoned with pesticides each year and up to 2 % of cases may prove fatal. At this juncture, biopesticides offer an alternative method of control that do not seem to provide the rapid development of resistance in the field, leave little, or no toxic residues and are generally harmless to beneficial insects and other non-target organisms.

Parasitoids and predators can be conserved, preserved and multiplied under laboratory conditions or in commercial production units for field release against target pests or diseases. A major benefit in the use of biopesticides is that they are safe for use by human beings and there are no reports regarding hazards caused due to the use of bio-pesticides, while there are innumerable instances on poisoning due to or non-target effects of chemical pesticides.

Developing countries would be highly benefited through development, exploitation and use of parasitoids and predators. The production, sale and use of biopesticides in a developing country can provide local employment opportunities, reduce health risks and costs due to chemical poisoning and environmental damage, improve export earnings through reducing chemical residue levels on export commodities; and in addition to this there are the benefits obtained through the extra control achieved by preserving natural enemies in crop systems and by maintaining indigenous biodiversity.

Biointensive pest management modules have been developed by the Project Directorate of Biological Control (PDBC) (now NBAIR) for management of pests on cotton, sugarcane, rice, citrus and several other crops (Singh 1996; Rabindra and Ballal 2002). These modules lay emphasis on release of biocontrol agents like parasitoids, predators and pathogens and reducing chemical pesticide applications to the possible extent. The pre-requisite of any biocontrol programme is to have a large-scale supply of beneficial agents. Today, in our country, there is a great demand for biocontrol agents, but the major problem is with respect to availability of good quality bioagents at the required place and time.

In India, several parasitoids and predators have been identified, evaluated and recommended for field releases against agricultural pests (Singh 2001). Technologies are available for the production and use of parasitoids and predators (Rabindra et al. 2003; Singh 2002). There are several success stories in the field of biosuppression of crop pests in our country

(Singh 1996). Biological control has gained maximum acceptance amongst sugarcane farmers of India through use of *Trichogramma* species

Biological control is now considered as the most important component of an integrated pest management strategy and typically involves an active human role. Predators, parasitoids and pathogens are utilized in biological control attempts against insect pests and in this chapter we deal with the utility of parasitoids and predators in biological control. Parasitoids have been used in biological control more than any other type of agent. A successful parasitoid should have a high reproductive rate, good searching ability, host-specificity, be adaptable to different environmental conditions and be synchronized with its target host (pest). No parasitoid has all these attributes, but the search should be for one with several of the above characteristics. Parasitoids are generally utilized in three overlapping types of biological control: conservation, classical biological control (introduction of natural enemies to a new locale) and augmentation. Conservation of natural enemies is probably the most important, readily available, generally simple and cost-effective. The role played by natural enemies in nature becomes evident when insecticide use is stopped or reduced. To tackle exotic pests (at times even native pests), we may have to turn to classical biological control. Unfortunately, classical biological control does not always work, the reasons for failure may include the release of too few individuals, poor adaptation of the natural enemy to environmental conditions at the release location and lack of synchrony between the life cycle of the natural enemy and the pest. The third type of biological control involves the supplemental release of natural enemies which could be inoculative (relatively few natural enemies released at a critical time of the season) or inundative (millions may be released). Habitat manipulation could be a useful approach, wherein the cropping system may be modified to favour or augment the natural enemies. Now potential parasitoids which are amenable to mass production are being reared and marketed by several insectaries, both Government and Private. These are being released against several crop pests. Success with such releases requires appropriate timing, dosage and sufficient number of releases.

With the introduction of exotic pests like the scolytid coffee berry borer, *Hypothenemus hampei* (Ferrari), coconut mite, *Aceria guerreronis* and serpentine leaf-miner, *Liriomyza trifolii* and also some of the indigenous pests like *Helicoverpa armigera* and *Spodoptera litura* becoming increasingly more serious, biological control should serve as a major component of IPM.

Indigenous Parasitoids

A successful parasitoid should have a high reproductive rate, good searching ability, host specificity, be adaptable to different environmental conditions and be synchronized with its host (pest). No parasitoid has all these attributes, but those with several of the above characteristics will be more important for use in suppressing pest populations.

In nature, several parasitoids been observed to be potential bio-agents of serious crop pests. Over a dozen parasitoids have been recorded on the citrus mealybug *Nipaecoccus viridis*. However, *Anagyrus dactylopii* was dominant, parasitizing up to 90% in the field (Ali 1957; Subba Rao et al. 1965).

On cabbage, cauliflower and other cole crops, diamondback moth (DBM), *Plutella xylostella* is a major pest. At Anand, Yadav et al. (1975) recorded up to 72% parasitism by *Cotesia plutellae*. In Karnataka and Tamil Nadu, *C. plutellae* was known to cause up to 80% parasitism (Jayarathnam 1977; Nagarkatti and Jayanth 1982). In the Nilgiris, *Diadegma semiclausum* provided parasitism ranging from 2.32 to 68% (Chandramohan 1994).

The potential indigenous larval parasitoids recorded on *H. armigera* in the pigeonpea and chickpea ecosystems are the ichneumonid early larval parasitoids *Campoletis chlorideae*, *Eriborus argenteopilosus* and tachinid late larval/larval-pupal parasitoids, *Goniophtalmus halli*, *Senometopia* (*Carcelia*) *illota* and *Palexorista laxa* (Bilapate et al. 1988). Indigenous natural enemies play an important role in the integrated pest management of rice.

Indigenous Predators

In India, several predators have been identified as potential biocontrol agents. For instance, more than 60 arthropod species have been recorded as predators of *Helicoverpa armigera* (Hübner). The important predators found feeding on *H. armigera* in India are chrysopids, anthocorids, ants, coccinellids and spiders (Manjunath et al. 1989; Duffield 1993, 1995). Chrysopids form an important group of predators. A number of studies have been conducted on the biology, population ecology, feeding potential and rearing of the potential ones such as *Chrysoperla zastrowi sillemi* (Esben-Petersen), *Mallada boninensis* (Okamoto), *Mallada astur* (Banks) and *Apertochrysa sp.* (Krishnamoorthy and Nagarkatti 1981; Patel et al. 1988; Singh et al. 1994; Bakthavatsalam et al. 1994).

The important indigenous coccinellids include *Coccinella septepunctata* Linnaeus, *Scymnus coccivora* Ayyar, *Chilocorus nigrita* Fabricius, *Cheilomenes sexmaculata* (Fabricius) and *Brumoides suturalis* (Fabricius). Amongst syrphids, the important ones include *Ischiodon scutellaris* (Fabricius), *Paragus serratus* (Fabricius) and *Paragus yerburiensis* Stuckenberg.

Aphidophagous coccinellid, *C. septempunctata* is more abundant in areas with low average temperature viz., northern parts of India. It plays important role in natural suppression of aphids like *Myzus persicae* (Sulzer), *Brevicoryne brassicae* (Linnaeus) and *Lipaphis erysimi* (Kaltenbach) infesting *rabi* oilseeds and cole crops. Similarly, syrphids like *I. scutellaris* and *Paragus* spp. are also found in very high numbers feeding on these aphids. *C. sexmaculata*, on other hand, is more abundant in warmer areas of southern India and it keeps aphid like *Aphis craccivora* Koch, infesting groundnut and pulses, at lower ebb during summer and *kharif* season.

Mass production techniques for these coccinellids (Joshi et al. 2003) and syrphids (Joshi et al. 1998) have been developed at the Project Directorate of Biological Control, Bangalore and are being multiplied throughout the year. However, there is need to evaluate these natural enemies on large-scale, either in open fields or at glasshouses.

Amongst indigenous coccidophagous coccinellids, *C. nigrita* has been utilized through inundative release, not only against *Melanaspis glomerata* (Green) but also on several other diaspine scales including red scale of citrus (Singh 1994). Other important coccinellids in this group are *Pharoscymnus horni* (Weise) and *S. coccivora*. These two play important role of assisting two major coccinellids viz., *C. nigrita* and *C. montrouzieri*, respectively in different fruit crops. By virtue of their small size, they are able to enter leaf sheath and crevices of bark, where crawlers of coccids generally reside, and feed on them at early stage of crop infestation.

For leaf and plant hoppers, colonization of mirid predator *Cyrtorhinus lividipennis* for which now rearing technique is available (over 1500 predators could be reared on 1 cc *Corcyra* egg), has proved to be effective, if releases are carried out @ 100 mirid bugs or 50–75 eggs/m^2 at 10 day interval. Weeds like *Cyperus sp.* help in off-season survival of mirid bug through harbouring plant hoppers. Predation by mirid bug was more on BPH resistant rice variety PTB 33. The presence of any combination of 3 nos./hill of spider *Lycosa preudoannulata, Oxyopus javanus* and *Tetragnatha sp.* checked the population of BPH and WBPH.

The coccinellid predator, *Cryptolaemus montrouzieri* though exotic has established well. It has proved to be very effective against the grape mealy bug, *Maconellicoccus hirsutus* (Singh 1989). Release of 10 beetles per vine could effectively suppress the grape mealybugs in about 75 days of release (Mani and Thontadarya 1988a). Dichlorvos, chlorpyriphos and all the commonly used fungicides at recommended concentrations are safe to all the stages *of C. montrouzieri*, thus allowing the combined use of *Cryptolaemus* with the above pesticides in the pest management programmes (Mani and Thontadarya 1988b; Babu 1986).

Amongst the different anthocorid predators recorded in other countries, *Orius* spp. appear to be the most promising, especially against thrips; examples being *Orius sauteri, Orius majusculus, Orius laevigatus* and *Orius insidiosus*. In India, anthocorids have been recorded as potential bioagents of different species of thrips in various ecosystems. But, systematic work is lacking in our country on the seasonal occurrence of the different potential anthocorid predators in our country. Information is lacking on the extent of control of thrips exerted by the natural populations of anthocorid predators in the different agroecosystems. *Orius* spp. are the most common anthocorids which have been collected from different crop ecosystems. *Orius tantillus* and *O. maxidentex* are the most common species collected. (Ballal and Gupta 2011)

Production techniques

Production techniques are also available for some potential parasitoids like Trichogrammatids, *Leptomastix dactylopii, Copidosoma koehleri, Telenomus remus*, etc. and predators like *C. z. sillemi, Scymnus coccivora, Pharoscymnus horni, Curinus coeruleus, Coccinella septempunctata, Cheilomenes sexmaculata, Chilocorus nigrita, Brumoides surturalis, Cardiastethus exiguus*, etc. (Singh et al. 2001b; Ballal et al. 2012; Ballal and Gupta 2011).

Conservation Biological Control

The conservation of natural enemies is probably the most important and readily available biological control practice available to growers. Natural enemies occur in all production systems, from the backyard garden to the commercial field. They are adapted to the local environment and to the target pest, and their conservation is generally simple and cost-effective. With relatively little effort the activity of these natural enemies can be observed. For example, parasitized aphid mummies are almost always present in aphid colonies. These natural controls are important and need to be conserved and considered when making pest management decisions. In many instances the importance of natural enemies has not been adequately studied or does not become apparent until insecticide use is stopped or reduced. Often, the best we can do is to recognize that these factors are present and minimize negative impacts on them. If an insecticide is needed, every effort should be made to use a selective material in a selective manner.

Conservation biological control practices such as refuges for natural bioagents, conserving weed plants harbouring predators and egg parasitoids, use of safer pesticides, judicious and selective use of non-persistent pesticides, strip treatment, spot treatment, etc. have been found to be effective conservation techniques in several crop ecosystems (Singh 2002). Use of kairomones, synomones, pheromones, adjuvants, etc. to increase the searching ability and retention of parasitoids, build-up population of biocontrol agents by providing artificial structures, food, alternate host, suppression of ants, etc., provision of grain sorghum in cotton plot, which serves as a source for natural enemies, etc are some conservation techniques.

Habitat manipulation techniques (to improve the population and performance of natural enemies) are easily incorporated into home gardens and even small-scale commercial plantings, but are more difficult to accommodate in large-scale crop production. There may also be some conflict with respect to pest control because of the difficulty in targeting the pest species as the refuges may be used by the pest insects as well as natural enemies. Habitat manipulation involves altering the cropping system to augment or enhance the effectiveness of a natural enemy. Many adult parasitoids benefit from sources of nectar and the protection provided by refuges such as hedgerows, cover crops and weedy borders. Mixed plantings and the provision of flowering borders can increase the diversity of habitats and provide shelter and alternative food sources. They are easily incorporated into home gardens and even small-scale commercial plantings, but are more difficult to accommodate in large-scale crop production. There may also be some conflict with pest control because of the difficulty of targeting the pest species as the refuges could be used by the pest insects as well as natural enemies.

Natural enemies may be conserved by using insecticides or formulations which are least harmful and by timing applications to reduce the impact on beneficial arthropods. Natural enemy populations may be enhanced by increasing the diversity of plant species in the vicinity of the crop, changing cultural practices to ensure continuous availability of hosts and by providing alternative food sources (Pawar 1986). Ballal and Singh (2001) reported that non-intervention and thus conservation of natural enemies to be the best strategy for *Helicoverpa armigera* management in the sunflower ecosystem.

Classical Biological Control

Biological control agents that are not host specific may pose threats to at-risk species and constraints have been applied to the types of organisms that may be used. The requirement for increased host specificity means exotic polyphagous predators are less appropriate for introduction and more research emphasis has been placed on parasitoid species (Goldson et al. 1994).

In many instances, the complex of natural enemies associated with an insect pest may be inadequate. This is especially evident when an insect pest is accidentally introduced into a new geographic area without its associated natural enemies. To obtain the needed natural enemies, we turn to classical biological control.

Classical biological control has proved its potential in our country with respect to some of the introduced natural enemies. The exotic natural enemies which have proved effective through augmentation include egg parasitoid *Telenomus remus* (Origin: Papua New Guinea) against *S. litura* infesting tobacco, the egg larval parasitoids *Chelonus blackburni* (Origin: Hawaii) and *Copidosoma koehleri* (Origin: Australia) against potato tuber moth *Phthorimaea operculella*. Not all introduced parasitoids were successful in managing the target pests.

Exotic parasitoids that have successfully established in our country include the encyrtids *Encarsia perniciosi* and *Aphytis diaspidis* for control of San Jose scale, *Quadraspidiotus perniciosus* and similarly, *Leptomastix dactylopii* against citrus mealybugs.

L. dactylopii introduced from the West Indies in 1983 is a fairly specific parasitoid of *Planococcus citri*, possessing excellent host searching ability. Field release of *Leptomastix* resulted in its establishment in mixed plantations of citrus and coffee, and also in citrus orchards in several parts of Karnataka, resulting in control of *P. citri* within 3–4 months. No insecticidal sprays were required subsequently for the control of *P. citri* in the following season (Manjunath 1985; Krishnamoorthy and Singh 1987; Nagarkatti et al. 1992).

Three strains of *E. perniciosi* viz., Californian, Russian and Chinese, were introduced for the control of *Q. perniciosus*. In addition, *A. diaspidis* (origin: Japan) was introduced from California. All the strains could establish and the Russian strain of the parasitoid gave 89% parasitism in Himachal Pradesh. *A. diaspidis* in combination with *E. perniciosi* gave 86.5% parasitism. In Kashmir, the Russian and Chinese strains appeared to be superior. American and Chinese strains of *E. perniciosi* were also released in the Kumaon hills of Uttar Pradesh; the population of the pest was reduced by about 95%. In Kashmir, releases of *E. perniciosi* and *Aphytis proclia* resulted in an increase of parasitism from 8.9 to 64.3%. Studies on the biology of *E. perniciosi* revealed that the multiplication rate of the parasitoid was over 10 times. In apple, release of *E. perniciosi* or *A. proclia* @ 2000/ infested tree gave effective control of San Jose scale (Rao et al. 1971; Singh 1989).

In the tobacco, cabbage and cauliflower ecosystems, the exotic parasitoid *T. remus* Nixon has proved to be potential parasitoid for the management of *S. litura.* Tobacco IPM has been successfully field demonstrated in farmers' field in Andhra Pradesh (PDBC-ICAR, 1999–2000).

Combination of exotic parasitoids, *C. blackburni* and *C. koehleri* with *Bt* products and granulosis virus has been found effective in managing the potato tuber moth, *P. operculella* in potato fields and in storage (Singh 1994).

For the management of the coffee berry borer, *H. hampei,* the bethylid parasitoids, *Prorops nasuta* and *Cephalonomia stephanoderis* were imported from Mexico in 1995 and the eulophid parasitoid, *Phymastichus coffea* and another consignment of *P. nasuta* from Colombia in 1999. The first test field releases were made in January–February, 1996. Though recoveries of *P. nasuta* and *C. stephanoderis* could be made, *P. nasuta* could not establish both in the laboratory and in the field in spite of repeated field releases. *C. stephanoderis* has established in several areas of Kodagu district, Wyanad and Lower Palanis. Carry over of the parasitoid from one season to the other has also been observed. More than 10,000 females of *P. coffea* have been released and establishment in small numbers has been observed (Sreedharan et al. 2001).

The spiraling whitefly, *Aleurodicus dispersus*, a native of the Caribbean region and Central America, probably came to India from Sri Lanka or the Maldives. It was first reported in 1993 from Kerala and later from other parts of peninsular India and the Lakshadweep islands. The pest is highly polyphagous and has been recorded on 253 host plants in India. Two aphelinid parasitoids, *Encarsia guadeloupae* and *E. sp.* nr. *meritoria*, have been fortuitously introduced together with the host into India. With the accidental introduction of both species of *Encarsia* into India, there has been a perceptible reduction in the population of *A. dispersus* (Ramani et al. 2002).

One of the most recent and significant success stories in the field of classical biological control is that of the excellent control of papaya mealy bug through introduction and field releases of exotic natural enemies. The papaya mealybug *Paracoccus marginatus* W & G was first recorded on papaya plants from Coimbatore in 2008 and later spread to different states viz. Kerala, Karnataka, Maharashtra and Tripura. Chemical pesticides could not give permanent relief and repeated use of chemical pesticides resulted in toxicity hazards, pollution and harmful effects on non-target beneficials. The natural enemies existing in nature like *Spalgis epius, Cryptolaemus montrouzieri* and *Scymnus coccivora* could not keep the papaya mealy bug population under check. NBAII imported three species of parasitoids *Acerophagus papaya, Pseudoleptomastix mexicana* and *Anagyrus loecki* (from USDA-APHIS at Puerto Rico), which are known to effectively suppress the papaya mealy bug in its native range. The parasitoids could be successfully multiplied and supplied to stake-holders all over the country. Inoculative releases of the parasitoids were also made in farmers' fields in different villages. The parasitoids could successfully establish in all the areas of release and suppress the papaya mealybug infestation on different crops (Shylesha et al. 2010). NBAII also trained entomologists/plant protection officials from SAUs, ICAR Institutes, KVKs, CIPMCs, Government Biocontrol Laboratories and Central Sericulture Research and Training Institute on the mass production, field release and conservation of the parasitoids.

Production and Utilization of Parasitoids and Predators

Besides introducing suitable exotic natural enemies, efforts should be made to develop more efficient and cost-effective production and utilization of indigenous natural enemies (Rabindra et al. 2003; Jalali et al. 2003). Success with field releases of natural enemies requires appropriate timing, release of the correct number of natural enemies per unit area or depending on pest density and release of quality bioagents. In many cases, the most effective release rate has not been identified as it will vary depending on crop type and target host density. Table 1 lists some of the parasitoids and predators, which could be released for the management of some major pests on different crop ecosystems.

Biological control through augmentation has gained maximum acceptance amongst sugarcane farmers of India. Use of *T. chilonis* has been effectively utilized for the management of sugarcane borers. Sugar mills have their own co-operative parasitoid production units and have contributed in a big way in adoption of biocontrol (PDBC 2000–2001; Singh et al. 2001a). Augmentation of the tachinid parasitoid *Sturmiopsis inferens* has decreased the population of shoot borers in Tamil

Table 1 Some biological control systems utilizing parasitoids

Crop/Pest	Biotic agents	Dosage per ha	Frequency of application
Sugarcane			
Chilo spp.	*Trichogramma chilonis*	50,000	Every 10 days, 8 times starting on 30-day-old crop for shoot borer and 60 days for other borers or during egg laying period
Pyrilla perpusilla	*Epiricania melanoleuca*	2–3 egg masses or 5–7 cocoons in 40 selected spots/ha	The releases are initiated before the onset of rainy season
Rice			
Scirpophaga incertulas & *Cnaphalocrocis medinalis*	*Trichogramma japonicum* *T. chilonis*	100,000	30, 37 and 44 days after transplanting (DAT)
Cotton			
Helicoverpa armigera, *Earias* spp., *Pectinophora gossypiella*	*T. chilonis*	150,000	Weekly 6 times starting from 40th day after planting or during the egg laying period
Tobacco			
Spodoptera litura	*Telenomus remus*	120,000	Five times at weekly interval
Coconut			
Opisina arenosella	*Goniozus nephantidis*	3000 adults	Need based or for each generation
	Cardiastethus exiguus	50 adults/tree	To coincide with egg or freshly hatched larval stage of the pest
Apple			
Eriosoma lanigerum	*Aphelinus mali*	1000 adults or mummies/ infested tree	Once, as soon as infestation is noticed
Quadraspidiotus perniciosus	*Encarsia perniciosi*	2000 adults/infested tree	Once, in spring
Cydia pomonella	*Trichogramma embryophagum*	2000 adults/tree	Releasing at weekly interval
Citrus			
Planococcus citri	*Leptomastix dactylopii*	3000 adults	Need based; under expert supervision
Tomato			
Helicoverpa armigera	*Trichogramma brasiliense* *T. pretiosum/T. chilonis*	50,000	Weekly interval/six times from 25th day after transplanting or during egg laying period

Nadu and the parasitoid permanently colonized in some pockets (Singh 1994). Similarly, inundative releases of *Isotima javensis* has given good results in the control of top borer, *Scirpophaga excerptalis* in north India.

There are several potential parasitoids in nature which are important mortality factors of major pests. On citrus butterfly *Papilio demoleus* Linnaeus, egg parasitoid *Trichogramma chilonis* Ishii parasitized up to 76% and *Telenomus sp.* nr. *incommodus* 78% in February (Krishnamoorthy and Singh 1988; Jalali and Singh 1990). *Distatrix papilionis* is the dominant parasitoid of caterpillars and *T. chilonis*, *T. incommodus* and *D. papilionis* caused a cumulative parasitism of 88% (Krishnamoorthy and Singh 1988). *T. chilonis, Melalophacharops sp.* and *D. papilionis* could be utilized for the biological suppression of butterflies attacking citrus. The eggs of fruit sucking moth, *Othreis fullonia* are successfully parasitized by *T. chilonis*, which suggests the possibility of utilizing *T. chilonis* for the control of this pest (Dodia et al. 1986).

Notable success has been achieved in the biosuppression of the hopper *Pyrilla perpusilla* in some states by the colonization/redistribution of the lepidopteran parasitoid, *Epiricania melanoleuca*. Misra and Pawar (1984) reported that this parasitoid when released @ 400,000–500,000 eggs or 2000–3000 cocoons/ha in eastern UP, West Bengal, Orissa, Karnataka, Kerala, Maharashtra, Rajasthan, Andhra Pradesh and Madhya Pradesh gave complete control of the pest. Pawar (1979) reported that in July–September, if 20–60 % parasitism of nymphs and adults are recorded there is no need to panic even if outbreak like situation is noticed.

Indigenous parasitoids play a major role in the management of the coconut black-headed caterpillar in the coconut ecosystem. Field release of the three stage specific *Opisina arenosella* parasitoids *viz Goniozus nephantidis, Elasmus nephantidis* and *Brachymeria nosatoi* at fixed norms and intervals in a heavily infested coconut garden (2.8 ha) for a period of 5 years resulted in highly significant reduction in *Opisina* population (Sathiamma et al. 2000). Follow-up observations revealed that even after 3 years no build-up of the pest was noted in the released site. The anthocorid predator *Cardiastethus exiguus* and *G. nephantidis* have been observed to be highly amenable to mass production and they have also proved to be highly effective against the egg and larval stages of *O. arenosella* as indicated in the recent field trials conducted at Kerala and Karnataka (Venkatesan et al. 2008).

The sugarcane woolly aphid, *Ceratovacuna lanigera,* was observed as a serious pest of sugarcane and reported in outbreak proportions from western and southern India (Rabindra et al. 2002; Joshi and Viraktamath 2004). The parasitoids which were recorded on this pest in Nagaland included *Aphelinus desantisi*, *Encarsia falvoscutellum*, *Diaeretiella rapae*, *Anagyrus sp.* and *Antocephalus sp.* (Tripathi 1995). In Assam, Jorhat *Encarsia flavoscutellum* was observed in abundant numbers parasitising woolly aphids. The heavy incidence of this parasitoid could prevent the further spread of the woolly aphid population. *Dipha* and *Micromus* are potential predators of SWA in nature, which keep the pest population under control.

In the cotton ecosystem, the indigenous parasitoid *Bracon greeni* gave satisfactory control of spotted and spiny bollworms *Earias* spp. in Karnal, Haryana (Khan and Rao 1960) and for the control of *Pectinophora gossypiella* in Coimbatore (Swamiappan and Balasubramanian 1980).

In rice ecosystem, conservation and inundative release of the egg parasitoid *T. japonicum* and *T. chilonis* along with the predator *Cyrtorhinus lividipennis* have given promising results. Weekly releases of *T. japonicum* and *T. chilonis* @ 100,000/ ha starting after a month of transplanting is recommended for the control of stem borer, *Scirpophaga incertulas* and leaf roller, *Cnaphalocrocis medinalis*. A total of three releases for *Rabi* and *Kharif* crops are sufficient. The trials conducted at Tamil Nadu, Maharashtra, Punjab, Assam and Kerala proved that Biocontrol Based Integrated Pest Management (BIPM) was either at par or better than farmers' practice in all the places (PDBC-ICAR 2001–2002).

The BIPM schedule for pest management includes releases of *C. z. sillemi* for sucking pests. This schedule was successful in Karnataka, Maharashtra and Gujarat

At the erstwhile Project Directorate of Biological Control (now NBAIR), the two indigenous early larval parasitoids of *H. armigera*—*C. chlorideae* and *E. argenteopilosus* could be continuously reared on alternate laboratory hosts and several basic studies conducted (Venkatesan et al. 1995; Ballal et al. 2000; Ballal et al. 2001b).

Geographical strains of *C. chlorideae* were obtained from different parts of the country and their biological parameters and performance evaluated and it was found that the Sehore strain was most efficient (Ballal and Ramani 1994). Variations were observed in the performance of *C. chlorideae* populations collected from different crop ecosystems. The lab-reared parasitoids which were originally from the pigeonpea ecosystem could not efficiently parasitise *H. armigera* larvae from the cotton ecosystem, whereas the parasitoids from the cotton ecosystem were capable of parasitizing more than 40 % of the larvae of cotton ecosystem (Ballal et al. 2001a). The studies clearly indicated that the performance of *C. chlorideae* is largely governed by the host plants on which the pest is found. Bajpai et al. (2002) reported that on chickpea plants, the chemical cues released during feeding by the

H. armigera was essential for *C. chlorideae* to be attracted to the infested plants and to induce parasitism. Parasitism was also governed by host plant variety (Ballal and Gupta 2003). This was also true for *Trichogramma* spp. (Ballal and Singh 2003).

In a successful attempt to bridge the gap between research and commerce through a path breaking work involving studying for 325 generations, a strain of *T. chilonis* with physiological tolerance to 0.07 % of endosulfan has been developed for the first time. The strain is distinctly superior (98 % control) to endosulfan spray (72.0 %) in the control of cotton bollworm. This strain has been transferred to M/s.Excel Industries Limited, Mumbai. It is multiplied on a large scale and distributed to the farmers under the trade name "Endogram". Endogram technology is registering a gradual spread in different states. In three years, 29700 ha of cotton and vegetables crops were treated with endogram in six different states.

This strain has been further developed for multiple tolerances to the recommended dosages of monocrotophos (0.05 %) and fenvalerate (0.002 %). Multiple insecticides tolerant strain of *T. chilonis* is for use on cotton, vegetables and rice, etc. where several insecticides. The strain is tolerant to endosulfan, monocrotophos and fenvalerate. It also shows moderate to high cross-tolerance for other insecticides too.

A strain of *T. chilonis* which can tolerate a temperature of 36 °C has been developed. High temperature tolerant strain of *T. chilonis* and *T. japonicum* can be utilized during months when temperatures are more than 35 °C. The strain can tolerate the temperatures up to 38 °C. It is useful against the sugarcane shoot borer and top borer and also others pests on cotton and vegetable crops during hot months. High host searching strain of *T. chilonis*, *T. japonicum*, *T. achaeae* and *T. bactrae* can be used on several pests as this strain has better host searching ability and higher fecundity. These strains can be used on a number of crops in milder climatic conditions (Jalali and Singh 1993; Jalali et al. 2006; Ballal et al. 2009b).

Mealy bugs like the common mealy bug (*Planococcus citri*), grape mealy bug (*Maconellicoccus hirsutus*), mango mealy bug (*Rastrococcus iceryoides*), spherical mealy bug (*Nipaecoccus viridis*), striped mealy bug (*Ferrisia virgata*), oriental mealy bug (*Planococcus lilacinus, P. pacificus, P. robustus*) and pineapple mealy bug (*Dysmicoccus brevipes*) cause serious damage and decrease the productivity and marketability of the produce. Some mealybugs have also been able to develop resistance to insecticides.

Cryptolaemus montrouzieri was introduced from Australia into India in June, 1898 for the control of soft green scale *Coccus viridis.* It could not establish on soft green scale. Later, it was reported as a common predator of many species of mealy bugs and to some extent on scale insects in Karnataka (Rao et al. 1971). In 1977, an insectory was established at Central Horticultural Experiment Station, Chethalli, Kodagu, Karnataka for its multiplication. This coccinellid can now be successfully mass produced and field released (Joshi et al. 2003). The production cost of 100 beetles in some private companies is Rs 70 (As per the 2000 price index). This beetle can also be reared using a semi-synthetic diet. A single grub is known to feed about 1500 eggs or 880 nymphs or 30 adult females of *M. hirsutus* (Mani et al. 2014). Now commercial insectaries are also producing and supplying *C. montrouzieri* to the growers. In fruit and plantation crops, the beetles are released @ 5–50 per plant, depending upon the severity of infestation and crop canopy. On each mealy bug infested plant of coorg mandarin, robusta coffee, arabica coffee and san–ramon coffee release of 10, 5, 3 and 2 beetles per plant resulted in reduction of mealy bug population and by 5th week the pest population reduced to negligible level. Beetles were released in 13 mixed planted orchards (citrus and coffee) and satisfactory results obtained. Field releases of *C. montrouzieri* @ 20 adults per tree gave excellent control of *F. virgata, M. hirsutus* and *P. lilacinus* on guava within 50 days in the presence of other local natural enemies. It was also found to be highly effective in suppressing the populations of *M. hirsutus* in grapes within 75 days. The predator was found effective in suppressing the mealy bugs on citrus, guava, grapes, mulberry, coffee, mango, pomegranate, custard apple, ber etc. and

Table 2 Biological control of mealy bugs and scale insects with *Cryptolaemus montrouzieri*

Crop	Species	Place	Result
Araucaria	*Eriococcus araucariae*	Karnataka	Completely wiped out
Brinjal	*Coccidohystrix insolita*	Karnataka	Suppressed
Crotons	*Planococcus minor*	Karnataka	Brings down the population
Ficus	*Chloropulvinaria psidii*	Karnataka	Managed successfully
Hibiscus	*Aphis gossypii*	Karnataka	Suppressed successfully
Jacoranda	*Saissetia hemisphaerica*	Karnataka	Suppressed
Jasmine	*Pseudococcus longispinus*	Karnataka	Kept under check
Mulberry	–	Karnataka	–
Mussaenda	*Orthesia insignis*	Karnataka	Suppressed
Neem	*Chloropulvinaria maxima*	Karnataka	Kept under check
Sapota	*Coccus viridis*	Karnataka	Suppressed
	Planococcus citri	–	–
Tomato	*Planococcus citri*	Karnataka	Suppressed
Ber	*Nipaecoccus viridis, P. lilacinus, P. citri M. hirsutus* and *Drepanococcus chiton*	Karnataka	Suppressed
Chow-chow	*P. lilacinus*	Karnataka	Suppressed
Citrus	*Planococcus citri* and *Nipaecoccus viridis*	Karnataka	Suppressed successfully
Coffee	*Planococcus* spp.	Karnataka	Suppressed successfully
Custard apple	*M. hirsutus, P. citri, P. lilacinus, F. virgata* and *N. viridis.*	Karnataka and Andhra Pradesh	Suppressed
Grapevine	*Maconellicoccus hirsutus* and *Planococcus citri*	Karnataka and Andhra Pradesh	Suppressed successfully
Guava	*Chloropulvinaria psidii Aphis gossypii Drepanococcus chiton Ferrisia virgata Planococcus citri* and *P. lilacinus*	Karnataka and Tamil Nadu	Suppressed successfully
Mango	*Chloropulvinaria polygonata Ferrisia virgata Planococcus citri Rastrococcus iceryoides* and *R. invadens*	Karnataka	Suppressed
Pomegranate	*Siphoninus phyllireae Maconellicoccus hirsutus, Planacoccus citri, P. lilacinus, Ferrisia virgata* and *Nipaecoccus viridis*	Karnataka	Suppressed

green shield scale on sapota, mango, guava, brinjal and crotons in Karnataka. It did not seriously impair the efficiency of local biocontrol agents (Table 2).

The conventional pesticides such as dichlorvos. chlorpyrifos, dicofol, fish oil rosin soap and most of the botanicals and fungicides are safe to *C. montrouzieri*.

C. montrouzieri does not feed on the mealybugs mummified due to parasitisation by *Anagyrus dactylopii*. *C. montrouzieri* in the presence of the encyrtid parasitoid *Anagyrus dactylopii* gave excellent control of *M. hirsutus* in vineyards. Similar control of *Ferrisia virgata* was achieved on guava with *C. montrouzieri* in the presence of the encyrtid parasitoid *Aenasius advena* (Mani et al. 1990). In Karnataka, *C. montrouzieri* although was found on *Dactylopius opuntiae* did not seriously impair the effectiveness of *Dactylopius* in controlling the weed *Opuntia.*

Chrysopids In India, 65 species of Chrysopids belonging to 21 genera have been recorded from various crop ecosystems. Some species are distributed widely and are important natural enemies for aphids and other soft bodied insects. Amongst them, *C. z. sillemi*, *Mallada boninensis*, *Apertochrysa crassinervis* and *Mallada astur* are the most common. The first two have been used in cotton ecosystem for protection from aphids and other soft bodied insects. *C. z. sillemi* has been recorded on cotton, green gram, sorghum, maize, safflower, sunflower and pigeonpea, predating on the pest like safflower aphid, maggots of safflower fruit fly, eggs of pentatomid bugs on green gram, sorghum aphid, eggs of Pyrilla, cotton aphid and leaf hoppers. In Himachal Pradesh, *C. z. sillemi* feeds on woolly aphid *Eriosoma lanigerum* colonies and hibernates in cocoons as prepupae from first week of November to early March.

C. z. sillemi can be multiplied on the eggs of *C. cephalonica.* A monocrotophos tolerant strain of *C. z. sillemi* has been selected by Gujarat Agricultural University, Anand. *C. z. sillemi* is now used extensively all over the country. *C. z. sillemi* is multiplied for commercial use by adopting a two step rearing. In the first step larval rearing, 120 three-day old chrysopid eggs are mixed with 0.75 ml of UV-irradiated *Corcyra* eggs in a plastic container (group rearing). On hatching, the larvae start feeding and on the 3rd day the larvae are transferred to 2.5 cm cubical cells of plastic louvers as the second step individual rearing. Total quantity of *Corcyra* eggs required for rearing 100 chrysopid larvae is 4.25 ml. They can also be produced on semi synthetic diet, which includes the utilization of wastes from other insect production units.

The cost of production and application of *C. z. sillemi* @ 100,000/ ha came to Rs 744, which could be reduced when the production capacity was increased. Attempts are on to reduce the cost involved in field use of chrysopids through manipulation of the dosages. Normally, chrysopids are recommended for use against different crop pests @ 50,000 or 100,000 1st instar larvae/hectare, 4–6 larvae/plant or 10–20 larvae/fruit plant are released. Depending on the situation, two releases are recommended. They are released on the plants along with sawdust, or dropped from the corrugated paper strips.

Anthocorids In India, very few attempts have been made to rear the anthocorid predators. Mukherjee et al. (1971) tried a synthetic diet for the rearing of *X. flavipes* (Reut.). Mass rearing methods have been standardised for four potential anthocorid predators, *Cardiastethus exiguus* Poppius (Ballal et al. 2003a), *Blaptostethus pallescens* Poppius (Ballal et al. 2003b) and *Xylocoris flavipes* (Reuter) (Ballal et al. 2013) and *Orius tantillus* Motshulsky (Gupta and Ballal 2006).

Techniques were not available to mass rear *Orius* spp. in India till recently. At the NBAIR, Bangalore, methods have now been standardised to multiply *Orius tantillus* on different host eggs. Earlier studies had indicated that the progeny production by *Orius maxidentex* when reared on sorghum midge, was 35.10 per female and 23.61 per female when reared on thrips. There are problems associated with continuous multiplication of host insects like thrips and midges. Hence other alternate laboratory hosts eggs were tried. *O. tantillus* could be continuously multiplied for 12 generations on UV irradiated *C. cephalonica* eggs, however, the progeny production was very low, the mean value being 3.1 per female. It is clear that there is a need to improve the diet provided to improve the progeny production. The UV-irradiated eggs of *Sitotroga cerealella* was also tried. *O. tantillus* could be reared more efficiently on *S. cerealella* eggs than on *C. cephalonica* eggs. The optimum temperature regime for multiplication of *O. tantillus* was found to be 24 and 28 °C as progeny production was maximum at these two temperatures, the values being 28.8 and 26.2 per female, respectively.

Laboratory studies were conducted to check the feeding preference of *O. tantillus* on parasitized and un-parasitized eggs of *Helicoverpa armigera.* Results of choice and no-choice tests showed that there was significantly higher preference for un-parasitized eggs in comparison to parasitized eggs, thus indicating that it may be possible to integrate releases of anthocorids and trichogrammatids for biological control of lepidopteran pests/thrips in different crop ecosystems (Gupta and Ballal 2007).

The anthocorid species which have been commonly used for field releases are: *Anthocoris nemoralis* (Fabricius), and *Orius* spp. Anthocorids are now being commercially produced in several countries. *C. exiguus* has been field evaluated against *O. arenosella* and *B. pallescens* against onion thrips and two spotted spider mites on bhendi. Both the anthocorids have proved to be potential predators for field use (Lyla et al. 2006; Ballal et al. 2009a).

Climate Change and Biological Control Though clear evidence is lacking, it is strongly felt that climate change may alter the effectiveness of biological control. Successful biocontrol agents are highly specific to the invasive species they are targeted to control and changes in the climatic factors may alter these inter specific interactions. Tritrophic interactions between plants, herbivorous insects and their natural enemies (predators, parasitoids and pathogens) result from a long co-evolutionary process specific to a particular environment and relatively stable climatic conditions. These tri-trophic interactions would be affected by climatic conditions in diverse ways. Extreme temperatures can affect both pests and their natural enemies. A warmer climate would increase the metabolic rate of insects and their natural enemies. Studies show that metabolic rate and hence burning of resources increases monotopically with temperature, while activity is maximum at intermediate temperatures. Therefore, increased temperature could lead to reduction in longevity and realized fecundity of temperate insect parasitoids, in turn causing decrease in their efficacy. Besides, exposure to stressful temperatures could induce lethal and sub lethal damages to parasitoids, generally decreasing mobility, ability to orient themselves to attractive odours and learning capacities and increasing production of male progeny. The endosymbiont bacteria associated with the parasitoid and host could be suppressed by short exposure to high temperature.

While the effects of global atmospheric changes on vegetation and resulting insect populations ('bottom-up interactions') are being increasingly studied, how these gases modify interactions amongst insects and their natural enemies ('top-down interactions') is less clear. As natural enemy efficacy is governed largely by behavioural mechanisms, altered prey finding and prey defence may change insect population dynamics.

Long term studies on effect of climate change on pests or natural enemies have not been conducted in India. Likely, impacts of any change in climate on populations of pests are manifold. They range from expansion in the geographical range of pests, increased risk of invasion, changes in overwintering patterns, natural enemy-pest interactions, changes in population growth rates, change in crop-pest synchrony, pest control factors and finally changes in pest complexes on spatial and temporal bases. Results obtained through current modelling approaches do not account for all the factors operating. Moreover, it may not be possible to replicate the methodology of other countries in the Indian context, given the wide-ranging socioeconomic conditions and different agro-ecology, and a different approach is needed to tackle the problem. Consolidation of all existing studies done in various parts of the country may act as a foundation, which can be supplemented by incorporation of the vast resources of data from various government agencies (Sehgal et al. 2006)

In India, a number of short term studies have been conducted to investigate the effect of abiotic factors, especially temperature on pests or natural enemies. Laboratory studies have shown that a temperature of 35 °C was detrimental to the different biological stages and adult longevity of *Campoletis chlorideae*, which is a potential indigenous parasitoid of *H. armigera* (Teggeli et al. 2004). Earlier, studies have clearly indicated the adverse effect of temperature on parasitoids. Singh and Ali (2006) reported that minimum and maximum temperatures showed a negative correlation with parasitization of *H. armigera* by *C. chlorideae*. Field studies in *H. armigera* infested chickpea fields in Himachal Pradesh indicated that the activity of *C. chlorideae* ceased when the mean maximum temperature reached above 40 °C Gupta and Desh Raj (2003). In chickpea fields in eastern Uttar Pradesh, parasitic activity of *C. chlorideae* was highest (80.5 %) when maximum and minimum temperatures and relative humidity were 24.5 °C, 8.6 °C and 85 %, and was lowest (22.2 %) when the above parameters were 36.6 °C, 18.5 °C and 85 %, respectively (Pandey et al. 2005).

The abundance of whitefly (*Bemisia tabaci*) and its parasitoid (*Encarsia lutea*) was monitored under agroclimatic conditions of Haryana. The pest population had a positive correlation with temperature, while the population of parasitized pupae showed negative correlation with temperature (Sharma et al. 2004). The infestation of the Oak tasar silkworm, *Antheraea proylei* by two species of tachinid parasitoids viz., *Blepharipa zebina* Walker and *Exorista sorbillans* (causing considerable loss to the oak tasar industry in Manipur) starts from March and reaches a peak during May with rise in temperature. This study indicated that a fair prediction for uzi fly infestation can be made from the prevailing abiotic conditions (Venkatachalapathy et al. 2002).

Khan and Misra (2003) reported that the populations of both the spiders and hoppers in the upland rice ecosystem in eastern UP during kharif were negatively correlated with temperature. Predaceous coccinellids could have varying responses to high temperature. To avoid high temperature, they could enter into diapause during pupal or larval stages, or could hide in land crevices or migrate (Indu and Chatterjee 2006). The activity of predatory fauna (coccinellids, chrysopids and syrphids) on aphids (*A. gossypii*) infesting isabgol (*Plantago ovata*) were studied in a field experiment conducted in Gujarat. The predatory activity of *Coccinella septempunctata*, *C. transversalis* and *Cheilomenes sexmaculata* increased due to increase in temperature (Patel and Borad 2005). *Coccinella septempunctata* and *Ischiodon scutellaris* are the established predators of mustard aphid *Lipaphis erysimi*. Maximum temperature had a significant negative relationship with the aphid population, but was positive for *C. septempunctata* and *I. scutellaris* (Tripathi et al. 2005). The natural parasitism of the predator *C. septumpunctata* by *Tetrastichus coccinellae* (*Oomyzus scaposus*) and the number of parasitoids emerging per coccinellid increased with an increase in temperature, thus adversely affecting the performance of the predator (Singh and Singh 2003).

Commercial Production of Parasitoids and Predators

Standard techniques are now available for the successful production of several parasitoids and predators, which could be followed by commercial insectaries. India's first private insectary, Biocontrol Research Laboratory was established at Bangalore in 1981. Since then numerous companies have come up country-wide, which produce parasitoids, predators, entomopathogens, plant disease antagonists, weed killers, etc. The PDBC has compiled an infobase on bioagent producers in India (private and government). As per this infobase (Biswas et al. 2000) and Singh (2002), there are 128 organizations producing bioagents.

Biological control workers have to face several major direct technical constraints in the process of production. These problems get further compounded by artificial selection forces and the conflicting requirements for natural enemies in a mass production programme. These technical obstacles include: non-availability of long term storage techniques for the most important alternate laboratory host insect *Corcyra cephalonica* and also for Tricho cards and mechanized application technology of parasitoids and predators, not available, problems associated with male-biased sex-ratio in the laboratory cultures and maintenance of cultures during summer and winter due to unfavourable temperature and humidity conditions, cannibalizm in chrysopids and in some coccinellid larvae which necessitates individual rearing, in vivo rearing of predators as it necessitates continuous production of host insects and host plants, infestation by *Bracon* and mites in *Corcyra* culture, disease insect cultures, effective in vitro mass production techniques for natural enemies on artificial diets, need for rearing at relatively high prey densities in the case of predatory mites, leading to high costs, occurrence of microbial contaminants fungi, bacteria, viruses, protozoa and nematodes in insect cultures leading to high mortality, prolonged development, diminutive adults, wide fluctuations in the quality of insects and direct pathological effects, lack of techniques that prevent selection pressures leading to genetic deterioration of the mass-produced natural enemies and loss of effectiveness, lack of techniques that prevent behavioural changes and/or the loss of vigour through poor nutrition when reared on alternative hosts or artificial diets, non-availability of commercial artificial diet for rearing of entomophagous insects, lack of automation to produce low-cost products and lack of good standard measures for evaluating the performance of mass reared biological control agents.

Future Thrusts

- Population dynamics of the pest and the natural biological control agents to be studied in detail before introducing an exotic natural enemy
- Standard production procedures to be developed and followed by all insectaries
- Strict quality control protocols to be followed by all insectaries
- Uniform release and evaluation techniques to be followed by all biological control researchers, which would enable the comparison of results
- In-depth studies on tri-trophic interactions between the pest, parasitoid and host plant
- Large-scale field trials to evaluate potential parasitoids in different agro-climatic regions
- Studies on kairomonal interventions to improve the performance of parasitoids
- Development of superior strains of parasitoids (insecticide tolerant, high temperature tolerant, with high searching ability, etc.)
- Climatic pattern mapping and climate mapping of a region are important in terms of risk assessment of pest as well as for biocontrol introductions. Eco-climatic assessment can provide valuable insight into species distribution, in relation to relevant climate data, particularly relating to assessment of the potential establishment of a particular biocontrol species.
- Future biocontrol attempts must consider climate variables in evaluating long term effectiveness
- Future research should concentrate on: (a) The physiological adaptations or ecological implications of exposure of parasitoids/predators/microbials to extreme climatic conditions and on the relationship between physiological adaptation and integration of a species within an ecosystem. (b) The over wintering strategies in parasitoids in relation to climate change. (c) Effect of climate change on tri-trophic interactions.
- The outcome of our research should enable us to answer some pertinent questions: (a) Could we adjust the practice of biological control by changing release schedules to compensate for the effects of climate? (b) Will the effect of climate be stronger on parasitoids and predators than on the prey insects? (c) How changes in herbivore and plant quality (including semiochemical emissions) following a rapid climate change affect a parasitoid or predator's life history traits.

Suppression of insect pests is of paramount importance considering that they can cause about 15–20% loss in agricultural production. The present paper highlights the potential as well as proven technologies of biological control that can be commercialized and upscaled to reach farmers. The future of "insecticide-less" pest management will be driven by a bouquet of parasitoids and predators complimented by entomopathogens.

References

Ali MS (1957) Some biological studies on *Pseudococcus vastator* (Maskell). Indian J Entomol 19:54–58

Babu TR (1986) Population density of grape mealybug *Maconellicoccus hirsutus* (Green) and its control by a predator *Cryptolaemus montrouzieri* Muls. Ph.D. Thesis, APAU, Hyderabad

Bakthavatsalam N, Singh SP, Pushpalatha NA, Bhumannavar BS (1994) Life tables of four species of chrysopids (Neuroptera: Chrysopidae). J Entomol Res 18:357–360

Ballal CR, Ramani S (1994) Studies on four geographical strains of *Campoletis chlorideae* Uchida (Hymenoptera: Ichneumonidae). In: Goel SC (ed) Biological control of insect pests. U. P. Zoological Society, Muzaffarnagar, pp 87–90

Ballal CR, Singh SP (2001) Effect of insectcide applications on the populations of *Helicoverpa armigera* (Hübner) (Lepidoptera: Noctuidae) and natural enemies in sunflower ecosystem. In: Proceedings of the International Meet Entomocongress 2000: Perspectives for the New Millennium, held at Trivandrum from 5–8 Nov 2000. Entomon (Spl. Issue):37–42

Ballal CR, Gupta T (2003) Influence of chickpea varieties on oviposition by *Helicoverpa armigera* and parasitising efficiency of *Campoletis chlorideae*, p 81. In: Proceedings of the National Symposium on Bio-Management of Insect Pests 29th to 31st March 2003, Annamalai University, Annamalainagar, Tamil Nadu, p 121

Ballal CR, Gupta T (2011) Anthocorids as potential predators. In: Bhardwaj SC, Singh S, Bhatnagar A (eds) Plant protection through eco-friendly techniques. Pointer Publishers, Jaipur, pp 92 – 111, 272

Ballal CR, Singh SP (2003) The effectiveness of *Trichogramma chilonis* Ishii, *Trichogramma pretiosum* Riley and *Trichogramma brasiliense* (Ashmead) (Hymenoptera: Trichogrammatidae) as parasitoids of *Helicoverpa armigera* (Hubner) (Lepidoptera: Noctuidae) on sunflower (*Helianthus annuus* L.) and redgram (*Cajanus cajan* (L.) Millsp.). Biocontrol Sci Technol 13:231–240

Ballal CR, Joshi S, Rao NS (2000) Technique for segregating male and female *Campoletis chlorideae* at cocoon stage. Pest Manage Hortic Ecosyst 6:106–109

Ballal CR, Jalali SK, Sunil Joshi, Rao NS (2001a) Variations in populations of *Campoletis chlorideae* Uchida and *Helicoverpa armigera* (Hubner) from different crop ecosystems. In: Darshan Singh, Mahal MS, Sohi AS, Dilawari VK, Brar KS, Singh SP (eds) Proceedings of the symposium on biocontrol based pest management for quality crop protection in the current millennium. Punjab Agricultural University, Ludhiana, pp 11–12, 224, 18–19

Ballal CR, Joshi S, Rao NS (2001b) Bio-deterioration of laboratory cultures of Campoletis chlorideae Uchida, pp 210–213, In: Verghese A, Reddy PP (eds) IPM in Horticultural Crops: emerging Trends in the New Millennium; Proceedings of the Second National Symposium on Integrated Pest Management (IPM); New molecules, Bio-Pesticides and Environment; at Bangalore, pp 257, 17–19 Oct 2001

Ballal CR, Singh SP, Poorani J, Tripti Gupta (2003a) Biology and rearing requirements of an anthocorid predator, *Blaptostethus pallescens* Poppius (Heteroptera: Anthocoridae). J Biol Control 17(1):29–33

Ballal CR, Gupta T, Joshi S, Chandrashekhar K (2009a) Evaluation of an anthocorid predator *Blaptostethus pallescens* against two-spotted spider mite *Tetranychus urticae*. Integr Control Prot Crops. IOBC/WPRS Bull 49:127–132

Ballal CR, Srinivasan R, Jalali SK (2009b) Evaluation of an endosulfan tolerant strain of *Trichogramma chilonis* on cotton. Biocontrol 54(6):723–732

Ballal CR, Gupta T, Joshi S (2012) Production protocols for and storage efficacy of an anthocorid predator *Cardiastethus exiguus* Poppius. J Environ Entomol 34(1):50–56

Ballal CR, Gupta T, Joshi S (2013) Production and evaluation of the warehouse pirate bug *Xylocoris flavipes* (Reuter) (Hemiptera: Anthocoridae). Presented during the *International Congress on Insect Science 2013*, 14th to 17th February, GKVK, Bangalore

Bilapate GG, Mokat RB, Lovekar RC, Bagade DN (1988) Population ecology of *Heliothis armigera* (Hubner) and its parasites on pulses. J Maharashtra Agric Univ 13:299–302

Biswas SR, Tandon PL, Singh SP (2000) Biocontrol resource infobase, project directorate of biological control, Bangalore (CD)

Chandramohan N (1994) Seasonal Incidence of Diamondback moth, *Plutella xylostella* L. and its Parasitoids in Nilgiris. J Biol Control 8:77–80

Dodia JF, Yadav DN, Patel RC (1986) Management of fruit sucking moth, *Othreis fullonica* L. (Lepidoptera: Noctuidae) in citrus orchards at Anand (Gujarat) GAU Res J 11:72–73

Duffield SJ (1993) Distribution of *Trichogramma* adults and level of parasitism of *Helicoverpa* eggs on cloths in sorghum and short duration pigeonpea. *Int Pigeonpea Newsl* 18:30–31

Duffield SJ (1995) Crop-specific difference in the seasonal abundance of four major predatory groups on sorghum and short duration pigeonpea. ICPN 2:74–75

Goldson SL, Philips CB, Barlow ND (1994) The value of parasitoids in biological control. NZ J Zool 21:91–96

Gupta T, Ballal CR (2006) Biology and feeding potential of an anthocorid predator, *Orius tantillus* (Heteroptera: Anthocoridae) on Sitotroga cerealella. Indian J Plant Prot 34(2):168–172

Gupta T, Ballal CR (2007) Feeding preference of anthocorid predators for parasitised and un-parasitised eggs. J Biol Control 21:73–78

Gupta RK, Desh Raj (2003) Extent of parasitism and seasonal activity of *Campoletis chlorideae* Uchida in chickpea ecosystem of lower hills of Himachal Pradesh. Indian J Plant Prot 31(2):5–8

Indu Chatterjee VC (2006) Studies on the life cycle of the lady beetle, *Coccinella septempunctata* L. Ann Plant Prot Sci 14(1):102–107

Jalali SK, Singh SP (1993) Superior strain selection of the egg parasitoid *Trichogramma chilonis* Ishii—Biological parameters. J Biol Control 7:57–60

Jalali SK, Rabindra RJ, Rao NS, Dasan CB (2003) Mass Production of Trichogrammatids and Chrysopids. Technical Bulletin No. 33, Project Directorate of Biological Control, Bangalore, p 16

Jalali SK, Singh SP, Venkatesan T, Murthy KS, Lalitha Y (2006) Development of endosulfan tolerant strain of an egg parasitoid *Trichogramma chilonis* Ishii (Hymenoptera: Trichogrammatidae). Indian J Exp Biol 44:584–590

Jayarathnam K (1977) Studies on the population dynamics of the diamondback moth *Plutella xylostella* (Linnaeus) (Lepidoptera: Yponomeutidae) and crop loss due to the pest on cabbage. Ph.D. thesis, UAS, Bangalore, pp 215

Joshi S, Viraktamath CA (2004) The sugarcane woolly aphid, *Ceratovacuna lanigera* Zehntner (Hemiptera: Aphididae): its biology, pest status and control. Curr Sci 87:307–316

Joshi S, Ballal CR, Rao NS (1998) An efficient and simple mass culturing technique for *Ischiodon scutellaris* (Fabricus), an aphidophagous syrphid. Indian J Plant Prot 26(1):56–61

Joshi S, Prashanth Mohanraj, Rabindra RJ, Rao NS (2003) Production and use of coccinellid predators. Project directorate of biological control, Bangalore, India. Technical Bullectin No. 32. pp 26

Khan Q, Rao VP (1960) Insect and mite pests in cotton in India—A monograph, Indian central cotton committee, Bombay, pp 217–301

Khan AA, Misra DS (2003) Abundance of spider in relation to biotic and abiotic factors in upland rice ecosystem of Eastern Uttar Pradesh. Plant Prot Bull Faridabad 55(3/4):23–29

Kimberling DN (2004) Lessons from history: predicting successes and risks of intentional introductions for arthropod biological control. Biol. Invasions 6:301–318

Krishnamoorthy A, Nagarkatti S (1981) A mass rearing technique for *Chrysopa scelestes* Banks (Neuroptera: Chrysopidae). J Entomol Res 5:93–98

Krishnamoorthy A, Mani M (1996) Biosuppression of *Helicoverpa armigera* (Hubn.) on tomato using two egg parasitoids, *Trichogramma brasiliensis* (Ashm.) and T. pretiosum (Riley). J Entomol Res 20:37–41

Krishnamoorthy A, Singh SP (1987) Biological control of citrus mealubug with an introduced parasite, *Leptomastix dactylopii* in India. Entomophaga 32:143–148

Krishnamoorthy A, Singh SP (1988) Observational studies on the occurrence of parasitoids of *Papilio* spp in citrus. India J Plant Prot 16:79–81

Lyla KR, Pathummal Beevi, Ballal CR (2006) Field evaluation of anthocorid predator, *Cardiastethus exiguus* Poppius against *Opisina arenosella* Walker (Lepidoptera: Oecophoridae) in Kerala. J Biol Control 20(2):229–232

Mani M (2008) Polyhouse efficacy of *Cryptolaemus montrouzieri* Mulsant for the suppression of *Planococcus citri* (Risso) on grapes and *Ferrisia virgata* (Cockerell) on guava. J Insect Sci 21:202–204

Mani M, Thontadarya TS (1988a) Field evaluation of *Cryptolaemus montrouzieri* Muls. in the suppression of *Maconellicoccus hirsutus* (Green) on grapevine. J Biol Control 2:14–16

Mani M, Thontadarya TS (1988b) Response of *Cryptolaemus montrouzieri* Muls. (Coccinellidae, Coleoptera) to commonly used pesticides in vineyards. J Biol Control 2:17–20

Mani M, Krishnamoorthy A, Singh SP (1990) The impact of the predator, *Cryptolaemus Montrouzieri* Mulsant, on pesticide-resistant populations of the striped mealybug, *Ferrisia virgata* (Ckll.) on guava in India. Int J Trop Sci 11(2):167–170

Manjunath TM (1985) *Leptomastix dactylopii* in India. Biocontrol News Inf 6:297

Misra MP, Pawar AD (1984) Use of *Epipyrops melanoleuca* Fletcher (Lepidoptera: Epipyropidae) for the biocontrol of sugarcane pyrilla, *Pyrilla perpusilla* (Walker) (Hemiopetra: Fulgoridae) Indian J Agric Sci 54:742–750

Manjunath TM, Bhatnagar VS, Pawar CS, Sithanantham S (1989). Economic importance of *Heliothis* spp. in India and an assessment of their natural enemies and host plants. Proc. orkshop Biol. Control of Heliothis: Increasing the Effectiveness of Natural Enemies, ed. EG King, RD Jackson, pp 197–228. New Delhi: Far East. Reg. Res. Off., US Dep. Agric. 550 pp

Mukherjee AB, Chaudhury AKS, Bronniman H, Chu Y (1971) Artificial diets for rearing of *Xylocoris flavipes* (Reuter), a predator of some pests of stored cereals. Indian J Entomol 33:356–358

Nagarkatti S, Jayanth KP (1982) Population dynamics of major pests of cabbage and their natural enemies in Bangalore district (India). In Proceedings of the InternationalConference of Plant Protection of Tropics, Malaysian Plant Protection Society, Kula Lumpur, Malasyia. pp 325–347

Nagarkatti S, Singh SP, Jayanth KP, Bhumannavar BS (1992) Introduction and establishment of *Leptomastix dactylopii* How. against *Planococcus* spp. in India. Indian J Plant Prot 19:102–104

Pandey RK, Singh GR, Tripathi A (2005) Role of natural enemies on larval population of *Helicoverpa armigera* on chickpea sown on different dates. Shashpa 12(1):35–37

Patel HM, Borad PK (2005) Natural enemies of *Aphis gossypii* in isabgol. J Med Aromat Plant Sci 27(2):287–290

Patel AG, Yadav DN, Patel RC (1988) Improvement in mass rearing technique of green lacewing *Chrysopa scelestes* Banks (Neuroptera: Chrysopidae). Guj Agric Uni Res J 14:1–4

Pawar AD (1979) Sugarcane pyrilla and its biological control. Plant Prot Bull 13:157–163

Pawar AD (1986) Role of natural enemies in the integrated pest management of rice. Plant Prot Bull India 38:15–26

PDBC-ICAR (1999–2000) Annual report, Project Directorate of Biological Control, Bangalore, pp 217

PDBC-ICAR (2000–2001) Annual report, Project Directorate of Biological Control, Bangalore—560 024, pp 218

PDBC-ICAR (2001–2002) Annual report, Project Directorate of Biological Control, Bangalore—560 024, pp 218

Rabindra RJ, Ballal CR (2002) Mass production of bioagents and their use in integrated pest management—problems and prospects. In: Sarath Babu B, Varaprasad KS, Anitha K, Prasada Rao RDVJ, Chakrabarty SK, Chandurkar PS (eds) Resources management in plant protection vol I. pp 255–273

Rabindra RJ, Mohanraj P, Poorani J, Jalali SK, Joshi S, Ramani S (2002) *Ceratovacuna lanigera* Zehntner (Homoptera: Aphididae), a serious pest of sugarcane in Maharashtra and attempts at its management by biological means. J Biol Control 16:171–172

Rabindra RJ, Jalali SK, Ballal CR (2003) Recent advances in biological control and future prospects. pp 15–30, In Proceedings of the National Symposium on Bio-Management of Insect Pests 29th to 31st March, 2003, Annamalai University, Annamalainagar, Tamil Nadu, pp 121

Raja J, Rajendran B, Pappiah CM (1998) Management of egg plant shoot and fruit borer *Leucinodes orbonalisI* Guen. In: Proceedings of the International Symposium on Pest Management in Horticultural Crops, Bangalore, 15–17 Oct1997, pp 84–86

Ramani S, Poorani J, Bhumannavar BS (2002) Spiralling whitefly, *Aleurodicus dispersus*, in India. Biocontrol News Info 23(2):55–62

Rao VP, Ghani MA, Sankaran T, Mathur KC (1971) A review of the biological control of insects and other pests in South East Asia and the pacific region. CIBC Technical communication. No. 6, pp 149

Sangwan HS, Verma SN, Sharma VK (1974) Possibility of integration of exotic parasite, *Trichogramma brasiliensis* Ashmead for the control of cotton bollworm. Indian J Entomol 34:69–70

Sathiamma B, Chandrika Mohan, Murali Gopal (2000) Biological potential and its exploitation in coconut pest management. In: Upadhyay RK, Mukerji KG, Chamola BP (eds) Biocontrol potential and its exploitation in sustainable agriculture, vol–2: Insect pests. Kluwer Academic/Plenum Publishers, New York, pp 421

Sehgal M, Das S, Chander S, Gupta NC, Kalra N (2006) Climate studies and insect pests: implications for the Indian context. Outlook Agric 35(1):33–40

Shylesha AN, Joshi S, Rabindra RJ, Bhumannavar BS (2010) Classical biological control of the papaya mealybug. Folder, National Bureau of Agriculturally Important Insects, Bangalore, p 3

Sharma SS, Pala Ram, Saini RK (2004) Population dynamics of *Bemisia tabaci* (Gennadius) and its parasitoid, *Encarsia lutea* (Masi), on cotton. J Cotton Res Dev 18(1):102–103

Singh HS, Singh R (2003) Parasitisation of *Tetrastichus coccinellae* Kurd. on *Coccinella septumpunctata* L. under field condition in Eastern Uttar Pradesh. Indian J Plant Protect 31(1):143

Singh R, Ali S (2006) Seasonal incidence of *Helicoverpa armigera* and *Campoletis chlorideae* on chick pea. Ann Plant Protect Sci 14(1):234–235

Singh SP (1989) Achievements of AICRP on Biological Control. Technical Bulletin, No. 2. Biological Control Centre, NCIPM, Faridabad, pp 20

Singh SP (1991) Natural enemies of *Heliothis armigera* with particular reference to studies made in India. Paper presented at ICAR/1 RPM/USDA. Joint Project development group meeting on managing insecticides resistance with focus on *Heliothis* resistance management in India held at Directorate of Rice Research, Hyderabad, 15–17 Oct 1991

Singh SP (1994) Fifteen years of AICRP on Biological Control. Technical Bulletin No. 8, Project Directorate of Biological Control, pp 320

Singh SP (1996) Biological systems in India. In: Rajiv K, Upadhyay K, Mukherjee G, Rajak RL (eds) IPM systems in agriculture—biocontrol in emerging biotechnology, vol 2. Aditya Books Pvt. Ltd., New Delhi, pp 216–328 and 620 pp

Singh SP (2002) Progress and potential of Biological Control in India. Thirteenth Dr. C. P. Alexander Memorial Lecture 2001, Dept of Zoology, University of Delhi, Delhi, pp 32

Singh SP, Jalali SK (1993) Evaluation of Trichogrammatids against *Plutella xylostella*. Trichogramma News 7:27

Singh SP, Jalali SK, Bhumannavar BS, Bakthavatsalam N, Pushpalatha NA (1994) Production and use of Chrysopid predators. Project Directorate of Biological Control, Bangalore, Technical Bulletin No. 10, 14 p

Singh SP, Murphy ST, Ballal CR (eds) (2001a) Augmentative Biocontrol—Proceedings of the ICAR-CABI Workshop, June 29th to July 1st, 2001. Project Directorate of Biological Control, Bangalore, 250 pp

Singh SP, Joshi S, Ballal CR (2001b) Innovations in mass rearing technology and techniques for mass releases, transportation and storage of natural enemies. In: Singh SP, Bhumannavar BS, Poorani J, Singh D (eds) Biological control—symposium on biocontrol based pest management for quality crop protection in the current millennium (July 18th to 19th, 2001). pp 1–33 and 181 pp

Sreedharan K, Vinod Kumar PK, Prakasan CB (eds) (2001) Coffee berry borer in India. Central Coffee Research Institute, Coffee Research Station, Karnataka, pp 112

Subba Rao BR Sangwar HS Abbasi OA Singh Y Ksheer Sagar AM (1965) New records of hymenopterous parasites of *Nipaecoccus vastator* (Maskell) (Homoptera: Coccidae), a serious pest of *Citrus* spp. from India. Indian J Entomol 27:109–110

Swamiappan M, Balasubramanian M (1980) Studies on mass multiplication and potentiality of *Chelonus blackburni* Cam. a broconid parasite of cotton bollworms. Entomon 5:73–75

Teggelli RG, Rao KJ, Kulkarni KA (2004) Effect of temperature and relative humidity and their interactions on the development of *Campoletis chlorideae* Uchida. Karnataka J Agric Sci 17(3):476–478

Tripathi GM (1995) Record of parasitoid and predator complex of sugarcane woolly aphid, *Ceratovacuna lanigera* Zehntner in Nagaland. Indian Sugar 44:883–885

Tripathi A, Mishra DS, Sharma RC (2005) Role of biotic and abiotic factors on mustard aphid, *Lipaphis erysimi* (Kalt.). Farm Sci J 14(1):84–85

Venkatesan T, Singh SP, Joshi S, Rao NS (1995) Breeding of *Campoletis chlorideae* Uchida (Hymenoptera: Ichneumonidae) on *Corcyra cephalonica* Stainton. J Biol Control 9:128–129

Venkatachalapathy M, Singh NI, Prasad B (2002) Role of abiotic factors on population build up and seasonal incidence of tachinid parasitoids infesting oak tasar silkworm, *proylei* J. (Saturniidae: Lepidoptera) in Manipur. Indian J Seric 41(2):106–111

Venkatesan T, Ballal CR, Rabindra RJ (2008) Biological control of Coconut Black-Headed Caterpillar *Opisina arenosella* using *Goniozus nephantidis* and *Cardiastethus exiguus* Technical Bulletin No. 39, Project Directorate of Biological Control, Bangalore, pp 14

Yadav DN, Patel RC, Manjunath TM (1975) Seasonal activity of *Apanteles plutellae* (Kurdj.), a larval parasite of *Plutella xylostella* (L.) at Anand (Gujarat, India). Indian J Plant Prot 3:111–115

Development, Characterization and Field Assessment of Multiple Insecticides and High Temperature Tolerant Strain of an Egg Parasitoid, *Trichogramma chilonis* Ishii Against Crop Pests

T. Venkatesan and S. K. Jalali

Abstract

The approach of integrated pest management (IPM) is to minimize the use of insecticides, there by biological control and other environmental friendly approaches can gain momentum. However, the use of chemical insecticides continues to be widely adopted and remains mainstay of insect pest control. In such crop scenario where insecticides are frequently used, releases of susceptible strain of a natural enemy may not give any appreciable control. The efficacy of trichogrammatids is also largely dependent on temperature conditions. In India, most of the crops are grown during monsoon, June–November, when the temperature is high (up to 40 °C). Many of the insecticides are also used frequently during the season, thereby reducing efficacy of trichogrammatids drastically at such stresses. In the present chapter, development, characterization and field evaluation of multiple insecticide and temperature tolerant strain of an egg parasitoid *Trichogramma chilonis* Ishii on crops is discussed.

Keywords

Characterization · Insect pests · *Trichogramma chilonis* · Temperature tolerant · Multiple insecticides tolerant

Introduction

The insect pest attack on cotton and vegetable crop causes yield loss about 11.0 and 7.5%, respectively, in the World (Wittwer 1979) and in Asia, 72% of potential yield loss occurs in cotton and vegetables. Insect pests cause up to 30% yield loss under modern agriculture compared to 5–10% under traditional agriculture in India. Several commercial crops, vegetable, and fruit crops are subjected to intensive plant protection measures. These crops receive about 80–90% of the total pesticide usage in India. Such usage has resulted in high level of insecticide resistance, necessitating repeated application of insecticides. This also often results in frequent outbreaks of

T. Venkatesan (✉) · S. K. Jalali
National Bureau of Agricultural Insect Resources, Post Bag No. 2491, H. A. Farm Post, Bellary Road, Hebbal, Bangalore, Karnataka, 560024, India
e-mail: tvenkat12@gmail.com

A. K. Chakravarthy (ed.), *New Horizons in Insect Science: Towards Sustainable Pest Management*,
DOI 10.1007/978-81-322-2089-3_29, © Springer India 2015

sucking pests and borers, DBM/bollworms pests. *Trichogramma* are released to suppress different caterpillar pests attacking maize, rice, sugarcane, cotton, vegetables, fruits, etc. (Singh and Jalali 1994). However, their use in crops where insecticides are most frequently used is diminishing due to high insecticidal usage and hot weather conditions found in some part of our country. Release of trichogrammatids in crops like cotton, tomato, cabbage, etc. are not giving satisfactory results due to high application of insecticides. Jalali et al. (2002) reported that the enemies were absent on cotton throughout India in sprayed fields and recorded trichogrammatids parasitizing only in four states, *viz.*, Punjab, Andhra Pradesh, Tamil Nadu, and Madhya Pradesh, out of nine cotton-growing states in India. Jalali and Singh (1993) observed trichogrammatids to be highly susceptible to all insecticides and even residues were found to effect parasitizing ability up to 21 days on cotton. Adult *Trichogramma* are quickly killed by broad-spectrum insecticides applied to cotton (Jalali and Singh 1993). The trichogrammatids are susceptible to a broad spectrum of insecticides and reduced parasitism had been reported in *T. exiguum* and *T. pretiosum* and *H. zea* and *Manduca* sp. in plots treated with pyrethroids (Campbell et al. 1991). Similarly, the drift of pesticide even a mile away and single application of pesticides in cotton reduced the efficacy of trichogrammatids (Stinner et al. 1974; Bull and House 1983; Bull and Coleman 1985).

Realizing the need of integrated pest management (IPM) approach to minimize the use of insecticide, biological control, and other environmental friendly approaches has gained momentum. However, the use of chemical insecticides continues to be widely adopted and remains mainstay of insect pest control. In such a crop scenario where insecticides are frequently used, releases of susceptible strain of *T. chilonis* will not give any appreciable supression.

Preliminary results revealed that multiple insecticides tolerant strain of *T. chilonis* gave 63 % more parasitism against cotton bollworms compared to susceptible laboratory strain in five places in three states. Again 11.2 % higher yield against existing IPM practice and 63.4 % higher yield was obtained to untreated check (Anonymous 2003). The efficacy of trichogrammatids is largely dependent on temperature conditions. Earlier studies have suggested that temperature >35 °C prevents adult emergence (López and Morrison 1980; Singh and Jalali 1994; Scholler and Hassan 2001), thereby reducing efficacy of trichogrammatids drastically at higher temperatures. In India, most of the crops are grown during March–November when the temperature is high and much of insecticides are used frequently during the period. Therefore, the strain of *T. chilonis* having insecticide and high temperature tolerance could be utilized against lepidopterous pests of crops in different parts of the country.

Material and Methods

Working out Initial LC_{50} Values for Field and Laboratory Collected Population of *T. chilonis*

The parasitoid, *Trichogramma chilonis* was collected from tomato fields in and around Bangalore where several rounds of insecticides were sprayed during crop period. This provided initial stock culture for work on genetic improvement. The experiment was carried out by serial dilution of three insecticides, *viz.*, endosulfan, monocrotophos, and fenvalerate by 1/2 seven times from field recommended dosages to work out LC_{50} values. The testing was carried out with laboratory and field collected parasitoids, after raising parasitoids in sufficient numbers in F_2 generation. The data obtained on mortality were subjected to probit analysis by statistical program SPSS version 8.0. The data were transformed to log base 10 before probit analysis and antilog of calculated values gave actual LC_{50} and LC_{90}. The fiducial limits slope and χ^2 values were also calculated.

Working out Initial Temperature Response for Field Collected and Laboratory Populations of *T. chilonis*

The experiment was carried out in growth chambers set at 32, 36, 40 and 45 °C, and at variable range of 32–38 °C. About 100 parasitized eggs (by *T. chilonis*) were kept in the glass vial for emergence. After emergence, egg card was provided @ 50 eggs/female. The vials were kept in above temperature to record percent mortality and percent parasitism after 6 and 24 h of exposure. The data on mortality and percent parasitism were analyzed by one-way ANOVA and means were separated by LSD values.

Development of Population of *T. chilonis* Having Multiple Insecticide and High Temperature Tolerance

The experiment was conducted with exposing field-collected population to three insecticides, *viz.*, endosulfan, monocrotophos, and fenvalerate as well as to variable high temperature (32–38 °C) simultaneously. The induction of tolerance in the multiple insecticides tolerant strain of *T. chilonis* to high temperature (32–38 °C) was carried out in the laboratory in growth chamber maintained at variable temperature for its ability to survive and parasitize *Corcyra cephalonica* (Stainton) eggs. Both strains were mixed together and allowed to mate for 24 h before exposure to insecticides and thereafter to higher temperature. The insecticides used were endosulfan (2.0 ml/L), monocrotophos (1.5 ml/L), and fenvalerate (0.4 ml/L). Parasitoids thus obtained were treated with same concentration till fixed parameters were achieved. The parameters fixed were $\leq 30\,\%$ mortality after 6 h of constant exposure and $\geq 90\,\%$ parasitism in sprayed condition. After 50 generations, concentrations were increased to double of field recommended that is, endosulfan (4.0 ml/L), monocrotophos (3.0 ml/L), and fenvalerate (0.4 ml/L). The mortality and parasitism data were tabulated for each generation.

LC_{50} Values to Determine Increased Tolerance to Insecticides and Temperature

The experiment was carried out by serial dilution method. The pesticide solution was prepared by taking dosages higher than field recommended dosages and serial dilution by reducing dilution by 1/2. About 100 adults were released in each vial in each concentration. The mortality was recorded after 6 and 24 h of constant exposure and percent parasitism was recorded after 6 days of exposure. The data obtained on mortality were subjected to probit analysis by statistical program SPSS version 8.0 and LC_{50}, LC_{90}, fiducial limits, slope, and χ^2 values were calculated.

Sequencing of Heat Shock Protein (HSP) for Determination of High Temperature Tolerance in Multiple Insecticides and High Temperature Tolerant Strain (MIHTTS) of *T. chilonis*

Heat tolerant adults were heat shocked at 40 °C for 30 min and allowed to recover at room temperature for an hour. The adults were freeze killed in liquid nitrogen. Total mRNA was extracted using Biogene RNA extraction kit as per manufacturers' instruction. The first strand synthesis was carried out by reverse transcriptase enzyme on total RNA. The amplification reaction was carried out in 200 µl PCR tubes using gradient thermal cycler. The PCR amplified DNA samples were electrophoresed on an Agarose gel along with the marker DNA to check the presence of heat shock protein (HSP).

Biochemical Characterization

GST-Activity

The conjugative activity of GST was assayed with CDNB as a substrate using standard procedure.

Carboxylesterase Activity

Five milligrams of *T. chilonis* adults were homogenized in a micro pestle and mortar with 0.2 ml of 50 mM ice-cold buffer of pH 7.5, containing 0.5 % (w/vol) TritonX-100. The homogenates were centrifuged at 4 °C at 12,000 g for 10 min. Supernatant was collected. This was repeated for two more extractions and supernatants were pooled. The standard procedure was followed to determine carboxylesterase activity (CE).

Isoenzyme Analysis

CE activity was determined for the different strains of *T. chilonis* in nondenaturing PAGE. Homogenates from each strain stained for 1-Napthyl acetate. Native PAGE was carried out using mini vertical gel system and 7.5% acrylamide (wt: vol) gel at pH 8.5 and stacking gel of 3% acrylamide (wt: vol). Km values were calculated for the lab susceptible and tolerant strains for the substrate 1-Napthyl acetate, the substrate concentrations 0.025 to 1 mM were used.

Microsomal Cytochrome C Reductase and p-Nitroanisole O-demethylase

Sigma reagent assay kit used for the preparation of microsomes. Sixty milligram each strains of *T. chilonis* adults were homogenized in a micro pestle and mortar with 2 ml of PBS pH 7.5 containing 0.1 mM EDTA.

Microsomal Cytochrome p-Nitroanisole O-demethylase

This assay measures the reduction of cytochrome c by NADPH—cytochrome c reductase in the presence of NADPH. The absorption spectrum of cytochrome c changes with oxidation/reduction state. Upon reduction a sharp absorption peak is monitored by the increase of cytochrome c absorbance at 550 nm. Cytochrome c reductase assay was performed by following the manufacture protocol.

Field Efficacy of the Multiple Insecticides and High Temperature Tolerant (MIHTTS) and Susceptible Strain of *T. chilonis* on Cotton, Tomato, and Cabbage Pests in Comparison to Insecticides Alone

Cotton

First Year A field trial was conducted at Sirsa, Haryana and crop was raised during *Kharif* (monsoon) season of 2007. The variety sown was CICR—2 in 10.0 acre with MIHTTS. The comparison was made with susceptible strain and farmer's practice in one acre each. Each treatment was separated by 200 m of barren land at all the locations. In tolerant strain plots imidacloprid, neem oil, and lambda-cyhalothrin was sprayed during the season, in susceptible strain released plots confidor, monocrotophos + cypermethrin, fenvalerate, lambda-cyhalothrin (two sprays), and neem oil was sprayed, in farmers' practice plot, insecticides sprayed were imidacloprid, monocrotophos + cypermethrin, fenvalerate, lambda-cyhalothrin (two sprays), endosulfan, and neem oil were sprayed once in a week and tabulated the results.

Second Year A field trial was conducted at Sirsa, Haryana, in collaboration with National Centre for Insect Pest Management (NCIPM), New Delhi during 2008. The variety sown was CICR—2 in 10.0 acre area. The comparison was made with susceptible and farmer's practice in one acre each. Each treatment was separated by 200 m of barren land at all the locations. The insecticides sprayed were imidacloprid, lambda-cyhalothrin, monocrotophos + cypermethrin, fenvalerate, endosulfan, and neem oil were sprayed once in a week and tabulated the results.

Trichogramma chilonis (MIHTTS and susceptible strain) was released in treatments T_1 and T_2 @ 150,000/ha/release in the form of parasitized eggs to cover the egg laying period of *H. armigera*. The release of parasitoids commenced with moth capture in pheromone traps. Eight days old parasitized cards were released at weekly interval from August to September 2007. The harvesting continued till mid November 2007.

Larval population was recorded at each place in 10 subplots in each treatment. Twenty plants in each subplot at each location were observed for the larval population. Thus, 200 plants were observed each time to record larval population per treatment. The percent boll damage, good and bad open bolls, and yield were recorded at harvest. Data on larval population, percent boll damage and yield data were subjected to one-way ANOVA and means were separated by CD values at 5%, wherever ANOVA was significant.

Tomato

First Year Field trial was conducted for the evaluation of MIHTTS of *T. chilonis* against *H. armigera* at Malur, Karnataka in collaboration with University of Agricultural Sciences (UAS), Bengaluru during 2007. Tomato was raised during summer 2007. The varieties sown were Shaktiman and Abhinava in 8.0 acre. Comparative evaluation was done in susceptible strain released plots (1 acre), farmer practice (1 acre), and in untreated control (200 m). Each treatment was separated by 200 m of barren land at all the locations. In tolerant strain plots confidor, neem oil, and lambda-cyhalothrin were sprayed during the season, in susceptible strain released plots imidacloprid, monocrotophos + cypermethrin, fenvalerate, lambda-cyhalothrin (two sprays), and neem oil was sprayed, in farmers' practice plot insecticides sprayed were imidacloprid, monocrotophos + cypermethrin, fenvalerate, lambda-cyhalothrin (two sprays), endosulfan and neem oil were sprayed once in a week.

Second Year Field trial was conducted for the evaluation of MIHTTS of *T. chilonis* against *H. armigera* at Malur, Karnataka during *Kharif* 2008. Tomato was raised during winter 2008–2009. The variety sown was 618—improved in 1.0 acre. Comparative evaluation was done in susceptible strain released plots, farmer practice and in untreated control (200 m). Each treatment was separated by 200 m of barren land at all the locations. In tolerant strain plots Dhanush, quinolphos (2 sprays), M—45 (3 sprays), copper oxychloride (2 sprays), Acrobat (2 sprays), Sectin (1 spray), Blue cop (1 spray) were sprayed during the season, in susceptible strain released plots Dimethomorph, Metacid, quinolphos, M—45, copper oxychloride, Fenamidome, 10 % mancozeb, Blue cop were sprayed, in farmers' practice plot monocrotophos, acephate, Rimon, Dash, acetamiprid, Combident Bloom, and Charm were sprayed once in a week.

T. chilonis (new insecticides tolerant strain and susceptible strain) was released in treatments T_1 and T_2 @ 50,000/ha/release in the form of parasitized eggs to cover the egg laying period of *H. armigera*. The release of parasitoids commenced with moth capture in pheromone traps in all the four areas. Eight days old parasitized cards were released at weekly interval from April to June, 2007. The harvesting continued till mid July 2007.

The eggs parasitism was recorded at each treatment in 10 subplots. Twenty plants in each subplot at each location were observed for the eggs, number of larvae and number of fruits bored. Thus, 200 plants were observed each time per treatment. The yield data were recorded at the end of trial. Field evaluation data on egg parasitism, number of larvae/plant and percent fruits bored and yield data were subjected to one-way ANOVA, and means were separated by CD values at 5 %, wherever ANOVA was significant.

Cabbage

First Year Field trial was conducted for the evaluation of MIHTTS of *T. chilonis* against diamond back moth (DBM) *Plutella xylostella* (L.) at Malur, Karnataka in collaboration with UAS. The cabbage var. Shristi was raised during summer season of 2007 in 4 acre. The comparison was made with susceptible strain in 0.5 acre and farmers' practice 1.0 acre and 200 m for untreated control. Each treatment was separated by 200 m of barren land at all the locations. In tolerant strain plots methomyl, thiodicarb, emamectin benzoate, and acephate were sprayed twice the season, in susceptible strain released plots methomyl, thiodicarb, emamectin benzoate, malathion, acephate acetamiprid and imidacloprid were sprayed twice, in farmers' practice plot methomyl, thiodicarb, emamectin benzoate, malathion, acephate and acetamiprid, and imidacloprid were sprayed in a week.

Second Year Field trial was conducted for the evaluation of MIHTTS of *T. chilonis* against DBM at Malur during *Kharif* 2008. The cabbage (variety—Maharani) was raised during winter season of 2008–2009 in 2 acre. The comparison was made with susceptible strain, farmers' practice and 200 m for untreated control. Each treatment was separated by 200 m of barren land at all the locations. In tolerant strain plots Takuni

(1 spray), Helicide (Ha NPV—2 sprays), Padson (1 spray), cypermethrin and chlorpyriphos (4 sprays), acephate and imidochloprid (1 spray) were sprayed during the season, in susceptible strain released plots Fame, Sumo powder, Confidor and acephate were sprayed twice, in farmers' practice plot insecticides sprayed were Dom, Ankur, Bancip, Lancer gold, cypermethrin, chlorpyriphos, acephate, endosulfan and imidacloprid twice in a week.

Trichogramma chilonis (new insecticides tolerant strain and susceptible strain) was released in treatments T_1 and T_2 @ 50,000/ha/release in the form of parasitized eggs to cover the egg laying period of *P. xylostella*. The release of parasitoids commenced with moth capture in pheromone traps in all the area. Eight days old parasitized cards were released at weekly interval from July to August, 2007.

The eggs parasitism was recorded at each treatment in 10 subplots. Twenty plants in each subplot at each location were observed for the egg masses, number of larvae and number of feeding punctures. Thus, 200 plants were observed each time per treatment. The yield data were recorded at the end of trial. Data on egg parasitism, no. of larvae/plant and feeding punctures, and yield data were subjected to one-way ANOVA and means were separated by CD values at 5 %, wherever ANOVA was significant.

Results and Discussion

Initial LC_{50} Values for Field and Laboratory Collected *T. chilonis*

The results are presented in Table 1. The LC_{50} values obtained for three insecticides, *viz*., endosulfan, monocrotophos, and fenvalerate after 6 h of constant exposure is indicated in Table 1. LC_{50} values of laboratory population was 0.08, 0.003, and 0.01 compared to 1.07, 0.70, and 0.04 for the field-collected population after 6 h of constant exposure to endosulfan, monocrotophos, and fenvalerate, respectively (Table 1). The significant χ^2 value for all the three tests indicated heterogeneity in the test.

Initial Temperature Response for Field and Laboratory Populations of *T. chilonis*

The results on initial evaluation of laboratory and field-collected populations exhibited differential response to higher temperature during 6 and 24 h exposure period. At 32 and 36 °C, and at variable temperature of 32–38 °C, indicated very low mortality in both populations. However at 40 and 45 °C, significantly high mortality was recorded in the laboratory population compared to the field-collected population. At 32 °C, no mortality was recorded upto 6 h exposure, indicating that this temperature is not higher threshold temperature for survival of *T. chilonis*. However at 40 °C, mortality of the adults was 59.7 % in laboratory population compared to 0.0 % in the field-collected population and it differed significantly amongst the populations (LSD = 5.92, $P = 0.05$). The laboratory population was found to be highly prone to next higher temperature of 45 °C as 96.1 % adults died within 6 h as compared to 9.2 % in the field-collected population. The low mortality recorded in the field-collected population suggests that adaptation to higher temperature is necessary to enhance potential of *T. chilonis* in high temperature (Table 2).

The results of exposure to high temperature for 24 h exhibited different results than 6 h exposure. The mortality in laboratory population (SS) was 47.7, 96.9, 100.0, 100.0, and 98.5 % as compared to 45.6, 77.7, 90.7, 97.1, and 57.1 % in 32, 36, 40, 45 °C and 32–38 °C, respectively. The different response of field-collected and laboratory populations to high temperature originates from its physiological adaptation to extreme temperature. The heat hardening is a well-known form of acclimatization in many invertebrates where exposure to high but sub-lethal temperature protects against subsequent heat induced death and heat hardening enhanced adult fitness in the field under hot conditions. The selection of *Trichogramma* lines for improvement of parasitization at constant low, medium, or high temperature indicated that a change in performance at one temperature concurrently resulted in opposite changes at distant temperature. Thus, genetic

Table 1 Dose mortality response (LC) of *T. chilonis* to three insecticides

Insecticide	Period for exposed (h)	LC_{50}	95% Fiducial limit		LC_{90}	95% Fiducial limit		Slope ± SE	χ^2
			Lower	Upper		Lower	Upper		
Laboratory population to									
Endosulfan	6	0.08	0.05	0.11	0.19	0.14	0.42	3.51 ± 0.36	18.8
Monocrotophos	6	0.003	–	–	0.01	–	–	2.06 ± 0.36	0.2
Fenvalerate	6	0.01	0.01	0.02	0.04	0.02	0.07	3.15 ± 0.24	17.6
Field-collected population to									
Endosulfan	6	1.07	0.65	1.64	3.71	2.21	16.46	2.38 ± 0.19	23.1
Monocrotophos	6	0.70	0.48	0.95	1.76	1.22	4.84	3.19 ± 0.24	23.9
Fenvalerate	6	0.04	–	–	0.003	–	–	1.07 ± 0.27	185.1

Table 2 Response of field-collected and laboratory-reared *T. chilonis* to high temperature and variable temperature regime

Temperature (°C)	Mortality (%)						Parasitism (%)		
	6 h		Mean (B)	24 h		Mean (B)			Mean (B)
	LP	FP		LP	FP		LP	FP	
32	0.0 (1.3)	0.0 (1.3)	0.0 (1.3)	47.7 (43.7)	45.6 (42.5)	46.6 (43.1)	54.0	46.7	50.4
36	4.1 (11.7)	0.9 (1.9)	2.5 (9.1)	96.9 (79.9)	77.7 (61.8)	97.3 (80.5)	6.7	45.0	25.9
40	59.7 (50.6)	0.0 (0.0)	29.8 (33.1)	100.0 (90.0)	90.7 (72.2)	95.3 (77.5)	0.0	18.3	9.2
45	96.1 (78.6)	9.2 (17.7)	52.7 (46.6)	100.0 (90.0)	97.1 (80.2)	98.6 (78.9)	0.0	2.3	1.2
32–38	1.8 (7.7)	0.0 (0.0)	0.9 (5.4)	98.5 (83.0)	57.1 (49.1)	77.8 (61.9)	0.0	63.3	36.1
Mean (A)	32.3 (34.6)	2.0 (8.1)		88.6 (70.3)	73.6 (59.1)		12.1	35.2	
	A factor	B factor	A x B	A factor	B factor	A x B	A factor	B factor	A x B
SEM±	1.26	2.00	2.83	3.40	5.38	7.61	3.34	4.52	7.62
LSD (P=0.05)	3.74	5.92	8.37	7.60	15.80	24.40	NS	14.8	19.4

Data in parentheses represent arcsine transformed values; *LP* laboratory population, *FP* field-collected population

trade-off in performance at different temperature and phenotypic plasticity in maternal selection may con-population evolution of the thermal niche in *Trichogramma.* To avoid genetic trade-off, a variable temperature of 32–38 °C was taken as one of the treatments.

The results of percent parasitism by laboratory population and field-collected population are presented in the Table 2. The result indicated that in general efficacy of *T. chilonis* was reduced in the high temperature. At 32 °C, parasitism by laboratory population was slightly higher compared to the field-collected population, but at all other higher temperature of 36, 40, 45 °C and at variable temperature of 32–38 °C percent parasitism by field-collected population was significantly more. At 40 and 45 °C, very low parasitism was recorded mainly due desiccation of the host eggs as humidity recorded was <20% at these temperature. Most surprisingly at variable temperature, percent parasitism by field-collected population was >63.3% compared to no parasitism by laboratory population. The results indicated that though lab population is capable of surviving in the temperature at 32–38 °C but it fails to parasitize its host. Thus, if releases of *Trichogramma* are to be considered during hotter months, ordinary lab population may not give any appreciable control of the pest.

Trichogrammatids are known to perform better in temperature range 20–32 °C. The effective pest control is determined by many factors including the quality and fitness of the parasitoid and complex interactions between the parasitoid, the target pest, the crop and environmental conditions. Considering the present scenario of very low levels of parasitism achieved during hotter months (35–40 °C, parasitism ≤25 %), it is absolutely necessary to go for the parasitoids that are field collected to high-temperature conditions. Hence, high-temperature field-collected population of *T. chilonis* will survive better in hostile crop environment and control lepidopteran pests efficiently.

The efficacy of trichogrammatids is largely dependent on temperature conditions. Earlier studies have suggested that temperature >35 °C prevents adult emergence (López and Morrison 1980; Singh and Jalali 1994; Scholler and Hassan 2001), thereby reducing efficacy of trichogrammatids drastically at higher temperature. Scott et al. (1997) examined the consequences of acclimation for survival and the fitness components of *T. carverae* and reported that acclimating wasps by rearing at constant temperature influenced parasitism rates at those temperature. They also observed that heat hardening at 33 °C during development also resulted in significant increases in survivorship of adults after exposure to 40 °C. Thomson et al. (2001) reported that heat hardening is a well known form of acclimatization in many invertebrates where exposure to high but sublethal temperature protects against subsequent heat induced death and heat hardening enhanced adult fitness in the field under hot conditions. Carrière and Boivin (2001) reported that selection of *Trichogramma* lines for improvement of parasitism at constant low, medium, or high temperature indicated that a change in performance at one temperature concurrently resulted in opposite changes at distant temperature. Thus, genetic trade-off in performance at different temperature and phenotypic plasticity in maternal selection may constrain evolution of the thermal niche in *Trichogramma.*

The efficacy of trichogrammatids is largely dependent on temperature conditions. Singh and Jalali (1994) found that the temperature >35 °C affects emergence, thereby reducing efficacy of trichogrammatids drastically. Abraham and Pradhan (1976) made an attempt to select strain of *T. chilonis* for tolerance to high temperature (32–35 °C) from heterogeneous populations of the parasitoids developed by interbreeding strains from Ambajipet (Andhra Pradesh). Cuddalore (Tamilnadu), Delhi, Lucknow, Ludhiana, and Mandya (Karnataka) but were unsuccessful. Ludhiana, Delhi, and Ambajiper strains, however, were more tolerant to high temperature and low humidity. Significant improvements in adult emergence, fecundity, and progeny production were made after rearing Ludhiana and Delhi strains for 32–33 generations at progressively increasing temperature from 30 to 33 °C and decreasing humidity from 60 to 10 %.

Development of *T. chilonis* Population with Multiple Insecticide and Temperature Tolerance

The results on the development process of a population of *T. chilonis* field-collected to three insecticides and high temperature revealed that percent mortality ranged from 94.0 to 100.0 % during 1st exposure. In endosulfan treatment, mortality remained >80.0 % for 36 generations after 6 h of constant exposure. In F_{45} generation, mortality recorded was 80.0 %. The concentration was increased to double the field recommended dosages, that is, 4.0 ml/L. Increasing concentration increased the mortality to 95.0 % and after constant selection for 10 generations mortality declined to 70.0 % and remained so throughout the period of development. The mortality declined due to constant exposure in each generation to the insecticides and high temperature, thus indicating that adaptation to conditions can induce tolerance in the parasitoid.

In monocrotophos treatment, mortality was 100.0 % in 1st generation in high temperature and sprayed situation after 6 h of constant exposure. The higher mortality obtained in monocrotophos indicated that this insecticide was more toxic than other insecticides. The mortality declined due to constant exposure after 45th generation and it was 80.0 %. The increase in concentration to double the field recommended dosages, that is, 3.0 ml/L increased mortality to 100.0% and after constant selection for 10 generations mortality declined to 80.0% and after 71 generations of selection, it further declined to 75.0%. The results indicated that exposing parasitoids continuously to insecticides enhances its surviving ability. In fenvalerate treatment, initial mortality was 94.0% and after 36 generations of selection it declined to 75.0% and it remained consistent till 45th generation. The increase in concentration, i.e., 0.4 ml/lit increased mortality to 95.0% and after constant selection for 10 generations mortality declined to 60.0% and after 71 generations of selection, it further declined to 57.0%. The results indicated that exposing parasitoids continuously to insecticides enhances its surviving ability. The result of parasitizing ability of *T. chilonis* when adults were exposed to sprayed situation is presented in Fig. 1. The initial egg parasitism obtained was 45.0, 35.0 and 50.0% when exposed to endosulfan, monocrotophos and fenvalerate treatments at high temperature. The percent egg parasitism ranged from 45.0 to 60.0% up to 12 generations after 24 h of constant exposure and increased to 75.0–90.0% and remained high till 50th generation. Increase in concentration resulted in decrease in parasitism. The parasitism ranged from 55.0–70.0% for another 10 generations and from 66th generation onwards, egg parasitism increased to 80.0–95.0% and remained ≥90.0% throughout the selection period. The results thus indicated usefulness of adapting multiple insecticide tolerant strain to high temperature to obtain a combined population, which is tolerant to all three insecticides and also to variable high temperature.

LC_{50} Values to Determine Increased Tolerance to Insecticides and Temperature

The initial LC_{50} values of *T. chilonis* existing population at the start of the work to three insecticides *viz.*, endosulfan, monocrotophos, and fenvalerate were 1.07, 0.70 and 0.04 ml/l, respectively. However, after exposure for 81 generations to three insecticides, LC_{50} values increased by 5.16, 6.27, and 6.0 times as compared to immediately after collecting parasitoids from the field. The formula applied to know percent increase in tolerance levels (T-C x 100/T) revealed that tolerance level to endosulfan, monocrotophos and fenvalerate increased by 98.2, 94.98 and 25.0% over laboratory population. Therefore continuous exposure to insecticides starting from low selected dosages induced tolerance in population (Table 4).

In India cotton alone consumes maximum pesticide usage (44.5%) followed by paddy-22.8% and fruits and vegetables-7% of the total pesticide consumption (Dudani and Sengupta 1991). *Trichogramma* sp. is used as a bioagent on lepidopterous pests attacking cotton, cereals, vegetables and fruits. The amenability of the parasitoid for mass rearing and release has contributed to its greater utility. However, greater reliance on synthetic pesticides had altered the ecological niche of the bioagent and reduced its potential. The development of resistance to insecticides has been more aggressive in insect pest of commercial crops like cotton, vegetables and fruits, where the onslaught of insect pests and also the consumption of pesticides is more. For the suppression of borers in the above mentioned crops, trichogrammatids have been widely used as biological control agent and of which particularly *H. armigera P. xylostella* L. have developed resistance to major group of pesticides. Many of the pesticides are extremely toxic to bioagents and this has warranted the development of tolerant strain of parasitoid. The trichogrammatids are susceptible to a broad spectrum of insecticides and reduced parasitism has been reported in *T exiguum* and *T. pretiosum* against *H. zea* and *Manduca* sp. in plots treated with pyrethroids (Campbell et al. 1991) (Table 3).

Similarly, the drift of pesticide even a mile away and single application of pesticides in cotton reduced the efficacy of trichogrammatids (Stinner et al. 1974; Bull and House 1983; Bull and Coleman 1985; Jalali and Singh 1993). Artificial selection in the laboratory improved resistance in *Aphytis melinus* generating a carbaryl-tolerant line with LC_{50} 5.13-time as great as the corresponding base colony and 19.7-time

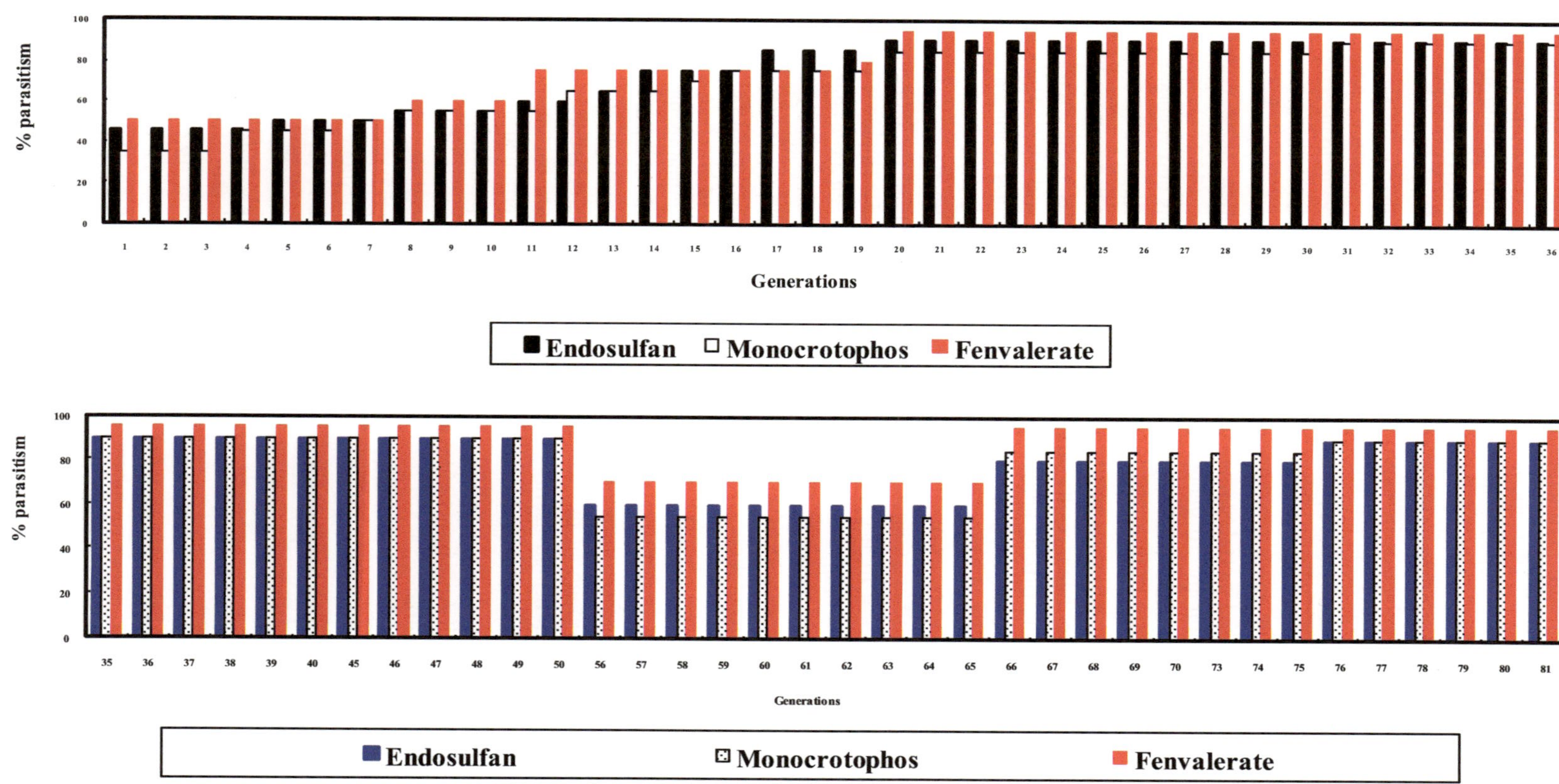

Fig. 1 Percent parasitism by T. chilonis during generations in various insecticides and high temperature (32–38 °C)

Table 3 Development of multiple insecticides and temperature tolerant strain of *T. chilonis*

Generations/temperature and RH	Endosulfan % mortality after 6 h	Monocrotophos % mortality after 6 h	Fenvalerate % mortality after 6 h
01/32–38° and 60%	95.0	100.0	94.0
36/32–38° and 60%[a]	80.0	85.0	75.0
37/32–38° and 60%	80.0	85.0	75.0
38/32–38° and 60%	80.0	85.0	75.0
39/32–38° and 60%	80.0	85.0	75.0
40/32–38° and 60%	80.0	85.0	75.0
41/32–38° and 60%	80.0	85.0	75.0
42/32–38° and 60%	80.0	85.0	75.0
43/32–38° and 60%	80.0	85.0	75.0
44/32–38° and 60%	80.0	85.0	75.0
45/32–38° and 60%	80.0	85.0	75.0
46/32–38° and 60%	95.0	100.0	94.0
47/32–38° and 60%	95.0	100.0	94.0
48/32–38° and 60%	95.0	100.0	91.0
49/32–38° and 60%	95.0	100.0	94.0
50/32–38° and 60%	95.0	100.0	94.0
51/32–38° and 60%[b]	95.0	100.0	95.0
52/32–38° and 60%	95.0	100.0	95.0
53/32–38° and 60%	95.0	100.0	95.0
54/32–38° and 60%	95.0	100.0	95.0
55/32–38° and 60%	95.0	100.0	95.0
56/32–38° and 60%	95.0	100.0	95.0
57/32–38° and 60%	95.0	100.0	95.0
58/32–38° and 60%	95.0	100.0	95.0
59/32–38° and 60%	85.0	95.0	90.0
60/32–38° and 60%	85.0	95.0	90.0
61/32–38° and 60%	70.0	80.0	60.0
62/32–38° and 60%	70.0	80.0	60.0
63/32–38° and 60%	70.0	80.0	60.0
64/32–38° and 60%	70.0	80.0	60.0
65/32–38° and 60%	70.0	80.0	60.0
66/32–38° and 60%	70.0	80.0	60.0
67/32–38° and 60%	70.0	80.0	60.0
68/32–38° and 60%	70.0	80.0	60.0
69/32–38° and 60%	70.0	80.0	60.0
70/32–38° and 60%	70.0	80.0	60.0
71/32–38° and 60%	70.0	75.0	57.0
72/32–38° and 60%	70.0	75.0	57.0
73/32–38° and 60%	70.0	75.0	57.0
74/32–38° and 60%	70.0	75.0	57.0
75/32–38° and 60%	70.0	75.0	57.0
76/32–38° and 60%	70.0	75.0	57.0
77/32–38° and 60%	70.0	75.0	57.0
78/32–38° and 60%	70.0	75.0	57.0
79/32–38° and 60%	70.0	75.0	57.0
80/32–38° and 60%	70.0	75.0	57.0
81/32–38° and 60%	70.0	75.0	57.0

[a] 36 generations to develop strain
[b] increasing insecticide concentration

Table 4 Dose mortality response (LC) of *T. chilonis* to three insecticides after 6 h of constant exposure of tolerant population

Insecticide	LC_{50}	95 % Fiducial limit		LC_{90}	95 % Fiducial limit		Slope ± SE	χ^2
		Lower	Upper		Lower	Upper		
After F_{81} generations								
Endosulfan	5.53	3.64	11.6	81.8	29.7	543.3	1.19±1.17	2.60
Monocrotophos	4.39	–	–	87.8	–	–	0.99±0.22	37.7
Fenvalerate	0.24	0.0	0.0	0.43	0.0	0.0	0.0	0.0

that of a relatively susceptible natural population (Rosenheim et al. 1989) (Table 4).

Sequencing Heat Shock Protein (HSP) for Determination of High Temperature Tolerance in Multiple Insecticides and High Temperature Tolerant Strain (MIHTTS) of *T. chilonis*

Sequencing of HSP PCR Products

A 300 bp HSP PCR product was gel eluted and sequenced using HSP-specific primers. Sequencing was done using both forward and reverse primers to check the accuracy of the sequencer. The similarity of the obtained sequence was found out using Blast program. Sequence showed homology with other HSP-70 sequences submitted in the database.

Biotin Labeling of HSP 300 bp Product

The 300 bp PCR was gel eluted and was labeled by random oligonucleotide DNA labeling method. The tube was incubated in water bath for 5–10 min and was cooled on ice. Biotin labeling mix 5 µl was added to klenow pol 1 µl and again incubated at 37 °C in a water bath over night. The biotin labeled probe was detected by spot hybridization using biotin chromogenic detection kit. The biotin labeled probe was detected by spot hybridization using biotin chromogenic detection kit (Fig. 2).

Which resulted intense bright spot. The PCR product was labeled with high efficiency.

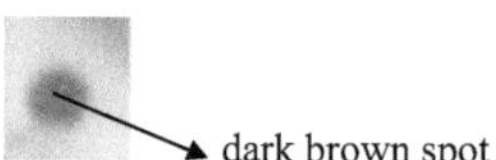

Fig. 2 Biotin labeled probe was detected by spot hybridization using biotin chromogenic detection kit

Biochemical Characterization

GST-Activity

GST—Conjugative activity in the susceptible was significantly lower than that of insecticide tolerant strains (endosulfan, monocrotopho and fenvalerate). Highest activity measured in MIHTTS followed by endosulfan, monocrotophos and fenvalerate tolerant strains. MIHTS shows 2.13-fold increase in activity, endosulfan, monocrotophos and fenvalerate tolerant strains show 1.7-fold increases in activity compared with the susceptible strain. However, there were no significant differences in activity among individual insecticides tolerant strains (Table 5).

Carboxylesterase (CE) Activity

Varied levels of CE activity found in the susceptible and insecticides tolerant strains were treated with endosulfan, monocrotophos and fenvalerate for 30 and 120 min. Table 6 shows that 2.92-fold decreased specific activity found in the susceptible strain was treated with endosulfan for 30 and 120 min, respectively. However, 1.25- and 1.5-decreased specific activity found in endosulfan tolerant strain was treated with endosulfan for 30 and 120 min, respectively. A 9.12- and 9.9-fold decreased specific activity found in susceptible strain treated with monocrotophos for 30 and 120 min, respectively. Similarly a 5.6-fold and 10-fold decreased specific activity found in monocrotophos strain treated with monocrotophos for 30 and 120 min, respectively. A 2.92-fold decreased specific activity found in the susceptible strain was treated with fenvalerate for 30 and 120 min. A. 1.25- and 1.4-fold decreased specific activity found in fenvalerate tolerant strain was treated with fenvalerate for 30 and 120 min, respectively (Table 6).

Michaelis constant (Km) values were calculated for the lab susceptible and tolerant strains for the substrate 1-Napthyl acetate, the substrate concentrations 0.025—1 mM were used. The Km value for the susceptible, endosulfan tolerant, fenvalerate tolerant, monocrotophos tolerant, and MIHTTS was found to be 0.02, 0.1, 0.032, 0.0735, and 0.02 mM, respectively. In terms of respective insecticides tolerance, endosulfan, fenvalerate and monocrotophos tolerant strains showed negative correlation toward esterase activity. Endosulfan tolerant strain showed fivefold higher Km value compared to lab susceptible strain. Monocrotophos tolerant strain showed 3.68-fold higher Km value compared to lab susceptible strain. Fenvalerate tolerant strain showed 1.6-fold higher Km value compared to susceptible strain (Table 7).

Isoenzyme Analysis

The results of native PAGE of esterase electromorphs from susceptible, tolerant strains, susceptible strain treated with endosulfan, monocrotophos and fenvalerate separately and tolerant strains treated with respective insecticides are shown in the Fig. 3. Nondenaturing PAGE of homogenates from each strain stained for 1-Napthyl acetate showed difference in composition of esterase isoenzyme. Esterase bands were designated E1, E2 and E3; E1 as the slowest migrating esterase and E3 as the fastest. E1, E2 and E3 bands were present in fenvalerate tolerant strain treated with fenvalerate. E1 and E3 bands were present in susceptible untreated and susceptible strain treated with monocrotophos. E2 and E3 bands were present in endosulfan tolerant, fenvalerate tolerant, monocrotophos tolerant strains and also in endosulfan tolerant treated with endosulfan and fenvalerate tolerant strain treated with fenvalerate. However, monocrotophos tolerant strain treated with monocrotophos showed only one E3 band. The differences in electromorphs may be related to individual isoenzyme activity to the substrate used and the genetic variation among strains. The results provide baseline information for further research on the involvement of esterases in the insecticides resistance mechanism of *T. chilonis*.

Microsomal Cytochrome C Reductase and p-Nitroanisole O-demethylase

This activity was determined by measuring the production of p-nitrophenol spectrophotometrically. The activity of fenvalerate tolerant strain was 5.65-fold more compared to susceptible strain (Table 8).

Microsomal Cytochrome p-Nitroanisole O-demethylase

One unit will reduce 1.0 micro mole of oxidized cytochrome c in the presence of 100 micro molar NADPH per minute at pH 7.8 at 25°C. The activity of resistant strain to fenvalerate was 10.5-fold more (Table 9).

Wu and Jiang (2003) reported insecticide resistance was associated with insensitivity to acetyl cholinesterase (AchE) in *P. xylostella* and *D. rapae* based on the kinetic parameters. Again in *D. rapae* insecticide resistance is due to AchE insensitivity and detoxification enzymes. Shusheng et al. (2003) found that fenvalerate resistance in *Cotesia plutellae* is positively related to mixed function oxidase (MFO), but not related to carboxylesterase (CE) and esterase (Es) activity. Perez-Mendoza (2000) studied the biochemical mechanism of resistance in a malathion tolerant strain of the *Hebrobracon hebetor*. It was found that malathion resistance in *H. hebetor* is associated with both an increased activity of the esterase E3 and null alleles of the esterases El and E2.

Table 5 GST activity in susceptible insecticide tolerant and MIHTT strain

Strain	Activity (nmol CDNB conjugated mg sample$-^{1}$ min$-^{1}$)
Susceptible	11.5
Endosulfan tolerant	20.0
Monocrotophos tolerant	20.5
Fenvalerate tolerant	20.3
MIHTTS	24.5

Table 6 Carboxylesterase activity in lab susceptible and insecticide tolerant strains treated with insecticides

Strain	Insecticide treated	Pesticide treated for (minutes)	Specific activity
Susceptible	Nil	Nil	180
Endosulfan tolerant	Nil	Nil	179
Susceptible	Endosulfan	30	61.4
Endosulfan tolerant	Endosulfan	30	143
Susceptible	Endosulfan	120	61.27
Endosulfan tolerant	Endosulfan	120	120
Monocrotophos tolerant	Nil	Nil	200
Susceptible	Monocrotophos	30	19.72
Monocrotophos tolerant	Monocrotophos	30	35.53
Susceptible	Monocrotophos	120	18.15
Monocrotophos tolerant	Monocrotophos	120	20
Fenvalerate tolerant	Nil	Nil	164
Susceptible	Fenvalerate	30	61.5
Fenvalerate tolerant	Fenvalerate	30	131
Susceptible	Fenvalerate	120	61.27
Fenvalerate tolerant	Fenvalerate	120	117

Table 7 Michaelis constant (Km) values of Carboxyl-esterase activity for 1-NA for susceptible and insecticide tolerant strains

Strain	Km (mM)
Lab susceptible	0.02
Endosulfan tolerant	0.1
Fenvalerate tolerant	0.032
Monocrotophos tolerant	0.0735
MIHTTS	0.02

Field Efficacy of the Multiple Insecticides and High Temperature Tolerant and Susceptible Strains of *T. chilonis* on Cotton, Tomato and Cabbage Pests in Comparison with Insecticides Alone

Cotton

The larval population recorded in tolerant *T. chilonis* plots was 0.5 larva/plant compared to 0.8 in laboratory population and 1.2 larvae/plant in farmers practice plots. The significant reduction in larval population in tolerant *T. chilonis* plots indicated tolerant *T. chilonis* survival in the sprayed cotton field and their ability to parasitize the eggs in greater numbers than the laboratory population.

During 1st year of trial, the mean boll damage was significantly lower in MIHTTS of *T. chilonis* plots compared to susceptible strain and farmers practice plots. The bad open bolls recorded

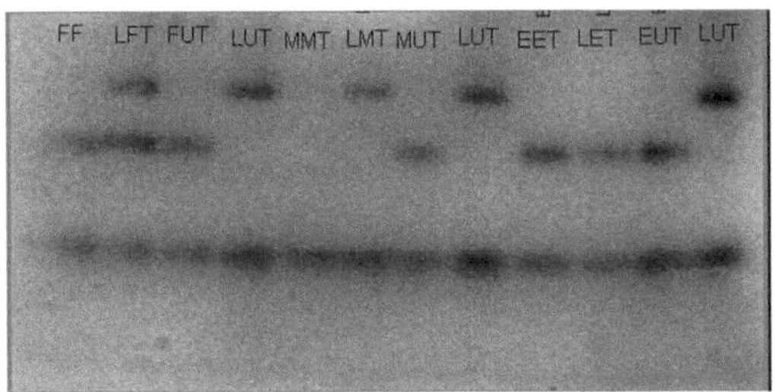

Fig. 3 Esterase isoenzymes banding pattern on native PAGE (Lane 1(*FF*)- Fenvalerate tolerant strain exposed to fenvalerate for 30 min; Lane 2 (*LFT*) Susceptible strain exposed to fenvalerate for 30 min; Lane 3 (*FUT*) Fenvalerate tolerant strain; Lane 4 (*LUT*) Susceptible strain; Lane 5 (*MMT*) Monocrotophos tolerant strain exposed to monocrotophos for 30 min; Lane 6 (*LMT*) Susceptible strain exposed to monocrotophos for 30 min; Lane 7 (*MUT*) Monocrotophos tolerant strain; Lane 8 (*LUT*) Susceptible strain; Lane 9 (*LUT*) Endosulfan tolerant strain exposed to endosulfan for 30 min; Lane 10 (*LUT*) Susceptible strain exposed to endosulfan for 30 min; Lane11 (*LUT*) Endosulfan tolerant strain; Lane 12 (*LUT*) Susceptible strain)

Table 8 Microsomal p-Nitroanisole O-demethylase activity

Strain	Units: mole min^{-1} mg^{-1} sample
Susceptible	0.02
Fenvalerate tolerant	0.113

Table 9 Cytochrome C-reductase activity

Strain	Units/ml
Susceptible	0.010
Fenvalerate tolerant	0.105

were 2.4/plant in field-collected population compared to 3.7 in laboratory population and 4.5 in farmers' practice plots, good open bolls recorded were 15.7, 11.0 and 12.0/plant, respectively. Percent boll damage generally in plots receiving MIHTTS of *T. chilonis* was significantly lower 2.2 compared to 3.3 and 3.5/plant in susceptible strain and farmers practice plots, respectively. The results showed that release of MIHTTS of *T. chilonis* was useful in reducing egg hatch, which resulted in lesser boll damage (Table 14). Plots that received the combination of MIHTTS of *T. chilonis* + insecticides recorded highest yield of 15.0 q/ha compared to 12.6 q/ha in susceptible strain of *T. chilonis* + insecticides and 14.1 q/ha in farmers practice plots. Besides, significant reduction in insecticide sprays was also observed in MIHTTS of *T. chilonis* released plots.

The larval population recorded in MIHTTS of *T. chilonis* plots was 0.3 larva/plant compared to 0.9 in susceptible strain of *T. chilonis* and 1.3 larvae/plant in farmers' practice plots. The significant reduction in larval population in tolerant strain plots indicated that tolerant strain's ability to survive in the sprayed cotton field and their ability to parasitize the eggs in greater numbers than the susceptible strain (Table 10).

During 2nd year of field evaluation, the mean boll damage was significantly lower in MIHTTS of *T. chilonis* plots compared to susceptible strain of *T. chilonis* and farmers' practice plots. The bad open bolls recorded were 2.7/plant in field-collected population compared to 4.2 in susceptible strain of *T. chilonis* and 5.6 in farmers' practice plots, good open bolls recorded were 18.3, 10.0, and 14.3/plant, respectively. Percent boll damage generally in plots receiving MIHTTS plots was significantly lower—3.5 compared to 4.8 and 4.1 in susceptible strain and farmers' practice plots, respectively. The results showed that release of MIHTTS strain was useful in reducing egg hatch, which resulted in lesser boll damage (Table 11).

Plots that received the combination of field-collected population + insecticides recorded highest yield of 14.8 q/ha compared to 12.4 q/ha in laboratory population + insecticides and 14.3 q/ha in farmers' practice plots. Besides, significant reduction in insecticide sprays was also observed in field-collected population released plots.

The mean boll damage was significantly lower in MIHTTS plots compared to susceptible strain and farmers' practice plots. The bad open bolls recorded were 2.7/plant in MIHTTS plots compared to 4.2 in susceptible strain and 5.6 in farmers' practice plots, good open bolls recorded were 18.3, 10.0, and 14.3/plant, respectively. Percent boll damage generally in plots receiving MIHTTS plots was significantly lower 3.5 compared to 4.8 and 4.1 in susceptible strain and farmers' practice plots, respectively. The results showed that release of MIHTTS was useful in

Table 10 Effect of field release of MIHTTS of *T. chilonis* on its efficacy against cotton bollworms and yield at Sirsa (Haryana) during 1st year trial

Treatments	No. of larvae/Plant	BOB	GOB	% boll damage	Yield (q/ha)
MIHTTS of *T. chilonis*	0.5	2.4	15.7	2.2	15.0
Susceptible strain	0.8	3.7	11.0	3.5	12.6
Farmers' practice	1.2	4.6	12.0	3.9	14.1
Untreated control	0.8	3.6	12.9	3.2	13.9
SEM. (±)	0.12	0.13	0.55	0.17	0.54
LSD (0.05)	0.27	0.30	1.27	0.39	1.25
LSD (0.01)	0.39	0.43	1.84	0.57	1.83
CV (%)	21.9	5.67	6.74	8.37	6.19

Table 11 Effect of field release of MIHTTS of *T. chilonis* on its efficacy against cotton bollworms and yield at Sirsa (Haryana) during 2nd year trial

Treatments	No. of larvae/ Plant	BOB	GOB	% boll damage	Yield (q/ha)
MIHTTS of *T. chilonis*	0.3	2.7	18.3	3.5	14.8
Susceptible strain	0.9	4.2	10.0	4.8	12.4
Farmers' practice	1.3	5.6	14.3	4.1	14.3
Untreated control	0.22	0.18	0.65	0.17	0.74
SEM. (±)	0.48	0.43	1.82	0.39	1.52
LSD (0.05)	0.73	0.82	2.45	0.57	2.12
LSD (0.01)	32.4	8.43	8.56	8.37	7.48

BOB bad open bolls, *GOB* good open bolls

Table 12 Effect of field release of MIHTTS of *T. chilonis* on percent egg parasitism and larval population of *H. armigera* on tomato in Malur (Karnataka) during 1st year trial

Treatments	Egg parasitism (%)	No. of larvae/Plant	Fruit bored (%)	Yield (q/ha)
MIHTRS of *T. chilonis*	80.0	1.5	0.78	495.0
Susceptible strain	13.3	4.8	4.05	407.1
Farmers' practice	5.0	3.7	0.91	421.5
Untreated control	20.4	5.9	11.5	257.4
SEM. (±)	3.3	0.66	0.77	1.35
LSD (0.05)	7.26	1.44	1.67	2.94
LSD (0.01)	10.18	2.02	2.34	4.12
CV (%)	17.75	26.36	28.12	0.54

reducing egg hatch, which resulted in lesser boll damage (Table 11).

Plots that received the combination of MIHTTS + insecticides recorded highest yield 14.8 q/ha compared to 12.4 q/ha in susceptible strain + insecticides and 14.3 q/ha in farmers' practice plots. Besides, significant reduction in insecticide sprays was also observed in field-collected population released plots.

Tomato

During 1st year field evaluation, egg parasitism by *Trichogramma* in various plots before initiation of releases was very low ranging from 1.0–4.0%. The post-treatment egg parasitism in MIHTTS of *T. chilonis* plots was significantly higher than egg parasitism by susceptible strain plots. In general, in farmers' practice plots egg parasitism was least, even lesser than untreated control (Table 12). The releases of *T. chilonis* commenced with moth capture in pheromone traps and coincided with period of egg laying

Table 13 Effect of field release of MIHTTS of *T. chilonis* on percent egg parasitism and larval population of *H. armigera* on tomato in Malur (Karnataka) during 2nd year trial

Treatments	Egg parasitism (%)	No. of larvae/plant	% fruit bored	Yield (q/ha)
MIHTTS of *T. chilonis*	65.5	1.8	1.17	502.0
Susceptible strain	27.2	3.6	7.11	364.0
Farmers' practice	8.8	1.9	2.20	419.0
Untreated control	12.6	4.8	14.65	236.0
SEM. (±)	1.9	0.85	1.12	3.39
LSD (0.05)	5.47	1.88	1.95	12.56
LSD (0.01)	9.36	3.14	3.34	23.48
CV (%)	14.44	18.52	19.78	12.45

of *H. armigera* at all the plots. The MIHTTS released indicated its ability to survive in higher proportion than laboratory population and that of natural parasitoid population in farmers practice plots and untreated control. The higher parasitism by MIHTTS compared to susceptible strain plots could be due to its tolerance to various insecticides in sprayed situation in the field.

The mean percent damage of fruits increased significantly in different treatments. The percent fruit damage in MIHTTS plots ranged from 0.78 compared to 4.05 per plant in susceptible strain and 0.91 per plant in farmers' practice plots and damage was highest in untreated control where it was 11.5 per plant. The larval population per plant also increased with increase in load of eggs, however, larval population was significantly less in MIHTTS plots + insecticide treated plots compared to susceptible strain + insecticides, farmers' practice plots and untreated control. The yield recorded was highest in MIHTTS plots + insecticides plots 495.0 compared to 407.1, 421.5, and 257.4 q/ha in susceptible strain plots, farmers' practice plots and untreated control, respectively (Table 12).

During 2nd year trial, egg parasitism by *Trichogramma* in various plots before initiation of releases was very low ranging from 0.0 to 1.0%. The post treatment egg parasitism in MIHTTS plots (65.5%) was significantly higher than egg parasitism by susceptible strain plots (27.2%), farmers' practice (8.8%) and untreated control (12.6%). In general in farmers practice plots egg parasitism was least, even lesser than untreated control (Table 13). The releases of *T. chilonis* commenced with moth capture in pheromone traps and coincided with period of egg laying of *H. armigera* at all the plots. The MIHTTS plots released indicated its ability to survive in higher proportion than susceptible strain plots and that of natural parasitoid population in farmers' practice plots and untreated control. The higher parasitism by MIHTTS compared to susceptible strain could be due to its tolerance to various insecticides in sprayed situation in the field.

The mean percent damage of fruits increased significantly in different treatments. The larval population per plant also increased with increase in load of eggs, however, larval population was significantly less in MIHTTS plots + insecticide treated plots (1.8/plant) compared to susceptible strain + insecticides (3.6/plant), farmers' practice plots (1.9/plant)) and untreated control (4.8/plant). The percent fruit damage in MIHTTS plots ranged from 1.17%, susceptible strain 7.11%, farmers' practice 2.2%, and damage was highest in untreated control where it was 14.65%. The yield recorded was highest in MIHTTS + insecticides plots (502 q/ha) compared to 364 q/ha in susceptible strain + insecticides, 419.0 q/ha in farmers' practice plots and untreated control (236 q/ha), respectively (Table 13).

Cabbage

In the field evaluation studies during 1st year, no egg parasitism was recorded by *Trichogramma* in any treatments. The post treatment egg parasitism in MIHTTS of *T. chilonis* plots was significantly higher (35.0%) than egg parasitism by susceptible strain (1.2%), farmers' practice (0.0%) and untreated control (2.0%). The MIHTTS released, indicated its ability to survive in higher proportion than susceptible strain. The higher parasitism by MIHTTS compared to susceptible strain could be due to its tolerance to various insecticides in sprayed situation in the field (Table 14).

Table 14 Effect of field release of MIHTTS of *T. chilonis* on its efficacy against diamondback moth of cabbage at Malur (Karnataka) during 1st year of trial

Treatments	Egg parasitism of stem borer (%)	No. of larvae/plant	Feeding punctures/plant	Yield (q/ha)
MIHTTS of *T. chilonis*	35.0	0.8	2.4	295.0
Susceptible strain	1.2	2.5	39.0	230.0
Farmers' practice	0.0	6.2	35.0	265.0
Untreated control	2.0	20.3	80.1	110.0
SEM. (±)	0.53	0.12	0.69	4.89
LSD (0.05)	1.16	0.27	1.50	10.66
LSD (0.01)	1.62	0.38	2.11	14.94
CV (%)	8.78	2.61	2.78	3.44

Table 15 Effect of field release of MIHTTS of *T. chilonis* on its efficacy against diamondback moth of cabbage at Malur (Karnataka) during 2nd year trial

Treatments	Egg parasitism of stem borer (%)	No. of larvae/plant	Feeding punctures/plant	Yield (q/ha)
MIHTTS of *T. chilonis*	46.53	0.33	10.90	315.0
Susceptible strain	0.80	5.75	94.70	228.0
Farmers' practice	0.0	2.10	15.10	278.0
Untreated control	3.70	12.75	134.16	128.0
SEM. (±)	0.78	0.32	1.12	6.45
LSD (0.05)	2.78	0.95	1.78	12.36
LSD (0.01)	5.17	2.23	3.12	19.56
CV (%)	15.99	8.79	12.65	8.86

The number of larvae also reduced drastically due to higher parasitism in MIHTTS plots. The larvae per plant was significantly less 0.8 in MIHTTS plots compared to 2.5, 6.2, and 20.3/plant in susceptible strain, farmers' practice and untreated control plots. Feeding punctures were also significantly less in MIHTTS released plots as compared to other treatments and in untreated control feeding punctures recorded were 80.1/plant, which was significantly higher than all treatments. This indicates that it may not be possible to raise cabbage without control measures. The yield recorded was highest in MIHTTS + insecticides plots 295.0 compared to 230.0, 265.0, and 110.0 q/ha in susceptible strain + insecticides plots, farmers' practice plots and untreated control, respectively (Table 14).

During the 2nd year trial, no egg parasitism was recorded by *Trichogramma* in any treatments. The post treatment egg parasitism in MIHTTS plots was significantly higher (46.53 %) than egg parasitism by susceptible strain (0.80 %), farmers' practice (0.0 %) and untreated control (3.70 %). The MIHTTS released indicated its ability to survive in higher proportion than susceptible strain. The higher parasitism by MIHTTS compared to susceptible strain could be due to its tolerance to various insecticides in sprayed situation in the field (Table 15).

The number of larvae also reduced drastically due to higher parasitism in MIHTTS plots. The larvae per plant was significantly less 0.33 in MIHTTS plots compared to 5.75, 2.10, and 12.75/plant in susceptible strain, farmers' practice and untreated control plots, respectively. Feeding punctures were also significantly less in the MIHTTS released plots (10.9/plant) as compared to susceptible strain plot (94.7/plant), farmers' practice (15.10/plant) and in untreated control feeding punctures recorded were 134.16/plant, which was significantly higher than all treatments. This indicates that it may not be possible to raise cabbage without control measures. The yield recorded was highest in MIHTTS + insecticides plots 315.0 q/ha compared to 228.0, 278.0, and 128.0 q/ha in susceptible strain + insecticides plots, farmers practice plots, and untreated control, respectively (Table 15).

Torre (2012) observed that simultaneous use of biological and chemical control is one of the most important goals of IPM, but has rarely been achieved. One explanation for this failure may be the limited number of evaluations of field populations of natural enemies for pesticide tolerance or resistance. Earlier work done has suggested that insecticide-tolerant natural enemies have proved effective in sprayed as compared to laboratory-bred populations. Jalali et al. (2006a) developed an endosulfan tolerant strain of *T. chilonis* and reported 55 % parasitism under sprayed condition compared to less than 5 % by a susceptible strain. Later on a multiple insecticides tolerant strain of *T. chilonis* was evaluated against American bollworm, *Helicoverpa armigera* on cotton all over India and proved efficacy of parasitoid under sprayed situation (Jalali et al. 2006b). Ballal et al. (2009) evaluated an endosulfan tolerant strain of *T. chilonis* against *H. armigera* cotton plants and reported that tolerant strain in conjunction with endosulfan spray resulted in significantly higher pest mortality in comparison to the endosulfan treatment or release of *T. chilonis* alone.

References

Abraham CC, Pradhan S (1976) Studies on developing races of *Trichogramma australicum* Girault suitable for high temperature-low humidity conditions. Madras Agri J 63:550–556

Anonymous (2003) Annual Report for the year 2002–2003. Project Directorate of Biological Control, Bangalore, India

Ballal CR, Srinivasan R, Jalali SK (2009) Evaluation of an endosulfan tolerant strain of *Trichogramma chilonis* on cotton. BioControl 54:723–732

Bull DL, Coleman RJ (1985) Effects of pesticides on *Trichogramma* spp. Southwestern Entomol 8:156–168

Bull DL, House VS (1983) Effects of different insecticides on parasitism of host eggs by *Trichogramma pretiosum* Riley. Southwest Entomol 8:46–53

Campbell CD, Walgenbach JF, Kennedy GG (1991) Effect of parasitoids on lepidopterous pests in insecticide treated and untreated tomatoes in western North Carolina. J Econ Entomol 84:1662–1667

Carrière Y, Boivin G (2001) Constraints on the evolution of thermal sensitivity of foraging in *Trichogramma*: genetic trade-offs and plasticity in maternal selection. Am Nat 157:570–581

Dudani AT, Sengupta S (1991) Status of banned and bannable pesticides. Voluntary Health Association of India, New Delhi

Jalali SK, Singh SP (1993) Susceptibility of various stages of *Trichogrammatoidea armigera* Nagaraja to some pesticides and effect of residues on survival and parasitising ability. Biocontrol Sci Technol 3:21–27

Jalali SK, Singh SP, Venkatesan T (2002) Natural enemies of cotton bollworms in sprayed fields in different states of India. Nat J Plant Improv 5:31–33

Jalali SK, Singh SP, Venkatesan T, Murthy KS, Lalitha Y (2006a) Development of endosulfan tolerant strain of an egg parasitoid *Trichogramma chilonis* Ishii (Hymenoptera: Trichogrammatidae). Indian J Exp Biol 44:584–590

Jalali SK, Venkatesan T, Murthy KS, Rabindra RJ, Lalitha Y, Udikeri SS, Bheemanna M, Sreenivas AG, Balagurunathan R, Yadav DN (2006b) Field efficacy of multiple insecticides tolerant strain of *Trichogramma chilonis* Ishii against American bollworm, *Helicoverpa armigera* (Hübner) on cotton. Indian J Plant Prot 34:173–180

López JD Jr, Morrison RK (1980) Overwintering of *Trichogramma pretiosum* in central Texas. Environ Entomol 9:75–78

Perez-Mendoza J, Fabrick JA, Zhu KY, Baker JE (2000) Alterations in esterases are associated with malathion resistance in *Habrobracon hebetor* (Hymenoptera: Braconidae). J Econ Entomol 93:31–37

Rosenheim JA, Hoy MA, Hoy A, Gorden J, Stewart JR (1989) Selecting for insecticide resistance in a California red scale parasitoid. Calif Agric 1989:17–18

Scholler M, Hassan SA (2001) Comparative biology and life tables of *Trichogramma evanescens* and *T. cacoeciae* with *Ephestia elutella*. Ent Exp Appl 98:35–40

Scott M, Berrigan D, Hoffmann AA (1997) Costs and benefits of acclimation to elevated temperature in *Trichogramma carverae*. Ent Exp Appl 85:211–219

Shu-sheng L, Yuan-xi Li, Zhen-hua T (2003) Host resistance to an insecticide favors selection of resistance in the parasitoid, *Cotesia plutellae* (Hymenoptera: Braconidae). Biol Control 28:137–143

Singh SP, Jalali SK (1994) Trichogrammatids. Tech Bull, No 7, pp 95 Project Directorate of Biological Control, Bangalore, India

Stinner RE, Ridgway RL, Coppedge JR, Morrison RK, Dickerson WA Jr (1974) Parasitism of *Heliothis* eggs after field releases of *Trichogramma pretiosum* in cotton. Environ Entomol 3:497–500

Thomson LJ, Robinson M, Hoffmann AA (2001) Field and laboratory evidence for acclimation without costs in an egg parasitoid. Functional Ecol 15:217–221

Torre JB (2012) Insecticide resistance in natural enemies-seeking for integration of chemical and biological controls. J Biofert Biopest 3:e104. doi:10.4172/2155-6202.1000e104

Wittwer SH (1979) Future technology advances in agriculture and their impact on regulatory environment. BioScience 29:603–610

Wu G, Jiang SR (2003) Resistance mechanisms to methamidosphos and methomyl in the parasitoid *Diaeretiella rapae* (M'Intosh). Acta Entomol Sin 46:292–298

Reproductive Alterations by *Wolbachia* in the Braconid *Cotesia vestalis* (Haliday)

K. Srinivasa Murthy, T. Venkatesan, S. K. Jalali and S. L. Ramya

Abstract

The endosymbiotic bacteria *Wolbachia*, associated with a number of hymenopteran parasitoids play role in the metabolism, physiology, and reproduction of their hosts. The impact on female progeny due to *Wolbachia* infection in the braconid *Cotesia vestalis* was investigated in different geographic populations of the parasitoid. The populations cured of *Wolbachia* recorded a reduction in male progeny compared to those infected. The sex ratio skewed toward males in the *Wolbachia* eliminated populations, altered toward higher females, when there was infection. There was 36.6% increase in female progeny over the males. The exploitation of *Wolbachia* for the biological manipulations of the parasitoid for effective pest management is discussed.

Keywords

Cotesia vestalis · Endosymbiont · Fitness benefits · *Wolbachia*

Introduction

Symbiotic bacteria have been associated with a number of parasitoids, known to inflict types of metabolic, physiological, and reproductive alterations, with the sex regulatory bacteria being the most frequent ones. Among the sex regulators, *Wolbachia* infects a large number of species including parasitic hymenoptera. About 2/3rd of all insect species are infected with *Wolbachia* (Werren and Windsor 2000). To date *Wolbachia* have been detected in 31 genera and 70 species of parasitic hymenoptera, as well as three dipeteran parasitoids, 26% of parasitoid wasps have *Wolbachia* (Iturbe and O'Neill 2007). *Wolbachia* spp. is maternally inherited obligate intracellular bacteria belonging to the α proteobacteria. The bacteria infect the reproductive tissues (ovaries and testes) of arthropods and are transmitted through the egg cytoplasm and alter reproduction in their hosts. *Wolbachia* arthropod relationships have variously been described as mutualistic (Girin and Bouletreau 1995), parasitic

K. S. Murthy (✉) · T. Venkatesan · S. K. Jalali ·
S. L. Ramya
National Bureau of Agriculturally Important Insects,
Bellary Road, Bangalore 560024, Karnataka, India
e-mail: ksm239@rediffmail.com

A. K. Chakravarthy (ed.), *New Horizons in Insect Science: Towards Sustainable Pest Management*,
DOI 10.1007/978-81-322-2089-3_30, © Springer India 2015

(Wade and Chang 1995; Werren et al. 1995; Stolk and Stouthamer 1996), pathogenic (Min and Benzer 1997), and symbiotic (James and Bollard 2000; Shoemaker et al. 2002). In arthropods, they have been implicated in several host reproductive modifications, including cytoplasmic incompatability, parthenogenesis, feminization, and male killing (Werren et al. 2008). These symbionts are the reproductive manipulators that promote their own spread in a population by encouraging the production of female progeny or reducing reproduction of uninfected females (Delgado and Cook 2009). These have been investigated for their potential use to host control because of their ability to modulate the sex ratio.

Cotesia vestalis (Haliday) (= *Cotesia plutellae* (Kurdjumov) (Hymenoptera: Braconidae), a solitary larval endoparasitoid, is one of the most important biological control agents of the diamondback moth, *Plutella xylostella* (Linnaeus) (Lepidoptera: Plutellidae), regarded as the most significant pest of *Brassica* crops. Parasitoids have developed a natural arsenal and a number of physiological mechanisms, to enable them to successfully colonize the host and regulate host development to their own benefit. One of these is through their association with symbionts. Obligate symbionts are required for the successful parasitism and suppression of the host immune system, as well as for inducing physiological alterations in the parasitized host (Consoli and Elliot 2006). Molecular evidence for the presence of endosymbiotic bacteria Wolbachia in *Cotesia* populations has been well documented (Rattan et al. 2011). The impact of *Wolbachia* infection on the parasitoid *C. vestalis* was investigated to rationalize the use of the parasitoid in pest management programmes.

Material and Methods

Seven populations of *C. vestalis* collected from different geographic locations of the country (Bangalore, Hoskote, Malur and Kolar, Bhubaneshwar, Varanasi, Salem, Shillong, Tirupathi, and Hyderabad) were considered for the study. Individual cocoons of *C. vestalis* from *P. xylostella* larvae were collected from field-grown cauliflower plants. The colony of *P. xylostella* was maintained on potted (0.2 × 0.3 m) mustard seedlings, *Brassica juncea* L. Czern for oviposition, in ventilated oviposition cages for the development of larval stages. Host larvae at early L3 stage were exposed to *C. vestalis* on mustard seedlings in ventilated cages and maintained on the plant until cocoon formation. Cocoons were collected and held in wooden-plastic cages (0.3 m^3) until adult emergence. Adult wasps were fed on honey.

Ten adult parasitoids each from different populations were surface sterilized in a series of double-distilled water and 70 % ethanol washes, then were freeze-killed at − 80 °C and transferred to an Eppendorf tubes and homogenization was done by crushing the adult in 20 µl of 5 % Chelex 100 MB DNA extraction buffer (BIO-RAD) using a DNA free disposable polypropylene pestle. This was followed by incubation for 3 h at 56 °C and then at 100 °C for 10 min. Eight microliters of 2.5 mg/ml Proteinase K solution were added to the tubes. Solutions were incubated at 55 °C for 1 h, heated twice to 90 °C for 15 min, and centrifuged for 2 min at 14,000 rpm. The supernatant was refreshed by a 1 min 14,000 rpm centrifugation. The supernatant was collected and gently mixed with 0.2 volume of Na-acetate (3 mM, pH 5.2) and 2 volumes of 100 % ethanol. After precipitation for 2 h at −20°C, the DNA was washed with 70 % ethanol, air dried and finally resuspended in 20 µl double-distilled water. DNA sample of 0.3 µl was used for PCR assays.

A molecular diagnostic approach was adapted for the detection of *Wolbachia* infection, since *Wolbachia* cannot be cultured. The assay was based on PCR mediated amplification of, and sequence determination of 16S rRNA gene. The presence of *Wolbachia* was verified by a PCR method based on the *Wolbachia* surface protein (wsp). Diagnostic PCR using the *Wolbachia*-specific primer set (forward:5′-CAT ACC TAT TCG AAG GGA TAG-3′; reverse: 5′-AGA TTC GAG TGAAAC CAA TTC-3′) was performed to determine the *Wolbachia* infection status of adult wasps. The PCR reaction was performed in a 500 µl PCR tube with a 25 µl reaction mixtures, each containing 1 mM dNTPs mix (3 µl), 5 ng/µl specific primer

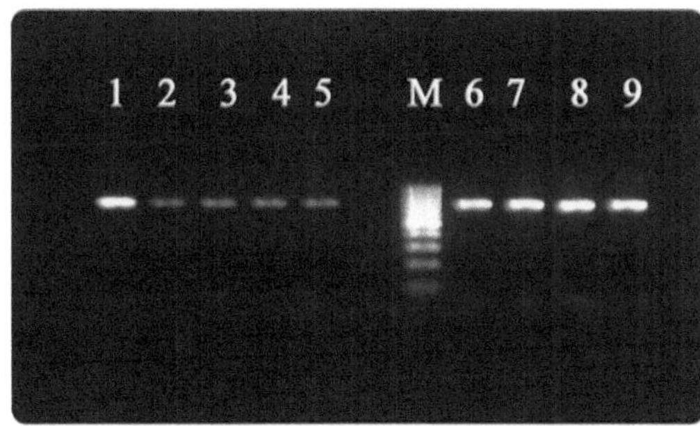

Fig. 1 PCR amplification of *wsp* gene of *Wolbachia* from *Cotesia vestalis* Lane *M*: 100 bp ladder, Lanes *1*, *2* and *3*: *C. vestalis* Bangalore (Hoskote, Malur & Kolar), Lanes *4*: *C. vestalis* (Varanasi), Lane *5*: *C. vestalis* (Tirupathi), Lane *6*: *C. vestalis* (Shillong), Lane *7*: *C. vestalis* (Bhubaneshwar), Lane *8*: *C. vestalis* (Salem) and Lane *9*: *C. vestalis* (Hyderabad)

(5 µl), 2.0 U Taq Polymerase (MBI, Fermentas), and 2 µl of template DNA solution (30 ng) in Taq reaction buffer. The reaction was set in the Thermal Cycler (Biorad Laboratories). The temperature profile for *Wolbachia*-specific PCR was a predenaturing step of 2 min at 94 °C, followed by 38 cycles of 30 s at 94 °C, 45 s at 55 °C, and 90 s at 72 °C, with a final extension step of 10 min at 72 °C. The amplified PCR products were resolved by horizontal gel electrophoresis in 1.8% Agarose gel with a low range ladder (Fermentas Mass Ruler 1000 bp), visualized under UV transilluminator and size of the amplified *Wolbachia*-specific bands was estimated by comparison with a co-migrating molecular weight standard. The wsp gene fragments from *Wolbachia* bacteria in *C. vestalis were* sequenced.

To determine the role of *Wolbachia* in the fitness attributes of the parasitoid, curing of *Wolbachia* with antibiotic, and feeding of *Wolbachia* to the populations free of *Wolbachia* was done. Heat and antibiotic treatments are both estimated methods of producing *Wolbachia* free individuals (Grieiner et al. 2002). Antibiotic Tetracycline (0.02%) a potent inhibitor of DNA dependent RNA polymerase of bacteria was used to produce *Wolbachia* free hosts. Tetracycline treatment of adults was accomplished by introducing a solution of tetracycline (0.02%) dissolved in 50% honey solution and fed to adult parasitoids. Feeding of the antibiotic was done for more than ten generations and each generation was checked for presence of *Wolbachia* by molecular methods until no detectable levels of *wsp* gene was amplified.

Wolbachia was isolated and pelleted from infected parasitoids. The protocol prescribed by Iturbe et al. (2010) was followed to obtain pure *Wolbachia.* Approximately 100 adults of the parasitoid were collected, surface sterilized for 3 min in 70% ethanol followed by sterile water. The insects were homogenized using a 40 ml cold SPG buffer (218 mM Sucrose, 3.8 mM KH_2PO_4, 7.2 mM KH_2PO_4, 4.9 mM L-Glutamate, pH 7.2). The extract was split in to four Falcon tubes containing another 20 ml SPG buffer each and centrifuged at 3200 × g for 15 min. The supernatant was subsequently filtered through syringe filters and *Wolbachia* were pelleted at 18,000 × g for 20 min in a Oakridge tube and resuspended in 4 × 750 µl cold SPG Buffer in eppendorf tubes. Intact *Wolbachia* were treated with 20 µl DNAse 1 for 30 min at 37 °C to remove host DNA contamination without disrupting the cells and resuspended in eppendorf tubes. The pellet was fed to the cured population of *C. vestalis* by mixing the pellet with 50% honey and feeding. The feeding was done for over ten generations. Presence of *Wolbachia* was detected by assays for *wsp* genes. The PCR using *Wolbachia*-specific primers for the wsp gene was performed to detect *Wolbachia* infection in the populations of *C. vestalis* from Bangalore, Bhubaneshwar, Hyderabad, Salem Shillong, Tirupathi, and Varanasi. All the populations revealed the presence of *Wolbachia* (Fig. 1).

Results and Discussion

Infection with *Wolbachia* resulted in greater female progeny production. Observations indicated that the sex ratio that was skewed toward males in the population, free of infection altered toward females when there was infection. There was 36.6% increase in female progeny over the males. (Table 1). *Wolbachia* can alter the normal pattern of sex determination in their host. The bacterium distorts host sex ratio via male killing, parthenogenesis induction or feminization. *Wol-*

Table 1 Impact of *Wolbachia* on sex ratio (% females) in different populations of *C.vestalis*

Population	% of females		Sex ratio (Female: Male)[a]	
	Wolbachia fed	*Wolbachia* cured	*Wolbachia* fed	*Wolbachia* cured
Bhubaneshwar	68.9	55.86	1.79:1	1:1.45
Bangalore	70.9	60.60	1.65:1	1:1.41
Shillong	74.07	66.22	1.51:1	1:1.35
Tirupathi	74.07	58.17	1.69:1	1:1.35
Varanasi	70.42	61.72	1:62:1	1:1.42
Salem	72.99	68.44	1.44:1	1:1.37
Hyderabad	72.99	70.92	1.41:1	1:1.37
Control	68.3	71.42	1:1.4	1:1.2

[a] Mean of 10 replications
CD (P=0.01 %) Populations (P) 16.4 NS
Treatments (T) 6.6 Sig
P x T 2.9 NS

bachia cause genetic male embryos to develop phenotypically as functional females. *Wolbachia* induces feminization by blocking the formation of androgenic glands, which produces the androgenic hormone responsible for male differentiation (Johanowicz and Hoy 2008).

Weeks et al. (2002) indicated that the shift toward females will occur without the elevated mortality of males, if nuclear genes or meiotic drive genes are involved in the distortion. Sex reversion, changing genetic males into functional neo females might occur due to reallocation of maternal resources from dead male embryos to their sisters, provides a direct physiological mechanism through which fitness compensation could favor male killing by cellular endosymbionts as in Harlequin beetle, *Acroeinus longimennus* (Zeh et al. 2005). Sex ratio distortion in transfected strain of Mediterranean flour moth *Ephestia kuehniella* was due to male killing (Hurst et al. 2002; Fujii et al. 2001; Dong et al. 2006). The removal of *Wolbachia* resulted in the recovery of 1:1 sex ratio. In *Drosophila bifasciata,* the sex ratio distortion was a result of reduced male hatching rate compared to uninfected females, due to the arrest of male embryos during the stage of development (mitotic abnormalities occurring during blastoderm formation and gastrulation) and chromosome segregation (chromatin remodeling) defects within the spindle in male embryos (Dodeine et al. 2001; Jia et al. 2009).

Histone-modifying enzymes or chromatin-remodeling complexes can be targeted to specific promoters by gene-specific or general transcription factors, *Wolbachia* may interfere with any of the transcription pathways that regulate some of these processes (Dobson 2003). Male killing is thought to benefit sibling females by eliminating competition. These illustrate the mutualism where both the host and bacteria are benefited (Narita et al. 2007). Because of the higher production of female progeny the infected females are predicted to be more efficacious in pest control. The feasibility of such utilization depends heavily upon how the transmission of wolbachia and the genes being driven in to a population occurs (Iturbe and O'Neill 2007). Maternally inherited endosymbionts spread through populations by increasing relative fitness of infected females. They achieve this by increasing the fecundity and or survival of infected females relative to uninfected females through metabolic processes (Doughlas 1994), thereby providing benefit to both the host and symbiont in a mutualistic association.

The detection of *Wolbachia* in the parasitoid populations of *C. vestalis* may prove to be useful for biological manipulations of the parasitoid as possible transgene drivers (Fry et al. 2004). These bacteria can drive particular mtDNA haplotypes through populations and alter reproductive biology. They can be utilized as vectors for spreading desirable genetic modifications in pest populations or as microbial agents to enhance productivity of natural parasitoids (Saiful Islam 2007). Infections can be manipulated by elimination, transfection or genetic modifications. However, effective exploitation of *Wolbachia* in pest management requires an in depth exploring of its role in insect biology and population dynamics.

References

Consoli FL, Elliot WK (2006) Symbiofauna associated with the reproductive system of *Cotesia flavipes* and *Doryctobracon areolatus* (Hymenoptera: Braconidae). Braz J Morphol Sci 23:463–470

Delgado AM, Cook JM (2009) Effects of a sex ratio distorting endosymbiont on mtDNA variation in a global insect pest. BMC Evol Biol 9:9–49

Dobson SL (2003) Reversing *Wolbachia*-based populations. Rev Biol 22:283–314

Dodeine F, Vavre F, Fleury F, Loppin B, Hochberg ME, Bouletreau M (2001) Removing symbiotic *Wolbachia* bacteria specifically inhibits oogenesis in a parasitic wasp. Proc Natl Acad Sci U S A 98:6247–6252

Dong P, Wang JJ, Zhao ZM (2006) *Wolbachia* infection and its influence on the reproduction of stored product psocid. *Liposcelis tricolor* (Pscoptera:Liposcelididae). J. Insect Sci 6:1–7

Doughlas AE (1994) Symbiotic interactions.Oxford University Press, Oxford

Fry AJ, Palmer MR, Rand DM (2004) Variable fitness effects of *Wolbachia* infection in *Drosophila melanogaster*. Heredity 93:379–389

Fujii Y, Kageyama D, Hoshizaki S, Ishiawa H, Sasaki T (2001) Transfection of *Wolbachia* in Lepidoptera: the feminiser of the adzuki bean borer, *Ostrinia scapulalis* causes male killing in the Mediterranean flour moth *Ephestia kuhieniella*. Proc Royal Soc Lond Series B Biol Sci 268:855–859

Girin C, Bouletreau M (1995) Microorganisms associated variation in host infestation efficiency in a parasitoid wasp *Trichogramma bourarachae* (Hymenoptera: Trichogrammatidae). Experientia 51:398–401

Grenier S, Silvia MG, Pintureau B, Francois L, Bolland P (2002) Use of Tetracycline in larval diet to study the effect of *Wolbachia* on hsot fecundity and clarify taxonomic status of *Trichogramma* species in cured bisexual lines. J Invertebr Pathol 80:13–21

Hurst GDD, Jiggins FM, Pomiankowski A (2002) Which way to manipulate host reproductions? *Wolbachia* that causes cytoplasmic incompatibility are easily invaded sex ratio distorting mutants. Am Soc Nat 160:360–373

Iturbe OI, O'Neill SL (2007) *Wolbachia*-host interactions: connecting phenotype to genotype. Curr Opin Microbiol 10:221–224

Iturbe IO, Woolfit M, Rances E, Duplouy A, O'Neill SL (2010) A simple protocol to obtain highly pure *Wolbachia* endosymbiont DNA for genome sequencing. J Microbiol Methods 84:134–136

James AC, Ballard JW (2000) Expression of cytoplasm incompatibility in *Drosophila simulans* and its impact on infection frequencies and distribution of *Wolbachia pipientis*. Evolution 54:1661–1672

Jia FX, Yang MS, Yang WJ, Wang JJ (2009) Influence of continuous high temperature conditions on *Wolbachia* infection frequency and the fitness of *Liposcelis tricolor* (Pscoptera:Liposcelididae). Environ Entomol 38:1365–1372

Johanowicz DL, Hoy MA (2008) The manipulation of arthropod reproduction by *Wolbachia* endosymbionts. Fla Entomol 81:310–317

Min KT, Benzer S (1997) *Wolbachia*, normally a symbiont of *Drosophila*, can be virulent, causing degeneration and early death. Proc Natl Acad Sci U S A 94:10792–10796

Narita S, Kageyama D, Nomura M, Fukatsu T (2007) Unexpected mechanism of symbiont—induced reversal of insect sex: feminizing *Wolbachia* continously acts on the butterfly *Eurena hecabe* during larval development. Appl Environ Microbiol 73:4332–4341

Rattan RS, Ashok BH, Reinke A, Gupta PR, Claus PWZ (2011) Molecular evidence for the presence of endosymbiotic bacteria *Wolbachia* in *Cotesia* populations (Hymenoptera: Braconidae). J Asia-Pacific Entomol 14:183–185

Saiful Islam S (2007) *Wolbachia* mediated reproductive alterations in invertebrate hosts and biocontrol implications of the bacteria: an update. Univ J Zool, Rajshahi Univ 26:1–19

Shoemaker DD, Machado CA, Molbo D, Werren JH, Windsor DM, Herre EA (2002) The distribution of *Wolbachia* in wasps: correlations with host phylogeny, ecology and population structure. Proc Royal Soc, London B 269:2257–2267

Stolk C, Stouthamer R (1996) Influence of a cytoplasmic incompatability-inducing *Wolbachia* on the fitness of the parasitoid wasp *Nasonia vitripennis*. Ent Expt et Appl 7:33–37

Wade M, Chang N (1995) Increased male fertility in *Tribolium confusum* beetles after infection with the intracellular parasitic *Wolbachia*. Nature 373:72–74

Weeks AR, Reynolds KT, Hoffmann AA, Mann H (2002) *Wolbachia* dynamics and host effects. Trends Ecol Evol 17:257–262

Werren JH, Windsor DM (2000) *Wolbachia* infection frequencies in insects: evidence of a global equilibrium. Proc Royal Soc Lond Series B Biol Sci 267:1277–1285

Werren JH, Windsor DM, Guo LR (1995) Distribution of *Wolbachia* among neotropical arthropods. Proc Royal Soc Lond. Series B. Biol Sci 262:197–204

Werren JH, Baldo L, Clark ME (2008) *Wolbachia*: master manipulators of Invertebrate biology. Nat Rev Microbiol 6:741–751

Zeh DW, Zeh JA, Bonilla MM (2005) *Wolbachia*, sex ratio bias and apparent male killing in the harlequin beetle. Heredity 95:41–49

Influence of Herbivores on the Biology of *Chrysoperla carnea* Stephens

P. N. Magar, N. S. Satpute, S. S. Madankar and S. P. Bhopale

Abstract

A laboratory study was conducted in the Biocontrol laboratory at Department of Entomology, Dr. Panjabrao Deshmukh Krishi Vidyapeeth, Akola, during *Kharif* season of 2010–2011, which assessed the influence of different cotton plant herbivores on the biological parameters of *Chrysoperla carnea.* The results revealed that the highest feeding potential was found on cotton aphids, followed by *Helicoverpa* (eggs and neonates), and nymphs of jassids 193.87, 179.53, and 142.74 numbers, respectively. The shortest larval duration was recorded on cotton aphids followed by *Helicoverpa* (eggs and neonates) as 9.40 and 10.80 days, respectively. Prolonged larval duration, pupal period, and incubation period were recorded on cotton jassids as 14.47, 6.80, 2.66 days, respectively. The highest fecundity was recorded on cotton aphids whereas, lowest fecundity on cotton jassids as 418.00 and 342.40 eggs/female, respectively. The longest adult longevity was found on cotton aphids.

Keywords

Chrysoperla carnea · Herbivores · *Helicoverpa* · Host suitability

Introduction

The green lacewing, *Chrysoperla carnea* (Stephens) (Neuroptera: Chrysopidae) is predominant among the predacious insect species. It is active throughout the year in India, feeding on insect pests on crops. The larvae are predacious, which feed on the eggs and neonate lepidopterous larvae, nymphs, and adults of whitefly, aphids, thrips, scale insect, mealy bug, mites, etc. (Singh and Narasimhan 1992; Tesfye and Gautam 2002). They feed on aphids in most of the agricultural and horticultural ecosystems including plantations as well as ornamental crops. The role of many chrysopids as predators of pests has been required over. Resistance to insecticides commonly used against the crop pests has led

P. N. Magar (✉) · N. S. Satpute · S. S. Madankar · S. P. Bhopale
Department of Entomology, Dr. Panjabrao Deshmukh Krishi Vidyapeeth, Akola, India
e-mail: pramodmagar08@gmail.com

A. K. Chakravarthy (ed.), *New Horizons in Insect Science: Towards Sustainable Pest Management,*
DOI 10.1007/978-81-322-2089-3_31, © Springer India 2015

to an interest toward utilizing them in integrated pest management (IPM) programs.

Chrysopids also have a great potential to be used as a bioagent against citrus aphids, whiteflies, citrus psylla, and citrus mealy bugs (Balsubramani and Swaminappn 1994). For achieving the effective management of sucking pests as well as early stages of lepidopteran pests, the study on biological parameters and predatory potential of *Chrysoperla carnea* (Stephens) against different pests was undertaken.

Materials and Methods

A laboratory experiment was conducted in Biocontrol laboratory at Department of Entomology, Dr. Panjabrao Deshmukh Krishi Vidyapeeth, Akola, during *Kharif* 2010–2011 with three treatments replicated five times in CRD.

Host Plant

Untreated cotton seeds of cultivar PKV Hy-2 were planted in field as well as in pots in the laboratory to collect pest population, i.e., aphids, jassids, and *Helicoverpa armigera* from fields as well as pots.

Insect Pest

The pest population, i.e., Aphids, *Aphis gossypii,* and jassids, *Amrasca biguttula biguttula* Ishida on cotton plants were collected from field, brought to laboratory and were offered to natural enemy green lacewing, *C. carnea.* The host suitability for *C. carnea* among herbivores, viz., aphids, jassids, and *Helicoverpa* was studied.

Natural Enemy

Chrysoperla carnea, initial culture of *C. carnea* was procured from Entomology section, College of Agriculture, Nagpur, and was reared in laboratory on eggs of rice moth, *C. cephelonica.*

a. *Feeding of C. carnea* Clean plastic vials were used for rearing *C. carnea* on different preys. A set of three plastic vials was used for experiment as per treatment replicated five times. In each glass vial, a single egg of *C. carnea* with known age was transferred. After hatching, individual larva was provided with known number of fresh prey daily. The preys were provided twice, once during morning at 9 h and thereafter in the evening at 17.00 h. The number of preys consumed were recorded daily. Adults of *C. carnea* were fed with honey, Protinex, and yeast mixture diet with the help of sponge. The sponge was replaced daily.
b. *Observations* A set of three larvae for each treatment was subjected for recording observations. Observations on biological parameters were recorded daily. Larval period was recorded daily instar wise. The larval instar was confirmed by the presence of exuviae. The instar wise larval period and consumption of prey food (instar wise) was noted. The total larval period was counted from hatching of eggs till pupation on each prey. A known quantity of individual prey was provided to *Chrysoperla* larva. The number of preys utilized as food by each larva was recorded by counting utilized and unutilized number of preys and thus, the feeding potential during larval period was noted. Larvae were allowed to pupate inside the same plastic vials. The period taken from initiation of pupation till the emergence of adults was noted and computed to determine the pupal duration.

Adult longevity for both male and female was noted. The period from the emergence of adult from pupae up to the death of adult was considered as adult longevity. The total number of eggs laid by female during the entire oviposition period was considered as the fecundity of female. The fecundity was determined by working out the average number of eggs laid by a female. The eggs of same age were transferred singly in separate plastic vials. The period taken till the hatching of egg was considered as the incubation period. The data on all relevant observations, thus

Table 1 Incubation period of *C. carnea*

Treatment	Incubation period in days					
	R1	R2	R3	R4	R5	Mean
Cotton aphids	2.67	3.00	2.67	3.33	3.00	2.934
Cotton jassids	3.67	3.67	3.00	3.33	3.67	3.468
Helicoverpa eggs and neonate	3.33	3.00	3.33	2.67	3.00	3.066
'F' test						Sig.
SE(m) ±						0.132
CD at 5%						0.398

Table 2 Total larval period of *C. carnea*

Treatment	Total larval period of Chrysoperla					
	R1	R2	R3	R4	R5	Mean
Cotton aphids	9.33	9.67	9.33	9.33	9.33	9.40
Cotton jassids	14.67	15.33	14.33	15.67	15.00	15.00
Helicoverpa eggs and neonate	10.00	10.67	10.33	10.00	10.00	10.20
'F' test						Sig.
SE(m) ±						0.152
CD at 5%						0.491

obtained were subjected to CRD analysis as per Gomez and Gomez (1984).

Incubation Period of *C. carnea* The minimum incubation period of 2.93 days was observed when *Chrysoperla* was provided with nymphs and adults of cotton aphids. It was followed by 3.06 days on *Helicoverpa* (eggs and neonates). Maximum incubation period of 3.46 days was recorded on cotton jassids. The present findings are in accordance with earlier reports by Mangrule (2002) and Balakrishnan et al. (2005).

Larval Period of *C. carnea*

The larval stage of *Chrysoperla* had three instars and the total larval period of *Chrysoperla* on different preys ranged between 9.40 and 15.00 days, minimum being on cotton aphids found at par with *Helicoverpa* (eggs and neonates) on which 10.20 days were required to complete the larval duration. The instar wise larval duration is given in Table 1–7. However, the larval period was observed to be prolonged when *Chrysoperla* was provided with cotton-fed nymphs of jassids and recorded as 15.00 days. This might be due to the unsuitability of jassids as prey for *Chrysoperla* and aphids being the most suitable prey among the three tested preys. Saminathan et al. (1999) reported 11.37 days of larval duration on neonates of *H. armigera.* Similarly, Bansod and Sarode (2000) reported that the total larval period was 9.87 and 12.57 days on nymphs of *A. gossypii* and neonates of *H. armigera.* (Table 1)

Feeding Potential of *C. carnea*

The predatory potential of *Chrysoperla* larvae was significantly influenced by different preys.

Significantly, higher predatory potential was observed with the cotton-fed nymphs and adults of aphids recording 198.1 number of preys/*Chrysoperla* larva with the highest host suitability and feeding potential. It was followed by the eggs and neonates of *Helicoverpa* recording 181.1 preys per larva, respectively. Kapadia and Puri (1992) reported that single larva of *C. carnea* consumed 765.51 nymph of *A. gossypii.* Kabissa et al. (1995) reported that the total consumption of *Chrysoperla* larvae on eggs of *H. armigera* and nymphs of *A. gossypii* were 169.80 and 171.80, respectively.

Table 3 Feeding potential of *C. carnea*

Treatment	No. of prey consumed by *Chrysoperla* larvae					
	R1	R2	R3	R4	R5	Mean
Cotton aphids	195.00	195.67	198.67	205.67	195.33	198.1
Cotton jassids	152.33	144.67	146.00	149.67	146.33	147.8
Helicoverpa eggs and neonate	175.67	180.00	180.33	185.67	184.00	181.1
'F' test						Sig.
SE(m) ±						1.735
CD at 5%						5.344

Table 4 Pupal period of *C. carnea*

Treatment	Pupal period of *Chrysoperla* in days					
	R1	R2	R3	R4	R5	Mean
Cotton aphids	6.33	6.00	6.00	6.33	6.33	6.2
Cotton jassids	6.33	7.00	6.67	6.67	6.67	6.67
Helicoverpa eggs and neonate	6.33	6.67	6.00	6.33	6.67	6.40
'F' test						Sig.
SE(m) ±						0.107
CD at 5%						0.328

Table 5 Adult longevity of male *Chrysoperla*

Treatment	Adult male longevity in days					
	R1	R2	R3	R4	R5	Mean
Cotton aphids	25.00	26.00	27.33	27.33	27.00	26.53
Cotton jassids	24.00	23.00	23.67	23.67	24.67	23.80
Helicoverpa eggs and neonate	24.33	24.67	26.00	27.00	25.00	25.40
'F' test						Sig.
SE(m) ±						0.415
CD at 5%						1.279

Table 6 Longevity of adult *C. carnea* female

Treatment	Adult female longevity in days					
	R1	R2	R3	R4	R5	Mean
Cotton aphids	37.33	39.67	38.67	38.00	37.33	38.20
Cotton jassids	34.67	37.00	36.67	33.33	34.33	35.20
Helicoverpa eggs and neonate	37.00	34.67	35.67	37.33	36.67	36.26
'F' test						Sig.
SE(m) ±						0.557
CD at 5%						1.711

Table 7 Fecundity of *C. carnea*

Treatment	No. of eggs laid/ female Chrysoperla					
	R1	R2	R3	R4	R5	Mean
Cotton aphids	417.33	414.00	416.00	416.33	413.67	415.47
Cotton jassids	340.67	344.33	344.67	340.67	342.33	342.53
Helicoverpa eggs and neonate	366.00	357.67	362.00	360.33	366.00	362.40
'F' test						Sig.
SE(m) ±						1.135
CD at 5%						3.499

Influence of Different Hosts on Pupal Period of *C. carnea*

The pupal period of *C. carnea* was significantly influenced by different preys. The pupal period recorded on different preys, viz., cotton aphids, *Helicoverpa* (eggs and neonates), and jassids was 6.2, 6.40, and 6.67 days, respectively, indicating the prey preference in that order (Table 2).

Balakrishnan et al. (2005) recorded 8.15, 8.00, and 8.40 day pupal period when *Chrysoperla* was provided with *A. gossypii*, *A. craccivora*, and neonates of *H. armigera*, respectively.

Influence of Different Hosts on Longevity of Adult *C. carnea*

Adult *C. carnea* males were identified from their short and broader abdomen and the data on longevity of adult male *Chrysoperla* revealed significant differences among the test preys. Significantly, longer male longevity of 26.53 days was observed on cotton aphids. The next best superior host observed was *Helicoverpa* eggs and neonates recording 25.40 days longevity. The shortest longevity of 23.80 days was found on cotton jassids (Table 3).

These results are in accordance with studies of Geethalakshmi et al. (2000), Dhepe (2001) and Mangrule (2002).

Influence of Different Hosts on Longevity of Adult *C. carnea*

The longevity of *C. carnea* adult female was found to be significantly influenced by different preys, which recorded significant longer female longevity of 38.20 days on cotton aphids. It was followed by 36.26 days when *Chrysoperla* was reared on eggs and neonates of *Helicoverpa*. (Table 4, 5, 6, 7)

Influence of Different Hosts on Fecundity of *C. carnea*

The maximum egg laying capacity of *Chrysoperla* was recorded as 415.47 eggs laid/female when reared on the nymphs and adults of cotton aphids. It was followed by eggs and neonates of *Helicoverpa* and the recorded reproductive potential was 362.40 eggs laid/female. However, significantly minimum fecundity of 342.40 eggs/female was observed on jassid nymphs fed on cotton.

References

Balakrishnan N, Baskaran RKM, and Mahadevan NR (2005) Development and predatory potential of green lacewing *Chrysoperla carnea* (Stephens) (Neuroptera:Chrysopidae) on different prey insects. Agri Sci Dig 25(3):194–197

Balsubramani V, Swamiappan M (1994) Development and feeding potential of the green lacewing *Chrysoperla carnea* (Stephens) on different insects pest of cotton. Anz Schadlingskundc, Pflanzenschutz, Umweltschutz 67(8):165–167

Gomez KA, Gomez A (1984) Statistical procedure for agricultural research, pp 1–68

Mangrule M (2002) Determination of suitable hosts for rearing of *Chrysoperla carnea* (Stephens). M. Sc. (Agri.) Thesis submitted to Dr. PDKV, Akola, MS, India

Saminathan VR, Baskaran RKM, Mahadevan NP (1999) Biology and predatory potential of green lacewing *Chrysoperla carnea* (Neuroptera: Chrysopidae) on different insects hosts. Ind J Agri Sci 502–505

Singh SP, Narasimham U (1992) Indian Chrysopidae. Biological control center, National centre for Integrated Pest Management, Bangalore, 34 p

Tesfye A, Gautam RD (2002) Feeding potential of green lacewing *Chrysoperla carnea* on non rice moth prey. Indian J Entomol 64(4):457–464

Status and Management of Three Major Insect Pests of Coconut in the Tropics and Subtropics

A. D. N. T. Kumara, M. Chandrashekharaiah, Subhash B. Kandakoor and A. K. Chakravarthy

Abstract

More than 900 species of pests are associated with cultivated and wild coconut palm. This number includes both invertebrates and vertebrates. Of these, red palm weevil, (*Rhynchophorus ferrugineus* Olivier) rhinoceros beetle (*Oryctes rhinoceros* L.), and coconut black-headed caterpillar (*Opisina arenosella* Walker) are the most important devastating insect pests of coconut in major coconut-growing areas of the world. These three insect pests are distributed wherever coconut palm occurs. Current status, bioecology, and the management of the three pests are reviewed and discussed in the light of the changing scenario on coconut and other palms.

Keywords

Black-headed caterpillar · Coconut · Red palm weevil · Rhinocerus beetle

Introduction

Coconut, *Cocus nucifera* L. (Palmaceae) is an important crop mainly in the tropical and subtropical regions of the world, and millions of people depend on this crop directly or indirectly. It is popularly called "kalpavriksha," "tree of life," due to multiple uses and it is one of the top ten most useful trees in the world, providing food for millions (Duke 1983). Across the world, several people are employed in coconut-based industries like coconut oil, dry coconut powder, tender coconut, coir, toiletries, etc. Coconut is grown in 93 countries mainly in Indonesia, Philippines, India, and Sri Lanka together accounting for 78 % of the total world production.

The coconut palm is attacked by a number of insect pests all around the year (Thampan 1975). Coleoptera is the most numerous among them, and a total of 323 species are associated with coconut palm (Child 1974). Most species of beetles feed on leaves, roots, or bores in plant buds. Curculionidae, Chrysomelidae, and Scarabaeidae mainly cause serious damage, resulting

A. D. N. T. Kumara (✉)
Crop Protection Division, Coconut Research Institute, Lunuwila, 61150, Sri Lanka
e-mail: adnthissakumara@yahoo.com

M. Chandrashekharaiah · S. B. Kandakoor · A. K. Chakravarthy
Department of Agricultural Entomology, Gandhi Krishi Vignan Kendra (GKVK), University of Agricultural Sciences, Bangalore, Karnataka 560065 India
e-mail: chakravarthyakshay@gmail.com, chandu.raiah@gmail.com

A. K. Chakravarthy (ed.), *New Horizons in Insect Science: Towards Sustainable Pest Management*,
DOI 10.1007/978-81-322-2089-3_32, © Springer India 2015

in loss of fronds and damage to palms (Howard et al. 2001). Large numbers of Lepidopterans are also recorded as major devastating pests of coconut, mainly feeding on leaves and inflorescence. The coconut mite, *Aceria guerreronis* Keifer, is a serious mite pest and damages immature nuts causing serious yield losses. The coconut beetle, *Brontispa longissima* Gestro is a serious chrysomelid pest of coconut in Southeast Asia. Interestingly, it does not occur in India and Sri Lanka. This is probably because the trade of coconut and planting materials takes in a sea route connecting the Maldives, Malaysia, Indonesia, Vietnam, and other East Asian countries. Red palm weevil (RPW) is widely considered the most devastating insect pest of palms in South and Southeast Asia and in the Middle East (Sivapragasam et al. 1990; Faleiro and Satarkar 2003). The detection of pest infestation is difficult because the grub starts feeding from inside the palm and never comes outside till the adult emergence. The rhinoceros beetle (RB) feeds on the growing portion of the palm leading to ragged appearance. The heavily attacked palm dies or gets exposed to damage by secondary pests (Thampan 1975). The coconut black-headed caterpillar (BHC) is a defoliating pest of coconut, attacking the coconut gardens in patches but heavily.

The tall nature of coconut palm creates difficulties to adopt pest management practices straightforwardly. A variety of cultural, biological, and chemical control measures have been employed to manage the pests. The difficulties in the detection of the correct time to manage the pest, hidden habitat of the most coconut pests, and the availability of suitable foods throughout the year create serious pest threats to the coconut palm worldwide (Kumara 2007). Root feeding, trunk injection and foliar spraying of synthetic pesticide, application of botanical pesticides, improving the palm vigor by the application of organic fertilizer, use of biological control methods such as predators and parasitoids, mass trapping of pests by using sex pheromone and aggregation pheromone, as well as adopting several physical and mechanical practices are some of the management practices widely used to combat the three coconut pests. These practices are employed singly, by combinations, or as an integrated package. In this chapter, the current status and the management practices of three important insect pests of coconut, viz., the RPW, the RB, and the BHC are discussed (Fig. 1).

Red Palm Weevil (RPW)

Biosystematics

RPWs, commonly known as Asian red palm weevils, are large, polyphagous insects (usually greater than 25 mm long) belonging to Rhynchophorinae, a subfamily within Curculionidae (Coleoptera) (Borror et al. 1964). The Asian species include, *Rhynchophorus ferrugineus*, *Rhynchophorus vulneratus* (Panzer), *Rhynchophorus distinctus* (Wattanapongsiri) *Rhynchophorus lobatus* (Ritsema), and *Rhynchophorus bilineatus* (Montrouzier) (Murphy and Briscoe 1999).

Distribution and Host Range

The RPW is native to southern Asia and Melanesia. It was first identified in the early twentieth century in South and Southeast Asia (Lefroy 1906; Brand 1917). In the later part of the twentieth century, the RPW spread to Middle East Asia, North Africa, Europe, and Australia (Buxton 1920; Abraham et al. 1998; Al-Ayedh 2008; Li et al. 2009; Faleiro 2006). In 1985, the RPW was first recorded from the northern United Arab Emirates in the Middle East and has become widespread in that area (Ferry and Gomez 2002). The pest was reported from the Savaran region in Iran in 1990 (Faghih 1996) and Egypt in 1993 (Cox 1993). In the same year, the weevil crossed into Europe, at first into southern Spain (Cox 1993; Barranco et al. 1995) and a decade later into Italy (Longo and Tamburino 2005), many southern European countries, and Turkey (Malumphy and Moran 2007). Recently, the RPW was detected in the Dutch Antillies and California, USA (Fig. 1; Ferry 2010; Nisson et al. 2010).

In the Mediterranean region, the RPW severely damages *Phoenix canariensis*. Currently,

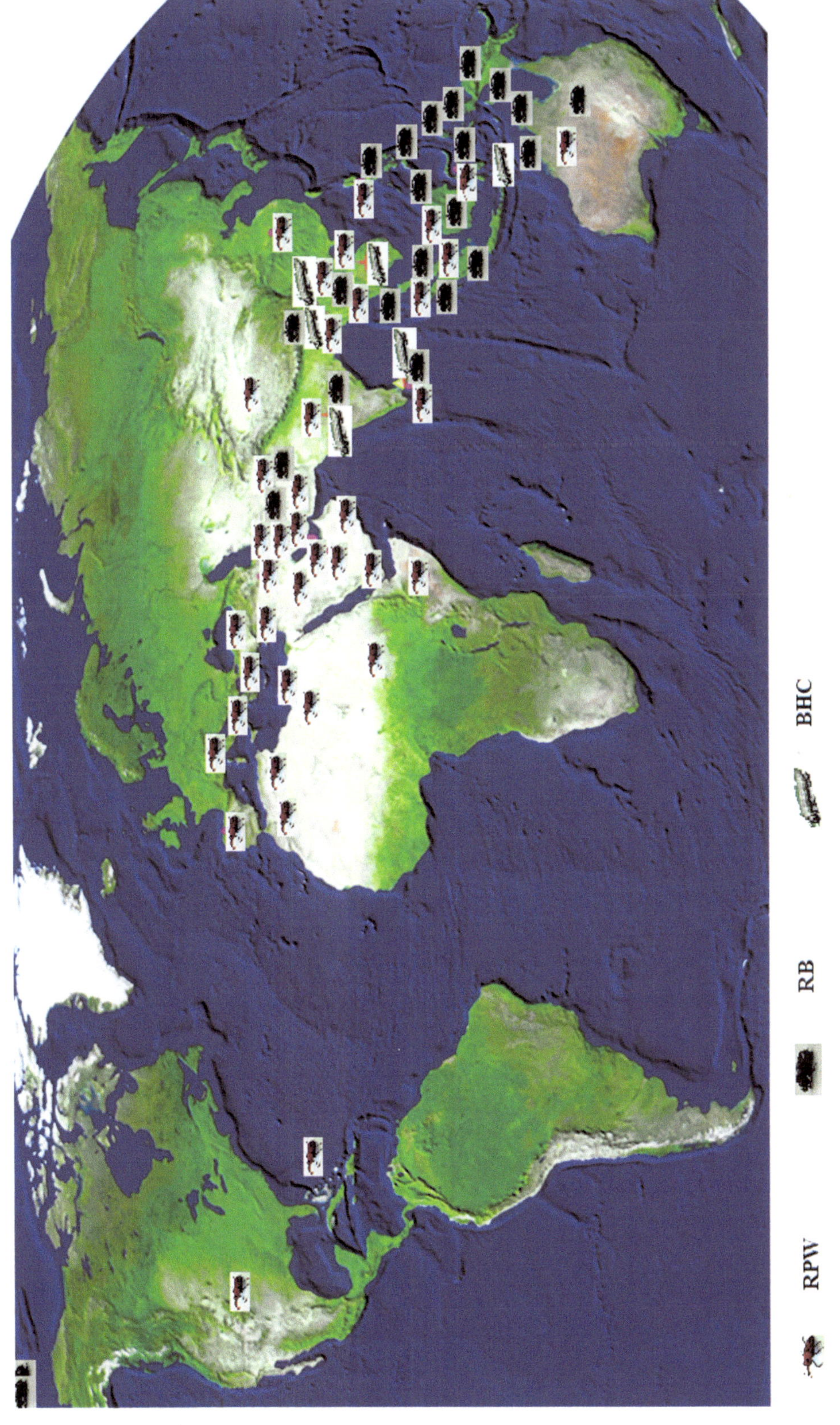

Fig. 1 Geographical distribution of the RPW, RB, and BHC

Table 1 Three major insect pests of coconut

Common name	Scientific name	Family and order	Number of species (approximately)	Geographical distribution	Pest status
RPW	*Rhynchophorus ferrugineus* Olivier	Curculionidae, Coleoptera	05	South Asia	Major
				Southeast Asia	Major
				Middle East	Major
				North Africa	Major
				Australia	Invasive
				Europe	Invasive
				North America	Invasive
RB	*Oryctes rhinoceros* L	Dynastidae Coleoptera	03	South Asia	Major
				Southeast Asia	Major
				Pacific Area	Major
				Africa	Minor
				Australia	Invasive
BHC	*Opisina arenosella* Walker	Lepidoptera: Oecophoridae)	01	South Asia	Major
				Southeast Asia	Invasive

the pest is reported in almost 15% of the global coconut-growing countries and in nearly 50% of the date palm-growing countries (Faleiro 2006). It prevails wherever palms are cultivated. The geographical spread of the RPW is mainly due to human intervention, by transporting infested young or adult date palm trees and offshoots from contaminated to uninfected areas (Alhudaib 1998; Gomez and Ferry 1999; Al-Saqer and Hassan 2011). Menon and Pandalai (1960) suggested that *R. ferrugineus* is a serious pest of coconut palms in India. It damages 34% of the coconut groves in Cochin, India. In Sri Lanka, it is the most serious pest causing fatal damage to young coconut palms of 3–10 years old (Brand 1917; Kirthisinghe 1960). It has been estimated that 10% of young coconut palms in the country is lost annually due to its attack (Mahindapala 1993). Accordingly, in 2000–2005, nearly 200,000 young palms have been killed by the RPW resulting in US$ 1,800,000 loss (Siriwardana et al. 2010). The RPW was first reported on coconut from Southeast Asia, and its host range included 19 palm species worldwide (Malumphy and Moran 2007).The list of known hosts include: *Cocos nucifera* (coconut palm), *Phoenix canariensis, Areca catechu, Arenga pinnata, Borassus flabellifer, Caryota maxima, Caryota cumingii, Corypha gebanga, Elaeis guineensis, Corypha elata, Livistona decipiens, Metroxylon sagu, Oreodoxa regia, Phoenix dactylifera* (date palm), *Phoenix sylvestris, Sabal umbraculifera, Trachycarpus fortunei*, and *Washingtonia* spp. However, sugarcane and *Agava americana* are used for laboratory rearing although their infestation in the field is not clear (Abraham et al. 1998).

Bioecology

The life cycle of the RPW varies between 45 and 139 days depending on environmental and geographical conditions and is spent inside the palm itself (Faleiro 2006; Esteban-Duran et al. 1998; Murphy and Briscoe 1999). The four stages of life cycle (Fig. 3) vary depending on egg, larva, pupa, and adult. The duration of each life stage and their numbers varies on host substrate and climatic factors (Table 1).The female RPW lays eggs, ranging from 58 to 531 in quantity, in the cracks, wounds, or crevices on the trunk of the tree. The light-yellow eggs (2.5 mm long) are laid close to the surface of the incision or wound. The grubs are white-yellow, and the larvae hatch from eggs in 4–6 days. They start feeding themselves by chewing the tissues of the plant and start moving toward the interior of the palm. The chewed up palm tissues and the thick brown fluid are oozing out from the tunnels of the trunk usually visualized after feeding of grubs. The grub lives

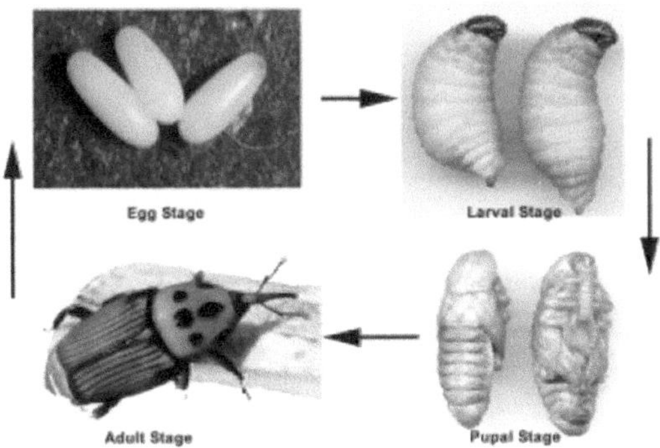

Fig. 2 Lifecycle of the RPW (Source: Prabahu and Patel 2009)

for 25–105 days, and it becomes a pupa in a cocoon made up of chewed-up tissues of the plant. The developmental state of the pupa usually takes 11–45 days (Abraham et al. 1998; Faleiro 2006). The adult RPW emerges out, or mostly the RPWs complete several generations inside the infested palm. Therefore, the RPW infestation is detected at such a late stage that it is not possible to save the infested palm. The RPW stays in the infested plant till it is hollow from inside and dead. After the death of the infested palm, the RPW moves to the neighboring palms. The rate of multiplication of the RPW is high as the female lays eggs continuously throughout the year (Fig. 2).

Management

Detection

The most critical factor in the management of the RPW is the detection of the damage at an early stage before severe damage to the internal tissues of the palm. The female weevil lays eggs on wounded tissues of the palm and the grub bores and begins its life in the palm, and normally never comes outside. Therefore, neither the grub nor the damage caused by it can be readily seen. Sometimes, a few small holes occur in the crown or on the soft stem. In many cases, the drying up of the young heart leaves or splitting of the petioles near the area of attack can be observed. But most often the attack by the weevil is noticed only when palms have been fatally infested and are beyond recovery (Menon and Pandalai 1960). The damage caused by the pest is severe, and once the weevil gets access to the palm, the final death of the palm is more or less certain. The first indication is the presence of holes on the stem with chewed fibrous material, sometimes protruding out (Child 1974). The RPW is a concealed tissue borer and all of its life stages are found inside the palm. Damage symptoms are indicated by the presence of tunnels in the trunk, oozing of thick yellow to brown fluid from the palm, the appearance of chewed-up plant tissue in and around an opening in the trunk, the presence of a fermented odor from the trunk, or toppling of the crown (Kaaheh et al. 2001).

Several detection methods are employed to detect the infected palms for treatment. The field staff surveys susceptible fields and regularly checks the symptoms of the infested palms. When the larvae are present in the palm, they produce sounds, due to chewing of palm tissue, crawling, emission, and quick oscillation (Pinhas et al. 2008).The sound generated by the chewing grub can be detected by endoscope or by placing the ear on stem (Hamad and Faith 2004). In the Middle East region, the infested palms produce a typical fermented odor, detected by sniffer dogs (Nakash et al. 2000). However, this may be possible if a considerable number of larvae are present, and at this stage the palm may be moderately to extensively damaged. Hence, the detection of infestations at an early stage is important to save the palms. Utilization of sound methodology to detect RPW-infested date palms has been attempted (Soroker et al. 2004), and an electronic device to detect infested coconut palms, although with less reliability and efficiency, has been developed in Sri Lanka (Fernando *pers. comm.*). Currently, digital signal-processing techniques are also used to identify the RPW in the palms (Al-Manie and Alkanhal 2005). Pinhas et al. (2008) developed a prototype that detected larvae of the RPW in offshoots of palms, which could be used in inspection of horticultural and ornamental palms traded between countries. However, this device is not portable. Siriwardana et al. (2010) developed and evaluated the portable RPW acoustic detector, which can be used for detecting the early stage of the damage efficiently and is able to detect 97% of affected palms. Recently, the image-processing method was developed by Al-Saqer and Hassan (2011).

Management

Because it is difficult to detect the damage by the RPW during the early stages of infestation, emphasis is generally focused on preventive measures relying on chemical applications. Control methods against RPW range from dusting of the leaf axils with insecticides after pruning, or spraying of the palm trunk, to localized direct injections of chemicals into the trunk (Faleiro 2006). All these treatments are often complemented with cultural and sanitary methods that include early destruction of the infested palm (Kurian and Mathen 1971) and prophylactic treatment of cut wounds (Pillai 1987). In newly spreading areas, preventive measures should be important including plant quarantine and plant certification, mass trapping using ferrugineol-based food-baited traps (Hallett et al. 1993), crop and field sanitation, preventive chemical treatments of gases, filling frond axils of young palms with a mixture of insecticides and curative treatments of infested palms in the early stages of attack, eradicating severely infested palms. These palms should be removed and destroyed by shredding (Dembilio and Jacas 2012). For avoiding infection, different precautionary measures have been considered, including avoiding mechanical damage or wounding the palm, application of repellents for the wound of palm trunks, containment/destruction of infested plants, field sanitation methods.

Insecticides

The most common and practical measure in chemical control is mainly based on the repeated applications of large quantities of synthetic insecticides employed in a range of preventive and curative procedures designed to contain the infestation. These procedures have been developed and refined since the 1970s in India, when work on application of organophosphates and carbamates ensured these chemicals to become the mainstay of the chemical approach to control RPW (Murphy and Briscoe 1999). In Sri Lanka, 20 to 30 ml of monocrotophos trunk injection to the affected palm at 2-monthly intervals twice has been recommended. (Fernando 2005). In Spain, a minimum of 8 preventive treatments with chlorpyrifos, imidacloprid, phosmetand, thiamethoxam per season (from March to November) are recommended to be applied as spray on the stipe, injected into the trunk, or as a drench, (Dembilio and Jacas 2012). Radiant (spinosad), Pyriproxyfen (IGRs) and Neemazal (plant extracts) were evaluated in laboratory against the RPW, and acute toxicity was recorded by high percents after treatment by Radiant followed by Pyriproxyfen while Neemazal did not exhibit acute toxicity. All tested insecticides exhibited lethal effect in the treated larvae and in the resulted pupae and adults from the treatment. Radiant was consistently the most toxic insecticide to the RPW based on LC50 recorded for general mortality (Hamadah and Tanani 2013).

The formulation, Imidacloprid SL, was successfully tested by Kaakeh (2005), in laboratory and semi-field assays against *R. ferrugineus*. Furthermore, high efficacy insecticides and botanical pesticides, biological control methods like entomopathogenic nematode, fungus and sterile insect techniques can be used as a package for preventive control of RPW. However, the systemic insecticide application through root feeding and stem injection are the only methods successfully reduced the RPW population in affected areas in initial stage of the infection (Prabhu et al. 2009; Khalifa et al. 2004; Abbas 2010). Efforts to develop biological management of RPW are in early stages (Abdullah 2009). Preliminary field trials suggest that an entomopathogenic fungus, *Beauveria bassiana*, partially controls RPW (Dembilio et al. 2010a). Combination trials of imidacloprid and entomopathogenic nematode *Steinernema carpocapsae* Weiser and the use of entomopathogenic nematode were initiated in countries like India and Saudi Arabia (Dembilio et al. 2010b).

Pheromone

The successful integrated pest management (IPM) tactics for the management of RPW is the use of pheromone traps. It can be used as pest-monitoring and large-scale mass-trapping program. The RPW male produced aggregation pheromone "rhynchophorol" identified as (2E)-6-methyl −2-hepten −4-ol for *Rhynchophorus palmarum* by Rochat et al. (1991). Subsequently, Hallett et al. (1993) identified and synthesized the "ferrugineol" (4-methyl −5-nonanol) by another male-produced aggregation pheromone. From that time, the pheromone technology has been widely used to manage both *R. palmarum* in oil palm and *R. ferrugineus* on coconut and date palm. Sri Lankan researchers, while studying the electroantennogram (EAG) response of male and female adults to 16 terpenes, reported that *R. ferrugineus* was sensitive to the size and the position of oxygen function, degree of unsaturation and degree of olefinic bonds in the molecules (Gunawardena 1994). Further, workers from the same laboratory reported the synthesis of ferrugineol by using Grignard reaction with butyl magnesium bromide and 2-methyl −1-pentanal. The activity of ferrugineol could be enhanced by combining it with n-pentanol, which is a major constituent of coconut sap, while decanol elicited the lowest EAG response (Gunawardena and Bandarage 1995a, b).

In 1994, the first pheromone trap evaluation started to monitor the RPW in the Middle East. The male-produced aggregation pheromones, chemically known as ferruginol (4-methyl −5-nonanol and 2. 4-methyl −5-nonunion), were available in various kinds of traps. In Saudi Arabia, inverted-bucket and upright-bucket traps are commonly used. In the United Arab Emirates (UAE), fabricated plastic traps are used while in India and Sri Lanka, bucket traps are mostly used (Faleiro 2006; Faleiro et al. 1998). The fundamental trap design is to have several windows in a container that allows the RPW to enter. The outside of the container is usually made rough by wrapping it with rough material. The RPW is lured by the pheromone and palm stems, leaf petiole pieces, pineapple pieces; fermented sugar solution or toddy are used synergistically to attract weevils. Insecticides or soap solutions are used as bait in the traps.

Pheromone for RPW - the aggregation pheromone is 4-methyl-5-nonanol and 2. 4-methyl-5-nonanone in 9:1 ratio

OH

4-methyl-5-nonanol

O

4-Methyl-5-nonanone

- Shelf Life: 60 months from manufacture date
- Solubility: Insoluble in water; soluble in alcohol
- Pheromone's Target Insect: Coconut weevil *Rhynchophorus vulneratus* and West Indian sugarcane weevil *Metamasius hemipterus*
  Source: http://www.bedoukian.com/products/product.asp?id=P4710-98

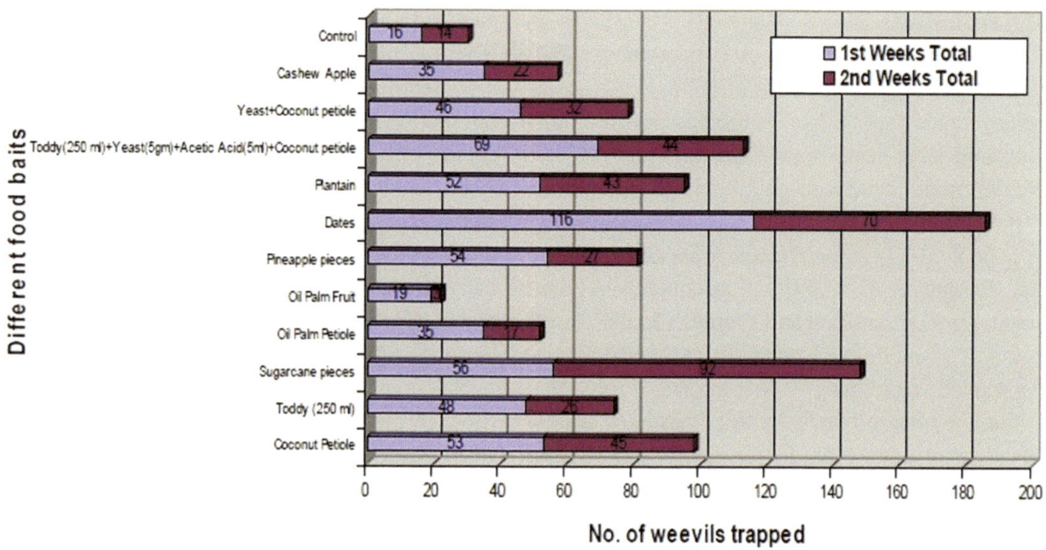

Fig. 3 Comparative weevil catches in red palm weevil pheromone traps using food baits (Source: Faleiro 2006)

The trap density varies from 0.5 to 10 ha (Faleiro 2006). Normally, it at least 1 or 2 traps/ha are recommended for use. In Saudi Arabia, pheromone traps brought down the infestation by 6.6 and 2.5% in 1993 and 1997, respectively (Vidyasagar et al. 2000). In Oman, the infestation was brought down by 24% in 1998 while obtaining 3% in 2003 (Al-khatri 2004) and the UAE gained 64% reduction in 2 years and 71% reduction in during one year (El-Ezaby et al. 1998; Oehlschager 2006). While India reduced infestation from 5% to zero within 1 year and from 2.4 to 0.2% during 1.5 years (Faleiro 2005a; Sujatha et al. 2006) and during 1997, Sri Lanka gained significant reduction using pheromone trap at 5 trap/ha, aggregation pheromone with fermented yeast sugar solution (Rajapakse 1998). In Sri Lanka it was revealed that the small-scale use of pheromone did not effectively reduce the pest infestation, and they suggest that using it as a large-scale area-wide mass-trapping program is most suitable for the management of pest (Fernando pers. comm.).

Chakravarthy et al. (2014) evaluated commercially available formulations at four concentrations to find the effective dosage for attracting maximum weevils. Pheromone 250, 500, 750, and 1000 mg were impregnated in wooden blocks (0.025 × 0.025 m) for 1 min, packed in a plastic cover (0.075 × 0.075 m), and sealed airtight in trilaminated pouches (0.075 × 0.075 m). These bucket traps (2 L) wrapped in gunny cloth having four holes of 4.5–cm diameter were used for trapping the weevils and beetles. Several types of food bait, such as ripened pineapple fruit pieces, sugarcane stem pieces, ripe dates, toddy solution, palm leaf petiole, palm stem pieces, etc., are used for weevil attraction as shown in Fig. 3 (Flairo 2006). The trap consisting of 1 L water and carbofuron 3G at 5 g and other chemicals as bait was used in each trap for the weevils (Fig. 4). The pheromone lure was tied to the inner surface of the bucket trap lid. The results of the study conducted by Falerio (2005b) were reconfirmed by Chakravarthy et al. (2014) and suggested that the pheromone compound 800–1000mg with food baited trap is the better for RPW mass trapping (Table 4).

In new areas and probable spreading zones, it is better to take strengthened quarantine measures to avoid the pest invasions. Avoiding the use of planting materials from infected areas, properly testing plant materials before transport from infected areas to uninfested areas, use of certified planting materials are essential for preventing the

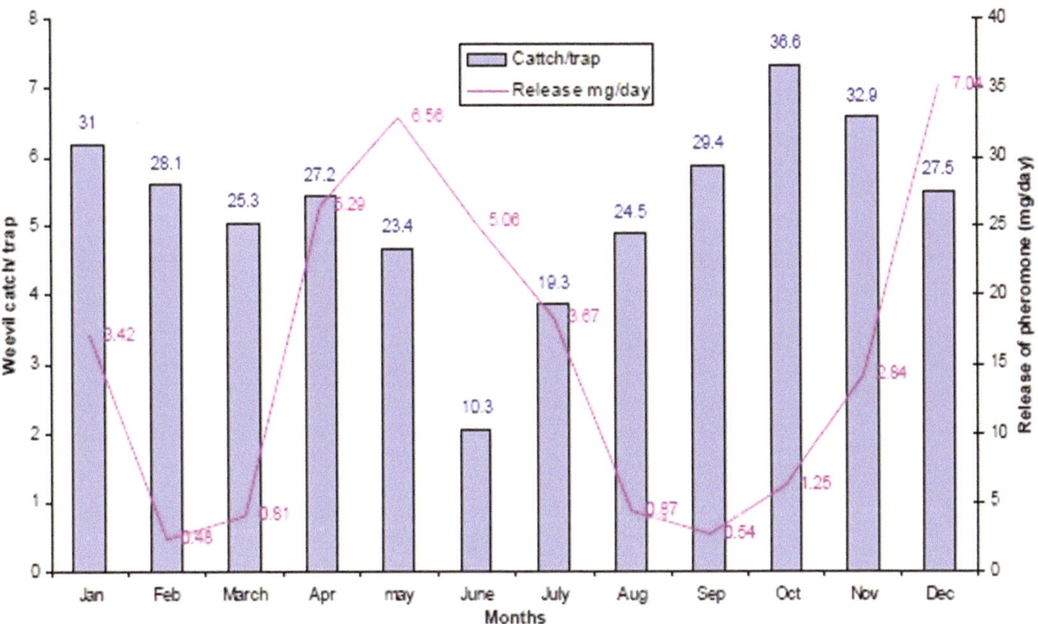

Fig. 4 Monthly RPW mass trapping using pheromone and the pheromone release rate (Source: Faleiro 2006)

Table 2 Host range of three major pests of coconut

Pest	Important host plants and region				
	Asia	Middle East	Europe	Africa	America
RPW	Coconut, date palm, oil palm, ornamental palms	Date palm, coconut	Date palm and ornamental palms	Coconut and date palm	Coconut, ornamental palms, and date palm
RB	Coconut, oil palm and date palm	Date palm, Coconut	–	Coconut and ornamental palm	Coconut
BHC	Coconut, other palms	–	–	–	–

pest introduction. The most successful way of controlling and managing the RPW is the use of the IPM program. It includes monitoring and taking care of the palm in susceptible age regularly; trapping adult RPWs using pheromone trap baited with synthetic pheromone and with synergists such as yeast-fermented sugar solution, ripened pineapple pieces, sugarcane stem pieces, or coconut petiole pieces; treating cuts and infections in palms; detecting the RPW at early stage; treating the plant in early stages with systemic insecticide if infected with RPW; eradicating and properly disposing infested palm or its parts; proper cutting of fronds incorporated with proper agricultural practices and training and educating farmers and Agriculture Department officers.

Rhinoceros Beetle

Biosystematics

Oryctes rhinoceros L, Scarabaeidae (Coleoptera), Dynastidinae. *O. monoceros, O. agamemnon, O. elegans* are other related species of the RB. Synonyms of RB are *Oryctes stentor* Castelnau, (1840) and *Scarabaeus rhinoceros* Linnaeus. Several common names are used for the RB around the world: Asiatic RB, bebete coco (French-Reunion (La Réunion)), black beetle, coconut black beetle, coconut palm RB, coconut RB (English), date palm beetle, dung beetle, escarabajo rinoceronte Asiático (Spanish), fruit

Table 3 Development time and number of instars for the RPW. (Source: Dembilio and Jacas 2012)

Authors	Feeding substrate	Development time (days)				Instars
		Egg	Larvae	Pupae	Adult	
Shahina et al. 2009	Honey in cotton	4–5	–	–	–	4
Shahina et al. 2009	Sugarcane lumps	4–5	50–80	20–30	74–115	9
Shahina et al. 2009	Apple slices	4–5	–	–	–	4
Abe et al. 2009	Apple slices	–	–	–	–	12
Salama et al. 2009	Banana slices	5	90	16–20	111–115	5
Salama et al. 2009	Sugarcane lumps	5	128	25–29	158–162	5
Salama et al. 2009	Squash fruit	5	83	20–24	108–112	5
Salama et al. 2009	Apple slices	5	103	16–18	124–126	5
Salama et al. 2009	Palm crown lumps	5	69	16–19	90–93	5
Kaakeh 2005	Sugarcane lumps	3–4	82	19	108	–
Kaakeh 2005	Palm heart lumps	3–4	86	21	124	–
Kaakeh 2005	Palm leaf base	3–4	84	18	119	–
Kaakeh 2005	Artificial diet	3–4	70–102	16–23	93–131	–
Martín-Molina 2004	Sugarcane lumps	3–4	88	25	116	11–17
Martín-Molina 2004	Artificial diet	3–4	93	30	128	7–12
Martín-Molina 2004	Palm lumps	–	–	–	–	8–15
Salama et al. 2002	Banana slices	–	–	13–22	–	–
Jaya et al. 2000	Sugarcane lumps	–	81	–	89	7
Esteban-Duran et al. 1998	Sugarcane lumps	–	76–102	19–45	139	–
Avand Faghih 1996	Palm lumps	1–6	41–78	–	–	–
Kranz et al. 1982	NS	2–3	60	14–21	76–84	–
Kalshoven 1981	Sago palm pith	–	–	–	105–210	–
Butani 1975	Sugarcan lumps	2–4	24–61	18–34	44–100	–
Rahalkar et al. 1972	Sugarcane lumps	3–4	32–51	15–28	50–82	–
Nirula 1956	Coconut slices	2–5	36–67	12–21	54–120	3
Viado and Bigornia 1949	Coconut slices	3	35–38	11–19	49–70	9
Lepesme 1947	NS	3	60	15	90–180	–
Dammerman 1929	NS	3	60–120	14	74–134	–
Leefmans 1920	Palm lumps	–	60	13–15	73–75	–
Ghosh 1912, 1923	Palm lumps	3–4	25–61	18–33	48–82	–

NS not specified

Table 4 Numbers of rhinoceros beetle and RPW trapped at four pheromone concentrations. (Source: Chakravarthy et al. 2014)

Treatments (mg)	Mean no. of beetle and weevil/trap	
	RB	RPW
250	33.00 (5.78)^{c}	78.00 (8.85)^{c}
500	42.00 (6.51)^{b}	264.00 (16.26)^{b}
750	98.00 (9.91)^{a}	695.80 (26.38)^{a}
1000	108.00 (10.41)^{a}	789.00 (28.10)^{a}
CD	0.34	0.27
± SEm	0.11	0.09

Mean of 10 traps/concentration, values with the same letter as superscript are nonsignificant at 5 % ($P < 0.05$)

stalk borer, Indischer Nashornkäfer (Dutch), Indischer Nashornkäfer (German), klappertor (Dutch), kumbang badak (Indonesia), kumbang tanduk (Indonesia), oryctes du cocotier (French), Palmen-Nashornkäfer (German), RB, rhinoceros du cocotier (French), scarab du cocotier (French) (Chandrika Mohan 2005).

Distribution and Host Range

The RB is one of the most damaging insects to coconut palm and African oil palm in South and Southeast Asia and the western Pacific Islands. The adult RBs feed on the growing point of the palm producing eventually ragged appearance of mature palm leaves. A severely attacked palm will die or be damaged by secondary-attack pests (Thampan 1975). The RB is distributed throughout Asia and the western Pacific. Thought to be native to the southern Asiatic region, the RB was introduced throughout the Pacific primarily as a result of the increased sea traffic during World War II. Floating logs containing larvae in tunnels might spread the pest to new areas (Bedford 1980; Howard et al. 2001; Gressitt 1957). Bedford (1980) reviewed the historical account of this species, "In Burma the pest first appeared in the extreme south of the peninsula. It probably entered from Malaysia about 1895 and worked its way north throughout the coconut growing areas of lower Burma over the following 15 years". It was accidentally introduced to a number of coconut growing areas of the Pacific and Indian Oceans. It is believed to have been introduced in rubber seedling potted plants from Sri Lanka to the Pacific island of Upolu, Western Samoa in 1909; from there it spread to the neighbouring island of Savail and to Tutuila in American Samoa. In 1921 the beetle was recorded in Keppel Island in the Kingdom of Tonga, but it was successfully eradicated in a campaign from 1922 to 1930. Wallis Island, about 320 km west of Samoa, became infested in 1931. RB introduced Palau Islands in about 1942 (Gressitt 1957), New Britain in 1942, and West Irian. Further establishments occurred in Vavau (Tonga), 1952; New Ireland, 1952; Pak Island and Manus Island (New Guinea), 1960; Tongatapu (Tonga), 1961; and the Tokelau Islands, 1963. The beetle was found at Suva on Viti Levu (main island of the Fiji group) early in 1953, and it has spread to at least 42 islands of the group, including all the important copra-producing ones, despite an intensive quarantine program to prevent this. An infestation of the beetle was reported from Guam in September 2007. In the Indian Ocean the island of Diego Garcia was infested during the First World War, possibly by beetles carried on troop ships (Orian 1959). Specimens were collected in the Cocos (Keeling) Islands in 1940. In 1962 it was found in Mauritius (Vinson 1963) and in 1978 in La Réunion (Chandrika Mohan 2005).

O. rhinoceros attacks the developing fronds of coconut, oil palm, and other palms in tropical Asia, and a number of Pacific Islands. Other species of RB such as *O. monoceros* L. African are associated with African palm species including coconut and other palms in the African region. Damaged fronds show typical triangular cuts. The beetle kills the palms (particularly seedlings and newly planted ones) when the growing point is destroyed during feeding on coconut, oil palm, betel nut, sago palm, and dates. They can also feed on *Pandanus* and other fleshy plants (Vargo 2000).

The larvae do not damage crops but instead grow in dead, decaying trunks and organic matter. The RB breeds in dead standing coconut palms killed by pest/disease/lightning and decaying organic materials like compost and sawdust heaps. (Bedford 1980). Decaying *Pandanus* trunk in Palau (Gressitt 1957) and heaps of decaying cocoa pod shells in New Ireland (Bedford 1976a) are also reported as breeding sites. In India (Nirula et al. 1956) and Mauritius, heaps of cattle dung were the most important breeding sites; in Burma, dead coconut stems, heaps of rotting paddy straw, and farm yard manure were most important (Ghosh 1923). In Sri Lanka, coconut logs and places near rafter mills (coconut sawdust), coir dust pits, and organic heaps were the most important breeding sites (Suwandharathne and Kumara 2007). Floating logs containing lar-

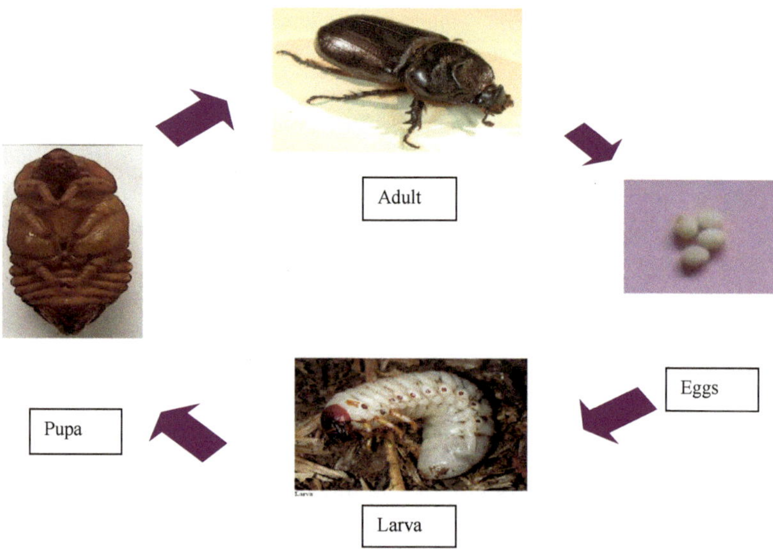

Fig. 5 Life cycle of the RB

vae in tunnels might spread the pest to new areas (Bedford 1980).

Bioecology

Female beetles lay eggs in rotting vegetation, especially in the trunks of rotting palms. The larvae bore and damage in the tunnel constructed by feces and silk among the spikes of flowers. The life cycle lasts from 4 to 9 months allowing more than one generation/year (Chen 1988). The beetle breeds in dead standing coconut palms that were killed by pest, disease, lightning, decaying organic materials like compost and sawdust heaps. Floating logs containing larvae in tunnels might spread the pest to new areas (Bedford 1980; Howard et al. 2001 cited Gressitt 1957). The female lays around 70–140 eggs on the organic substrate; after hatching between 11 and 20 days the emerged larvae remain and feed on the organic matter around 80–130 days. Pupation takes place in the breeding site, and organic matter cocoon made around the pupae and adult emergence occur after 14–30 days. The life cycle is completed within nearly 9 months and adult longevity about 4–6 months (Fig. 5) (Suwandarathne, pers. comm.).

The adult beetle bores into the soft tissue of the bud by cutting and chewing the tender unopened leaves and inflorescences. In the process, the leaves and inflorescences are severely damaged. The affected leaves, on the emergence, will give a characteristic fan-like appearance where the leaflets are cut off in the same place on both sides of the leaf stalk. When the attack is on the unopened spathe, the inflorescence gets destroyed. Sometimes the beetles have also been found boring into the soft tissues of the tender nuts. Though death is not common in the grown-up trees, the beetle may cause death of the young palms by boring into the growing point and destroying it, and repeated attacks may cause death (Thampan 1975). In oil palm, the RB bores into the base of the cluster of spears, causing wedge-shaped cuts in the unfolded fronds. In younger palms, the effect of damage can be much more severe (Wood 1968). Attack by adults may reduce yield and kill seedlings. They may provide entry points for lethal secondary attacks by RPW or pathogens (Bedford 1980).

Management

Mechanical and cultural tools

The management of the RB includes destroying breeding sites and collection and destruction of bio-stages of the beetle from the manure heaps or pits. When applying organic manure to the palms, it should be covered with thick (>0.15 m) soil layer or applied as a thin layer (<0.1 m). Other possibilities are: regular examination of seedlings and removing the beetle physically using a metal hook; application of repellents like neem seed cake or powder 150 g with sand (1:2), or Carbofuran 3G 40 g with sand (1:1), Naphthalene balls (10.5 g) covered with fine sand at 45-day intervals, pongamia seed cake with sand (1:2) and filled with Sevidol 8G (25 g with 200 g fine sand) thrice in April, September, and December into the bases of the three innermost leaf petiole gaps between stem and leaf petiole; spraying 0.01% Carbaryl (50WP) in breeding sites; application of used engine oil at the basal area of leaf petioles (CPCRI 2012; Jayanth et al. 2009; Fernando pers com). Phorate (10%) granules is reported to give protection for up to 60 days when applied at 5 g/palm. Application of naphthalene balls in the leaf axil at the base of spindle leaf at 12 g/palm provides good protection against the pest in Malaysia and India. This treatment gives 45–60 days protection to the palm. Application of oil cakes of neem (*Azadirachta indica* A. Juss., Meliaceae) or marotti (*Hydnocarpus wightiana* Bl., Bixaceae) in powder form at 250 g mixed with equal volume of sand, thrice a year to the base of the spindle leaf of the coconut palm is an effective prophylactic method against the RB (Chandrika Mohan 2005).

Biological Control

Under biological control, mainly two biological agents were used: entomopathogenic fungus and the *Oryctes* virus. For the management of immature stages entomopathogenic fungi, *Metarhizium anisopliae,* were used. They were of two types: long or short spored, with the long-spored (*M. anisopliae* var. *major*) varieties isolated from *Oryctes* spp. being more virulent (Ramle et al. 1999). Malaysian isolates killed all third instar larvae after 14 days (Ramle et al. 1999, 2006). The infective ability of a strain may become reduced by culturing on media but restored considerably following the infection of a host (Fargues et al. 1983). *M. anisopliae* infection is most frequent when rainfall and humidity are high; the spores should be applied to breeding sites or can be practiced preparing artificial breeding sites and applying fungus as a biopesticide. The fungus can be mass-produced on the broken maize seed grains and methods of mass-producing and on-farm production techniques have been developed and reported from Asian Pacific countries. Fresh spores or dry spores with mycelium or as a powder formulations were used and incorporated into the breeding ground (Dangar et al. 1991; Ramle et al. 2007; Subaharan 2004; Tey 1995).

Use of *Oryctes rhinoceros* Nudivirus (OrNV) is a key factor to control in areas where the pest is nonendemic. Virus inoculum mixed with sugar solution and it inoculated by pouring to the mouth of adult beetle or allowing adult beetle to swim the solution for infection. Infected beetles were released to field for spreading the virus among the population. The virus virulence of different geographical genomic isolates is different, and their efficacy varied from place to place (Crawford et al. 1986). Hence, the dosages of virions administered varied. With this caveat in mind, an isolate from Leyte Island, Philippines caused more larval mortality than did isolates from other locations in the country or from Samoa (Zelazny 1979). On the Malay Peninsula, the isolate Ma07, extracted from adult midguts from the west coast, caused higher mortality in larvae and adults and was deemed more virulent than the widespread isolate PV505 from the Philippines; the isolate from Sabah caused the lowest mortality (Ramle et al. 2005). In some areas like the Andaman Islands (Jacob 1996), the Minicoy Island (Mohan and Pillai 1993), the Maldive Islands (Zelazny et al. 1992; Zelazny et al. 1990), and Oman (Kinawy 2004), after introducing OrNV, the RB damage subsequently cut down significantly. The OrNV is effectively used in Malaysia, India, the Philippines, and Samoa to reduce the damage (Babjan et al. 19951996; Ramle et al. 2005; Zelazny and Alfiler 1991; Marschall and JIoane 1982).

Pheromone

In the Philippines, the *O. Rhinoceros* pheromone, ethylcrysanthemumate, was first tested and trapped in 7–25 % of the adult population during 3 years (Ragoussis 2007). The male producing aggregation pheromone compound 4-methyloctanoate identified from Indonesia and it was reported that it is 10 times more effective than ethylcrysanthemumate (Wood 1968). Subsequently, ethyl 4-methyloctanoate pheromone produced a commercial synthesis (Munoz et al. 2009; Ragoussis et al 2007) and is spreading rapidly. Trapping with this pheromone as a management component is now applied in Malaysia and Indonesia (Norman and Basri 2004; Oehlschlager 2007) and is used as a tool in ecological studies. Several trap designs of differing costs and effectiveness are used: pheromone emitted from a sachet dispenser has been tested (Desmier de Chenon et al. 2001); Plastic bucket trap; parabolic trap; single- or double-vane traps (Oehlschlager 2007); and PVC tube trap. The latter is a tube, 16 cm in diameter and 2 m in height with two-side openings or windows in the upper part and an open top, which stands in a bucket and simulates an upright coconut trunk. It catches more beetles than the bucket trap and others. Adding empty oil palm fruit stalk material or coir dust cow dung mixture enhances the catch (Morin et al. 2001). In India, different doses of 4-methyloctanoate were field evaluated and found significantly higher beetle catches from 750 and 1000mg doses at the rate of 1 trap/ha (Table 4) (Chakravarthy et al. 2014). A nanomatrix and polymer composite was developed to load the RB pheromone, ethyl 4-methyloctanoate. The pheromone loaded to the nanomatrix showed extended duration of release when subjected to thermal gravity analysis.

Pheromone for RB – the structure of aggregation pheromones

Ethylchrysanthemate

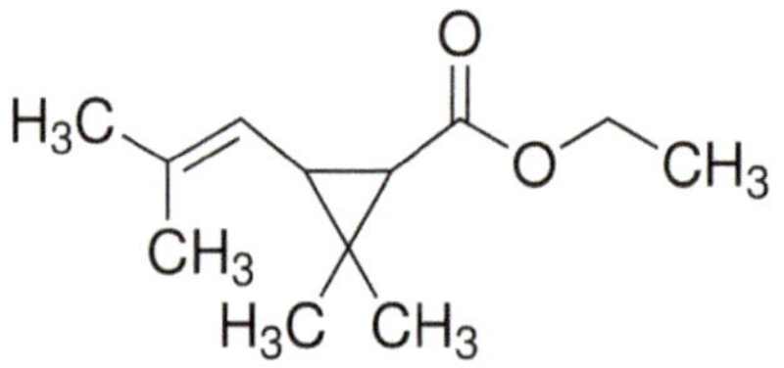

H_3C, O, O, CH_3, CH_3, H_3C, CH_3

Source :http://www.sigmaaldrich.com/catalog/product/aldrich/w506109?lang

Ethyl 4-methyloctanoate (Main pheromone component)

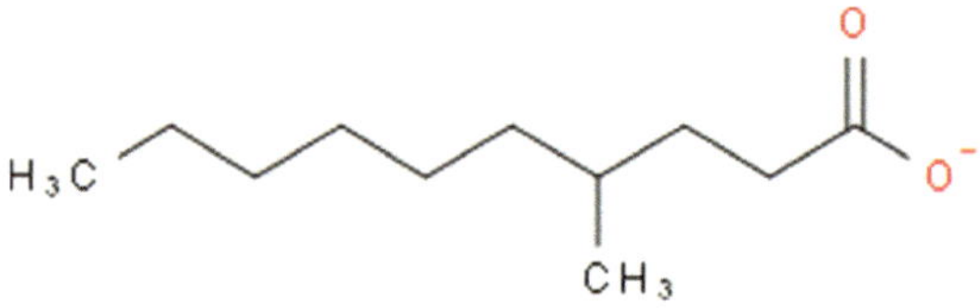

O, H_3C, O^-, CH_3

Ethyl 4-methyloctanoate

- Pheromone target insect: Coconut rhinoceros beetle, *Oryctes rhinoceros*
- Appearance: colorless clear liquid

Source: http://www.chemfrog.com/getPSupply/HNMFGH/

In field trapping experiments, (4 S)-ethyl 4-methyloctanoate and the racemic mixture were equally attractive and ten times more effective in attracting beetles than ethyl chrysanthemumate. Ethyl 4-methylheptanoate was as attractive as ethyl chrysanthemumate and more attractive than the 4-methyloctanoic acid, but further studies are required before it can be classified as an aggregation pheromone (Chakravarthy et al. 2014). Compared to ethyl 4-methyloctanoate alone, combinations of the three male-produced compounds did not increase attraction, whereas addition of freshly rotting oil palm fruit bunches to pheromone-baited traps significantly enhanced attraction. With increasing dose, captures of *O. rhinoceros* increased, but doses of 6, 9, and 18 mg/day were competitive with 30 mg/day lures. Newly designed vane traps were more effective in capturing beetles than were barrier or pitfall traps. The results of this study indicated that there is potential for using ethyl 4-methyloctanoate in operational programs to suppress *O. rhinoceros* in oil palm plantations (Hallett et al. 1995).

The total number of RB caught in pheromone traps varied with concentrations. Nearly 98 and 108 beetles were caught in pheromone traps baited with pheromone lure with 750 and 1000 mg, respectively, and were found significantly superior to other treatments. Records of 250 and 500 mg proved inferior with only 33 and 42 beetles, respectively, in the pheromone traps (Chakravarthy et al. 2014). The peak activity of beetles was found during the 29th to 53rd week after the installation of the traps, i.e., from July to March. The beetles caught in the traps ranged from 0.0 to 4.5 beetles/trap/week recorded from the 29th to the 53rd week after the installation of the traps in India. Beetles caught in the trap varied from 0.25 to 1.75 beetles/trap/week recorded from the 29th to the 53rd week after trap installation. Rhinolure is an aggregation pheromone effective in mass trapping both males and females of the coconut RB (Chakravarthy et al. 2014). The active compound serving as an attractant, ethyl 4-methyloctanoate is supplied as a bubble formulation in sachets, and the chemical is suspended in lure. The trap should be installed at about 0.8 m from the ground and is effective at 1/ha. The septa can be suspended in the upper lid of the bucket, taking care to avoid direct sunlight as it would be affecting the performance of the lure. The bucket should contain holes and rough corrugations on the lateral sides just below the upper lid so that the beetles that are attracted alight on this rough surface before entering the holes.

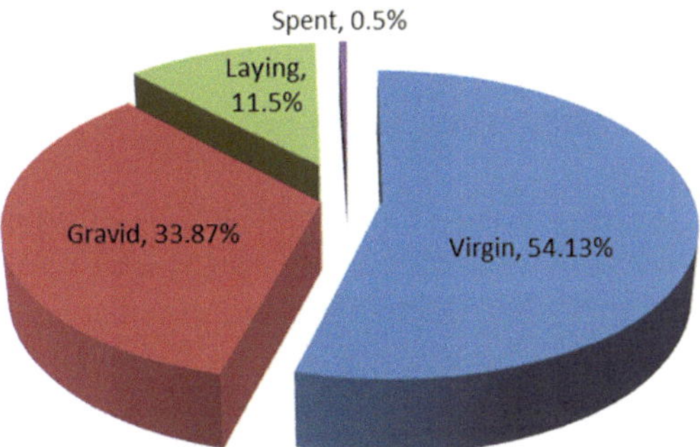

Fig. 6 Reproductive status of RB females in pheromone traps (Source: Jayanth et al. 2009)

Field evaluation of traps revealed that nanomatrix-loaded pheromone (240 mg) trapped 18.0 beetles/trap/month, followed by the commercial lure (containing 800 mg pheromone), which trapped 12 beetles/trap/month. Studies on the longevity of pheromone lures indicated that the commercial lure was exhausted in 3 months whereas in the nanomatrix, it remained active up to 8 months (CPCRI 2012). Jayanth et al. (2009) evaluated the reproductive status of RB females captured in aggregation pheromone traps and revealed that the majority of females captured were virgin or gravid females (Fig. 6). Hence, the mass trapping of RB is particularly more effective due to attract female beetles that had not started reproductive activity.

The control of the RB is difficult once they invade the new area; therefore, it is better to implement quarantine measures in those countries that are found in the probable spreading zones. If new infections are noticed, immediate control measures should be taken to stop the spreading of the pest. Once they spread, integrated management measures, such as mechanical and biological use of pheromone and chemical control tactics, should be taken to manage the pest.

Coconut Black-Headed Caterpillar

Biosystematics

The black-headed caterpillar (BHC) was first described by Walker and was placed under the family Cryptophasidae (Lepidoptera). Meyrick (1905) described the same as *Nephantis serinopa* under the family Xyloryctidae. In 1981, Baker compared the holotype of *O. arenosella* by Walker and *N. serinopa* by Meyrick and found that both are conspecific. Hence, the name *Opisina arenosella* is being used again with *N. serinopa* as a synonym.

Distribution and Host Range

The BHC is considered a serious defoliating pest on coconut. Its natural range extends from India to Sri Lanka (Perera 1987), Burma (Ghosh 1923), Bangladesh (Alam 1962), and, recently reported, Thailand (Bao-qian et al. 2013). Damage to coconut in India was first recorded from Andhra Pradesh in 1909 (Rao et al. 1948), and this pest menace is frequently noticed in South India. The distribution within the region showed patchy or spot distribution and recorded frequent outbreaks (Perera 1987; Sundararaju 1985; Nadarajan and Channabasavanna 1980).

Coconut Palmyra and some ornamental palms were recorded in Sri Lanka and India as hosts, i.e., Palmyra (*Borassus*), *Corypha*, *Hyphaene*, *Phoenix*, and *Roystonia* (Rao et al. 1948; Nirula et al. 1951b). Butani (1975) recorded it as a minor pest of date palm. Manjunath (1985) recorded that larvae feed on banana leaves during an outbreak in India. However, from laboratory studies it was concluded that fan palm (*Livistonia chinensis*), wild date palm (*Phoenix sylvestris*), and date palm (*P. dactylifera*) are suitable host plants, but banana is not.

Bioecology

Females of *O. arenosella* lay their eggs in small groups on the undersurface of coconut leaflets. Eggs are generally deposited in the vicinity of feeding larvae, and it has been suggested that this results in the slow spread of outbreaks to peripheral, uninfested trees (Perera 1987). The larvae usually have five instars and feed on the undersurface of coconut leaves, at first gregariously, then singly, consuming the lower epidermis and mesophyll but leaving the upper epidermis intact. The upper surface of the leaf has a characteristic scorched appearance where caterpillars have fed. The larvae construct a gallery of silk and frass, into which they retreat if disturbed. Pupation takes place within the larval gallery. Adults are frequently found during the day resting on the undersurface of the leaves of palms damaged by the larvae. The moth flies at night, but little is known of its dispersal abilities. The egg stage lasts on average 3 days and the five larval instars last six, seven, seven, five, and ten days, respectively (Perera, 1987). Although there are normally five instars, up to eight have been recorded in the laboratory when the larvae are stressed (the supernumerary instars would be indistinguishable from the fifth instar in the field). The pupal stage lasts for an average of 8 days, and the total length of the preadult life cycle is thus approximately 46 days. Adult longevity is 7–9 days, during which the female lays about 152 eggs (Perera 1987). In Sri Lanka, (Perera et al. 1988) BHCs have shown to follow partially discrete generation cycles during outbreaks. Ramkumar et al. (2006) studied the population cycles of the BHC and found that there are discrete generation cycles. Further, they suggested that the possible reason for the discrete generation may be protandry.

The BHC is an important and the most destructive pest in many commercial and subsistence coconut-cultivating areas, because moderate- to high-density populations of the BHC do indeed cause considerable yield loss. The pest infestation is mainly confined to the lower fronds, and in severe infestation, several hundreds to thousands of larvae can be observed on a palm. The caterpillar feeds on chlorophyll by scraping lower epidermis of leaflets and constructs galleries of silk and frass. The infested fronds give burnt-up appearance and the affected palms often take several years to

Table 5 Summary of classical biological control of the BHC. (Source: Cock and Perera 1987)

Parasitoid	Release country	Recommendation
Antrocehalus pandens Walker	Sri Lanka	No field recoveries, not suitable for field releases
Bessa remota Aldrich	Sri Lanka, India	No field recoveries, not suitable for field releases, fields are unable to reach the larval galleries
Bracon brevicornis Wesmeal	Sri Lanka, India	Field recoveries occur, suitable for releases
Elamus nephantidis Rohwer	Sri Lanka	No field recoveries
Eriborus trocanteratus Morley	India	Recoveries occur after adding the parasitoid complex
Stomatomyia bessiana Baranoff	India	Colony did not persist
Tetrastichus Israeli Mani and Kurian	Sri Lanka	No recoveries
Trichogamma brasiliensis Ashmead	Sri Lanka	No recoveries
Trichogamma minutum Riley	Sri Lanka, India	Not recoveries

recover completely. Further, BHC attack results in heavy yield loss (>50%), and the infested palms can regain the normal yield potential during the fourth year following the pest attack, provided the pest infestation is brought under control (Chandrika Mohan et al. 2010).

Management

Cultural and Chemical Control

The early stage of the pest, artificial defoliation such as cutting and destruction of lower fronds can delay the infestation (Perera 1989). Originally, management of the species was accomplished by removing the infested fronds of the coconut palms or using light traps in order to physically remove the infestation from the palm. However, frond removal reduces the palm's yield drastically, and does not guarantee to resolve the infestation (Cork and Hall 1998). Chemical insecticides are of course used in the control. Trunk injection of Monocrotophos of tall palms at 3–6 g of active ingredient per palm was translocated and accumulated in the leaves in quantities sufficient to kill the BHC, and experimental palms indicated that the insecticidal effect persists for about 6 months (Kanagarathnam and Pinto 1985); therefore, it is suggested that this treatment would be adequate to control a succession of larvae hatching out of eggs over a period of time. However, alternative methods to chemicals have been sought out in order to reduce the chemical residues on the produced fruit, as well as maintain the health of predatory animals and beneficial parasitoids.

Another method is use of botanicals and biopesticides against O. arenosella and are come from different forms. Biopesticides have proven to be as effective as chemical pesticides in many cases. The control of the BHC has been accomplished with the use of both garlic- and neem-based biopesticides in India.A commercial formulation of soluneem was evaluated and recommended to use as application through root feeding. These treatments act as poison to the species and are administered as *O. arenosella* consume the leaves. In studies, reduction of all stages of larvae as well as pupae were observed and drastically reduced the damage incurred by the palms. Application of organic farming practices also resulted in the reduction of pest infestation in applied gardens. Chakravarthy et al. (2012) evaluated the combination of organic manure and biopesticide application with synthetic pesticides and they found that less number of infested palms and low defoliation in organic manure biopesticide applied fields.

Biological Control

Naturally, a large number of predatory animals like birds, spiders, anthocoridae, reduvidae, and carabidae are recorded (Cock and Perera 1987). A large number of parasitoids associated with the BHC were recorded, and they caused the suppression of the population. BHC-associated parasitoids include Braconidae, Eulopidae, Chalisidae, Bethylidae, Ichneumonidae, Elasmidae, Tachinidae, Phoridae, Stenomidae, and Eupelmidae(Cock and Perera 1987). The classical biological control method was tested both in India and Sri Lanka; however, it was not successful in both countries (Table 5).

The augmentative releases of parasitiods, *Trichogramma minimum* Riley, *Goniozus nephantidis* Muesebeck and *Bracon brevicornis* Wesmeal, are all known parasitoids of the species, and work by parasitizing larvae at different instars (Venkatesan et al. 2009). *G. nephantidis* and *B. brevicornis* both parasitoid wasps, parasitize third to seventh instar larvae, leading to the eventual shriveling and death of the organism. Wasps have been observed parasitizing up to 57 % of the resident larvae, which would reduce the population of the BHC significantly. While *G. nephantidis* proves to be the dominant parasitizing species over *B. brevicornis* due to more developed parental care in *B. brevicornis* and therefore reduced number of parasitization, they both act as effective species in controlling the coconut BHC (Venkatesan et al. 2009).

Use of Pheromone

The presence of female sex pheromone in the BHC has been revealed from studies in Sri Lanka and India (Murthy et al. 1995). Studies were conducted to ascertain the attraction of virgin females of the BHC to conspecific males. One- to two-day-old virgin females were individually confined in net cages and fixed to the sticky traps. The traps were placed horizontally in the canopy of the infested palms for two consecutive nights. The results indicated that the number of male moths trapped in baited traps was significantly higher than in unbaited traps. Hence, the results revealed the attractiveness of virgin females to conspecific males of *O. arenosella* due female sex pheromone released from caged virgin females (Fernando and Chandrasir 1997). Work conducted in the 1980s by the Natural Resource Institute using insects from Sri Lanka demonstrated the presence of four electro-physiologically active compounds in extracts from female moths (Cork and Hall, 1998). The putative pheromone components were identified and field tested in Sri Lanka but high catches in unbaited traps hampered the field work (Cork and Hall 1998).

Srinivasa and Muralimohan (2009) expressed the scope for utilizing sex pheromones for the management of *O. arenosella*. The behavioral attributes like that the adults emerge restricted to 10–15 days over 30 successive days in a generation, that populations are protandrous, and the reproductive biology of *O. arenosella* shows less probability of trapping an unmated male in mass trapping. Further, they suggested that the sex pheromones can directly contribute toward downsizing populations of *O. arenosella* if they are used for disrupting natural mating.

Female-produced sex pheromone (Z, Z,Z) 3,6,9-Tricosatriene released by the female was identified, artificially synthesized, and field-tested in India (Bhanu et al. 2009). Bhanu et al. (2011) made an attempt of GC-MS analyses of volatiles collected from virgin females. *O. arenosella* confirmed the structure of the *O. arenosella* pheromone as (Z, Z,Z)-3,6,9-tricosatriene (Z3Z6Z9–23Hy). Further, field trials indicated that using wing traps PVC vial dispensers with 100 µg, the pheromone loading was significantly superior in attracting the male moths. The trap catches of the BHC male moths in ten pheromone traps were able to indicate moth emergence peaks.

(Z,Z,Z)-3,6,9-Tricosatriene
Formula: C23H4
Source:http://www.pherobase.com/database/synthesis/synthesis-detail-Z3Z6Z9-23Hy.php

Chandrashekharaiah et al. (2012) demonstrated the calling behavior of the female BHC, the male response toward sex pheromone, and the behavioral responses of male and female *O. arenosella* to female pheromone. Studies on exploitation of pheromone traps as a surveillance and monitoring tool in IPM of *O. arenosella* were also conducted. Nearly 50% of population reduction was found in mass trapping of male BHC moths. Chandrashekharaiah (2013a) confirmed the distinct moth emergence periods of BHC males using pheromone traps and nearly five generations of BHC/year. The duration of moth emergence and nonemergence periods of moth varied from 34 to 45 days and 44 to 56 days, respectively. The maximum number of days of moth emergence and nonemergence was recorded during March, April, and May. From these studies, it is inferred that an average moth emergence period lasted for 41.50 days (SD=±3.93) followed by 48.16 (SD=±4.35) days of nonemergence period. The pheromone traps placement should not be universal or the same for all the places. But initially, traps should be installed based only on the visual observation on stage of the pest; in subsequent generations, the traps can be placed 70–80 days after initial set up.

Chandrashekharaiah et al. (2013b) demonstrated that the optimum dosage pheromone per lure was 0.1 mg. Studies on the standardization of lure type revealed that the commercial plastic vial type (65.75 moths/trap/generation) was more effective than vial with cap, black septa, and red septa (31.50, 56.25, 41.50, and 26.25 moths/trap/generation). The pheromone release rate profile of field-installed dispensers quantified at different day intervals using gas chromatography with flame ionization detectors. The results suggested that, initially, i.e., up to 10 days, the release rate was maximum (60 and 75% pheromone released within 10 and 20 days, respectively) and later, it was further reduced as the days advanced. The studies on trap-type standardization indicated that the commercial cross vane trap (153.40 moths/trap/generation) was more effective than a wing vane trap (29.66 moths/trap/generation), wing vane trap with open side (64.60 moths/trap/generation), wing vane trap with large size (122.66 moths/trap/generation), and funnel trap (5.40 moths/trap/generation). The comparison of the trap with and without pheromone lure indicated that the trap with lure proved more effective (77.66 moths/trap/generation) than the trap without lure (23.87 moths/trap/generation).

Chandrashekharaiah et al. (2013c) demonstrated the robustness of mass trapping technology using sex pheromone traps. The study was conducted in 14.2 ha with 1700 infested coconut trees. Nearly 661 (I and II generation, respectively) and 836 (III generation) cross vane traps baited with lure were installed in the study area uniformly. Nearly 73,739, 52,392, and 7953 moths were trapped in I, II, and III generation, respectively, with the larval (2.97±0.63 (Mean±SD) larvae/leaflet before mass trapping) reduction of 34.27 (1.93±0.64 larvae/leaflet), 88.76 (0.12±0.37 larvae/leaflet) and 93.97 % (0.09±0.03 larvae per leaf) in II, III, and IV generations, respectively; whereas in the control plot (2.5 ha), the larval population was increased continuously up to III generation (1.83±0.22 to 5.27±2.12 larvae/leaflet in I and III generation, respectively) and reduced in the subsequent generations. Reduction in larval numbers was achieved in treated plot due to continuous trapping of male moths using pheromone traps.

Chandrashekharaiah (2013a) showed that the integration of larval parasitoids with pheromone traps have a cumulative effect on BHC population. Further, larval parasitoids can effectively integrate with pheromone traps without having any adverse effects. From the above studies, it is confirmed that pheromone traps are more effective in suppressing BHC population than the release of parasitoids and other methods. Further, they are eco-friendly, without causing any environmental hazards and safe to nontarget organisms. The pheromone traps and lures immediately after mass trapping can be easily collected and buried in the soil or it can be recycled for further use. This technology can also be easily combined with other management practices without any adverse effect. With these results, the pheromone traps can be recommended for the management of BHC either alone or in combination with the release of larval parasitoid, *G. nephantidis*.

References

Abbas MST (2010) IPM of the red palm weevil, *Rhynchophorus ferrugineus*. In: Ciancio A, Mukerji KG (eds) Integrated management of arthropod pests and insect borne diseases, integrated management of plant pests and diseases 5. Springer, NY, pp 209–233

Abdullah MAR (2009) Biological control of the red palm weevil, *Rhynchophorus ferrugineus* (Olivier) (Coleoptera: Curculionidae) by the parasitoid mite, *Rhynchopolipus rhynchophori* (Ewing) Acarina: Podapolipidae). J Egypt Soc Parasitol 39:679–686

Abraham VA, Al-Shuaibi M, Faleiro JR, Abozuhairah RA, Vidyasagar PSPV (1998) An integrated approach for the management of red palm weevil *Rhynchophorus ferrugineus* Olivier-A key pest of date palm in the Middle East. Sultan Qaboos University. J Sci Res Agric Sci 3:77–83

Alam M (1962) A list of insects and mites of eastern Pakistan. Report of the Agriculture Department, Dhaka, pp 107

Al-Ayedh H (2008) Evaluation of date palm cultivars for rearing the red date palm weevil, *Rhynchophorus ferrugineus* (Coleoptera: Curculionidae. Florida Entomologist Soc 91:353–358

Al-Manie MA, Alkanhal MI (2005) Acoustic detection of the red date palm weevil. World Acad Sci Eng Technol 2:160–163

Babjan B, Devi KS, Dangar TK, Sathiamma B (1995) Biological suppression of *Oryctes rhinoceros* by re-release of *Baculovirus oryctes* in an infected contiguous area. J Plant Crops 23:62–63

Baq-Qian L, Zhen Y, Jin QI-AN, Hai-Bo W, Yue-Guan FU, Wei-Dong LI Zheng-Qiang PENG (2013) Exotic pest alert: *Opisina arenosellaï* (Lepidopteraï: Oecophoridae). J Biosafety 22(1):17–22

Barranco P, DeLa Pena J, Cabello T (1995) Un Nuevo curculio'nido tropical para la fauna Europa, *Rhynchophorus ferrugineus* (Olivier 1790), (Curculionidae: Coleoptera). Boletin de la Asociaction Espanola de Entomologia 20:257–258

Bedford GO (1980) Biology, ecology and control of palm Rhinoceros beetle. Ann Rev Entomol 26:213

Bhanu KRM, Hall DR, Mathew T, Malvika C, Prabhakara MS, Awalekar RV, Jayanth KP (2009) Female sex pheromone of coconut black headed Caterpillar, *Opisina arenosella* Walker (Lepidoptera: Oecophoridae). 25th annual meeting. Int Soc Chemical Ecol. Neuchatel, August 23–27, pp 160

Bhanu KRM, Hall DR, Mathew T, Malvika C, Prabhakara MS, Awalekar RV, Jayanth KP (2011) Monitoring of Opisina arenosella by using female sex pheromones. APACE, pp 39.

Borror DJ, Delong DJ, Triplehorn CA (1964) An introduction to the study of insects, 4th edn. Holt, Rinehart & Winston, New York, 852 pp

Brand E (1917) Coconut red weevil. Some facts and fallacies. Tropical Agriculture Magazine of the Ceylon Agricultural Society 49:22–24

Butani DK (1975) Insect pests of fruit crops and their control date palm. Pesticides 9:40–42

Buxton PA (1920) Insect pests of dates and the date palm in Mesopotamia and elsewhere. Bull Entomol Res 11:287–303

Chakravarthy AK, Chandrashekharaiah M, Kandakoor SB, Nagaraj DN (2014) Efficacy of aggregation pheromone in trapping red palm weevil (Rhynchophorus ferrugineus Olivier) and rhinoceros beetle (Oryctes rhinoceros L.) from infested coconut palms. J Environ Biol 35:479–484

Chandrashekharaiah M, Bhanu KRM, Chakravarthy AK (2012) Pheromone Mass Trapping Technology for management of Coconut black-headed caterpillar, *Opisina arenosella* Walker (Lepidoptera: Oecophoridae) in Southern Karnataka, India. ICE –12: The 12th International Course In: Chemical Ecology; Frontiers Of Chemical ecology; Max Planck Institute For Chemical Ecology Jena, Germany.

Chandrashekharaiah M (2013a) Pheromone mass trapping technology for management of coconut black-headed caterpillar, *Opisina arenosella* walker (Lepidoptera: Oecophoridae) in southern Karnataka. Ph D Dissertation, University of Agricultural Sciences, Bangalore, India, pp 88–117

Chandrashekharaiah M, Bhanu KRM, Prabhakar MS, Mallik B, Muralidharan K, Chakravarthy AK (2013b) Standardisation of coconut black headed caterpillar, *Opisina arenosella* Walker monitoring and mass trapping by pheromone. Proceedings of International Conferences on Insect Science, ICIS 2013, Feb 14–17, Bangalore, India, p 46. (Eds. Chakravarthy AK, Ashok Kumar CT, Abraham Varghese and Thyagaraj NE)

Chandrashekharaiah M, Bhanu KRM, Prabhakar MS, Mallik B, Muralidharan K, Chakravarthy AK (2013c) Pheromone Mass Trapping For The Management Of Discrete Generations Of The Lepidopteron Pest, *Opisina arenosella* Walker (Lepidoptera: Oecophoridae) In Coconut Ecosystems. In: International Chemical Ecology Conference, Melbourne in 17–21 August 2013, p. 107.

Chandrika M (2005) Rhinoceros beetle, CPCRI India & IUCN/SSC Invasive Species Specialist Group (ISSG) report. http://www.issg.org/database/species/reference_files/oryrhi/IPMpackage.pdf

Chandrika M, Radhakrishnan Nair CP, Kesavan Nampoothiri C, Rajan P (2010) Leaf-eating caterpillar (*Opisina arenosella*) induced yield loss in coconut palm. Int J Trop Insect Sci 30(3):132–137

Chen ZC (1988) The ecology and control of major pests on *Cocos nucifera*. Chinese J Entomol Special Publ 2:81–96

Child R (1974) Coconuts. 2nd ed. Longmans, Green and Co., London, England.CPCRI, 2012, Annual Report 2011–12, Central Plantation Crops Research Institute, Kasaragod, 671124, Kerala, India, 128 p

Cork MJW, Perera PACR (1987) Biological control of *Opisina arenosella* Walker (Lepidoptera Oecophoridae) Biocontrol news and inf 8:283–310.

Cork A, Hall DR (1998) Application of pheromones for crop pest management in the Indian sub-continent. J Asia-Pacific Entomol 1(1):35–49

Cox ML (1993) Red palm weevil, *Rhynchophorus ferrugineus* in Egypt. FAO Plant Protection Bulletin 41:30–31

CPCRI (2012) Annual Report 2010–2011, Central Plantation Crops Research Institute Kasaragod, 671124, Kerala, India, 146 p

Crawford AM, Zelazny B, Alfiler AR (1986) Genotypic variation in geographical isolates of *Oryctes* baculovirus. J Gen Virol 67:949–952

Dangar TK, Geetha L, Jayapal SP, Pillai GB (1991) Mass production of entomopathogen Metarhizium anisopliae in coconut water wasted from copra making industry. J Plant Crops 19:54–69

Dembilio O, Quesada-Moraga E, Santiago-Álvarez C, Jacas JA (2010a) Potential of an indigenous strain of the entomopathogenic fungus, Beauveria bassiana as a biological control agent against the Red Palm Weevil, *Rhynchophorus ferrugineus*. J Invertebrate Pathology 104:214–221

Dembilio O, Llácer E, Altube MM, Jacas JA (2010b) Field efficacy of imidacloprid and *Steinernema* a chitosan formulation against the red palm weevil *Rhynchophorus ferrugineus* (Coleoptera: Curculionidae) in *Phoenix canariensis*. Pest Manag Sci 66:365–370

Desmier de Chenon R, Asmady Sudharto PS (2001) New improvement of pheromone traps for the management of the rhinoceros beetle in oil palm plantations. In Proc. PIPOC Int. Palm Oil Congr., Kuala Lumpur, Malaysia, 20–22 August, pp 624–632

Duke JA (1983) Handbook of energy crops (unpublished). http://www.hort.purdue.edu/newcrop/duke_energy/Cocos_nucifera.html

Esteban-Duran J, Yela JL, Beitia Crespo F, Jimenez A (1998) Biology of red palm weevil, *Rhynchophorus ferrugineus* (Olivier) (Coleoptera: Curculionidae: Rhynchophorinae), in the laboratory and field life cycle, biological characteristics in its zone of introduction in Spain, biological method of detection and possible control. Boletin Sanidad Vegetal Plagas 24:737–748

Faghih AA (1996) The biology of red palm weevil, *Rhynchophorus ferrugineus* Oliv. (Coleoptera, Curculionidae) in Savaran region (Sistan province, Iran). Applied Entomological Phytopathology 63:16–86

Faleiro JR (2005a) Biology and Management of the Red Palm Weevil: India.

Faleiro JR (2005b) Pheromone technology for the management of red palm weevil *Rhynchophorus ferrugineus* Olivier (Coleoptera: Rhynchophoridae)—A key pest of coconut, Technical Bulletin No.4, ICAR Research Complex for Goa. 40 pp.

Faleiro JR (2006) A review of the issues and management of the red palm weevil, *Rhynchophorus ferrugineus* (Coleoptera: Rhynchophoridae) in coconut and date palm during the last one hundred years. Int J Tropical Insect Sci 26:135–154

Faleiro JR, Abraham VA, Al-Shuaibi MA (1998) Role of pheromone trapping in the management of red palm weevil. Indian Coconut J Coconut Development Board 29:1–3

Faleiro JR, Satarkar V (2003) Age and fecundity of female red palm weevils, *Rhynchophorus ferrugineus* Oliver (Coleoptera: Curculionidae) Captured by Pheromone Traps in Coconut Plantations of India. Elsevier Sci Ltd., Oxon, England

Fargues JF, Robert PH (1983) Effects of passaging through scarabeid hosts on virulence and host specificity of two strains of the entomopathogenic hyphomycete *Metarhizium anisopliae*. Can J Microbiol 29:576–583

Fernando LCP (2005) Report of the crop protection division. In: Jayasekara C, Evarad JMDT, Appuhamy PAHN and Bandaranayake EMT (Eds) Annual Reoprt of Coconut Research Institute, Sri Lanka

Fernando LCP, Chandrasirl KAS (1997) More evidence on the attraction of males to the female sex pheromone of *Opisina arenosetta* Walker. Cocos 12:79–81

Ferry M (2010) The red palm weevil in California. Palms 54:203

Ferry M, Gomez S (2002) The red palm weevil in the Mediterranean area. Palms 46:172–178

Ghosh CC (1923) The palm beetles in Burma, with notes on other palm pests. In: Proceedings of the V Entomology Meeting, PUSA, pp 99–103

Gomez VS, Ferry M (1999) Attempts at biological control of date palm pests recently found in Spain, pp 121–125. In Proc. of the First Regional Symposium for Applied Biological Control in Mediterranean Countries, Cairo, 25–29 October 1998, Imprimerir Sacco, Toulouse, France

Gressitt JL (1957) Hispine beetles of the South Pacific. Nova Guinea 8:205–324

Gunawardane NE, Bandarage UK (1995a) A 4 methyl 1,5- nonanole (ferruginol) as an aggregation pheromone of the coconut pest, *Rhynchophorus ferrugineus* (Coleoptera: Curculionidae) synthesis and used in a preliminary field assay. J Natl Couns Sri Lanka 23:71–79

Gunawardane NE, Bandarage UK (1995b) Enhancement of the activity of ferruginol by N- pentanol in an attractant baited trap for the coconut pest, *Rhynchophorus ferrugineus* (Coleoptera: Curculionidae). J Natl Couns Sri Lanka 23:81–86

Hallett RH, Gries G, Borden JH, Czyzewska E, Oehlschlager AC, Pierce HD Jr, Angerilli NPD, Rauf A (1993) Aggregation pheromones of two Asian palm weevils, *Rhynchophorus ferrugineus* and R. vulneratus. Naturwissenschaften 80:328–331

Hallett RH, Perez AL, Gries G, Gries R, Pierce HD Jr, Yue J, Oehlschlager AC, Gonzalez LM, Borden JH (1995) Aggregation pheromone of coconut rhinoceros beetle, *Oryctes rhinoceros* (L.) (Coleoptera: Scarabaeidae). J Chem Ecol 21(10):1549–1570

Hamad SAS, El Faith MM (2004) Endoscope: a potential diagnostic tool for red palm weevil infestation. Proceeding of the Date Palm Regional Workshop on

Ecosystem-Based IPM for Date Palm in Gulf Countries, March 2004, pp 89–93

Hamadah SH, Tanani MA (2013) Laboratory studies to compare the toxicity for three insecticides in the red palm weevil *Rhynchophorus ferrugineus*. IJ Biol Pharm Allied Sci 2(3):506–519

Howard FW, Moore RM, Giblin-Davis RG, Abad (2001) Insects on Palms. CABI Publisher. London, England. Ho, C.T and Toh P.Y. 1982. Some investigations into the control of Oryctes rhinoceros L. in coconut plantings. Planter 58:492–506

Jacob TK (1996) Introduction and establishment of baculovirus for the control of rhinoceros beetle *Oryctes rhinoceros* (Coleoptera: Scarabaeidae) in the Andaman Islands (India). Bull Entomol Res 86:257–262

Jayanth KP, Mathew MT, Narabenchi GB, Bhanu KRM (2009) Reproductive status of Oryctus rhinoceros females captured in aggregation pheromone traps. Indian Coconut J 6:17–20

Kaakeh W (2005) Longevity, fecundity, and fertility of the red palm weevil, *Rynchophorus ferrugineus* Olivier (Coleoptera: Curculionidae) on natural and artificial diets. Emir J Agric Sci 17:23–33

Kaahkeh W, Aboul-Nour MM, Khamis AA (2001) The red Palm Weevil: the most destructive agricultural pest. United Arab Emirates University Printing Press, UAE

Kanagarathnam P, Pinto JLJG (1985) Effect of Monocrotophos on the leaf eating caterpillar *Opisina arenosella* Walker, when injected into the trunk of the coconut palm. Cocos 3:9–15

Kinawy MM (2004) Biological control of the coconut palm rhinoceros beetle (*Oryctes rhinoceros* L. Coleoptera: Scarabaeidae) using *Rhabdionvirus oryctes* Huger in Sultanate of Oman. Egypt J Biol Pest Control 14:113–118

Kirthisinghe JKF (1960) Red weevil and its control. Ceylon Cocon Plts Re 4(2):34–38

Kumara ADNT (2007) Movement pattern of *Neoseiulus baraki*, a predatory mite of coconut mite Aceria guerreronis in the field. M. Sc. Thesis, University of Ruhuna Sri lanka, pp 1–6

Lefroy HM (1906) The more important insects injurious to Indian Agriculture. Memoirs Department Agriculture India I:139–v

Li Y, Zhu ZR, Ju R, Wang LS (2009) The Red palm weevil *Rhynchophorus ferrugineus* (Coleoptera: Cucculionidae), Newly reported from zhejiang, china and update of geographical distribution. Florida Entomol. Soc., 92:386–387

Mahindapala R (1993) Integrated management of red palm weevil. CARP GTZ Agricultural Research Management Project Discussion document 2:35–41

Malumphy MH (2007) Red palm weevil, *Rhynchophorus ferrugineus*. Plant Pest Notice, Central Science Laboratory 50:1–3

Marschall KJ, Ioane I (1982) The effect of rerelease of *Oryctes rhinoceros* baculovirus in the biological control of rhinoceros beetles in Western Samoa. J Invertebr Pathol 39:267–276

Menon KPV, Pandalai KM (1960) The coconut palm-a monograph. Indian Central Coconut Committee, Ernakulam, India

Mohan KS, Pillai GB (1993) Biological control of Oryctes rhinoceros (L.) using an Indian isolate of *Oryctes baculovirus*. Insect Sci Appl 14:551–558

Morin JP, Sudharto PS, Purba RY, Desmier DChenonR, Kakul T (2001) A new type of trap for capturing *Oryctes rhinoceros* (Scarabaeidae, Dynastinae), The main pest in young oil palm and coconut plantings. Coconut Res De 17:13–22

Munoz L, Bosch MP, Rosell G, Guerrero A (2009) Asymmetric synthesis of ®- and (S)-4-methyloctanoic acids. A new route to chiral fatty acids with remote stereocenters. Tetrahedron Asymmetr 20:420–424

Murphy ST, Briscoe BR (1999) The red palm weevil as an alien invasive: biology and the prospects for biological control as a component of IPM. Biocontrol News Inform 20:35–45

Murthy KS, Gour TB, Reddy DDR, Babu TR, Zaheruddeen SM (1995) Pheromone baits for coconut black beaded caterpillar, *Opisina arenosella* Walker. Insect Envt 1:3–4

Nadarajan L, Channabasavanna GP (1980) Population dynamics of coconut black headed caterpillar, *Nephantis serinopa* Meyrick (Lepidoptera: Cryptophasidae) and its parasites. Mysore J Agric Sci 11(4):533–541

Nakash J, Osam Y, Kehat M (2000) A suggestion to use dogs for detecting red palm Weevil (*Rhynchophorus ferrugineus*) infestation in date palm in Israel. Phytoparasitica 28:153–154

Nirula KK (1956) Investigation on the pest of coconut palm, Part-IV. *Rhynchophorus ferrugineus*. Indian Coconut J 10:28–44

Nirula KK, Antony J, Menon KVP (1951b) Investigations on the pests of coconut palm. Part II control with DDT sparay. Indian Coconut J 4:225–223

Nisson N, Hodel D, Hoddle MS (2010) Red Palm Weevil, *Rhynchophorus ferrugineus* (Olivier). Center for Invasive Species Research, University of California

Oehlschlager C (2007) Optimizing trapping of palm weevils and beetles. Acta Hortic 736:347–368

Perera PACR (1987) Studies on *Opisina arenosella* (Walker) and its natural enemies in Sri Lanka. Ph.D. thesis, University of London, UK

Perera PACR, Hussell MP, Godfray CJ (1988) Population dynamics of the coconut caterpillar, *Opisina arenosella* Walker (Lepidoptera Xyloryctidae), in Sri Lanka. Bull Entomol Res 78:479–492

Pinhas J, Soroker V, Hetzroni A, Mizrach A, Teicher M, Goldberger J (2008) Automatic acoustic detection of the red palm weevil. Comput Electron Agric 63:131–139

Prabhu ST, Patil RS (2009) Studies on the biological aspects of red palm weevil, *Rhynchophorus ferrugineus* (Oliv.). Karnataka J Agric Sci 22:732–733

Ragoussis V, Giannikopoulos A, Skoka E, Grivas P (2007) Efficient synthesis of (+/−)-4-methyloctanoic acid, aggregation pheromone of rhinoceros beetles of the genus Oryctes (Coleoptera: Dynastidae, Scarabaeidae). J Agric Food Chem 55:5050–5052

Rajapakse CNK, Gunawardena NE, Perera KFG (1998) Pheromone baited trap for the management of red palm weevil *Rhynchophorus ferrugineus* (Coleoptera: Curculionidae) population in coconut plantations. Cocos 13:54–65

Ramkumar K, Muralimohan K, Kiranmayi L, Srinivasa YB (2006) Discrete generation cycles in the tropical moth *Opisina arenosella*. Res Commun Curr Sci 91(6):811–816

Ramle M, Basri MW, Norman K, Mukesh S, Siti Ramlah AA (1999) Impact of *Metarhizium anisopliae* (Deuteromycotina: Hyphomycetes) applied by wet and dry inoculum on oil palm rhinoceros beetles, *Oryctes rhinoceros* (Coleoptera: Scarabaeidae). J Oil Palm Res 11:25–40

Ramle M, Wahid MB, Norman K, Glare TR, Jackson TA (2005) The incidence and use of *Oryctes* virus for control of rhinoceros beetle in oil palm plantations in Malaysia. J Invertebr Pathol 89:85–90

Ramle M, Basri MW, Norman K, Siti Ramlah AA, Hisham H (2006) Research into the commercialization of *Metarhizium anisopliae* (Hyphomycetes) for biocontrol of oil palm rhinoceros beetle, *Oryctes rhinoceros* (Scarabaeidae), in oil palm. J Oil Palm Res Spec Issue:37–49

Ramle M, Norman K, Ang BN, Ramlah Ali AS, Basri MW (2007) Application of powder formulation of Metarhizium anisopliae to control *Oryctes* rhinoceros in rotting oil palm residues under leguminous cover crops. J Oil Palm Res 19:319–331

Rao YR, Cherian MC, Ananthanarayanan KP (1948) Infestation of *Nephantis serinopa* Meyrick in South India, and their control by the biological method. Indian J Entomol 10:205–247

Al-Saqer SM, Ghulam MH (2011) Red Palm Weevil (*Rynchophorus Ferrugineous*, Olivier) Recognition by image processing techniques. Am J Agric Biol Sci 6(3):365–376

Siriwardena KAP, Fernando LCP, Nanayakkara N, Perera KFG, Kumara ADNT, Nanayakkara T (2010) Portable Acoustic Device for Detection of Coconut Palms infested by *Rynchophorus ferrugineus* (Coleoptera: Curculinoidae). J Crop Prot 29:25–29

Sivapragasam A, Arikiah A, Ranjit CA (1990) The Red stripe weevil, *Rhynchophorus schach* Olivier (Coleoptera: Curculionidae): an increasing menace to coconut palms in Hilir Perak. Planter 66:113–123

Soroker VY, Nakash U, Landau A, Mizrach A, Hetzroni A, Gerling D (2004) Utilization of sounding methodology to detect infestation by *Rhynchophorus ferrugineus* on palm offshoots. Phytoparasitica 32:6–8

Srinivasa YB, Muralimohan K (2009) Scope for managing age-structured populations of the coconut black headed caterpillar, *Opisina arenosella* (Walker) using sex pheromones. Pest Manage Hortic Ecosyst 15(2):89–92

Subaharan K (2004) Green muscardine fungus for coconut rhinoceros beetle control. Planter 80:717–720

Sujatha A, Chalapathirao Rao NBV, Rao DVR (2006) Field evaluation of two pheromone lures against red weevil, (*Rhynchophorus ferrugineus* Oliv.) in coconut gardens in Andhra Pradesh. J Plant Crops 34(3):414–416

Sundararaju D, Rodrigues OF (1985) Survey on incidence of coconut leaf eating caterpillar, *Opisina arenosella* (Walker) and its natural enemies in Goa. Indian Coconut J 7:3–6

Tey CC, Ho CT (1995) Infection of *Oryctes rhinoceros* L. by application of *Metarhizium anisopliae* (Metsch.) Sorokin to breeding sites. Planter 71:563–567

Thampan PK (1975) The Coconut Palm and Its Product. Green Villa Publishing House, India

Venkatesan T, Jalali SK, Srinivasamurthy K (2009) Competitive interactions between *Goniozus nephantidis* and *Bracon brevicornis*, parasitoids of coconut pest *Opisina arenosella*. Int J Pest Manag 55(3):257–263

Vidhyasagar PSPV, Al-saihati AA, Al-Mohanna OE, Subbei AI, Absul Mohsin AM (2000) Management of Red palm weevil, *Rhynchophorus ferrugineus* Olivier A serious pest of date palm in Al-Qatif, Kingdom of Saudi Arabia. J Plant Crops 28(1):35–43

Wood BJ (1968) Pests of Oil Palm in Malaysia and Their Control. Incorporated Society of Planters Kuala Lumpur, Malaysia

Zelazny B (1979) Virulence of the baculovirus of Oryctes rhinoceros from ten locations in the Philippines and in Western Samoa. J Invertebr Pathol 33:106–107

Zelazny B, Alfiler AR (1991) Ecology of baculovirus-infected and healthy adults of *Oryctes rhinoceros* (Coleoptera: Scarabaeidae) on coconut palms in the Philippines. Ecol Entomol 16:253–259

Zelazny B, Lolong A, Crawford AM (1990) Introduction and field comparison of baculovirus strains against *Oryctes rhinoceros* (Coleoptera: Scarabaeidae) in the Maldives. Environ Entomol 19:1115–1121

Zelazny B, Lolong A, Pattang B (1992) *Oryctes rhinoceros* (Coleoptera: Scarabaeidae) populations suppressed by a baculovirus. J Invertebr Pathol 59:61–68

Area-Wide Integrated Pest Management in Pigeonpea

B. V. Bhede, B. B. Bhosle, O. P. Sharma, S. T. Shinde and A. G. Badgujar

Abstract

Insect pests are the major constraints for pigeonpea yield reduction. Large-scale integrated pest management (IPM) of pigeonpea was undertaken on 1000 ha in Parbhani in farmers' participatory mode. The pest population was less in IPM fields than non-IPM fields. The population of pod borer larvae was 0.39/plant in IPM, and 0.64/plant in non-IPM. Plume moth larvae were 0.16 and 0.25/plant in IPM and non-IPM fields, respectively. The webbings by spotted pod borer (*Maruca vitrata* Gey.) were 0.08 and 0.19/plant in IPM and non-IPM fields, respectively. The pod damage due to pod borer complex was less in IPM (7.71 %) than non-IPM (21.52 %) fields. The immature stages (maggot and pupa) of pod fly were 2.14 and 7.99/50 pods in IPM and non-IPM fields, respectively. The population of natural enemies like coccinellids (0.21/plant in IPM and 0.09/plant in non-IPM), chrysopids (0.03/plant in IPM and 0.01/plant in non-IPM) and spider (0.36/plant in IPM and 0.24/plant in non-IPM) was more in IPM fields. The parasitisation of pod borer larvae was 3.63 % in IPM and 1.78 % in non-IPM. The impact of the IPM practices resulted in increased yield (10.18 q/ha in IPM and 8.41 q/ha in non-IPM) with a net profit of ₹ 7710/ha. IPM module insect pests effectively managed and conserved natural enemies.

B. V. Bhede (✉) · B. B. Bhosle · S. T. Shinde ·
A. G. Badgujar
Department of Entomology, Marathwada Krishi Vidyapeeth, Parbhani, Maharashtra 431402, India
e-mail: bvb1093@gmail.com, bvb1093@rediffmail.com

O. P. Sharma
Plant Protection, NCIPM, New Delhi, India

A. K. Chakravarthy (ed.), *New Horizons in Insect Science: Towards Sustainable Pest Management*,
DOI 10.1007/978-81-322-2089-3_33, © Springer India 2015

Keywords

Area-wide · IPM pigeonpea · Parasitisation

Introduction

Pigeonpea, *Cajanus cajan* (L.) is one of the most important pulse crops in India. Insect pests feeding on flowers and pods are the most important biotic constraints affecting pigeonpea yield. In India, pigeonpea is attacked by nearly 250 species of insect pests (Sachan et al. 1994). The pod borers are the key impediments for the low productivity. In Maharashtra, pod borer *Helicoverpa armigera* (Hubner), plume moth *Exelastis atomosa* (Walsingham), pod fly *Melanagromyza obstusa* (Malloch), spotted pod borer *Maruca vitrata* Geyer are considered significant in incurring economic losses. These pod borers' damage amounts to 57 % pods and 35 % seeds with final yield loss up to 28 % (Saho 2002). The farmers use only synthetic insecticides for pest management. Increasing insecticide resistance has lead to a greater risk of control failure (Armes et al.1992).

Material and Methods

The large-scale integrated pest management (IPM) programme was demonstrated and evaluated on 1000 ha of farmers' field in Parbhani district (Maharashtra) during 2010–2011. This programme was implemented in Chinchtakali, Mardasgaon and Gopa villages. The farmers were provided with IPM components including chemical insecticides. All agronomic practices were followed as per the recommendation of MKV, Parbhani. The IPM module included the following practices. Deep ploughing in summer; use of high-yielding and disease-resistant varieties, viz. BSMR 736, BSMR 853, BDN 708; installation of pheromone traps, at the rate of 5/ha, to monitor pod borer *H. armigera*; installation of bird perches, at the rate of 50/ha; intercropping with greengram, blackgram, mung, udid, soybean and sorghum; in case of sole crops, mixing of 1 % sorghum or pearl millet seeds with pigeonpea seeds; hand collection of big *H. armigera* larvae; spraying 5 % NSKE or Azadirachtin 3000 ppm at bud initiation stage; spraying HaNPV at the rate of 450 LE/ha (2×10^9 POB/ml) for early instar *H. armigera* larvae; spraying of emamectin benzoate 5SC at the rate of 200 g/ha and monocrotophos 36 SL at the rate of 1000 ml/ha.

In non-IPM fields, only the recommended agronomic practices were followed and no IPM inputs were used. These select farmers were mostly dependent on chemical insecticides. Two fields of IPM and non-IPM were selected from each village for periodical observations. Observations on incidence of pests and their natural enemies were made by following the standard procedures. The observations on larva of pod borer *H. armigera* and plume moth *E. atomosa*, immature stages of pod fly *M. obtusa*, number of webbings by spotted pod borer *M. vitrata*, pod damage, number of coccinellids, chrysopids, spiders and pod borer larval parasitisation were recorded at weekly intervals. Yield was recorded for both IPM and non-IPM fields, and economics worked out.

Results and Discussion

Pest Status

Population of Pod Borer

The results of IPM demonstration showed that population of pod borer larvae was less in IPM fields than non-IPM fields throughout the season. The average larval population of *H. armigera* was 0.39/plant in IPM, whereas 0.64/plant in non-IPM fields (Table 1). The incidence of pod borer larvae was noticed during 44th MW (25–31 October 10). The peak population was recorded in 48th MW (22–28 November 10) in IPM (0.76 larvae/plant) and non-IPM (0.88 larvae/plant).

Population of Plume Moth

The incidence of plume moth was initiated in 45th MW (1–7 November 10) in both IPM and non-IPM fields. The population ranged from 0.08

Table 1 Incidence of pod borers on pigeonpea in IPM and non-IPM fields

MW	Duration	No. of *Helicoverpa armigera* larvae/plant		No. of *Exelastis atomosa* larvae/plant		No. of webbings by *Maruca vitrata* larvae/plant	
		IPM	Non-IPM	IPM	Non-IPM	IPM	Non-IPM
43	18–24 October10	0.00	0.00	0.00	0.00	0.00	0.00
44	25–31 October10	0.03	0.10	0.00	0.00	0.00	0.08
45	01–07 Novober10	0.46	0.60	0.11	0.21	0.13	0.30
46	08–14 Novober10	0.55	0.79	0.08	0.18	0.06	0.16
47	15–21 Novober10	0.62	0.83	0.41	0.58	0.11	0.19
48	22–28 Novober10	0.76	0.88	0.20	0.30	0.10	0.20
49	29 Novober10–05 December 10	0.58	0.73	0.00	0.01	0.05	0.16
50	06–12 December10	0.53	0.70	0.11	0.35	0.00	0.15
51	13–19 December10	0.36	0.76	0.40	0.45	0.20	0.35
52	20–26 December10	0.25	0.83	0.25	0.33	0.26	0.41
1	27 December10–02 January 11	0.15	0.85	0.20	0.31	0.01	0.05
	Mean	0.39	0.64	0.16	0.25	0.08	0.19

to 0.41 larvae/plant in IPM and from 0.01 to 0.58 larvae/plant in non-IPM fields. Overall, the population was minimal in IPM fields (0.16/plant) as compared to non-IPM fields (0.25/plant). The maximum population was recorded in 47th MW (15–21 November 10) in IPM and non-IPM fields (Table 1).

Incidence of Spotted Pod Borer

The number of webbings on plant due to larvae of spotted pod borer was minimal in all IPM fields (0.08/plant) compared to that of non-IPM (0.19/plant). The peak incidence was recorded in 52nd MW (20–26 December 10) in both IPM and non-IPM fields (Table 1).

Pod Damage Due to Pod Borer Complex

The percentage of pod damage was less in IPM (7.71 %) as compared to non-IPM fields (21.52 %). The pod damage was observed from 48th MW (22–28 November 10), and thereafter increased till the end of season (Table 2). The highest pod damage was noticed in 52nd MW (20–26 December 10).

Immature Stages of Pod Fly

The immature stages (maggot and pupa) of pod fly in pods of pigeonpea were noticed in 50th MW (6–12 December 10). The range was from 0.66 to 5.66 immature stages/50 pods in IPM, and from 4.33 to 17.66 immature stages/50 pods in non-IPM fields. The immature stages records increased till the end of season. The average of the immature stages was 2.14 and 7.99/50 pods in IPM and non-IPM, respectively.

Grain Damage due to Pod Fly

At harvest, the observations on grain damage due to pod fly indicated that the grain damage in IPM was 5.74 % whereas it was 12.20 % in non-IPM fields.

Status of Natural Enemies on Pigeonpea in IPM and Non-IPM

Predators

The population of natural enemies like coccinellids (0.21 in IPM and 0.09/plant in non-IPM), chrysopids (0.03 in IPM and 0.01/plant in non-IPM) and spiders (0.36 in IPM and 0.24/plant in non-IPM) was more in IPM fields. The coccinellid population was 0.03–1.01/plant in IPM, and 0.01–0.41/plant in non-IPM (Table 3). The chysopids population was 0.02–0.10/plant in IPM and 0.01–0.04/plant in non-IPM.

Table 2 Pod damage due to pod borer complex in IPM and non-IPM fields

MW	Duration	Pod damage (%)		No. of immature stages of pod fly (maggot and pupa)/50 pods	
		IPM	Non-IPM	IPM	Non-IPM
47	15–21 November 10	0.00	0.00	0.00	0.00
48	22–28 November 10	2.33	10.33	0.00	0.00
49	29 November 10–05 December10	5.66	15.33	0.00	0.00
50	06–12 December 10	10.33	23.33	0.66	4.33
51	13–19 December 10	11.66	33.33	2.33	15.66
52	20–26 December 10	13.66	35.66	6.33	18.33
1	27 December 10–02 January 11	10.33	32.66	5.66	17.66
	Mean	7.71	21.52	2.14	7.99

Table 3 Population of natural enemies in IPM and non-IPM fields

MW	Duration	No. of spiders/plant		Parasitisation of pod borer larvae (%)	
		IPM	Non IPM	IPM	Non IPM
43	18–24 October 10	0.51	0.34	0.00	0.00
44	25–31 October 10	0.56	0.40	0.00	0.00
45	01–07 November 10	0.57	0.50	4.00	2.50
46	08–14 November 10	0.58	0.44	4.50	1.00
47	15–21 November 10	0.56	0.43	7.00	3.50
48	22–28 November 10	0.48	0.30	8.60	5.00
49	29 November 10–05 December 10	0.05	0.01	5.00	2.00
50	06–12 December 10	0.30	0.06	4.80	1.00
51	13–19 December 10	0.15	0.08	2.00	0.00
52	20–26 December 10	0.15	0.10	2.00	1.00
1	27 December 10–02 January 11	0.00	0.00	1.50	0.00
	Mean	0.36	0.24	3.63	1.78

Table 4 Economics of IPM and non-IPM in pigeonpea fields

Particulars	IPM	Non-IPM
Yield (q/ha)	10.18	8.41
Number of sprayings	3.83	5.67
Cost of spraying (₹/ha)	2320	2950
Net income (₹/ha)	40,720[a]	33,640
Increase in yield (q/ha)	1.77	–
Additional profit due to increase in yield (₹/ha)	7080	–
Additional profit due to savings of sprayings (₹/ha)	630	–
Net profit due to IPM (₹/ha)	7710	–

[a]Price ₹ 4000/q

Parasitisation of Pod Borer Larvae

The percentage of parasitized larvae was more in IPM (3.63 %) than non-IPM (1.78 %). The range of parasitized larvae was 1.50–8.60 % in IPM and 1.00–5.00 % in non-IPM. The ichneumonids, *Eriborus argenteopilosus* (Cameron) and *Campoletis chlorideae* (Uchida), were major parasitoids recorded on *H. armigera*.

Economics

IPM fields recorded the highest yield compared to non-IPM fields. The average yield was 10.18 q/ha in IPM and 8.41 q/ha in non-IPM fields. The insecticide sprayings required 3.83 in IPM fields and 5.67 in non-IPM fields. The adoption of IPM resulted in savings of ₹ 630/ha due to less cost of insecticide sprayings in IPM. Also, the additional profit of ₹ 7080/ha is recorded due to IPM. The IPM crop increased the net profit by ₹ 7710/ha (Table 4).

Babriya et al. (2010), Sharma et al. (2011) and Chaudhary et al. (2008) reported effectiveness of chemical insecticides in management of pod borer using similar strategies. Mandal and Mishra (2003) and Sharma et al. (2011) reported successful management of pod fly with timely detection and use of chemical insecticides. Bisane et al. (2008) recorded similar parasitisation reports of *H. armigera*. The present evaluation of pigeonpea IPM benefited the farmers who participated in this programme.

References

Armes NJ, Jadhav DR, Bond GS, King ABS (1992) Insecticide resistance in *Helicoverpa armigera* (Lepidoptera: Noctuidae) in South India. Pestic Sci 16:131–144

Babriya PM, Kabaria BB, Patel VN, Joshi MD (2010) Chemical control of gram pod borer (*Helicoverpa armigera* Hubner) infesting pigeonpea. Legum Res 33(3):224–226

Bisane KD, Borkar SL, Lande GK, Bhalkare SK (2008) Parasitization of *Helicoverpa armigera* (Hubner) on pigeon pea. J Biol Control 22(2):455–461

Chaudhary RG, Saxena H, Dhar V, Prajapati RK (2008) Evaluation and validation of IPM modules against wilt, Phytophthora blight, pod borer and pod fly in pigeonpea. J Food Legum 21(1):58–60

Mandal SMA, Mishra BK (2003) Bioefficay of insecticides and biopesticides against pod borers and pod fly on pigeonpea. J Appl Zoo Res 14(10):42–43

Sachan JN, Yadava CP, Ahmad R, Katti G (1994) Insect pest management in pulse crops. In: Dhaliwal GS, Arora R (eds) Trends in agricultural insect pest management. Commonwealth Publishers, New Delhi, p 308

Saho BK (2002) Comparative efficacy of synthetic insecticides and plant products against the incidence of pigeonpea pod borers. Indian J Ent 64(2):195–201

Sharma OP, Bhosle BB, Kamble KR, Bhede BV, Seeras NR (2011) Management of pigeonpea pod borers with special reference to pod fly (*Melanagromyza obtusa*). Indian J Agric Sci 81(6):539-543

Sustainable Management of Tea Mosquito Bug *Helopeltis antonii* Signoret (Miridae: Hemiptera) on Cashew

C. Manja Naik, A. K. Chakravarthy, Timmanna and N. E. Thyagaraj

Abstract

A series of observations were recorded on natural enemies and their role in suppression of tea mosquito bug (TMB) *Helopeltis antonii* Signoret in the cashew plantations of Zonal Agricultural Research Station, Brahmavar and Pethri village, Udupi district and in maidan (plains) tracts of Chintamani, Karnataka during 2006–2008. The ant, *Oecophylla smaragdina* Fabricius was the most effective predator against TMB. Cashew trees fully colonized by *O. smaragdina* received the least TMB damage of 11.40 + 1.87 and 8.71 + 1.23 % during 2006–2008 in coastal and maidan Karnataka, respectively. TMB damage colonized with *O. smaragdina* was 23.40 + 2.16 in coastal and 20.12 + 2.05 in maidan habitat. The unsprayed cashew trees with no ant recorded maximum damage of 47.42 + 3.71 and 52.36 + 3.86 % in coastal and maidan tracts, respectively. Higher nut yield of 3.70 kg/tree in coastal and 2.43 kg/tree in maidan tract was recorded from the trees fully colonized with *O. smaragdina*. The yellow crazy ant (*Anoplolepis gracilipes* F Smith) was a deterring factor for the spread, establishment and effectiveness as a predator on colonies of *O. smaragdina* in cashew plantations in Karnataka. *Telenomus* species was the predominant egg parasitoid on TMB. Maximum egg parasitisation of 16.80, 15.35 and 12.70 % was recorded on TMB eggs in Brahmavar, Pethri and Chintamani cashew plantations, respectively, in December, 2006–2008. The sequential sprays of monocrotophos (0.05 %)—λ cyhalothrin (0.005 %)—carbaryl (0.10 %) registered the least percentage TMB damage, higher nut yield and higher C to B ratio.

C. M. Naik (✉) · A. K. Chakravarthy · Timmanna · N. E. Thyagaraj
Department of Entomology, University of Agricultural Sciences, GKVK, Bangalore 560 065, India
e-mail: naik_196710@yahoo.com

A. K. Chakravarthy
Department of Agricultural Entomology, Gandhi Krishi Vignan Kendra (GKVK), University of Agricultural Sciences, Bangalore, Karnataka, India

A. K. Chakravarthy (ed.), *New Horizons in Insect Science: Towards Sustainable Pest Management*, DOI 10.1007/978-81-322-2089-3_34, © Springer India 2015

Keywords

Botanicals · Cashew · Management practices · Tea mosquito bug

Introduction

Cashew (*Anacardium occidentale* L.) is an important cash crop. With the rapid expansion of cashew crop acreage, the insect pest problems also increased in Karnataka, South India. This is quite apparent from the fact that the production is not in pace with the increase in area. In spite of the adoption of recommended package of practices, decline in production is mainly due to tea mosquito bug (TMB) *H. antonii* Signoret. More than 50 insect pests are recorded on cashew in India (Devasahayam and Nair 1986). Amongst them, TMB alone causes 30–100% yield loss (Abraham and Nair 1981) in outbreak situations (Sundararaju and Sundarababu 1999). TMB causes serious damage to the tender leaf, growing tip, inflorescence, apple and the nut. Cashew grows wild in nature; and in Karnataka, South India, the Horticulture Department and the government encourage cashew plantation in vast stretches of wastelands/degraded/government lands. Private owners of the lands are also encouraged to cultivate cashew. Wherever cashew grows, people tend not to use insecticides to suppress pests. Therefore, eco-friendly management practices are desired for the TMB and other pests.

Material and Methods

During 2006, six field surveys on pests were conducted in cashew plantations of coastal, maidan (plane landscape) and other habitats of Karnataka to find activity of ants, species compositions, flushing shoots damaged by *H. antonii* and other pests and number of healthy nuts. Attempts were made to establish ant colonies on cashew trees in Brahmavara, Puttur and Chintamani cashew plantations of 2 ha each at each location. *O. smaragdina* species of ants were released on five randomly selected cashew trees after the release of ants. Observations were recorded for 1.5 years at periodic intervals (2–3 months duration). The natural ant colonies of *O. smaragdina* on existing trees with terminal branches with ants were incised, and ants were held in perforated polythene bags and released on the labelled cashew trees and periodic observations were recorded for 3 years. Studies on seasonal activity of egg parasitoid of *Telenomus* sp. on TMB in Coastal Karnataka was investigated using field surveys. Data were collected from two study sites at Brahmavar and in Pethri village cashew plantation. The cashew plant parts (shoot, petioles, midrib and panicles) containing eggs of *H. antonii* were collected at monthly intervals and counted under a zoom stereomicroscope. They were treated with carbendazim 0.1% solution for 10 min. After treatment, samples were dried to remove the dampness of carbendazim solution placed in a plastic container completely wrapped in a black paper (Geisberger 1993) to record the emergence of parasitoids and percentage of TMB egg parasitization.

Predatory ants associated with cashew plantations and their effect on *H. antonii* population was assessed at 20 locations in west coast and maidan Karnataka. Observations were recorded on insecticide-sprayed and non-sprayed cashew plantations in October–February, 2006–2007 and 2007–2008, to determine the differences in TMB infestation and occurrence of natural enemies. The ant specimens were identified by Mr. T. M. Mushtak Ali, Department of Agricultural Entomology, UAS, GKVK, Bangalore. Abundance and effectiveness of predatory ants were determined on their predominance and predation of TMB on cashew in coastal Karnataka. Data on the ant colonisation and damage on reproductive parts of the tree and yield (kg/tree) were also recorded.

Indices for assessing abundance of predatory ant *O. smaragdina* were made based on quick examination of an individual cashew tree (Way and Khoo 1991) as follows.

+	(Few) less than 20 workers/tree; no trails; very few or no homopterans; no nests
++	(Moderate number) > 20–50 *Oecophylla smaragdina*
+++	(Common) > 50–500 *Oecophylla smaragdina* usually some distinct trails on tree trunk or canopy but rarely on ground
++++	(Abundant) > 500–1000 *Oecophylla smaragdina*; well-defined trails in canopy and occasionally on trunk and along ground
+++++	(Very abundant) > 1000 *Oecophylla smaragdina* or with strong trails; interconnecting or virtually all trees across their canopies and or along ground

The effectiveness of biopesticides and insecticides against TMB infestation were tested at Zonal Agricultural Research Station, Brahmavara on cashew cultivar, V-4. Three sprays were given in sequence on new flush, panicle emergence and nut and fruit developmental stages of the trees. Treatments were replicated thrice in a randomized block design.

Treatment	ml/l or %
Pongamia oil–pongamia oil–pongamia oil	2–2–2
PSKE–PSKE–PSKE	2–2–2
Neem oil–neem oil–neem oil	5–5–5
NSKE–NSKE–NSKE	2–2–2
Beauveria bassiana–Beauveria bassiana–Beauveria bassiana	2–2–2
Monocrotophos–λ-cyhalothrin–carbaryl	0.05–0.005–0.10
PSKE–λ cyhalothrin–carbaryl	2–0.05–0.10
Beauveria bassiana–λ-cyhalothrin–carbaryl	2–0.005–0.10
NSKE–λ cyhalothrin–carbaryl	2–0.05–0.10
Monocrotophos–endosulfan–carbaryl	0.05–0.05–0.10
Control	0.10

t-test was adopted for comparison of two treatment means

PSKE pongamia seed kernel extract, *NSKE* neem seed kernel extract

Results and Discussion

Field surveys revealed that of 1520 cashew trees across different habitats of Karnataka, predatory green ants were detected in 40 % cashew trees. However, nesting colonies of the ant were found only on 18 % cashew trees. Ant fauna of cashew trees comprised six species, *O. smaragdina* being the dominant species. The number of flushing shoots damaged by *H. antonii* where ants were found was less (14.50 %) compared to trees where green ants were absent (37 %). Undisturbed patches with native vegetation where branches of the cashew trees were touching each other or other trees/bushes held the ant colonies of *O. smaragdina*.

Predatory Ants on TMB

The abundance of different ant species foraging on cashew trees was observed at 20 locations of coastal Karnataka and Chintamani (Table 1). In coastal Karnataka, the ground ant species *Diacamma rogasum* Loguill and *Camponotus compresus* Fabricius were observed to be less abundant and nesting on the cashew tree was low to negligible. The arboreal species, *O. smaragdina*, was observed at all the locations surveyed.

In non-sprayed plantations, *O. smaragdina* population was abundant and occupied the entire cashew tree canopy. The other species of ants were absent on the trees colonized with *O. smaragdina*. It has well-defined colonies that it defends aggressively against other ant species. The trees colonized by *O. smaragdina* were interconnected with well-defined trails of worker ants invariably using aerial connections, where the canopies overlapped. Trails of *O. smaragdina* were observed on the ground where canopies were not overlapping. The incompatibility of *O. smaragdina* and the other ants was confirmed by spatial separation at the edges of their territory.

In the maidan tract of Karnataka, eight ant species foraging in cashew ecosystems were observed. *O. smaragdina* was abundant and nesting on the cashew tree branches was recorded. The other species of ants were not competing with the predominant species, *O. smaragdina*. Other ant species were found foraging at the edges of the tree canopy. *O. smaragdina* makes nests by spinning silk between leaves with moderately sized, densely spaced cashew leaves, around 5–6, medium sized nests were recorded per tree. Each nest consisted of leaflets with the ventral side wound

Table 1 Composition of ant species on cashew, Karnataka, South India

Location	Coastal/maidan part	Sub family	Arboreal/ground	Abundance	Sprayed/unsprayed
Mundur	*D. rugosum*	Ponerinae	Ground	+	Sprayed
Ullal	*O. smarigdina*	Formicinae	Arboreal	+	Sprayed
Konaje	*O. smarigdina*	Formicinae	Arboreal	+++	Non-sprayed
Brahmavar	*O. smarigdina*	Formicinae	Arboreal	+	Sprayed
Pethri	*O. smarigdina*	Formicinae	Arboreal	+	Sprayed
Korgi	*O. smarigdina*	Formicinae	Arboreal	+	Sprayed
Gujjadi	*O. smarigdina*	Formicinae	Arboreal	+	Sprayed
Hebri	*O. smarigdina*	Formicinae	Arboreal	+	Sprayed
	O. smarigdina	Formicinae	Arboreal	+++++	Non-sprayed
	D. rugosum	Ponerinae	Ground	+	Non-sprayed
	C. compresus	Formicinae	Ground	+	Non-sprayed
	Crematogaster wroughtonii (Forel)	Myrmicinae	Arboreal	+	Non-sprayed
Belvai	*O. smarigdina*	Formicinae	Arboreal	+	
Bantakal	*O. smarigdina*	Formicinae	Arboreal	+++++	Non-sprayed
	O. smarigdina	Formicinae	Arboreal	+++++	Non-sprayed
	D. rugosum	Ponerinae	Ground	+	Non-sprayed
	C. compresus	Formicinae	Ground	+	Non-sprayed
	C. wroughtonii	Myrmicinae	Arboreal	+	Non-sprayed
Cherkadi	*O. smarigdina*	Formicinae	Arboreal	+	
Kalanja	*C. compresus*	Formicinae	Ground	+	
Guthiger	*O. smarigdina*	Formicinae	Arboreal	+	
Savanur	*D. rugosum*	Ponirinae	Ground	+	
Bellare	*O. smarigdina*	Formicinae	Arboreal	+	
Kairapalli	*C. compresus*	Formicidae	Ground	+	
	O. smarigdina	Formicinae	Arboreal	+	Sprayed
Cherkadi	*O. smarigdina*	Formicinae	Arboreal	+	
	C. wroughtonii	Myrmicinae	Arboreal	+	Sprayed
ARS Chintamani	*O. smarigdina*	Formicinae	Arboreal	++++	Non-sprayed
	C. wroughtonii	Myrmicinae	Arboreal	+	Non-sprayed
	Crematogaster sp	Myrmicinae	Arboreal	+	Non-sprayed
	C. rothnyi (Mayer)	Myrmicinae		+	Non-sprayed

+ very scarce; ++ very scarce; +++ less abundant; ++++ abundant; +++++ very abundant

up with silk or several overlapping leaflets spun together.

The other species of ants, *D. rugosum* and *C. wroughtonii* nests were recorded, whether arboreal or on the ground. Ground nests, which sometimes occurred abundantly in a thick layer of cashew leaves and close to the trunks of trees, were observed on borders of plantation near bunds and roads. The activity of ants on cashew trees sprayed with insecticides was low, and few or no ants were observed on such trees. It was abundant (++++) to very abundant (+++++) in non-sprayed plantations. Among the different ant species recorded in cashew plantations, *O. smaragdina* appeared to be a predominant arboreal ant in cashew plantations at both the locations observed.

Predatory ant, *O. smaragdina* workers captured insect preys and carried them away to the nests by operating singly or in groups. In the study sites, the distribution of ant species indicated that single species of *O. smaragdina* occupied 80 % of the cashew plantation observed, and was found to be very abundant (+++++) (>1000 workers occupying a tree canopy of 15 m radius).

Cashew trees fully colonized by the predatory ant *O. smaragdina* exhibited the least percentage damage 11.40+1.87 and 8.71+1.23 during

Table 2 Egg parasitisation of tea mosquito bug by *Telenomus* sp. in field. (pooled data 2006–2008)

Location (percentage egg parasitisation by TMB) (%)			
Month	Brahmavar	Pethri	Chintamani
January	7.80	15.35	4.64
February	8.50	7.40	5.17
March	2.10	5.30	3.14
April	4.20	3.00	2.23
May	1.30	0.60	0.92
June	1.82	0.30	1.03
July	0.40	0.90	0.67
August	0.90	0.80	2.02
September	4.40	3.63	4.23
October	4.80	5.50	5.24
November	5.70	7.42	7.14
December	16.80	5.70	12.70

2006–2008 in coastal and maidan Karnataka, respectively. Percentage damage on the trees partly colonized with *O. smaragdina* was 23.40+2.16 in coastal and 20.12+2.05 in maidan tract. The unsprayed cashew trees with no ant recorded maximum percentage damage 47.42+3.71 and 52.36+3.86 in Brahmavar and Chintamani, respectively during the survey. Higher nut yield of 3.70 and 2.43 kg/tree was recorded in the trees fully colonized with *O. smaragdina* in coastal and maidan Karnataka, respectively. Based on the results obtained, it was observed that *O. smaragdina* appeared as a predominant predatory ant on *H. antonii* in cashew-growing regions of Karnataka. Its role as a key predator is of potential significance in keeping the TMB population under check leading to increased yield. Increasing the activities of *O. smaragdina* could further reduce the *H. antonii* population. Similar observations were made by Chong (1987) and Chin et al. (1988). *O. smaragdina* significantly reduced damage of *H. antonii* on cashew. This is in concurrence with the findings of Peng et al. (1995) and Wijetunga et al. (2003).

Observations for 4 years (2007–2010) on cashew plantations at Puttur and Chintamani revealed that colonies of *O. smaragdina* did not establish in the presence of the yellow crazy ants (*Anoplolepis gracilipes* F. Smith). *O. smaragdina* ants' activity levels got gradually decreased and wiped out from the plantation within a year. Efforts to establish colonies of *O. smaragdina* failed at Puttur and Chintamani mainly because of the occurrence of crazy ants. Baiting has proved the most effective method to manage crazy ants by the fish-based protein—Fipronil. It has been proved that concentrations below 0.5 g/ha a. i. has no effect on reptiles, birds or mammals, and as Fipronil does not dissolve in water therefore does not adversely affect water supplies (Fishes 2007). Solutions to crazy ant problem can be found in prevention which will only be achieved through increased monitoring of ants and products into cashew plantations and by increasing public awareness.

Effect of *Telonomus* of TMB

The egg parasitoid *Telenomus* species was the predominant parasitoid on *H. antonii.* Maximum mean egg parasitisation of 16.80, 15.35 and 12.70% were recorded from host eggs collected in Brahmavar, Pethri and Chintamani cashew plantations, respectively, during December (Table 2) 2006–2008. Similar observations on the effectiveness of *Telenomus* species on TMB eggs were reported by Sundararaju (1993) and CIBC (1983). Cashew trees fully colonized by *O. smaragdina* recorded the least percentage damage of 11.40+1.87 and 8.71+1.23 during 2006–2008 in coastal and maidan tracts of Karnataka respectively (Table 3). TMB damage on trees partly colonized with *O. smaragdina* ants were 23.40+2.16 in coastal and 20.12+2.05 in maidan

Table 3 Effect of *O. smaragdina* colonization on infestation of *H. antonii* on cashew

Zone	Sprayed/ non-sprayed plantation	Colonization of *Oecophylla smaragdina*	No. of *Oecophylla smaragdina* workers/tree	No. of trees observed	Percentage damage by TMB	Nut yield
Coastal Karnataka	Sprayed	None	(Few) <20	52	12.30+2.31	4.30
	Non-sprayed	None	(Moderate) <20	32	47.42+3.71	1.76
	Non-sprayed	Part	Moderate >20–50	21	23.40+2.16	2.18
	Non-sprayed	Full	Abundant >500–1000	35	11.40+1.87	3.70
Maidan Karnataka	Sprayed	None	(Few) <20	28	8.35+1.62	2.60
	Non-sprayed	None	(Moderate) <20	17	52.36+3.86	1.02
	Non-sprayed	Part	Moderate >20–50	18	20.12+2.05	1.52
	Non-sprayed	Full	Abundant >500–1000	16	8.71+1.23	2.43

None cashew trees with no ants, *Part* part of the cashew tree canopy colonized by ants, *Full* entire tree canopy colonized by ants

tract. The unsprayed cashew trees without ants recorded maximum percentage TMB damage of 47.42+3.71 and 52.36+3.86 in coastal and maidan tracts, respectively. Higher nut yield of 3.70 and 2.43 kg/tree was recorded on trees fully colonized with *O. smaragdina.*

Effect of Biopesticides and Insecticide on TMB

Data on effectiveness of treatments were recorded at 10, 20 and 30 days after treatment. Monocrotophos (0.05 %) treated trees at flushing registered significantly less TMB damage (14.82 %). Maximum flush damage of 26.73 % was recorded with microbial pesticide *B. bassiana* at 30 DAT. Among botanicals, *Pongamia* seed kernel extract (2 %) was significantly superior over neem products. These observations are in concurrence with the findings of Thirumalaraju et al. (1997; Table 3).

Endosulfan (0.05 %) recorded 16.72 % damage on panicle at 30 DAT, and it was significantly superior over the botanicals and entomopathogen *B. bassiana.* The new molecule, λ-cyhalothrin (0.005 %), was at par with endosulfan (0.05 %). These two insecticidal treatments were significantly superior over other treatments. The first spray of monocrotophos (0.05 %) on the new flush and λ-cyhalothrin (0.005 %) on the panicle were at par with the recommended sprays of monocrotophos and endosulfan at the respective stages of the crop. The botanicals and the microbial pesticide alone were not effective against TMB.

During nut and fruit developmental stages, carbaryl (0.10 %) in the sequential spray schedule of monocrotophos (0.05 %)—λ-cyhalothrin (0.005 %) recorded the least percentage TMB damage of 4.36 % at 30 DAT, and it was at par with the recommended spray schedule: monocrotophos (0.05 %)—endosulfan (0.05)—carbaryl (0.10 %). The undamaged nuts appeared spotless, healthy and larger at nut developmental stage, 30 days after treatment. Botanicals, viz. pongamia oil, neem oil, NSKE and the microbial *B. bassiana*, alone were not effective against TMB. However, PSKE (2 %) was found effective in combating the menace.

It is evident that the sequential sprays of monocrotophos (0.05 %)—λ-cyhalothrin (0.005 %)—carbaryl (0.10 %) at flushing, panicle and nut and fruit developmental stages were effective in combating TMB. The effectiveness of monocrotophos (0.05 %)—λ-cyhalothrin (0.005 %)—carbaryl (0.10 %) was in confirmation with the findings of Bhat and Raviprasad (2007). Highest mean nut yield of 1106.57 kg/ha was recorded in sequential spray schedules: monocrotophos (0.05 %)—λ-cyhalothrin (0.005 %)—carbaryl (0.10 %), followed by monocrotophos (0.05 %)—endosulfan (0.05 %)—carbaryl (0.10 %; 879.91 kg/ha; Table 4). Lower nut yield in botanicals and *B. bassiana* treated trees may be due to high percentage damage by TMB on flushing, panicle and nut and fruit developmen-

Table 4 Effect of botanicals and insecticides against tea mosquito bug in field (pooled)

Treatments	TMB damage (%)		
	New flush	Panicle	Nut and fruits
	30 DAT	30 DAT	30 DAT
Pongamia oil–pongamia oil–pongamia oil	9.64 (18.12)	15.13 (22.85)	2.56 (9.16)
PSKE–PSKE–PSKE	7.50 (15.82)	12.13 (20.35)	1.90 (7.98)
Neem oil–neem oil–neem oil	17.07 (24.34)	21.42 (27.54)	5.46 (13.50)
NSKE–NSKE–NSKE	15.40 (23.08)	19.20 (25.98)	3.50 (10.66)
Beauveria bassiana–Beauveria bassiana–Beauveria bassiana	20.22 (26.73)	24.73 (29.82)	5.98 (13.75)
Monocrotophos–λ-cyhalothrin–carbaryl	5.00 (12.87)	7.0 (15.30)	0.60 (4.36)
PSKE–λ cyhalothrin–carbaryl	6.61(14.82)	8.30 (16.72)	0.98 (5.63)
Beauveria bassiana–λ-cyhalothrin–carbaryl	7.53 (15.92)	7.56 (15.95)	0.76 (5.01)
NSKE–λ cyhalothrin–carbaryl	19.10 (25.91)	12.40 (20.55)	1.75 (7.18)
Monocrotophos–endosulfan–carbaryl	6.12 (14.29)	17.92 (25.01)	1.06 (5.87)
Control	22.67 (28.43)	27.92 (31.88)	7.87 (16.28)
CV	4.77	4.71	13.56
CD (0.05)	1.63	1.84	2.09

Values given in the parentheses are Arc sine percentage
DAT days after treatment

Table 5 Nut yield and economics for the treatments evaluated against tea mosquito bug in field

Treatments	Nut yield (kg/tree) treatments (₹)	Cost of control (₹/ha)	Net returns over	Benefit over control (₹)	C:B
Pongamia oil–pongamia oil–pongamia oil	398.72	1200	18,736	4486	01:03.7
PSKE–PSKE–PSKE	477.04	1200	22,652	8402	01:07.0
Neem oil–neem oil–neem oil	393.97	1850	17,849	3599	01:01.9
NSKE–NSKE–NSKE	439.07	1570	20,384	6134	01:03.9
Beauveria bassiana–Beauveria bassiana–Beauveria bassiana	309.72	1500	13,986	−264	01:00.0
Monocrotophos–λ-cyhalothrin–carbaryl	1106.57	2278	53,048	38,798	01:17.0
PSKE–λ cyhalothrin–carbaryl	879.91	2278	41,718	27,468	01:12.1
Beauveria bassiana–λ-cyhalothrin–carbaryl	684.71	1913	32,323	18073	01:09.4
NSKE–λ cyhalothrin–carbaryl	510.86	1913	23,630	9380	01:04.9
Monocrotophos–endosulfan–carbaryl	488.31	2013	22,403	8148	01:04.0
Control	285	–	14,250	–	–

Cost of raw cashew nut at the rate of ₹ 50/kg, the cost of treatments include cost of insecticides and labour

tal stages. These findings are in agreement with Sundararaju (2004).

The C to B ratio indicated that monocrotophos (0.05 %)—λ-cyhalothrin (0.005 %)—carbaryl (0.10 %) treated trees recorded higher net gain 1:17.03 (Table 5). Among the botanicals PSKE (2%) registered higher C to B ratio of 1:7.0. Lower C to B ratio in botanicals and entomopathogen may be due to their low effectiveness in suppressing the pest leading to high percentage damage, as also reported by Thirumalaraju et al. (2002).

Sustainable practices comprising *B. bassiana, O. smaragdina, Telenomus* sp., pongamia seed kernel extract and λ-cyhalothrin for the *H. antonii* management on cashew in coastal Karnataka, South India, inferred that the integration of all these possible pest management strategies, help in keeping this dollar-earning crop free from the deadly insect pest. With the presence of *O. smaragdina* ant colonies, negligible or no application of insecticides, the population of other natural enemies, viz. coccinellids, chrysopids, entomopathogens, mantids, spiders and others can also be sustained in the cashew ecosystem. This will help in realizing higher cashew nut yields and residue free kernels. Utilization of the predatory ant *O. smaragdina* against TMB on cashew needs group participation.

References

Abraham CC, Nair GM (1981) Effective management of the tea mosquito bugs for breaking the yield barriers in cashew. Cashew Causerie 3(1):6–7

Bhat PS, Raviprasad TN (2007) Management of tea mosquito bug *Helopeltis antonii* sign. With newer insecticide/products. National Seminar on Research Development and Marketing of cashew, 20–21st Nov 2007 p 56

Chin PK, Sipat AB, Khoo KC (1988) Studies on the predator-prey relationship between *Oecophylla smaragdina* and *Helopeltis theobromae* using the radiotracer techniques. In Proc Int Symp on Mod Insect Control Nucl Tech Biotechnol 427–435

CIBC (Common Wealth Institute of Biological Control) (1983) Possibility for the use of natural enemies in the control of *Helopeltis* spp. (Miridae). Biocontrol News Inf 4:7–11

Chong KK (1987) The cocoa mirid in Peninsular Malaysia and its management. Planter 63(5):16–520

Devasahayam M, Nair CPR (1986) The tea mosquito bug, *Helopeltis antonii* Signoret on cashew in India. J Plant Crops 14(1):1–10

Giesberger G (1983) Biological control of *Helopeltis* pest of cocoa in Java, with special reference to the role of black cocoa ant *Dolichoderus bituberculatus* Mayr. In the biological control system. Arch Cocoa Res 2:91–180

Manja Naik (2008) Bioecology and management of tea mosquito bug, *Helopeltis antonii* Signoret (Heteroptera: Miridae) on cashew in coastal Karnataka, PhD thesis UAS, Bangalore pp 224

Peng RK, Christian K, Gibb K (1995) The effect of the green ant, *Oecophylla smaragdina* (Hymenoptera: Formicidae), on insect pests of cashew trees. Australia Bull Ent Res 85:279–284

Sundararaju D (1993) Studies on the parasitoids of tea mosquito bug, *Helopeltis antonii* Sign. (Heteroptera: Miridae) on cashew with special reference to *Telenomus* sp. (Hymenoptera: Scelionodae). J Biol Control 7(1):6–8

Sundararaju D (2004) Evaluation of certain insecticides against tea mosquito bug on cashew. Cashew 18(2):22–26

Sundararaju D, Sundarababu PC (1999) *Helopeltis* spp. (Heteroptera: Miridae) and their management in plantation and horticultural crops of India. J Plantation Crops 27:155–174

Thirumalaraju GT (2002) Ecology and management of tea mosquito bug *Helopeltis antoniii* Signoret (Hemiptera: Miridae) on cashew. Ph. D. Thesis, University of Agricultural Sciences, Bangalore, Karnataka, India, pp 115

Thirumalaraju GT, Shankaranarayana V, Lingaiah HB, Vishnuvardhana, Chandregowda M (1997) Evaluation of different insecticides for control of tea mosquito bug (*Helopeltis antonii* Sign) in cashew under maidan parts of Karnataka. Cashew 11(4):37–40

Way MJ, Khoo KC (1991) Colony dispersion and nesting habits of the ants, *Dolichoderus thoracicus* and *Oecophylla smaragdina* (Hymenoptera: Formicidae) in relation to their success as biological control agents on cocoa. Bull Ent Res 81:341–350

Wijetunga PMAPK, Ahangama D, Ranaveera B (2003) Biology of the cashew pest *Helopeltis antonii* sign and its predators. Trop Agric Res 15:188–198

Integrated Pest Management (IPM) for Reducing Pesticide Residues in Crops and Natural Resources

G. V. Ranga Rao, B. Ratna Kumari, K. L. Sahrawat and S. P. Wani

Abstract

Investigation on the pesticide residues during 2006–2009 in various crops and natural resources (soil and water) in the study village (Kothapally, Telangana State (TS)) indicated the presence of a wide range of insecticidal residues. Pooled data of the 80 food crop and cotton samples, two rice grain samples (3 %) showed beta endosulfan residues, and two (3 %) soil samples showed alpha and beta endosulfan residues. In vegetables of the 75 tomato samples, 26 (35 %) were found contaminated with residues of which 4 % had residues above MRLs. Among the 80 brinjal samples, 46 (56 %) had residues, of these 4 % samples had residues above MRLs. Only 13 soil samples from vegetable fields were found contaminated. The frequency of contamination in brinjal fields was high and none of the pulses and cotton samples revealed any pesticide contamination. IPM fields showed substantial reduction sprays which in-turn reflected in lower residues. Initial studies on water analysis indicated the presence of residues in all water sources with higher in bore wells compared to open wells, however, by 2009 the water bodies reflected no residues above the detectable level.

Keywords

IPM · Natural Resources · Residues

Introduction

Ever increasing demand for food, feed, and fiber, due to increased population, requires increased productivity on a sustained basis. With the advent and adoption of improved technologies such as high-yielding crop varieties and the use of fertilizers and pesticides, considerable progress has been achieved in boosting agricultural production (Foley 2011). However, during this process of enhancing productivity, the use of agrochemicals became an integral part of the present day agriculture. Globally, approximately 2.5 million tons of pesticides are used annually in agriculture. Latest information on pesticide use across

G. V. Ranga Rao (✉) · B. R. Kumari · K. L. Sahrawat · S. P. Wani
International Crops Research Institute for the Semi-Arid Tropics, Patancheru, Telangana State, 502 324 India
e-mail: g.rangarao@cgiar.org

A. K. Chakravarthy (ed.), *New Horizons in Insect Science: Towards Sustainable Pest Management*,
DOI 10.1007/978-81-322-2089-3_35, © Springer India 2015

the world clearly indicated an increase from about US$7 billion to US$12 billion from 2000 to 2012 with a similar trend across the globe (Plumer 2013).

Worldwide, approximately 9000 species of insects and mites, 50,000 species of plant pathogens, and 8000 species of weeds damage crops. Insect pests cause an estimated loss of 14 %, plant pathogens cause 13 % loss, and weeds cause another 13 % loss (Pimentel 2009). Pesticides use is indispensable in agricultural production. About one-third of the agricultural products are produced by using pesticides. Without pesticide application, the loss of fruits, vegetables, and cereals from pest injury would reach 78, 54, and 32 %, respectively 2008). In view of the world's limited croplands and growing population; it is necessary to take all measures to increase crop production in order to ensure food safety (Zhang et al. 2011). On the other hand, Knutson and other researchers pointed out that if the consumption of pesticides is prohibited, the food production in the USA would drop sharply and the food prices would soar.

Drivers of food security and crop protection issues are discussed relative to food losses caused by pests. Insect pests globally consume food estimated to feed an additional one billion people. Key drivers include rapid human population increase, climate variability, loss of beneficial on-farm biodiversity, reduction in per capita cropped land, and water shortages. The use of integrated pest management (IPM) in agriculture is urgently needed, and is also being widely adopted globally. IPM offers a 'toolbox' of complementary crop- and region-specific crop protection solutions to address these rising pressures. IPM aims for more sustainable solutions by using complementary technologies. The applied research challenge now is to reduce selection pressure on single solution strategies, by creating additive/synergistic interactions between IPM components. IPM is compatible with organic, conventional, and genetically modified (GM) cropping systems and is flexible, allowing regional fine-tuning. It reduces the pest levels below economic thresholds utilizing key 'ecological services', particularly bio-control. Landscape scale 'ecological engineering', together with genetic improvement of new crop varieties, will enhance the durability of the pest-resistant cultivars (conventional and GM). The IPM will also promote compatibility with the use of semio-chemicals, bio-pesticides, precision pest monitoring tools, and rapid diagnostics. These combined strategies are urgently needed; and are best achieved via multi-disciplinary research, including complex spatio-temporal modeling at the farm and landscape scales. Integrative and synergistic use of existing and new IPM technologies will help meet the future food needs more sustainably in the developed and developing countries. The aim of this chapter is to provide further evidence to show that IPM indeed can reduce pesticide use without sacrificing the yields of the major crops studied.

Status on Pesticide Related Issues

There have been many studies on determining the ill effects of pesticide exposure (McCauley et al. 2006). The World Health Organization and the UN Environment Programme estimate that each year, 3 million farm workers in the developing world experience severe pesticide poisoning of whom about 18,000 were fatal (Miller 2004). A study with 23 school children who were shifted to organic food from normal diet, a dramatic reduction in the levels of organo-phosphorus pesticides in their system was observed (Lu et al. 2006).

Excessive and non-judicious use of insecticides has led to the degradation of environmental quality, pest resistance, pest resurgence and the contamination of agricultural products and natural resources. Most of the studies on pesticides conducted in Asia reflect the presence of pesticide residues in significant amounts in food and agricultural commodities, and pesticide pollution does exist in the country; and is a cause of concern for public health (Kumari et al. 2002, 2003, 2004, 2005, 2006). Pesticides applied to the soil or that eventually end in the soil in agricultural areas can contribute to the contamination of surface and ground waters (Gilliom et al. 2006; McMahon et al. 2006).

Information from India showed that about 51 % of the food material is contaminated with residues in comparison to 21 % worldwide, of which 20 % were above MRL prescribed by FAO standards (Anon 1999). The contaminated food is generally not discarded in the developing countries, but enters the food chain out of ignorance, innocence and equally importantly out of lack of affordability by the consumers. Lack of awareness of the consequence of pesticide- contaminated food could be one of the reasons for increased incidences of cancers in developing world. Besides the damage to human health, an indiscriminate use of chemical pesticides adversely affects the natural bio-diversity that results in the reduction of natural enemies (Ranga Rao et al. 2005).

Exposure of humans to the hazardous chemicals directly in the fields and indirectly through contaminated diet resulted in the occurrence of residues of organo-chlorines in human blood (3.3–6.3 mg per L) and milk (3.2–4.6 mg per L) samples from lactating women. High levels of pesticide residues (15–605 times) were observed in blood samples of cotton farmers from four villages in Punjab (Anon, 2005). In the past few decades with the benefits of synthetic pesticides being clearly recognized, the usage has steadily increased from 2.2 g ha^{-1} active ingredient (a.i.) in 1950 (David 1995) to 381 g ha^{-1} by 2007 i.e., about 270- fold increase (Anon 2009).

Various inappropriate practices in the use of pesticides cause possible poisoning symptoms generally among farmers who do not wear protective clothing (Ntow et al. 2006). Perceptions by farmers of pesticide efficacy were found to play a major role in farmers' behavior towards the use of pesticides and the adoption of alternative methods of pest control such as IPM (Hashemi and Damalas 2010). For example, pesticide use on any crop depends on the farmer's attitude whether to enhance the productivity to meet the market demands in search of enhanced income or subsistence farming for livelihood (Erbaugh et al. 2000).

For maintaining the quality of a commodity, it is essential to keep the produce free of pesticide residues. A zero level residue in the finished product is not only desired but also needed for eco-preservation and human health as well. The necessity of pesticide residue analysis in various agro-based commodities has become more relevant in the present context. Implementation of IPM strategies will help to reduce the dependence on toxic pesticides associated with agriculture to enhance productivity of healthy products and profitability.

The chemical residues from the soil find their way to the aquatic systems or get accumulated in the plant products (grain, root, stem etc.). Farmer field schools organized in India on cotton situation brought out the importance of IPM in reducing pesticide-induced risks at the farm level without sacrificing the yields (Mancini 2006). The constraints in the adoption of protective clothing in tropics were discussed by Kishi (2005).

Integrated Pest Management (IPM)

Globally, there is an increasing pressure on the agriculture sector to produce more food to meet increased demand of the growing populations all around the world. This has increased the need for intensive plant protection with increased use of pesticides, leading to complex environmental implications. Several national and international agencies and nongovernmental organizations are presently engaged in supporting research and the use of eco-friendly approaches for crop protection practices for the sustainable environment.

The basic concept of IPM is the containment of pests below economically damaging levels, using a combination of control measures. Two fundamental principles are: (1) that as individual pest control methods are often not successful alone and (2) that pests only need to be managed when present at populations high enough to cause economic damage. The IPM relies on the integration of various plant protection options with a selective use of insecticides in a regulated program. This refers to an active program of monitoring pest and natural enemy population levels. Four primary components of IPM include: host plant resistance, manipulation of the farming system, enhanced bio-control, and selective use of biorational and/or synthetic pesticides.

IPM is the most environment-friendly approach of crop-protection and prescribes the use of chemical pesticides as the last resort. However, most of the farming communities in India are not much educated. Therefore, they are averse to adopt the program. Implementation of the IPM strategies reduces toxic pesticides in agriculture to enhance productivity of healthy products and profitability. The inclusion of eco-friendly IPM packages in the plant protection measures is the need of the hour to save the crop losses from the biotic stresses and to sustain and improve the agricultural production, soil health, and overall environmental quality. Insecticide residues in non-IPM vegetable fields were higher than those recorded for the IPM fields (Arora and Singh 2004; Sardana et al. 2005). The insecticide residues in the IPM-managed vegetable (tomato and cucumber) fields ranged from 0.004 to 0.027 mg kg^{-1}, while the residues ranged from 0.005 to 0.106 mg kg^{-1} in the non-IPM fields (Ranga Rao et al. 2009a).

On-Farm Experience

Under integrated watershed management program and bio-intensive pest management (BIPM) technologies were initiated in farmer participatory approach during 2000 in Kothapally village of TS to alleviate the plant protection problems in crops like cotton, pigeonpea, and chickpea. During 2000–2001, pigeonpea BIPM farmers applied one spray each of neem fruit extract and HNPV, followed by manual shaking (3–5 times) and did not apply any chemicals. Non-IPM farmers sprayed 3–4 times with chemicals. During the 2001–2002 season, BIPM farmers used one spray each of neem and HNPV followed by manual shaking (2–4 times), while the non-IPM farmers used 2–3 rounds of chemical sprays. In chickpea, during the post rainy season 2000–2001, the BIPM plots received 1–3 sprays of HNPV, while the non-IPM farmers did not take any plant protection measures for their crops. During 2001–2002, BIPM farmers applied one spray of neem fruit extract and two sprays of HNPV, while non-IPM farmers used two sprays of chemicals.

The larval population in BIPM pigeonpea plots was always found lower than those of non-IPM plots, where farmers applied 3–4 sprays of chemicals. BIPM interventions resulted in the substantial decrease in borer damage to pods and seeds with 34 % and 21 % pod and seed damage compared to 61 and 39 % pod and seed damage in non-IPM plots. This lower pod borer damage in the BIPM plots also reflected in higher yield of 0.77 t ha^{-1} compared to 0.53 t ha^{-1} in farmer practice treatment. The observations on egg and larval population during 2001–2002 indicated similar trend as in the previous season. The BIPM interventions resulted in 33 and 55 % reduction in pod and seed damage, respectively. The BIPM plots yielded 0.55 t ha^{-1} compared to 0.23 t ha^{-1} yield in non-IPM plots, even although the overall yield levels were low (Ranga Rao et al. 2007).

In chickpea, egg and larval population during 2000–2001 indicated the onset of the pests during the first fortnight of November when the crop was around 30 d old (with one egg $plant^{-1}$), and the number continued to increase until the first fortnight of December when the crop attained podding stage and later declined by the end of January. The difference in plant protection practices between BIPM and non-IPM plots was clearly reflected in the lower larval population in BIPM fields throughout the vulnerable phase of the crop. The BIPM farmers also harvested three times higher yields of 0.78 t ha^{-1} compared to 0.25 t ha^{-1} in non-IPM fields, which was primarily due to an effective pest management and the adoption of improved variety (ICCV 37) developed at the International Crops Research Institute for the Semi-Arid-Tropics (ICRISAT).

Chemical Usages on Different Crops

Detailed crop surveys on the use of chemicals on different crops during 2005–2006 in India brought out the following proportion of pesticide inputs in various crops: cotton: (51 %), rice (10 %), pigeonpea (6 %), maize (2 %), chickpea

Table 1 Quantity of common used pesticides, used by farming community and the recommended doses

Chemical (No. of farmers)	Chemical group	Quantity of chemical used (ml ha^{-1})		
		Mean	Range	Recommended
Endosulfan (185)	Organochlorine	1580	375–5000	1000
Monocrotophos (251)	Organophosphate	1590	250–3750	750
Indoxacarb (169)	Chloro-nicotil	418	63–1250	250
Spinasod (133)	Microbial	213	50–500	125
Cypermethrin (82)	Pyrethroid	1753	250–2500	500
Imidacloprid (51)	Neonicotinoid	305	63–750	125

(1 %), groundnut (2 %), and chilly (28 %) of the total pesticides usage in the selected project locations (World bank DM ICRISAT, final Report Anon 2007). In Asian agriculture, about 80 % of the plant protection chemicals utilized were in cotton and vegetables, although the area was only about 5 % of the total. Similar trend was also noticed in India with 75 % of the chemical use in these crops covering only 5 % of the cultivated area (Vasantharaj David 1995). Of these, chilly was found to be highly intensive crop with 15–20 sprays in a 6-month period, contributing to heavy residues on the products, hindering its export. Results from Table 1 clearly show the use of excess dosage of plant protection chemicals by farmers. This could be due to their ignorance, low confidence on the efficacy of chemicals, lack of effectiveness due to the occurrence of insecticidal resistance in key species, and inappropriate application. Since intensive plant protection in a limited area was responsible for major residues and environmental issues those areas should be given the priority to reverse the ill effects caused by the use of chemicals.

The studies related to pesticide use the following implementation of IPM in 17 selected villages, indicated substantial reduction in pesticide application from 11 sprays to 4 sprays in cotton, 2.1 to 1.6 in rice, 2.9 to 2.2 in pigeonpea, and 2.9 to 2.3 in chickpea during 2005 and 2007 (Table 2). This impact was due primarily to the periodic farmer researcher interactions, training imparted to the farmers and their keenness on judicious use of chemical pesticides. Mancini (2006) also described similar results with about 75 % reduction in pesticide use in contact villages compared to 28 % in the noncontact villages without compromising crop yields through farmer field schools.

The crop samples analyzed for pesticide residues in 15 contact (41 samples) and 5 noncontact (15 samples) villages revealed presence of pesticide residues in all samples of which 38 samples had residues below 0.001 ppm (Anon 2007). However, one sample each of *Dolichos* and tomato only had residues of monocrotophos and chlorpyriphos above the maximum residue limits (MRLs) prescribed by the FAO. According to Peter Melchett (2008), the level of pesticide residues in juice drinks in the UK was on an average 34 times more than those permitted in drinking water and sometimes up to 300 fold. Studies conducted by Yaong Bai et al. (2006) in vegetables in the Shaanxi area of China revealed the occurrence of residues of five organophosphorus pesticides ranging from 0.004 to 0.257 ppm; and in 18 of 200 samples, the residue levels exceeded MRLs. The occurrence of pesticide residues in the in samples in the study clearly indicated the status of residues and the need for developing strategies for their management.

Bio-Rationals

The term covers a range of alternatives to synthetic chemical pesticides of biological origin. Their main feature is specificity to avoid nontarget mortality and associated problems. The use of bio-pesticides is an important component of IPM strategy for all major crops. The best-known examples are the neem-based products, which have shown to be effective against a number of pests, NPV being used for the control of important

Table 2 Comparison of pesticide use on selected crops in villages before and after the implementation of IPM

Village (No. of farmers)	No. of insecticidal sprays											
	Cotton			Paddy			Pigeonpea			Chickpea		
	2005	2007	Reduction (%)	2005	2007	Reduction (%)	2005	2007	Reduction (%)	2005	2007	Reduction (%)
Daulatabad (11)	–	–	–	2.0	1.7	15.0	3.3	3.3	0.0	–	–	–
Mudireddypalli (19)	–	–	–	2.3	2.1	8.7	3.1	3.3	−6.5	–	–	–
Peddaravelli (11)	7.9	2.2	72.2	1.5	0.8	46.7	2.0	1.3	35.0	–	–	–
Pullagiri (14)	6.9	3.6	47.8	2.6	2.3	11.5	2.7	1.8	33.3	–	–	–
Indrakal (17)	7.5	4.1	45.3	2.3	2.1	8.7	2.7	2.0	25.9	–	–	–
Musapet (9)	–	–	–	1.8	0.8	55.6	–	–	–	–	–	–
Addakal (11)	–	–	–	2.5	2.1	16.0	–	–	–	–	–	–
Chandapur (16)	16.5	6.8	58.8	2.7	1.7	37.0	3.0	2.3	23.3	2.9	2.4	17.2
Kamalpally (15)	9.5	3.1	67.4	–	–	–	2.8	2.5	10.7	2.7	2.6	3.7
Gundlamachnur (17)	13.7	3.6	73.7	2.2	1.7	22.7	2.9	1.7	41.4	3.0	2.6	13.3
Lingapur (18)	10.3	4.0	61.2	2.1	1.6	23.8	2.5	1.6	36.0	2.7	1.8	33.3
Kyasaram (21)	14.7	4.2	71.4	2.4	2.1	12.5	3.1	2.3	25.8	2.7	2.4	11.1
Alirajpet (15)	10.9	3.3	69.7	2.1	1.7	19.0	3.0	2.4	20.0	2.9	2.2	24.1
Kukunurpally (16)	16.4	3.2	80.5	1.7	1.4	17.6	3.0	1.9	36.7	–	–	–
Vattimeena-pally(16)	8.1	3.4	58.0	–	–	–	2.9	2.1	27.6	3.6	2.6	27.8
Medipallyka-lam (20)	15.5	3.9	74.8	1.8	1.5	16.7	3.5	2.9	17.1	2.8	2.0	28.6
Kummera (15)	9.9	3.4	65.7	1.8	0.5	72.2	2.6	1.6	38.5	2.6	1.9	26.9
Mean	11.4	3.8	65.1	2.1	1.6	25.6	2.9	2.2	24.3	2.9	2.3	20.7

Absence of crop in the village; Obtained from Ranga Rao et al. 2009; Obtained from ICRISAT World Bank DM project final report 2007

pests like *Helicoverpa armigera* and *Spodoptera* spp. In addition, *Bacillus thuringiensis* (Bt) has gained importance in suppressing pest populations in crops like cotton and vegetables.

There are several bio-pesticides commercially available for use by farmers. There were approximately 175 registered bio-pesticide active ingredients in India and 700 products globally (Ranga Rao and Goplakrishnan 2009). Awareness of the need for safer agents has grown with an increasing concern for the toxicity of synthetic pesticides. Hence, biorational pesticides have immense potential. A number of neem-based formulations are being produced by small-scale formulators and marketed as insecticides. Most of them are made from neem oil and contain varying amounts of Azadirachtin. There have, however, been problems with the maintenance of consistent quality. To overcome this, farmers are encouraged to procure neem seeds and prepare their own spray containing 5 % neem-fruit-powder extract using the prescribed procedure.

Hence, several integrated pest management (IPM) programs have adopted neem as one of the prime options for creating greater stability and sustainability in crop production. In the present IPM module, the use of neem during the vegetative phase, followed by the application of Helicoverpa Nucleo Polyhedrosis virus (HNPV), a popular insect pathogen at flowering and need-based application of chitin inhibitors (novaluron, flufenoxuron) instead of conventional insecticide (endosulfan) during pod formation phase in pest management would be of immense help in augmenting the natural enemies in the chickpea ecosystem (Ranga Rao et al. 2008).

Effect of IPM Options on Soil Inhabiting Natural Enemies

Studies to assess the effects of select treatments on soil inhabiting natural enemies during 1998–2000 postrainy seasons revealed that their population started building up during the vegetative phase (302 trap^{-1}) and attained the peak during the flowering phase (455 trap^{-1}) and subsequently there was a gradual decline during pod formation and preharvested phases of the crop. Observations on the effects of various treatments on soil inhabiting natural enemies at vegetative phase revealed that plots treated with endosulfan had significantly lower populations (107.7 trap^{-1}) with 64 % reduction compared to the control (302.3 trap^{-1}). The plots treated with HNPV showed minimum disturbance to natural enemies with a catch of 267.1 trap^{-1}, on par with the control (Ranga Rao et al. 2008).

These studies clearly indicated the population dynamics of soil inhabiting natural enemies and their potential in suppressing the pod borer. Considering the preference by insect pests and their associated natural enemies live and feed on chickpea than other legume crops (Ranga Rao and Shanower 1999), it is necessary to integrate safer and effective pest management options in the chickpea IPM programs in order to obtain maximum advantage from the natural enemies. Hence, one should be cautious in the selection and sequencing of control measures to maintain the ecological balance and healthy environment. The results from these investigations have provided further insight to the earlier studies on the effective use of IPM options in the management of key pests and their natural enemies with less deleterious effects on natural enemies.

Effect of IPM Options on Aerial Natural Enemies in the Chickpea Canopy

Using a De Vac® at 22, 54, 76, and 99 DAS during the 1998–1999 season assessed the impact of various IPM options on aerial natural enemies. The results from these studies at 22 DAS revealed lower number of natural enemies in plots treated with endosulfan (39.5) compared to plots treated with HNPV (69.7), IPM (51.0) and control (87.1). Observations at 54 DAS 2 days after the third spray suggested a similar trend with a significant reduction (58 %) in the number of natural enemies in the plots treated with endosulfan. However, there was no significant reduction in the number of natural enemies in the plots treated with either neem (20.8) or HNPV (21.5) compared to the control (23.8). Perusal of the data at 76 DAS revealed that the plots treated with endosulfan recorded the less number of aerial natural enemies (18.0) while neem, HNPV, and IPM treatments had populations of 25.3, 28.8, and 27.3, respectively, compared to control (32.2). At 99 DAS, the natural enemy populations in plots treated with endosulfan were found significantly low (9.5) and the other treatments were on par with each other. The overall effect of endosulfan, neem, and HNPV indicated 52, 29, and 14 % reduction in population of aerial natural enemies, respectively, over control. (Ranga Rao et al. 2008)

Effect of IPM options on larval parasitoids of H. armigera. During the study period, the larval parasitization *of H. armigera* was mainly by *Campoletis chlorideae*. Apart from *C. chlorideae,* the other larval-pupal parasitoid, *Carcelia illota* Curron, a tachinid was recorded only in control plots, however, its incidence was only 2 %. Two years study during 1998–2000 at ICRISAT fields, the overall effect of endosulfan, neem, HNPV, and IPM treatments indicated 35, 20, 16, and 21 % reduction, respectively.

In subsequent studies during 2003–2004, post-rainy season in chickpea revealed the overall effect in two samples of larval collections (at 26 and 56 DAS) lower parasitization in plots treated with endosulfan (2.3 %) with 60 % reduction over control. The larval parasitization from plots treated with neem fruit extract (4.7 %) and neem oil (5.2 %) indicated 17 and 11 % reduction in population, respectively, over control. The bio-pesticide HNPV-treated plot recorded higher number of parasitized *H. armigera* larvae (5.7 %) with 2.8 % reduction in population, which was on par with control.

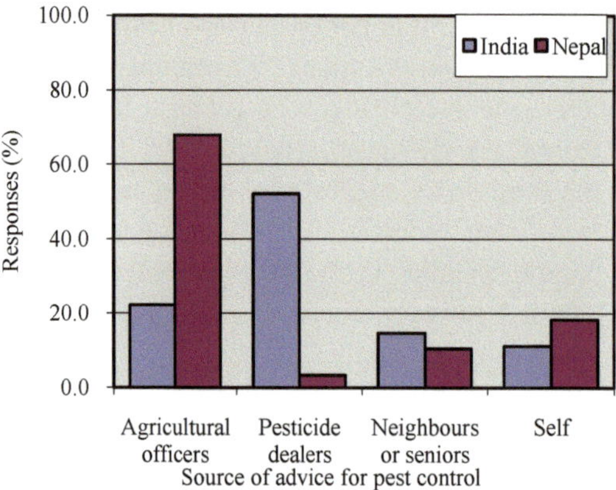

Fig. 1 Sources of advice to farmers in pest control in India and Nepal. (Obtained from Ranga Rao et al. 2009b)

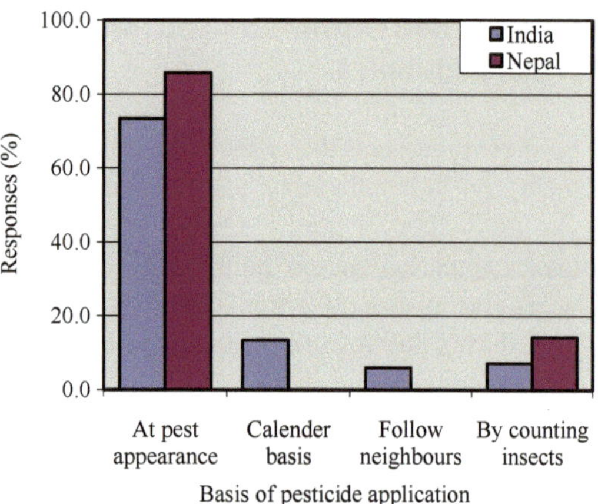

Fig. 2 Basis of pesticide application by farmers in India and Nepal. (Obtained from Ranga Rao et al. 2009b)

Farmer Perception of Plant Protection

Participatory rural appraisal (PRA) was undertaken in 70 villages in India and Nepal, covering 1185 farmers to generate baseline information on the current plant protection practices. The study revealed that 93 % of the farmers in India and 90 % in Nepal had adopted chemical control for the management of various insect pests in different crops. However, less than 20 % of the farmers expressed confidence on the efficacy of the current plant protection measures. In India, 52 % farmers get their plant protection advice from pesticide dealers. While in Nepal, majority of the farmers (69 %) make their plant protection decisions through agricultural officers (Fig. 1). A majority of the farmers (73 % in India and 86 % in Nepal) initiate the plant protection based on the first appearance of the pest, irrespective of their population, crop stage, and their damage relationships (Fig. 2). About 50 % of the farmers in India and 20 % in Nepal were not using any protective clothing while spraying. Health problems associated with the application of plant protection chemicals were reported by farmers. The cost of plant protection on various crops ranged from 7 to 40 % of the total crop production cost. Although IPM has been advocated for the past two decades, only 32 % in India and 20 % of farmers in Nepal were aware of the IPM practices. IPM implementation in selected villages brought 20–65 % reduction in pesticide use on different crops (Ranga Rao et al. 2009b).

Knowledge on Integrated Pest Management (IPM)

Though IPM has been advocated for over two decades, only 32 % farmers in India and 20 % in Nepal were aware of IPM practices. Among the various bio-pesticides, majority of the farmers (76 % in India and 93 % in Nepal) have adopted neem in their pest management programs. Though the farmers in India and Nepal were aware of biopesticides and natural enemies, their integration into the IPM was only 32 % in India and 20 % in Nepal. This low adoption of IPM in various crops was primarily due to the non-availability of IPM inputs at the farm level, the complexity of the IPM modules for different crops, lack of information on the ill effects of toxic chemicals and the existing insufficient extension networks.

Insecticide Residue Monitoring: A Case Study

Pesticide residue monitoring was taken up at Kothapally and Enkepally villages of Ranga Reddy district, TS in food crops (rice, maize, pigeon-

Table 3 Pesticide residues in vegetable samples collected from farmers' fields, Kothapally village, Ranga Reddy district during 2007

Crop (No. of samples)	Range of pesticide residue level (mg kg^{-1})			
	Monocrotophos	Chlorpyrifos	Endosulfan	Cypermethrin
Brinjal (10)	0.003 (<0.001–0.007)	0.008 (<0.001–0.040)	0.019 (<0.001–0.089)	0.052 (<0.001–0.283)
Cucumber (10)	0.004 (0.001–0.011)	0.066 (0.001–0.330)	0.019 (0.002–0.030)	0.010 (0.001–0.034)
Okra (10)	0.013 (<0.001–0.044)	0.605 (0.001–5.154)	0.130 (0.001–0.784)	0.025 (<0.001–0.112)
Ridgegourd (6)	0.015 (<0.001–0.041)	0.050 (0.001–0.223)	0.021 (0.002–0.061)	0.086 (0.001–0.352)
Tomato (23)	0.005 (<0.001–0.025)	0.035 (<0.001–0.151)	0.032 (<0.001–0.466)	0.024 (<0.001–0.141)

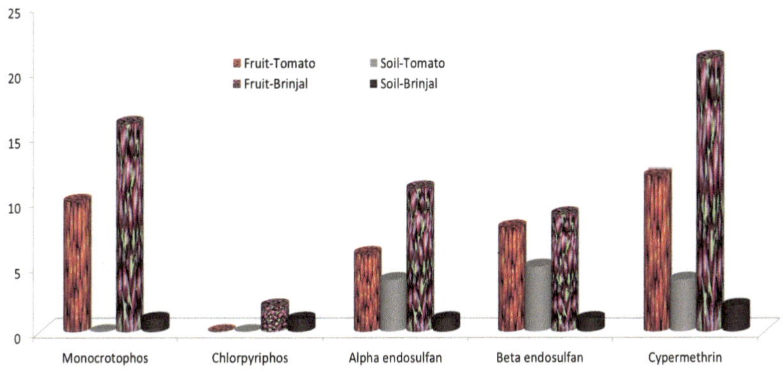

Fig. 3 Frequency distribution of insecticide residues in vegetable crops and soil samples from their respective fields

pea), vegetables (tomato and brinjal), and cotton besides soil and water during 2006 and 2009 seasons. The pesticide residue analysis during 2006 and 2007 revealed the presence of residues of chlorpyriphos and cypermethrin above MRL in 10% of the samples of brinjal and tomato. In fact, most of the water samples from bore as well as open wells showed considerable level of residues though they are below the MRLs (Table 3).

Pesticide Residues in Food Grains and Cotton

Analysis of food grains, cotton, and soil samples showed that out of all grain samples analyzed, one sample of rice grain was contaminated with beta endosulfan (0.5 μg g^{-1}). Alpha (0.02 μg g^{-1}) and beta endosulfan (0.02 μg g^{-1}) residues were detected in one soil sample collected from maize field during 2008 season. Only two samples contained beta endosulfan residue— one rice grain sample (0.008 μg g^{-1}) and one soil sample collected from rice field (0.03 μg g^{-1}) during the 2009 season. However, none of the pigeonpea grain and cotton lint samples were contaminated with insecticide residues. The presence of endosulfan residues in rice grain and soil from rice field could be attributed to the fact that farmers used endosulfan for pest control in various fields (Fig. 3). Detection of endosulfan residues in maize cultivated fields and cobs was in consonance with the study conducted by Singh et al. (1992). Senapati et al. (1992) reported the absence of residues in pigeonpea grain at harvest. Samant et al. (1997) and Nayak et al. (2004) also reported nondetectable levels of chlorpyriphos and endosulfan in the black gram and green gram seeds. The nondetection of residues in soils from pigeonpea fields are in agreement with the results of Tanwar and Handa (1998). A shift in cotton cultivation from traditional varieties to Bt varieties, which requires less number of sprays according to our survey, might be one of the reasons for nondetectable residues in cotton lint. Suganya Kanna et al. (2007) also did not observe any resi-

Table 4 Insecticide residues in tomato and brinjal and in respective soil samples from the fields observed in Kothapally and Enkepally villages during 2008–2009 and 2009–2010 cropping seasons

No. of samples analyzed/ contaminated	Insecticides detected	Frequencies	Residue range (μg g^{-1})	MRL (μg g^{-1})*
In Tomato				
Fruit 75 (26)	Monocrotophos	10	0.006–0.2	0.2
	Alpha endosulfan	5	0.01–0.2	2.0
	Beta endosulfan	8	0.008–0.07	2.0
	Cypermethrin	11	0.06–0.5	0.5
Soil 40 (13)	Monocrotophos	–	–	–
	Alpha endosulfan	4	0.05–0.8	–
	Beta endosulfan	3	0.02–0.2	–
	Cypermethrin	3	0.01–0.3	–
Brinjal				
Fruit 80 (46)	Monocrotophos	17	0.01–0.2	0.2
	Chlorpyriphos	2	0.009–0.01	0.2
	Alpha endosulfan	15	0.009–1.0	2.0
	Beta endosulfan	10	0.006–3.0	2.0
	Cypermethrin	21	0.01–0.2	0.2
Soil 40 (5)	Monocrotophos	1	0.06	–
	Chlorpyriphos	1	0.03	–
	Alpha endosulfan	1	0.1	–
	Beta endosulfan	1	0.01	–
	Cypermethrin	1	0.02	–

*Maximum residue limit

dues of imidacloprid and acetamiprid in cotton lint.

Insecticide Residues in Vegetables and Soil

Studies organized on the pesticide residues in vegetable (brinjal, cucumber, okra, ridge gourd, and tomato) and water samples collected from Kothapally Adarsha watershed in Rangareddy district, TS, India during 2007 revealed the presence of monocrotophos (range 0.001–0.044 mg kg^{-1}), chlorpyrifos (0.001–5.154 mg kg^{-1}), cypermethrin (0.001–0.352 mg kg^{-1}) and endosulfan (0.001–0.784 mg kg^{-1}). The residues of monocrotophos and endosulfan were below MRL in all the 59 vegetable samples, while the residues of chlorpyrifos were above MRL in four samples and cypermethrin in two samples.

The data on insecticide residues in tomato fruits and soil are presented in Table 4. Out of the 15 tomato fruit samples analyzed during the 2008 summer season from two villages, eight (53 %) samples were found to be contaminated with all the insecticide groups under study, except for chlropyriphos; and the residue concentration ranged from 0.01 to 0.3 μg g^{-1}. However, one sample showed monocrotophos residue above the MRL. During the *Kharif* 2008 season, 40 % of the samples (6 out of 15) were contaminated (0.006 to 0.3 μg g^{-1}). One (0.07 μg g^{-1}) out of the 15 samples contained insecticide residues during the *Rabi* 2008 season. During the 2009 summer season, low concentrations of residues in 7 out of 15 samples (47 %) were detected showing monocrotophos as the major insecticide. Four samples out of 15 contained residues during the 2009 *Kharif* season, however they were below MRLs. (Table 4). Out of the 10 soil samples 3 (33 %) contained cypermethrin residues (ranging from0.1 to 0.3 μg g^{-1}) in the 2008 summer season. Alpha and beta endosulfan residues (0.02 to 0.07 μg g^{-1}) in 5 out (55 %) of 10 samples were detected during 2008 *Kharif* season. During the 2008 *Rabi*, only 1 out of 10 soil samples

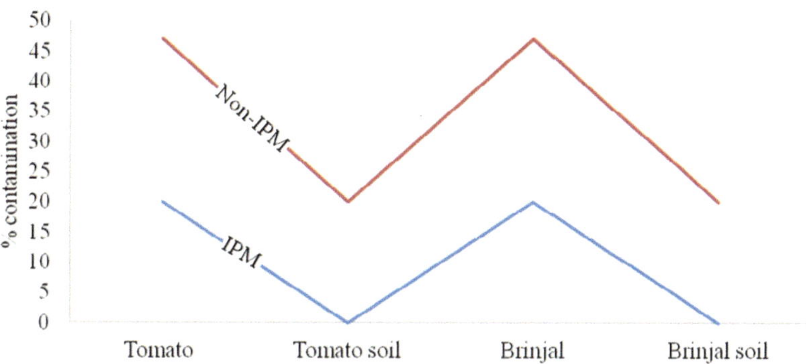

Fig. 4 Impact of IPM in reducing insecticides residues in tomato and brinjal crops and soils

contained beta endosulfan residue (ranging from 0.03–0.2 μg g^{-1}). Three out of ten soil samples contained alpha endosulfan and cypermethrin residues (ranging from 0.04 to 0.8 μg g^{-1}) during the 2009 *Kharif* season.

In brinjal during 2008 summer season, the frequency of contamination with cypermethrin (0.009 to 3.0 μg g^{-1}) was higher, and it was in 9 out of 16 brinjal fruit samples. Beta endosulfan was present in greater concentration (3.0 μg g^{-1}) and was above the MRL (Table 6). A contamination level of 69 % (11 out of 16) with monocrotophos and cypermethrin as the main contaminants (residue concentration ranging from0.006 to 0.2 μg g^{-1}). In 7 out of 16 samples, residues of monocrotophos, alpha endosulfan and cypermethrin (44 % contamination) were detected during 2008 *Rabi* season, (0.009 to 0.1 μg g^{-1}). Sixty nine per cent (11 out of 16) of the samples were found contaminated during 2009 summer season, and the residue concentration ranged from 0.006 to 0.2 μg g^{-1}. In 8 out of 16 samples (0.01–2.0 μg g^{-1}) insecticide residues were detected during the 2009 *Kharif* season. The results of soil analysis are shown in Table 3. Monocrotophos (0.06 μg g^{-1}) and chlorpyriphos (0.03 μg g^{-1}) residues were detected in the samples collected in 2008 summer season. During the 2008 *Kharif* season, insecticide residues were not detected in the samples. One out of the eight (13 %) samples collected contained the residues of different insecticides (ranging from 0.01 to 0.1 μg g^{-1}) during the 2008 *Rabi*, 2009 summer, and 2009 *Kharif* seasons. The presence of monocrotophos in selected vegetable samples in concentrations above the MRL probably was due to unauthorized sale by pesticide dealers and their use by farmers, although this insecticide was banned for use on vegetables as per the Insecticide Act, 1968 as on 28th December, 2006 (Sharma 2007). The contamination of soil samples with insecticide residues from the field planted with brinjal was lower as compared to the samples from the field planted with tomato. This could be attributed to greater canopy cover under brinjal and longer duration of the crop as suggested by Jayashree and Vasudevan (2007) in paddy canopy and the movement of residues to the soil and in the runoff water.

Considering overall all samples, of the 80 food crop and cotton samples, only two rice grain samples (3 %) showed beta endosulfan residues and two (3 %) out of 80 soil samples showed alpha and beta endosulfan residues. In vegetables, of the 75 tomato samples, 26 (35 %) were found contaminated with residues and 4 % had residues above MRLs. In soil samples (Fig. 5), 13 samples (26 %) out of the 50 samples from tomato fields had residues. Among the 80 brinjal samples, 46 (56 %) had residues; and out of these 4 % samples had residues above MRLs. Only 13 % of the soil samples from brinjal fields were contaminated (Fig. 3 and 4).

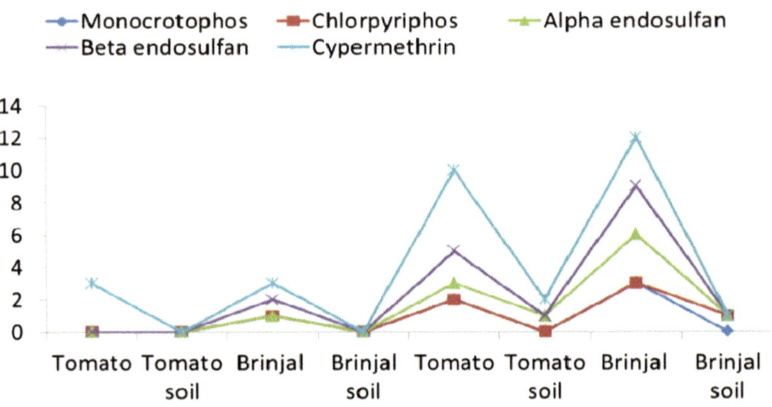

Fig. 5 Frequency distribution of insecticide residues in crops and soil samples taken from the IPM and Non-IPM fields planted with vegetables

Table 5 Pesticide residues in two vegetable samples collected from IPM and farmers practice plots, Kothapally village, Ranga reddy district, TS, 2007

Crop	Treatment (No. of samples)	Residue levels (mg kg^{-1})			
		Monocrotophos	Chlorpyrifos	Endosulfan	Cypermethrin
Tomato	IPM (18)	0.005	0.034	0.012	0.023
Tomato	Non-IPM (5)	0.005	0.041	0.101	0.028
Cucumber	IPM (5)	0.004	0.027	0.011	0.009
Cucumber	Non-IPM (5)	0.005	0.106	0.026	0.012

Insecticide Residues in Water

The pesticide residue analysis during 2006 and 2007 revealed the presence of residues of chlorpyriphos and cypermethrin in most of the water samples from bore as well as open wells showed considerable level of residues though they are below the MRLs. During 2006–2007, residues of all the four pesticides were found higher in bore well water compared to open well samples (Table 6). Residues of endosulfan were higher by 300 %, cypermehrin by 89 %, monocrotophos by 50 %, and chlorpyrifos by 9 % in bore wells compared to samples collected from the open wells. The total residue concentrations of all the four pesticides were high in water samples from bore wells (0.036 mg kg^{-1}) than water samples from the open wells (0.023 mg kg^{-1}). Low levels of residues in open wells could be due to greater exposure to the environment thereby more scope for degradation. These studies brought about the status of selected conventional pesticides used for farming activities. Though the levels of toxicity in several samples were below MRL's considering their occurrence in all samples one should critically look into the eco system to make sure the crops and the agro ecosystem were free from the toxicants.

Water analysis during 2009 from food crop fields and vegetable fields did not reveal any insecticide residues. According to the WHO (2004), most of the organochlorine pesticides are practically insoluble in water. Our results are in agreement with the findings of Jagdishwar Reddy et al. (1997) who reported no insecticide residues in river, tank and canal water. However, most of the documented review on pesticide residues in water in India indicated the presence of highly persistent organochlorines like DDT, HCH, lindane, and heptachlor and endosulfan in different water sources. The suspended residues were probably quickly decomposed by sunlight through photo degradation reaction and hence pyrethroids did not persist longer on the surface or sub-surface

Table 6 Pesticide residue levels in water samples collected from open and bore wells of Kothapally village, Ranga Reddy district during different phases of IPM (2006-09)

Source of water sample	Residue levels (mg kg^{-1})[a]				Total
	Monocrotophos	Chlorpyrifos	Endosulfan	Cypermethrin	
Initial phase of IPM upto 2006					
Bore well	0.003 (<0.001–0.004)	0.012 (<0.001–0.018)	0.004 (<0.001–0.005)	0.017 (<0.001–0.029)	0.036
Open well	0.002 (<0.001–0.002)	0.011 (0.004–0.017)	<0.001 (<0.001)	0.009 (<0.001–0.009)	0.023
During 2009					
Bore well	ND	ND	ND	ND	ND
Open well	ND	ND	ND	ND	ND

ND Not detected

[a] Mean of four open and two bore wells (Values in the parenthesis denote the range)

water samples (Awasthi 1997; Nwankwoala and Osibonjo 1992) studied the organochlorine pesticide residues in surface waters in Ibadan (Nigeria). This may be due to the indiscriminate use of chemicals and perhaps could be contamination from local as well as upstream areas.

Impact of Integrated Pest Management in Minimizing Insecticide Residues

To understand the impact of IPM modules in the reduction of insecticide residues, samples of crop, soil, and water were monitored from selected IPM farmers and the results compared with the samples collected from the nonIPM farmers from two villages, *viz.*, Kothapally and Enkepally of Ranga Reddy district, Andhra Pradesh. As vegetables are the major source of chemical use, tomato and brinjal were covered in this study. Five tomato and five brinjal farmers were selected from Kothapally village and IPM schedule was given to them. studies organized on the pesticide residues in vegetable (brinjal, cucumber, okra, ridgegourd, and tomato) and water samples collected from Kothapally Adarsha watershed in Rangareddy district, Andhra Pradesh, India during 2007 revealed the presence of monocrotophos (range 0.001–0.044 mg kg^{-1}), chlorpyrifos (0.001 to 5.154 mg kg^{-1}), cypermethrin (0.001 to 0.352 mg kg^{-1}) and endosulfan (0.001 to 0.784 mg kg^{-1}). The residues of monocrotophos and endosulfan were below MRL in all the 59 vegetable samples while the residues of chlorpyrifos were above MRL in four samples and cypermethrin in two samples. The water samples also revealed the presence of pesticide residues but were below MRLs (Table 6). Among the food crops and cotton analyzed for the insecticide residues (monocrotophos, chlorpyriphos, alpha endosulfan, beta endosulfan, and cypermethrin), one rice grain sample (0.5 μg g^{-1}) out of five samples collected from Kothapally was contaminated and among the soil samples, residues were detected in one soil sample (0.02 μg g^{-1}) collected from maize field during 2008 in Enkepally. Only two samples were contaminated—one rice grain sample (0.008 μg g^{-1}) and one soil sample (0.03 μg g^{-1}) collected from rice field during 2009 from Enkepally. Out of the total 45 tomato fruit samples analyzed from Kothapally for insecticide residues over a period of five seasons in 2008 and 2009, 11 samples (24%) were found to contain residues. In Enkepally, the residues were observed in 50% of samples (15 out of 30 samples) during this period. However, none of the samples from Kothapally and 7% of contaminated samples from Enkepally had residues above MRLs. Overall, out of the 30 soil samples collected from tomato fields during 2008 and 2009, only six samples (20%) contained insecticide residues compared to 35% in Enkepally. Among the 40 brinjal samples analyzed during 2008 and 2009 seasons, 17 (43%) samples from Kothapally and 29 (73%) samples from the Enkepally contained insecticide residues. The overall residue levels in brinjal during the study period indicated 7% of samples in Enkepally above MRLs.

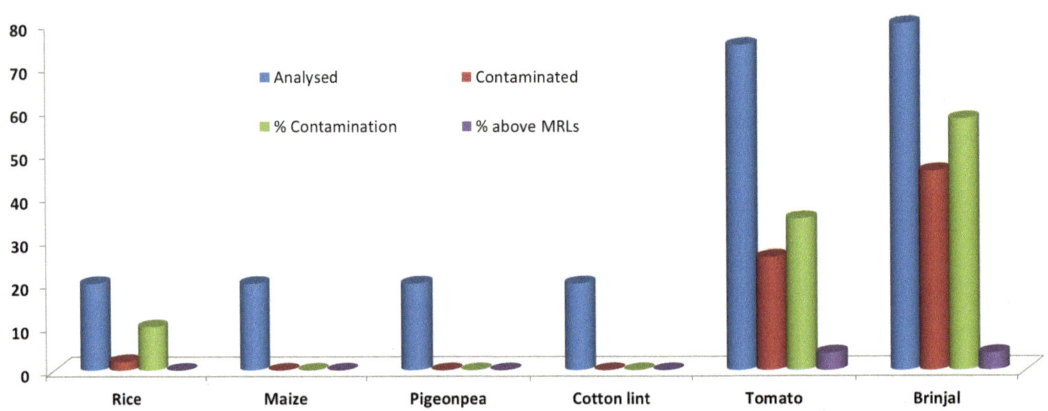

Fig. 6 Percent contaminated samples of various food, fiber, and vegetable crops from Kothapally village during 2008–2010

Soil analysis in five various seasons showed that only 10 and 15 % of the samples collected from brinjal fields were contaminated in Kothapally and Enkepally, respectively; and none of the water samples collected from food crops, cotton, and vegetable crops were contaminated (Fig. 6 and Table 4).

As a result of close interactions with researchers and farmers covering various activities on natural resources and crop improvement, the farmers are familiar and adopting the good agricultural practices. The awareness in farmers on various aspects, particularly efficient use of water, the importance of improved cultivars and plant protection practices has increased substantially and most of the senior farmers are presently at the forefront in spreading the technologies to others.

With the introduction of transgenic cotton in this village during 2005, the adoption presently is 100 %, which has facilitated farmers in reducing the pesticide use; for example, from 20 (while using traditional varieties) to at present 3–4 sprays. Though pesticides are still in use in this village (mostly on vegetables), the farmers are quite aware of the bio-pesticides such as neem, vermiwash, and HNPV; and they strictly follow the need-based application of plant protection options. The data obtained in 2008–2009 on pesticide residues clearly indicated a down word trend in the occurrence of beta endosulfan, monocrotophos and cypermethrin in only 4 % of brinjal and tomato samples. After thorough implementation of IPM, the water samples from various fields in Kothapalli village were found free from residues. This clearly emphasizes the impact of intensive implementation of the IPM in this village during the past one decade (Figs. 4 and 5). This is one example in which there is a remarkable turnaround from a bad situation which was rectified, through a greater level of education followed by adoption of eco- friendly approaches.

Thus, by adopting the IPM strategies in their village (Kothapally), senior farmers including Mr Narayana Reddy, Mr Narsimha Reddy, and Ms Laxmi are very comfortable in sharing their knowledge in the use of BIPM approaches in addressing the environmental and health issues. In this process, now the whole farming community of the village adopted the protective clothing and took the oath that they see no one sprays any plant protection chemicals without a protective gear. At present, this village is a role model for sustainable improvement of natural resources with improved productivity and environment.

The world has long produced enough calories, around 2700 per day per human, more than enough to meet the United Nations projection of a population of nine billion by 2050, up from the current seven billion. There are hungry people not because food is lacking, but because not all of those calories go to feed humans (a third go to feed animals, nearly 5 % are used to produce biofuels, and as much as a third is wasted, all along the food chain, Mark Bittman 2013).

The Way Forward

An adequate support for plant protection research is essential to meet the challenges of producing healthy food from the available land with minimal adverse effect on the environment. Technologies such as developing resistant varieties, enhancing natural enemies, improving the cultural control, judicious use of chemical pesticides and IPM will have a significant role to play in the future. Indeed most operational IPM systems have a relatively simple, yet effective beginning. In this way, even where resources may be quite limited, an effective IPM system can often be developed and adopted to suit the local situations. Biological control of pests through the use of natural enemies is an important component of the IPM strategy due to its environmental soundness and wide acceptability. Interest in biological control of pests in agricultural crops is increasing. Apart from the harm from the chemicals to human health and environment, pesticides can easily disrupt the natural control of pests and diseases by killing their natural enemies. Without these beneficial organisms, farmers become more dependent on the use of pesticides. Without the progress in the recent plant protection research, the hunger and poverty alleviation would have been worst but need to be taken further. This cannot be achieved through the individual research agenda of any one organization; and hence appropriate research partnerships including the international organizations, national institutes, non-governmental agencies and farmers should work together to make the dream of safe food and safe environment true.

References

Anon (1999) Pesticide Action Network Asia and Pacific "Warning: pesticides are dangerous to your health"

Anon (2005) Very high levels of pesticides found in blood samples in Punjab, India. The Hindu Daily June 8

Anon (2007) World Bank—ICRISAT—DM Project: traditional technology with a modern twist project completion report June 2005—May 2007

Anon (2009) Indian pesticides industry, Vital for ensuring food security. http://www.bharatbook.com

Arora S, Singh DK (2004) Residues of insecticides from IPM and Non-IPM fields of Okra and Brinjal. Pesticide Res J 16(2):68–70

Awasthi MD (1997) Degradation and persistence of synthetic pyrethroids in tropical soil and aquatic environment. Indian J Envi Tox 7(1):36–38

Bai Y, Zhou L, Wang J (2006) Organophosphorus pesticide residues in market in Shaanxi area, China. Food Chemistry 98(2):240–242

Bittman M (2013) http://www.nytimes.com/2013/10/15/opinion/how-to-feed-the-world.html?ref=markbittman

Erbaugh JM, Semana AR, Adipala E, Isubikalu P (2000) The influence of farmer perception on pesticide usage for management of cowpea field pests in eastern Uganda African. Crop Sci J 8 (3):317–326

Foley JA, Ramankutty N, Brauman KA, Cassidy ES, Gerber JS, Johnston M, Mueller ND, O'Connell C, Ray DK, West PC, Balzer C, Bennett EM, Carpenter SR, Hil J, Monfreda C, Polasky S, Rockström J, Sheehan J, Siebert S, Tilman D, Zaks DPM (2011) Solutions for a cultivated planet. Nature Analysis 478:337–342

Gilliom RJ, Barbash CG, Crawford PA, Hamilton JD, Martin N, Nakagaki LH, Nowell JC, Scott PE, Stackelberg GP, Thelin A, Wolock DM (2006) The quality of our nation's waters-Pesticides in the Nation's streams and ground water, 1992-2001. Circ. 1291. USGS, Washington, DC. Available at http://ca.water.usgs.gov/pnsp/pubs/circl1291/

Hashemi MS, Damalas CA (2010) Farmers' perceptions of pesticide efficacy: reflections on the importance of pest management practices adoption. J. Sust Agricul 35(1):69–85

Jagdishwar Reddy D, Narasimha Rao B, Narasimha Reddy K, Sultan MA, Ramesh Babu T (1997) Monitoring of pesticide residues in river, tank and canal water. Pesticide Res J 9(1):97–100

Jayashree R, Vasudevan N (2007) Organochlorine pesticide residues in ground water of Thiruvallur district, India. Environmental Monitoring and Assessment 128:209–215

Kishi M (2005) The health impacts of pesticides: what do we know, what can be done? In: Pretty J (ed), The pesticide detox. Earthscan, London, pp. 23–38

Kumari B, Madan VK, Kumar R, Kathpal TS (2002) Monitoring of seasonal vegetables for pesticide residues. Environmental Monitoring and Assessment 74:263–270

Kumari B, Gulati R, Kathpal TS (2003) Monitoring of pesticidal contamination in honey. The Korean J Apic 18(2):155–160

Kumari B, Madan VK, Singh J, Singh S, Kathpal TS (2004) Monitoring of pesticidal contamination of farmgate vegetables from Hisar. Environmental Monitoring and Assessment 90:65–77

Kumari B, Singh J, Singh S, Kathpal TS (2005) Monitoring of butter and ghee (clarified butter fat) for pesticidal contamination from cotton belt of Haryana, India. Environmental Monitoring and Assessment 105:111–120

Kumari B, Madan VK, Kathpal TS (2006) Monitoring of pesticide residues in fruits. Environmental Monitoring and Assessment 123:407–412

Lu C, Toepel K, Irish R, Fenske RA, Barr DB, Bravo R (2006) Organic diets significantly lower children's dietary exposure to organophosphorus pesticides. Environ Health Perspect 114(2):260–263

Mancini F (2006) Impact of integrated pest management farmer field schools on pesticide use and farmers' ecological knowledge on cotton farming in India. Dissertation

McCauley LA, Anger WK, Keifer M, Langley R, Robson MG, Rohlman D (2006) Studying health outcomes in farm worker populations exposed to pesticides. Envi Health Perspectives 114(3):953–960

McMahon PB, Dennehy KF, Bruce BW, Bohlke JK, Michel RL Grudak JJ, Hurlbut DB (2006) Storage and transit time of chemicals in thick unsaturated zones under rangeland and irrigated cropland, High Plains, United States. Water Resources Research 42:W03413. doi:10.1029/2005WR004417

Melchett P (2008) More pesticides in UK juice drinks. http://www.timesonline.co.uk/tol/life_and_style/food_and_drink/article5375565.ece

Miller GT (2004) Sustaining the Earth, 6th edition. Thompson Learning, Inc. Pacific Grove, California. 9:211-216. http://www.worstpolluted.org/projects_reports/display/82

Nayak SK, Chhibber RC, Ujagir R (2004) Residue of endosulfan in harvested grains of Blackgram (*Vigna mungo* L.) protected during different growth stages. Insect Envi 10(2):68

Ntow WJ, Gijzen HJ, Kelderman P, Drechsel P (2006) Farmer perceptions and pesticide use practices in vegetable production in Ghana. Pest Manag Sci 62(4):356–365

Nwankwoala AU, Osibanjo O (1992) Baseline levels of selected organochlorine pesticide residues in surface waters in Ibadan (Nigeria) by electron capture gas chromatography. Sci Total Environ 119:179–190

Pimentel D (2009) Integrated pest management: innovations development process. Chapter 3 pp 83–87. springer.com/chapter/10.1007%2F978-1-4020-8992-3_3

Plumer B (2013) http://www.washingtonpost.com/blogs/wonkblog/wp/2013/08/18/the-world-uses-billions-of-pounds-of-pesticides-each-year-is-that-a-problem

Ranga Rao GV, Shanower TG (1999) Cultural practices in Insect Management with reference to groundnut. Indian J. of Plant Protection 27(1–2):87–108

Ranga Rao GV, Gopalakrishnan S (2009) Bio-pesticides research at ICRISAT: a consortium model. Pages 17 In proceedings of Expert Consultation on Biopesticides and Biofertilizers for Sustainable Agriculture, Organized by Asia-Pacific Association of Agricultural Research Institutions (APAARI) & Council of Agriculture, Taipei (COA), Taiwan Agricultural Research Institute, Taichung, 27–29 October 2009

Ranga Rao GV, Rameswar Rao V, Reddy YVR, and Murthy KSRK (2005) Status of safety in pesticide use in Asia: Options for improvement. pp 129–136 in proceedings of the National Seminar on Pesticide Residues and their Risk Assessment organized by Food and Drug Toxicology Rsearch Centre, National Institute of Nutrition, Hyderabad, January 20–21 2005.

Ranga Rao GV, Rupela OP, Wani SP, Rahman SJ, Jyothsna JS, Rameshwar Rao V, Humayun V (2007) Bio-intensive pest management reduces pesticide use in India. Pesticide News 76:16–17

Ranga Rao GV, Visalakshmi V, Suganthy M, Vasudeva Reddy P, Reddy YVR, Rameshwar Rao V (2008) Relative toxicity of neem on natural enemies associated with chickpea ecosystem: a case study. Int J Trop Insect Sci 27(3/4):229–235

Ranga Rao GV, Sahrawat KL, Srinivasa Rao CH, Binitha D, Kirankumar Reddy K, Bharath BS, Rameshwar Rao V, Murthy KVS, Wani SP (2009a) Insecticide residues in vegetable crops grown in Kothapally watershed, Andhra Pradesh, India: A case study. Indian J Dry land Agricul Res Development 24(2):21–27

Ranga Rao GV, Rameshwar Rao V, Prasanth VP, Khannal NP, Yadav NK, Gowda CLL (2009b) Farmers' perception on plant protection in India and Nepal: a case study. Int J Tropic Insect Sci 29(3):158–168

Samant P, Senapati HK, Pal AK, Pattnaik MR, Swain N (1997) Persistence of chlorpyriphos in greengram and blackgram. Pesticide Res J 9(2):261–262

Sardana HR, Bambawale OM, Singh DK, Kado LN (2005) Monitoring of insecticide residues in IPM and non-IPM fields of okra and brinjal. Indian J Plant Protec 33(2):197–201

Senapati HK, Sahoo BK, Pattnaik MR, Pal AK (1992) Persistence of some common pesticides in pigeonpea. Orissa J Agricul Res 5(1–2):100–103

Sharma KK (2007) Pesticides registered/banned/restricted in India. Pesticide Residue Analysis Manual, pp 171–176

Singh YP, Gangwar SK, Thakur NSA (1992) Persistence of endosulfan in/on maize at medium high altitude hills of Meghalaya. Indian J Plant Protec 20(2):195–197

Suganya Kanna S, Karuppuchamy P, Kuttalam S, Chandrasekaran S (2007) Harvest time residues of Acetamiprid 20 SP in Cotton. Pestology 31(8):28–30

Tanwar RS, Handa SK (1998) Persistence, translocation and metabolism of endosulfan residues in Pigeonpea (*Cajanus cajan* L. Mill sp.). Pesticide Res J 10(1):73–79

Vasantharaj David B (1995) The pesticides industry. Kothari's Desk Book series. pp 464

World Health Organization (2004) Heptachlor and heptachlor epoxide in drinking water. Background document for development of WHO Guidelines for Drinking-Water Quality. WHO/SDE/WSH/03.04/99. WHO, Geneva.

Zhang WJ, Jiang FB, Ou JF (2011) Global pesticide consumption and pollution: with China as a focus. Proceedings of the International Academy of Ecology and Environmental Sciences 1(2):125–144

Biological Consequences of Climate Change on Arthropod Biodiversity and Pest Management

Hari C. Sharma

Abstract

Global warming and climate change will trigger major changes in geographical distribution and population dynamics of insect pests, insect–host-plant interactions, activity and abundance of natural enemies, and efficacy of crop protection technologies. Changes in geographical distribution and incidence will affect both crop production and food security. Insect pests presently confined to tropical and subtropical regions will move to temperate regions along with a shift in the areas of production of their host plants, while distribution and relative abundance of some insect species vulnerable to high temperatures in the temperate regions may decrease as a result of global warming. The relative efficacy of pest control measures such as host-plant resistance, natural enemies, biopesticides, and synthetic chemicals is likely to change as a result of global warming and climate change. There is an urgent need to assess the efficacy of various pest management technologies under diverse environmental conditions and develop appropriate strategies for pest management to mitigate the adverse effects of climate change.

Keywords

Climate change · Host-pant interaction · Population dynamics

Introduction

Insect pests cause an estimated annual loss of 13.6% globally, and the extent of losses in India has been estimated to be 17.5% (Dhaliwal et al. 2010). The pest-associated losses are likely to increase as a result of changes in crop diversity and climate change. Climate change and its variability will have major effect on water availability, forest cover, biodiversity, crop production, and food security (Fig. 1). Changes in rainfall pattern are of greater importance for agriculture than the annual changes in temperature, especially in regions where lack of rainfall may be a limiting factor for crop production. Geographical

H. C. Sharma (✉)
International Crops Research Institute for the Semi-Arid Tropics (ICRISAT), Patancheru, Telangana 502324, India
e-mail: H.Sharma@cgiar.org

A. K. Chakravarthy (ed.), *New Horizons in Insect Science: Towards Sustainable Pest Management*,
DOI 10.1007/978-81-322-2089-3_36,

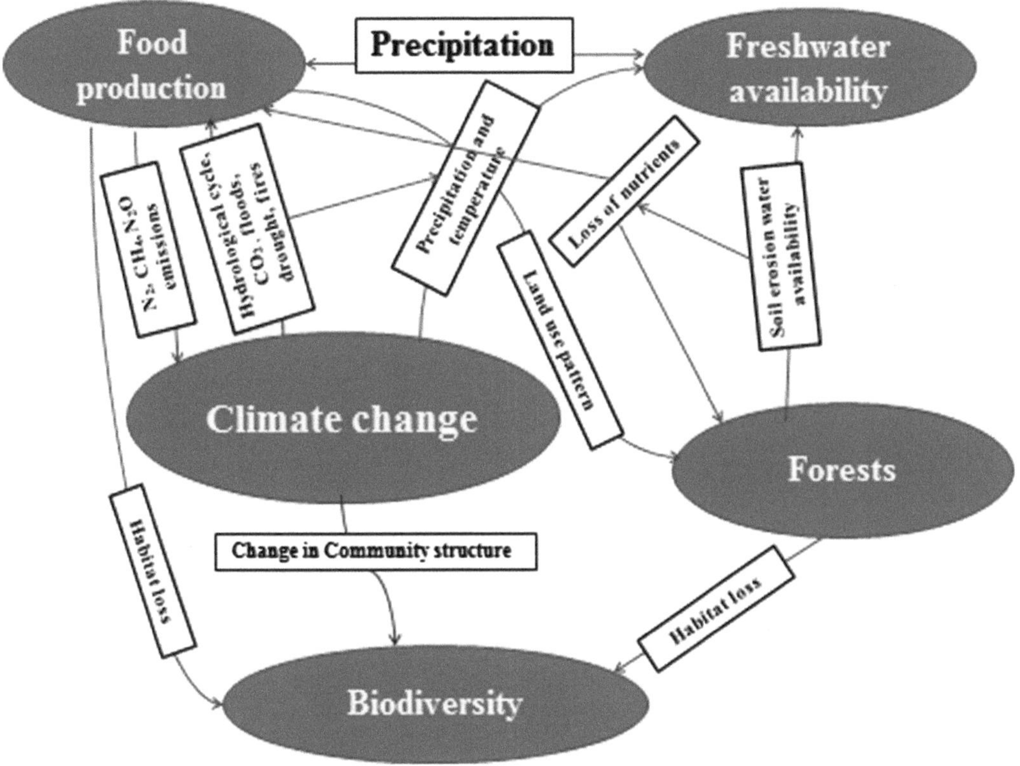

Fig. 1 Climate change effects on water availability, forest cover, biodiversity, and food security

distribution of tropical and subtropical insect pests will extend along with shifts in the areas of production of their host plants, while distribution and relative abundance of some insect species vulnerable to high temperatures in the temperate regions may decrease. High mobility and rapid population growth will increase the extent of losses due to insect pests. Current estimates of changes in climate indicate an increase in global mean annual temperatures of 1 °C by 2025 and 3 °C by the end of the next century, and the date at which an equivalent doubling of CO_2 will be attained is estimated to be between 2025 and 2070, depending on the level of emission of greenhouse gasses (IPCC 1990; Crowley 2000; Fig. 2). Mean annual temperature changes between 3 and 6 °C are estimated to occur across Europe, with greatest increases occurring at high latitudes.

Effect of Global Warming on Arthropod Diversity and Extinction of Species

Arthropods (insects, spiders, and mites) are the most diverse group of organisms, which can serve as useful indicators of the effect of global warming and climate change on different agro-ecosystems (Table 1). Arthropods are the most diverse component of terrestrial ecosystems and occupy a wide variety of functional niches and microhabitats (Kremen et al. 1993). Responses of arthropods to pollution depend on both temperature and precipitation in such a way that ecosystem-wide adverse effects are likely to increase under predicted climate change (Zvereva and Kozlov 2010). Consequences of temperature increases of 1–2 °C will be comparable in magnitude to the currently seen climate change in the Antarctic

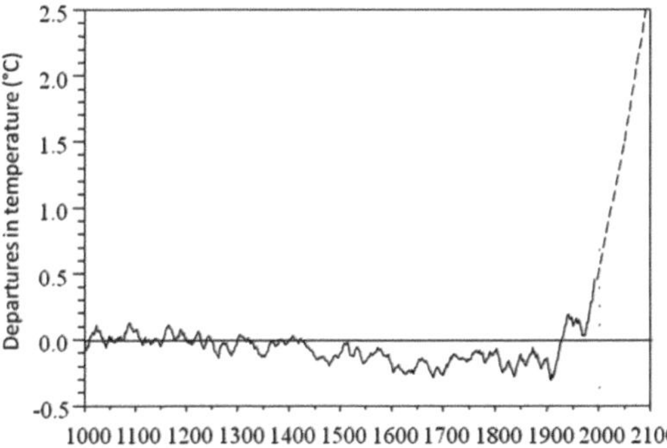

Fig. 2 Likely increase in temperature over the next 100 years (Crowley 2000)

(Bokhorst et al. 2008). Speciation takes between 100 and 1,000,000 years, providing between 10 and 10,000 new species per year, and 99.9% of all species that ever lived have become extinct. We are now living through the sixth extinction spasm, which is largely driven by human activities. The current extinction rates are 100–1000 times greater than what has happened earlier. Nearly 45 and 275 species are going extinct everyday as a result of human activities (Bokhorst et al. 2008).

Effect on Geographic Distribution and Population Dynamics of Insect Pests

Low temperatures are often more important than high temperatures in determining geographical distribution of insect pests. Increasing temperatures may result in a greater ability to overwinter in insect species limited by low temperatures at higher latitudes, extending their geographical range (EPA 1989; Hill and Dymock 1989). Spatial shifts in distribution of crops under changing climatic conditions will also influence the distribution of insect pests in a geographical region (Parry and Carter 1989). However, whether or not an insect pest would move with a crop into a new habitat will depend on other environmental conditions such as the presence of overwintering sites, soil type, and moisture, e.g., populations of the corn earworm (*Heliothis zea* (Boddie)) in North America might move to higher latitudes/altitudes, leading to greater damage in maize and other crops (EPA 1989). For all the insect species, higher temperatures, below the species' upper threshold limit, will result in faster development, resulting in rapid increase of pest populations as the time to reproductive maturity is reduced. In addition to the direct effects of temperature changes on development rates, increases in food quality due to plant stress may result in dramatic increases in growth of insect pest populations, while the growth of certain insect pests may be adversely affected (Maffei et al. 2007). Pest outbreaks are more likely to occur in stressed plants as a result of the weakening of the plants' defensive system, and thus, increasing the level of susceptibility to insect pests. Global warming will lead to earlier infestation by *Helicoverpa armigera* (Hub.) in North India (Sharma 2010), resulting in increased crop loss. An increase of 1 and 2 °C in temperature will cause northward shifts in the potential distribution of the European corn borer, *Ostrinia nubilalis* (Hub.) of up to 1220 km, with an additional generation in nearly all regions where it is currently known to occur (Porter et al. 1991). Overwintering of insect pests will increase as a result of climate change, producing larger spring populations as a base for a buildup in their numbers in the following season. Many insects such as *Helicoverpa* spp. are migratory, and therefore, may be well adapted to exploit new opportunities by moving rapidly into new areas as a result of climate change (Sharma 2005).

Table 1 Species diversity among different groups of organisms

Organisms	Number of species
Viruses, algae, protozoa, etc.	80,000
Bacteria	4000
Fungi	72,000
Plants	270,000
Animals: invertebrates (insects)	1,360,000
Animals: vertebrates	48,500
Total	*1,834,500*

Effects on Pollinators/Scavengers

- Altered profiles of pollinators/scavengers
- Extinction and/or emergence of new pollinators/scavengers

- Changes in composition of pollinators
- Asynchrony in pollinator activity and plant phenology
- Landscape changes due to change in pollinators and scavengers

Effects of Climate Change on Pest Management

Effects on Expression of Resistance to Insect Pests

Host-plant resistance to insects is one of the most environmental friendly components of pest management. However, climate change may alter the interactions between insect pests and their host plants (Sharma et al. 2012b). Resistance to sorghum midge, observed in India, breaks down under high humidity and moderate temperatures in Kenya (Sharma et al. 1999). There will be increased impact on insect pests which benefit from reduced host defenses as a result of the stress caused by the lack of adaptation to suboptimal climatic conditions. Problems with new insect pests will occur if climatic changes favor the introduction of non-resistant crops or cultivars. The introduction of new crops and cultivars to take advantage of the new environmental conditions is one of the adaptive methods suggested as a possible response to climate change (Parry and Carter 1989).

Insect–host-plant interactions will change in response to the effects of CO_2 on nutritional quality and secondary metabolites of the host plants. Increased levels of CO_2 will enhance plant growth, but may also increase the damage caused by some phytophagous insects (Coviella and Trumble 1999). The effects of increased atmospheric CO_2 on herbivory will not only be species-specific, but also specific to each insect–plant system. Increased CO_2 may also cause a slight decrease in nitrogen-based defenses (e.g., alkaloids) and a slight increase in carbon-based defenses (e.g., tannins). Lower foliar nitrogen due to CO_2 causes an increase in food consumption by herbivores up to 40 %, while unusually severe drought increases the damage by insect species such as spotted stem borer, *Chilo partellus* (Swin.) in sorghum (Sharma et al. 2005).

Effect on Effectiveness of Transgenic Crops

Environmental factors such as soil moisture, soil fertility, and temperature have strong influence on the expression of Bt toxins in transgenic plants (Sachs et al. 1998). Cotton bollworm, *Heliothis virescens* (F.) destroyed Bt cottons due to high temperatures in Texas, USA (Kaiser 1996). Similarly, *H. armigera* destroyed the cotton crop in the second half of the growing season in Australia because of reduced production of Bt toxins in the transgenic crops. Possible causes of the failure of insect control may be inadequate production of the toxin protein, effect of environment on transgene expression, locally resistant insect populations, and development of resistance due to inadequate management (Sharma and Ortiz 2000). It is therefore important to understand the effects of climate change on the efficacy of transgenic plants for pest management.

Activity and Abundance of Natural Enemies

Relationships between insect pests and their natural enemies will change as a result of global warming, resulting in both increases and decreases in the status of individual pest species. Changes in temperature will also alter the timing of diurnal activity patterns of different groups of insects, and changes in interspecific interactions could also alter the effectiveness of natural enemies for pest management (Hill and Dymock 1989). Quantifying the effect of climate change on the activity and effectiveness of natural enemies will be a major concern in future pest management programs. The majority of insects are benign to agro-ecosystems, and there is much evidence to suggest that this is due to population control through interspecific interactions among insect pests and their natural enemies (pathogens, parasites, and predators). Oriental armyworm,

Mythimna separata (Walk.) populations increase during extended periods of drought (which is detrimental to natural enemies), followed by heavy rainfall (Sharma et al. 2002). Aphid abundance increases with an increase in CO_2 and temperature; however, parasitism rates remain unchanged in elevated CO_2. Temperature not only affects the rate of insect development but also has a profound effect on fecundity and sex ratio of parasitoids (Dhillon and Sharma 2009). The interactions between insect pests and their natural enemies need to be studied carefully to devise appropriate methods for using natural enemies in pest management.

Biopesticides and Synthetic Insecticides

There will be an increased variability in insect damage as a result of climate change. Higher temperatures will make dry seasons drier, and conversely, may increase the amount and intensity of rainfall, making wet seasons wetter than at present. Current sensitivities on environmental pollution, human health hazards, and pest resurgence are a consequence of improper use of synthetic insecticides (Sharma 2012a). Natural plant products, entomopathogenic viruses, fungi, bacteria, nematodes, and synthetic pesticides are highly sensitive to the environment. Increase in temperatures and UV radiation and a decrease in relative humidity may render many of these control tactics less effective, and such an effect will be more pronounced on natural plant products and biopesticides. Therefore, there is a need to develop appropriate strategies for pest management that will be effective under situations of global warming in future. Farmers will need a set of pest control strategies that can produce sustainable yields under climatic change.

The relationship between the inputs costs and the resulting benefits will change as a result of changes in insect–plant interactions. This will have major bearing on economic thresholds, as greater variability in climate will result in variable impact of pest damage on crop production. Increased temperatures and UV radiation and low relative humidity may render many of these control tactics less effective, therefore, there is a need to address these issues on an urgent basis for sustainable crop production and food security.

References

Bokhorst S, Huiskes A, Convey P, van Bodegom PM, Aerts R (2008) Climate change effects on soil arthropod communities from the Falkland Islands and the Maritime Antarctic. Soil Biol Biochem 40:1547–1556

Coviella CE, Trumble JT (1999) Effects of elevated atmospheric carbon dioxide on insect-plant interactions. Conserv Biol 13:700–712

Crowley TJ (2000) Causes of climate change over the past 1000 years. Science 289:270–277

Dhaliwal GS, Jindal V, Dhawan AK (2010) Insect pest problems and crop losses: changing trends. Indian J Ecol 37:1–7

Dhillon MK, Sharma HC (2009) Temperature influences the performance and effectiveness of field and laboratory strains of the Ichneumonid parasitoid, *Campoletis chlorideae*. BioControl 54:743–750

Environment Protection Agency (EPA) (1989) The potential effects of global climate change on the United States, Vol 2: National studies. Review of the Report to Congress, US Environmental Protection Agency, Washington DC

Hill MG, Dymock JJ (1989) Impact of climate change: agricultural/horticultural systems. DSIR Entomology Division Submission to the New Zealand Climate Change Program. Department of Scientific and Industrial Research, Auckland

IPCC (1990) The potential impacts of climate change on agriculture and forestry. Intergovernmental Panel on Climate Change. World Meteorological Organization and UN Environment Program, Geneva

Kaiser J (1996) Pests overwhelm Bt cotton crop. Nature 273:423

Kremen C, Colwell RK, Erwin TL, Murphy DD, Noss RF, Sanjayan MA (1993) Terrestrial arthropod assemblages: their use in conservation planning. Conserv Biol 7:796–808

Maffei ME, Mithofer A, Boland W (2007) Insects feeding on plants: rapid signals and responses proceeding induction of phytochemical release. Phytochemistry 68:2946–2959

Parry ML, Carter TR (1989) An assessment of the effects of climatic change on agriculture. Clim Change 15:95–116

Porter JH, Parry ML, Carter TR (1991) The potential effects of climate change on agricultural insect pests. Agric Forest Meteor 57:221–240

Sachs ES, Benedict JH, Stelly DM, Taylor JF, Altman DW, Berberich SA, Davis SK (1998) Expression and segregation of genes encoding Cry1A insecticidal proteins in cotton. Crop Sci 38:1–11

Sharma HC (2005) *Heliothis/Helicoverpa* management: emerging trends and strategies for future research. Oxford & IBH, New Delhi (Science Publishers, USA)

Sharma HC (2010) Effect of climate change on IPM in grain legumes. In: 5th International Food Legumes Research Conference (IFLRC V), and the 7th European Conference on Grain Legumes (AEP VII), 26–30th April 2010, Anatalaya, Turkey

Sharma HC (2012a) Climate change effects on activity and abundance of insects: implications for crop protection and food security. In: Kang MS, Banga SS (ed) Combating climate change: an agricultural perspective. Taylor and Francis, Boca Raton

Sharma HC (2012b) Effect of global warming on insect–host plant–environment interactions. In: 24th International Congress of Entomology, 19–24 Aug 2012, Daegu, South Korea

Sharma HC, Ortiz R (2000) Transgenics, pest management and the environment. Curr Sci 79:421–437

Sharma HC, Mukuru SZ, Manyasa E, Were J (1999) Breakdown of resistance to sorghum midge, *Stenodiplosis sorghicola*. Euphytica 109:131–140

Sharma HC, Sullivan DJ, Bhatnagar VS (2002) Population dynamics of the Oriental armyworm, *Mythimna separata* (Walker) (Lepidoptera: Noctuidae) in South-Central India. Crop Prot 21:721–732

Sharma HC, Dhillon MK, Kibuka J, Mukuru SZ (2005) Plant defense responses to sorghum spotted stem borer, *Chilo partellus* under irrigated and drought conditions. Int Sorghum Millets Newsl 46:49–52

Zvereva EL, Kozlov MV (2010) Responses of terrestrial arthropods to air pollution: a meta-analysis. Environ Sci Poll Res Inter 17:297–311

Temperature-Based Phenology Modeling and GIS-Based Risk Mapping: A Tool for Forecasting Potential Changes in the Abundance of Mealybug *Phenacoccus solenopsis* Tinsley (Hemiptera: Pseudococcidae)

Babasaheb B. Fand, Henri EZ Tonnang, Mahesh Kumar, Ankush L. Kamble and SK Bal

Abstract

Phenacoccus solenopsis Tinsley (Hemiptera: Pseudococcidae) is a highly invasive and a polyphagous pest of worldwide importance. Its recent outbreak and rapid spread in Indian cotton growing belt caused large scale devastation. A study was undertaken with a basic assumption that the future distribution and abundance of *P. solenopsis* will be affected seriously by temperature alterations due to global climate change, which might further aggravate the yield losses. The population growth potential of *P. solenopsis* was estimated at six constant temperatures ranging from 15 to 40 °C. The phenology models established using best fitting functions in a rate summation and cohort up-dating approach were employed in a geographic information system for mapping population growth potentials according to real-time or interpolated temperature data, for both current and future climate to predict the impact of climate change. The risks for population establishment and survival, average numbers of generations and potential population increase/year were computed using interpolated daily minimum and maximum temperatures at a spatial resolution of 10 arc minutes obtained from worldclim database (www.worldclim.org). The real-time weather station data from two selected locations across India were used to analyze within-year variation of pest population increase due to seasonal climate fluctuations. The model predicted favorable temperature range for *P. solenopsis* development, survival, and reproduction within a

B. B. Fand (✉) · M. Kumar · A. L Kamble · SK Bal
National Institute of Abiotic Stress Management (ICAR), Malegaon, Baramati, Pune, Maharashtra 413115, India
e-mail: babasahebfand@gmail.com

H. EZ Tonnang
Crop Management and Production Systems Division, International Potato Center (CIP), Lima 12, Peru

A. K. Chakravarthy (ed.), *New Horizons in Insect Science: Towards Sustainable Pest Management*, DOI 10.1007/978-81-322-2089-3_37,

range of 20–35 °C with maximum population growth potential and shorter generation length at 30 °C. The findings revealed significant changes in *P. solenopsis* activity under climate change scenario, including expansion of a geographical distribution range at higher latitudes and altitudes, marked increase in the number of generations/year and increased abundance and damage activity in present distribution range in India. The study generated knowledge on temperature-dependent population dynamics and growth potential of *P. solenopsis* crucial for undertaking agroecoregion specific management strategies.

Keywords

Climate change · Invasive pests · Phenology modeling · Risk mapping

Introduction

Agricultural productivity is substantially limited by the pest damage at various stages of crop growth; with worldwide average yield loss of 18 % despite timely crop protection measures (Oerke et al. 1994, 2006). In India, pest damage varies considerably in different agroclimatic regions across the country mainly due to differential impacts of several abiotic factors such as temperature, humidity, and rainfall (Reed and Pawar 1981), which indicates further intensification of yield losses due to increased incidence of insect pests in future climate change scenario. The global climate change, a well-established fact and reality has been predicted to raise the mean surface temperature of earth by 1.5–5.8 °C by the end of this century (Govindasamy et al. 2003; Hijmans et al. 2005; IPCC 2007). Insects being poikilotherms, abiotically stressful environment in changing climate is expected to impact negatively their diversity and abundance; and ultimately the extent of damage caused in economically important agricultural crops (Porter et al. 1991; Sutherst 2000; Bale et al. 2002; Ward and Masters 2007). This may affect perilously the agricultural production and the livelihood of farmers especially in tropical and subtropical countries where larger proportion of work force is directly depending on climate sensitive sectors such as agriculture (IPCC 2007; Chahal et al. 2008). The sound forecasting tools based on detailed analysis of pest species' life table under varied environmental conditions will be crucial in undertaking agroecoregion specific management strategies against aggravating pest problems in the context of climate change.

The solenopsis mealybug, *Phenacoccus solenopsis* Tinsley (Hemiptera: Pseudococcidae) is considered to be one of the most devastating pests of cotton in India, where crop losses range from 30 to 60 % (Dhawan et al. 2007; Jhala et al. 2008; Nagrare et al. 2009). The mealybug species native to North America has spread throughout the country causing devastation (Nagrare et al. 2011; Vennila et al. 2010, 2011). Because of its invasiveness and rapid spread across India within a short period of time, *P. solenopsis* has been the major focus of researchers. This chapter describes a temperature-dependent simulation model predicting population growth potential of *P. solenopsis* under varied environmental conditions.

Material and Methods

Culture of *P. solenopsis*

Mealybugs collected from infested plants in the field were cultured in laboratory on sprouted potato tubers at 27 ± 2 °C temperature and 65 ± 5 %

relative humidity (RH) (Gautam 2008). About 2–3 gravid females of *P. solenopsis* were released on the sprouted potato tubers kept in plastic jars (10 cm diameter) with camel hair brush. The jars were covered with clean black muslin cloth tied with rubber band. The mealybug culture developed after 10–12 days was then used for subsequent studies.

Life Table Studies

The effects of temperature on development, survival, and reproduction of *P. solenopsis* were studied on cohorts of single life stages in controlled incubation chambers at six constant temperatures, i.e., 15, 20, 25, 30, 35, and 40 °C. A group of 25 newly emerged crawlers (0–24-h old) of *P. solenopsis* was transferred into a small plastic jar (10 × 7.5 cm) containing a medium sized sprouted potato tuber. The jar was covered with a muslin cloth tied with rubber band. The experimental setup was replicated six times with a total of 300 individuals tested at each temperature. The development and survival of insects were recorded daily by observing them under stereomicroscope. The similar procedure was followed for second- and third-instar mealybugs. The development time and mortality of nymphs during each instar was recorded daily. In *P. solenopsis*, the immature intended to develop as males construct a loosely woven silky filamentous cocoon after second moult and undergoes two moults inside with a prepupal stage, before emerging as winged adults. Hence, 20 newly woven cocoons were separated and observed for emergence at each test temperature in ten replications. For recording fecundity and longevity, newly moulted six adult females (0–24-h old) from the colony of *P. solenopsis* were confined separately to a plastic jar (10 × 7.5 cm size) containing a sprouted potato tuber covered with muslin cloth. The experiment was repeated five times. Similarly, freshly emerged winged adults from puparia were observed for their longevity at respective temperatures. The survival time was recorded separately for males and females.

Mode Fitting and Analysis

For simulating temperature-dependent life table parameters of *P. solenopsis*, we used Insect Life Cycle Modeling Software (Version 3.0), developed by International Potato Center, Lima, Peru (Sporleder et al. 2012). Overall phenology models were established using best fitting functions in a rate summation and cohort up-dating approach. Cumulative frequency of development time of each life stage and temperature was plotted against normalized development times, i.e., time/median development time by fitting a lognormal distribution curve. Sharpe and DeMichele model (Sharpe and DeMichele 1977; Schoolfield et al. 1981) was used for describing temperature dependence of mean development rates of immature stages and adult senescence. The equation of model is:

$$r(T)=\frac{RH025\times\dfrac{T}{298.16}\times\exp\left[\dfrac{\Delta Ha}{R}\left(\dfrac{1}{298.16}-\dfrac{1}{T}\right)\right]}{1+\exp\left[\dfrac{\Delta Hl}{R}\left(\dfrac{1}{Tl}-\dfrac{1}{T}\right)\right]+\exp\left[\dfrac{\Delta Hh}{R}\left(\dfrac{1}{Th}-\dfrac{1}{T}\right)\right]},$$

where r (T) is the development rate at temperature T (°K), R is the universal gas constant (1.987 cal degree^{-1} mol^{-1}), RHO_{25} is the development rate at 25 °C temperature (298.16 °K), assuming no enzyme inactivation, ΔH_a is the enthalpy of activation of reaction catalyzed by enzyme (cal mol^{-1}), ΔH_l is the change in enthalpy at low temperature (cal mol^{-1}), ΔH_h is the change in enthalpy at high temperature (cal mol^{-1}), T_l is the low temperature at which enzyme is half active and T_h is the high temperature at which enzyme is half active.

A second-order polynomial function [m (T) = aT^2 + bT + c, where m (T) is the mortality or fecundity at temperature T (°C)], was fitted to describe temperature dependence of immature mortality and female fecundity. Age-related oviposition rate was described by exponential model [$y = (1-\exp(-(ax+bx^2+cx^3)))$, where y is the cumulative oviposition rate, and x is the normalized female age].

Spatial Simulation and Risk Mapping in Geographic Information System (GIS)

Climate Data

Climate data required for spatial simulations at current climatic scenario were obtained from Worldclim database (http://www.worldclim.org/). The current data comprised interpolated monthly minimum and maximum temperatures obtained from long-term time series (1950–2000) from a global network (Hijmans et al. 2005). For simulating *P. solenopsis* risk under climate change scenario, we used downscaled data (Ramirez and Jarvis 2010) (freely downloadable at http://gisweb.ciat.cgiar.org/GCMPage) of the SRES-A_1B emission scenario for the year 2050 (IPCC 2007). The climate data were used at a resolution of 10 arc minutes. For simulating within year variability of *P. solenopsis* population increase at different locations, i.e., point-by-point analyses, we used actual temperature data obtained from two weather stations in India, viz., Ludhiana (Punjab State) and Akola (Maharashtra State).

Temperature Simulations

Daily minimum and maximum temperature data are used by insect life cycle modeling (ILCYM) for simulating pest population growth potential for each 15 min time step using a cosine function for half-day temperature predictions. For the first half-day predictions, the following equation is used:

$$Ti = \frac{(Max - Min)}{2} \times \cos\left(\frac{\pi \times (i - 0.5)}{48}\right) + \frac{(Min + Max)}{2},$$

where *Ti* is the temperature (°C) of time step *i* ($i = 1, 2, 3, \ldots 48$) and Min and Max are the daily minimum and maximum temperatures.

The same procedure was then repeated to estimate *Ti* for the second half day using minimum temperature of the next day in the equation. As worldclim database only include monthly aggregates of temperature data, ILCYM replaces the daily minimum and maximum temperatures in above equation with monthly averages for half-day temperature predictions in each 15 min time step. In this case, a value of *Ti* remains constant for all days of the month, except for the last day that employs minimum temperature of first day of next month to calculate *Ti* for second half day (for details see Sporleder et al. 2008; Kroschel et al. 2013).

Estimation of Life Table Parameters

Based on the simulated temperatures and model outputs, the life table parameters, viz., net reproductive rate (R_0), mean generation time (T), intrinsic rate of natural increase (r_m), finite rate of increase (λ), and doubling time (D_t) were estimated at range of constant temperatures using life table simulation function of ILCYM (Sporleder et al. 2008, Kroschel et al. 2013). The estimated life table parameters were plotted against Julian days to visualize variability in pest potentials due to seasonal climate variation.

Estimation of Pest Risk Indices

Based on the estimated life table parameters and simulated temperatures, following three indices for pest risk at each data point were calculated.

a. **Establishment risk index (ERI)or survival risk index**

This index characterizes the geographical areas having risk of survival and establishment of the pest insect. The survival index is calculated by the following formula:

$$\text{ERI} = (1 - x\text{Nymph 1}) \times (1 - x\text{Nymph 2}) \times (1 - x\text{Nymph 3}),$$

where

$$x = \frac{\sum \text{Number of days, a specific immature life stage does not survive}}{365}.$$

b. **Generation index (GI)**

This index represents the mean number of generations that a given pest can produce within a year. It is calculated by averaging the sum of estimated

Table 1 Mean development time and survival of *P. solenopsis* life stages at constant temperatures (Replications: nymphs = 12; male pupae = 10; adults = 5; total sample size: nymphs = 300; male pupae = 100; adults = 30)

Temperature (°C)	Nymph1		Nymph2		Nymph3		Male pupae		Female	Male
	Mean development time (days)	Survival (%)	Mean development time (days)	Survival (%)	Mean development time (days)	Survival (%)	Mean development time (days)	Survival (%)	Survival time (days)	Survival time (days)
15	14.44 (0.11)	41.65 (1.82)	16.51 (0.12)	77.00 (1.67)	13.89 (0.12)	87.00 (1.13)	18.98 (0.10)	71.00 (2.07)	50.54 (0.59)	3.78 (0.06)
20	11.98 (0.10)	47.67 (1.95)	12.18 (0.10)	86.00 (1.21)	10.02 (0.10)	91.00 (0.83)	14.81 (0.08)	81.00 (1.91)	44.44 (0.51)	2.95 (0.05)
25	6.72 (0.06)	71.33 (2.40)	7.78 (0.06)	93.00 (0.87)	5.83 (0.06)	94.00 (0.85)	9.05 (0.06)	90.00 (0.91)	29.47 (0.40)	2.67 (0.05)
30	5.45 (0.05)	76.67 (1.60)	6.32 (0.05)	95.00 (0.86)	5.56 (0.06)	96.30 (0.91)	6.96 (0.05)	93.00 (0.81)	30.87 (0.35)	1.83 (0.04)
35	3.89 (0.05)	55.33 (1.46)	4.94 (0.05)	75.30 (1.51)	3.30 (0.04)	89.70 (1.01)	3.85 (0.04)	84.00 (1.05)	23.80 (0.19)	1.79 (0.04)
40	3.0 (0.04)	29.33 (1.50)	3.54 (0.04)	54.00 (1.28)	3.01 (0.05)	85.00	3.46 (0.00)	79.00	18.44 (0.11)	1.41 (0.01)

Numbers in parentheses are standard errors

generation lengths for each Julian day as shown in the following formula:

$$GI = \frac{\sum_{x=1}^{365} 365/Tx}{365},$$

where Tx is the predicted generation length in days at Julian day x ($x = 1, \ldots, 365$)

c. **Activity index (AI)**

This index represents the finite rate of increase and hence population growth potential of given pest species. It is calculated by taking a log of the products of the estimated finite rates of increase for each Julian day as shown in the following formula:

$$AI = \log_{10} \prod_{x=1}^{365} \lambda x,$$

where λx is the finite rate of increase at Julian day x ($x = 1, \ldots, 365$).

GIS Mapping of Risk Indices

The maps showing the pest risk potentials were visualized by employing the estimated risk indices in "potential population analysis and mapping" module of ILCYM software. The module facilitates spatial simulations of pest populations through grid based within a defined area according to grid-specific monthly temperatures interpolated from available databases (worldclim in this case). The estimated indices were organized in matrices of 12×12 dimensions using longitude as columns and latitude as rows, representing 12 matrices each for ERI, GI, and AI. The generated ASCII files (.asc) were converted to grid format by importing in DIVA-GIS, an open-source geographic information system (downloadable at http://www.diva-gis.org) for visualizing the risk maps.

Results and Discussion

Effect of Temperature on *P. Solenopsis*

Development of all the immature life stages of *P. solenopsis* was significantly affected by the temperature. Development time decreased significantly with increasing temperatures within the evaluated temperature range (Table 1). The development of immature stages at temperature extremes (15 and 40 °C) was best described by nonlinear model, Sharpe and DeMichele function (for all stages, df = 4, 13; $P \leq 0.001$; Nymph 1, F = 173.14; Nymph 2, F = 195.57; Nymph 3, F = 71.10; Male pupa, F = 182.77) (Figs. 1a, 1b,

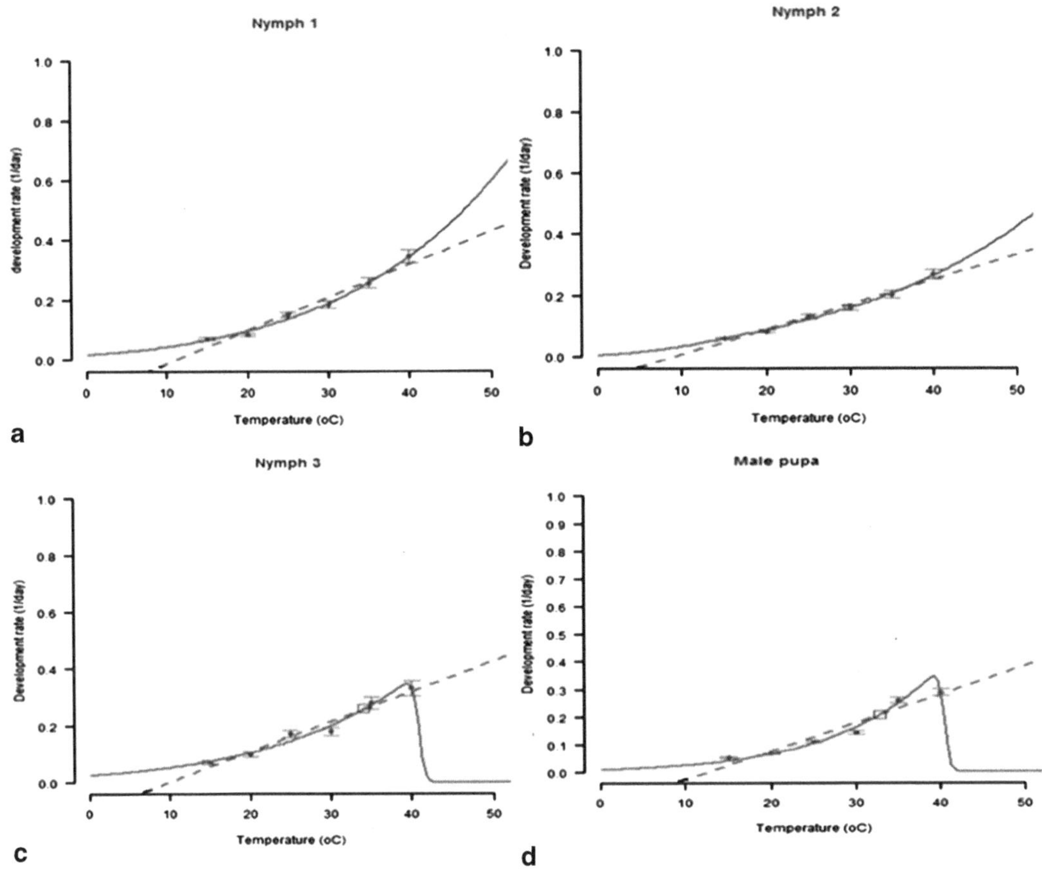

Fig. 1 Temperature-dependent developmental rates (1/day) for immature stages of *P. solenopsis*: nymph 1 (**a**), nymph 2 (**b**), nymph 3 (**c**), and male pupae (**d**) (Sharpe and DeMichele model). Bars represent standard deviation of the median

1c, and 1d). The first-instar nymphs were highly sensitive to the temperature extremes; however, later instars had a better survival at all the temperatures within the evaluated range (Table 1). Longevity of adults (both male and female) decreased significantly with increasing temperature due to increased senescence rates (for both sexes, df=3, 2; male: F=17.47, P=0.055; female: F=12.75, P=0.074) (Table 1; Figs. 2a and 2b). The oviposition showed curvilinear response with maximum fecundity at 30 °C and dropping off at temperatures below and above it (second-order polynomial function, F=4.16; df=2, 3; P=0.136) (Fig. 3). The model predicted favorable temperature range for *P. solenopsis* development, survival, and reproduction between 20 and 35 °C with maximum population growth potential and shorter generation length at 30 °C (Fig. 4).

Risk Mapping of *P. solenopsis*

Establishment Risk

Figure 5a indicates regions in India where *P. solenopsis* population theoretically could establish according to temperature conditions at current climate scenario (worldclim data 1950–2000). Regions with ERI 1 indicate that at least certain proportion of pest population is always expected to survive and complete life cycle throughout the year. The likelihood of permanent establishment of *P. solenopsis* is reduced considerably in areas on a map where ERI is less than one. More than 80 % of area in India, barring extreme Northern Himalayas and North-Eastern high hills is predicted with high establishment risk. Besides its present distribution range, the pest is predicted

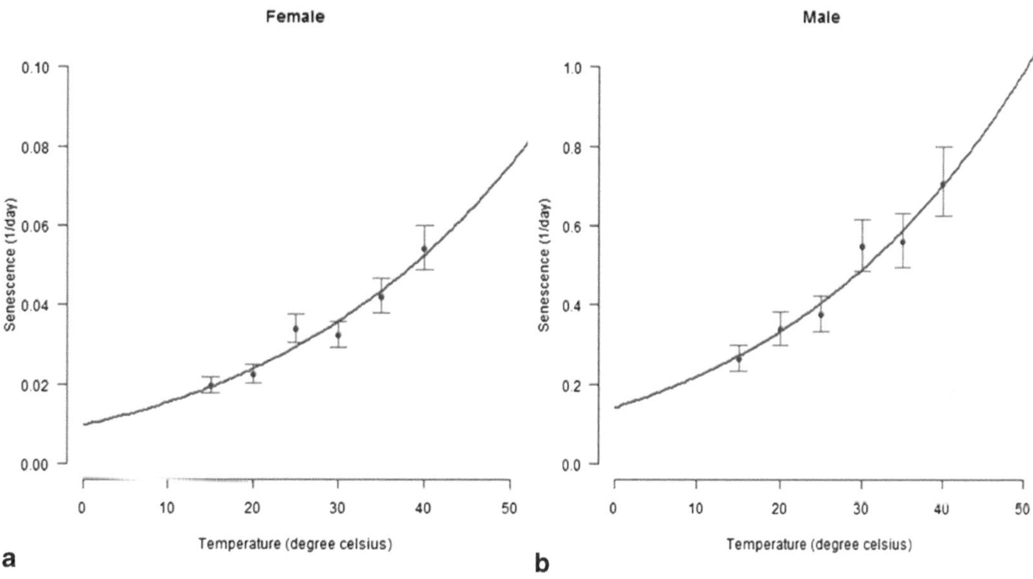

Fig. 2 Temperature-dependent survival rates (1/day) for adults of *P. solenopsis*: female (**a**) and male (**b**). Fitted curves, Sharpe-DeMichele model. Bars represent standard deviation of the mean

to extend its geographic distribution and spread, especially at higher latitudes and altitudes due to future climate change (Fig. 5b).

Generation Index

Average numbers of generations that can theoretically be produced per year by *P. solenopsis* under current temperature conditions are visualized in Fig. 6a. According to the model predictions, under current climate conditions *P. solenopsis* can produce maximum of 10–11 generations per year. The regions included West- and South-east coasts of India. Under future climate scenario of the year 2050, a significant increase in the number of *P. solenopsis* generations with a maximum increase of 2–3 generations is predicted for India, particularly in the regions of Gangetic plains, Western Deserts, Deccan plateau, and Eastern Ghats where warmer conditions prevail (Fig. 6b).

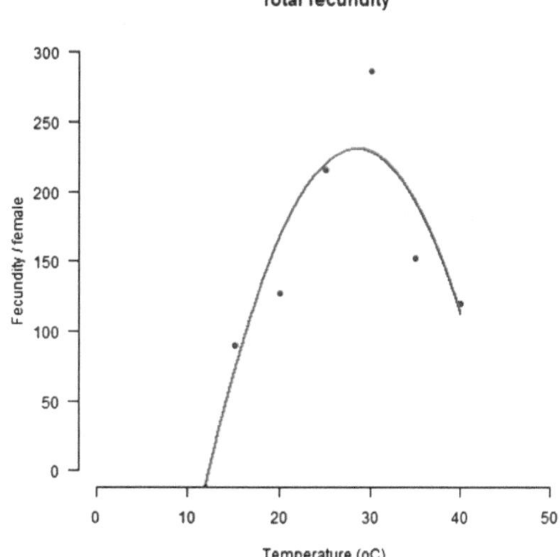

Fig. 3 Temperature-dependent oviposition curve (second-order polynomial regression model)

Damage Potential

AI explains the temperature-dependent finite rate of population increase within a year. Every increase of the index by a value of 1 indicates a tenfold increase of the pest population. Under present climatic conditions, Westerns semiarid region, Deccan plateau and Eastern Ghats of India were predicted optimal for *P. solenopsis* activity (Fig. 7a). Similarly, under climate change scenario drastic change in *P. solenopsis* population growth and abundance is predicted throughout the country, barring Northern and North-Eastern Himalayas where low temperatures limit the pest survival (Fig. 7b).

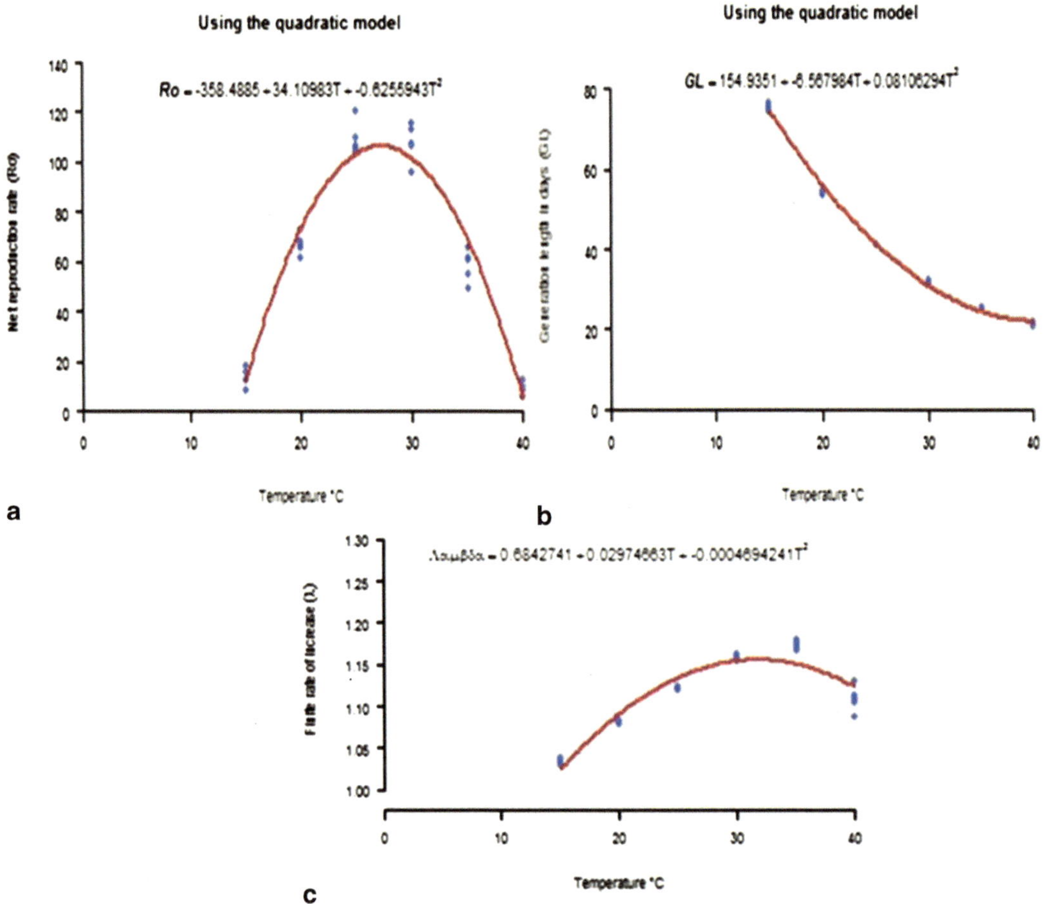

Fig. 4 Life table parameters of *P. solenopsis* estimated through model prediction over a range of temperatures: **a** net reproduction rate (*R0*), **b** mean generation time (*T*), and **c** finite rate of increase (λ)

Within Year Variability of Pest Population

Daily weather station data from two locations across the cotton growing zones of India, viz., Ludhiana (Punjab) and Akola (Maharashtra) were used to predict seasonal variability in *P. solenopsis* population growth and abundance. In India, cotton is sown around May–June and harvested around October–December in different parts of the country; however, wide range of flora present abundantly throughout the year supports the carryover and perpetual of *P. solenopsis* during off seasons (Vennila et al. 2010, 2011). According to the analysis, in both selected locations *P. solenopsis* populations potentially might increase with a maximum finite rate of about 1.16 during the total cropping season and reduced to 1.0 during off season and cool winter months (Fig. 8a and 8b).

However, the population growth potential varied considerably among the seasons (cropping and off seasons) and across the agroclimatic zones, which might be due to larger within year temperature fluctuations and also differences in sowing dates of the crop. According to the model predictions, at Ludhiana, *P. solenopsis* activity is considerably high throughout the entire crop season with peak infestation during June–August (Fig. 9a). Contrarily, the pest is predicted to have maximum finite rate of increase during of April–May at Akola (Fig. 9b). The temperature increase

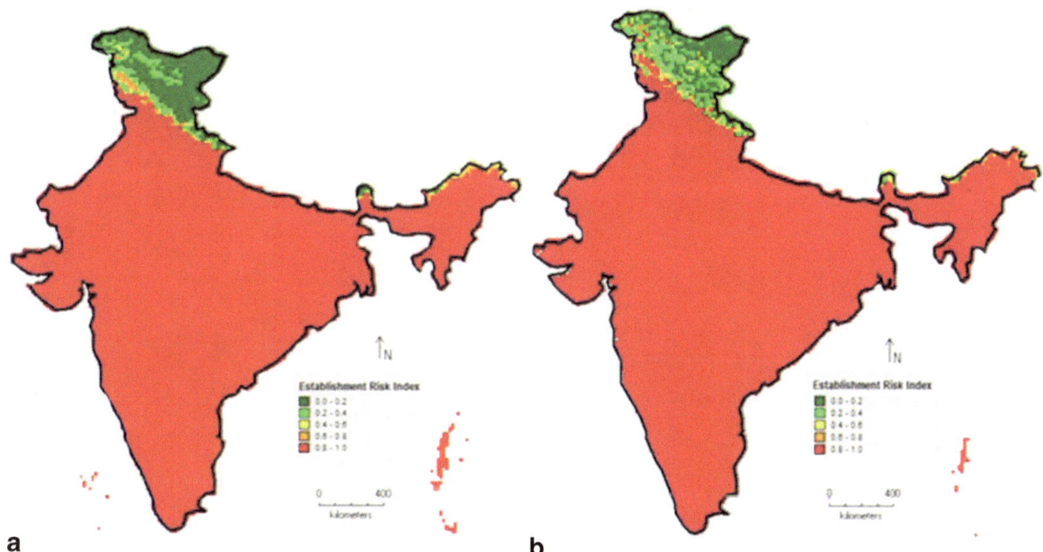

Fig. 5 Establishment and distribution of *P. solenopsis* in India at current (1950–2000) (**a**) and future (SRES A1B scenario 2050) (**b**) climatic conditions simulated using interpolated minimum and maximum temperature data from worldclim database. *P. solenopsis* is rarely established in the geographical regions on a map where the index value is lower than 0.8

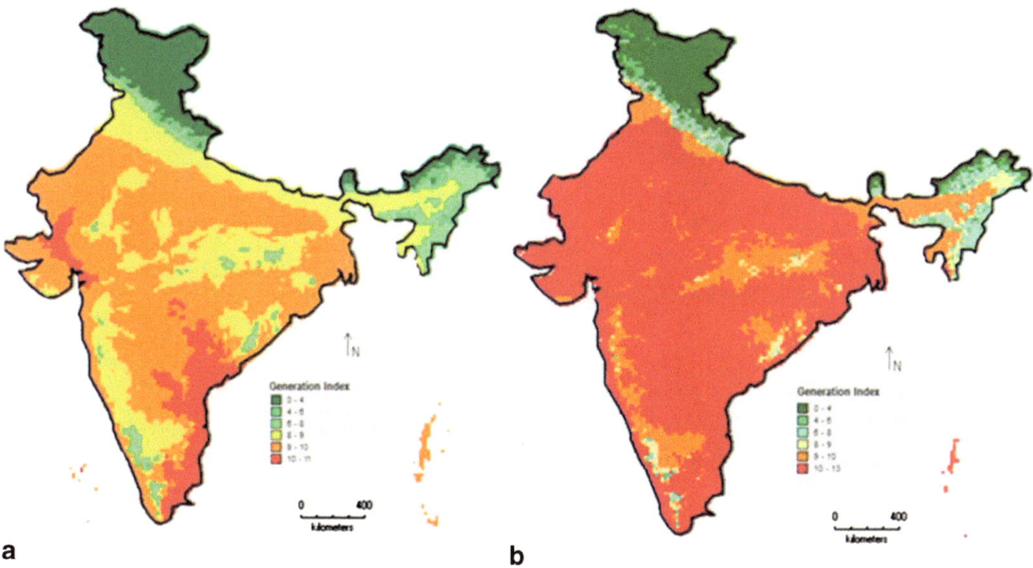

Fig. 6 Change in the generation index (GI) of *P. solenopsis* in India at current (1950–2000) (**a**) and future (SRES A1B scenario 2050) (**b**) climatic conditions simulated using interpolated minimum and maximum temperature data from worldclim database. The index represents mean number of generations; a *P solenopsis* population can produce within a year under current (**a**) and future (**b**) climatic conditions

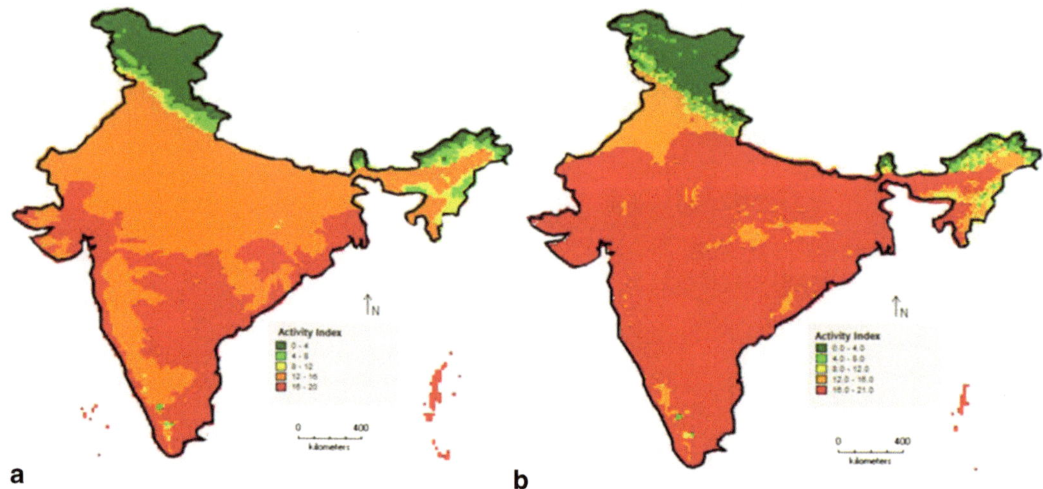

Fig. 7 Activity index (AI) for *P. solenopsis* in India at current (1950–2000) (**a**) and future (SRES A1B scenario 2050) (**b**) climatic conditions simulated using interpolated minimum and maximum temperature data from worldclim database. An index value of 1 represents tenfold potential population increase within a year at present and future climatic conditions

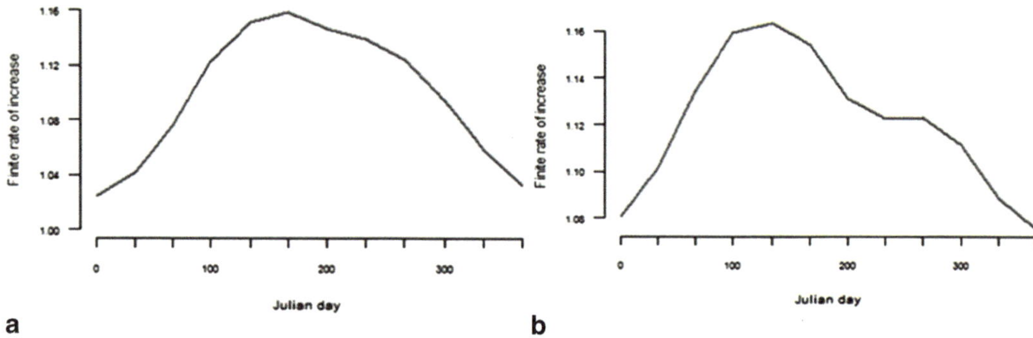

Fig. 8 Simulated finite rate of increase of *P. solenopsis* throughout the year at present climatic conditions for Ludhiana (**a**) and Akola (**b**)

due to climate change in these areas is predicted to increase the finite rate of increase of *P. solenopsis* throughout the year with maximum during the cotton cropping season (Fig. 10a and 10b).

The temperature influenced significantly the development, survival, and reproduction of *P. solenopsis*. The development times of the immature stages of *P. solenopsis* reported in this study are in conformity with earlier reports (Fand et al. 2010; Nikam et al. 2010; Vennila et al. 2010; Ali et al. 2012; Asifa et al. 2012; Prasad et al. 2012). Relatively, lower mortality rates reported by Prasad et al. (2012) in first-instar nymphs of *P. solenopsis* are slightly deviating from this study, which largely may be because of differences in rearing procedures and food source/host plant used for rearing of test insect, that seriously affect survival of test organisms. A curvilinear response for fecundity with a maximum at 30 °C (286.43 eggs/female) and dropping off at temperatures below and above this are in conformity with reports of Nagrare et al. (2009, 2011) and Prasad et al. (2012). The life table parameters estimated from the outputs of *P. solenopsis* simulation model are in agreement with the previous reports (Nagrare et al. 2009, 2011; Fand et al. 2010; Prasad et al. 2012).

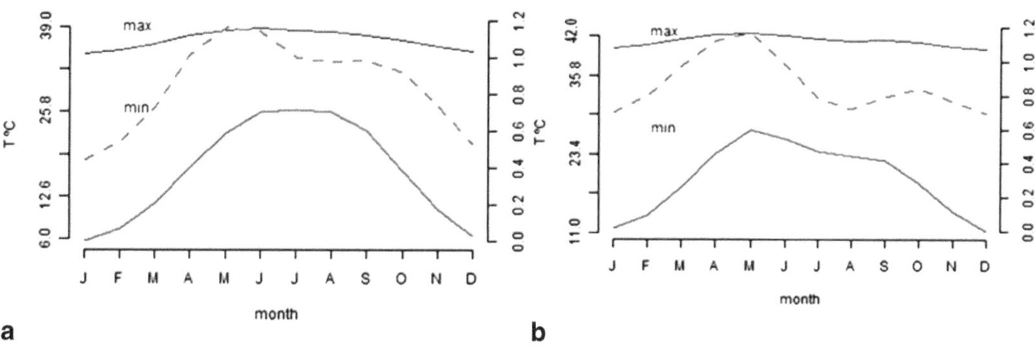

Fig. 9 Simulated finite rate of increase of *P. solenopsis* plotted against daily minimum and maximum temperatures recorded at two weather stations across India: Ludhiana (**a**) and Akola (**b**)

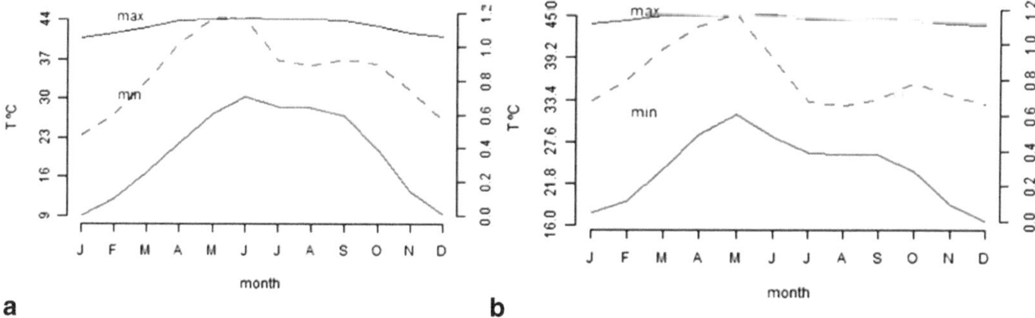

Fig. 10 Simulated finite rate of increase of *P. solenopsis* plotted using interpolated daily minimum and maximum temperatures at 10 arc minutes resolution (SRES A1B scenario for the year 2050) obtained from worldclim database for two locations of India: Ludhiana (**a**) and Akola (**b**)

In India, *P. solenopsis* is reported from areas where the map shows an ERI of 0.8 and above, but this does not necessarily means that it is permanently established in those areas. This suggests that almost 80% of area in India except Northern Himalayas and North-Eastern high hills region are highly suitable habitats for *P. solenopsis* establishment and survival. Present findings regarding geographical habitat suitability for *P. solenopsis* establishment are in conformity with the predictions of MAXENT model (Fand 2012). The generation indices simulated for India in this study, gave reasonably good predictions when compared with the literature data. Tanwar et al. (2011) reported that a female of *P. solenopsis* can produce as many as 15 generations per year in field conditions. Based on the laboratory life table studies at 26±1 °C, *P. solenopsis* is reported to complete ten overlapping generations in a year (Arif et al. 2012). A significant increase in the number of generations per year under future climate conditions has been predicted by this study. Ultimately, the potential for population abundance and damage by the pest species will also increase. The increased abundance of *P. solenopsis* under future climatic conditions indicates that it may pose serious threat to agriculture in India. The impact of climate change, especially temperature increase on the geographic distribution and spread of *P. solenopsis* in India has been reported (Fand 2012).

Analyses of the within-year variation in population increase contribute to better understanding of temporal variation in *P. solenopsis* growth and development in response to changing temperature conditions. A peak activity of *P. solenopsis* during the months of June–August predicted for the north zone of India (Ludhiana) is in conformity with earlier reports (Vennila et al. 2010, 2011; Nagrare et al. 2011). Dhawan et al. (2007) recorded highest field infestation in the months of July–August (30–34 standard meteorological weeks) in Bathinda, Muktsar, Ferozepur, and Faridkot districts of Punjab State. In Tamil Nadu, *P. solenopsis* population was maximum during June, decreased slowly during September and there was no incidence up to February (Suresh et al. 2010).

The model predicted favorable temperature range for *P. solenopsis* development, survival, and reproduction within a range of 20–35 °C with maximum population growth potential and shorter generation length at 30 °C. The knowledge on temporal variation in *P. solenopsis* growth and development in response to changing temperature conditions help in undertaking effective management strategies. The findings revealed significant changes in *P. solenopsis* activity under climate change scenario, which included expansion of a geographical distribution range at higher altitudes, marked increase in the number of generations per year and increased abundance and damage activity in present distribution range in India. The present predictions on the future distribution, survival, and abundance of *P. solenopsis* clearly indicate that the invasiveness of this pest will be aggravated and intensified under projected climate changes.

Acknowledgments This study is a part of the research project "Abiotic stresses affecting crop-insect pest interactions in the context of global climate change (Project Code-IXX08575)" of the School of Atmospheric Stress Management, National Institute of Abiotic Stress Management (NIASM), Baramati, Pune, Maharashtra (India). The authors gratefully acknowledge financial support from Indian Council of Agricultural Research, New Delhi. They are thankful to the Director, NIASM for providing all the necessary facilities and extending his cooperation and support to carry out present investigations. We are grateful to the Heads of the Department of Agrometeorology, Punjab Agricultural University, Ludhiana (Punjab) and Dr. Panjabrao Deshmukh Krishi Vidyapeeth, Akola (Maharashtra) for providing temperature data.

References

Ali A, Asifa Hameed, Saleem M, Khalil N, Saeed M, Malik NA (2012) Effect of temperature and relative humidity on the biology of cotton mealy bug (*Phenacoccus solenopsis* Tinsley). J Agric Res 50:89–101

Arif MI, Rafiq M, Wazir S, Mehmodd N, Ghaffar A (2012) Studies on Cotton Mealybug, *Phenacoccus solenopsis* (Pseudococcidae: Homoptera), and its natural enemies in Punjab, Pakistan. Int J Agric Biol 14:557–562

Asifa Hameed Aziz MA Aheer GM (2012) Impact of ecological conditions on biology of cotton mealybug, *Phenacoccus solenopsis* (Hemiptera: Pseudococcidae) in laboratory. Pakistan J Zool 44:685–690

Bale JS, Masters GJ, Hodkinson ID, Awmack C, Bezemer TM, Brown VK, Butterfield J, Buse A, Coulson JC, Farrar J, Good JEG, Harrington R, Hartley S, Jones TH, Lindroth RL, Press MC, Symrnioudis I, Watt AD, Whittaker JB (2002) Herbivory in global climate change research: direct effects of rising temperature on insect herbivores. Global Change Biol 8:1–16

Chahal SK, Bains GS, Dhaliwal LK (2008) Climate change: mitigation and adaptation. International Conference on Climate change, biodiversity and food security in the South Asian Region, 3–4 November, Punjab State Council for Science and Technology, Chandigarh and United Nations Educational, Scientific and Cultural Organization, New Delhi

Dhawan AK, Singh K, Saini S, Mohindru B, Kaur A, Singh G, Singh S (2007) Incidence and damage potential of mealybug, *Phenacoccus solenopsis* Tinsley on cotton in Punjab. Indian J Ecol 34:110–116

Fand BB (2012) Modeling the impact of climate change on potential geographic distribution of polyphagous mealybug *Phenacoccus solenopsis* in India pp 37–38 in Proceedings of the IVth National Symposium on Plant protection in horticultural crops- emerging challenges and sustainable pest management, organised by Association or the Advancement of Pest Management in Horticultural Ecosystems, Bangalore, 25–28 April. Indian Institute of Horticultural Research, Bangalore

Fand BB, Gautam RD, Subhash Chander, Suroshe SS (2010) Life table analysis of the mealybug, *Phenacoccus solenopsis* Tinsley (Hemiptera: Pseudococcidae) under laboratory conditions. J Entomol Res 34:175–179

Gautam RD (2008) Biological pest suppression, 2nd edn. Westville Publishing House, New Delhi

Govindasamy B, Duffy PB, Coquard J (2003) High-resolution simulations of global climate, part 2: effects of increased greenhouse gases. Clim Dyn 21:391–404

Hijmans RJ, Cameron SE, Parra JL, Jones PG, Jarvis A (2005) Very high resolution interpolated climate surfaces for global land areas. Int J Climatol 25:1965–1978

IPCC (2007) Climate change- impacts, adaptation and vulnerability. In: Parry ML, Canziani OF, Palutikof JP, van der Linden PJ, Hanson CE (eds). Cambridge University Press, Cambridge, p 976

Jhala RC, Bharpoda TM, Patel MG (2008) *Phenacoccus solenopsis* (Hemiptera: Pseudococcidae), the mealybug species recorded first time on cotton and its alternate host plants in Gujarat, India. Uttar Pradesh J Zool 28:403–406

Kroschel J, Sporleder M, Tonnang HEZ, Juarez H, Carhuapoma P, Gonzales JC, Simon R (2013) Predicting climate-change- caused changes in global temperature on potato tuber moth *Phthorimaea operculella* (Zeller) distribution and abundance using phenology modeling and GIS mapping. Agricultural and Forest Meteorology 170:228–241.

Nagrare VS, Kranthi S, Biradar VK, Zade NN, Sangode V, Kakde G, Shukla RM, Shivare D, Khadi BM, Kranthi KR (2009) Widespread infestation of the exotic mealybug species, *Phenacoccus solenopsis* Tinsley (Hemiptera: Pseudococcidae), on cotton in India. Bull Entomol Res 99:537–541

Nagrare VS, Kranthi S, Rishi Kumar, Dhara Jothi B, Amutha M, Deshmukh AJ, Bisane KD, Kranthi KR (2011) Compendium of cotton mealybugs. Central Institute for Cotton Research, Nagpur, p 42

Nikam ND, Patel BH, Korat DM (2010) Biology of invasive mealy bug, *Phenacoccus solenopsis* Tinsley (Hemiptera: Pseudococcidae) on cotton. Karnataka J Agric Sci 23:649–651

Oerke EC (2006) Crop losses to pests. J Agric Sci 144: 31–43

Oerke EC, Debne HW, Schonbeck F, Weber A (1994) Crop production and crop protection. Elsevier, Amsterdam

Porter JH, Parry ML, Carter TR (1991) The potential effects of climatic change on agricultural insect-pests. Agric Fort Meteorol 57:221–40

Prasad YG, Prabhakar M, Sreedevi G, Ramachandra Rao G, Venkateswarlu B (2012) Effect of temperature on development, survival and reproduction of the mealybug, *Phenacoccus solenopsis* Tinsley (Hemiptera: Pseudococcidae) on cotton. Crop Prot 39:81–88

Ramirez J, Jarvis A (2010) Downscaling global circulation model outputs: the delta method decision and policy analysis. Working Paper 1. International Center for Tropical Agriculture, CIAT

Reed W, Pawar CS (1981) Heliothis: a global problem. In: ICRISAT Proceedings of the International workshop on heliothis management, pp 9–14. November 15–20, International Crops Research Institute for the Semi-Arid Tropics, Patancheru, Andhra Pradesh, India

Schoolfield RM, Sharpe PJH, Magnuson CE (1981) Non-linear regression of biological temperature-dependent rate models based on absolute reaction-rate theory. J Theor Biol 88:719–731

Sharpe PJH, DW DeMichele (1977) Reaction kinetics of poikilothermic developement. J Theor Biol 64:649–670

Sporleder M, Simon R, Juarez H, Kroschel J (2008) Regional and seasonal forecasting of the potato tuber moth using a temperature-driven phenology model linked with geographic information systems. In: Kroschel J, Lacey L (eds) Integrated pest management for the potato tuber moth, *Phthorimaea operculella* Zeller—a potato pest of global importance. tropical agriculture 20 advances in crop research 10. Margraf Publishers, Weikersheim

Sporleder M, Simon R, Gonzales JC, Carhuapoma P, Juarez H, De Mendiburu F, Kroschel J (2009) ILCYM—Insect life cycle modelling (version 2.1): a software package for developing temperature-based insect phenology models with applications for regional and global pest risk assessments and mapping (User Manual). p 79 International Potato Center, Lima, Peru

Sporleder M, Tonnang HEZ, Juarez H, Carhuapoma P, Gonzales JC, Simon R, Kroschel J (2012) ILCYM—Insect life cycle modeling (version 3.0): a software package for developing temperature-based insect phenology models with applications to regional and global analysis of insect population and mapping. (User Manual). 124, International Potato Center, Lima, Peru

Suresh SR, Jothimani P, Sivasubramanian P, Karuppuchamy R, Samiyappan, Jonathan EI (2010) Invasive mealybugs of Tamil Nadu and their management. Karnataka J Agric Sci 23:6–9

Sutherst R (2000) Climate change and invasive species: a conceptual framework. In: Mooney H, Hoobs R (eds) Invasive species in a changing world. Island Press, Washington DC, pp 211–240

Tanwar RK, Jeyakumar P, Singh A, Jafri AA, Bambawale OM (2011) Survey for cotton mealybug, *Phenacoccus solenopsis* (Tinsley) and its natural enemies. J Env Biol 32:381–384

Vennila S, Ramamurthy VV, Deshmukh A, Pinjarkar DB, Agarwal M, Pagar PC, Prasad YG, Prabhakar M, Kranthi KR, Bambawale OM (2010) A treatise on mealybugs of Central Indian cotton production system. Technical Bulletin No. 24, New Delhi: National Centre for Integrated Pest Management

Vennila S, Prasad YG, Prabhakar M, Rishi Kumar, Nagrare V, Amutha M, Dharajyothi, Agarwal M, Sreedevi G, Venkateswarlu B, Kranthi KR, Bambawale OM (2011) Spatio-temporal distribution of host plants of cotton mealybug, *Phenacoccus solenopsis* Tinsley in India, Technical Bulletin No. 26, New Delhi: National Centre for Integrated Pest Management

Ward NL, Masters GJ (2007) Linking climate change and species invasion: an illustration using insect herbivores. Global Change Biol 13:1605–1615